AF796464

SolidWorks for Designers

Release 2005

Sham Tickoo

Professor

Department of Mechanical Engineering Technology

Purdue University Calumet

Hammond, Indiana

U.S.A.

Contributing Authors

Gurpreet Singh

Sr. CADD Engineer

Deepak Maini

Sr. CADD Engineer

CADCIM Technologies

CADCIM Technologies

(www.cadcim.com)

U.S.A.

CADCIM Technologies

SolidWorks for Designers: Release 2005
Sham Tickoo

Published by CADCIM Technologies, 525 St Andrews Drive, Schererville, IN 46375 USA.
Copyright © CADCIM Technologies. All rights reserved. No part of this publication may be reproduced or distributed in any form or by any means, or stored in the database or retrieval system without the prior permission of CADCIM Technologies.

ISBN 1-932709-04-5

Cover designer: *CADCIM Technologies*
Technical editors: *Sanjib Sahu, Monalisa Bhol*
Copy editor: *Pragya Katariya and Manoj Bachheti*
Cover illustration: *Created by Deepak Maini and Gurpreet Singh using SolidWorks, PhotoWorks, and Photoshop software*
Typeface: *10/12 New Baskerville Bt*

NOTICE TO THE READER

Publisher does not warrant or guarantee any of the products described in the text or perform any independent analysis in connection with any of the product information contained in the text. Publisher does not assume, and expressly disclaims, any obligation to obtain and include information other than that provided to it by the manufacturer.

The reader is expressly warned to consider and adopt all safety precautions that might be indicated by the activities herein and to avoid all potential hazards. By following the instructions contained herein, the reader willingly assumes all risks in connection with such instructions.

The Publisher makes no representation or warranties of any kind, including but not limited to, the warranties of fitness for particular purpose or merchantability, nor are any such representations implied with respect to the material set forth herein, and the publisher takes no responsibility with respect to such material. The publisher shall not be liable for any special, consequential, or exemplary damages resulting, in whole or part, from the reader's use of, or reliance upon, this material.

www.cadcim.com

DEDICATION

To teachers, who make it possible to disseminate knowledge to enlighten the young and curious minds of our future generations

To students, who are dedicated to learning new technologies and making the world a better place to live

THANKS

To the faculty and students of the MET department of Purdue University Calumet for their cooperation

To Santosh Tickoo, Sanjib Sahu, Monalisa Bhol, and Pragya Katariya of CADCIM/CADSoft Technologies for their valuable help

Free Teaching Aids for Faculty

The following teaching aids are available for free to faculty.

1. *All Part files, Assembly files, and Drawing files used for tutorials, exercises, and illustrations in this book.*
2. *PowerPoint presentations for every chapter of the book.*
3. *Instructor's Guide with answers to review questions and solution to exercises.*
4. *Course outlines.*
5. *Students projects.*
6. *Free online technical support by contacting* **techsupport@cadcim.com**.

To access the Web site that contains these teaching aids, please contact the author, Prof. Sham Tickoo, at the following address.

stickoo@calumet.purdue.edu
or
tickoo@cadcim.com

For Students: *You can download solid modeling exercises, tutorials, and special topics by accessing the author's Web site* **http://technology.calumet.purdue.edu/met/tickoo/students/students.htm** *or the publisher's Web site* **www.cadcim.com**

For more information, please visit www.cadcim.com

Table of Contents

Chapter 2: Editing and Modifying the Sketches

Chapter 3: Adding Relations and Dimensions to the Sketches

Chapter 4: Advanced Dimensioning Techniques and Base Feature Options

Chapter 5: Creating Reference Geometries

Chapter 6: Advanced Modeling Tools-I

Chapter 7: Advanced Modeling Tools-II

Chapter 12: Working With Drawing Views-I

Chapter 13: Working With Drawing Views-II

About the Author

Sham Tickoo is a professor of Mechanical Engineering Technology at Purdue University Calumet. He has been in the education and training since 1981 and joined Purdue University in the year 1987. Since then he has been teaching drafting and design, AutoCAD, AutoLISP, 3ds max, 3ds viz, Mechanical Desktop, Pro/Engineer, Autodesk Inventor, SolidWorks etc.

Prof. Tickoo is also one of the best known authors of CAD/CAM books in the world market. He has authored a number of books on the latest CAD/CAM technology. He is one of the very few authors in the world whose books are translated and sold in the countries like Italy, China, Philippines, and Russia. The list of some of his books is given below:

In addition to his academic career, he has been employed as a design engineer, quality control engineer, software developer, and has also worked as a consultant for a number of companies in US and Canada.

Prof. Tickoo has developed a software package called "SMLayout" that has been registered by Autodesk Inc. as a third party software product. This software package is currently being used by some companies in Canada, United States, and Columbia. SMLayout is written in the AutoLISP programming language. It generates flat layout drawing of some complicated geometric shapes like: transitions, cones, cone-cylinder intersections, sphere-cylinder intersections, and cylinder-cylinder intersections. This software package addresses the needs of steel fabricators and manufacturers of sheet metal products. The programs also provide a dimensioning option for automatic dimensioning of the drawing.

In March 2000, Prof. Tickoo was issued a US patent for a product called "Self Adjusting Cargo Organizer" that he designed for vehicles.

Preface

SolidWorks 2005

SolidWorks, developed by SolidWorks Corporation, is one of the world's fastest growing solid modeling softwares. It is a parametric feature-based solid modeling tool and it not only unites the three-dimensional (3D) parametric features with two-dimensional (2D) tools, but also addresses every design-through-manufacturing process. SolidWorks is also known as standard in 3D. The latest in the family of SolidWorks, SolidWorks 2005 includes a number of customer requested enhancements. This implies that SolidWorks is completely tailored to the customer's needs. Based mainly on the users feedback, this solid modeling tool is remarkably user-friendly and it allows you to be productive from day one.

The 2D drawing views of the components are easily generated in the **Drawing** mode. The drawing views that can be generated include detailed, orthographic, isometric, auxiliary, section, and so on. You can use any predefined standard drawing document to generate the drawing views. Besides displaying the model dimensions in the drawing views or adding reference dimensions and other annotations, you can also add the parametric Bill Of Materials (BOM) and balloons in the drawing view. If a component in the assembly is replaced, removed, or a new component is assembled, the modification will be automatically reflected in the BOM placed in the drawing document. The bidirectional associative nature of this software ensures that any modification made in the model is automatically reflected in the drawing views and any modification made in the dimensions in drawing views automatically updates the model.

SolidWorks for Designers Release 2005 is a text book written with an intent of helping the users who are interested in the 3D design. This book is written with the tutorial point of view with learn-by-doing as the theme. Real world mechanical engineering industry examples and tutorials have been used to ensure that the user relates his knowledge of the book with the actual mechanical industry designs. The main features of the book are as follows.

- **Tutorial Approach**

  The author has adopted the tutorial point-of-view with learn-by-doing as the theme throughout the book. This approach will guide the users through the process of creating the model in the tutorial.

- **Real-world Mechanical Engineering Projects as Tutorials**
  The author has used the real-world mechanical engineering projects as tutorials in this book so that the reader can correlate the tutorials in this book with the real-time models in the mechanical engineering industry.

- **Coverage of major SolidWorks Modes**
  All major modes of SolidWorks are covered in this book, which include the **Part** mode, the **Assembly** mode, and the **Drawing** mode.

- **Tips and Notes**
  Additional information related to the topics is provided to the users in the form of tips and notes.

- **Learning Objectives**
  The first page of every chapter provides in brief the topics that will be covered in that chapter. This will help the users to easily refer to a topic.

- **Tools Section**
  Every chapter begins with the tools section that provides the detailed explanation of the SolidWorks tools and commands.

- **Self-Evaluation Test, Review Questions, and Exercises**
  Every chapter ends with a Self-Evaluation Test so that the users can assess their knowledge of the chapter. The author has given the answers of the Self-Evaluation Tests so that the users can compare their answers with the correct answers. The Review Questions and Exercises are also given at the end of each chapter and can be used by the Instructors as test questions and exercises in the classroom.

- **Heavily Illustrated Text**
  The text in this book is heavily illustrated with the help of around 700 line diagrams and 800 photos that support the tools sections and tutorials.

Features of the Text

Learning Objectives

Each chapter begins with Learning Objectives, which are the points that in brief explain that tools and options that the reader will be able to work with after completing the chapter.

Tutorials

These are real-world mechanical, automobile engineering projects that are explained in a step-by-step method.

Self-Evaluation Test

These are the questions given at the end of each chapter. Answer to these questions are also given at the end of the chapter so that users can verify their answers with the correct answers.

Review Questions

These are the questions that are also given at the end of the chapter. However, the answer to these questions are not given in the book and so the instructor can use these questions as the test questions. The teaching faculty will get answers to these questions in the Instructor's Guide that is provided by the author.

Note

The author have provided a lot of additional information to the users about the topic being discussed in the form of notes.

Tip

Special information and techniques are provided in the form of tip that allow the users to increase their efficiency.

New

This indicates the new commands or tools introduced in SolidWorks 2005.

Enhanced

This indicates the existing commands and tools that are enhanced in SolidWorks 2005.

Introduction

SolidWorks 2005

Welcome to the world of Computer Aided Designing (CAD) with SolidWorks. If you are a new user of this software package, you will be joining hands with thousands of users of this parametric feature-based, and one of the most user-friendly software package. If you are familiar with the previous releases of this software, you will be able to upgrade your designing skills with the tremendous improvement in this latest release.

SolidWorks, developed by SolidWorks Corporation, USA, is a feature-based, parametric solid-modeling mechanical design and automation software. SolidWorks, also known as the standard in 3D, is the first CAD package to use the Microsoft Windows graphic user interface. The use of the drag-drop (DD) functionality of Windows makes this CAD package extremely easy to learn. The Windows graphic user interface makes it possible for the mechanical design engineers to innovate their ideas and implement them in the form of virtual prototypes or solid model, large assemblies, subassemblies, and detailing and drafting.

SolidWorks is one of the products of SolidWorks Corporation, which is a part of Dassault Systemes. SolidWorks also works as the platform software for a number of softwares. This implies that you can also use other compatible softwares within the SolidWorks window. There are a number of softwares provided by the SolidWorks Corporation, which can be used as Add-Ins with SolidWorks. Some of the softwares that can be used on SolidWorks's working platform are listed below.

1. SolidWorks Animator
2. PhotoWorks
3. FeatureWorks
4. COSMOS/Works
5. COSMOS/Motion
6. COSMOS/Flow
7. eDrawings
8. SolidWorks Piping
9. CAMWorks
10. Toolbox
11. Mold Base

As mentioned earlier, SolidWorks is a parametric, feature-based and easy to use mechanical design automation software. It enables you to convert the basic 2D sketch into a solid model using simple but highly effective modeling tools. SolidWorks does not restrict you to 3D solid output, but it extends to the bidirectional associative generative drafting. It also enables you to create the virtual prototype of a sheet metal component and its flat pattern, which helps you in complete process planning for designing and creating a press tool. SolidWorks helps you to extract of core and cavity of a model that has to be molded or casted. With SolidWorks, you can also create complex parametric shapes in the form of surfaces. Some of the important modes of SolidWorks are discussed next.

Part Mode

The **Part** mode of SolidWorks is a feature-based parametric environment in which you can create solid models. You are provided with three default planes named as the **Front Plane**, **Top Plane**, and the **Right Plane**. First, you need to select the sketching plane to create the sketch for the base feature. When you select the sketching plane, you enter the sketching environment. The sketches for the model are drawn in the sketching environment using easy to use tools. After drawing the sketches, you can dimension them and apply the required relations in the same sketching environment. The design intent is captured easily by adding relations and equations, and using the design table in the design. You are provided with the standard hole library known as the hole wizard in the **Part** mode. You can create simple holes, tapped holes, counterbore holes, countersink holes, and so on using this wizard. The holes can be of any standard such as ISO, ANSI, JIS, and so on. You can also create complicated surfaces using the surface modeling available within the **Part** mode. Annotations such as weld symbols, geometric tolerance, datum references, and surface finish symbols can be added to the model within the **Part** mode. The standard features that are used frequently can be saved as library features and can be retrieved when needed. The palette feature library is also provided in SolidWorks, which contains a number of standard mechanical parts and features. You can also create the sheet metal components in the **Part** mode of SolidWorks using the related tools. Besides, you can also analyze the part model for various stresses applied to the model in the real physical conditions. COSMOSXpress is a very easy and user-friendly tool for stress analysis in SolidWorks. It helps you to reduce the cost and time in testing your design in real physical testing conditions (destructive tests). You can analyze the component during modeling in the SolidWorks windows. In addition, you can work with the weld modeling within the **Part** mode of SolidWorks by creating steel structures and adding weld beads. All standard weld types and welding conditions are available for your reference. You can extract the core and cavity in the **Part** mode by using the mould design tools.

Assembly Mode

In the **Assembly** mode, you can assemble components of the assembly with the help of the required tools. There are two types of methods of assembling the components:

1. Bottom-up assembly
2. Top-down assembly

In the bottom-up assembly approach, the assembly is created by assembling the previously created components and maintaining their design intent. In the top-down approach, the components are created in the assembly mode. You may begin with some ready-made parts and then create other components in the context of the assembly. You can refer to the features of some components of the assembly to drive the dimensions of the other components. The **SmartMates** option allows you to assemble all the components using a single button. While assembling the components of an assembly, you can also animate it by dragging. Besides you can also check the working of your assembly. Collision detection is one of the major features of the SolidWorks assembly. Using this feature, you can rotate and move the components and detect the collision between the components that you have assembled. You can see the realistic motion of the assembly by using physical dynamics. Physical simulation is used to simulate the assembly with the effects of motors, springs, and gravity on the assemblies.

Drawing Mode

The **Drawing** mode is used for the documentation of the parts or the assemblies created earlier, in the form of drawing views and the detailings in the drawing views. There are two types of drafting done in SolidWorks:

1. Generative drafting
2. Interactive drafting

Generative drafting is the process of generating the drawing views of the part or assembly created earlier. The parametric dimensions and annotations added to the component in the **Part** mode can be generated in the drawing views. Generative drafting is bidirectional associative in nature. Automatic BOMs and balloons can be added while generating the drawing views of an assembly.

In interactive drafting you have to create the drawing views by sketching them using the normal sketching tools, after which you have to add dimensions to them.

SYSTEM REQUIREMENTS

The system requirements to ensure the smooth running of SolidWorks on your system are as follows:

- Microsoft Windows XP Professional or Windows 2000 Professional.
- Intel Pentium, Xeon, or AMD Athlon-based computer.
- 256 MB RAM minimum (512 MB to 1 GB or higher recommended).
- Graphic card OpenGL (recommended).
- Mouse or any other compatible pointing device.
- Internet Explorer version 5.5 or higher recommended.

GETTING STARTED WITH SolidWorks

Install SolidWorks on your system and then start it by choosing the **Start** button available on the lower left corner of the screen. Choose **All Programs** to display the **Program** menu. Choose **SolidWorks 2005** to display the cascade menu and then choose **SolidWorks 2005**, as shown in Figure I-1.

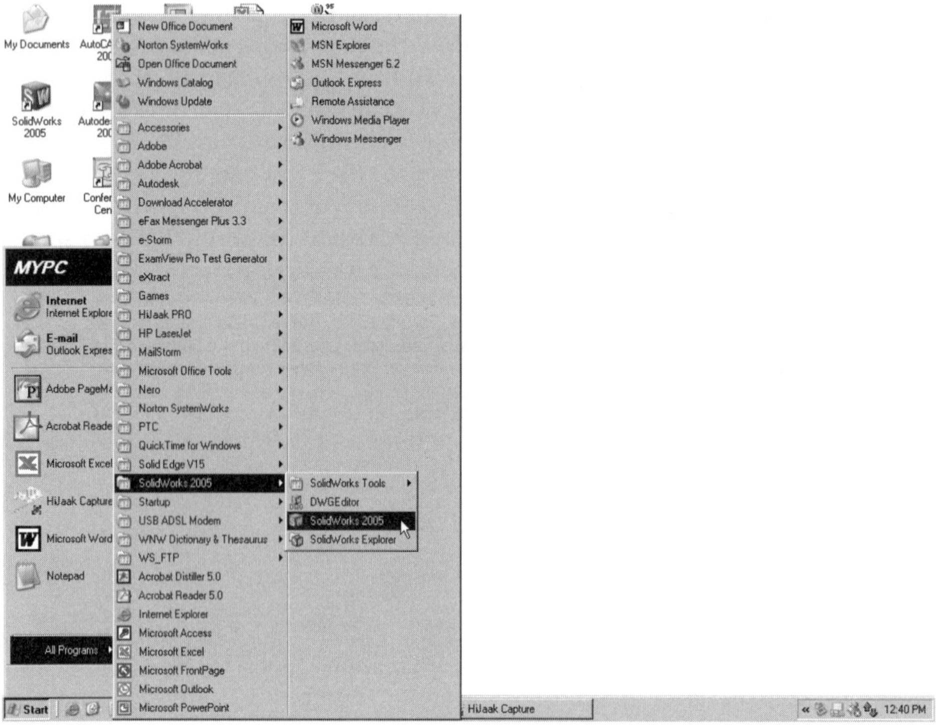

Figure I-1 *Starting SolidWorks using the taskbar shortcut*

The system will now prepare to start SolidWorks. The SolidWorks window will be displayed on the screen. On opening SolidWorks for the first time, the **SolidWorks License Agreement** dialog box will be displayed, as shown in Figure I-2; choose the **Accept** button from this dialog box.

Tip. *You can also start **SolidWorks 2005** by double-clicking the **SolidWorks 2005** shortcut icon from the desktop of your computer. You need to create the shortcut icon of **SolidWorks 2005 if it** is not created by default. The shortcut icon of **SolidWorks 2005** is created by choosing the **Start** button available on the lower left corner of the screen. Choose **Programs** to display the **Program** menu and from it, choose **SolidWorks 2005** to display the cascade menu. Move the cursor on **SolidWorks 2005** and right-click to display the shortcut menu. Move the cursor to the **Send To** option; the cascade menu is displayed. Choose the **Desktop (create shortcut)** option from the cascade menu. The **SolidWorks 2005** shortcut icon will be placed on the desktop of your computer.*

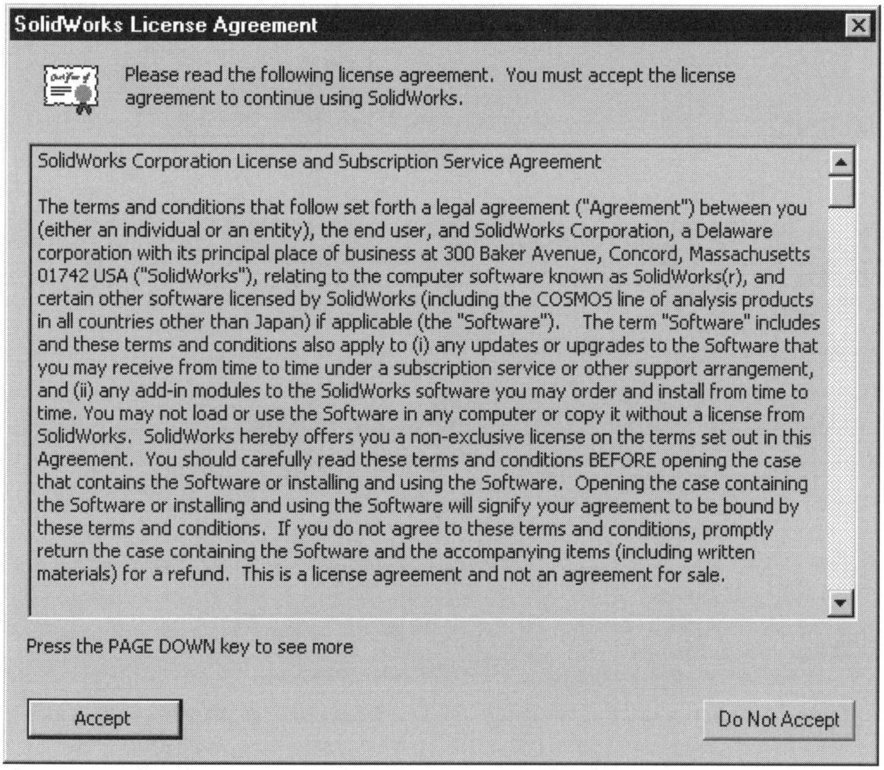

***Figure I-2 SolidWorks License Agreement** dialog box*

The SolidWorks 2005 window is opened and the **SolidWorks Resource Task Pane** is displayed on the right, as shown in Figure I-3. This window can be used to open a new file or an existing file, and use the on-line tutorials. This window can also be used to visit the Web site of the SolidWorks partners.

Choose the **New Document** button from the **Getting Started** group in the **SolidWorks Resources Task Pane** to open a new file. Alternatively, you can also choose the **New** button from the **Standard** toolbar. The **New SolidWorks Document** dialog box is displayed, as shown in Figure I-4.

Tip. *In the earlier releases of SolidWorks, the **Tip of the Day** dialog box was displayed by default when you started a new SolidWorks session. This release onward, the tip of the day will be displayed at the bottom of the **Task Pane**. You can choose **Next Tip** to view additional tips. These tips are extremely useful in helping you efficiently use SolidWorks. It is recommended that you view at least 2 to 3 tips every time you start a new session of SolidWorks 2005.*

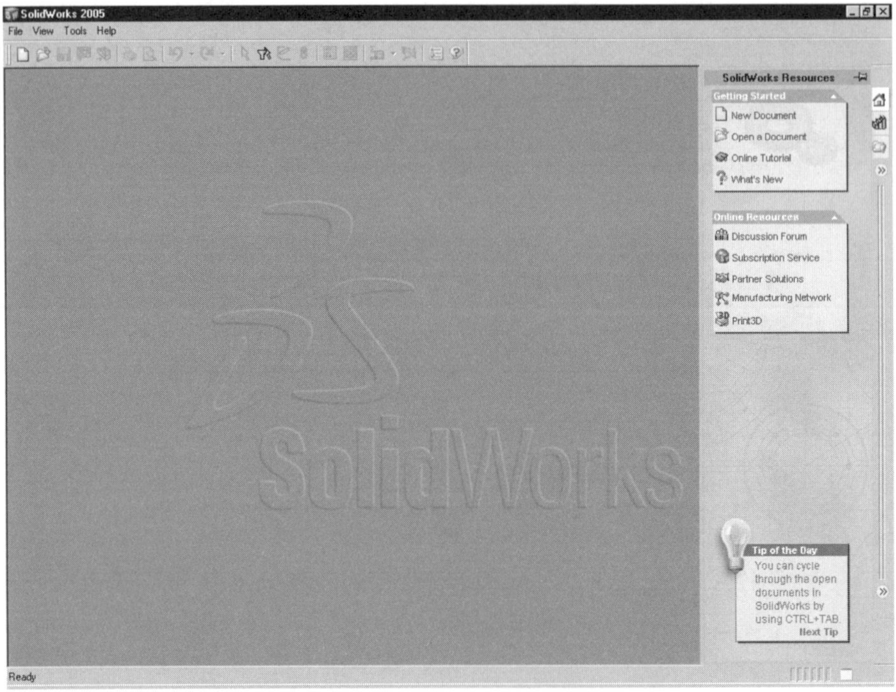

Figure I-3 SolidWorks 2005 window and the ***SolidWorks Resource Task Pane***

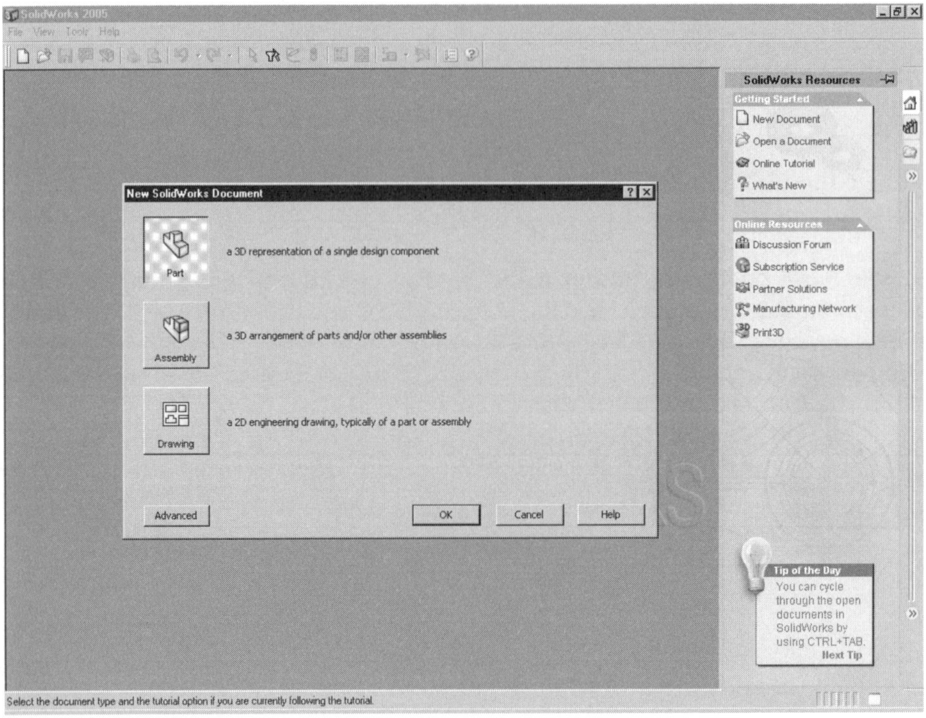

Figure I-4 *The* ***New SolidWorks Document*** *dialog box*

Choose the **Part** button for creating a part model and then choose **OK** from the **New SolidWorks Document** dialog box to enter the **Part** mode of SolidWorks. Note that the **Task Pane** is automatically closed if you start a new file using the **Standard** toolbar. The initial screen display when you start a new part file of SolidWorks, using the **New** button in the **Standard** toolbar, is shown in Figure I-5.

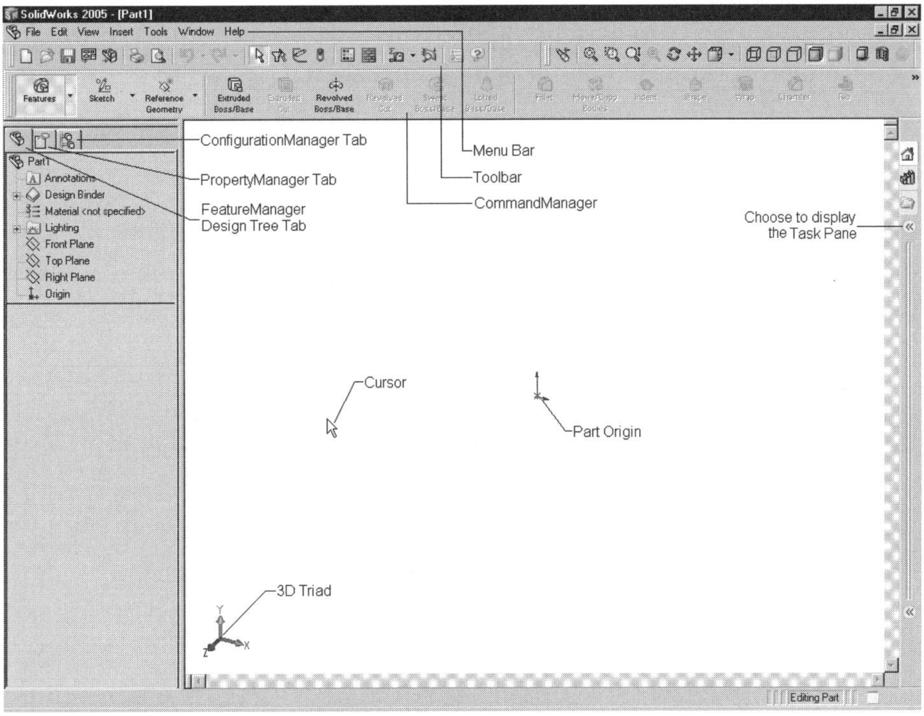

***Figure I-5** Components of new part document*

It is evident from the screen that SolidWorks is a very user-friendly solid modeling tool. Apart from the default CommandManager shown in Figure I-5, you can also invoke a number of other CommandManagers by moving the cursor on the CommandManager and right-clicking to display the shortcut menu. Choose the **Customize CommandManager** option from the shortcut menu; a cascading menu is displayed. Using this cascading menu, you can select the CommandManager that you need to display.

COMMANDMANAGER

There are three options to invoke a tool in SolidWorks: the CommandManager, the use of the menu bars available on the top of the screen, and the shortcut menu. The menu bar and the shortcut menu are discussed later. While working with CommandManager you will realize that this is the most convenient method to invoke a tool. Different types of CommandManagers are used for different design environments. The CommandManagers available in various modes of SolidWorks are discussed next.

Part Mode CommandManagers

A number of CommandManagers can be invoked in the **Part** mode. The CommandManagers that are extensively used during the designing process in this environment are described next.

Sketch CommandManager

This CommandManager is used to enter and exit the 2D sketching environment and the 3D sketching environment. The tools available in the CommandManager are used to draw sketches for the features. This CommandManager is also used to add relations and smart dimensions to the sketched entities. The **Sketch CommandManager** is shown in Figure I-6.

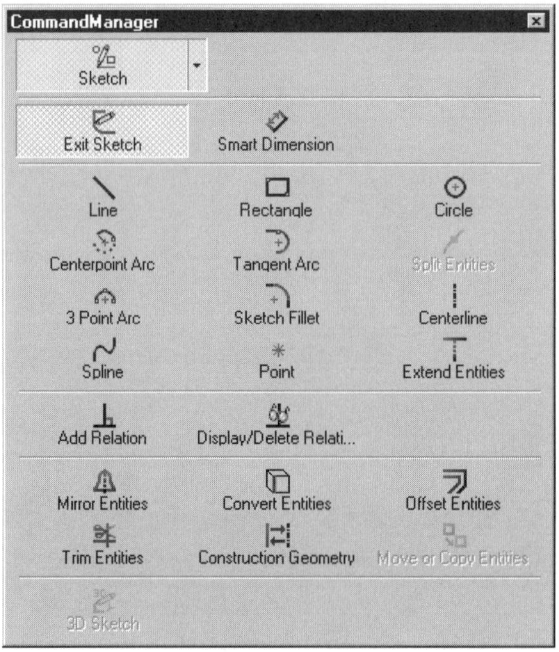

Figure I-6 Sketch CommandManager

Dimension/Relations CommandManager

This CommandManager is used to add various dimensions such as horizontal dimension, vertical dimension, ordinate dimension, and so on. The **Dimension/Relation CommandManager** is displayed in Figure I-7.

Features CommandManager

This is one of the most important CommandManagers provided in the **Part** mode. Once the sketch is completed, you need to convert the sketch into a feature using the modeling tools. This CommandManager provides all the modeling options that are to be used for feature-based solid modeling. The **Features CommandManager** is shown in Figure I-8.

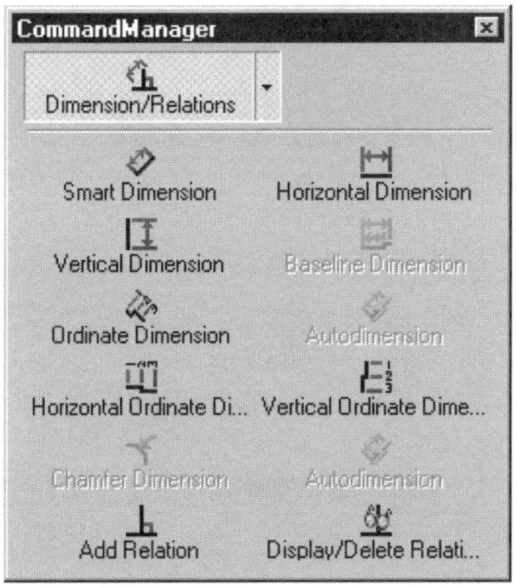

Figure I-7 Dimension/Relations CommandManager

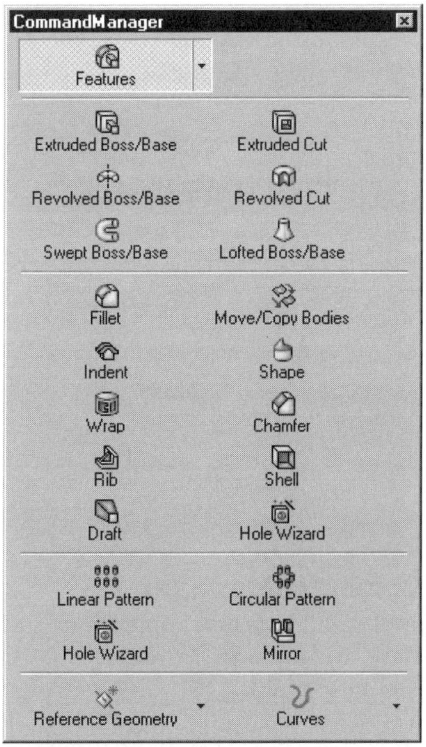

Figure I-8 Features CommandManager

Reference Geometry CommandManager

This CommandManager is used to create the reference geometry features like planes, axes, coordinate systems, points, and mate references. The **Reference Geometry CommandManager**, along with all its options, is shown in Figure I-9.

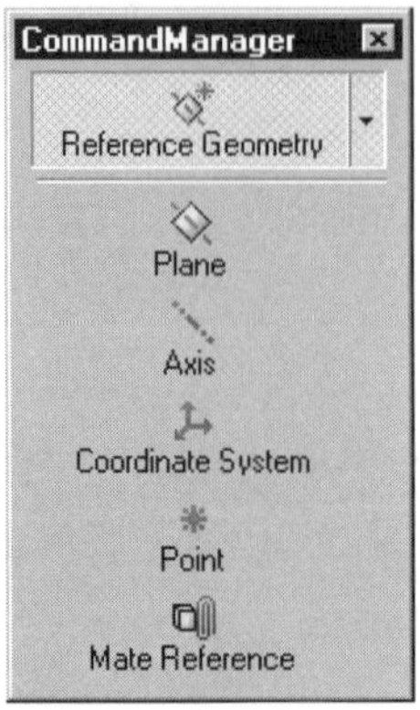

Figure I-9 Reference Geometry CommandManager

Sheet Metal CommandManager

This CommandManager provides you the tools that are used in creating the sheet metal parts. As mentioned earlier, in SolidWorks, you can create sheet metal parts while working in the **Part** mode. This is done with the help of the **Sheet Metal CommandManager** shown in Figure I-10.

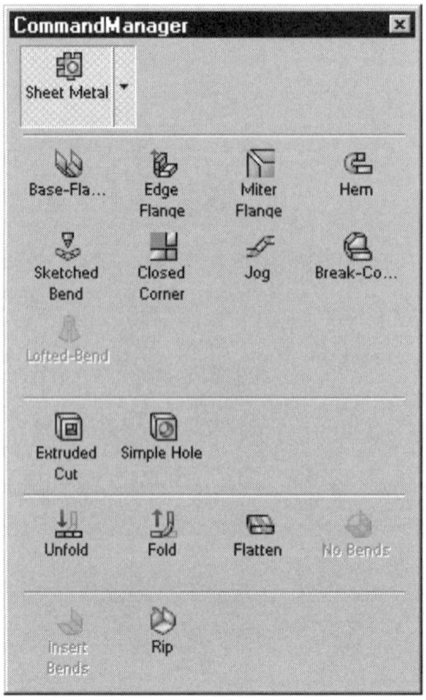

Figure I-10 Sheet Metal CommandManager

Surfaces CommandManager

The options available in this CommandManager are used to create complicated surface features. These surface features are also converted into the solid features. The **Surfaces CommandManager** is shown in Figure I-11.

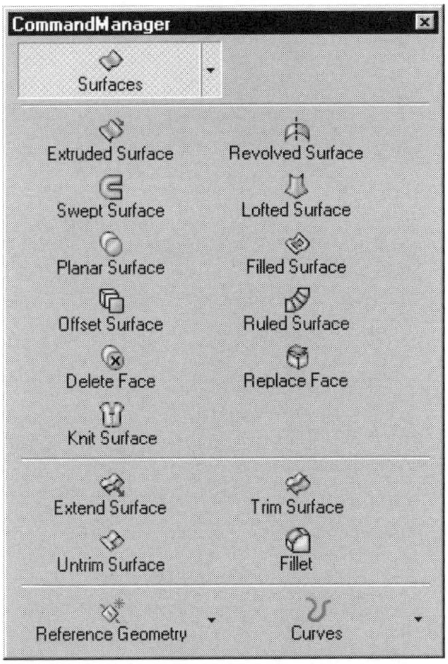

Figure I-11 Surfaces CommandManager

Curves CommandManager

The options available in this CommandManager are used to create the curves, such as the projected curve, split line, curve through points, helical curve, and so on. The **Curves CommandManager** is shown in Figure I-12.

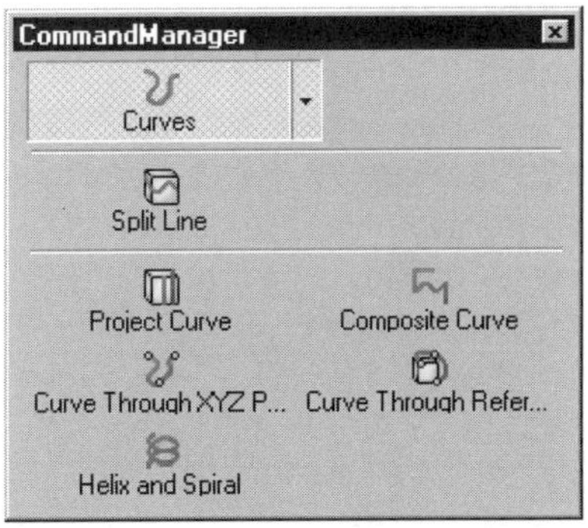

Figure I-12 Curves CommandManager

Molds CommandManager

The options in this CommandManager are used to design the mold and extract its core and cavity. The **Molds CommandManager** is shown in Figure I-13.

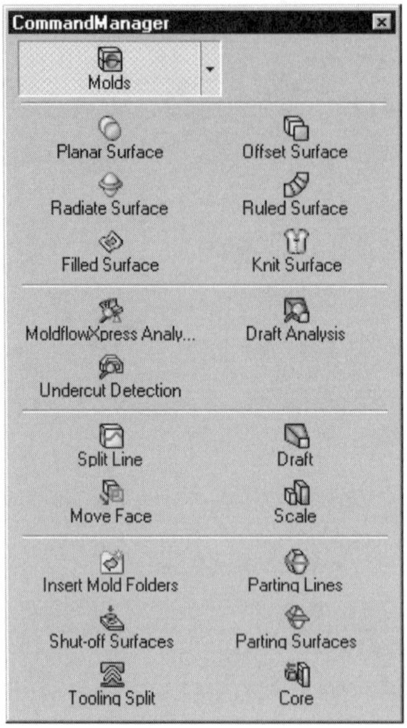

Figure I-13 Molds CommandManager

View CommandManager

This CommandManager is used to zoom, pan, and orient the model. The shading mode of the solid model can also be changed using this toolbar. You can also create the section view of a model using the option available in this CommandManager. The **View CommandManager**, along with all its buttons, is shown in Figure I-14.

Standard Views CommandManager

This CommandManager is used to orient the model using the standard view options. The **Standard Views CommandManager** is shown in Figure I-15.

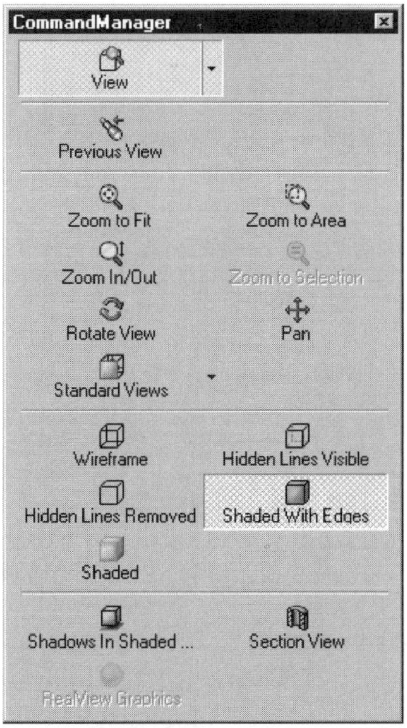

Figure I-14 View CommandManager

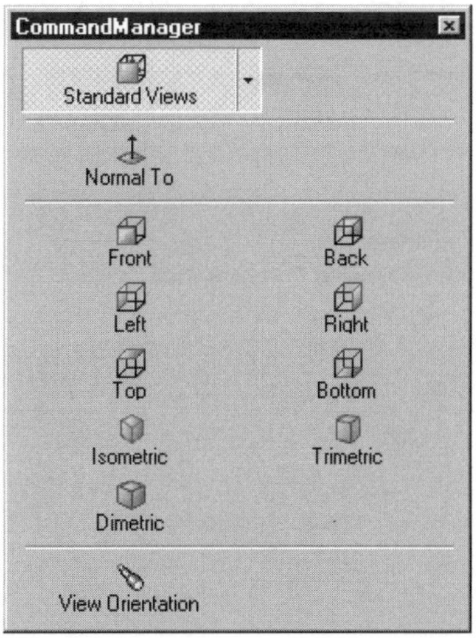

Figure I-15 Standard Views CommandManager

Tools CommandManager

This CommandManager is used to measure the distance between two entities, to add the equations in the design, to calculate the mass properties of a solid model, and so on. The **Tools CommandManager** is shown in Figure I-16.

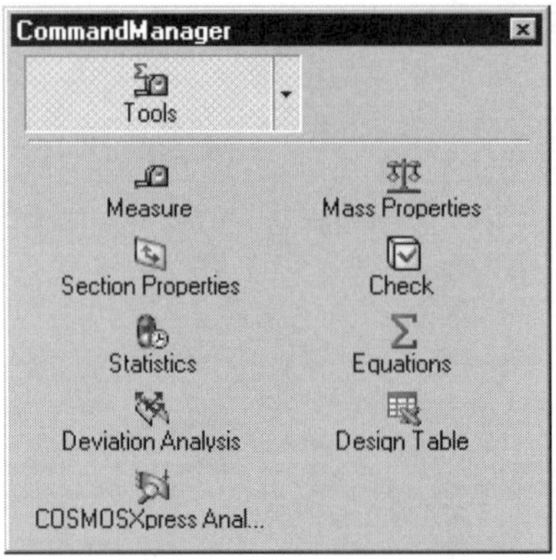

Figure I-16 Tools CommandManager

Assembly Mode CommandManagers

The CommandManager in the **Assembly** mode is used to assemble the components, create an explode line sketch, and simulate the assembly. This CommandManager is discussed next.

Assemblies CommandManager

This CommandManager is used to apply various types of mates to the assembly. Mates are the constraints that can be applied to the components to restrict their degree of freedom. You can also move and rotate the component in the assembly, change the hidden and the suppression state of the assembly and individual component, edit the component of assembly, and so on. The **Assemblies CommandManager** is shown in Figure I-17.

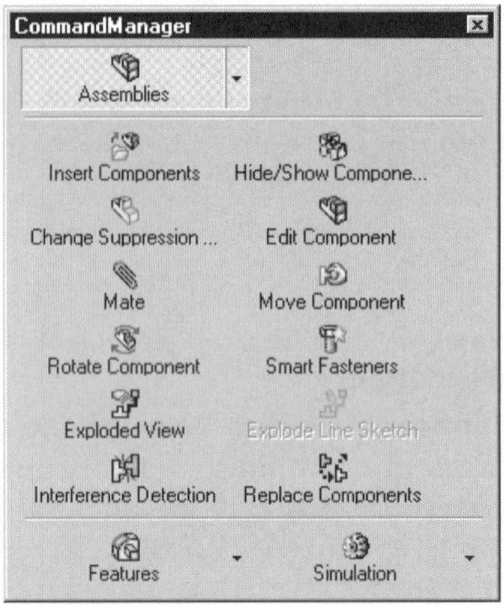

Figure I-17 Assemblies CommandManager

Drawing Mode CommandManagers

A number of CommandManagers can be invoked in the **Drawing** mode. The CommandManagers that are extensively used during the designing process in this mode are described next.

Drawings CommandManager

This CommandManager is used to generate the drawing views from an existing model or an assembly. The views that can be generated using this CommandManager are model view, three standard views, projected view, section view, aligned section view, detail view, crop view, relative view, auxiliary view, and so on. The **Drawings CommandManager** is shown in Figure I-18.

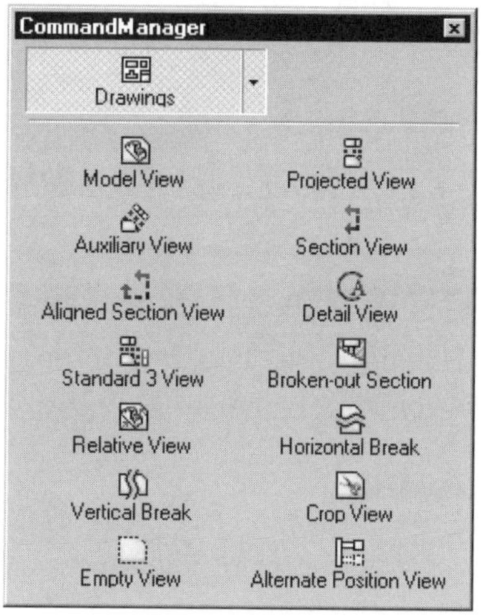

Figure I-18 Drawings CommandManager

Tip. *You can also create 2D drawings and drawing views using the normal sketching tools available in the **Sketch Tools** toolbar. The **Sketch Tools** toolbar is also available in the **Drawing** mode.*

Annotations CommandManager

The **Annotations CommandManager** is used to generate the model items and to add the notes, balloons, geometric tolerance, surface finish symbols, and so on to the drawing views. The **Annotations CommandManager** is shown in Figure I-19.

DIMENSIONING STANDARD AND UNITS

While installing SolidWorks on your system, you can specify the units and the dimensioning standard for dimensioning the model. You are provided with many dimensioning standards such as ANSI, ISO, DIN, JIS, BSI, and GOST. You are also provided with various units such as millimeters, centimeters, inches, and so on. This book follows millimeters as the units and ISO as the dimensioning standard. Therefore, it is recommended that you install SolidWorks with ISO as the dimensioning standard and millimeters as units.

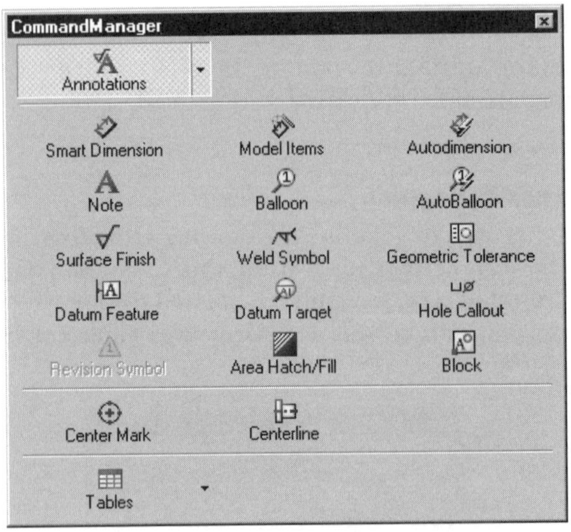

Figure I-19 Annotations CommandManager

IMPORTANT TERMS AND THEIR DEFINITIONS

Before you proceed further in SolidWorks, it is very important for you to understand the following terms. These terms have been widely used in this book.

Feature-based Modeling

Feature is defined as the smallest building block that can be modified individually. In SolidWorks, the solid models are created by integrating a number of these building blocks. A model created in SolidWorks is a combination of a number of individual features with each feature related to the other, directly or indirectly. These features understand their fit and function properly, and therefore can be modified any time during the design process. If the proper design intent is maintained while creating the model, these features automatically adjust their values to any change in their surrounding. This provides greater flexibility to the design.

Parametric Modeling

The parametric nature of a software package is defined as its ability to use the standard properties or parameters in defining the shape and size of a geometry. The main function of this property is to derive the selected geometry to a new size or shape without considering its original size or shape. You can change or modify the shape and size of any feature at any stage of the design process. This property makes the designing process very easy. For example, consider the design of the body of a pipe housing shown in Figure 1-20.

In order to change the design by modifying the diameter of the holes and the number of holes on the front, top, and the bottom face, you need to simply select the feature and change the diameter and the number of instances in the pattern. The modified design is shown in Figure 1-21.

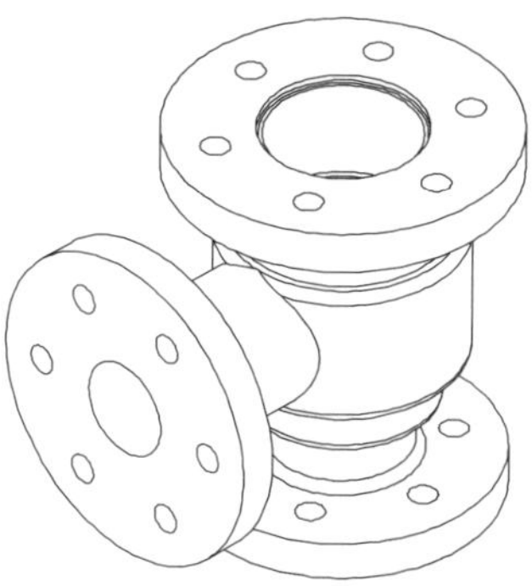

Figure I-20 Body of pipe housing

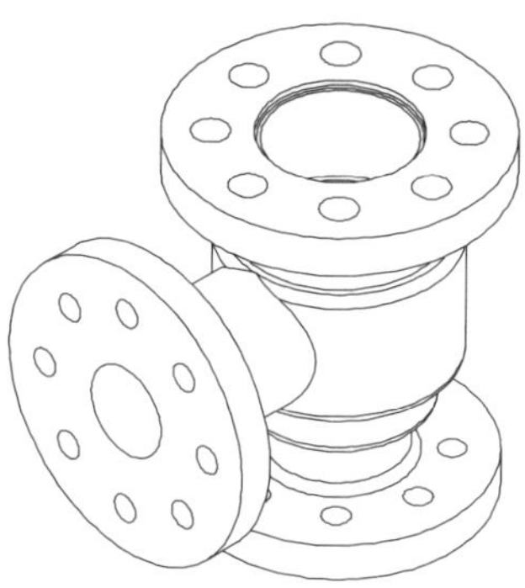

Figure I-21 Design after modifications

Bidirectional Associativity

As mentioned earlier, SolidWorks has different modes such as the **Part** mode, **Assembly** mode, and **Drawing** mode. There exists a bidirectional associativity between all these modes, which ensures that any modification made in the model in any one of the modes of SolidWorks is automatically reflected in the other mode immediately. For example, if you modify the dimension of a part in the **Part** mode, the change will be reflected in the **Assembly** and the **Drawing** modes. Similarly if you modify the dimensions of a part in the drawing views generated in the **Drawing** mode, the changes will be reflected in the **Part** and **Assembly** modes. Consider the drawing views shown in Figure I-20. These are the drawing views of the body of the pipe housing shown in Figure I-22. Now, when you modify the model of the body of pipe housing in the **Part** mode, the changes will be reflected in the **Drawing** mode automatically. Figure I-23 shows the drawing views of the pipe housing after increasing the diameter and the number of holes.

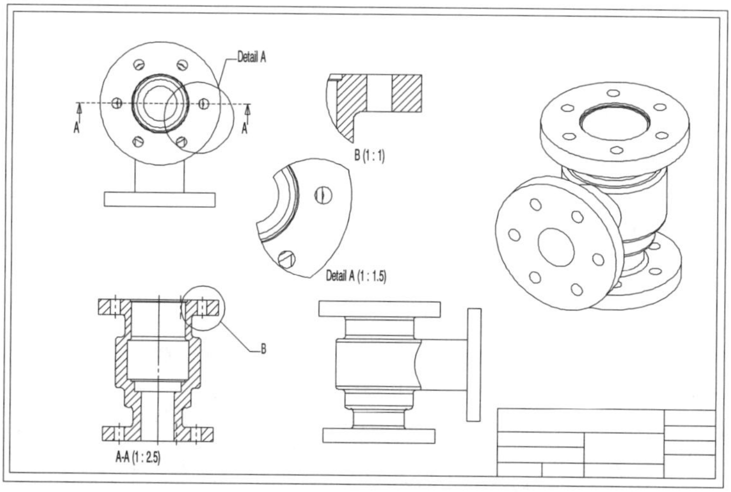

Figure I-22 Drawing views of the body part before making the modifications

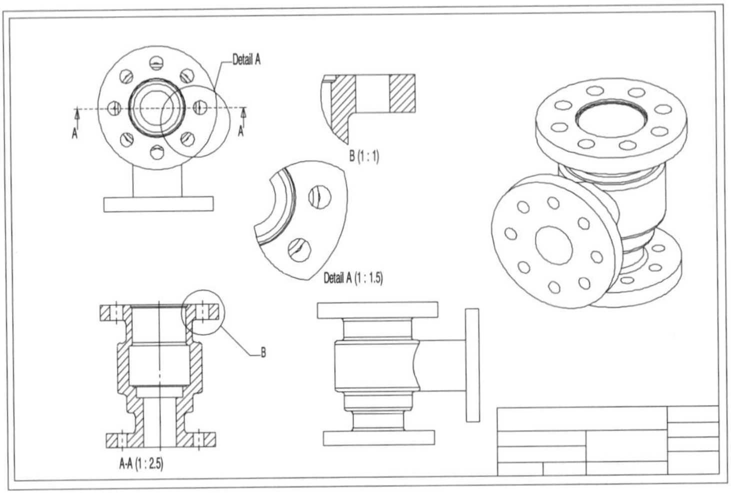

Figure I-23 Drawing views after modifications

Windows functionality

SolidWorks is the first Windows-based 3D CAD package. It uses the graphical user interface of Windows, drag and drop functionality, and the copy paste functionality of Windows. For example, consider a case where you had created a hole feature on the front planar surface of a model. Now, to create another hole feature on the top planar surface of the same model, select the hole feature and press CTRL+C (copy) on the keyboard. Next, select the top planar surface of the base feature and press CTRL+V (paste). You can also drag and drop the standard features from the feature pallet library on the face of the model on which the feature has to be added. Therefore, SolidWorks is easiest to learn 3D CAD package for mechanical engineers.

Geometric Relations

Geometric relations are the logical operations that are performed to add a relationship (like tangent or perpendicular) between the sketched entities, planes, axes, edges, or vertices. When adding the relations, one entity can be a sketched entity and the other entity can be a sketched entity, or an edge, face, vertex, origin, plane, and so on. There are two methods to create the geometric relations:

1. Automatic Relations
2. Add Relations

Automatic Relations

The sketching environment of SolidWorks has been provided with the facility of autorelations. This facility ensures that the geometric relations are applied to the sketch automatically while creating it. Automatic relations are also provided in the **Drawing** mode while working with interactive drafting.

Add Relations

Add relations is used to add the geometric relations manually to the sketch. The 16 types of geometric relations that can be manually applied to the sketch are as follows:

Horizontal

This relation forces the selected line segment to become a horizontal line. You can also select two points and force them to be aligned horizontally.

Vertical

This relation forces the selected line segment to become a vertical line. You can also select two points and force them to be aligned vertically.

Collinear

This relation forces the two selected entities to be placed in the same line.

Coradial

This relation is applied to two selected arcs, circles, or an arc and a circle to force them to become equi-radius and at the same time share the same centerpoint.

Perpendicular

This relation is used to make the selected line segment perpendicular to another selected segment.

Parallel

This relation is used to make the selected line segment parallel to another selected segment.

Tangent

This relation is used to make the selected line segment, arc, spline, circle, or ellipse tangent to another arc, circle, spline, or ellipse.

Note

In case of splines, the relations are applied to their control points.

Concentric

This relation forces two selected arcs, circles, a point and an arc, a point and a circle, or an arc and a circle to share the same centerpoint.

Midpoint

This relation forces a selected point to be placed on the midpoint of a line.

Intersection

This relation forces a selected point to be placed at the intersection of two selected entities.

Coincident

This relation is used to make two points or a point and a line or a point and an arc coincident.

Equal

The equal relation forces the two selected lines to become equal in length. This relation is also used to force two arcs, or two circles, or an arc and a circle to have equal radii.

Symmetric

The symmetric relation is used to force the selected entities to become symmetrical about a selected centerline, so that they remain equidistant from the centerline.

Fix

This relation is used to fix the selected entity to a particular location with respect to the coordinate system of the current sketch. The endpoints of the fixed line, arc, spline, or an elliptical segment are free to move along the line.

Pierce

This relation forces the sketched point to be coincident to the selected axis, edge, or curve where it pierces the sketch plane. The sketched point in this relation can be the endpoint of the sketched entity.

Merge

This relation is used to merge two sketched points or endpoints.

Library Feature

Generally, in a mechanical design, some features are used frequently. In most of the other solid modeling tools, you need to create these features whenever you need them. However, SolidWorks allows you to save these feature in a library so that you can retrieve them whenever you want. This saves a lot of designing time and effort of a designer.

Feature Palette

The feature palette is also a very unique feature of SolidWorks. Using the feature palette library you can access some ready-to-use drop-down standard mechanical parts such as shafts, keyways, forming parts for a sheet metal, and so on.

Design Table

Design tables are used to create a multiinstance parametric component. For example, some components in your organization have the same geometry but different dimensions. Instead of creating each component of the same geometry with a different size, you can create one component and then, using the design table, create different instances by changing the dimension as per your requirement. Using the design table, you can access all those components with the same geometry having different sizes in a single part file.

Equations

Equations are the analytical and numerical formulae applied to the dimensions during the sketching of the feature sketch or after sketching the feature sketch. The equations can also be applied to the placed features.

Collision Detection

Collision detection is used to detect the interference and collision between the parts of an assembly when the assembly is in motion. While creating the assembly in SolidWorks, you can detect the collision between the different parts of the assembly by moving and rotating its components.

What's Wrong? functionality

While creating a feature of the model or after editing a feature created earlier, if the geometry of the feature is not compatible and the system is not able to construct that feature, then the **What's Wrong?** functionality is used to detect the possible error that occurs while creating the feature.

2D Command Line Emulator

The 2D command emulator is an Add-In of SolidWorks. You can activate this by choosing **Tools > Add-Ins** from the menu bar. On choosing this option, the **Add-Ins** dialog box is displayed. Select the **SolidWorks 2D Emulator** check box and choose **OK** from the **Add-Ins** dialog box. A command section is displayed at the bottom of the graphics area. This 2D Command line emulator is useful for invoking the commands by typing them. You can type the commands in the 2D Command line emulator.

COSMOSXpress

In SolidWorks 2005, you are provided with COSMOSXpress, which is an analysis tool in SolidWorks to execute the static or stress analysis. In COSMOSXpress you can only execute the linear static analysis. Using the linear static analysis, you can calculate the displacement, strain, and stresses applied on a component with the effect of material, various loading conditions, and restraint conditions applied on a model. A component fails when the stress applied to it reaches a certain permissible limit. The Static Nodal stress plot of the crane hook designed in SolidWorks and analyzed using COSMOSXpress is shown in Figure I-24.

Physical Dynamics

Physical Dynamics is used to observe the motion of the assembly. With this option selected, the component dragged in the assembly applies a force to the component that it touches. As a result, the other component is moved or rotated within it allowable degrees of freedom.

Physical Simulation

Using **Physical Simulation** you can simulate the assemblies created in the assembly environment of SolidWorks. You can assign and simulate the effect of different simulation elements such as linear, rotary motors, and gravity to the assemblies. After creating a simulating assembly, you can record and replay simulation.

Seed Feature and Seed Component

The original feature that is used as the parent feature to create any type of pattern or mirror feature is known as the seed feature. You can edit or modify only a seed feature. You cannot edit the instances of the pattern feature.

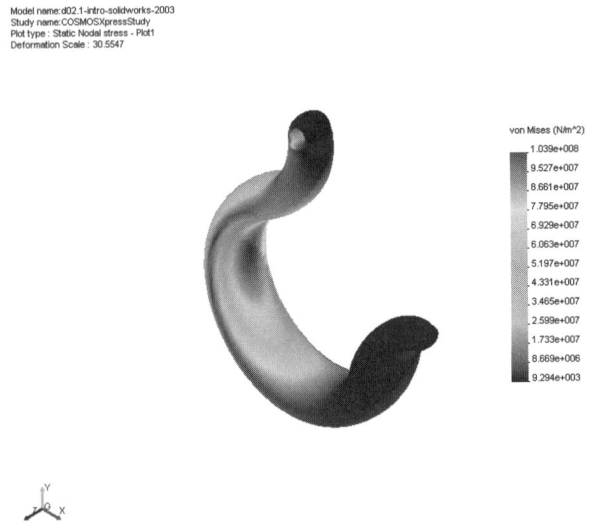

Figure I-24 *The crane hook analyzed using COSMOSXpress*

The original component that is used to create a derived pattern, local pattern, or a mirrored component is known as the seed component.

FeatureManager Design Tree

The **FeatureManager Design Tree** is one the most important component of SolidWorks screen. It contains the information about the default planes, material, lights, and all the features that are added to the model. When you add features to the model using various modeling tools, the same are also displayed in the **FeatureManager Design Tree**. You can easily select and edit the features using the **FeatureManager Design Tree**. When you invoke any tool to create a feature, the **FeatureManager Design Tree** is replaced by the respective **PropertyManager**. At this stage, the **FeatureManager Design Tree** is displayed in the drawing area.

Absorbed Features

Features that are directly involved in creating other features are known as absorbed features. For example, sketch of an extruded feature is an absorbed feature of the extrude feature.

Child Features

The features that are dependent on their parent and whose existence is not possible without their parent features are known as child features. For example, consider a cube with filleted edges. If you delete the cube, the fillet feature will also be deleted because its existence is not possible without its parent feature.

Dependent Features

Dependent features are those that depend on their parent feature but can still exist without the parent feature with some minor modifications. If the parent feature is deleted, then by specifying other references and modifying the feature, you can retain the same feature.

AUTO-BACKUP OPTION

SolidWorks also allows you to set the option to save the SolidWorks document automatically after a regular interval of time. While working on a design project, if the system crashes, you may loose the unsaved design data. If the auto-backup option is turned on, your data is saved automatically after regular intervals. To set this option on choose **Tools > Options** from the menu bar; the **System Options - General** dialog box is displayed. Select the **Backups** option from the display area provided on the left of this dialog box. Now, choose the **Save auto recover info every** check box. The spinner provided on the right of the check box is enabled. Using this spinner, you can set the number of changes after which the document would be saved automatically. By default, the backup files are saved in *X:\Documents and Settings\Admisitrator <name of your machine>\Local Settings\TempSWBackupDirectory* location (where *X* is the drive in which you have installed SolidWorks 2005 and the *Local Settings* folder is a hidden folder). You can also change the path of this location. To change this path, select the button provided on the right of the edit box; the **Browse For Folder** dialog box is displayed. Using this you can specify the location of the folder to save the backup files. If you need to save the backup files in the current folder, then select the **Save backup files in the same location as the original** check box. You can set the number of backup files that you need to save using the **Number of backup copies per document** spinner. After setting all the options, choose the **OK** button from the **System Options - Backup** dialog box.

SELECTING HIDDEN ENTITIES

Sometimes, while working on a model, you need to select an entity that is either hidden behind another entity or is not displayed in the current orientation of the view. SolidWorks allows you to select these entities using the **Select Other** option. For example, consider a case where you need to select the back face of a model, which is not displayed in the current orientation. In such case, move the cursor over the visible face such that the cursor is also in line with the back face of the model. Now, right-click and choose **Select Other** from the shortcut menu; the cursor changes to the select other cursor and the **Select Other** list box is displayed. This list box displays all entities that can be selected. The item on which you move the cursor in the list box is highlighted in the drawing area. You can select the hidden face using this box.

COLOR SCHEME

SolidWorks allows you to use various color schemes as the background color for the screen, color and display style of **FeatureManager Design Tree**, and for displaying the entities on the screen. Note that the color scheme used in this book is neither the default color scheme nor the predefined color scheme. To set this color scheme, choose **Tools > Options** from the menu bar; the **System Options - General** dialog box is displayed. Select the **Colors** option from the left of this dialog box. The option related to the color scheme is displayed in the dialog box and the name of the dialog box is changed to **System Options - Colors** dialog box. In the **System colors** area, the **Viewport Background** option available in the scroll down area is selected by default; choose the **Edit** button. Select white color from the **Color** dialog box and choose the **OK** button. Select the **Blue** option from the **PropertyManager Color** drop-down list available in the **System colors** area and choose the **None** option from the **PropertyManager Skin** drop-down list. Now, clear the **Match graphics area and**

FeatureManager backgrounds check box. After setting the color scheme, you need to save it so that next time if you need to set this color scheme, you do not need to configure all settings. You just need to select the name of the saved color scheme from the **Current Color Scheme** drop-down list. Choose the **Save As Scheme** button; the **Color Scheme Name** dialog box is displayed. Enter the name of the color scheme as **SolidWorks 2005** in the edit box provided in the **Color Scheme Name** dialog box and choose the **OK** button. Now, choose the **OK** button from the **System Options - Color** dialog box.

Note

The colors of the entities that are mentioned in the description or tutorials in this book are those that are displayed if the operating system is Windows XP. However, if you use Windows 2000, the colors of the entities that you get on the screen may be different from those mentioned in this book.

Chapter 1

Drawing Sketches for the Solid Models

Learning Objectives

After completing this chapter you will be able to:

- *Understand the requirement of the sketching environment.*
- *Open a new part document.*
- *Understand various terms used in the sketching environment.*
- *Work with various sketching tools.*
- *Use the drawing display tools.*
- *Delete the sketched entities.*

THE SKETCHING ENVIRONMENT

Most of the products designed using SolidWorks are a combination of sketched, placed, and derived features. The placed and derived features are created without drawing a sketch, but the sketched features require a sketch to be drawn first. Generally, the base feature of any design is a sketched feature and is created using a sketch. Therefore, while creating any design, the first and foremost point is to draw the sketch for the base feature. Once you have drawn the sketch, you can convert it into the base feature and then add the other sketched, placed, and derived features to complete the design. In this chapter, you will learn how to create the sketch for the base feature using various sketcher entities.

In general terms, a sketch is defined as the basic contour for the feature. For example, consider a spanner shown in Figure 1-1.

Figure 1-1 *Solid model of a spanner*

This spanner consists of a base feature, a cut feature, a mirror feature (cut on the back face), fillets, and an extruded text feature. The base feature of this spanner is shown in Figure 1-2. It is created using a single sketch drawn on the **Front Plane**, as shown in Figure 1-3. This sketch is drawn in the sketching environment using various sketching tools. Therefore, to draw the sketch of the base feature, you first need to invoke the sketching environment where you will draw the sketch.

Note

Once you are conversant with various options of SolidWorks, you can also use a derived feature or a derived part as the base feature.

The sketching environment of SolidWorks can be invoked any time in the **Part** mode or **Assembly** mode. You just have to specify that you want to draw the sketch of a feature and then select the plane on which you want to draw the sketch.

Note

You will learn how to invoke the sketching environment in SolidWorks 2005 later in this chapter.

Figure 1-2 Base feature of the spanner

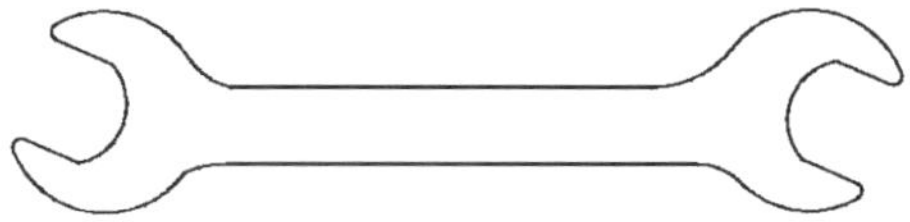

Figure 1-3 Sketch for the base feature of the spanner

STARTING A NEW SESSION OF SolidWorks 2005

To start a new session of SolidWorks 2005, choose **Start > Programs > SolidWorks 2005 > SolidWorks 2005** from the **Start** menu or double-click on the **SolidWorks 2005** icon placed on the desktop of your computer; the SolidWorks 2005 window will be displayed. If you are starting SolidWorks application for the first time after installing it, the **Welcome to SolidWorks** dialog box will also be displayed, as shown in Figure 1-4. This dialog box welcomes you to SolidWorks and helps you customize SolidWorks installation. The options available in this dialog box are discussed next.

Help customization Area

The options available in this area are used to define the type of help you need to invoke while working with SolidWorks. By default, the **I am new user. Show Quick Tips to help me to started** radio button is selected in this area. With this radio button selected, you are provided with quick tips that will guide you through the process of starting as a new user. Keep this radio button selected if you are a new user of SolidWorks.

If you are an existing user of SolidWorks, select the **I am new to this version of SolidWorks. Show me Interactive What's New help** radio button. This ensures that the **Interactive What's New** help is displayed while you are working with the tools that are enhanced or introduced in this release of SolidWorks.

If you do not want to invoke any dynamic help topic, select the **Do not show me any dynamic help** radio button.

Work flow customization Area

The options available in the **Work flow customization** area of this dialog box are used to customize the default visibility of toolbars and menu bars. This customize the toolbars and menu bars on the basis of the area of your expertise, such as product design, machine design, and mold design. Select the check box that belongs to the area of your expertise.

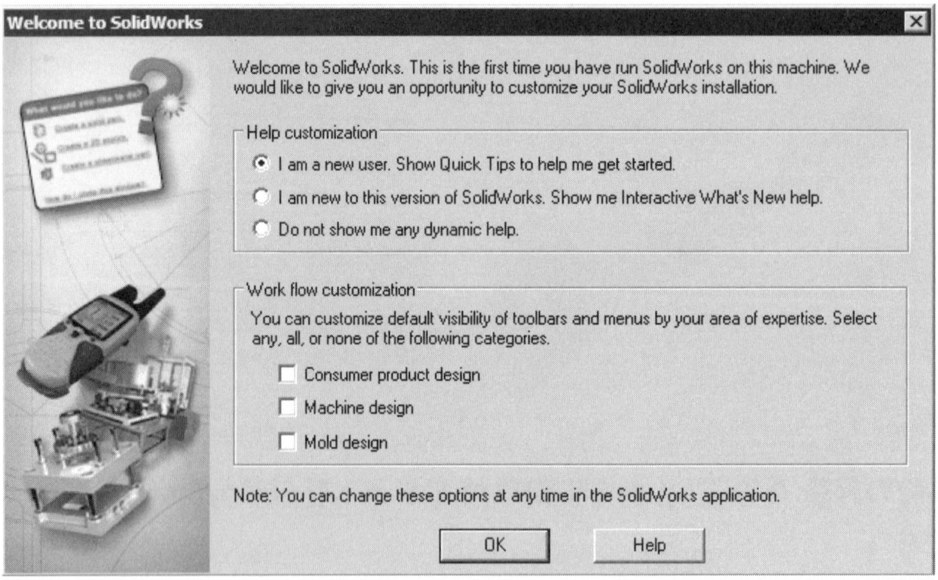

Figure 1-4 *The* ***Welcome to SolidWorks*** *dialog box*

This textbook follows the beginners point of view. Therefore, you need to keep the default radio button selected in the **Help customization** area and clear all check boxes in the **Work flow customization** area. Next choose the **OK** button from the **Welcome to SolidWorks** dialog box. The **Welcome to SolidWorks** dialog will disappear and you can view the **SolidWorks 2005** window as shown in Figure 1-5.

You will notice that **Task Pane** is displayed on the right of the SolidWorks 2005 window. This **Task Pane** is provided with various options that are used to start a new file, open an existing file, browse the related links of SolidWorks, and so on. The options available in the **Task Pane** are discussed next.

SolidWorks Resources Task Pane

By default, the **SolidWorks Resources Task Pane** is displayed when you start the SolidWorks session. Different options provided in the groups available in this **Task Pane** are discussed next.

Getting Started

The options available in this group are used to start a new document, open an existing document, and invoke the interactive help topics.

Online Resources

The options available in this group are used to invoke the discussion forum of SolidWorks, subscription services, solution partners, manufacturing network, and print 3D Web sites.

Tip of the Day

The **Tip of the Day** group provides you with a useful tip that will enhance your performance

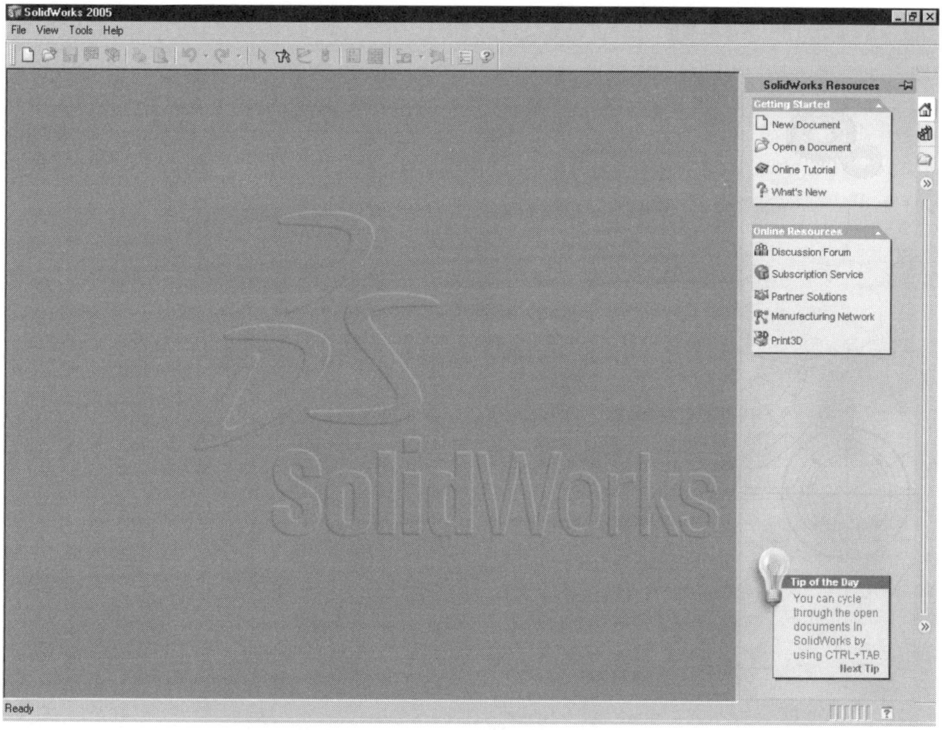

***Figure 1-5** The SolidWorks 2005 window*

of using SolidWorks. Choose the **Next Tip** provided on the lower right corner of the **Tip of the Day** dialog box to switch to the next tip.

Design Library

The **Design Library Task Pane** is invoked by choosing the **Design Library** tab from the **Task Pane**. This **Task Pane** is used to browse the default **Design Library** available in SolidWorks, browse the toolbox components, and access the **3D ContentCentral** Web site. To access the toolbox components, Toolbox Add-in needs to be installed on your computer. To access the **3D ContentCentral** Web site, your computer needs to be connected to the Internet.

File Explorer

The **File Explorer Task Pane** is used to explore the files and folders that are saved on the hard disk of your computer.

STARTING A NEW DOCUMENT IN SolidWorks 2005

To start a new document in SolidWorks 2005, choose the **New Document** option from the **Getting Started** group of the **SolidWorks Resources Task Pane**; the **New SolidWorks Document** dialog box will be displayed, as show in Figure 1-6. You can also invoke this dialog box by choosing the **New** button from the **Standard** toolbar. The options provided in this dialog box are discussed next.

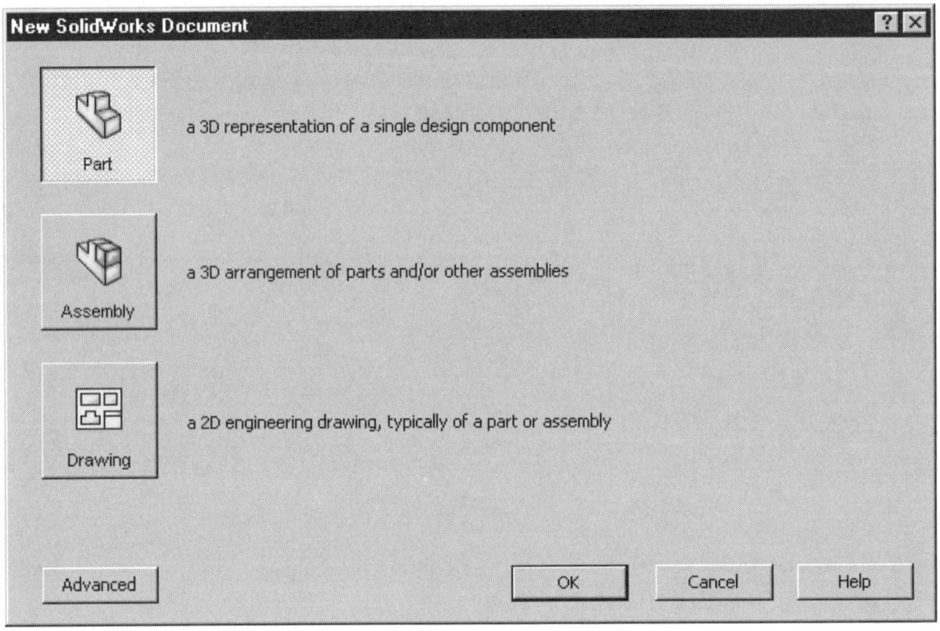

Figure 1-6 ***New SolidWorks Document*** *dialog box*

Part

The **Part** button is chosen by default in the **New SolidWorks Document** dialog box. Choose the **OK** button to start a new part document to create solid models or sheet metal components. When you start a new part document, you will enter the **Part** mode.

Assembly

Choose the **Assembly** button and then the **OK** button from the **New SolidWorks Document** dialog box to start a new assembly document. In an assembly document, you can assemble the components created in the part documents. You can also create components in the assembly document.

Drawing

Choose the **Drawing** button and then the **OK** button from the **New SolidWorks Document** dialog box to start a new drawing document. In a drawing document, you can generate or create the drawing views of the parts created in the part documents or the assemblies created in the assembly documents.

Note

*When you start a new session of SolidWorks, the **What would you like to do?** window is displayed, which assists you in working with SolidWorks. To close this window, choose the (X) button available close to the lower right corner of the SolidWorks window. The (X) button changes into the (?) button, which you can choose to display this window again.*

Tip. *If you invoke the **New SolidWorks Document** dialog box using the **Task Pane** and start a new document, the **Task Pane** will remain expanded even when new document is started. If you invoke the **New SolidWorks Document** dialog box using the toolbar or menu bar, the **Task Pane** will collapse and remain collapsed after starting the new document. You will learn more about expanding and collapsing the **Task Pane** later in this chapter.*

THE SKETCHING ENVIRONMENT

Whenever you start a new part document, by default you are in the part modeling environment. But, you need to start the design by first creating the sketch of the base feature in the sketching environment. You can invoke the sketching environment using the **Sketch** tool available in the **Standard** toolbar. You can also choose the **Sketch** button from the **CommandManager** (Figure 1-7) to invoke the **Sketch CommandManager**.

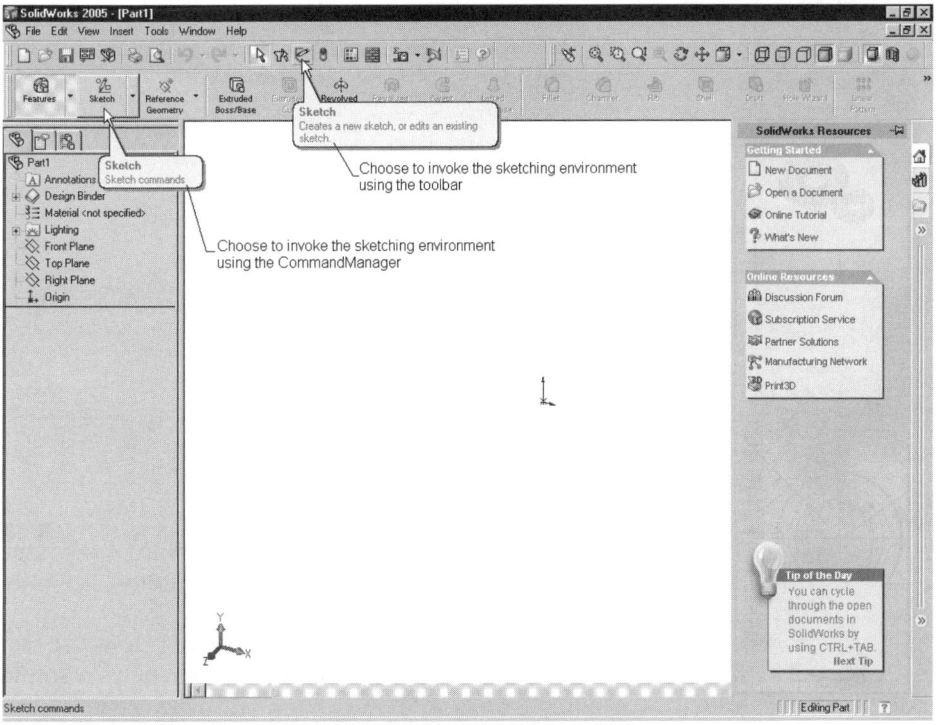

***Figure 1-7** Different methods of invoking the sketching environment in SolidWorks 2005*

When you choose the **Sketch** button from the **Standard** toolbar or choose any tool from the

Tip. *You will notice that **Task Pane** is still available on the right of the drawing area. This decreases the drawing area; to increase the drawing area you need to collapse the **Task Pane**. To collapse the **Task Pane**, click anywhere in the drawing area, choose any one of the arrow symbols » available on the right of the **Task Pane**, or click on the bar attached to these two arrows. You can also collapse the **Task Pane** by clicking once in the drawing area.*

*To expand the **Task Pane** at any stage of the design cycle, choose any one of the arrows provided on the right of the **Task Pane**.*

Sketch CommandManager, the **Edit Sketch PropertyManager** is displayed and you are prompted to select the plane on which the sketch will be created. Also, the three default planes available in SolidWorks 2005 (**Front Plane**, **Right Plane**, and **Top Plane**) are temporarily displayed on the screen, as shown in Figure 1-8.

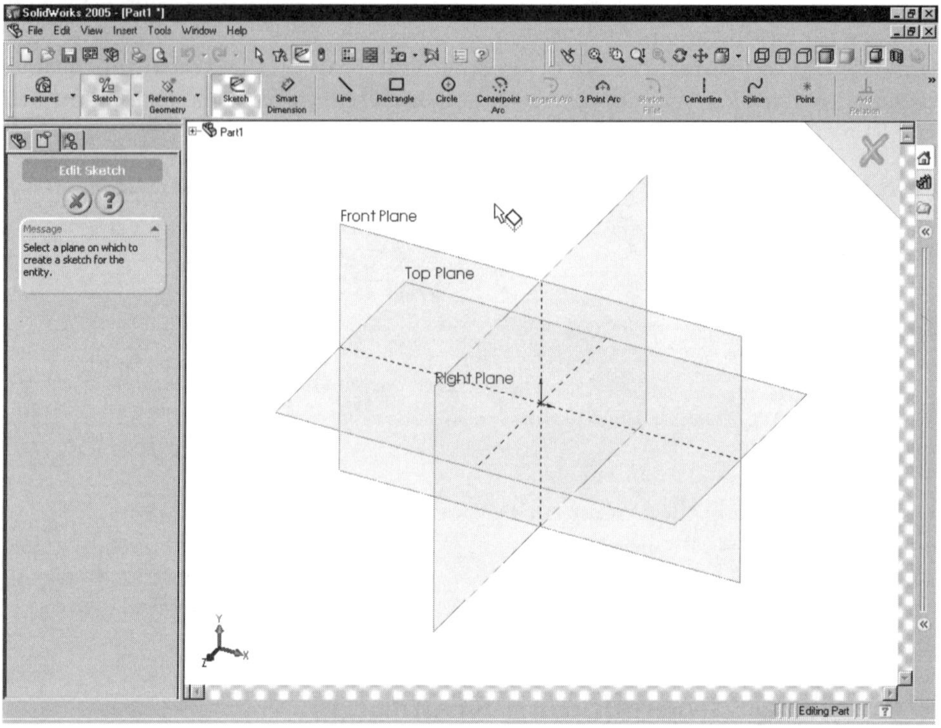

Figure 1-8 Selecting the plane to create the sketch of the base feature

Depending on the requirement of the design, you can select any plane to draw the sketch of the base feature. The selected plane is automatically oriented normal to the view so that you can easily create the sketch. Also, the **CommandManager** now displays various sketching tools to draw the sketch.

The default screen appearance of a SolidWorks part document in the sketching environment is shown in Figure 1-9.

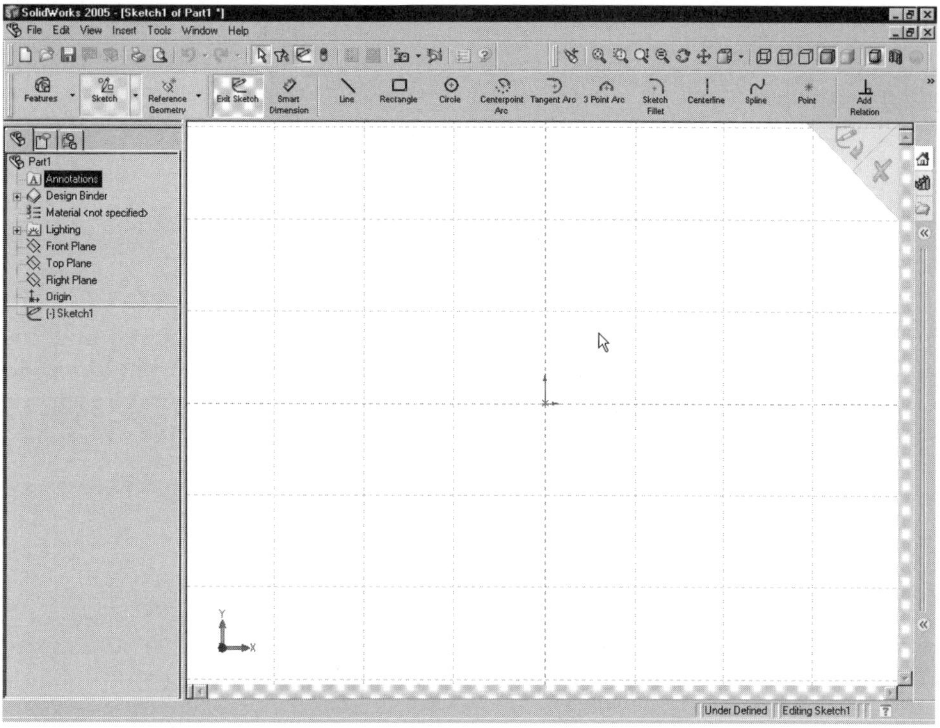

Figure 1-9 Default screen display of a part document in the sketching environment

SETTING UP THE DOCUMENT OPTIONS

When you install SolidWorks on your computer, you will be prompted to specify the dimensioning standards and units for measuring linear distances. The settings specified at that time are the default settings and whenever you start a new SolidWorks document, it will use those settings. However, if you want to modify these settings for a particular document, you can easily do so using the **Document Properties** dialog box. To invoke this dialog box, choose **Tools > Options**. When you invoke this option, the **System Options** dialog box will be displayed. In this dialog box, choose the **Document Properties** tab; the name of this dialog box will be changed to **Document Properties** dialog box. Setting the options for the current document using this dialog box is discussed next.

Modifying the Dimensioning Standards

To modify the dimensioning standards, invoke the **System Options** dialog box and then choose the **Document Properties** tab. You will notice that the **Detailing** option is selected by default in the area on the left to display the detailing options, as shown in Figure 1-10.

The default dimensioning standard that was selected while installing SolidWorks will be selected in the drop-down list provided in the **Dimensioning standard** area. You can select the required dimensioning standard from this drop-down list. The standards that are available are ANSI, ISO, DIN, JIS, BSI, GOST, and GB. You can select any one of these dimensioning standards for the current document.

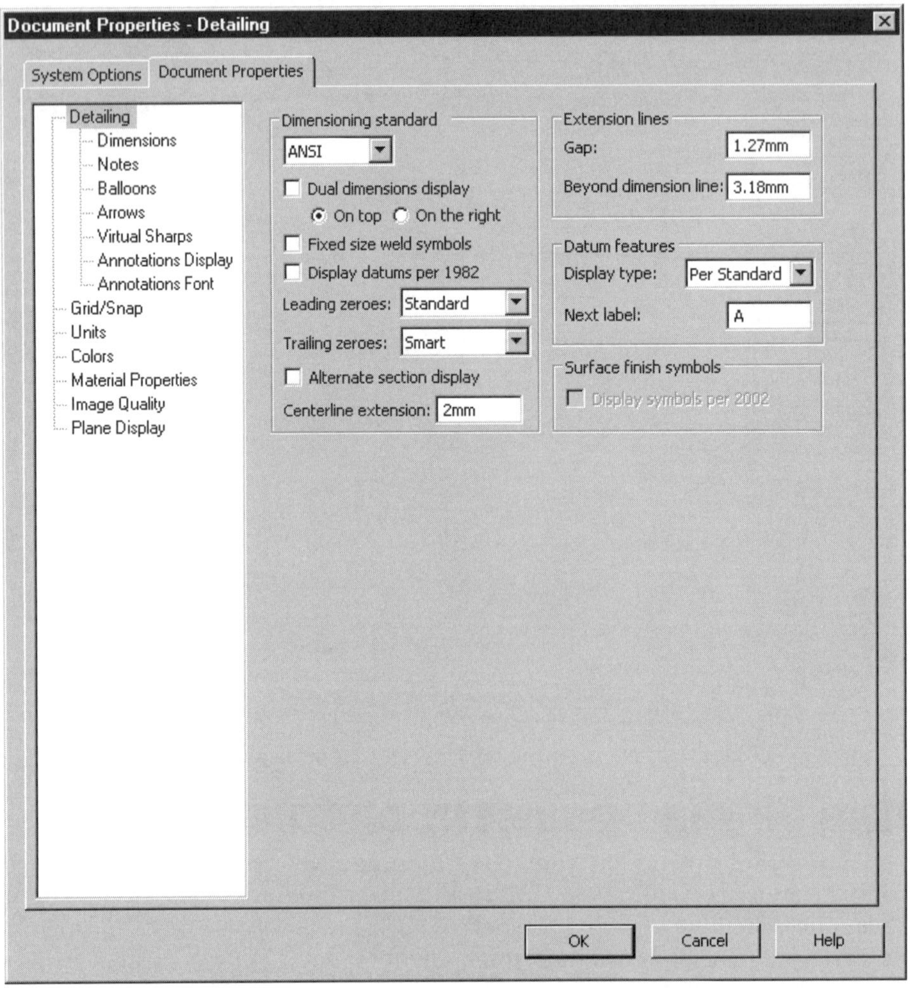

Figure 1-10 Setting the dimensioning standards

Modifying the Linear and Angular Units

To modify the linear and angular units, invoke the **System Options** dialog box and then choose the **Document Properties** tab. In this tab, select the **Units** option from the area on the left to display the options related to linear and angular units, as shown in Figure 1-11. The default option for measuring the linear distances that was selected while installing SolidWorks will be provided in the drop-down list provided in the **Length units** area. You can set the units from the options provided in the **Unit system** area. If you need to specify some units other than those provided in this area, choose the **Custom** radio button to invoke the options available in the **Length units** area. You can set the units using the drop-down list provided in this area. The units that can be selected are angstroms, nanometers, microns, millimeters, centimeters, meters, microinches, mils, inches, feet, and feet & inches. You can also change the units for angular dimensions by selecting them from the drop-down list in the **Angular units** area. The angular units that can be selected are Degrees, Deg/Min, Deg/Min/Sec, and Radians.

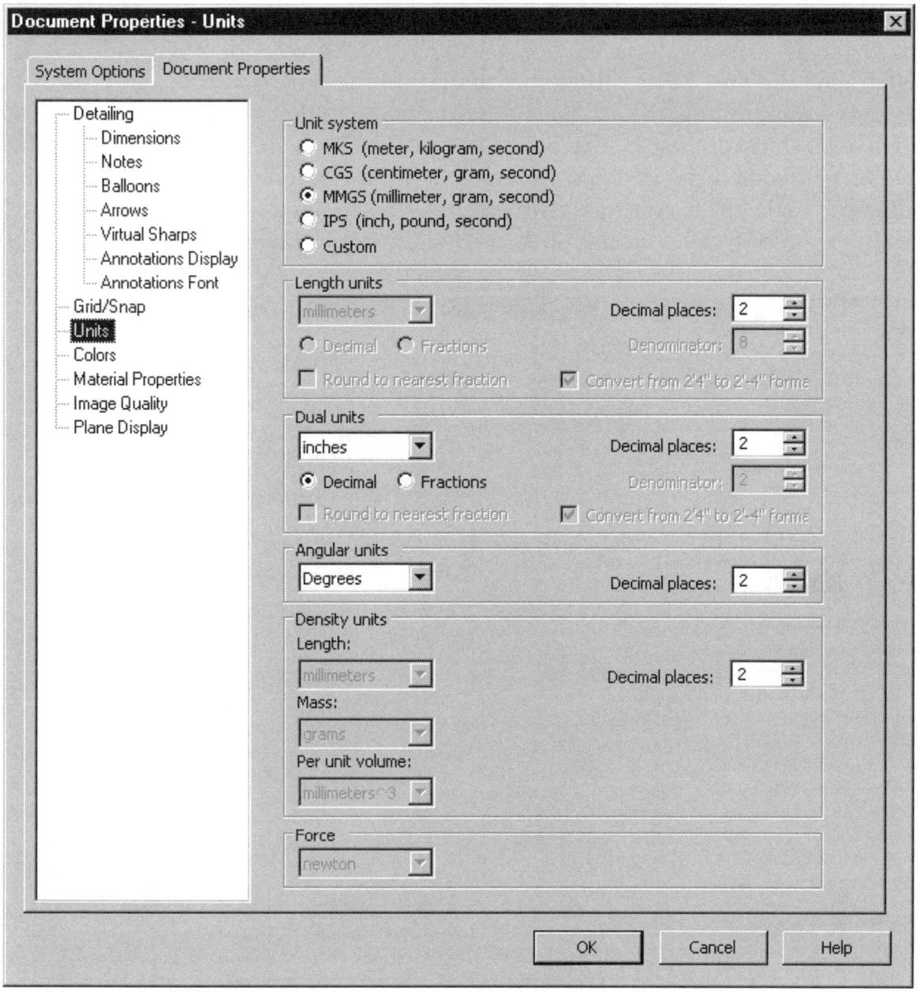

Figure 1-11 Setting the dimensioning units

Modifying the Snap and Grid Settings

In the sketching environment of SolidWorks, you can make the cursor jump through a specified distance while creating the sketch. Therefore, if you draw a sketched entity, its length will change in the specified increment. For example, while drawing a line if you make the cursor jump through a distance of 10 mm, the length of the line will be incremented by a distance of 10 mm. To do this, choose **Tools > Options** from the menu bar to display the **System Options** dialog box. To ensure that the cursor jumps through the specified distance you need to invoke the snap option. Select the **Relations/Snap** branch of the **Sketch** option to display the related options. From the options available on the right, select the **Grid** check box and clear the **Snap only when grid is displayed** check box if it is selected. This check box when selected snaps the sketched entities only when grid is displayed.

Now, choose the **Document Properties** tab to invoke the **Document Properties** dialog box.

Select the **Grid/Snap** option to display the related options, see Figure 1-12. The distance through which the cursor will jump is dependent on the ratio between the values in the **Major grid spacing** and **Minor-lines per major** spinners available in the **Grid** area. For example, if you want that the coordinates should change in the increment of 10 mm, you will have to make the ratio of major and minor lines to 10. This can be done by setting the value of the **Major grid spacing** spinner to **100** and that of the **Minor-lines per major** spinner to **10**. Similarly, to make the cursor jump through a distance of 5 mm, set the value of the **Major grid spacing** spinner to **50** and that of the **Minor-lines per major** spinner to **10**.

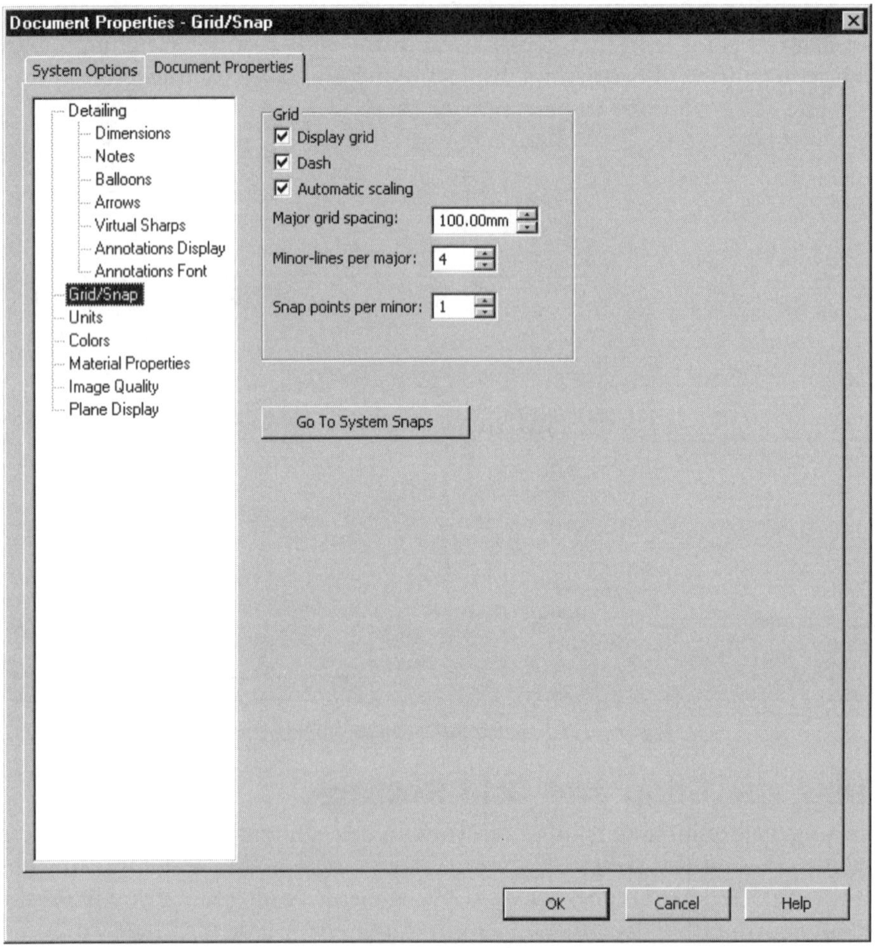

***Figure 1-12** Modifying the grid and snap settings*

Tip. *If you want to display the grid in the sketching environment, select the **Display grid** check box from the **Grid** area of the **Document Properties - Grid/Snap** dialog box.*

While drawing a sketched entity by snapping through grips, the grips symbol is displayed below the right of the cursor.

Note

Remember that these settings will only be for the current documents. When you open a new document, it will have the settings that were defined while installing SolidWorks.

LEARNING ABOUT SKETCHER TERMS

Before you learn about the various sketching tools, it is important for you to understand some terms that are used in the sketching environment. These terms are discussed next.

Origin

The origin is a red color icon that is displayed in the center of the sketching environment screen. This icon consists of two arrows displaying the X and Y axes directions of the current sketching plane. The point of intersection of these two axes is the origin point and the coordinates of this point are 0,0.

Inferencing Lines

The inferencing lines are the temporary lines that are used to track a particular point on the screen. These lines are dashed lines and are automatically displayed when you select a sketching tool in the sketching environment. These lines are created from the endpoint of a sketched entity or from the origin. For example, if you want to draw a line from the point where two imaginary lines intersect, you can use the inferencing lines to locate the point and then draw the line from that point. Figure 1-13 shows the use of inferencing lines to locate the point of intersection of two imaginary lines.

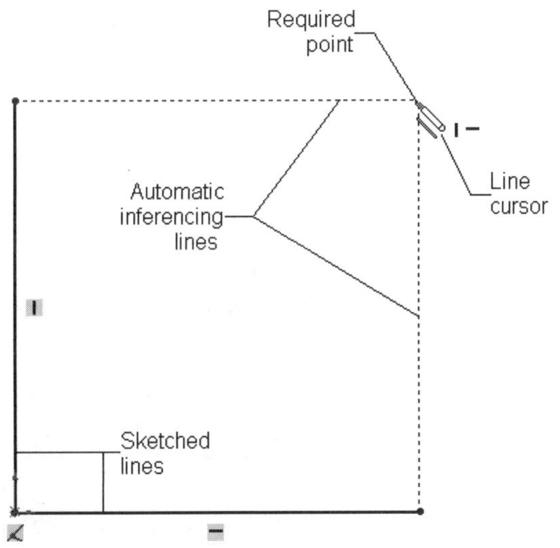

Figure 1-13 *Using inferencing lines to locate a point*

Figure 1-14 shows the use of inferencing lines to locate the center of an arc. Notice that the inferencing lines are created from the endpoint of the line and from the origin.

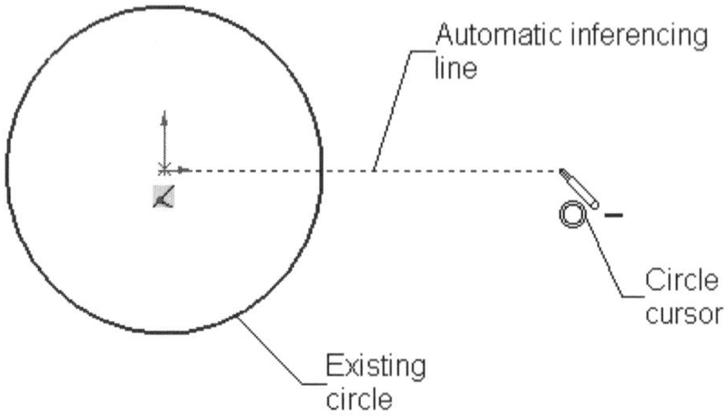

Figure 1-14 *Using inferencing lines to locate the center of a circle*

Note
The inferencing lines that are displayed on the screen will be either blue or yellow. The blue inferencing lines suggest that the relations are not added to the sketched entity and the yellow inferencing lines suggest that the relations are added to the sketched entity. You will learn about various relations in later chapters.

Select Tool

Toolbar:	Standard > Select

The **Select** tool is used to select a sketched entity or exit any sketching tool that is active. You can select the sketched entities by selecting them one by one using the left mouse button. You can also hold the left mouse button down and drag the cursor around the multiple sketched entities to define a box and select the multiple entities. There are two methods of selection, box selection and cross selection, which are discussed next.

Note
*You can also invoke the **Select** tool or exit a sketching tool by pressing the ESC key.*

Selecting the Entities Using the Box Selection

A box is a window created by pressing the left mouse button and dragging the cursor from left to right in the drawing area. A box has a property that all entities that lie completely inside it will be selected. The selected entities will be displayed in green.

Selecting the Entities Using the Cross Selection

When you press the left mouse button and drag the cursor from right to left in the drawing area, the cross selection method is invoked. The box drawn in the cross selection method consists of dotted lines. The cross selection method has a property that all entities that lie completely or partially inside the dotted box or the entities that touch the dotted box will be selected.

Now, you are familiar with the important sketching terms. Next you will learn about the sketching tools available in SolidWorks.

DRAWING LINES

CommandManager:	Sketch > Line
Menu:	Tools > Sketch Entities > Line
Toolbar:	Sketch > Line

Lines are one of the basic sketching entities available in SolidWorks. In general terms, a line is defined as the shortest distance between two points. As mentioned earlier, SolidWorks is a parametric solid modeling tool. This property allows you to draw a line of any length and at any angle and later force it to the desired length and angle. To draw a line in the sketching environment of SolidWorks, choose the **Line** tool from the **Sketch CommandManager**. You can also invoke **Line** tool by pressing L from the keyboard. You will notice that the cursor, which was an arrow earlier, is replaced by the line cursor. The line cursor is actually a pencil-like cursor with a small inclined line below the pencil. Also, the **Insert Line PropertyManager** is displayed, as shown in Figure 1-15.

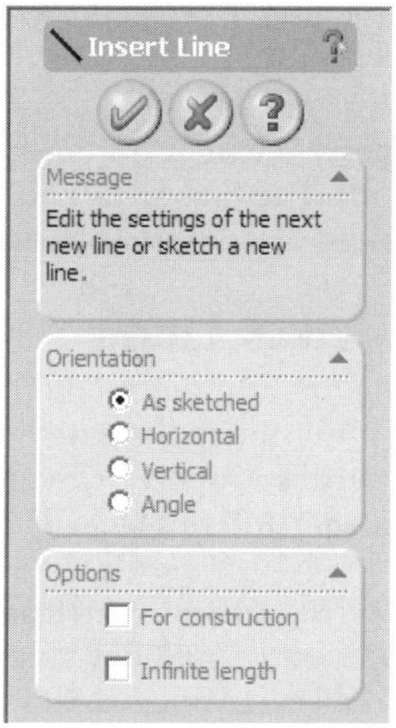

Figure 1-15** The **Insert Line PropertyManager

Tip. *The **What's New** symbol provided on the right of the **PropertyManager** implies that there is some enhancement in this tool or it is a new tool in this release of SolidWorks. If you click on this symbol, the **SolidWorks 2005 What's New** dialog box will be displayed.*

The **Message** rollout of the **Insert Line PropertyManager** informs you to edit the settings of the next line or sketch a new line. Using the options available in this **PropertyManager**, you can set the orientation and other sketching options of drawing a line. All these options are discussed next.

Orientation Rollout

The **Orientation** rollout is used to define the orientation of the line to be drawn. By default, the **As sketched** radio button is selected and you can draw the line in any orientation. If you select the **Horizontal** radio button, you can only draw horizontal lines. When you select this radio button, the **Parameters** rollout is displayed and you can specify the length of the line in the **Length** spinner provided in this rollout. If you select the **Add dimension** check box from this rollout, a dimension displaying the length of the line will be applied to the drawn line. You will learn more about dimensioning in the later chapters.

If you select the **Vertical** radio button, you can only draw vertical lines. The **Parameters** rollout is also displayed where you can set the parameters for drawing the vertical line.

The **Angle** radio button is selected to draw lines at a specified angle. When you select this radio button, the **Parameters** rollout is displayed, where you can set the values of the length of the line and the angle or orientation.

Options Rollout

The **For construction** check box available in this rollout is used to draw a construction line. You will learn more about construction lines later in this chapter. The **Infinite length** check box is used to draw a line of infinite length.

After setting the options in this **PropertyManager**, you need to draw the line. In SolidWorks, there are two methods of drawing lines. The first method is to draw continuous lines and the second method is to draw individual lines. Both these methods are discussed next.

Drawing Chain of Continuous Lines

This is the default method of drawing lines. In this method, you just have to specify the start point and the endpoint of the line using the left mouse button. As soon as you specify the start point of the line, the **Line Properties PropertyManager** will be displayed. The options in the **Line Properties PropertyManager** will not be available at this stage.

After specifying the start point, move the cursor away from the start point and specify the endpoint of the line using the left mouse button. A line will be drawn between the two points. You will notice that the line is green and has filled circles at the two ends. The line will be displayed in green because it is still selected.

Move the cursor away from the endpoint of the line and you will notice that another line is attached to cursor. The start point of this line is the endpoint of the last line and the length of this line can be increased or decreased by moving the cursor. Because this line stretches like a rubber band as you move the cursor, it is called a rubber-band line. The point that you specify next on the screen will be taken as the endpoint of the second line and a line will be drawn such that the endpoint of the second line is taken as the start point of the new line and the point you specify is taken as the endpoint of the new line. Now, a new rubber-band line is displayed starting from the endpoint of the last line. This is a continuous process and you can draw a chain of as many continuous lines as you need by specifying the points on the screen using the left mouse button.

You can exit the continuous line drawing process by pressing the ESC key from the keyboard, by choosing the **Select** tool, or by double-clicking on the screen. You can also right-click to display the shortcut menu and choose the **End Chain** option from the shortcut menu.

Figure 1-16 shows a sketch drawn using continuous lines. This sketch is started from the lower left corner and the horizontal line is drawn first. To close the loop using the last line, as soon as you move the cursor close to the start point of the first line, you will notice that a red circle is displayed at the start point. If you specify the endpoint of the line at this stage, the loop will be closed and no rubber-band line will be displayed now. This is because the loop is already closed and you may not need another continuous line now. However, the **Line** tool is still active and you can draw other lines.

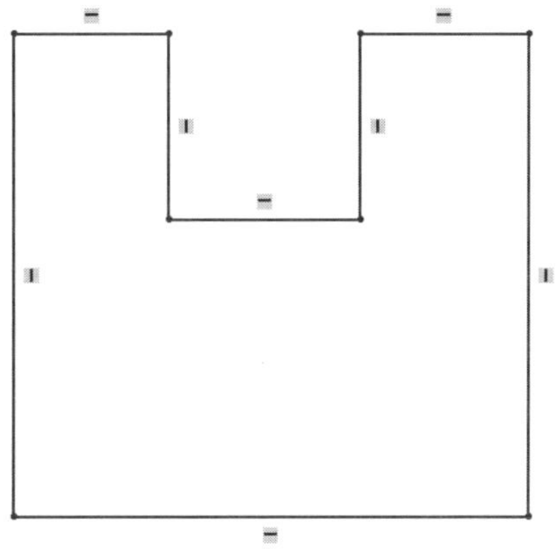

Figure 1-15 *Sketch drawn with the help of continuous lines*

Note

*When you exit the line drawing process by double-clicking on the screen or by choosing **End chain** from the shortcut menu, the current chain is ended but the **Line** tool is still active and you can draw other lines.*

Drawing Individual Lines

This is the second method of drawing lines. Using this method, you can draw individual lines and the start point of the next line will not necessarily be the endpoint of the last line. To draw individual lines, you need to press and hold the left mouse button down and drag the cursor from the start point of the line to the endpoint. Once you have dragged the cursor to the endpoint, release the left mouse button; a line will be drawn between the two points.

To make the process of sketching easy in SolidWorks easy, you are provided with the **PropertyManager**. The **PropertyManager** is a table that will be displayed on the left of the screen as soon as you select the first point of any sketched entity. The **PropertyManager** has all parameters related to the sketched entity such as the start point, endpoint, angle, length, and so on. You will notice that as you start dragging the mouse, the **Line Properties PropertyManager** is displayed on the left of the drawing area. All options in the **Line Properties PropertyManager** will be available when you release the left mouse button. Figure 1-17 shows a partial view of the **Line Properties PropertyManager**.

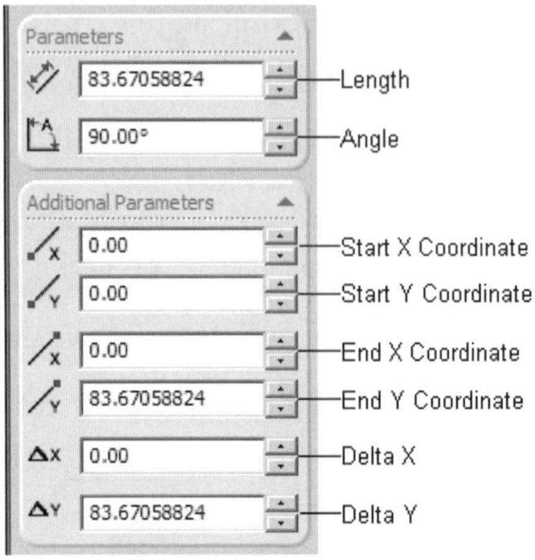

Figure 1-17** Partial display of the **Line Properties PropertyManager

Note

*The **Line Properties PropertyManager** will also display additional options about relations. You will learn more about relations in later chapters.*

After you have drawn the line, modify the parameters in the **Line Properties PropertyManager** to make the line to the desired length and angle. You can also dynamically modify the line by holding its endpoints and dragging them.

Line Cursor Parameters

When you draw lines in the sketching environment of SolidWorks, you will notice that a numeric value is displayed above the line cursor, see Figure 1-18. This numeric value indicates the length of the line you draw. This value is the same as that in the **Length** spinner of the **Line Properties PropertyManager**. The only difference is that in the **Line Properties PropertyManager**, the value will be displayed with more precision.

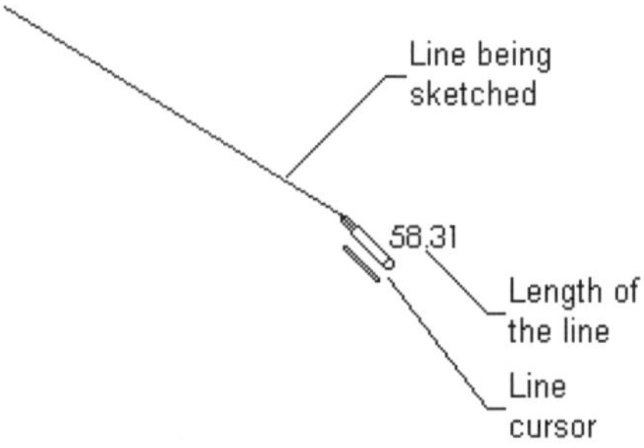

***Figure 1-18** The length of the line displayed on the screen while drawing the line*

The other thing that you will notice while sketching is that sometimes when you are drawing vertical or horizontal lines, a ▬ or ▮ symbol is displayed below the line cursor. These are the symbols of the **Vertical** and **Horizontal** relations. SolidWorks automatically applies these relations to the lines. These relations ensure that the lines you draw are vertical or horizontal and not inclined. Figure 1-19 shows the symbol of the **Vertical** relation on a line and Figure 1-20 shows the symbol of the **Horizontal** relation on a line.

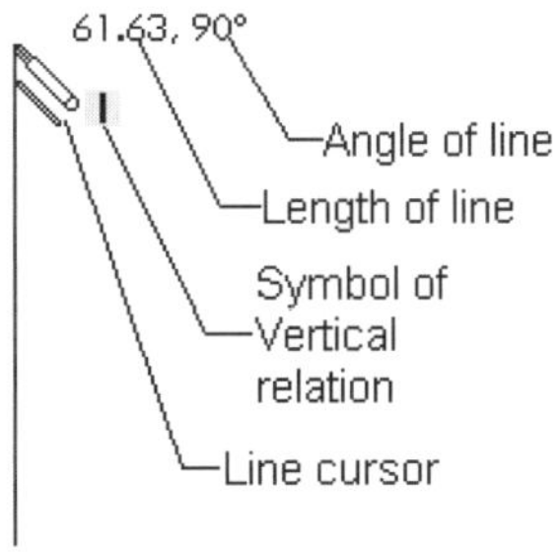

Figure 1-19 Symbol of the ***Vertical*** *relation*

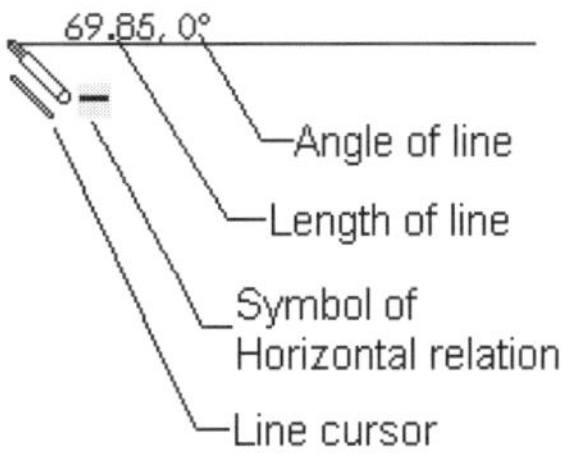

Figure 1-20 Symbol of the ***Horizontal*** *relation*

Note

In addition to the ***Horizontal*** *and* ***Vertical*** *relations, you can apply a number of other relations such as* ***Tangent, Concentric, Perpendicular, Parallel,*** *and so on. You will learn about all these relations in later chapters.*

The other options of the ***Line Properties PropertyManager*** *will be discussed in later chapters.*

Drawing Tangent or Normal Arcs Using the Line Tool

SolidWorks allows you to draw tangent or normal arcs originating from the endpoint of the line while drawing continuous lines. Note that these arcs can be drawn only if you have drawn at least one line, arc, or spline. To draw such arcs, draw a line by specifying the start point and the endpoint. Move the cursor away from the endpoint of the last line to display the rubber-band line. Now, when you move the cursor back to the endpoint of the last line, the arc mode is invoked and the line cursor is replaced by the arc cursor. The angle and the radius of the arc is displayed above the arc cursor. You can also invoke the arc mode by right-clicking and choosing **Switch to arc** from the shortcut menu or pressing the A key from the keyboard.

To draw a tangent arc, invoke the arc mode by moving the cursor back to the endpoint of the last line. Now, move the cursor through a small distance along the tangent direction of the line; a dotted line will be drawn. Next, move the cursor in the direction in which the arc should be drawn. You will notice that a tangent arc is drawn. Specify the endpoint of the tangent arc using the left mouse button. Figure 1-21 shows an arc tangent to an existing line. To draw a normal arc, invoke the arc mode. Now, move the cursor through a small distance in the direction normal to the line and then move it in the direction of the endpoint of the arc; the normal arc will be drawn, as shown in Figure 1-22.

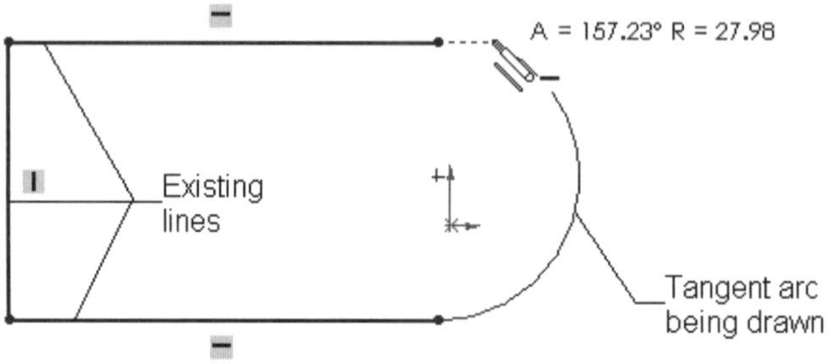

Figure 1-21 Drawing a tangent arc using the ***Line*** *tool*

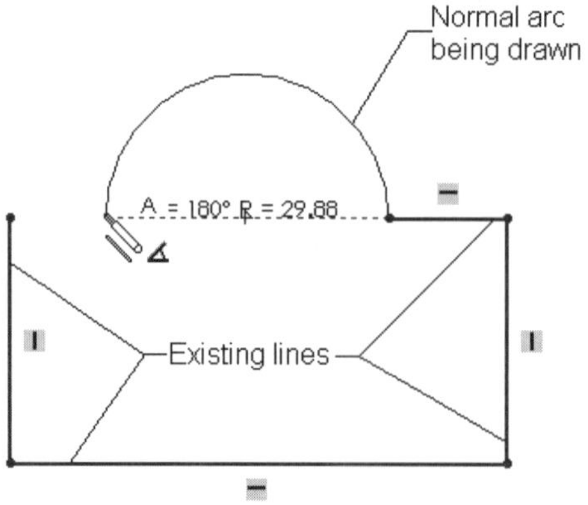

Figure 1-22 Drawing a normal arc using the ***Line*** *tool*

As soon as the endpoint of the tangent or normal arc is defined, the line mode will be invoked again. You can continue drawing lines using the line mode or move the cursor back to the endpoint of the arc to invoke the arc mode.

Note

If the arc mode is invoked by mistake while drawing lines, you can cancel the arc mode and invoke the line mode again by pressing the A key from the keyboard. Alternatively, you can right-click and choose ***Switch to Line*** *from the shortcut menu or move the cursor back to the endpoint and press the left mouse button to invoke the line mode.*

Drawing Construction Lines or Centerlines

CommandManager:	Sketch > Centerline
Menu:	Tools > Sketch Entities > Centerline
Toolbar:	Sketch > Centerline

Construction lines or centerlines are those that are drawn only for the aid of sketching. These lines are not considered while converting the sketches into features. You can draw a construction line similar to the sketched line by using the **Centerline** tool. You will notice that when you draw a construction line, the **For Construction** check box in the **Line Properties PropertyManager** is selected. You can also draw a construction line by sketching the line using the **Line** tool and then selecting the **For Construction** check box available in the **Line Properties PropertyManager**. When you select this check box, the line is turned into a centerline.

DRAWING CIRCLES

CommandManager:	Sketch > Circle
Menu:	Tools > Sketch Entities > Circle
Toolbar:	Sketch > Circle

In SolidWorks, there are two methods of drawing circles. The first method is by specifying the center point of a circle and then defining its radius. The second method is to draw a circle by defining three points that lie on its periphery. Both these methods are discussed next. To draw a circle, choose the **Circle** button from the **Sketch CommandManager**; the **Circle PropertyManager** will be displayed, as shown in Figure 1-23.

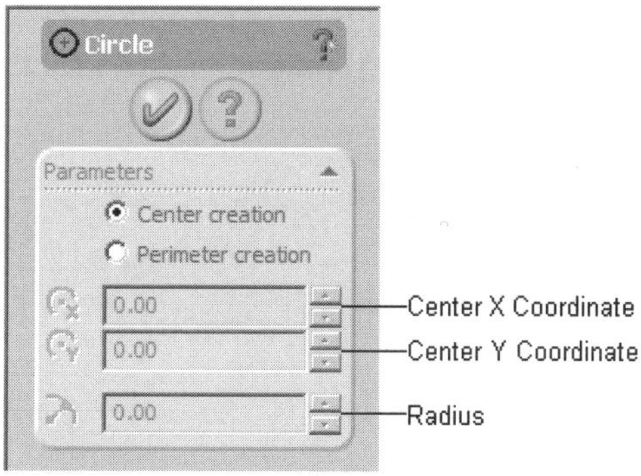

Figure 1-23** The **Circle PropertyManager

Drawing Circles by Defining its Centerpoint

When you invoke the **Circle PropertyManager**, the **Center creation** radio button is selected by default in the **Parameters** rollout of the **Circle PropertyManager**. This radio button is selected to draw a circle by specifying it center. You will notice that the arrow cursor is replaced by the circle cursor. The circle cursor consists of a pencil and two concentric circles below the pencil. Specify the center point of the circle and then move the cursor to define its radius. The current radius of the circle is displayed above the circle cursor. This radius will change as you move the cursor. The coordinates of the center point of the circle and the radius updated dynamically are shown in the **Circle PropertyManager**. You can define any arbitrary radius of the circle and then modify it to the desired value by using the **Circle PropertyManager**. Figure 1-24 shows a circle being drawn using the **Center creation** option of the **Circle** tool.

Drawing Circles by Defining Three Points

The **Perimeter creation** radio button is used to draw a circle by defining three points on it. To draw a circle using this option, invoke the **Circle PropertyManager** and select the **Perimeter creation** radio button from the **Parameters** rollout. The select cursor will be replaced by a three point circle cursor. Specify the first point of the circle in the drawing area. Now, specify

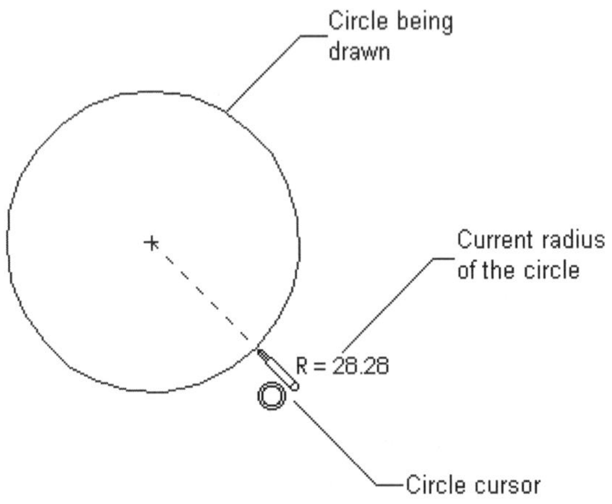

***Figure 1-24** Drawing a circle using the **Center creation** option*

the other two points on the circle. The resulting circle will be highlighted in green and you can set its parameters in the **Circle PropertyManager**. Figure 1-25 shows a circle being drawn by specifying three points on it.

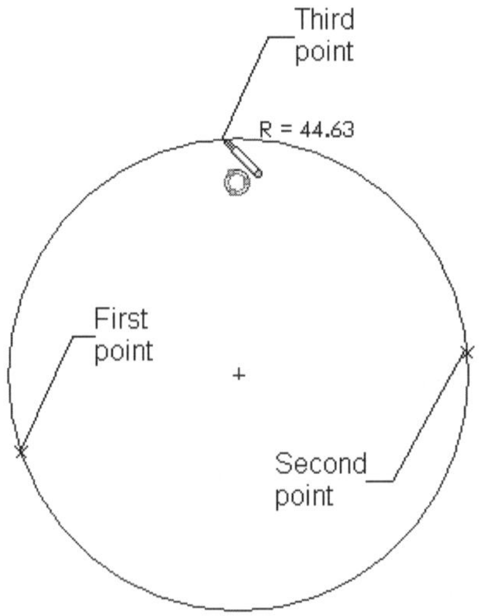

***Figure 1-25** Drawing a circle using the **Perimeter creation** option*

Drawing a Construction Circle

If you want to sketch a construction circle, draw a circle using the **Circle** tool and then select the **For construction** check box available in the **Circle PropertyManager**.

Tip. *To convert a construction entity back to the sketched entity, invoke the select tool and then select the construction entity. The entity will turn green and the **PropertyManager** will be displayed on the left of the drawing area. From the **PropertyManager**, clear the **For Construction** check box. The construction entity will again be changed into a sketched entity and will be displayed with continuous line.*

DRAWING ARCS

In SolidWorks, you can draw arcs using three methods: **Tangent/Normal Arc**, **Centerpoint Arc**, and **3 Point Arc**. All these methods can be invoked separately by choosing their respective buttons from the **Sketch Tools** toolbar. These methods are discussed next.

Drawing Tangent/Normal Arcs

CommandManager:	Sketch > Tangent Arc
Menu:	Tools > Sketch Entities > Tangent Arc
Toolbar:	Sketch > Tangent Arc

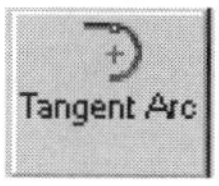

The tangent arcs are those that are drawn tangent to an existing sketched entity. The existing sketched entities include sketched and construction lines, arcs, and splines. The normal arcs are those that are drawn normal to an existing entity. You can draw tangent and normal arcs using the **Tangent Arc** tool.

To draw a tangent arc, invoke the **Tangent Arc** tool; the arrow cursor will be replaced by the arc cursor. Move the arc cursor close to the endpoint of the entity you want to select as the tangent entity. You will notice that a red circle is displayed at the endpoint. Also, a yellow symbol displaying two concentric circles appears below the pencil. Now, press the left mouse button once and move the cursor along the tangent direction through a small distance and then move the cursor to size the arc. The arc will start from the endpoint of the tangent entity and its size will change as you move the cursor. Note that the angle and the radius of the tangent arc are displayed above the arc cursor, see Figure 1-26.

To draw a normal arc, invoke the **Tangent Arc** tool. Move the arc cursor close to the endpoint of the entity you want to select as the normal entity. A red circle will be displayed at the endpoint and a yellow symbol displaying two concentric circles appears below the pencil. Now, press the left mouse button once and move the cursor along the normal direction through a small distance and then move the cursor to size the arc, refer to Figure 1-27.

While drawing the arc, as soon as you start moving the cursor after specifying the start point, the **Arc PropertyManager** will be displayed. However, the options in the **Arc PropertyManager** are not enabled at this stage. These options are enabled only after you have completed drawing the tangent or normal arc.

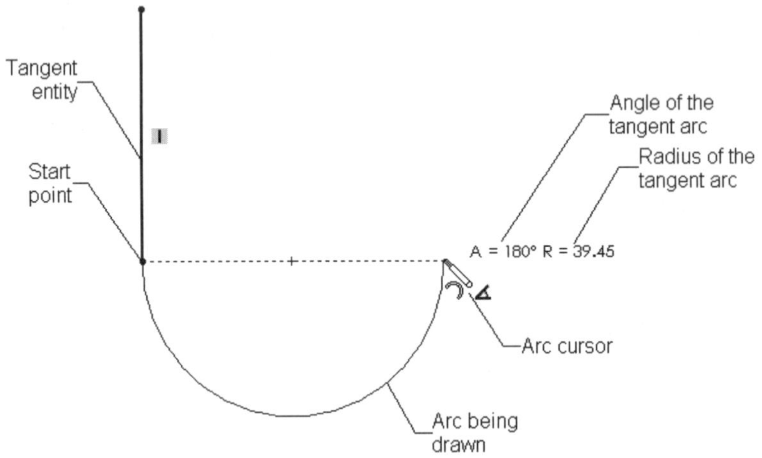

Figure 1-26 *Drawing tangent arc*

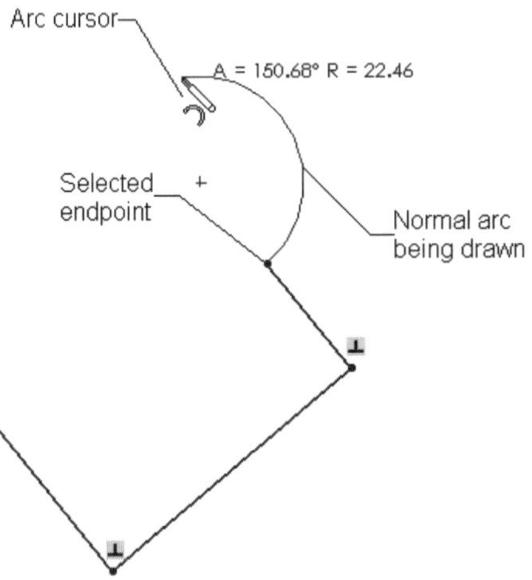

Figure 1-27 *Drawing normal arc*

You can draw an arbitrary arc and then modify its value using the **Arc PropertyManager**. Figure 1-28 shows a partial view of the **Arc PropertyManager**.

Note

When you select a tangent entity to draw a tangent arc, the ***Tangent*** *relation is applied between the start point of the arc and the tangent entity. Therefore, if you change the coordinates of the start point of the arc, the tangent entity will also be modified accordingly.*

Tip. *You can invoke the list of recently used commands by right-clicking in the drawing area. The cascading menu that is displayed when you choose the* ***Recent Command*** *option from the shortcut menu displays the eight most recent commands.*

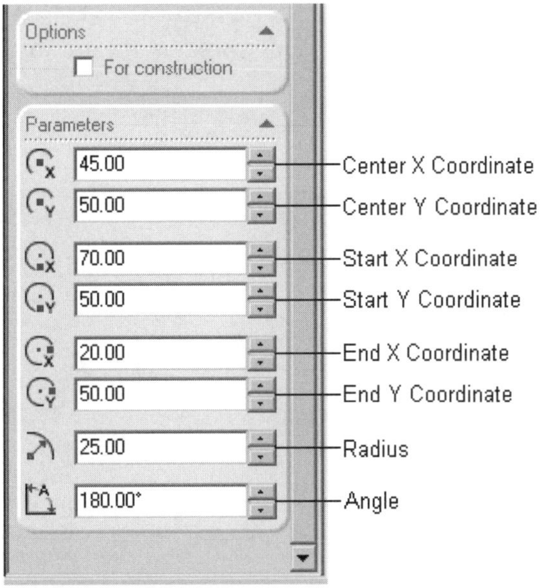

Figure 1-28** Partial display of the **Arc PropertyManager

Drawing Centerpoint Arcs

CommandManager:	Sketch > Centerpoint Arc
Menu:	Tools > Sketch Entity > Centerpoint Arc
Toolbar:	Sketch > Centerpoint Arc

The center point arcs are those that are drawn by defining the center point, start point, and endpoint of the arc. When you invoke this tool, the arrow cursor is replaced by the arc cursor. As mentioned earlier, an arc cursor consists of a pencil and an arc below the pencil.

To draw a center point arc, invoke the **Centerpoint Arc** tool and then move the arc cursor to the point that you want to specify as the center point of the arc. Press the left mouse button once at the location of the center point and then move the cursor to the point from where you want to start the arc. You will notice that a dotted circle is displayed on the screen. The size of this circle will modify as you move the mouse. This circle is drawn for your reference and the center point of this circle lies at the point that you specified as the center of the arc. Press the left mouse button once at the point that you want to select as the start point of the arc. Next, move the cursor to specify the endpoint of the arc. You will notice that the reference circle is no longer displayed and an arc is being drawn with the start point as the point that you specified after specifying the center point. Also, the **Arc PropertyManager**, similar to the one that is shown in the tangent arc, is displayed on the left of the drawing area. Note that the options in the **Arc PropertyManager** will not be available at this stage.

If you move the cursor in the clockwise direction, the resulting arc will be drawn in the clockwise direction. However, if you move the cursor in the counterclockwise direction, the resulting arc will be drawn in the counterclockwise direction. Specify the endpoint of the arc using the left mouse button. Figure 1-29 shows the reference circle that is drawn when you move the mouse button after specifying the center point of the arc and Figure 1-30 shows the resulting center point arc.

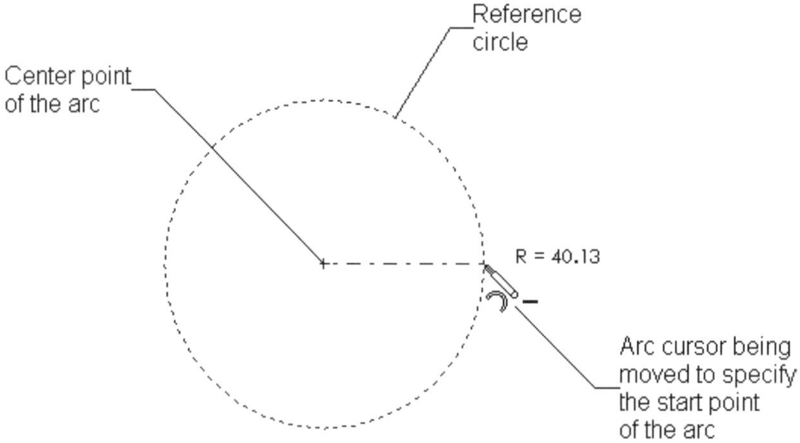

***Figure 1-29** Specifying the center point and the start point of the center point arc*

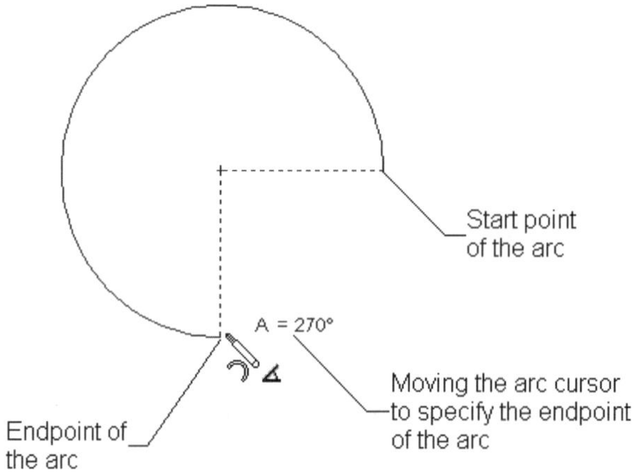

***Figure 1-30** Moving the cursor to specify the start point and the endpoint of the arc*

Drawing 3 Point Arcs

CommandManager:	Sketch > 3 Pt Arc
Menu:	Tools > Sketch Entities > 3 Point Arc
Toolbar:	Sketch > 3 Pt Arc

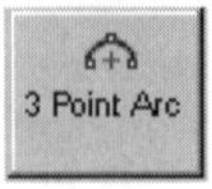

The three point arcs are those that are drawn by defining the start point and endpoint of the arc, and a point somewhere on the arc. When you invoke this tool, the arrow cursor is replaced by the arc cursor.

To draw a three point arc, invoke the **3 Pt Arc** tool and then move the arc cursor to the point that you want to specify as the start point of the arc. Press the left mouse button once at the location of the start point and then move the cursor to the point that you want specify as the endpoint of the arc. As soon as you start moving the cursor after specifying the start point, a reference arc will be drawn and the **Arc PropertyManager** will be displayed. However, the options in the **Arc PropertyManager** will not be available at this stage.

Using the left mouse button, specify the endpoint of the arc. You will notice that the reference arc is no longer displayed. Instead, a solid arc is displayed and the cursor is attached to the arc. As you move the cursor, the arc will also be modified dynamically. Using the left mouse button, specify a point on the screen to create the arc. The last point that you specify will determine the direction of the arc. The options in the **Arc PropertyManager** will be displayed once you draw the arc. You can modify the properties of the arc using the **Arc PropertyManager**. Figure 1-31 shows the reference arc that is drawn by specifying the start point and the endpoint of the arc and Figure 1-32 shows the resulting three point arc.

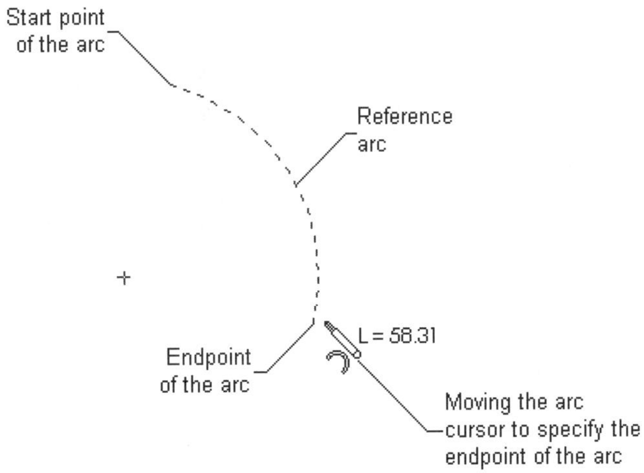

Figure 1-31 *Specifying the start point and the endpoint of the arc*

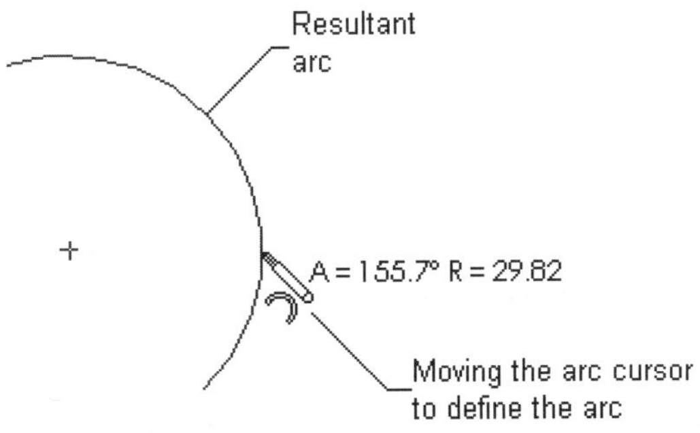

Figure 1-32 *Specifying a point on the arc to draw it*

DRAWING RECTANGLES

CommandManager:	Sketch > Rectangle
Menu:	Tools > Sketch Entities > Rectangle
Toolbar:	Sketch > Rectangle

In SolidWorks, the rectangles are drawn by specifying two opposite corners of the rectangle. To draw a rectangle, invoke the **Rectangle** tool; the arrow cursor will be replaced by the rectangle cursor. Move the cursor to the point that you want to specify as the first corner of the rectangle. Press the left mouse button once at the first corner and then move the cursor and specify the other corner of the rectangle using the left mouse button. You will notice that the length and width of the rectangle are displayed above the rectangle cursor. The length is measured along the X axis and the width is measured along the Y axis. Figure 1-33 shows a rectangle being drawn by specifying two opposite corners.

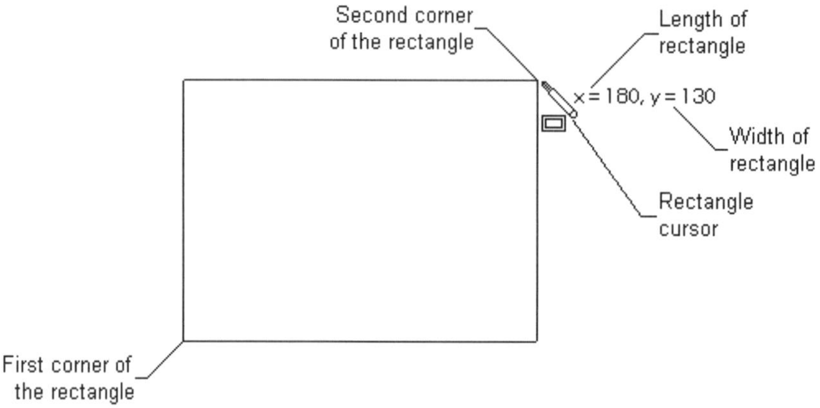

Figure 1-33 *Drawing a rectangle by specifying two opposite corners*

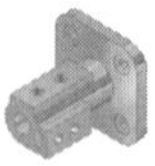

Note

When you draw a rectangle, the ***PropertyManager*** *will not be displayed. This is because a rectangle is considered as a combination of four individual lines. Therefore, after drawing the rectangle, if you select one of the lines of the rectangle using the* ***Select*** *tool, the* ***Line Properties PropertyManager*** *will be displayed. You can modify the parameters of the selected line using the* ***Line Properties PropertyManager****.*

Remember that because the relations are applied to all four corners of the rectangle, if you modify the parameters of one of the lines using the ***Line Properties PropertyManager****, the other three lines will also be modified accordingly.*

You can convert a rectangle into a construction rectangle by selecting all lines together using a window and then selecting the ***For Construction*** *check box from the* ***PropertyManager****.*

DRAWING PARALLELOGRAM

Menu:	Tools > Sketch Entities > Parallelogram

In SolidWorks, the **Parallelogram** tool can be used to draw a parallelogram and also to draw a rectangle at an angle. The methods used to draw both these entities are discussed next.

Drawing a Rectangle at an Angle

To draw a rectangle at an angle, invoke the **Parallelogram** tool from the menu bar. The cursor will be replaced by the parallelogram cursor. Move the cursor to the point you want to specify as the start point of one of the edges of the rectangle. Press the left mouse button once at this point and move the cursor to size the edge. You will notice that a reference line is being drawn. Depending on the current position of the cursor, the reference line will be horizontal, vertical, or inclined at some angle. The current length of the edge and its angle will be displayed above the parallelogram cursor. Using the left mouse button, specify the endpoint of the edge such that the resulting reference line is at an angle.

Next, move the cursor to specify the width of the rectangle. You will notice that a reference rectangle is drawn at an angle. Also, irrespective of the current position of the cursor, the width will be specified normal to the first edge, either above or below. Using the left mouse button, specify a point on the screen to define the width of the rectangle. The reference rectangle will be converted into a sketched rectangle. Figure 1-34 shows a rectangle drawn at an angle.

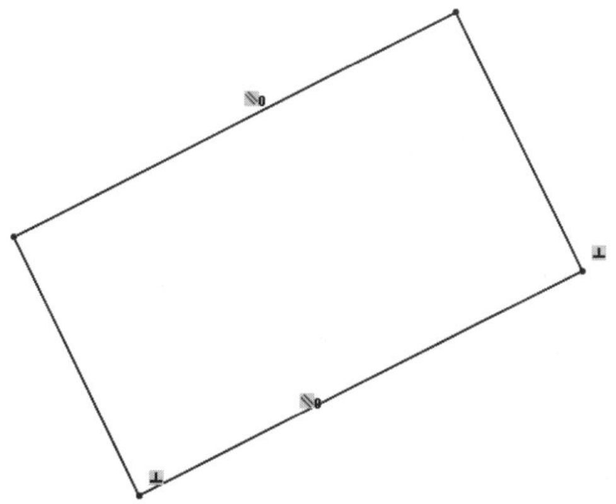

Figure 1-34 *Rectangle at an angle*

Drawing Parallelograms

To draw a parallelogram, invoke the **Parallelogram** tool from the menu bar. The cursor will be replaced by the parallelogram cursor. Specify two points on the screen to define one edge of the parallelogram. Next, press the CTRL key from the keyboard once and then move the mouse to define the width of the parallelogram. You will notice that the width is no longer added normal to the first edge. As you move the mouse, a reference parallelogram will be drawn. The size and shape of the reference parallelogram will depend on the current location of the cursor.

Specify a point on the screen to define the parallelogram. Figure 1-35 shows a parallelogram drawn at an angle.

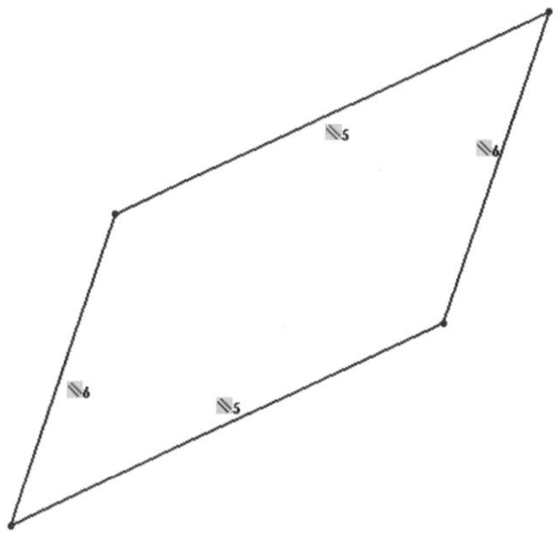

Figure 1-35 *Parallelogram at an angle*

Note
Similar to rectangles, each edge of a parallelogram is considered as a separate line. Also, in the case of parallelograms, the ***PropertyManager*** *is not displayed while you are drawing it.*

Tip. *You can also invoke the tools to draw lines, arcs, circles, or rectangles using the shortcut menu that is displayed when you right-click in the drawing area.*

DRAWING POLYGONS

Menu: Tools > Sketch Entities > Polygon

A polygon is defined as a multisided geometric figure in which the length of all sides and the angle between them are the same. In SolidWorks, you can draw a polygon with the number of sides ranging from 3 to 40. The dimensions of a polygon are controlled using the diameter of a construction circle that is either inscribed inside the polygon or circumscribed outside the polygon. If the construction circle is inscribed inside the polygon, the diameter of the construction circle will be taken from the edges of the polygon. If the construction circle is circumscribed about the polygon, the diameter of the construction circle will be taken from the vertices of the polygon.

To draw a polygon, invoke the **Polygon** tool; the **Polygon PropertyManager** will be displayed, as shown in Figure 1-36.

Set the parameters such as the number of sides, inscribed or circumscribed circle, and so on in the **Polygon PropertyManager**. You can also modify these parameters after drawing the

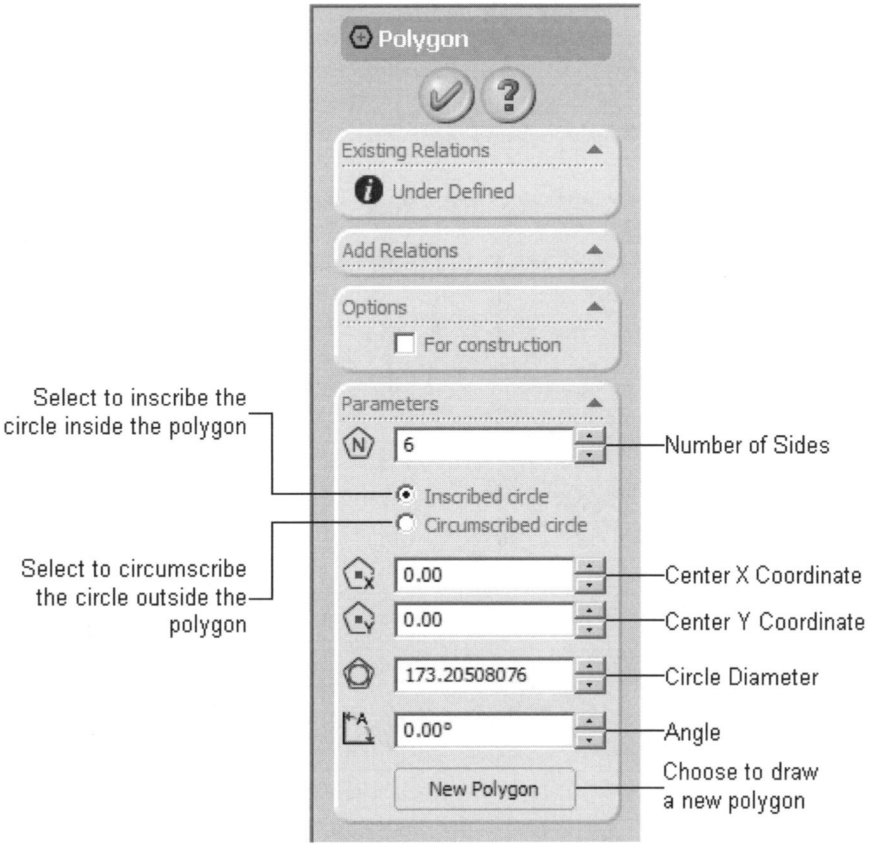

Figure 1-36 Polygon PropertyManager

polygon. When you invoke this tool, the arrow cursor will be replaced by the polygon cursor. Press the left mouse button at the point that you want to specify as the center point of the polygon and then move the cursor to size the polygon. The length of each side and the rotation angle of the polygon will be displayed above the polygon cursor as you drag it. Using the left mouse button, specify a point on the screen after you get the desired length and rotation angle of the polygon. You will notice that based on whether you selected the **Inscribed circle** or the **Circumscribed circle** radio button in the **Polygon PropertyManager**, a construction circle will be drawn inside or outside the polygon. After you have drawn the polygon, you can modify the parameters such as the center point of the polygon, the diameter of the construction circle, the angle of rotation of the polygon, and so on using the **Polygon PropertyManager**. If you want to draw another polygon, choose the **New polygon** button provided below the **Angle** spinner in the **Polygon PropertyManager**.

Figure 1-37 shows a six-sided polygon with the construction circle inscribed inside the polygon and Figure 1-38 shows a five-sided polygon with the construction circle circumscribed outside the polygon. Note that the reference circle is retained with the polygon. Remember that this circle will not be considered while converting the polygon into a feature.

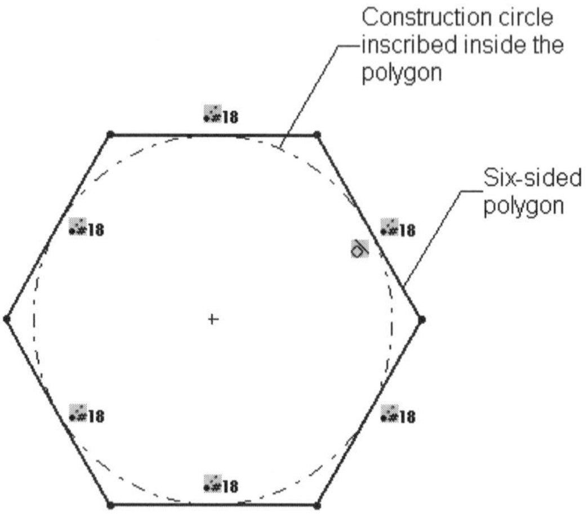

Figure 1-37 Six-sided polygon with construction circle inscribed inside the polygon

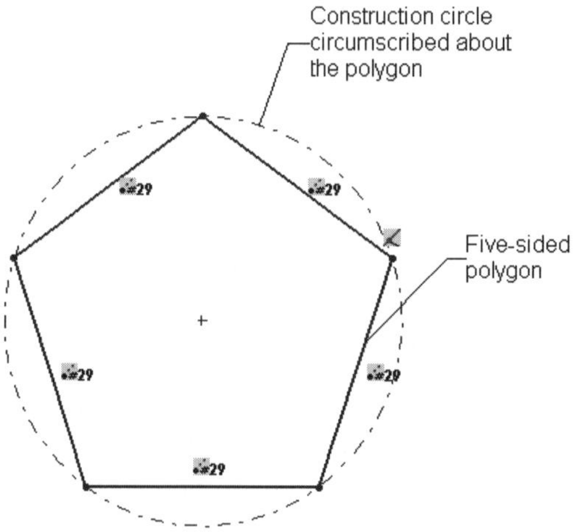

Figure 1-38 Five-sided polygon with construction circle circumscribed outside the polygon

DRAWING SPLINES

CommandManager:	Sketch > Spline
Menu:	Tools > Sketch Entities > Spline
Toolbar:	Sketch > Spline

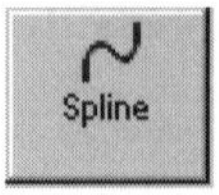

In SolidWorks, you can draw a spline by continuously specifying the endpoints of the spline segments using the left mouse button. This method of drawing splines is similar to that of drawing continuous lines. After specifying all points of the spline, right-click and choose **Select** to exit the **Spline** tool. Choosing **End Spline**

will exit the current spline but the **Spline** tool will still be active and you can draw another spline. Figure 1-39 shows a spline drawn with its start point at the origin.

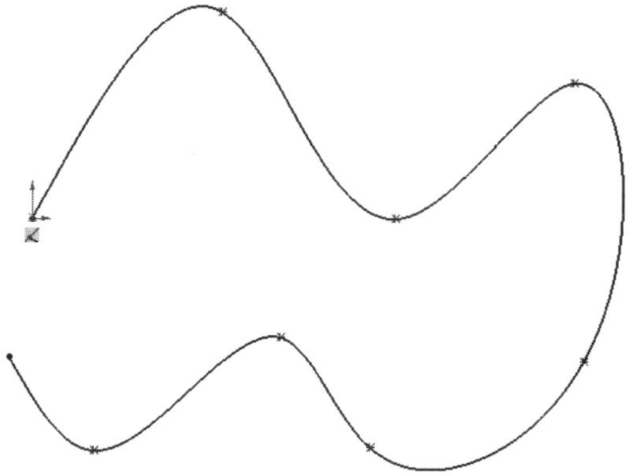

Figure 1-39 *Sketched spline with its start point at the origin*

Note

Handles are provided on the spline where points are specified to draw spline. These handles are displayed when you select the spline using the ***Select*** *tool. You will learn more about these handles in the later chapters while editing splines.*

Similar to drawing individual lines, you can also create individual spline segments by specifying the start point and then dragging the mouse to specify the endpoint.

Tip. *After creating the spline, when you select it using the* ***Select*** *tool, the* ***Spline PropertyManager*** *is displayed. The current handle will be displayed with a cyan filled square and its number and the corresponding X and Y coordinates will be displayed in the* ***Spline PropertyManager****. You can modify these coordinates to modify the selected spline. A double-sided arrow will also be displayed along with the handle. You will learn more about this in the later chapters.*

PLACING SKETCHED POINTS

CommandManager:	Sketch > Point
Menu:	Tools > Sketch Entities > Point
Toolbar:	Sketch > Point

To place a sketched point, choose the **Point** tool from the **CommandManager** and then specify the point on the screen where you want to place the sketched point. The **Point PropertyManager** will be displayed with the X and Y coordinates of the current point. You can modify the location of the point by modifying its X and Y coordinates in the **Point PropertyManager**.

DRAWING ELLIPSES

Menu: Tools > Sketch Entities > Ellipse

In SolidWorks, an ellipse is drawn by specifying its center point and then specifying the two ellipse axes by moving the mouse. To draw an ellipse, invoke the **Ellipse** tool from the menu bar; the arrow cursor will be replaced by the ellipse cursor. Move the cursor to the point that you want to specify as the center point of the ellipse. Press the left mouse button once at the center point of the ellipse and then move the cursor to specify one of the ellipse axes. You will notice that a reference circle is drawn and two values are displayed above the ellipse cursor, see Figure 1-40. The first value that shows R = * is the radius of the first axis that you are defining and the second value that shows r = * is the radius of the other axis. While defining the first axis, the second axis is taken equal to the first axis. This is the reason why a reference circle is drawn and not a reference ellipse.

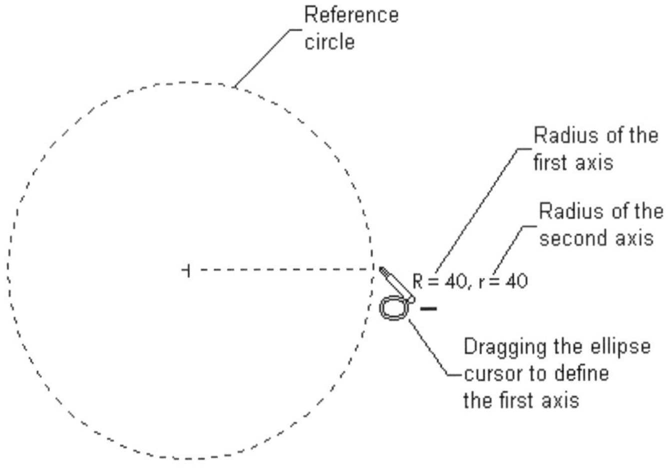

Figure 1-40 *Dragging the cursor to define the first axis*

Specify a point on the screen to define the first axis. Next, move the cursor to size the other ellipse axis. You will notice that the **Ellipse PropertyManager** is displayed. As you move the cursor, the second value above the ellipse cursor that shows r = * will change dynamically as you move the cursor on the screen. Using the left mouse button, specify a point on the screen to define the second axis of the ellipse, see Figure 1-41.

DRAWING ELLIPTICAL ARCS

Menu: Tools > Sketch Entities > Partial Ellipse

In SolidWorks, the process of drawing an elliptical arc is similar to that of drawing an ellipse. You will follow the same process of defining the ellipse first. The point that you specify on the screen to define the other axis of the ellipse is taken as the start point of the elliptical arc. You can define the endpoint of the elliptical arc by specifying a point on the screen as shown in Figure 1-42.

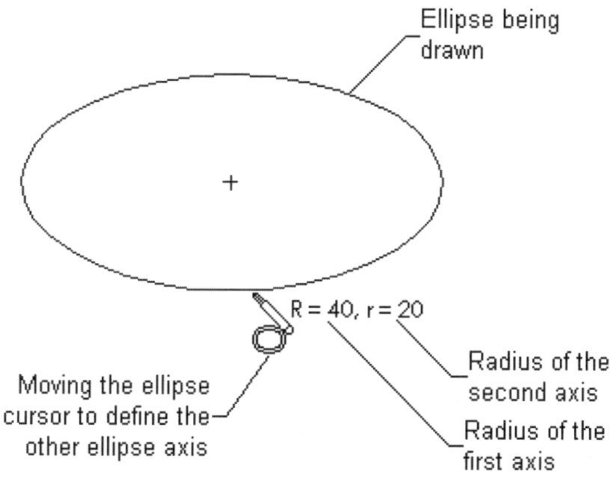

Figure 1-41 *Defining the second axis of the ellipse*

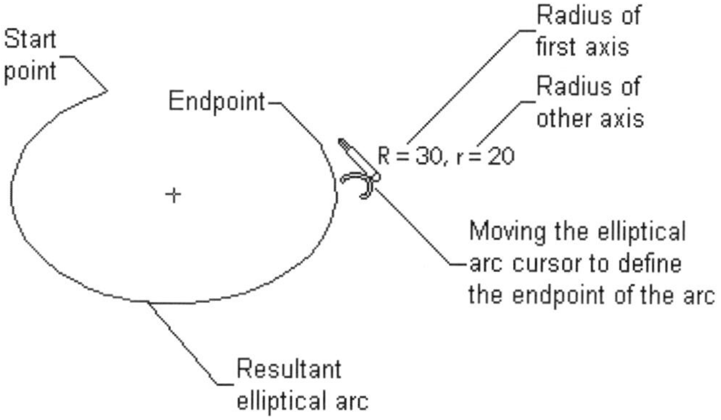

Figure 1-42 *Drawing the elliptical arc*

You can also set the parameters of the elliptical arc in the **Ellipse PropertyManager** shown in Figure 1-43.

DRAWING PARABOLIC CURVES

Menu:	Tools > Sketch Entities > Parabola

In SolidWorks, you can draw a parabolic curve by specifying the focus point of the parabola and then specifying two points on the guide of the parabolic curve. To draw a parabolic curve, invoke the **Parabola** tool from the menu bar; the cursor will be replaced by the parabola cursor. Move the cursor to the point that you want to specify as the focal point of the parabola. Press the left mouse button once at the focal point and then move the cursor to define the apex point and size the parabola. You will notice that a reference parabolic arc is displayed. As you move the cursor away from the focal point, the parabola will be flattened. After you get the basic shape of the parabolic curve, specify a point on the screen using the left mouse button. This point is taken as the apex of the parabolic curve. Next, specify two points on the screen with respect to the reference parabola to define the guide of the parabolic curve, see Figure 1-44.

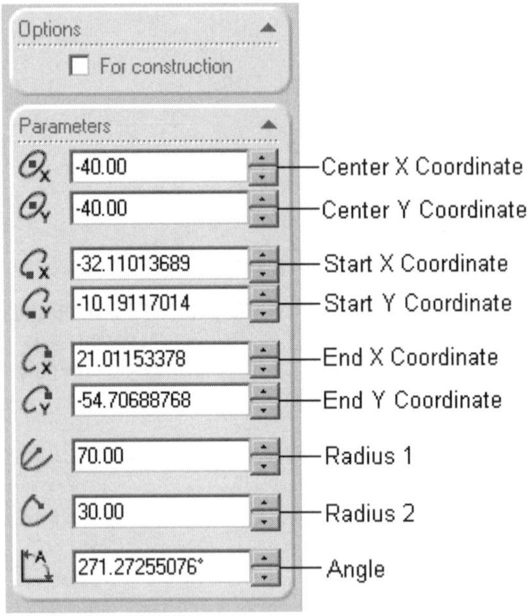

Figure 1-43 *Partial view of the* ***Ellipse PropertyManager***

Figure 1-44 *Drawing the parabola*

As you move the mouse after specifying the focal point of the parabola, the **Parabola PropertyManager** will be displayed. But the options in the **Parabola PropertyManager** will not be available. These options will be available only after you have drawn the parabola. Figure 1-45 shows a partial view of the **Parabola PropertyManager**.

DRAWING DISPLAY TOOLS

The drawing display tools are one of the most important tools provided in any of the solid modeling software. These tools allow you to modify the display of a drawing by zooming or panning it. Some of the drawing display tools available in SolidWorks are discussed in this chapter. The remaining tools are discussed in the later chapters.

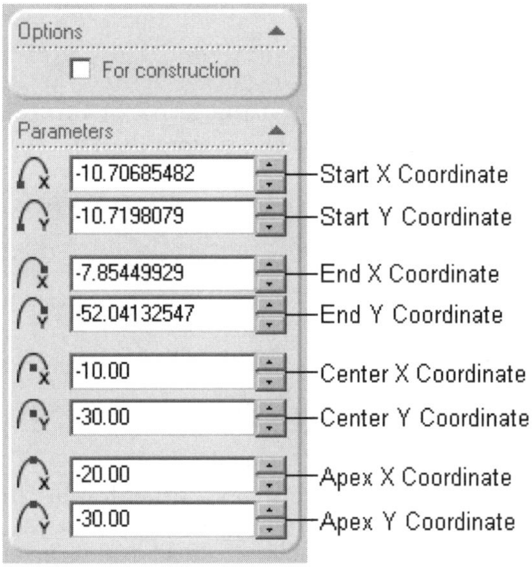

Figure 1-45 Partial view of the ***Parabola PropertyManager***

Zoom to Fit

Menu:	View > Modify > Zoom to Fit
Toolbar:	View > Zoom to Fit

The **Zoom to Fit** tool is used to increase or decrease the drawing display area so that all sketched entities or dimensions are fitted inside the current view.

Zoom to Area

Menu:	View > Modify > Zoom to Area
Toolbar:	View > Zoom to Area

The **Zoom to Area** tool is used to magnify a specified area so that the part of the drawing inside the magnified area can be viewed in the current window. The area is defined by a window that is created by dragging the cursor and specifying two opposite corners of the window. When you choose this tool, the cursor is replaced by a magnifying glass cursor. Press and hold the left mouse button down and drag the cursor to specify two opposite corners of the window. The area enclosed inside the window will be magnified.

Zoom In/Out

Menu:	View > Modify > Zoom In/Out
Toolbar:	View > Zoom In/Out

The **Zoom In/Out** tool is used to dynamically zoom in or out of the drawing. When you invoke this tool, the cursor is replaced by the zoom cursor. To zoom out of a drawing, press and hold the left mouse button down and drag the cursor in the downward direction. Similarly, to zoom in a drawing, press and hold the left mouse button down and drag the cursor in the upward direction. As you drag the cursor, the drawing display will be modified dynamically. After you get the desired view, exit this tool by choosing the **Select** tool from the **Sketch** toolbar. You can also exit this tool by right-clicking and choosing **Select** from the shortcut menu or by pressing the ESC key.

Tip. *You can also use the keyboard shortcuts to invoke some of the drawing display tools. For example, to invoke the* ***Zoom to Fit*** *tool, press the F key. Similarly, to zoom out of a drawing, press the Z key and to zoom in, press the SHIFT+Z keys.*

Zoom to Selection

Menu:	View > Modify > Zoom to Selection
Toolbar:	View > Zoom to Selection

The **Zoom to Selection** tool is used to modify the drawing display area such that the selected entity is fitted inside the current display. After selecting the entity, choose the **Zoom to Selection** button. The drawing display area will be modified such that the selected entity fits inside the current view.

Pan

Menu:	View > Modify > Pan
Toolbar:	View > Pan

The **Pan** tool is used to drag the view in the current display. This process is similar to changing the view by using the scroll bars available in the drawing area.

Tip. *You can also invoke the* ***Pan*** *tool using the CTRL key and the arrow keys on the keyboard. For example, to pan toward the right, press the CTRL key and then press the right arrow key a few times. Similarly, to pan upward, press the CTRL key and then press the up arrow key a few times.*

Redraw

Menu:	View > Redraw

The **Redraw** tool is used to refresh the screen. Sometimes when you draw a sketched entity, some unwanted elements remain on the screen. To remove these unwanted elements from the screen, choose this tool. The screen will be refreshed and all the unwanted elements will be removed. You can also invoke this tool by pressing the CTRL+R keys from the keyboard.

DELETING THE SKETCHED ENTITIES

You can delete the sketched entities by selecting them using the **Select** tool and then pressing the DELETE key from the keyboard. You can select the entities by selecting them individually or select more than one entity by defining a window or crossing around the entities. When you select the entities, they turn green. When they turn green, press the DELETE key from the keyboard. You can also delete the sketched entities by selecting them and choosing the **Delete** option from the shortcut menu that is displayed on right-clicking.

TUTORIALS

Tutorial 1

In this tutorial, you will draw the sketch of the model shown in Figure 1-46. The sketch is shown in Figure 1-47. You will not dimension the sketch. The solid model and the dimensions are given only for your reference. **(Expected time: 30 min)**

Figure 1-46 *Solid model for Tutorial 1*

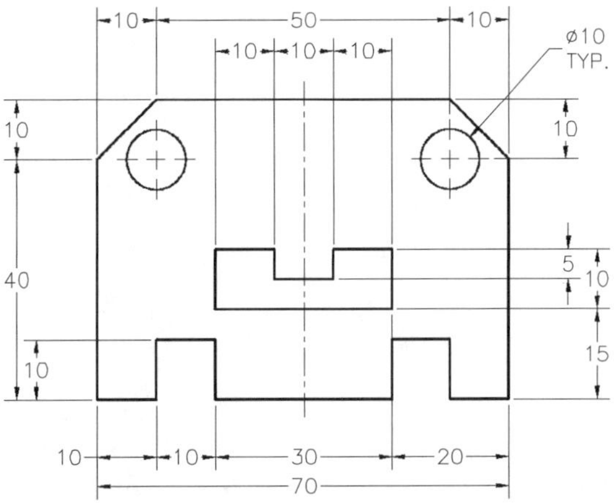

Figure 1-47 *Sketch of the model*

The steps that will be followed to complete this tutorial are listed below:

a. Start SolidWorks and then start a new part document.
b. Switch to the sketching environment.
c. Draw the sketch of the model using the **Line** and **Circle** tools, refer to Figures 1-50 through 1-52.
d. Save the sketch and then close the document.

Starting SolidWorks and Starting a New Part Document

1. Start SolidWorks by choosing **Start > Programs > SolidWorks 2005 > SolidWorks 2005** or by double-clicking on the shortcut icon of SolidWorks 2005 available on the desktop of your computer.

   The **SolidWorks 2005** window is displayed along with the **SolidWorks Resources Task Pane** on its right. The **Getting Started**, **Online Resources**, and **Tip of the Day** groups are displayed in this **Task Pane**.

   You can get many valuable tips from the **Tip of the Day** group. These tips are helpful to make the full utilization of this CAD package.

2. Choose the **New Document** option from the **Getting Started** group of the **SolidWorks Resources Task Pane** to display the **New SolidWorks Document** dialog box.

4. The **Part** button is chosen by default. Choose the **OK** button from the **New SolidWorks Document** dialog box as shown in Figure 1-48.

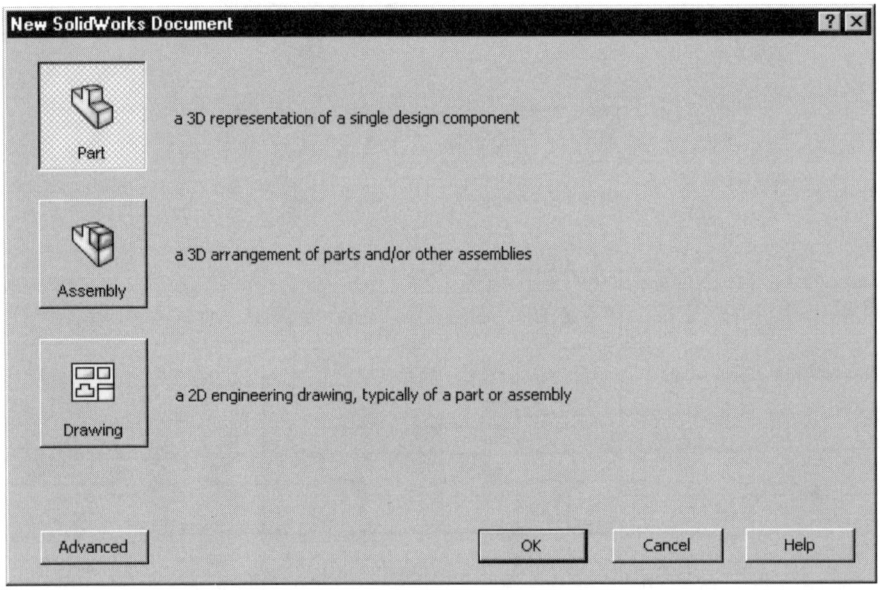

***Figure 1-48** **New SolidWorks Document** dialog box*

Tip. *If the shortcut icon of SolidWorks is not created automatically on the desktop of your computer when you install SolidWorks, you can create it manually. To create the shortcut icon on the desktop, choose **Start > Programs > SolidWorks 2005** to display the SolidWorks cascading menu. Right-click on **SolidWorks 2005** in the cascading menu and then choose **Send To > Desktop (create shortcut)** from the shortcut menu.*

A new SolidWorks part document will be started. When you start a new part document, the part modeling environment is active by default. As you first need to draw the sketch of the feature, you need to invoke the sketching environment.

5. Choose the **Sketch** button from the **Standard** toolbar; the **Edit Sketch PropertyManager** is displayed and you are prompted to select a plane on which you want to draw the sketch.

6. Select **Front Plane** from the drawing area; the sketching environment is invoked and the plane is oriented normal to the view. You will notice that a red origin is displayed in the center of the screen, indicating the sketching environment. The default screen appearance of the sketching environment of SolidWorks is shown in Figure 1-49.

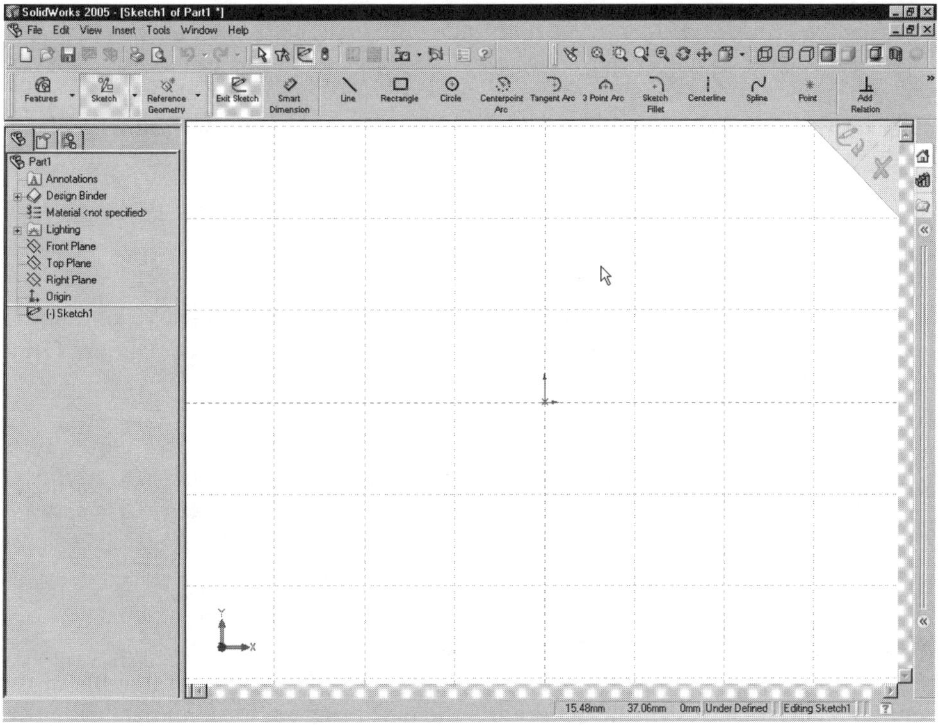

Figure 1-49 *Screen display in the sketching environment*

Setting the Units and Grid

It is assumed that while installing SolidWorks, you selected the option of measuring the length in millimeters. This is the reason why the length will be measured in millimeter in the current file. But if you selected some other unit, you need to make some initial settings of changing the linear and angular units before you proceed with drawing the sketch.

1. Choose **Tools > Options** from the menu bar to invoke the **System Options - General** dialog box.

2. Choose the **Document Properties** tab; the name of the dialog box is changed to **Document Properties - Detailing**.

3. Select the **Units** option from the area on the left to display the options related to linear and angular units.

4. Select the **MMGS (millimeter, gram, second)** radio button from the **Unit system** area if it is not selected by default. Also, select the **Degrees** option from the drop-down list provided in the **Angular units** area.

Note
*If you selected **Millimeters** as the unit while installing SolidWorks, you can skip the points discussed earlier in this section.*

5. Select **Grid/Snap** from the area on the left. Set the value of the **Major grid spacing** spinner to **100** and the value of the **Minor-lines per major** spinner to **20**.

6. Now, choose the **Go To System Snaps** button; the system options related to relations and snap are displayed.

7. Select the **Grid** check box from the **Sketch Snaps** area if it is cleared. Make sure that you clear the **Snap only when grid is displayed** check box if it is selected. Choose **OK** to exit the dialog box.

Tip. *If the grid is displayed on the screen when you invoke the sketching environment for the first time, then you can set the option to turn off the grid display. Right-click in the drawing area to display the shortcut menu. The **Display Grid** option has a check mark on its left, indicating that this option is chosen. Select this option again to turn the grids off.*

Drawing the Outer Loop of the Sketch

It is a good practice to draw the sketch on one side of the origin, preferably in the first quadrant. This is because while generating the part program for manufacturing the part, you will have a reference for work origin in advance.

The sketch of the model consists of an outer loop, two circles inside the outer loop, and a cavity. Therefore, it will be drawn using the **Line** and **Circle** tools. You will first draw the outer loop and then the inner entities. Note that in the sketching environment, the lower right corner of the SolidWorks window displays three areas. The first area displays the X, Y, and Z coordinates of the current location of the cursor. These coordinates will be modified as you move the cursor around the drawing area. You will use the coordinate display to draw the sketch of the model.

You will start drawing the sketch from the lower left corner of the sketch and the outer loop will be drawn using the continuous lines.

1. Choose the **Line** button from the **Sketch CommandManager** to invoke the **Line** tool; the arrow cursor is replaced by the line cursor.

2. Move the cursor in the first quadrant close to the origin; the coordinates of the point are displayed close to the lower right corner of the screen.

3. Press the left mouse button at the point whose coordinates are 10 mm 10 mm 0 mm and then move the cursor horizontally toward the right.

   You will notice that the symbol of the **Horizontal** relation is displayed below the line cursor and the length of the line is displayed above the line cursor.

   As the length of the first horizontal line at the lower left corner of the sketch is 10 mm, you will move the mouse until the length of the line above the line cursor is shown as 10.

4. Press the left mouse button when the length of the line that is displayed above the line cursor shows a value of 10. Make sure the **Horizontal** relation symbol is displayed below the cursor.

   The first horizontal line is drawn. Because you are drawing continuous lines, the endpoint of the last line is automatically selected as the start point of the next line.

5. Move the line cursor vertically upward. The symbol of the **Vertical** relation is displayed below the line cursor and the length of the line is displayed above the line cursor.

6. Press the left mouse button when the length of the line displayed above the line cursor shows a value of 10 and the **Vertical** relation symbol is displayed below the cursor.

   A vertical line of length 10 mm will be drawn and displayed in green. Also, as this is the line that is selected now, the previous line will no longer be highlighted and therefore will be displayed in blue.

7. Move the line cursor horizontally toward the right. Press the left mouse button when the length of the line above the line cursor shows a value of 10. This draws the next horizontal line of 10 mm length.

8. Move the line cursor vertically downward and press the left mouse button when the length of the line on the line cursor shows a value of 10.

Tip. *If by mistake you invoke the arc mode while drawing lines, move the cursor back to the endpoint of the previous line and press the left mouse button. The line mode will be invoked again.*

9. Move the line cursor horizontally toward the right and press the left mouse button when the length of the line on the line cursor shows a value of 30.

10. Move the line cursor vertically upward and press the left mouse button when the length of the line on the line cursor shows a value of 10.

11. Move the line cursor horizontally toward the right and press the left mouse button when the length of the line on the line cursor shows a value of 10.

12. Move the line cursor vertically downward and press the left mouse button when the length of the line on the line cursor shows a value of 10.

13. Move the line cursor horizontally toward the right and press the left mouse button when the length of the line on the line cursor shows a value of 10.

14. Move the line cursor vertically upward and press the left mouse button when the length of the line on the line cursor displays a value of 40.

    The next line that you will draw is an inclined line that makes an angle of 135-degree. To draw this line, you will move the cursor in a direction that makes an angle of 135-degree.

15. Move the line cursor such that the line is drawn at an angle of 135-degree and the length of the line displays a value of 14.14 above the cursor.

16. Press the left mouse button at this location to specify the endpoint of the inclined line.

17. Move the line cursor horizontally toward the left and press the left mouse button when the length of the line on the line cursor displays a value of 50.

    You will notice that some yellow inferencing lines are displayed when you move the cursor.

18. Move the line cursor in the direction diagonally downward where the value of the angle displays a value of 135-degree and the length of the lines displays a value of 14.14.

19. Press the left mouse button at this location.

20. Move the cursor vertically downward to the start point of the first line.

    You will notice that when you move the cursor close to the start point of the first line, a red circle is displayed. Symbols of the **Vertical** and **Coincident** relations are displayed on the right of the cursor. The length of the line shows a value of 40.

21. Press the left mouse button when the red circle is displayed. Right-click to display the shortcut menu and choose the **Select** option to exit the **Line** tool.

    This completes the sketch of the outer loop. Note that the display of the sketch is small. Therefore, you need to modify the drawing display area such that the sketch fits the screen. This is done using the **Zoom to Fit** tool.

22. Choose the **Zoom to Fit** button from the **View** toolbar to fit the current sketch on the screen. The outer loop of the sketch is completed and is shown in Figure 1-50. Note that in this figure, the display of grid is temporarily turned off for better visibility. To do so, clear the **Display grid** check box from the **Grid** area of the **Document Properties - Grid/Snap** dialog box.

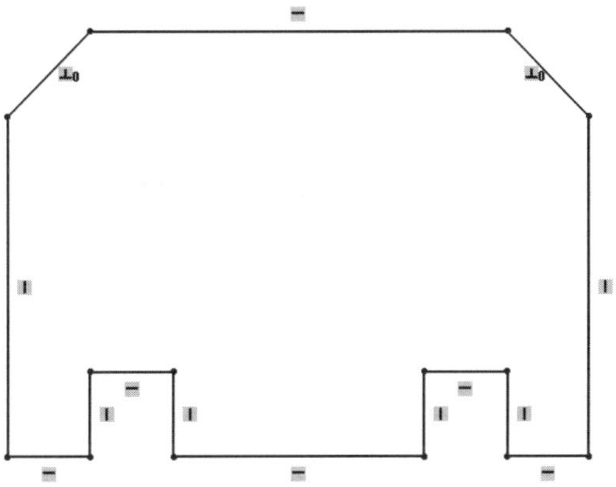

Figure 1-50 *Outer loop of the sketch*

Drawing Circles

The circles will be drawn using the **Circle** tool. You will use the inferencing line originating from the start points and endpoints of the inclined lines to specify the center point of the circles. At a given time you can either snap to grid or use inferencing lines to draw the sketches. Therefore, you need to turn off the snapping to grid.

1. Choose **Tools > Options** from the menu bar to invoke the **System Options - General** dialog box. Select **Relations/Snaps** from the left of this dialog box. Clear the **Grid** check box and choose the **OK** button from this dialog box.

2. Choose the **Circle** button from the **Sketch CommandManager** to invoke the **Circle** tool.

    As the **Select** tool was active earlier, the cursor earlier was the arrow cursor. But when you invoke the **Circle** tool, the arrow cursor will be replaced by the circle cursor.

4. Move the circle cursor close to the lower endpoint of the right inclined line and then move it toward the left. Remember that you will not press the left mouse button at this moment.

An inferencing line is displayed originating from the lower endpoint of the right inclined line. When you move the cursor toward the left, you will notice that at the point where the cursor is vertically in line with the upper endpoint of the right inclined line, another inferencing line is originated from the upper endpoint of the right inclined line. This inferencing line will intersect the inferencing line generated from the lower endpoint of the inclined line.

5. Press the left mouse button at the point where the inferencing lines from both the endpoints of the inclined lines intersect. Now, move the circle cursor toward the left to define a circle.

6. Press the left mouse button when the radius of the circle displayed above the circle cursor shows a value close to 5.

7. The **Circle PropertyManager** is displayed. Set the value of the **Radius** spinner to 5 from this **PropertyManager**.

8. Similarly, draw the circle on the left using the inferencing lines generating from the endpoints of the left inclined line. The sketch after drawing the two circles inside the outer loop is shown in Figure 1-51. In this figure also, the grids are turned off for clarity.

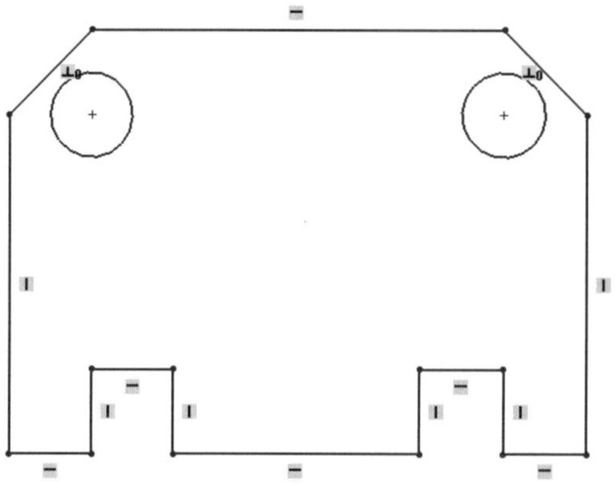

Figure 1-51 *Sketch after drawing the two inner circles*

9. Right-click in the drawing area and choose **Select** to exit the **Circle** tool.

Drawing the Sketch of the Inner Cavity

Next, you will draw the sketch of the inner cavity. To draw the sketch of the inner cavity, you will start drawing with the lower horizontal line. Before proceeding further, you need to invoke the snap to grid option.

1. Select the **Grid** option from the **System Options** dialog box. Now, invoke the **Line** tool by pressing the L key from the keyboard; the arrow cursor is replaced by the line cursor.

2. Move the line cursor to a location whose coordinates are 30 mm 25 mm 0 mm.

3. Press the left mouse button at this point and move the cursor horizontally toward the right. Press the left mouse button when the length of the line above the line cursor shows a value of 30.

4. Move the line cursor vertically upward and press the left mouse button when the length of the line on the line cursor displays a value of 10.

5. Move the line cursor horizontally toward the left and press the left mouse button when the length of the line on the line cursor displays a value of 10.

6. Move the line cursor vertically downward and press the left mouse button when the length of the line on the line cursor displays a value of 5.

7. Move the line cursor horizontally toward the left and press the left mouse button when the length of the line on the line cursor displays a value of 10.

8. Move the line cursor vertically upward and press the left mouse button when the length of the line on the line cursor displays a value of 5.

9. Move the line horizontally toward the left and press the left mouse button when the length of the line on the line cursor displays a value of 10.

10. Move the line cursor vertically downward to the start point of the first line. Press the left mouse button when the red circle is displayed. The length of the line at this point will show a value of 10.

11. Right-click and choose **Select** from the shortcut menu. This completes the sketch for Tutorial 1.

12. Choose the **Zoom to Fit** button from the **Standard** toolbar to fit the display of the sketch on the screen. The final sketch for Tutorial 1 is shown in Figure 1-52.

Saving the Sketch

It is recommended that you will create a separate folder for saving the tutorial files of this book. When you invoke the option to save a document, the default folder *\My Documents* will be displayed. You will create a folder with the name *SolidWorks* in the *My Documents* folder and then create the folder of each chapter inside the *SolidWorks* folder. As a result, you can save the tutorials of a chapter in the folder of that chapter.

1. Choose the **Save** button from the **Standard** toolbar to invoke the **Save As** dialog box. Create the *SolidWorks* folder inside the *\My Documents* folder and then create the *c01* folder inside the *SolidWorks* folder.

2. Enter the name of the document as *c01tut1* in the **File name** edit box and choose the **Save** button. The document will be saved in the *\My Documents\SolidWorks\c01* folder.

3. Close the document by choosing **File > Close** from the menu bar.

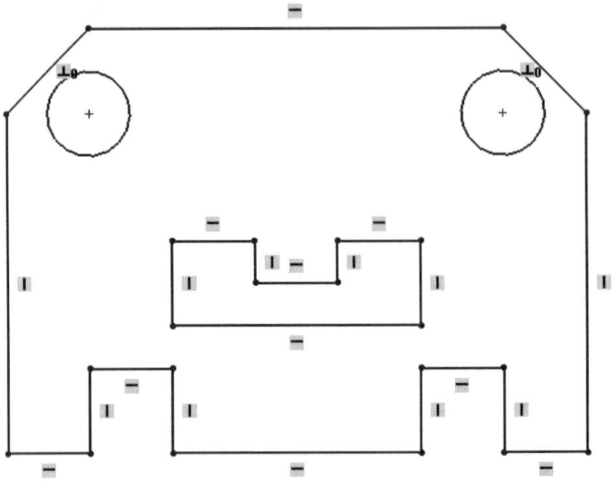

Figure 1-52 Final sketch for Tutorial 1

Tip. *If you open a document that was saved in the sketching environment, it will be opened in the sketching environment only and not in the part modeling environment.*

Tutorial 2

In this tutorial, you will draw the basic sketch of the revolved solid model shown in Figure 1-53. The sketch of the revolved solid model is shown in Figure 1-54. Do not dimension the sketch. The solid model and dimensions are given only for your reference.

(Expected time: 30 min)

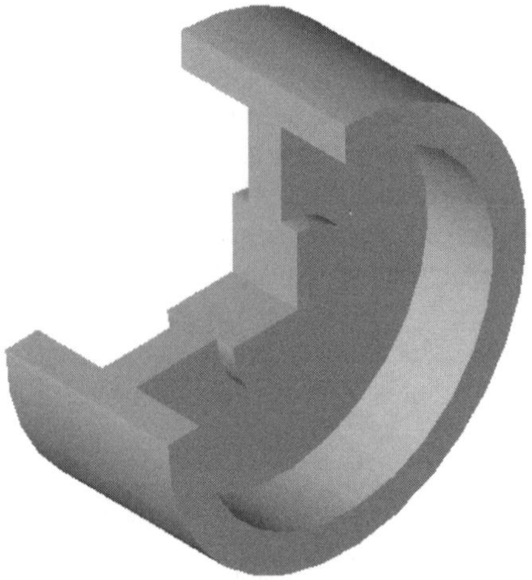

Figure 1-53 Revolved model for Tutorial 2

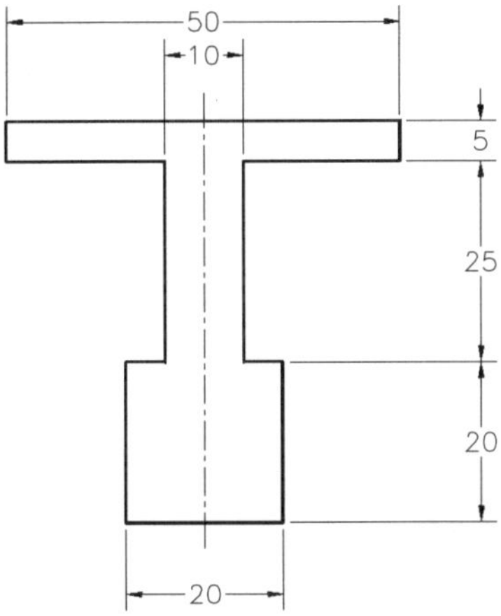

Figure 1-54 *Sketch for the revolved model*

The steps that will be followed to complete this tutorial are listed below.

a. Start a new part document.
b. Switch to the sketching environment.
c. Modify the settings of the snap and grid so that the cursor jumps through a distance of 5 mm instead of 10 mm.
d. Draw the sketch of the model using the **Line** tool, refer to Figure 1-55.
e. Save the sketch and then close the document.

Starting a New Document

1. Choose the **New** button from the **Standard** toolbar to invoke the **New SolidWorks Document** dialog box.

2. The **Part** button is chosen by default in the **New SolidWorks Document** dialog box. Choose **OK**.

   A new SolidWorks part document is started. As mentioned earlier, when you start a new part document, the part modeling environment is active by default. But, because you first need to draw the sketch of the revolved model, you need to invoke the sketching environment.

3. Choose the **Sketch** button from the **Standard** toolbar to display the **Edit Sketch PropertyManager**. Select **Front Plane**.

   A red origin and the **Sketch CommandManager** are displayed. Also, the confirmation corner is displayed with the **Exit Sketch** and **Delete Sketch** options on the upper right corner of the drawing area. This indicates that the sketching environment is activated.

Modifying the Snap and Grid Settings and the Dimensioning Units

Before you proceed with drawing the sketch, you need to modify the grid and snap settings so that you can make the cursor jump through a distance of 5 mm.

1. Choose **Tools > Options** from the menu bar to invoke the **System Options - General** dialog box. Choose the **Document Properties** tab.

2. Select the **Grid/Snap** option from the area on the left to display the options related to linear and angular units. Set the value of the **Major grid spacing** spinner to **50**. Make sure the value of the **Minor-lines per major** spinner is **10**.

   The coordinates close to the lower left corner of the SolidWorks window will show an increment of 5 mm instead of the default increment of 25 mm when you exit the dialog box.

   If you selected a unit other than millimeter while installing SolidWorks, you need to change the unit for the current drawing.

3. Select the **Units** option from the area on the left of the **Document Properties - Grid/Snap** dialog box.

4. Select the **MMGS (millimeter, gram, second)** radio button from the **Unit system** area.

5. Make sure the **Grid** check box is selected in the **System Options - Relation/Snaps** dialog box. After setting all parameters, choose the **OK** button from this dialog box.

Drawing the Sketch

As evident from Figure 1-52, the sketch will be drawn using the **Line** tool. You will start drawing the sketch from the lower left corner of the sketch.

1. Choose the **Line** button from the **Sketch CommandManager**; the arrow cursor is replaced by the line cursor.

2. Move the line cursor to a location whose coordinates are 40 mm 0 mm 0 mm.

3. Press the left mouse button down at this point and move the cursor horizontally toward the right. Press the left mouse button again when the length of the line above the line cursor shows a value of 20.

4. Move the line cursor vertically upward and press the left mouse button when the length of the line on the line cursor displays a value of 20.

5. Move the cursor horizontally toward the left and press the left mouse button when the length of the line on the line cursor displays a value of 5.

6. Move the line cursor vertically upward and press the left mouse button when the length of the line on the line cursor displays a value of 25.

7. Move the line cursor horizontally toward the right and press the left mouse button when the length of the line on the line cursor displays a value of 20.

8. Move the line cursor vertically upward and press the left mouse button when the length of the line on the line cursor displays a value of 5.

9. Move the line cursor horizontally toward the left and press the left mouse button when the length of the line on the line cursor displays a value of 50.

10. Move the line cursor vertically downward and press the left mouse button when the length of the line on the line cursor displays a value of 5.

11. Move the line cursor horizontally toward the right and press the left mouse button when the length of the line on the line cursor displays a value of 20.

12. Move the line cursor vertically downward and press the left mouse button when the length of the line on the line cursor displays a value of 25.

13. Move the line cursor horizontally toward the left and press the left mouse button when the length of the line on the line cursor displays a value of 5.

14. Move the line cursor vertically downward to the start point of the first line. Press the left mouse button when the red circle is displayed. The length of the line at this point will be 20 mm

15. Right-click and choose **Select** from the shortcut menu.

    The sketch is completed but does not fit the screen. Therefore, you need to modify the display area such that the sketch fits the screen.

16. Choose the **Zoom to Fit** button from the **View** toolbar to fit the sketch on the screen. The final sketch for Tutorial 2 is shown in Figure 1-55. In this figure, the display of grid is turned off for clarity.

Tip. *You will notice that the bottom horizontal line in the sketch is black and the remaining lines are blue. In the next chapter, you will learn about the reason why some entities in the sketch have a different color.*

Saving the Sketch

1. Choose the **Save** button from the **Standard** toolbar to invoke the **Save As** dialog box.

2. Enter the name of the document as *c01tut2* in the **File name** edit box and choose the **Save** button.

   The document will be saved in the *\My Documents\SolidWorks\c01* folder.

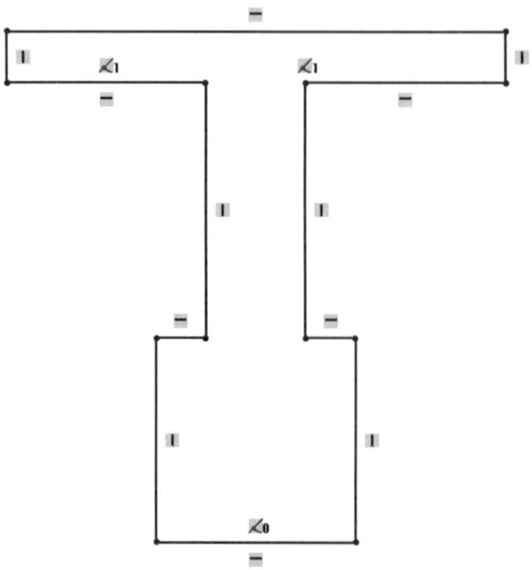

Figure 1-55 Final sketch for Tutorial 2

3. Close the document by choosing **File > Close** from the menu bar.

Tutorial 3

In this tutorial, you will draw the basic sketch of the model shown in Figure 1-56. The sketch to be drawn is shown in Figure 1-57. Do not dimension the sketch; the solid model and dimensions are given only for your reference. **(Expected time: 30 min)**

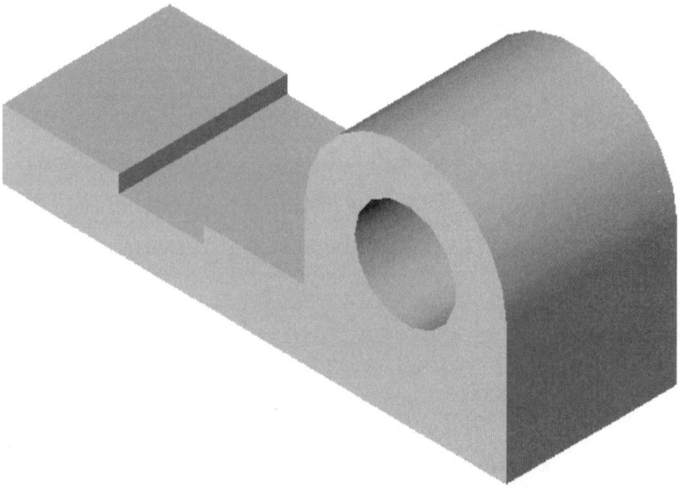

Figure 1-56 Solid model for Tutorial 3

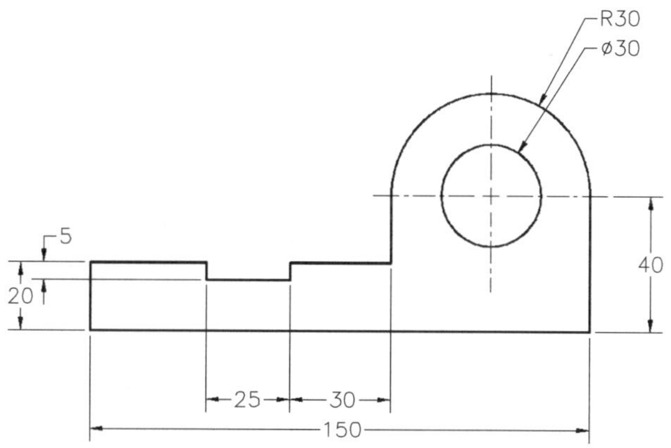

Figure 1-57 *Sketch for Tutorial 3*

The steps that will be followed to complete this tutorial are listed below.

a. Start SolidWorks and then start a new part file.
b. Switch to the sketching environment.
c. Modify the settings of the snap and grid so that the cursor jumps through a distance of 5 mm instead of 10 mm.
d. Draw the outer loop of the sketch using the **Line** tool, refer to Figure 1-58.
e. Draw the inner circle using the **Circle** tool, refer to Figure 1-59.
f. Save the sketch and then close the file.

Starting a New File

1. Choose the **New** button from the **Standard** toolbar to invoke the **New SolidWorks Document** dialog box.

2. The **Part** button is chosen by default in the **New SolidWorks Document** dialog box. Choose the **OK** button.

   A new SolidWorks part document will be started. As you first need to draw the sketch of the revolved model, you need to invoke the sketching environment.

4. Choose the **Sketch** button from the **Standard** toolbar. The **Edit Sketch PropertyManager** is displayed. Select **Front Plane**.

   A red origin is displayed and the **Sketch CommandManager** is displayed above the drawing area. Also, the confirmation corner with the **Exit Sketch** and **Delete Sketch** options is displayed on the upper right corner of the drawing area. This indicates that the sketching environment is activated.

Modifying the Snap and Grid Settings and Dimensioning Units

As the dimensions in the sketch are multiple of 5, you need to modify the grid and snap settings so that you can make the cursor jump through a distance of 5 mm instead of 10 mm.

1. Choose **Tools > Options** from the menu bar to invoke the **System Options - General** dialog box. Choose the **Document Properties** tab.

2. Select the **Grid/Snap** option from the area on the left to display the options related to linear and angular units. Set the value of the **Major grid spacing** spinner to **50**. Make sure the value of the **Minor-lines per major** spinner is **10**.

   The coordinates displayed close to the lower left corner of the SolidWorks window will show an increment of 5 mm instead of the default increment of 10 mm when you close the dialog box.

3. Choose **OK** to close the dialog box.

Drawing the Outer Loop

As evident from Figure 1-57, the sketch consists of an outer loop and an inner circle. Therefore, this sketch will be drawn using the **Line** and **Circle** tools. You will start drawing from the lower left corner of the sketch. As the length of the lower horizontal line is 150 mm, you need to modify the drawing display area such that the drawing area in the first quadrant is increased. This can be done using the **Pan** tool.

1. Choose the **Pan** button from the **View** toolbar; the arrow cursor is replaced by the pan cursor.

2. Press and hold the left mouse button down and drag the cursor toward the bottom left corner of the screen.

   You will notice that the origin also moves toward the bottom left corner of the screen, thus increasing the drawing area in the first quadrant.

3. After dragging the origin close to the lower left corner, release the left mouse button.

4. Choose the **Line** button from the **Sketch CommandManager**; the pan cursor is replaced by the line cursor.

5. Move the line cursor to a location whose coordinates are 40 mm 0 mm 0 mm.

6. Press and hold the left mouse button down at this point and move the cursor horizontally toward the right. Press the left mouse button again when the length of the line above the line cursor shows a value of 150.

7. Move the line cursor vertically upward and press the left mouse button when the length of the line on the line cursor displays a value of 40.

   The next entity that has to be drawn is a tangent arc. As mentioned earlier, you can draw a tangent arc using the **Line** tool also. Drawing arcs from within the **Line** tool is a recommended method when you need to draw a sketch that is a combination of lines and arcs. This increases the productivity by reducing the time taken in invoking the tools for drawing an arc and then invoking the **Line** tool to draw lines.

8. Move the line cursor away from the endpoint of the last line and then move it back to the endpoint.

   The arc mode is invoked and the line cursor is replaced by the arc cursor. Also, the **Arc PropertyManager** is displayed in place of the **Line Properties PropertyManager**.

9. Move the arc cursor vertically upward to a small distance.

10. When the dotted line is displayed, move the cursor toward the left.

    You will notice that a tangent arc is being drawn. The angle of the tangent arc and its radius are displayed above the arc cursor.

11. Press the left mouse button when the angle value above the arc cursor shows 180 and the radius shows a value of 30 to complete the arc.

    The required tangent arc is drawn. As mentioned earlier, the line mode is automatically invoked after you have drawn the arc using the **Line** tool. Therefore, the arc cursor will be replaced by the line cursor and the **Arc PropertyManager** will be replaced by the **Line Properties PropertyManager**.

12. Move the line cursor vertically downward and press the left mouse button when the length of the line on the line cursor displays a value of 20.

13. Move the line cursor horizontally toward the left and press the left mouse button when the length of the line on the line cursor displays a value of 30.

14. Move the line cursor vertically downward and press the left mouse button when the length of the line on the line cursor displays a value of 5.

15. Move the line cursor horizontally toward the left and press the left mouse button when the length of the line on the line cursor displays a value of 25.

16. Move the line cursor vertically upward and press the left mouse button when the length of the line on the line cursor displays a value of 5.

17. Move the line cursor horizontally toward the left and press the left mouse button when the length of the line on the line cursor displays a value of 35.

18. Move the line cursor to the start point of the first line. Press the left mouse button when the red circle is displayed.

    The length of the line at this point will be 20 mm

19. Right-click and choose **Select** from the shortcut menu to exit the **Line** tool.

20. Choose the **Zoom to Fit** button to fit the sketch on the screen. This completes the outer loop of the sketch. The sketch after drawing the outer loop is shown in Figure 1-58.

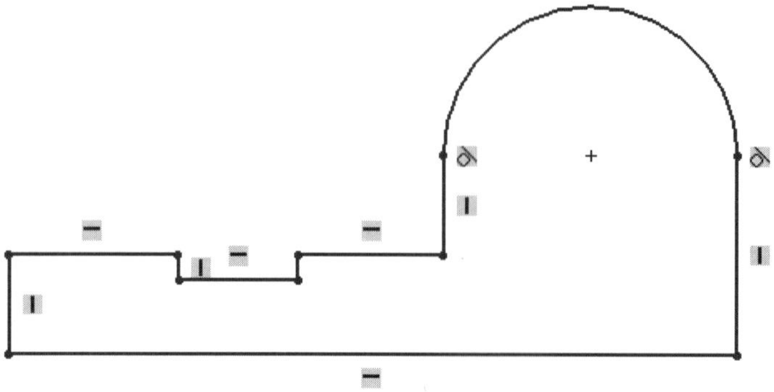

Figure 1-58 *Sketch after drawing the outer loop*

Drawing the Circle

The circle in the sketch will be drawn using the **Circle** tool. The center point of the circle will be the center point of the arc, which will be displayed by a plus sign. This plus sign is automatically drawn when you draw the arc. You can select this center point to draw the circle.

1. Choose the **Circle** button from the **Sketch CommandManager** to invoke the **Circle** tool; the arrow cursor is replaced by the circle cursor.

2. Move the circle cursor close to the center point of the arc and press the left mouse button when the red circle is displayed.

3. Move the cursor toward the left and when the radius of the circle above the circle cursor shows a value of 15, press the left mouse button. A circle of 15 mm radius is drawn.

4. This completes the sketch for Tutorial 3. Right-click and choose the **Select** option from the shortcut menu to exit the **Circle** tool.

The final sketch for Tutorial 3 is shown in Figure 1-59.

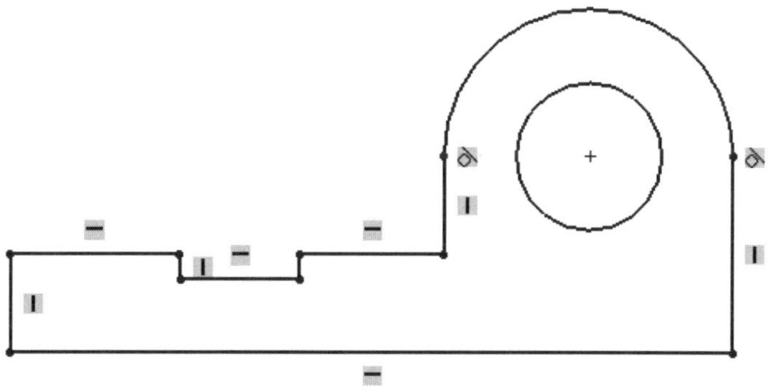

Figure 1-59 Final sketch for Tutorial 3

Saving the Sketch

1. Choose the **Save** button from the **Standard** toolbar to invoke the **Save As** dialog box.
2. Enter the name of the document as *c01tut3* in the **File name** edit box and choose the **Save** button.
3. Close the document by choosing **File > Close** from the menu bar.

Tutorial 4

In this tutorial, you will draw the sketch of the model shown in Figure 1-60. The sketch of the model is shown in Figure 1-61. Do not dimension the sketch. The dimensions and the solid model are given only for your reference. **(Expected time: 30 min)**

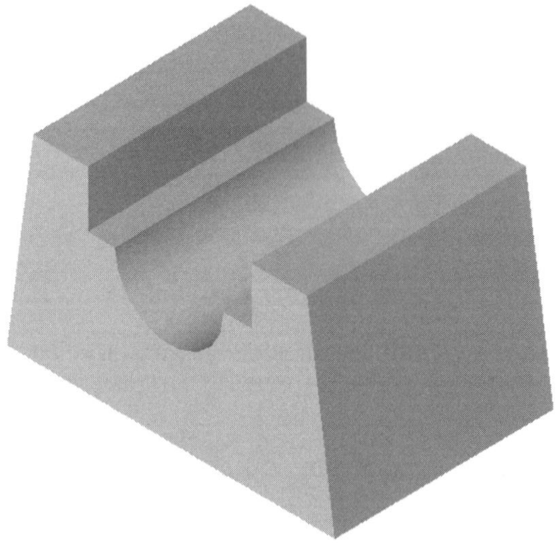

Figure 1-60 Model for Tutorial 4

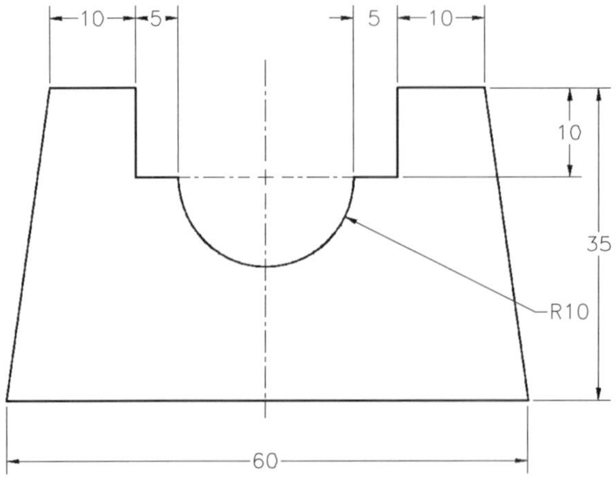

Figure 1-61 Sketch for Tutorial 4

The steps that will be followed to complete this tutorial are listed below.

a. Start SolidWorks and then start a new part document.
b. Switch to the sketching environment.
c. Modify the settings of the snap and grid so that the cursor jumps through a distance of 5 mm instead of 10 mm.
d. Draw the sketch using the **Line** tool, refer to Figure 1-62.
e. Save the sketch and then close the file.

Opening a New File

1. Choose the **New** button from the **Standard** toolbar to invoke the **New SolidWorks Document** dialog box.

2. The **Part** button is chosen by default in the **New SolidWorks Document** dialog box. Now choose the **OK** button.

   As you first need to draw the sketch of the model, you need to invoke the sketching environment.

4. Choose the **Sketch** button from the **Standard** toolbar and select the **Front Plane** to invoke the sketching environment.

Modifying the Snap and Grid Settings and Dimensioning Units

As evident in Figure 1-61, the dimensions in the sketch are multiple of 5. Therefore, you need to modify the grid and snap settings so that the cursor jumps through a distance of 5 mm instead of 10 mm.

1. Choose **Tools > Options** from the menu bar to invoke the **System Options - General** dialog box. Choose the **Document Properties** tab.

2. Select the **Grid/Snap** option from the area on the left to display the options related to linear and angular units. Set the value of the **Major grid spacing** spinner to **50** and the value of the **Minor-lines per major** spinner to **10**.

   The coordinates displayed close to the lower left corner of the SolidWorks window show an increment of 5 mm when you close this dialog box.

3. Choose **OK** to close the dialog box.

Drawing the Sketch

The sketch will be drawn using the **Line** tool. The arc in the sketch will also be drawn using the same tool. You will start drawing from the lower left corner of the sketch.

1. Invoke the **Line** tool by pressing the L key from the keyboard; the arrow cursor is replaced by the line cursor.

2. Move the line cursor to a point whose coordinates are 30 mm 0 mm 0 mm.

3. Press the left mouse button at this point and move the cursor horizontally toward the right. Press the left mouse button again when the length of the line above the line cursor shows a value of 60.

   The bottom horizontal line of 60 mm length is drawn.

4. Choose the **Zoom to Fit** button from the **View** toolbar to increase the display of the line that is drawn.

   As mentioned earlier, you can also invoke the drawing display tools while some other tool is active. After modifying the drawing display, the tool that was active before invoking the drawing display tool will be restored and you can continue using that tool. Therefore, after the drawing display area is modified, the **Line** tool will be restored and you can continue drawing lines.

5. Move the line cursor along a direction that makes an angle close to 98-degree with the positive X axis direction. The angle can be checked from the spinner below the **Length** spinner in the **Line Properties PropertyManager**.

6. Press the left mouse button when the length of the line displayed above the line cursor shows a value of 35.36.

7. Move the line cursor horizontally toward the left and press the left mouse button when the length of the line above the line cursor shows a value of 10.

8. Move the line cursor vertically downward and press the left mouse button when the length of the line above the line cursor shows a value of 10.

9 Move the line cursor horizontally toward the left and press the left mouse button when the length of the line above the line cursor shows a value of 5.

   Next, you need to draw the arc that is normal to the last line.

10. Move the line cursor away from the endpoint of the last line and then move it back close to the endpoint.

    The arc mode is invoked and the line cursor is replaced by the arc cursor. Also, the **Line Properties PropertyManager** is replaced by the **Arc PropertyManager**.

11. Move the arc cursor vertically downward up to the next grid point.

12. Move the arc cursor toward the left.

    You will notice that a normal arc is being drawn and the angle and radius of the arc is displayed above the arc cursor.

13. Press the left mouse button when the angle value on the arc cursor is 180 and the radius value is 10. An arc normal to the last line is drawn and the line mode is activated.

14. Move the line cursor horizontally toward the left and press the left mouse button when the length of the line on the line cursor shows a value of 5.

15. Move the line cursor vertically upward and press the left mouse button when the length of the line on the line cursor shows a value of 10.

16. Move the line cursor horizontally toward the left and press the left mouse button when the length of the line on the line cursor shows a value of 10.

17. Move the line cursor to the start point of the first line and press the left mouse button when the red circle is displayed.

18. Press the ESC key to exit the **Line** tool.

    This completes the sketch. However, you need to modify the drawing display area such that the sketch fits the screen.

19. Choose the **Zoom to Fit** button from the **View** toolbar to modify the drawing display area. The final sketch for Tutorial 4, without the grid display, is shown in Figure 1-62.

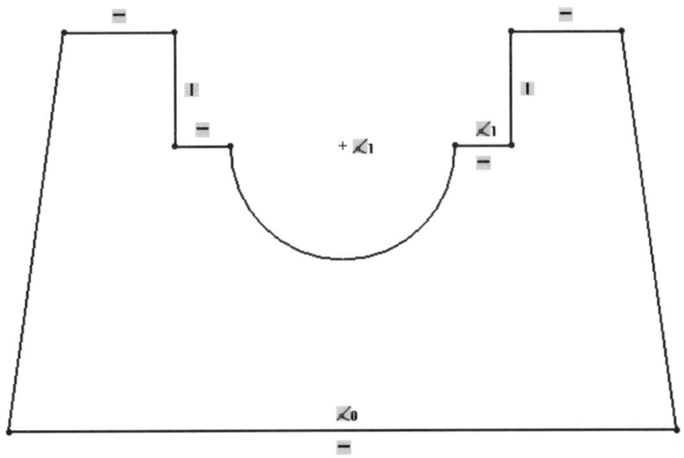

Figure 1-62 *Final sketch for Tutorial 4*

Saving the Sketch

1. Choose the **Save** button from the **Standard** toolbar to invoke the **Save As** dialog box.

2. Enter the name of the document as *c01tut4* in the **File name** edit box and choose the **Save** button.

3. Close the document by choosing **File > Close** from the menu bar.

SELF-EVALUATION TEST

Answer the following questions and then compare your answers with those given at the end of this chapter.

1. The base feature of any design is a sketched feature and is created by drawing the sketch. (T/F)

2. You can also invoke the **3Pt Arc** tool from within the **Line** tool. (T/F)

3. By default, the cursor jumps through a distance of 5 mm. (T/F)

4. When you save a file in the sketching environment, it is opened in the part modeling environment when you open it for the next time. (T/F)

5. You can convert a sketched entity into a construction entity by selecting the _________ check box provided in the **PropertyManager**.

6. To draw a rectangle at an angle, you need to use the _________ tool.

7. The _________ are the temporary lines that are used to track a particular point on the screen.

8. You can also invoke the _________ tool or exit a sketching tool by pressing the ESC key.

9. When you select a tangent entity to draw a tangent arc, the _________ relation is applied between the start point of the arc and the tangent entity.

10. The rectangle is considered as a combination of individual _________.

REVIEW QUESTIONS

Answer the following questions.

1. The three point arcs are those that are drawn by defining the start point of the arc, the endpoint of the arc, and a point on the arc. (T/F)

2. You can also delete the sketched entities by selecting them and choosing the **Delete** option from the shortcut menu, which is displayed on right-clicking. (T/F)

3. The origin is a blue icon that is displayed in the middle of the sketcher screen. (T/F)

4. In SolidWorks, circles are drawn by specifying the center point of the circle and then entering the radius of the circle in the dialog box that is displayed. (T/F)

5. When you open a new SolidWorks document, it is not maximized in the SolidWorks window. (T/F)

6. In SolidWorks, a polygon is considered as a combination of which of the following entities.

   (a) Lines (b) Arcs
   (c) Splines (d) None

7. Which one of the following options is not displayed in the **New SolidWorks Document** dialog box?

   (a) **Part** (b) **Assembly**
   (c) **Drawing** (d) **Sketch**

8. Which one of the following entities is not considered while converting a sketch into a feature?

   (a) Sketched circles (b) Sketched lines
   (c) Construction lines (d) None

9. When you select a line of the rectangle, which of the following **PropertyManager** will be displayed?

   (a) **Line Properties PropertyManager** (b) **Line/Rectangle PropertyManager**
   (c) **Rectangle PropertyManager** (d) None

10. While drawing an elliptical arc, which of the following **PropertyManager** will be displayed?

    (a) **Arc PropertyManager** (b) **Ellipse PropertyManager**
    (c) **Elliptical Arc PropertyManager** (d) None

EXERCISES

Exercise 1

Draw the sketch of the model shown in Figure 1-63. The sketch to be drawn is shown in Figure 1-64. Do not dimension the sketch. The solid model and dimensions are given only for your reference. **(Expected time: 30 min)**

Figure 1-63 Solid model for Exercise 1

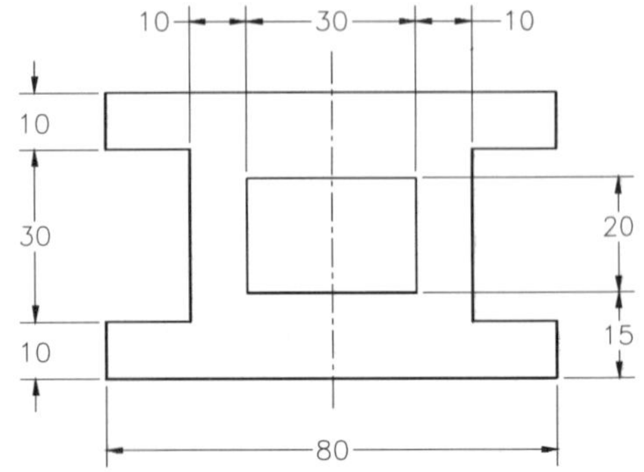

Figure 1-64 Sketch for Exercise 1

Exercise 2

Draw the sketch of the model shown in Figure 1-65. The sketch to be drawn is shown in Figure 1-66. Do not dimension the sketch. The solid model and dimensions are given only for your reference. **(Expected time: 30 min)**

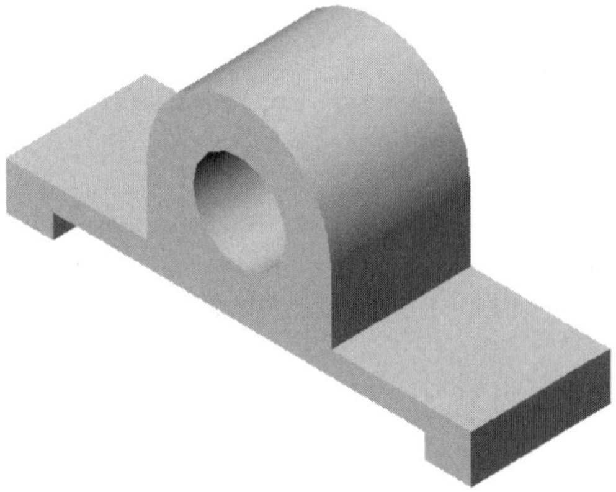

Figure 1-65 Solid model for Exercise 2

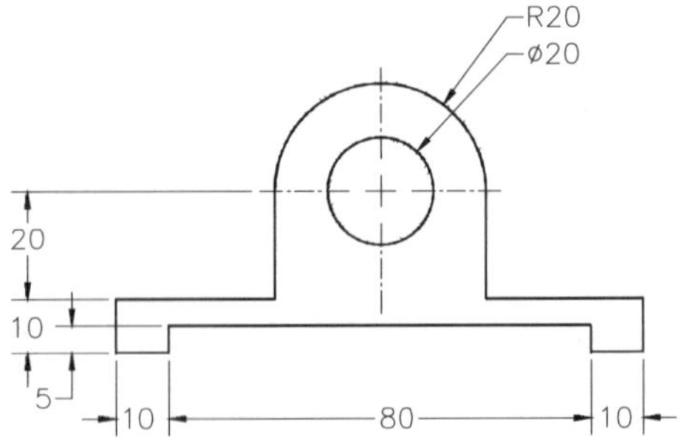

Figure 1-66 Sketch for Exercise 2

Answers to Self-Evaluation Test

1. T, **2.** T, **3.** F, **4.** F, **5. For Construction**, **6. Parallelogram**, **7.** inferencing lines, **8. Select**, **9. Tangent**, **10.** lines

Chapter 2

Editing and Modifying the Sketches

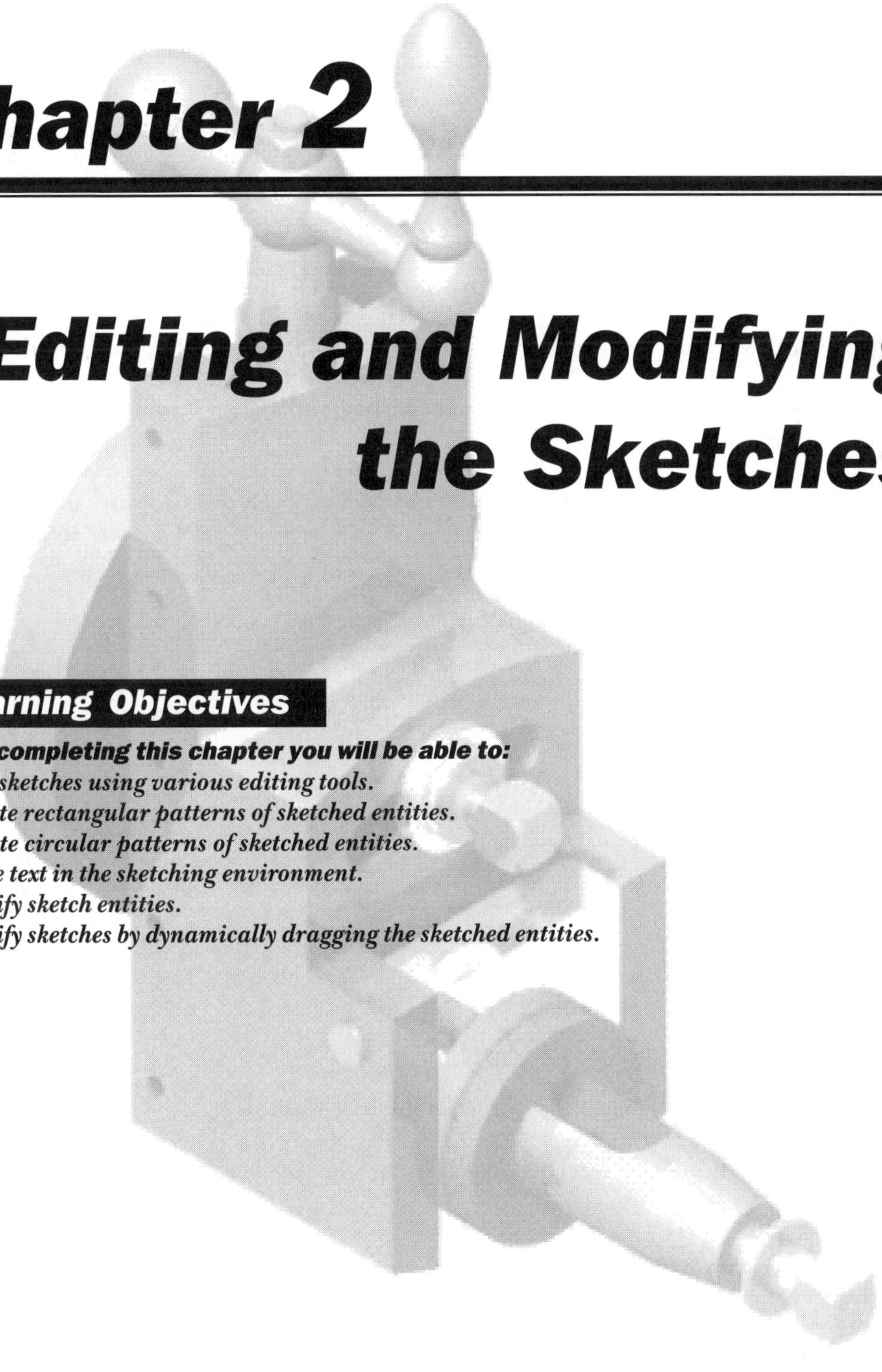

Learning Objectives

After completing this chapter you will be able to:
- *Edit sketches using various editing tools.*
- *Create rectangular patterns of sketched entities.*
- *Create circular patterns of sketched entities.*
- *Write text in the sketching environment.*
- *Modify sketch entities.*
- *Modify sketches by dynamically dragging the sketched entities.*

EDITING THE SKETCHED ENTITIES

SolidWorks provides you with a number of tools that can be used to edit the sketched entities. These options include trimming the sketched entities, extending the sketched entities, offsetting the sketched entities, mirroring the sketched entities, and so on.

Trimming the Sketched Entities

CommandManager:	Sketch > Trim Entities
Menu:	Tools > Sketch Tools > Trim
Toolbar:	Sketch > Trim Entities

The **Trim Entities** tool is used to trim the unwanted entities in a sketch. You can use this tool to trim a line, arc, ellipse, parabola, circle, spline, or centerline that is intersecting another line, arc, ellipse, parabola, circle, spline, or centerline. You can also extend the entities using the **Trim** tool. To use the trim option, choose the **Trim Entities** button from the **Sketch CommandManager**; the **Trim PropertyManager** will be displayed, as shown in Figure 2-1.

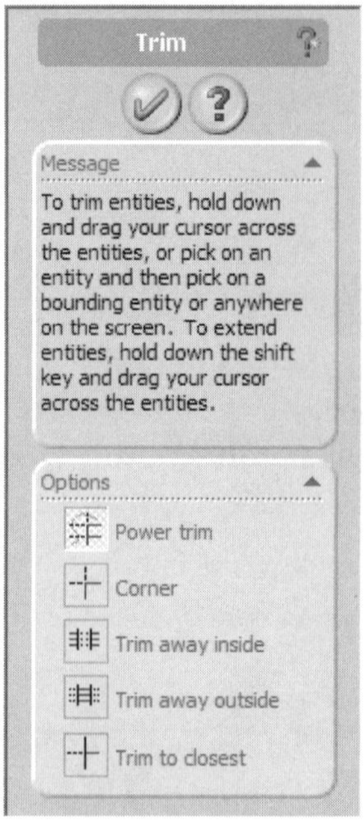

*Figure 2-1 The **Trim PropertyManager***

The options available in this **PropertyManager** to trim the sketched entities are discussed next.

Power trim

When the **Power trim** button is chosen in the **Options** rollout, the **Message** rollout provided in this **PropertyManager** will inform you about the procedure of trimming and extending the sketched elements using this option. To trim the unwanted portion of the sketch using this option, press and hold the left mouse button down and drag the cursor. You will notice that a gray-colored drag trace line is displayed along the path of the cursor. When you drag and place the cursor on an unwanted sketched entity, it will be trimmed and a small red-colored box will be displayed in its place. You can continue trimming entities by dragging the cursor on them. After trimming all unwanted entities, release the left mouse button.

To extend the sketched entities using this tool, press and hold the left mouse button down and press the SHIFT key from the keyboard. Now, drag the cursor up to the entity that needs to be extended. As the cursor reaches the entity, it will be extended to the nearest intersection. If there is no intersection on the path of the entity, it will not be extended. You can continue extending entities by dragging the cursor on the elements. After extending all required entities, release the left mouse button.

Corner

The **Corner** button available in the **Options** rollout is used to trim or extend elements in such a manner that the resulting elements form a corner. To trim unwanted elements using this option, choose the **Corner** button from the **Options** rollout; you will be prompted to select an entity. Select the entity from the geometry area. You need to make sure that you select the area of the entity that you want to retain. Now, you will be prompted to select another entity. Select the second entity from the geometry area, refer to Figure 2-2. The portion of the entities from where the selection was made will be retained and the other portion will be removed, resulting in a corner, as shown in Figure 2-3.

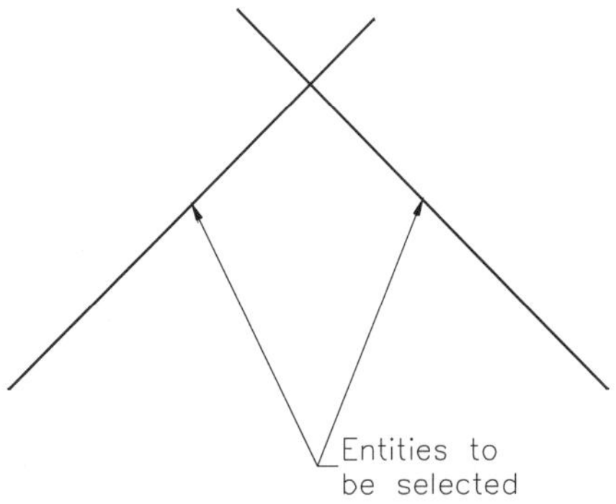

Figure 2-2 *Entities to be selected for trimming*

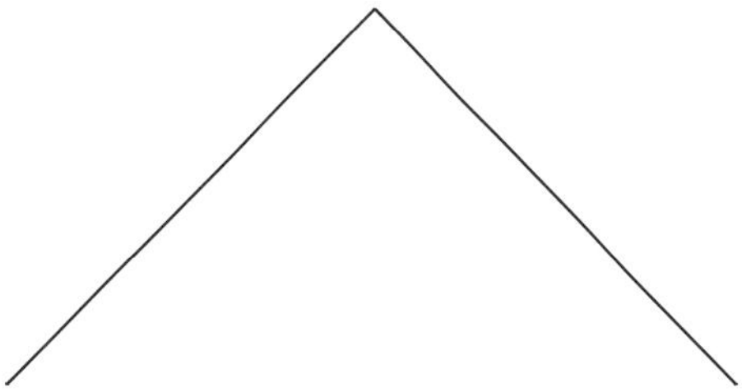

Figure 2-3 *Trimmed entities*

You can also extend the entities using this tool. To do so, choose the **Corner** button from the **Options** rollout. Select the entities to be extended; the selected entities will be extended to their apparent intersection, refer to Figures 2-4 and 2-5.

Trim away inside

The **Trim away inside** button available in the **Options** rollout is used to trim the inside portion of the selected entity. This portion is defined by two bounding entities. To trim the entities using this tool, invoke the **Trim PropertyManager** and choose the **Trim away inside** button from the **Options** rollout. The **Message** rollout will inform you to select two bounding entities, and then to select the entities to be trimmed. Select the bounding entities from the drawing area, refer to Figure 2-6. Now, select the entities to be trimmed from the drawing area. As you select an entity

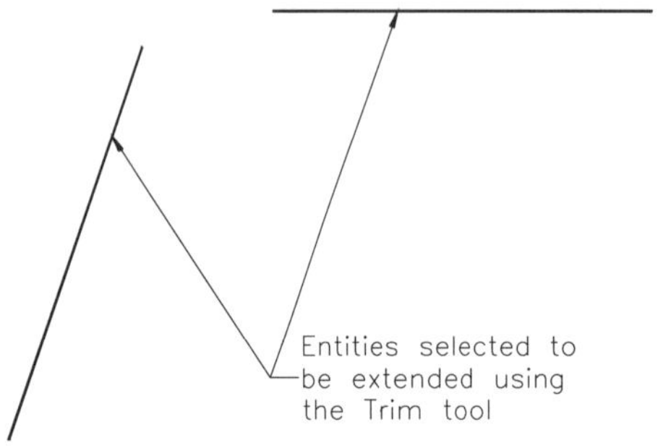

Figure 2-4 *Entities selected for extending*

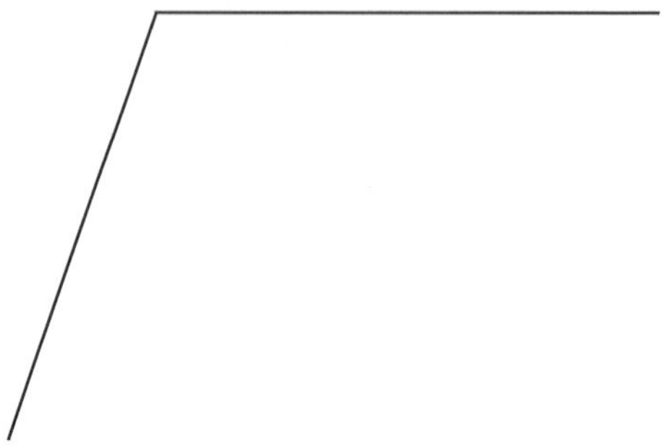

Figure 2-5 *Sketch after extending*

to be trimmed, the portion of the entity inside the bounding entities will be removed and the portion outside that will be retained, refer to Figure 2-6.

Trim away outside

The **Trim away outside** button available in the **Options** rollout is used to trim the outside portion of the selected entity. This portion is defined by two bounding entities. To trim the entities using this tool, invoke the **Trim PropertyManager** and choose the **Trim away outside** button from the **Options** rollout. The **Message** rollout will inform you to select two bounding entities, and then to select the entities to be trimmed. Select the bounding entities from the drawing area, refer to Figure 2-7. Now, select the entities to be trimmed from the drawing area. As you select an entity to be trimmed, the portion of the entity outside the bounding entities will be removed and the portion inside that will be retained, refer to Figure 2-7.

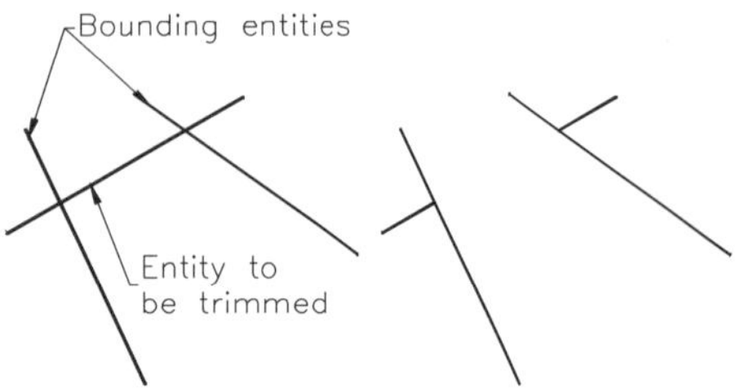

Figure 2-6 *Entities trimmed using the **Trim away inside** option*

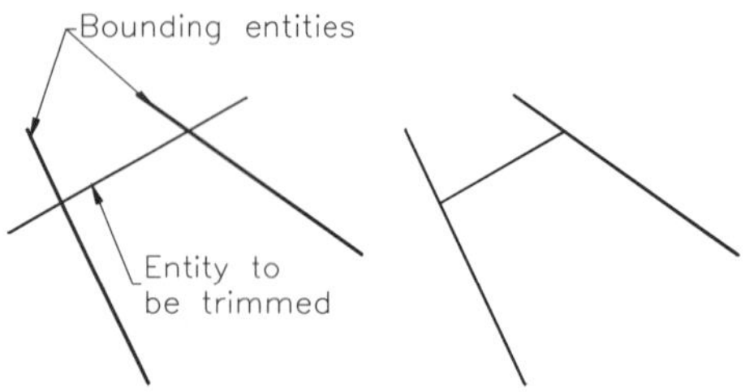

Figure 2-7 *Entities trimmed using the **Trim away outside** option*

Trim to closest

The **Trim to closest** tool is used to trim the selected entity to its closest intersection. To trim entities using this tool, invoke the **Trim PropertyManager** and choose the **Trim to closest** button from the **Options** rollout; the cursor will be replaced by the trim cursor. Move the trim cursor near to the portion of the sketched entity to be removed. The entity or the portion of the entity to be removed will be highlighted in red. Press the left mouse button to remove the highlighted entity. Figure 2-8 shows the entities to be trimmed and Figure 2-9 shows the sketch after trimming the entities.

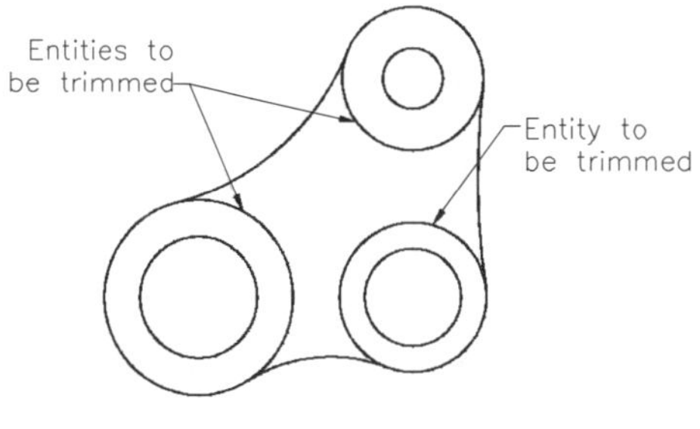

Figure 2-8 *Entities to be trimmed*

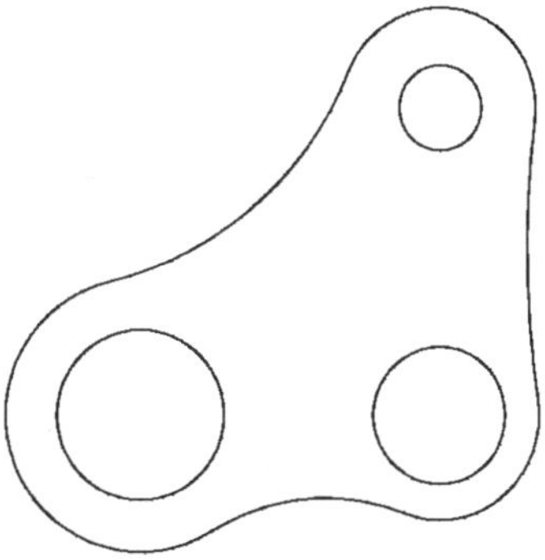

Figure 2-9 *Sketch after trimming the entities*

You can also use this option to extend the sketched entities. To do this, move the trim cursor to the entity to be extended. When the sketched entity turns red, drag the cursor; the entity to be extended will be displayed in green. Drag the cursor to the entity up to which it will be extended. You will notice the preview of the extended entity in red color. Release the left mouse button when the preview of the extended entity appears.

Tip. *In SolidWorks, all buttons are not displayed by default in toolbars or **CommandManagers**. You need to customize them and add buttons according to your need and specifications. The next topic is Extending the Sketched Entities. The **Extend** tool button is not available in the **Sketch CommandManager** by default. To insert the **Extend Entities** button in the **Sketch CommandManager**, you need to follow the procedure given below.*

*1. Choose **Tools > Customize** from the menu bar to display the **Customize** dialog box.*
*2. Choose the **Commands** tab from the **Customize** dialog box.*
*3. Select **Sketch** from the **Categories** area of the **Customize** dialog box.*
*4. Press and hold the left mouse button down on the **Extend Entities** button from the **Buttons** area of the **Customize** dialog box; the description of the **Extend Entities** button will be displayed in the **Description** area.*
*5. Drag the mouse to the **Sketch CommandManager** and release the left mouse button to place the **Extend Entities** button on this **CommandManager**.*
*6. Choose **OK** from the **Customize** dialog box.*

*You can insert the other tool buttons in the **Sketch CommandManager** or any other CommandManager by following the same procedure. To remove the tool button from a CommandManager, invoke the **Customize** dialog box and drag the tool button you want to remove from the CommandManager to the graphics area.*

*When you customize and add a tool to the **CommandManager**, it will automatically be added to the toolbar also.*

Extending the Sketched Entities

CommandManager:	Sketch > Extend Entities	*(Customize to Add)*
Menu:	Tools > Sketch Tools > Extend	
Toolbar:	Sketch > Extend Entities	*(Customize to Add)*

The **Extend Entities** tool is used to extend the sketched entity to intersect the next available entity. The tool is used to extend a line, arc, ellipse, parabola, circle, spline, or centerline to intersect another line, arc, ellipse, parabola, circle, spline, or centerline. The sketched entity is extended up to its intersection with another sketched entity or a model edge. Choose the **Extend Entities** button from the **Sketch CommandManager** and move the extend cursor close to the portion of the sketched entity to be extended. The entity to be extended will be displayed in red and its preview will also be displayed in red. Press the left mouse button to complete the extend operation. Figure 2-10 shows the sketched entities before extending and Figure 2-11 shows the sketched entities after extending.

Tip. *If the preview of the sketched entity to be extended is shown in the wrong direction, move the extend cursor to a position on the other half of the entity and observe the new preview.*

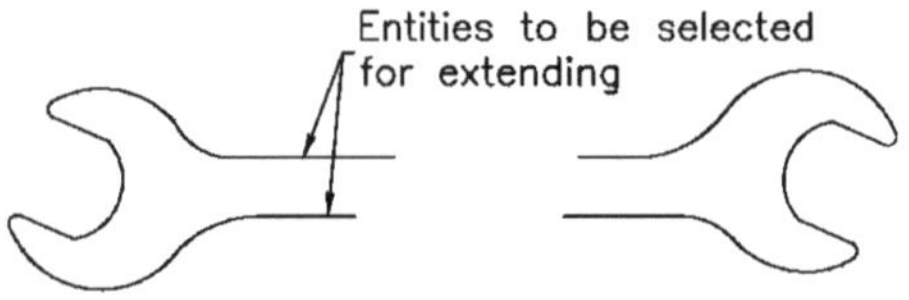

Figure 2-10 *Sketched entities before extending*

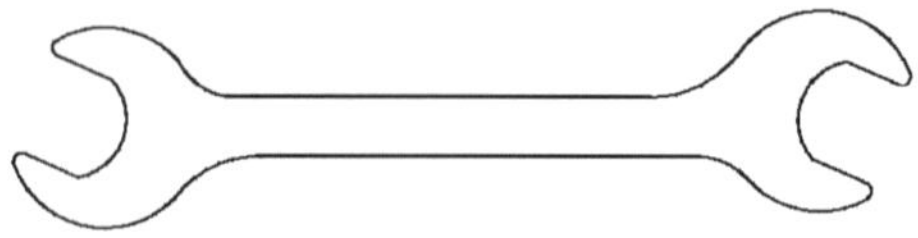

Figure 2-11 *Sketched entities after extending*

Tip. *You can toggle between the **Trim Entities** and **Extend Entities** tools using the shortcut menu that will be displayed on right-clicking when one of these tools is active.*

Filleting the Sketched Entities

CommandManager:	Sketch > Sketch Fillet
Menu:	Tools > Sketch Tools > Fillet
Toolbar:	Sketch > Sketch Fillet

A fillet creates a tangent arc at the intersection of two sketched entities. It trims or extends the entities to be filleted, depending on the geometry of the sketched entity. You can apply a fillet between two nonparallel lines, between two arcs, between two splines, between an arc and a line, between a spline and a line, and between a spline and an arc. The fillet between two arcs, or between an arc and a line depends on the compatibility of the geometry to be extended or filleted along a given radius. You can invoke the **Sketch Fillet** tool first and then select the entities to be filleted. You can also hold the CTRL key down and first select two entities to be filleted and then invoke this tool. When you invoke this tool, the **Sketch Fillet PropertyManager** will be displayed, as shown in Figure 2-12.

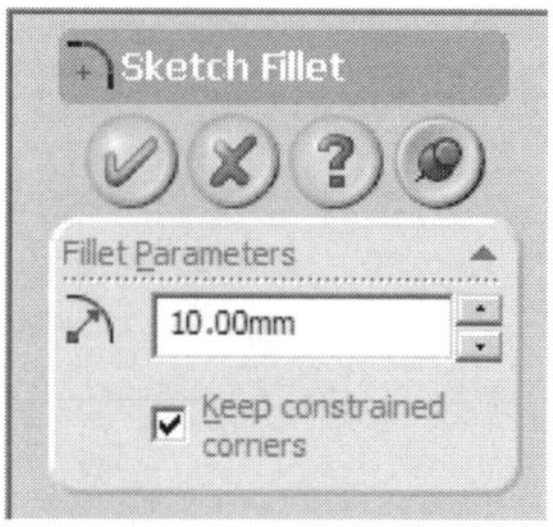

Figure 2-12 Sketch Fillet PropertyManager

Set the value of the **Fillet Radius** spinner and press ENTER or choose the **Apply** button from the **Sketch Fillet** dialog box. You can also select the sketched entities after invoking the **Sketch Fillet PropertyManager**. You can also select the intersection points of the entities to be filleted. You can also select the nonintersecting entities for creating a fillet. While filleting the nonintersecting entities the selected entities will be extended to form a fillet. If the **Keep constrained corners** check box is selected, the dimension and geometric relations applied to the sketch will not be deleted. If you clear the **Keep constrained corners** check box, you will be prompted to delete the relations applied to the corners of the sketched entities to be filleted. The **Undo** button will be displayed in the **Sketch Fillet PropertyManager** when you create at least one sketch fillet. Figures 2-13 and 2-14 show the sketched entities before and after applying the radius.

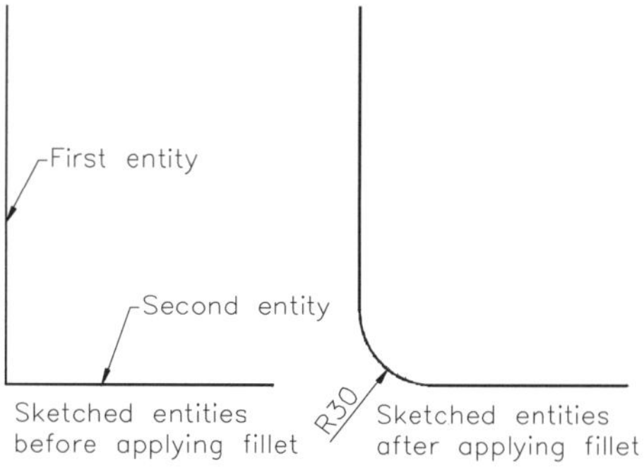

***Figure 2-13** Sketched entities before and after creating a fillet*

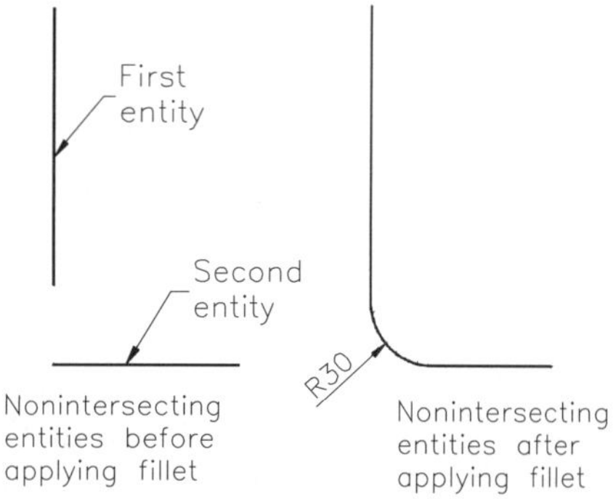

***Figure 2-14** Sketched entities before and after creating a fillet*

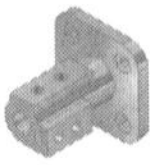

Note

The consecutive fillets with the same radius are not dimensioned individually; an automatic equal radii relation is applied to all fillets.

The fillet creation between two splines, between a spline and a line, and between a spline and an arc depends on the compatibility of the spline to be trimmed or extended.

Tip. *The other method of creating a sketched fillet between two entities is to drag a window around the entities to be filleted. To fillet the entities using this method, invoke the **Sketch Fillet PropertyManager**. Set the fillet radius spinner to the required value. Now, drag the cursor to create a window around the two entities to be filleted such that they are enclosed inside the window. As soon as you release the left mouse button, the fillet will be created between the two selected entities.*

You can also select a vertex to fillet the two entities that form that vertex.

Chamfering the Sketched Entities

CommandManager:	Sketch > Sketch Chamfer	*(Customize to Add)*
Menu:	Tools > Sketch Tools > Chamfer	
Toolbar:	Sketch > Sketch Chamfer	*(Customize to Add)*

The **Sketch Chamfer** tool is used to apply a chamfer to adjacent sketch entities. The chamfer can be specified by angle-distance or distance-distance options. You can apply a chamfer between two nonparallel lines; the lines may be intersecting lines or nonintersecting lines. The creation of a chamfer between two nonintersecting lines depends on the length of the lines and the chamfer distance. To create a chamfer, choose the **Sketch Chamfer** button from the **Sketch CommandManager** and select two entities to be chamfered. You can also select the two entities before invoking the **Chamfer** tool. When you invoke this tool, the **Sketch Chamfer PropertyManager** will be displayed, as shown in Figure 2-15. The options available in the **Sketch Chamfer PropertyManager** are discussed next.

Figure 2-15 Sketch Chamfer PropertyManager

Angle-distance

The **Angle-distance** radio button is selected to create the chamfer by specifying the angle and the distance, refer to Figure 2-16. When you select this radio button, the **Direction 1 Angle** spinner will be displayed below the **Distance 1** spinner. The **Distance 1** spinner is used to

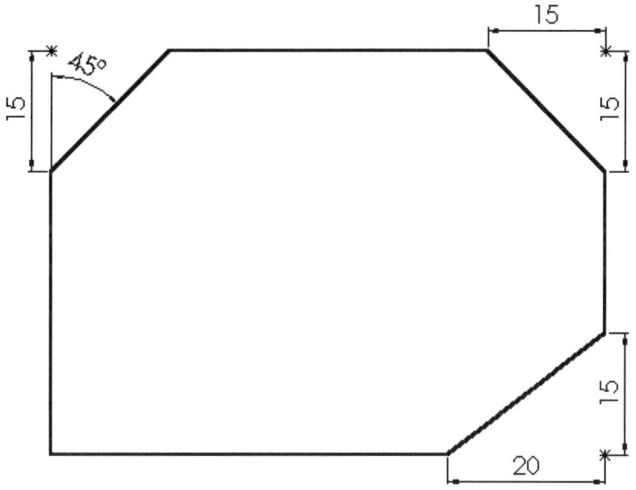

Figure 2-16 Creating various types of chamfers

specify the distance and the **Direction 1 Angle** spinner is used to specify the angle. Note that the angle will be measured from the first entity that you selected.

Distance-distance

The **Distance-distance** radio button is selected to create the chamfer by specifying the distances along both the selected lines, refer to Figure 2-16. When you invoke the **Sketch Chamfer PropertyManager**, the **Distance-distance** radio button and the **Equal distance** check box are selected by default. Therefore, you can create an equal distance chamfer between the selected entities. The distance value can be specified in the **Distance 1** spinner.

Equal distance

The **Equal distance** check box is selected to specify an equal distance in both directions for creating the sketched chamfer. This check box is selected by default. If you clear this check box, you can specify two different distances for creating the chamfer, refer to Figure 2-16. When you clear this check box, the **Distance 2** spinner will be displayed below the **Distance 1** spinner to set the value of the distance in the second direction. Note that the distance 1 value will be measured along the first entity that you selected and the distance 2 value along the second entity.

You can also use the shortcut menu that is invoked to use various methods of creating a chamfer.

Note

If you apply a sketch chamfer to the entities that are constrained using some relations or dimensions, the ***SolidWorks*** *warning box will be displayed. This window will show the message:* ***At least one sketch constraint is about to be lost, Chamfer anyway?*** *Choose the* ***Yes*** *button from this warning box. You will learn more about the relations and dimensions that constrain the sketch in later chapters.*

Offsetting the Sketched Entities

CommandManager:	Sketch > Offset Entities
Menu:	Tools > Sketch Tools > Offset Entities
Toolbar:	Sketch > Offset Entities

Offsetting is one of the easiest methods of drawing parallel lines or concentric arcs and circles. You can select the entire chain of entities as a single entity or select an individual entity to be offset. You can offset selected sketched entities, edges, loops, faces, and curves. When you choose the **Offset Entities** button from the **Sketch Tools** toolbar, the **Offset Entities PropertyManager** will be displayed, as shown in Figure 2-17. The options available in the **Offset Entities PropertyManager** are discussed next.

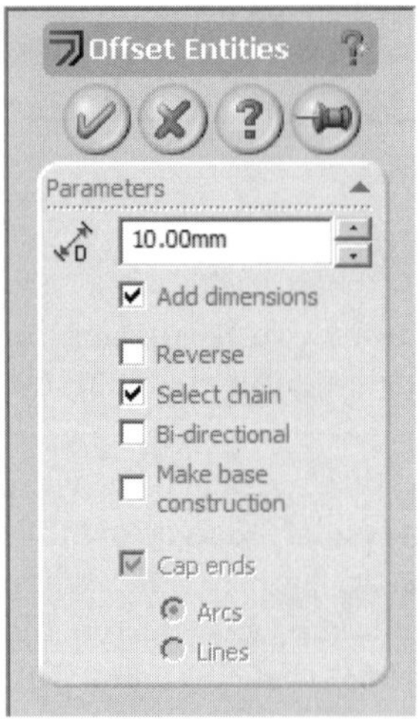

Figure 2-17 Offset Entities PropertyManager

Note

*While performing any kind of editing operation, if you want to clear the current selection set, right-click in the drawing area and choose **Clear Selections** from the shortcut menu.*

Offset Distance

The **Offset Distance** spinner is used to set the distance through which the selected entity needs to be offset. You can set the value of the offset distance using this spinner or by dragging the offset entity in the drawing area.

Add Dimension

The **Add dimension** check box is selected by default in the **Parameters** rollout. This check box when selected will add a dimension representing the offset distance between the parent entity and the resulting offset entity.

Reverse

The **Reverse** check box is selected to change the direction of offset. Note that while offsetting the entities by dragging, you do not need this check box. This is because you can change the direction of offset by dragging the entities in the required direction.

Select Chain

The **Select chain** check box is selected to select the entire chain of continuous sketched entities that are in contact with the selected entity. When you invoke the **Offset** tool, the **Select chain** check box will be selected by default. If you clear this check box, only the selected sketched entity will be offset.

Bi-directional

The **Bi-directional** radio button is selected to create the offset of the selected entity in both directions of the selected entity. If the **Bi-directional** check box is selected, the **Reverse** check box will not be available in the **Parameters** area.

Make Base Construction

The **Make base construction** check box is selected to convert the parent entity or the base entity into a construction entity.

Cap Ends

The **Cap ends** check box is available only if the **Bi-directional** check box is selected. This option when selected closes the ends of the bidirectionally offset entities. You can select the type of cap to close the ends. If you are offseting a closed entity on both directions, this option will not be available.

Figure 2-18 shows a new chain of entities created by offsetting the chain of entities and Figure 2-19 shows offsetting of a single entity.

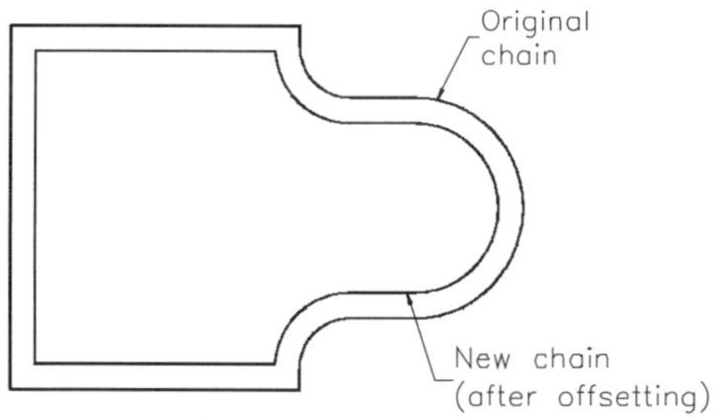

***Figure 2-18** Offsetting a chain of entities*

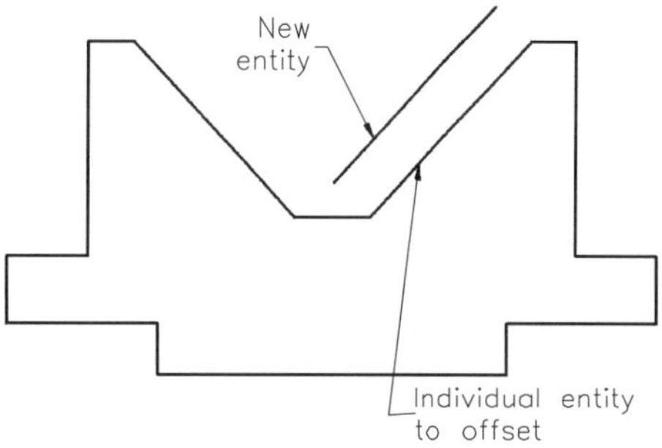

***Figure 2-19** Offsetting a single entity*

Mirroring the Sketched Entities

CommandManager:	Sketch > Mirror Entities
Menu:	Tools > Sketch Tools > Mirror
Toolbar:	Sketch > Mirror Entities

The **Mirror Entities** tool is used to create the mirror image of the selected entities. The entities are mirrored about a centerline. When you create the mirrored entity, SolidWorks applies the symmetric relation between the sketched entities. If you change the entity, its mirror image will also change. To mirror the existing entities, choose the **Mirror Entities** button from the **Sketch CommandManager**; the **Mirror PropertyManager** will be displayed, as shown in Figure 2-20.

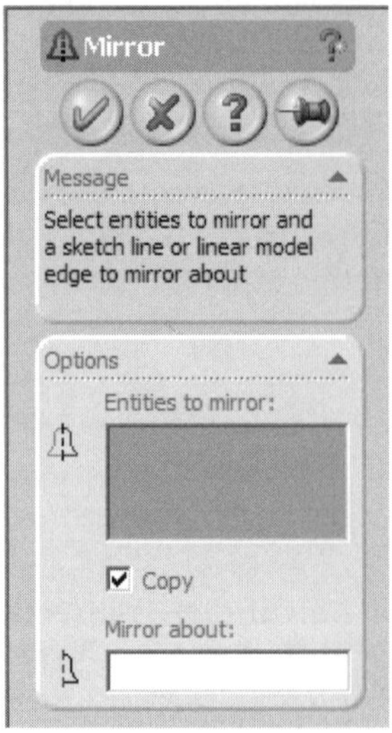

Figure 2-20** The **Mirror PropertyManager

You are prompted to select entities to mirror. Select the entities to mirror from the drawing area. The name of the selected entities will be displayed in the **Entities to mirror** selection area. After selecting all entities to mirror, click once in the **Mirror about** selection area. While selecting the entities, the  symbol will also be displayed. If you right-click, while this symbol is displayed below the cursor, the selection mode in the next selection area will be activated automatically. You will be prompted to select a line or a linear model edge to mirror about. Select a line or a centerline from the drawing area that will be used as the mirror line. The preview of the mirrored image will be displayed. You need to make sure that **Copy** check box is selected in the **Mirror PropertyManager**. If you clear this check box, the parent selected entities will be removed and only the mirror image will be retained when you mirror the sketched entities. Choose the **OK** button from the **Mirror PropertyManager**.

Figure 2-21 shows the sketched entities with the centerline and Figure 2-22 shows the resulting mirror image of the sketched entities.

Mirroring While Sketching (Dynamic Mirror Entities)

CommandManager:	Sketch > Dynamic Mirror Entities	*(Customize to Add)*
Menu:	Tools > Sketch Tools > Dynamic Mirror	
Toolbar:	Sketch > Dynamic Mirror Entities	*(Customize to Add)*

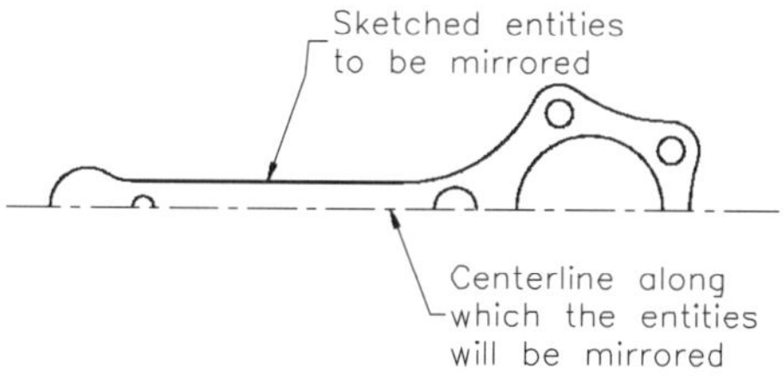

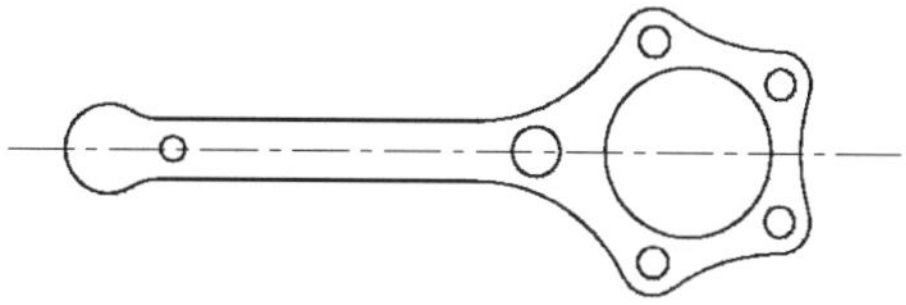

Figure 2-21 *Selecting the sketched entities and the centerline* *Figure 2-22* *Sketch after mirroring the geometry*

The **Dynamic Mirror Entities** tool is used to mirror the entities along a symmetry line while sketching. This tool is recommended when you are drawing symmetric sketches. To use this tool, choose the **Dynamic Mirror Entities** button from the **Sketch CommandManager**; the **Mirror PropertyManager** will be displayed, as shown Figure 2-23.

Figure 2-23 *The* ***Mirror PropertyManager***

The **Message** rollout of the **Mirror PropertyManager** informs you to select a sketch line or a linear model edge to mirror about. Select a line or a centerline from the drawing that will be used as symmetry line. The symmetry symbols appear at both ends of the centerline to indicate that automatic mirroring is activated, as shown in Figure 2-24. Now, start drawing the sketch. The sketched entity that you draw on one side of the centerline will automatically be created on the other side of the mirror line (centerline). As evident in Figure 2-25, the entities are mirrored automatically while sketching. Figure 2-26 shows the complete sketch with automatic mirroring.

Note

After completing the sketch using the automatic mirroring option, you can clear this option by choosing the ***Dynamic Mirror Entities*** *button from the* ***Sketch CommandManager****.*

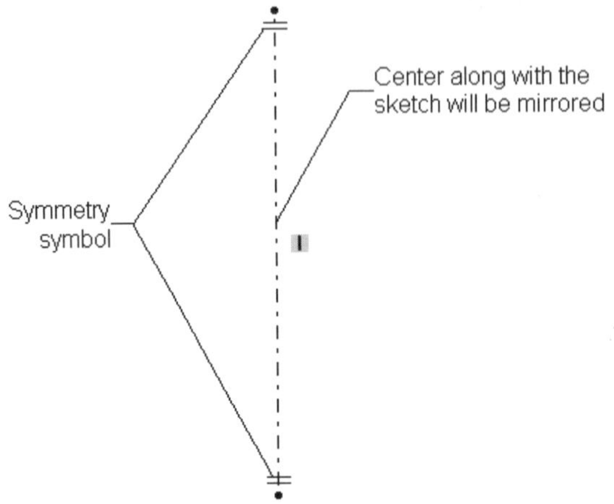

Figure 2-24 *The centerline and the symmetry symbols that indicate that automatic mirroring is activated*

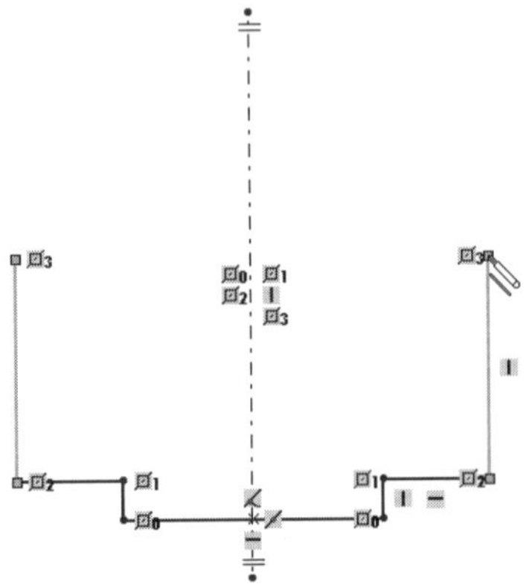

Figure 2-25 *Sketching using automatic mirroring active*

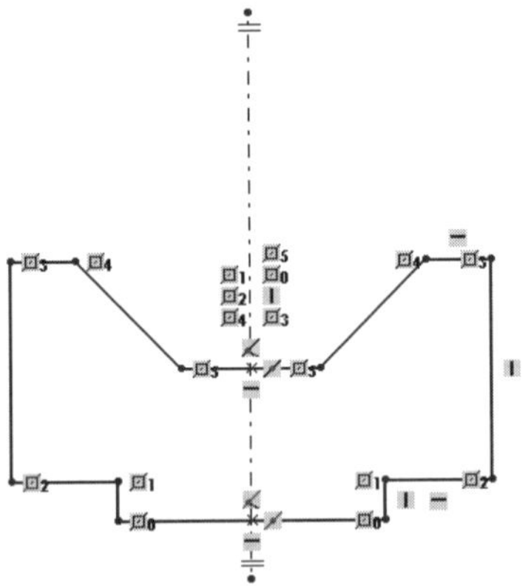

Figure 2-26 *Sketch drawn with automatic mirroring*

MOVING, ROTATING, AND SCALING THE SKETCHED ENTITIES

Command Manager:	Sketch > Move or Copy
Menu:	Tools > Sketch Tools > Move or Copy
Toolbar:	Sketch > Move or Copy

The **Move or Copy Entities** tool allows you to move, rotate, or scale the selected sketched entities. You can also make a copy of the sketched entities while moving, rotating, or scaling the entities. When you invoke this tool, the **Move or Copy PropertyManager** will be displayed. You can select the option to move, rotate, or scale the selected entities in this **PropertyManager**. However, the menu bar provides you separate options to invoke any of these three operations. You can choose any of these operations from the

Tools > **Sketch Tools** menu. You can also invoke any of these operations using the shortcut menu that will be displayed when you select the entities and right-click.

The operations that can be performed using the **Move or Copy Entities** tool are discussed next.

Moving or Copying the Sketched Entities

To move or copy the sketched entities, invoke this tool. Remember that this tool will be available only when at least one sketched entity is drawn. When you invoke this tool, the **Move or Copy PropertyManager** will be displayed, as shown in Figure 2-27, and you will be prompted to select sketch items or annotations. The options available in this **PropertyManager** are discussed next.

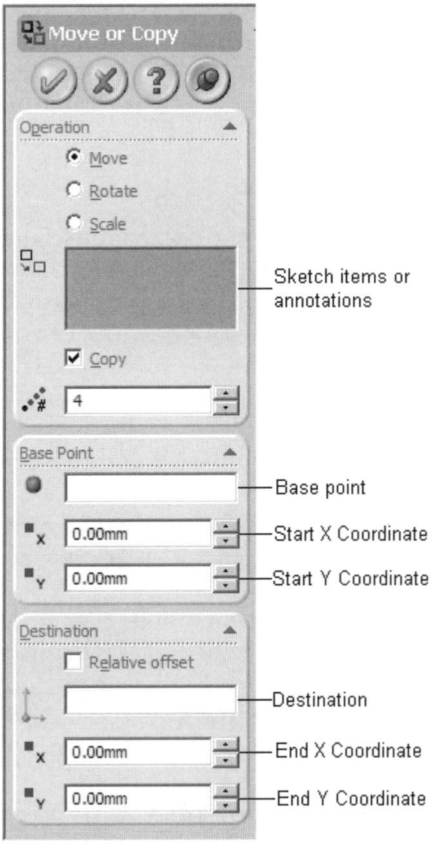

Figure 2-27 Move or Copy PropertyManager

Operation Rollout

The options in the **Operation** rollout are used to select the operation to be performed and to select the entities. You can first invoke the **Move or Copy PropertyManager** and then select the entities to be moved or copied. The selected entities will be displayed in the **Sketch items or annotations** selection area. After selecting all entities, right-click to accept the selection set.

You can select the **Copy** check box if you want to copy the selected entities. When you select this check box, the **Number of Copies** spinner will be displayed below this check box. You can specify the number of copies to be created using this spinner. You can select the **Keep relations** check box from this area to retain the relations applied to the selected entities even after performing this operation. This check box will not be available if you select the **Copy** check box.

Base Point Rollout

The options in this rollout are used to specify the base point for moving or copying the selected entities. After selecting the entities to be moved or copied, as soon as you right-click, you will be prompted to specify the base point of the move. Also, the base point icon will be attached to the arrow cursor. You can define the base point by specifying a point on the screen or by entering the coordinates of the base point in the **Start X Coordinate** and **Start Y Coordinate** spinners available in this area. If you want to use the spinners, set the values in the spinners and then press the TAB key to accept the values. The base point icon will be automatically fixed at the base point that you define. If you want to redefine the base point, right-click in the drawing area and choose **Define Base Point**. The previous selection will be cleared and you will be prompted to specify the base point again.

Destination Rollout

The options in this rollout are used to specify the destination where the entities will be placed after moving or copying. As soon as you define the base point, this rollout will be activated and you will be prompted to specify the destination of the move. Also, the destination icon will be attached to the arrow cursor. As you move the cursor on the screen, the selected entities will also move dynamically. You can define the destination point by specifying a point on the screen or use the **End X Coordinate** and **End Y Coordinate** spinners to enter the values.

If you select the **Relative offset** check box, the **Base Point** rollout will no longer be displayed and the **Destination** rollout also shows the **Delta X** and **Delta Y** spinners. By selecting this check box, you can define the displacement along the X and Y directions. All selected entities will be moved in the X and Y directions through the distances specified in the two spinners.

If you select the **Copy** check box in the **Operation** rollout, the preview of the specified number of copies will be displayed and the gap between the copies is defined by the difference between the base point and the destination point. Figure 2-28 shows the selected entities being moved using the **Move or Copy** tool. In this figure, the upper and right edges of a rectangle are selected to be moved.

Rotating or Copying the Sketched Entities

Command Manager:	Sketch > Rotate or Copy	*(Customize to Add)*
Menu:	Tools > Sketch Tools > Rotate or Copy	
Toolbar:	Sketch > Rotate or Copy	*(Customize to Add)*

To rotate or copy the sketched entities, invoke the **Rotate or Copy** tool. You can also select the entities and right-click to display the shortcut menu. In this menu, choose **Rotate or Copy** to invoke this tool. When you invoke this tool, the **Move or Copy**

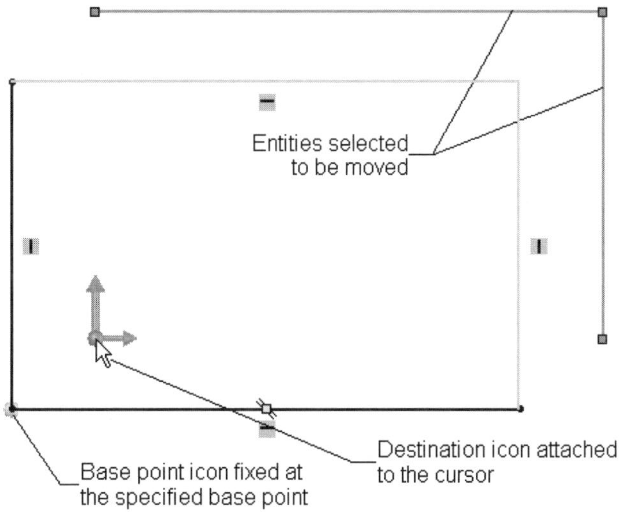

Figure 2-28 *Moving the selected entities*

PropertyManager will be displayed, as shown in Figure 2-19, and you will be prompted to select sketch items or annotations.

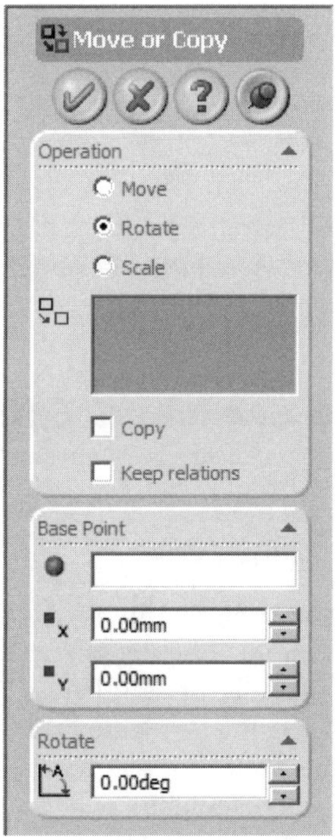

***Figure 2-29 Move or Copy PropertyManager** for the **Rotate** tool*

The entities that you select to rotate are displayed in the **Sketch items or annotations** selection area of the **Operation** rollout. After selecting the entities to rotate, right-click; you will be prompted to specify the base point about which to rotate. You can use the options in the **Base Point** rollout. These options are similar to those discussed while moving the entities.

As soon as you specify the base point, the **Angle** spinner in the **Rotate** rollout will be highlighted. You can specify the angle of rotation using this spinner. You can also drag the mouse on the screen to define the angle of rotation.

Figure 2-30 shows a rectangle being rotated by dragging the cursor to define the angle of rotation. The lower right vertex of the rectangle is taken as the base point of rotation.

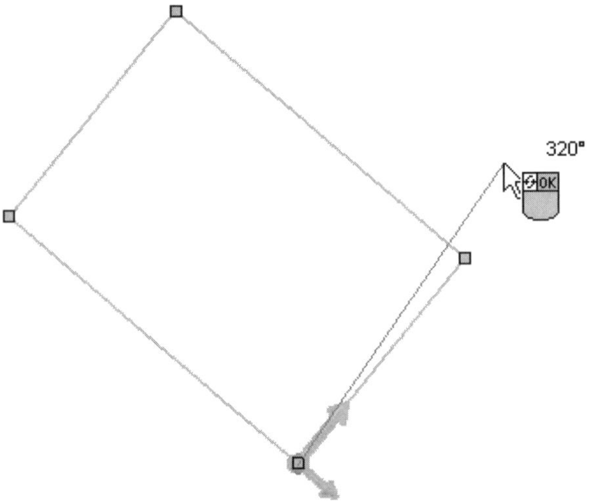

***Figure 2-30** Defining the rotation angle for the rectangle by dragging the mouse*

Scaling or Copying the Sketched Entities

Command Manager:	Sketch > Scale or Copy	*(Customize to Add)*
Menu:	Tools > Sketch Tools > Scale or Copy	
Toolbar:	Sketch > Scale or Copy	*(Customize to Add)*

To scale or copy the sketched entities, invoke the **Scale or Copy** tool using the menu bar or using the shortcut menu. The **PropertyManager** that is displayed is similar to the one that is displayed for rotating. The only difference is that the **PropertyManager** for scaling has the **Scale Factor** spinner in the **Operation** rollout. This spinner is used to specify the scale factor for scaling the selected entities. Note that in this case, the **Keep relations** check box is not available.

After you select the entities to be scaled, define the base point. Next, specify the scale factor using the spinner and then right-click to complete this operation.

CREATING PATTERNS

While sketching the base feature of a model, sometimes you may need to place the sketched entities in a particular arrangement such as along linear edges or around a circle. For example, refer to Figures 2-31 and 2-32, which show the base features with the slots. These slots are created with the help of linear and circular patterns of the sketched entities. The tools that are used to create linear and circular patterns of sketched entities are discussed next.

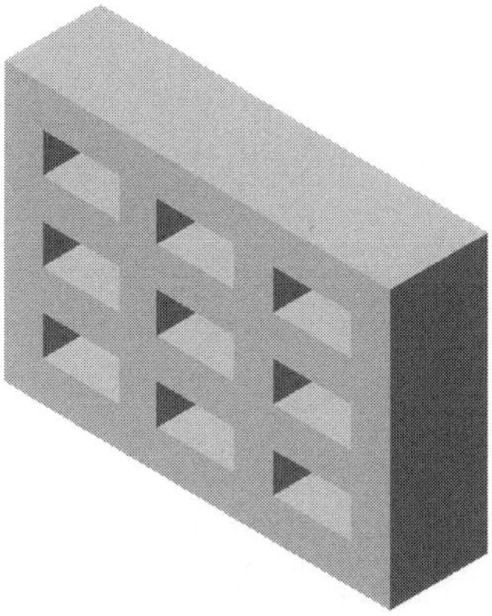

Figure 2-31 *Base feature with slots created about linear edges*

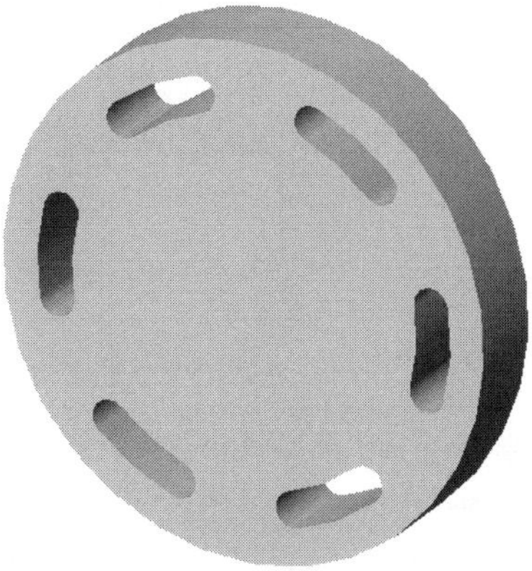

Figure 2-32 *Base feature with slots created around a circle*

Creating Linear Pattern of the Sketched Entities

CommandManager:	Sketch > Linear Sketch Step and Repeat *(Customize to Add)*
Menu:	Tools > Sketch Tools > Linear Sketch Step and Repeat
Toolbar:	Sketch > Linear Sketch Step and Repeat *(Customize to Add)*

In SolidWorks, the linear pattern of the sketched entities is created using the **Linear Sketch Step and Repeat** tool. To create the linear pattern, select the sketched entities using the **Select** tool and then choose this button from the **Sketch CommandManager**. The **Linear Sketch Step and Repeat** dialog box will be displayed as shown in Figure 2-33 and the preview of the linear pattern will be shown on the screen in the background. Also, the arrow cursor will be replaced by the linear pattern cursor. Note that if you have not selected the sketched entities to pattern before invoking this tool, you will have to select them one by one using the linear pattern cursor. You cannot define a window to select more than one entity using the linear pattern cursor.

The options available in this dialog box are discussed next.

Direction 1 Area

By default, the first direction of a linear pattern is taken along the direction of the positive X axis and is indicated in the preview of the pattern along direction 1. The options under this area are used to set the number of instances, spacing, and the angle along the first direction. These options are discussed next.

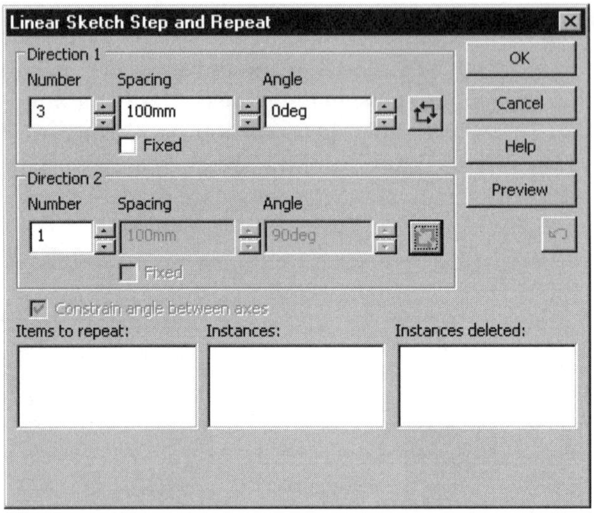

***Figure 2-33 Linear Sketch Step and Repeat** dialog box*

Number

The **Number** spinner is used to specify the number of instances in the linear pattern along the first direction. As you modify the value of this spinner, the number of instances in the preview of the pattern will also be modified. The default value of this spinner is 3.

Spacing

The **Spacing** spinner is used to specify the spacing between each instance along the first direction. Remember that the spacing is measured from the center of the entities selected for creating the pattern. Figure 2-34 shows the preview of the linear pattern that is displayed in the background. As evident in the figure, the number of instances along direction 1 is three and the distance is measured between the center of two consecutive entities.

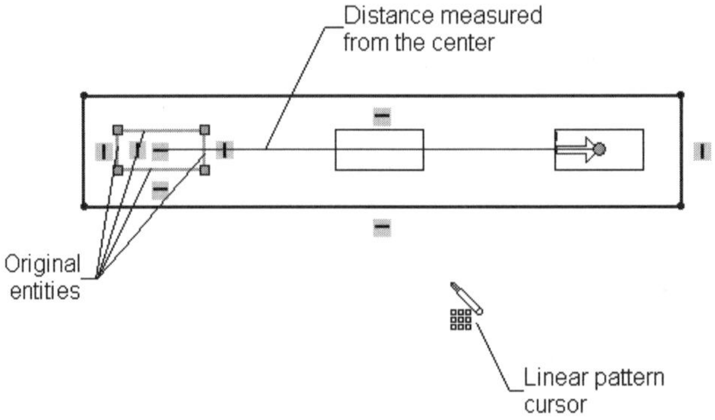

***Figure 2-34** Preview of the linear pattern with three instances along direction 1 and one instance along direction 2*

Angle

The **Angle** spinner is used to specify the angle of the linear pattern along the first direction.

Remember that the orientation of the entities selected to be patterned will not change as you enter an angle value for the first direction. The default value of the angle is 0-degree. Figure 2-35 shows a linear pattern at an angle of 30-degree.

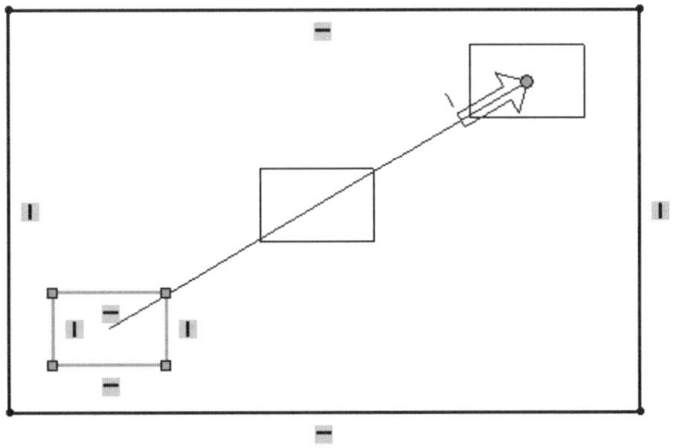

Figure 2-35 Preview of the linear pattern at an angle of 30-degree along direction 1

Reverse Direction 1

The **Reverse Direction 1** button is available on the right of the **Angle** spinner. This button is chosen to reverse direction 1.

Fixed

The **Fixed** check box is available below the **Spacing** spinner. If this check box is selected, the incremental spacing between the instances along direction 1 will be placed as a dimension between the first two instances after you create the pattern. Also, a centerline will be drawn between the first two instances. Note that this dimension will not be placed from center to center. It will be displayed from the first entity in the first instance in the pattern to the first entity in the second instance.

Direction 2 Area

By default, the second direction of a linear pattern is taken along the direction of the positive Y axis. The options under this area are used to set the number of instances, spacing, and the angle along the second direction. These options are discussed next.

Number

The **Number** spinner is used to specify the number of instances in the linear pattern along the second direction. The default value of this spinner is 1. This is the reason the other options in the **Direction 2** area are not available. As soon as you set the value of this spinner to more than 1, the other options in this area will be available.

Spacing

The **Spacing** spinner is used to specify the spacing between each instance along the second direction. As mentioned earlier, the spacing is measured from the center of the entities selected for creating the pattern.

Angle

The **Angle** spinner is used to specify the angle of the linear pattern along the second direction. This value will always be measured from the positive X axis direction. As mentioned earlier, the orientation of the entities selected to be patterned will not change as you enter an angle value for the pattern. The default value of the angle is 90-degree. Figure 2-36 shows a linear pattern with the angle of direction 1 as 0-degree and the angle of direction 2 as 60-degree.

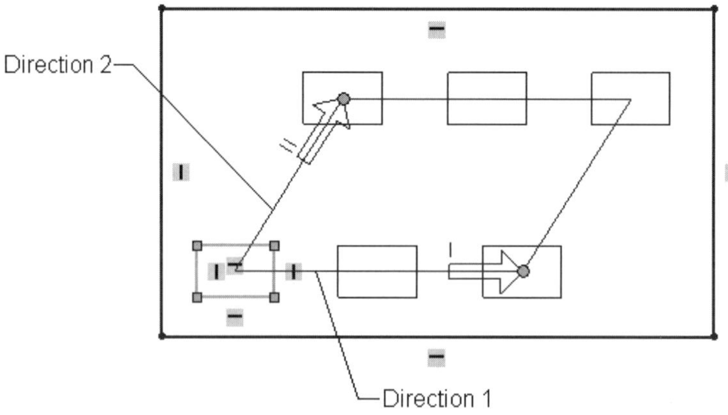

Figure 2-36 *Preview of the linear pattern at an angle of 0-degree along direction 1 and 60-degree along direction 2*

Reverse Direction 2

The **Reverse Direction 2** button is available on the right of the **Angle** spinner. This button is chosen to reverse direction 2.

Fixed

The **Fixed** check box is available below the **Spacing** spinner. If this check box is selected, the incremental spacing between the instances along direction 2 will be placed as a dimension between the first two instances after you create the pattern. Also, a centerline will be drawn between the first two instances.

Constrain angle between axes

The **Constrain angle between axes** check box will be available only if you set the value of the **Number** spinner in the **Direction 2** area to more than 1. This check box is selected by default. As a result, the angle between the two directions of the linear pattern will be constrained and placed as the dimension value when you create the pattern. However, if this check box is cleared, the angle between the two directions will not be displayed and will not be related to each other.

Items to repeat

The **Items to repeat** list box displays all entities selected to pattern. If you want to remove an entity from the selection set, select that entity from this box. The background color of the box will become reddish and the entity will be highlighted in blue. Now, press the DELETE key from

Tip. *You can also specify the spacing and angle value dynamically in the preview of the linear pattern. To do this, press the left mouse button on the control points displayed at the end of the arrow in the pattern preview and drag the cursor. After placing the arrow at the desired location, release the left mouse button. The new spacing and angle values will be displayed in their respective spinners.*

the keyboard. The removed entity will no longer be displayed in the preview of the pattern and will not be selected in the original entities.

Instances

The **Instances** list box displays all instances of the linear pattern, except the original instance. The instances are displayed in a matrix format with the first number indicating the instances along direction 1 and the second number indicating the instances along direction 2. Note that the matrix [1,1] is not displayed because it is the original instance. If you want to remove an instance from the pattern, select it in this list box and then press the DELETE key from the keyboard. You will notice that the deleted instance is no longer displayed in the preview of the pattern. Remember that this instance will not be deleted from the memory of the pattern. It will be displayed in the **Instances deleted** list box.

Tip. *You can also add entities in the current selection set or remove them from the current selection set by selecting them using the linear pattern cursor. As you add or remove the instances, the effect can be seen dynamically in the preview of the pattern.*

Instances deleted

The **Instances deleted** list box displays all instances that are deleted from the linear pattern. As mentioned earlier, the deleted instances are not removed from the memory of the pattern. They are stored in this list box. If you want to restore the deleted instance, select it in this box and then press the DELETE key. You will notice that the selected instance is again displayed in the preview and also in the **Instances** list box.

Undo

The **Undo** button is chosen to undo the operations performed in the **Linear Step Sketch and Repeat** dialog box. All operations will be undone one by one if you keep choosing this button and the original parameters will be restored. After everything is undone, this button will not be available.

Creating Circular Pattern of the Sketched Entities

CommandManager:	Sketch > Circular Sketch Step and Repeat *(Customize to Add)*
Menu:	Tools > Sketch Tools > Circular Sketch Step and Repeat
Toolbar:	Sketch > Circular Sketch Step and Repeat *(Customize to Add)*

In SolidWorks, the circular pattern of the sketched entities is created using the **Circular Sketch Step and Repeat** tool. To create the circular pattern, select the sketched entities using the **Select** tool and then choose this button from the **Sketch CommandManager**. The **Circular Sketch Step and Repeat** dialog box will be displayed, as shown in Figure 2-37, and the preview of the circular pattern will be shown on the screen in the background. Also, the arrow cursor will be replaced by the circular pattern cursor.

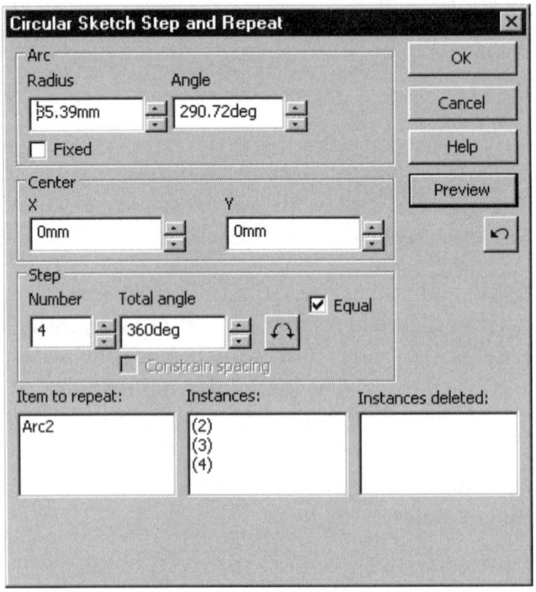

Figure 2-37 ***Circular Sketch Step and Repeat*** *dialog box*

The options provided in the **Circular Sketch Step and Repeat** dialog box are discussed next.

Arc Area

The options under the **Arc** area are used to specify the radius and angle of the imaginary circle around the circumference of which the instances will be arranged. These options are discussed next.

Radius

The **Radius** spinner is used to specify the radius of the imaginary circle around the circumference of which the instances of the circular pattern are arranged. As you modify this value, the center point of the imaginary circle will also change.

Angle

As you invoke the **Circular Sketch Step and Repeat** tool, a reference circle will be drawn with dashed lines around which the instances of the circular pattern will be arranged. Also, a reference line is drawn from the center of the entities selected to pattern and the center of the reference circle. The **Angle** spinner is used to modify the angle of this reference line drawn between the center of entities and the center of the reference circle. As you modify the value of this spinner, the center of the reference circle will also modify accordingly.

Fixed

If you select the **Fixed** check box, the radius of the reference circle will be placed as a linear dimension value after you create the circular pattern.

Center Area

Two spinners are available in the **Center** area: the **X** spinner and the **Y** spinner. These two spinners are used to specify the X and Y coordinates of the center point of the reference circle around which the instances of the circular pattern will be arranged. By default, the center of the reference circle is coincident with the origin. Therefore, the value of both these spinners will be 0 mm, 0 mm. You will notice that an arrow is displayed at the center point of the reference circle and the tip of the arrow coincides with the origin. You can define the new coordinates of the center of the reference circle in these spinners. You can also define the center of the reference circle by pressing the left mouse button on the control point provided on the tip of the arrow and dragging the arrow to the desired location. The values of the **Radius** and the **Angle** spinner in the **Arc** area and the **X** and **Y** spinners in the **Center** area will modify accordingly.

Note

If you know the location of the center point of the circular pattern, dragging the arrow at the center of the reference circle is the recommended method of defining the center of the circular pattern.

Step Area

The options in the **Step** area are used to specify the parameters such as the number of instances in the pattern, the angle between the instances, and so on. These options are discussed next.

Number

The **Number** spinner is used to specify the number of instances in the circular pattern. The default value of this spinner is **4**.

Total angle

The **Total angle** spinner is used to specify the total angle between all instances of the circular pattern. The default value of this spinner is 360-degree. You can enter the desired value of the angle in this spinner.

Reverse Rotation

The **Reverse Rotation** button is chosen to reverse the direction of the pattern. By default, the instances of the circular pattern are arranged in the clockwise direction. When you choose this button, the direction will be reversed to counterclockwise direction. To restore the counterclockwise direction, choose this button again. Figure 2-38 shows the preview of the circular pattern with four instances arranged in the counterclockwise direction.

Equal

The **Equal** check box is selected to specify the total angle between all instances of the circular pattern. If you clear this check box, the **Total angle** spinner will be replaced by the **Spacing** spinner. You can specify the incremental spacing between two successive instances of the circular pattern in this spinner. Figure 2-39 shows the preview of the circular pattern with 70-degree incremental spacing between two successive instances.

Constrain spacing

The **Constrain spacing** check box is selected to constrain the incremental spacing between

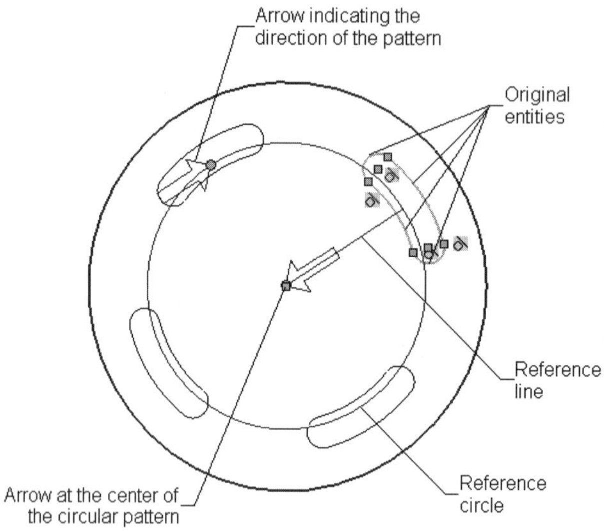

***Figure 2-38** Parameters associated with the circular pattern*

Tip. *You can modify the total angle between the instances by pressing the left mouse button on the tip of the direction arrow and dragging the cursor.*

the individual instances of the circular pattern. If this check box is selected, the incremental spacing will be placed as an angular dimension when you create the pattern. Figure 2-40 shows a circular pattern with the angle and radius values placed in the pattern.

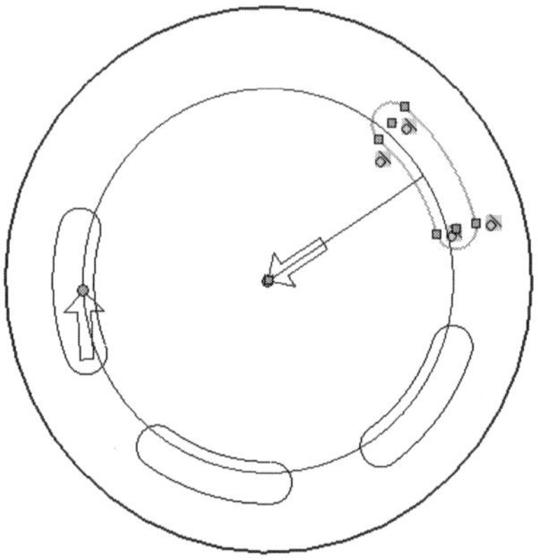

***Figure 2-39** Creating circular pattern by defining incremental angle between individual instances*

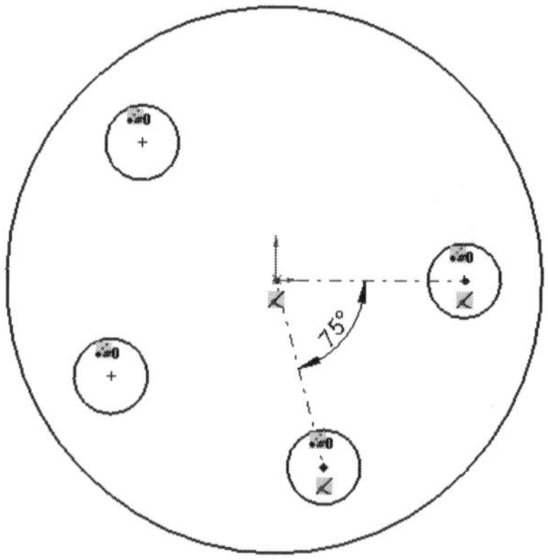

***Figure 2-40** Radius and angle values placed in the circular pattern*

Note
The other options in this dialog box are similar to those discussed in the linear pattern.

EDITING PATTERNS

You can edit the patterns of the sketched entities by using the shortcut menu that will be displayed when you right-click on any instance of the pattern. Depending on whether you right-clicked on the instance of the linear or the circular pattern, the **Edit Linear Step and Repeat** or the **Edit Circular Step and Repeat** option will be available in the shortcut menu. Figure 2-41 shows the partial view of the shortcut menu that will be displayed when you right-click one of the instances of a circular pattern.

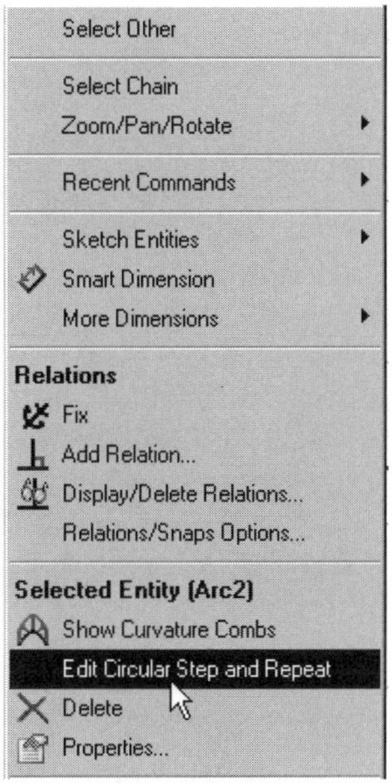

Figure 2-41 *The partial view of the shortcut menu displayed on right-clicking on an instance of a circular pattern*

Depending on whether you choose the option to edit a linear pattern or a circular pattern, the **Linear Sketch Step and Repeat** or the **Circular Sketch Step and Repeat** dialog box will be displayed. Note that only those options that can be edited will be available in these dialog boxes.

WRITING TEXT IN THE SKETCHING ENVIRONMENT

Command Manager:	Sketch > Text	*(Customize to Add)*
Menu:	Tools > Sketch Entities > Text	
Toolbar:	Sketch > Text	*(Customize to Add)*

You can also create text in the sketching environment of SolidWorks that can be later used to create join or cut features. To write the text, choose the **Text** button from the **Sketch CommandManager**; the **Sketch Text PropertyManager** will be displayed, as shown in Figure 2-42. Enter the text in the box available in the **Text** rollout. You will notice that by default, the text will be started from the sketch origin. However, the text will actually be placed at the location that you specify on the screen before or after

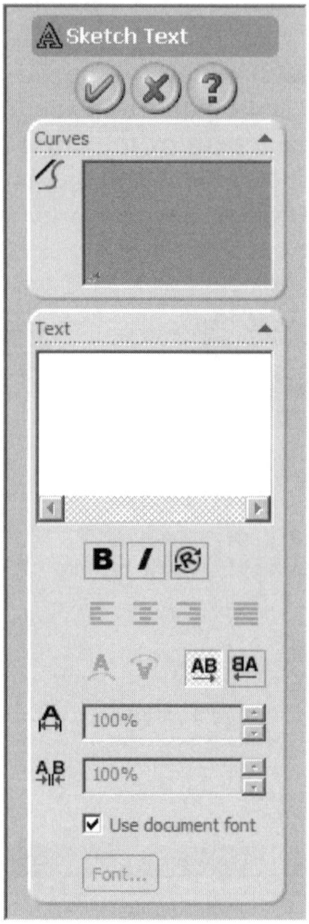

Figure 2-42 Sketch Text PropertyManager

writing the text. You can also change the format, font, justification, and so on using the options available in the **Text** rollout. Figure 2-43 shows the text created using the **Text** tool.

Figure 2-43 *Sketch created using the* ***Text*** *tool*

You can also create the text along a curve. To do so, you first need to create a curve. The curve can be an arc, a spline, a line, or a combination of line, arc, and spline. Next, invoke the **Sketch Text PropertyManager** and select the curve or curves along which you need to create the text. Now, enter the text in the **Text** edit box; you will observe that the text is created along the arc. You can use the **Flip Horizontal** and **Flip Vertical** options to modify the display of the text. Figure 2-44 shows the text created along an arc.

Figure 2-44 Text created along the arc

MODIFYING SKETCHED ENTITIES

Most of the sketches require modification at some stage of the design. Therefore, it is important for any designer to understand the process of modification in SolidWorks. Modification of various sketched entities is discussed next.

Modifying a Sketched Line

You can modify the sketched lines by using the **Line Properties PropertyManager**, which is displayed when you select it using the **Select** tool. Note that if the selected line is part of a rectangle, polygon, or a parallelogram, the entire object will be modified as you modify the line. This is because the relations are applied to all lines of a rectangle, polygon, and a parallelogram.

Similarly, you can also modify the centerlines using the **Line Properties PropertyManager**, which will be displayed when you select a centerline.

Modifying a Sketched Circle

To modify a sketched circle, select it using the **Select** tool to display the **Sketched Circle PropertyManager**. The coordinate values of the center point of the circle and the value of the radius will be displayed. You can modify the values that you want.

Tip. *The status of the sketched entity that you select for modification is displayed in the* ***Existing Relations*** *rollout of the* ***PropertyManager****, which is displayed. For example, if the selected sketched entity is fully defined, it will be displayed in the* ***PropertyManager*** *and if the entity is underdefined, the* ***PropertyManager*** *will display a message that the entity is underdefined.*

Modifying a Sketched Arc

To modify a sketched arc, select it using the **Select** tool. The **Arc PropertyManager** will be displayed with the coordinate values of the center point, the start point, and the endpoint. The values of the radius and the included angle will also be displayed. You can modify the values that you want.

Modifying a Sketched Polygon

To modify a sketched polygon, right-click on any one edge of the polygon to display the shortcut menu. Choose the **Edit polygon** option from the shortcut menu to display the **Polygon PropertyManager**. You can modify the selected polygon using the options in the **Polygon PropertyManager** or draw a new polygon.

Note

If you right-click on the reference circle that is automatically drawn when you draw a polygon, the ***Edit polygon*** *option will not be available in the shortcut menu.*

Modifying a Spline

Enhanced

You can perform five types of modifications on a spline. The first is the modification of the coordinates of the selected control point. The second is adding more control points on the spline. The third is modifying the spline using the Spline Handles. The fourth is adding more Spline Handles. The fifth is adding a curvature control symbol. These modifications are discussed in detail next.

Modifying the Coordinates of Control Points

To modify the control points of a spline, select it using the **Select** tool; the **Spline PropertyManager** will be displayed. Partial view of the **Spline PropertyManager** is displayed in Figure 2-45. All control points of the spline will be displayed and the current control point will be displayed with a square with a filled circle inside it. The number and the coordinates of the current handle will be displayed in the **Spline PropertyManager**. You can modify the coordinates using the spinner. You can select any other handle by using the **Spline Point Number** spinner.

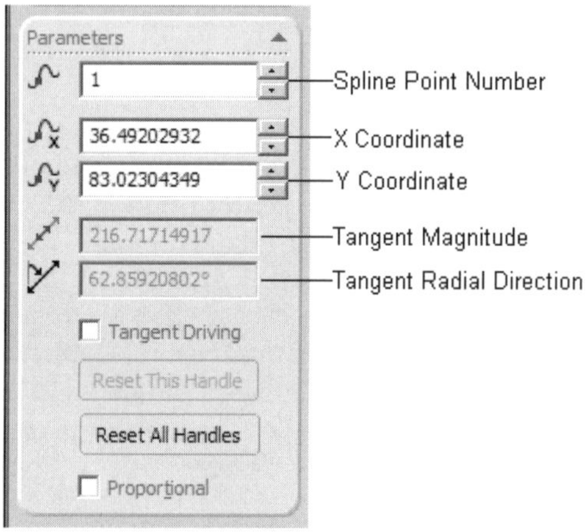

Figure 2-45 *Partial view of the* ***Spline PropertyManager***

Adding Control Points

To add control points to the spline, choose **Tools > Spline Tools > Insert Splint Point** from the menu bar. You can also select the spline and right-click to invoke the shortcut menu. In this menu, choose the **Spline Point** option. Now, select the points on the spline where you want to add the control point. A circle will be displayed at the point specified on the spline and the spline handle will also be displayed attached to the inserted point. You will learn more about the spline handles later in this chapter. Similarly, you can add as many control points as you want to the spline. After adding the required number of control points, invoke the **Select** tool and then select the spline again. You will notice that the boxes of control points are displayed on all points you specified on the spline. You can display the coordinates of a control point using the **Spline Point Number** spinner available in the **Spline PropertyManager**.

Modifying the Spline Using the Spline Handles

The spline handle is a line with arrows on its both endpoints. The spline handles will be displayed when you select the spline using the **Select** tool. The default spline handles will be displayed in gray, which implies that they are not selected. When you move the cursor over a spline handle, it will be highlighted in red and a double-sided arrow will be displayed below the cursor. Select the spline handle when it is displayed in red; the spline handle will be displayed in green. The number of respective control points, and their coordinates are displayed in the **Parameters** rollout. Drag the cursor to dynamically modify the magnitude of the tangent and the radial direction of the tangent. As you drag the cursor, the values of the magnitude of the tangent and the radial direction of the tangent will be modified dynamically in the **Tangent Magnitude** and **Tangent Radial Direction** spinners, respectively. The **Tangent Driving** check box is also selected automatically when you drag the cursor. After modifying the shape of the spline, click anywhere in the drawing area to exit the current selection set. You will notice that currently modified spline handle is displayed in blue, which implies that the default setting of this spline handle is modified. Similarly, you can edit other spline handles.

Adding/Deleting the Spline Handles

To add more spline handles, select the spline and right-click to invoke the shortcut menu. Choose the **Add Tangency Control** option from the shortcut menu; a blue-colored double-sided arrow will be displayed attached to the selected spline. Move the cursor to the location where you want to place the spline handle and click on that location. A spline control point will also be added along with the spline handle. You can also apply relations to the spline handles. You will learn more about relations in the later chapters. You can also delete the spline handles by selecting them and pressing the DELETE key from the keyboard.

Adding the Curvature Control

You can also add a curvature control symbol which is used to modify the curvature of the spline. To add a curvature control, select the spline and invoke the shortcut menu. Choose the **Add Curvature Control** option from it; the curvature control symbol  will be attached to selected spline. Move the cursor to the location where you want to place the curvature control symbol. A spline handle will also be placed tangent to the spline at the location where the curvature control symbol is attached. Now, select and drag the circle placed at the bottom of the this symbol. You will notice that the curvature of the spline is modified dynamically as you drag the cursor. After modifying the shape of the spline release the left mouse button. You can also delete the curvature control symbol by selecting it and pressing the DELETE key from the keyboard.

Modifying a Sketched Point

To modify a sketched point, select it using the **Select** tool; the **Sketched Point PropertyManager** will be displayed. You can modify the coordinates of the sketched point using this **PropertyManager**.

Modifying an Ellipse or an Elliptical Arc

To modify an ellipse or an elliptical arc, select it using the **Select** tool; the **Ellipse PropertyManager** will be displayed. You can modify the required parameters using the options available in this **PropertyManager**.

Modifying a Parabola

To modify a parabola, select it using the **Select** tool; the **Parabola PropertyManager** will be displayed. You can modify the parameters of the parabola from this **PropertyManager**.

Dynamically Modifying and Copying the Sketched Entities

In the sketcher environment of SolidWorks, you can modify or copy the sketched entities by dynamically dragging them using the left mouse button. For example, consider a case where you create a sketch of a rectangle and you want to increase the size of the rectangle. You simply have to select any of the lines of the rectangle or any of its vertices and hold the left mouse button down to drag the cursor. Drag the sketch according to your requirement and then release the left mouse button. If you choose **Tools > Sketch Setting > Detach Segment on Drag** from the menu bar and select a line of a rectangle to drag, the line segment will be detached from the rectangle. This option is not activated by default. As a result, the segments are not detached on dragging.

You can also copy the sketched entities dynamically. Select the sketched entity or entities to be copied using the CTRL key or by dragging a window around them. Press and hold the CTRL key down. Now, press and hold the left mouse button down and drag the selected entity or entities. The preview of the copied entities will be displayed. Release the left mouse button at the location where you want to place the new entities. You can make multiple copies of the selected entities by repeating this procedure.

Splitting the Sketched Entity

CommandManager:	Sketch > Split Entities	*(Customize to Add)*
Menu:	Tools > Sketch Tools > Split Curve	
Toolbar:	Sketch > Split Curve	*(Customize to Add)*

Using the **Split Entities** option you can split a sketched entity into two or more entities by specifying the split points. To split an entity, choose the **Split Entities** button from the **Sketch CommandManager**; the current cursor will be replaced by the split entities cursor. Move the cursor to an appropriate location where you want to split the sketched entity. When the cursor snaps to the entity, press the left mouse button to add a split point. Now, right click to display the shortcut menu and choose the select option from the shortcut menu. Using the select cursor, select the sketched entity. You will notice that the sketched entity is divided in two entities and a split point is added between the two sketched entities. You can add as many split points as you need. Remember that to split a circle, a full ellipse, or a closed spline, you need to split them at least at two points.

You can also delete the split points to convert the split entity into a single entity. To delete the split point, select the split point and press the DELETE key from the keyboard. You can also right-click on the split point to display the shortcut menu and choose the **Delete** option from the shortcut menu.

TUTORIALS

Tutorial 1

In this tutorial, you will create the base sketch of the model shown in Figure 2-46. The sketch of the model is shown in Figure 2-47. You will create the sketch of the base feature using the normal sketch tools and then modify and edit the sketch using various modifying options.

(Expected time: 30 min)

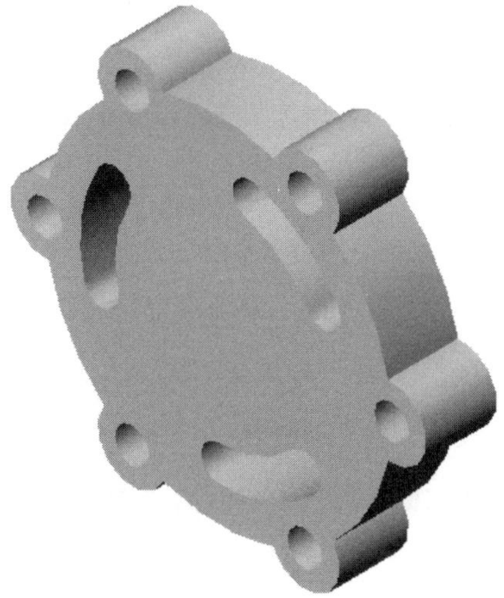

Figure 2-46 *Solid Model for Tutorial 1*

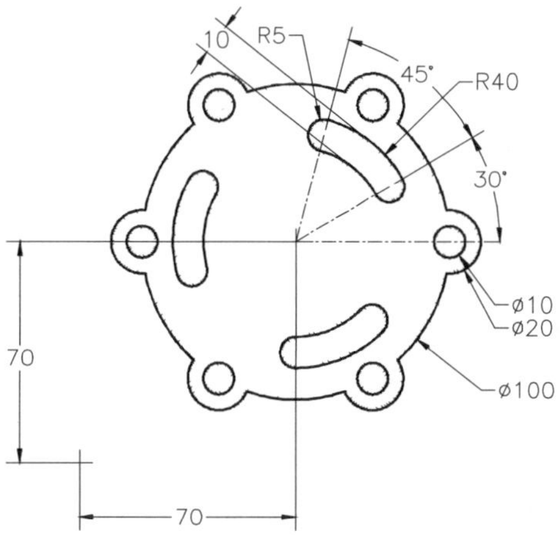

Figure 2-47 *The sketch of the base feature*

The steps that will be followed to complete this tutorial are listed next.

a. Start SolidWorks and then start a new part document.
b. Maximize the part file document and then switch to the sketching environment.
c. Draw the outer loop of the sketch of the given model, refer to Figures 2-48 and 2-49.
d. Create the inner cavity using the **Centerpoint Arc** and **Tangent arc** tools, refer to Figure 2-50.
e. Use the **Circular Sketch Step and Repeat** tool to create a circular pattern of the inner cavity, refer to Figure 2-51.
f. Complete the sketch by creating the circles that define the hole in the outer loop.
g. Using the **Circular Sketch Step and Repeat** tools, create the circles for the remaining holes in the outer loop, refer to Figure 2-52.

Starting SolidWorks and Starting a New SolidWorks Document

1. Start SolidWorks by choosing **Start** > **Programs** > **SolidWorks 2005** > **SolidWorks 2005** or by double-clicking on the shortcut icon of SolidWorks 2005 available on the desktop of your computer.

   The **SolidWorks Resources Task Pane** is displayed on the right of this window.

2. Choose the **New Document** option from the **Getting Started** group of the **SolidWorks Resources Task Pane** to display the **New SolidWorks Document** dialog box will be displayed.

3. The **Part** button is chosen by default. Choose the **OK** button from this dialog box.

   A new SolidWorks part document is started. The **SolidWorks Resources Task Pane** is also displayed and some of the display of the drawing area is consumed by it. Therefore, you need to exit this **Task Pane**.

4. Click anywhere in the drawing area to exit the **SolidWorks Resources Task Pane**.

   By default, when you start a new part document, the part modeling environment is active. But because you need to draw the sketch of the feature, you need to invoke the sketching environment.

5. Choose the **Sketch** button from the **Standard** toolbar; the **Edit Sketch PropertyManager** is invoked and you are prompted to select the plane to create the sketch.

6. Select the **Front Plane**; the sketching environment is invoked and the plane is oriented normal to the view.

   You will notice that red sketch origin is displayed along with the confirmation corner.

Modifying the Snap and Grid Settings and the Dimensioning Units

Before you proceed with drawing the sketch, you need to modify the grid and snap settings so that you can make the cursor jump through a distance of 10 mm.

1. Choose **Tools > Options** from menu bar and then choose the **Document Properties** tab to invoke the **Document Properties - Detailing** dialog box.

2. Choose the **Grid/Snap** option from the area on the left. Set the value of the **Minor-lines per major** spinner to **10** and the **Major grid spacing** spinner to **100**.

   The coordinates displayed close to the lower left corner of the SolidWorks window shows an increment of 10 mm when you move the cursor in the drawing area after exiting the dialog box.

3. If by default the grid is displayed when you invoke the sketching environment, you can remove the display of the grid. To remove the grid clear the **Display grid** check box from the **Grid** area.

4. Choose the **Go To System Snaps** button and select the **Grid** check box and clear the **Snap only when grid is displayed** check box, if it is selected.

   If you selected a unit other than millimeter to measure the length while installing SolidWorks, you need to select the units for the current drawing.

5. Invoke the **Document Properties** tab and select the **Units** option from the area on the left of the **Document Properties - Grid/Snap** dialog box.

6. Select the **MMGS (millimeter, gram, second)** radio button from the **Unit system** area and **Degrees** from the drop-down list available in the **Angular units** area.

7. Choose the **OK** button after making the necessary settings.

Drawing the Outer Loop of the Sketch

As evident from Figure 2-47, the sketch consists of the outer loop and inner cavities. It is recommended that for complex sketches, you should first create the outer loop of the sketch. Therefore, you need to create the outer loop of the sketch first. Then you can proceed with the inner cavities.

The origin of the sketcher environment is placed in the middle of the drawing area and you have to create the sketch in the first quadrant. Therefore, it is recommended that you modify the drawing area, which can be done using the **Pan** tool.

1. Choose the **Pan** button from the **View** toolbar to invoke the **Pan** tool; the select cursor is replaced by the pan cursor.

2. Press and hold the left mouse button down near the sketch origin and drag the cursor to the lower left corner of the drawing area. The sketch origin moves near the lower left corner of the drawing area.

Tip. *You can also pan using the CTRL+middle mouse button. To use this option, hold the CTRL key down and then press the middle mouse button and drag the cursor.*

3. Choose the **Circle** button from the **Sketch CommandManager**; the pan cursor is replaced by the circle cursor.

4. Move the cursor to a location where the value of the coordinates is 70 mm 70 mm 0 mm.

5. Specify the center point of the circle at this location and move the cursor horizontally toward the right. When the radius above the circle cursor shows a value 50, press the left mouse button.

6. Choose the **Zoom to Area** button from the **View** toolbar. Using the left mouse button, drag the cursor to define a window such that the sketched circle and the sketch origin are placed in the window.

   When you release the left mouse button the display area of the sketch is increased.

7. Choose the **Circle** button from the **Sketch CommandManager** and move the cursor to the right quadrant of the circle. All quadrants of the circle are displayed and the right quadrant on which you have placed the cursor is displayed in red. The symbol of coincident constraint is displayed below the cursor.

9. Using the left mouse button, specify the center point of the circle at this location and move the cursor horizontally toward the right. When the value of the radius above the circle cursor shows a value 10, press the left mouse button.

10. Choose the **Trim Entities** button from the **Sketch CommandManager** and choose the **Trim to closet** button from the **Trim PropertyManager**.

11. Trim the part of the sketch so that the sketch looks similar to that shown in Figure 2-48.

12. Choose the **Circular Sketch Step and Repeat** button from the **Sketch CommandManager**. You may have to customize this button to add in the **Sketch CommandManager**.

    The **Circular Sketch Step and Repeat** dialog box is displayed and the cursor is replaced by the circular pattern cursor.

13. Using the circular pattern cursor, select the smaller trimmed circle. The preview of the circular pattern with default setting is displayed in the drawing area.

    You will notice that the center of the circular pattern, which is displayed by an arrow, is placed at the origin. But because the origin is not the actual center of the circular pattern, you need to modify the center of the circular pattern. This can be done by entering the coordinates of the point in the **X** and **Y** spinners available in the **Center** area of this dialog box. But the recommended method of modifying the center of the circular pattern is by dragging the arrow displayed at the center of the pattern.

14. Move the circular pattern cursor at the control point available at the end of the arrow displayed at the origin.

15. Press and hold the left mouse button down at the control point and drag it to the center of the outer trimmed circle in the sketch. Release the left mouse button when a red circle is displayed at the center point of the outer trimmed circle.

    You will notice that both **X** and **Y** spinners in the **Center** area show 70 mm as the values. This is because the center of the outer trimmed circle is located at a distance of 70 mm along the X and Y axis directions.

16. Set the value of the **Number** spinner in the **Step** area to **6**. Accept all other default values and choose the **OK** button to create the pattern.

17. Trim the unwanted portion of the outer trimmed circle using the **Trim Entities** tool. You need to use the **Trim to closet** button for this trimming. After trimming, the sketch should look similar to that shown in Figure 2-49.

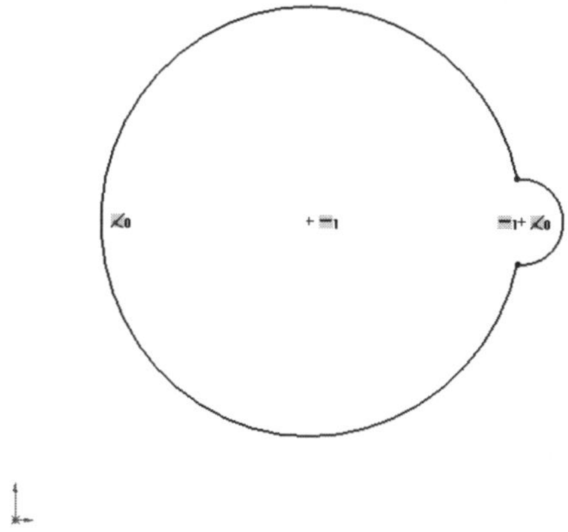

Figure 2-48 Sketch after trimming the unwanted entities

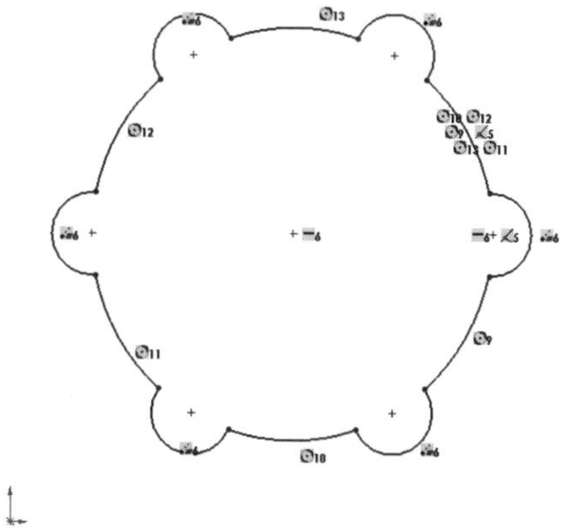

Figure 2-49 Outer loop of the sketch

Drawing the Sketch of the Inner Cavity

Next, you need to draw the sketch of the inner cavities. You will draw the sketch of one of the cavities and then create a circular pattern of this cavity. The number of items in the circular pattern will be 3.

Before creating the sketch of the inner cavity, you need to create a centerline that will act as a reference for creating the sketch of the inner cavity.

1. Invoke the **System Options - Relations/Snaps** dialog box. Make sure that the **Angle** check box is selected and set the value of the **Snap Angle** spinner to **30**.

2. Choose the **Centerline** button from the **Sketch CommandManager**. Move the cursor close to the center of the outer circle. When the red circle is displayed, specify the start point of the centerline. Move the cursor to a location where the angle of the line is shown as exactly 30-degree and the length is close to 35 in the **Parameters** rollout of the **PropertyManager**. At this point, specify the endpoint of the line.

3. Exit the **Centerline** tool and select the centerline; the **Line PropertyManager** is displayed. Set the value of the **Length** spinner to **35**.

4. Choose the **Circle** button from the **Sketch CommandManager**.

5. Move the cursor to the upper endpoint of the centerline. When the red circle is displayed, specify the center point. Move the cursor horizontally toward the right and draw a circle of radius close to 5 mm.

6. Set the value of the radius of the circle to **5** in the **Radius** spinner of the **Circle PropertyManager**.

7. Choose the **Circular Sketch Step and Repeat** button from the **Sketch CommandManager**; the **Circular Sketch Step and Repeat** dialog box is displayed and the circle is automatically selected to be patterned. This is because the circle was selected when you invoked this tool.

   Note that the center of the circular pattern is placed at the origin. You need to change the center of the circular pattern to the center of the outer trimmed circle.

8. Move the circular pattern cursor at the control point available at the end of the arrow displayed at the origin.

9. Press and hold the left mouse button down and drag the cursor to the center point of the outer trimmed circle. Release the left mouse when the red circle is displayed.

10. Set the value of the **Number** spinner to **2**.

11. Set the value of the **Total angle** spinner to 45-degree and choose the **Reverse Rotation** button.

12. Choose **OK** from the **Circular Sketch Step and Repeat** dialog box.

13. Choose the **Centerpoint Arc** button from the **Sketch CommandManager**.

14. Move the cursor to the center point of the outer trimmed circle and specify the center point

of the arc. Now, move the cursor to the intersection of the centerline and the circle. Specify the start point of the arc when the intersection symbol appears.

15. Move the cursor in the counterclockwise direction. When the value of the angle above the arc cursor shows a value of 45-degree, press the left mouse button to specify the endpoint of the arc. At this point, the second inner circle is highlighted in red.

    Next, you need to offset the arc created in the previous step through an offset distance of 10.

    The arc created earlier is displayed in green, indicating that it is already selected. If the arc is not selected, select the arc using the left mouse button before invoking the offset tool.

16. Choose the **Offset Entities** button from the **Sketch CommandManager**; the **Offset Entities PropertyManager** is displayed.

17. The default value of the **Offset Distance** spinner is **10**. Clear the **Reverse** check box, if selected. Choose the **OK** button from the **Offset Entities PropertyManager**.

18. Choose the **Trim Entities** button from the **Sketch CommandManager**. Trim the unwanted portion of the inner cavity and then exit this tool. If the sketch turns red and the **SolidWorks** warning window appears, choose **OK** to close the window. The sketch after completing the inner cavity is shown in Figure 2-50.

Creating the Pattern of the Inner Cavity

Next, you need to create the pattern of the inner cavity using the **Circular Sketch Step and Repeat** tool. The pattern of the inner cavity consists of three instances. The center point of the pattern lies at the center point of the outer trimmed circle.

Before invoking this option, exit the **Trim** tool if it is active. Next, press the CTRL key from the keyboard and using the left mouse button, select all entities of the inner cavity.

1. Choose the **Circular Sketch Step and Repeat** button from the **Sketch CommandManager** toolbar to invoke the **Circular Sketch Step and Repeat** dialog box.

2. Press and hold the left mouse button down at the control point of the arrow displayed at the origin. Drag it to the center of the outer trimmed circle in the sketch. Release the left mouse button when the red circle is displayed.

   You will notice that the **X** and **Y** spinners in the **Center** area show the value 70 mm. The default value of the number of items in the pattern is 4. But because you need three items in the pattern, you need to modify this value.

3. Set the value of the **Number** spinner in the **Step** area to **3**. Accept all other default values and choose the **OK** button to create the pattern. If the sketch turned red earlier, it will turn back to blue and the **SolidWorks** information box is displayed. Choose **OK** from this dialog box. The sketch after creating the pattern of the inner cavity is shown in Figure 2-51.

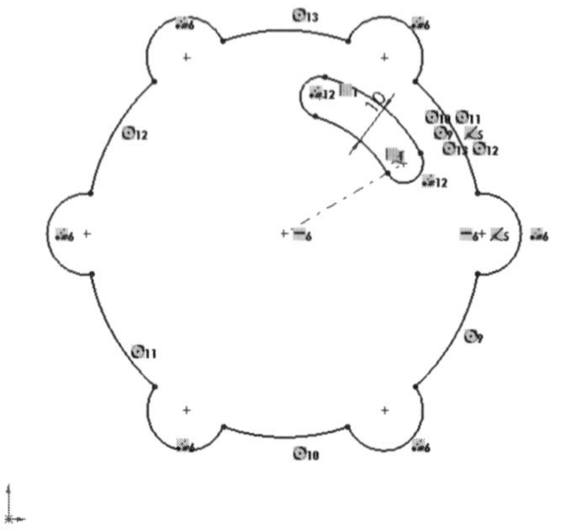

Figure 2-50 *The sketch of the outer loop and the inner cavity*

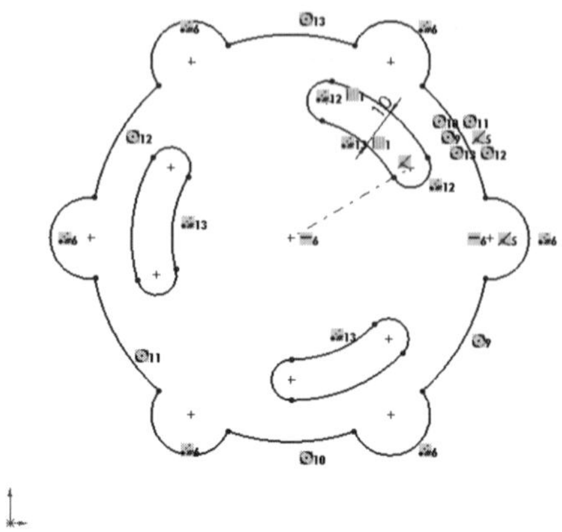

Figure 2-51 *Sketch after creating the pattern of the inner cavity*

Sketching the Holes

Next, you need to draw the sketch of the holes. As evident from Figure 2-47, you need to draw a total of six circles. After drawing the first circle, you need to draw the other five circles by creating a circular pattern of the parent circle.

1. Choose the **Circle** button from the **Sketch CommandManager**.

2. Select the center point of the trimmed circle at the left quadrant of the outer circle as the center point of the new circle.

3. Press and hold the CTRL key down and draw a circle of radius close to 5. Make sure you press the CTRL key so that the cursor does not snap to the points or grid.

4. Set the value of the **Radius** spinner to **5** in the **Circle PropertyManager**.

5. Choose the **Circular Sketch Step and Repeat** button from the **Sketch CommandManager** to invoke the **Circular Sketch Step and Repeat** dialog box.

6. Move the center of the circular pattern to the center point of the outer trimmed circle.

7. Set the value of the **Number** spinner to **6** in the **Step** area of this dialog box. Accept all other default values and choose **OK** in this dialog box.

The final sketch of Tutorial 1 is shown in Figure 2-52.

Saving the Sketch

As mentioned earlier, it is recommended that you create a separate folder for saving the tutorial files of each chapter. When you invoke the option to save the document, the default folder that is displayed is the one in which you saved the last document. Browse to the *My Documents/SolidWorks* folder. Now, you will create another folder inside the *SolidWorks* folder and save the document.

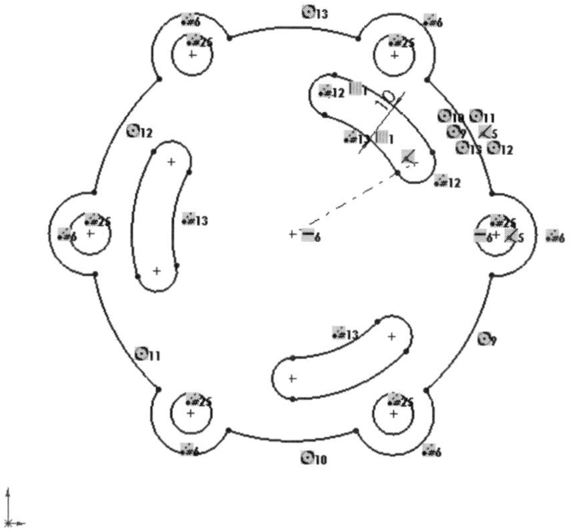

***Figure 2-52** Final sketch for Tutorial 1*

1. Choose the **Save** button from the **Standard** toolbar to invoke the **Save As** dialog box.

2. Choose the **Create New Folder** button from the **Save As** dialog box. Enter the name of the folder as *c02* and press ENTER.

3. Enter the name of the document as *c02tut1* in the **File name** edit box and choose the **Save** button.

   The document will be saved in the */My Documents/SolidWorks/c02* folder.

4. Close the file by choosing **File** > **Close** from the menu bar.

> **Tip**. *If you do not want to display the symbols of the relations that are displayed on the sketched entities, choose **View > Sketch Relations** from the menu bar to clear this option.*

Tutorial 2

In this tutorial, you will create the base sketch of the model shown in Figure 2-53. The sketch of the model is shown in Figure 2-54. You will create the sketch with mirror line and mirror tool. After creating the sketch, you will modify it by dragging the sketched entities.

(Expected time: 30 min)

Figure 2-53 Solid model for Tutorial 2

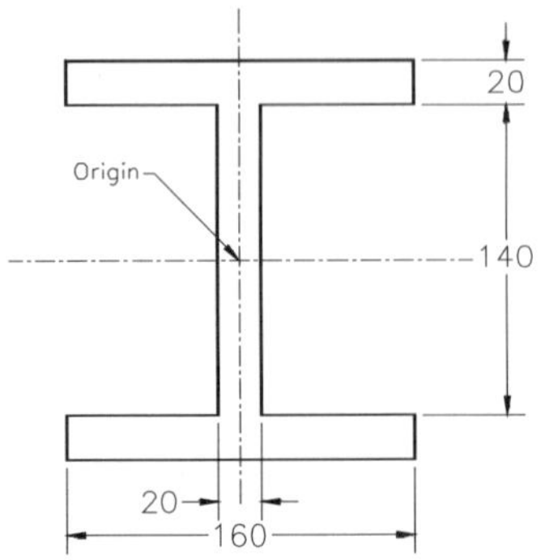

Figure 2-54 Sketch of the model

The steps that will be followed to complete this tutorial are listed below.

a. Start a new part document.
b. Maximize the part document and then switch to the sketching environment.
c. Create the centerlines and create a mirror line using one of the centerlines and the **Mirror Entities** tool.
d. Create the sketch in the third quadrant; the sketch will automatically be mirrored on other side, refer to Figure 2-55
e. Mirror the entire sketch along the second mirror line, refer to Figure 2-56.
f. Modify the sketch by dragging the sketched entities, refer to Figure 2-57.

Starting a New Document

1. Choose the **New** button from the **Standard** toolbar to invoke the **New SolidWorks Document** dialog box.

2. The **Part** button is chosen by default. Choose the **OK** button from the **New SolidWorks Document** dialog box to start a new SolidWorks part document.

   Next, you need to invoke the sketching environment.

3. Choose the **Sketch** button from the **Standard** toolbar and select the **Front Plane** to invoke the sketching environment.

   Next, you need to modify the snap and grid settings.

4. Invoke the **Document Properties - Grid/Snap** dialog box. Set the value of the **Major grid spacing** spinner to **100** and the **Minor-lines per Major** spinner to **10**.

   It is assumed that you have selected the unit of measurement as millimeter while installing SolidWorks. However, if you have selected any other unit while installation, you need to change it.

5. Select the **Units** option available below the **Grid/Snap** option. Select the **MMGS (millimeter, gram, second)** option from the **Unit system** area. Choose the **Degrees** option from the drop-down list available in the **Angular units** area. Choose **OK** to exit the dialog box.

Drawing the Center Lines and Converting a Center line Into a Mirror Line

In this tutorial, you need to draw the sketch of the given model with the help of a mirror line. The sketches that are symmetrical along any axis are recommended to be drawn using the mirror line. The mirror line is drawn using the **Centerline** and **Dynamic Mirror Entities** tools. After drawing a mirror line, when you draw the entities on one side of the mirror line, the same entities are also automatically drawn on the other side of the mirror line. A symmetrical relation is also applied to the entities on both sides of the mirror line. Therefore, if you modify an entity on one side of the mirror line, the same modifications are reflected to the mirrored entity, and vice versa. First, you need to draw a mirror line.

1. Choose the **Centerline** button from the **Sketch CommandManager**.

2. Move the line cursor to a location where the value of the coordinates is 0 mm 100 mm 0 mm. You may have to zoom the drawing to reach that point.

3. Specify the start point of the centerline at this point and move the cursor vertically downward and draw a line of 200 mm length. You may have to zoom and pan the drawing to draw a line of this length.

4. Now, double-click anywhere on the screen to end the current chain. You can also press the right mouse button anywhere on the drawing area to invoke the shortcut menu and choose the **End chain** option from the shortcut menu.

5. Move the line cursor to a location where the value of the coordinates is -100 mm 0 mm 0 mm.

6. Specify the start point of the centerline and move the cursor horizontally toward the right to draw a line of 200 mm length. Pan the drawing, if required.

7. Zoom to fit the drawing and then right-click to invoke the **Select** tool. The line cursor will be replaced by the select cursor.

8. Select the vertical centerline; it will be displayed in green.

9. Choose **Dynamic Mirror Entities** from the **Sketch CommandManager** to convert the selected centerline to a mirror line and activate the automatic mirror option.

You can confirm the creation of the mirror line and the activation of the automatic mirror option by observing the symmetrical symbol applied to both ends of the centerline.

Drawing the Sketch

Next, you need to draw the sketch of the base feature. You will draw the sketch in the third quadrant and the same sketch will be created automatically on the other side of the mirror line. The symmetrical relation is applied between the parent entity and the mirrored entity.

1. Press the L key from the keyboard to invoke the **Line** tool.

2. Move the line cursor to a location where the value of the coordinates is 0 mm -100 mm 0 mm.

3. Specify the start point of the line at this location and move the cursor horizontally toward the left.

4. Specify the endpoint of the line when the length of the line above the line cursor shows a value 90.

   You will notice that as soon as you specify the endpoint of the line, a mirror image is automatically created on the other side of the mirror line. This line drawn as the mirrored entity is merged with the line drawn on the left. Therefore, the entire line becomes a single entity. Remember that the lines will merge only if one of the endpoints of the line you draw is coincident with the mirror line.

5. Move the cursor vertically upward and specify the endpoint of the line when the length of the line above the line cursor shows a value 30.

   You will notice that as soon as you specify the endpoint of a line, a mirror image is automatically created on the other side of the mirror line.

6. Move the line cursor toward the right and specify the endpoint of the line when the length of the line above the line cursor shows a value 30. A mirror image is automatically created on the other side of the mirror line.

7. Move the line cursor vertically upward and specify the endpoint of the line when the line cursor snaps the horizontal centerline. Exit the **Line** tool. The sketch after creating the lines is shown in Figure 2-55.

Mirroring the Entire Sketch

After creating one half of the sketch, you need to mirror the entire sketch about the horizontal centerline. But first, you need to disable automatic mirroring.

1. As evident from the symmetric relation symbol on the vertical centerline, automatic mirroring is activated. Choose the **Dynamic Mirror Entities** button again to disable automatic mirroring. Now, the vertical centerline will not work as a mirror line.

2. Use the box selection method to select all lines sketched earlier and the horizontal centerline. Make sure you do not select the vertical centerline.

3. Choose the **Mirror Entities** button from the **Sketch CommandManager**. The entire sketch is mirrored about the horizontal centerline. The sketch after mirroring is shown in Figure 2-56.

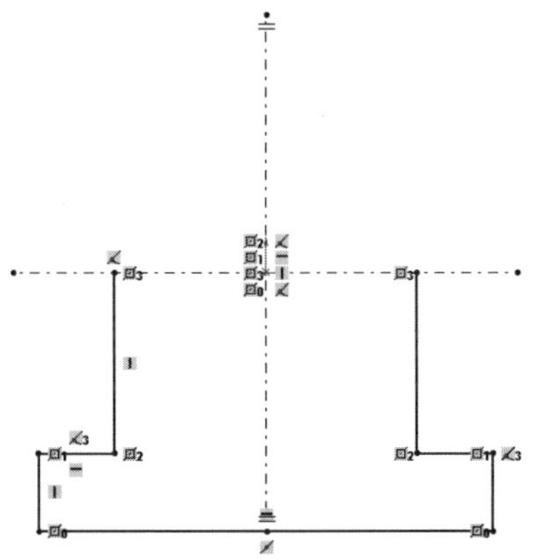

Figure 2-55 Sketch after drawing the lines

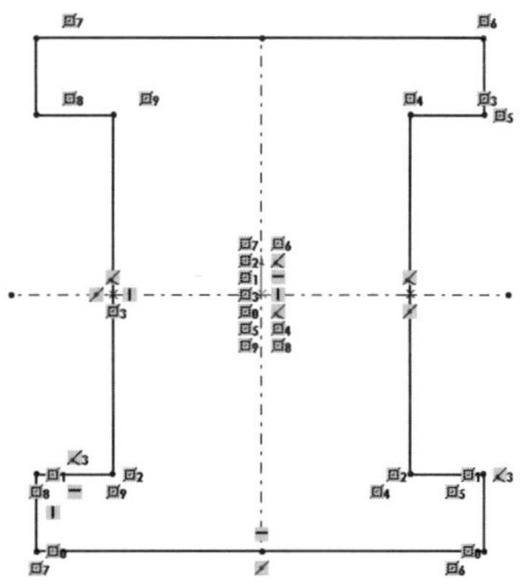

Figure 2-56 Sketch after mirroring the sketched entities

Modifying the Sketch by Dragging

Next, you need to modify the sketch by dragging. While dragging the entities, you will observe that the corresponding mirrored entity will also be modified.

1. Select the lower right vertical line. The **Line PropertyManager** will be displayed on the left of the drawing area.

   You will notice that the value of the length in the **Length** spinner shows a value **30**. But the required length of this line is 20. Therefore, you need to edit the length of this line. In this tutorial, you will edit the sketch by dragging. You will observe that when you drag one line, all dependent lines are also modified. This is because they are created as mirror images.

2. Select the top horizontal line and then move the cursor to the right endpoint of this line. When the red circle appears, drag the cursor vertically downward. Release the left mouse button when the cursor snaps to the next grid point.

   You will observe that all sketched entities that are related to the dragged entity are also modified.

3. Select the lower left vertical line and drag one of its endpoint horizontally toward the right. Release the left mouse button when the cursor snaps to the next grid point on the right.

4. Select the middle left vertical line and drag one of the endpoints of the line horizontally toward the right. Release the left mouse button when the cursor snaps to the grid point on the left of the vertical centerline.

   The final sketch after modifying the sketched entities by dragging is shown in Figure 2-57.

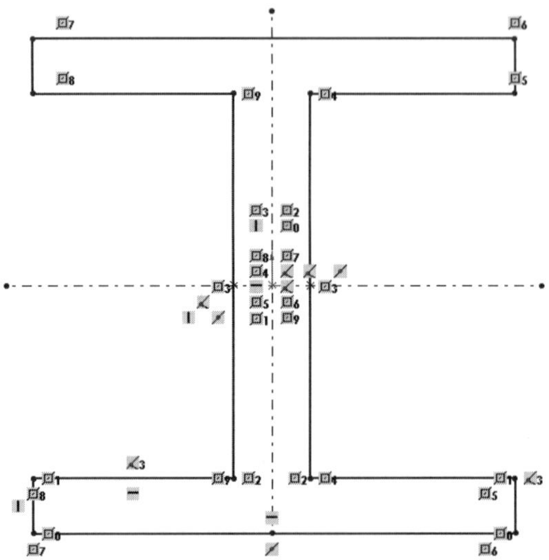

Figure 2-57 Final sketch for Tutorial 2

Saving the Sketch

1. Choose the **Save** button from the **Standard** toolbar to invoke the **Save As** dialog box.

2. Browse to the *SolidWorks/c02* folder if it is not displayed by default.

3. Enter the name of the document as *c02tut2* in the **File name** edit box and choose the **Save** button.

   The document will be saved in the */My Documents/SolidWorks/c02* folder.

3. Close the document by choosing **File > Close** from the menu bar.

Tutorial 3

In this tutorial you will create the basic sketch of the model shown in Figure 2-58. The sketch is shown in Figure 2-59. Do not dimension the sketch; the solid model and the dimensions are given only for your reference. **(Expected Time: 30 min.)**

Figure 2-58 *Solid model for Tutorial 3*

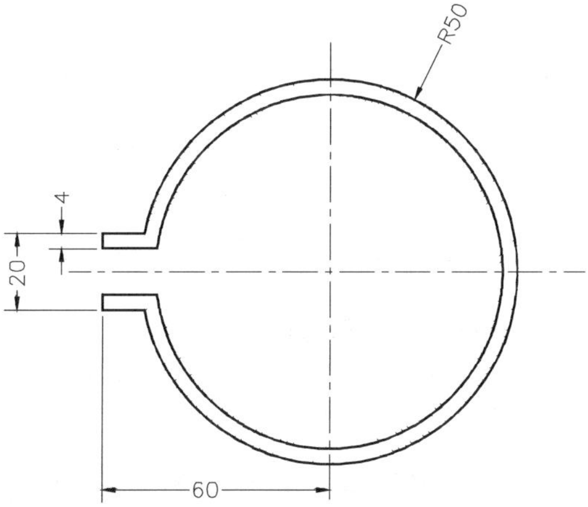

Figure 2-59 *Sketch for Tutorial 3*

The steps that will be followed to complete this tutorial are listed below.

a. Start a new part document.
b. Maximize the part file document and then switch to the sketching environment.
c. Create a centerline that will be used for reference.
d. Draw and edit the sketch using the **Mirror Entities** and **Trim Entities** tools, refer to Figure 2-60.
e. Offset the entire sketch, refer to Figure 2-61.
f. Complete the final editing of the sketch using the sketch extend and sketch trim tools, refer to Figures 2-62 and 2-63.

Starting a New Document

1. Choose the **New** button from the **Standard** toolbar to invoke the **New SolidWorks Document** dialog box.

2. The **Part** button is chosen by default. Choose the **OK** button from the **New SolidWorks Document** dialog box.

3. Choose the **Sketch** button from the **Standard** toolbar and select **Front Plane** to invoke the sketching environment.

4. Using the **Document Properties - Grid/Snap** dialog box, set the value of the **Major grid spacing** to **100** and the **Minor-lines per major** spinner to **10**.

Drawing the Centerline

Before drawing the sketch, you need to draw a centerline that will act as a reference for the other sketch entities. This centerline will also be used for mirroring.

1. Choose the **Centerline** button from the **Sketch CommandManager**.

2. Move the cursor to a location where the value of the coordinates is -70 mm 0 mm 0 mm.

3. Specify the start point at this location and move the cursor horizontally toward the right.

4. Specify the endpoint of the centerline when the value of the length of the centerline above the line cursor shows a value of 140.

5. Double-click anywhere in the drawing area to end line creation.

Drawing the Outer Loop of the Sketch

Next, you need to draw the outer loop of the sketch using the sketch tools. As evident from Figure 2-59, the sketch is drawn using the **Circle** and **Line** tools.

1. Choose the **Circle** button from the **Sketch CommandManager**; the line cursor will be replaced by the circle cursor.

2. Move the circle cursor to the origin and when the red circle is displayed, press the left mouse button to specify the center point of the circle.

3. Move the cursor horizontally toward the right and draw a circle of 50 mm radius.

4. Choose the **Zoom to Fit** button from the **View** toolbar to increase the display of the sketch.

5. Choose the **Line** button from the **Sketch CommandManager** and move the cursor to a location where the value of the coordinates is -60 mm 10 mm 0 mm. Specify the start point of the line at this location.

6. Move the cursor horizontally toward the right and press the left mouse button to specify the endpoint of the line when the line cursor snaps the circle. Exit the **Line** tool.

7. Press the CTRL key from the keyboard and using the left mouse button, select the horizontal line created in the last step and the centerline. Both selected entities are displayed in green color.

8. Choose the **Mirror Entities** button from the **Sketch CommandManager**; the mirror image of the lower horizontal line is created on the other side of the centerline.

9. Choose the **Line** button from the **Sketch CommandManager** and move the cursor to the left endpoint of the upper horizontal line. Specify the start point of the line when the red circle is displayed.

10. Move the cursor vertically downward. Specify the endpoint of the line when the cursor snaps to the left endpoint of the lower horizontal line.

11. Choose the **Trim Entities** button from the **Sketch CommandManager**; the line cursor is replaced by the trim cursor. Trim the unwanted portion of the circle, as shown in Figure 2-60.

Offsetting the Entities

After drawing the outer loop of the sketch, you need to draw the inner cavity. The first step in drawing the inner cavity of the sketch is offsetting the entire sketch inwards.

1. Choose the **Offset Entities** button from the **Sketch CommandManager**; the **Offset Entities PropertyManager** is displayed on the left of the drawing area.

2. Set the value of the **Offset Distance** spinner to **4**. Select any one entity of the sketch; the entire sketch is selected.

   When you select the sketch, the preview of the offset sketch is displayed in the drawing area. But the direction of the offset is outside the sketch. The required direction of offset should be inside the sketch. Therefore, you need to flip the direction.

3. Move the cursor inside the sketch and press the left mouse button to offset the sketch in the reverse direction. A dimension with a value of 4 is displayed with the sketch.

   The sketch after offsetting the outer loop is shown in Figure 2-61.

4. Choose the **Extend Entities** button from the **Sketch CommandManager**. The select cursor is replaced by the extend cursor.

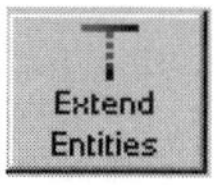

   If the **Extend Entities** button is not available in the **Sketch CommandManager**, you can invoke the extend option from **Tools > Sketch Tools > Extend**. As discussed earlier, you

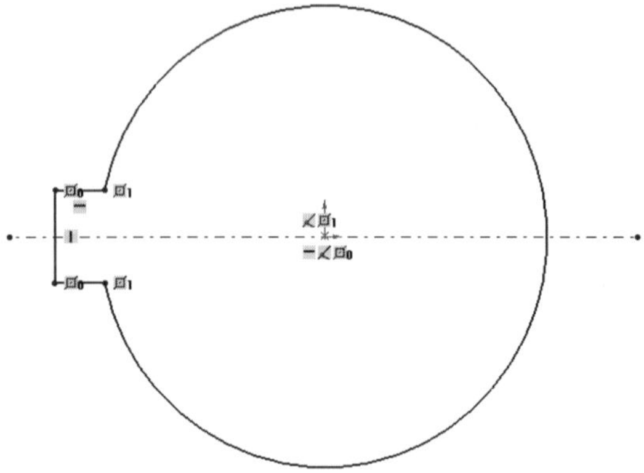

Figure 2-60 *Sketch of the outer loop*

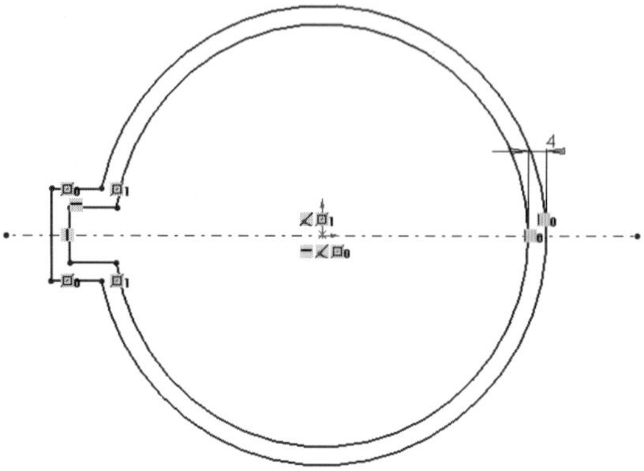

Figure 2-61 *Sketch after offsetting the outer loop*

can insert the **Extend Entities** button in the **Sketch CommandManager** using the **Customize** dialog box.

5. Move the cursor close to the left end of the lower horizontal line of the inner sketch. You can preview the extended line in red.

   You need to move the cursor a little toward the left if the preview of the extended line appears on the right.

6. Press the left mouse button to extend the line.

7. Similarly, extend the upper horizontal line of the inner sketch. The sketch after extending the lines is shown in Figure 2-62.

8. Choose the **Trim Entities** button from the **Sketch CommandManager** and the extend cursor is replaced by the trim cursor.

9. Using the left mouse button, trim the unwanted entities. The final sketch is shown in Figure 2-63.

Saving the Sketch

1. Choose the **Save** button from the **Standard** toolbar to invoke the **Save As** dialog box.

2. Enter the name of the document as *c02tut3* in the **File name** edit box and choose the **Save** button. Close the file by choosing **File > Close** from the menu bar.

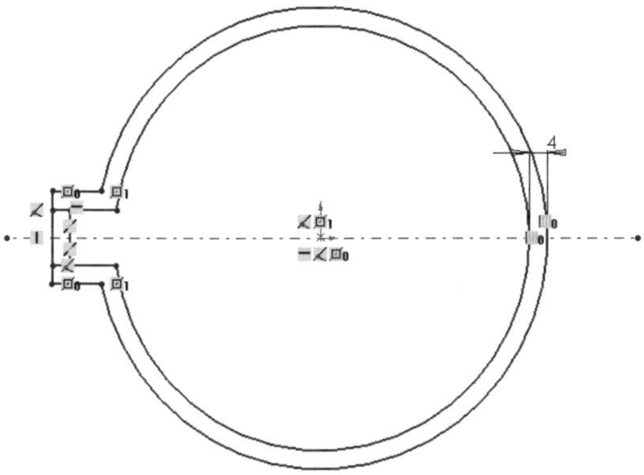

Figure 2-62 *Sketch after extending the lines*

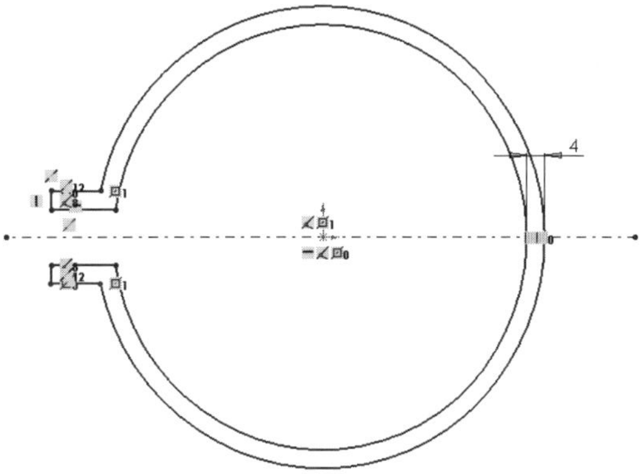

Figure 2-63 *Final sketch for Tutorial 3*

Tip. *Setting the snap and grid parameters for each document is a very tedious and time-consuming process. To overcome this problem you can create a template in which all snap, grid, and units are set. Whenever you have to create a part document you just have to choose that template from the* ***Template*** *tab of the* ***New SolidWorks Document*** *dialog box. To create and save a template, create a new document in the Part mode and set the parameters of units, snap, and grid from the* ***Document Properties*** *dialog box. Choose the* ***Save*** *button from the* ***Standard*** *toolbar or choose* ***File > Save*** *from the menu bar. The* ***Save As*** *dialog box will be displayed. Browse* ***X:/Program Files/SolidWorks/data/templates*** *location to save the file.* ***X:*** *is the drive in which you have installed SolidWorks. Select the* ***Part Templates (*.prtdot)*** *option from the* ***Save as type*** *drop-down list. Enter the name of the template file in the* ***File name*** *edit box. Choose the* ***Save*** *button from the* ***Save As*** *dialog box.*

Next time when you invoke the ***New SolidWorks Document*** *dialog box to start a new part document, choose the* ***Advanced*** *button to display the advanced version of this dialog box. The advanced version of the* ***New SolidWorks Document*** *dialog box will have the* ***Tutorial*** *tab that lists the template that you created, along with the other default templates. Using the same procedure, you can also create the user-defined templates for drawing and assembly documents. You will learn more about drawing and assembly documents in later chapters.*

Next time when you invoke the ***New SolidWorks Document*** *dialog box, the advanced version of this dialog box will be displayed. To restore the original version, choose the* ***Novice*** *button.*

SELF-EVALUATION TEST

Answer the following questions and then compare your answers with those given at the end of this chapter.

1. The **Trim** option is also used to extend the sketched entities. (T/F)

2. In the sketching environment, you can apply fillets to two parallel lines. (T/F)

3. You can apply a fillet to two nonparallel and nonintersecting entities. (T/F)

4. You cannot offset a single entity; you have to select a chain of entities to create an entity using the **Offset Entities** tool. (T/F)

5. You can choose **Insert** > **Customize** from the menu bar to display the **Customize** dialog box. (T/F)

6. The design intent is not captured in the sketch created using the mirror line. (T/F)

7. The __________ tool is used to create a linear pattern in the sketcher environment of SolidWorks.

8. The __________ tool is used to create a circular pattern in the sketcher environment of SolidWorks.

9. To modify a sketched circle, select it using the __________ tool to display the __________ **PropertyManager**.

10. The __________ tool is used to invoke the automatic mirroring option.

REVIEW QUESTIONS

Answer the following questions.

1. You cannot extend the sketched entity using the trim tool. (T/F)

2. The preview of the entity to be extended is displayed in red. (T/F)

3. You cannot apply a sketch fillet to two nonintersecting entities. (T/F)

4. The sketched entities can be mirrored without using a centerline. (T/F)

5. The **Fixed** check box in the **Linear Sketch Step and Repeat** dialog box is selected to specify the dimension between the first two instances along a specified direction after you create the pattern. (T/F)

6. Which **PropertyManager** is displayed when you choose the **Sketch Fillet** button from the **Sketch CommandManager**?

   (a) **Sketch Fillet** (b) **Fillet**
   (c) **Surface Fillet** (d) **Sketching Fillet**

7. Which **PropertyManager** is displayed on the left of the drawing area when you choose **Tools** > **Sketch Tools** > **Chamfer** from the menu bar?

   (a) **Sketch Chamfer** (b) **Sketcher Chamfer**
   (c) **Sketching Chamfer** (d) **Chamfer**

8. Which tool is used to create automatic mirror line?

   (a) **Dynamic Mirror Entities** (b) **Mirror**
   (c) **Automatic Mirror** (d) None

9. Which tool is used to break a sketched entity in two or more entities?

   (a) **Split Entities** (b) **Trim Sketch**
   (c) **Break Curve** (d) **Trim Curve**

10. Which tool is used to create a circular pattern in SolidWorks?

   (a) **Circular Pattern** (b) **Circular Sketch Step and Repeat**
   (c) **Array** (d) None

EXERCISES

Exercise 1

Create the sketch of the model shown in Figure 2-64. The sketch is shown in Figure 2-65. The solid model and dimensions are given only for reference. Create the sketch on one side and then mirror it on the other side. Make sure you do not use the mirror line option to draw this sketch. This is because if you draw the sketch using the mirror line, some relations are applied to the sketch. These relations interfere while creating fillets. **(Expected time: 30 min)**

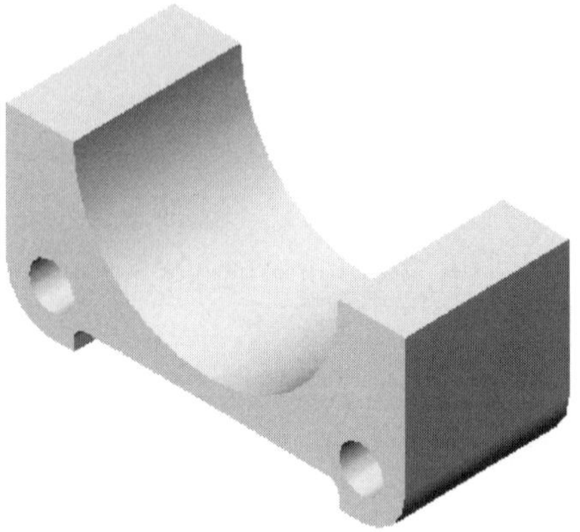

Figure 2-64 Solid model for Exercise 1

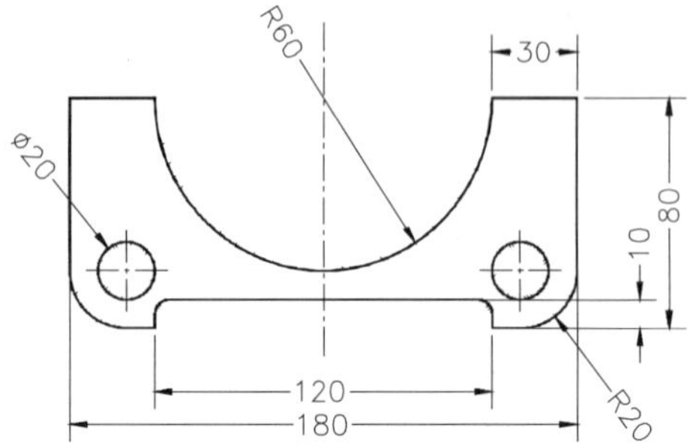

Figure 2-65 Sketch for Exercise 1

Exercise 2

Create the sketch of the model shown in Figure 2-66. The sketch is shown in Figure 2-67. The solid model and dimensions are given only for reference. Create the sketch using the sketching tools and then edit the sketch using the circular pattern tool and the trim tool.

(Expected time: 30 min)

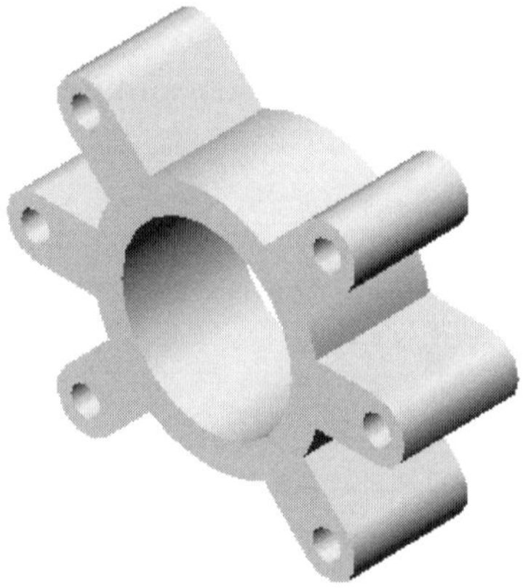

Figure 2-66 Solid model for Exercise 2

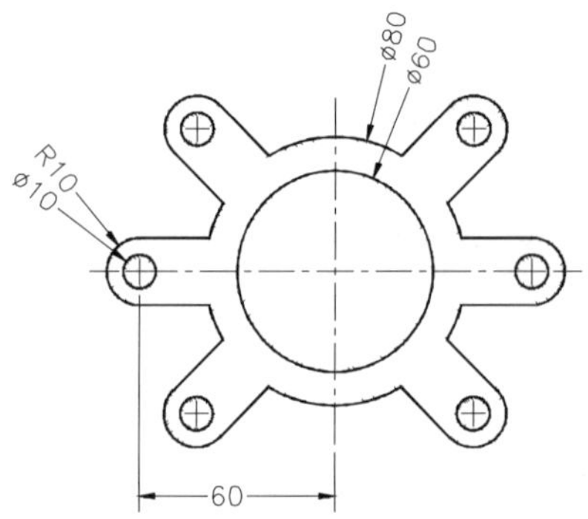

Figure 2-67 Sketch for Exercise 2

Exercise 3

Create the sketch of the model shown in Figure 2-68. The sketch is shown in Figure 2-69. This model is created using a revolved feature; therefore, you will create the sketch on one side of the centerline. The solid model and dimensions are given only for reference.

(Expected time: 30 min)

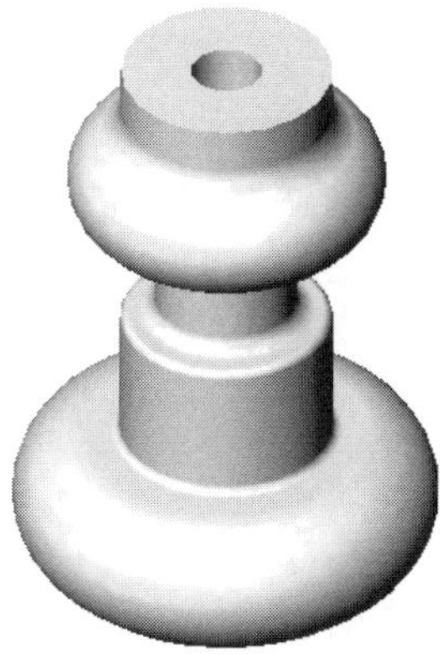

Figure 2-68 *Solid model for Exercise 3*

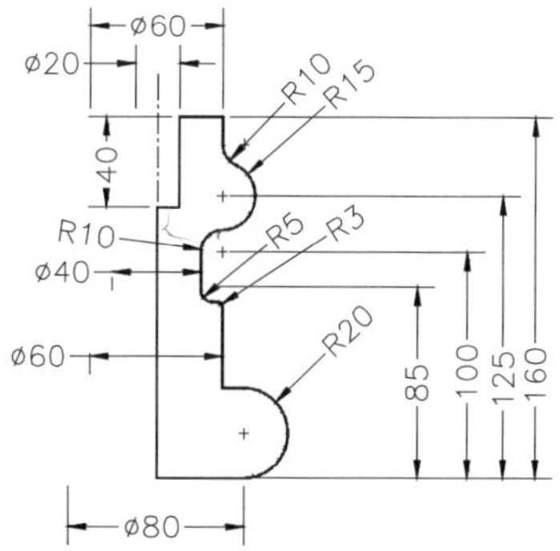

Figure 2-69 *Sketch for Exercise 3*

Exercise 4

Create the sketch of the model shown in Figure 2-70. The sketch is shown in Figure 2-71. This model is created using a revolved feature; therefore, you will create the sketch on one side of the centerline. The solid model and dimensions are given only for reference.

(Expected time: 30 min)

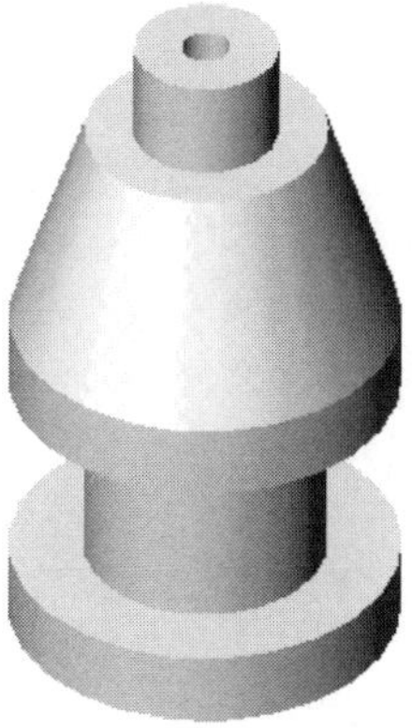

Figure 2-70 *Solid model for Exercise 4*

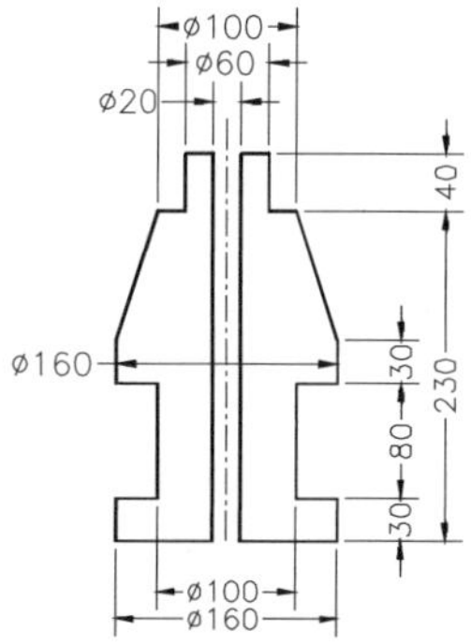

Figure 2-71 *Sketch for Exercise 4*

Answers to Self-Evaluation Test

1. T, **2.** F, **3.** T, **4.** F, **5.** F, **6.** F, **7. Linear Sketch Step and Repeat**, **8. Circular Sketch Step and Repeat**, **9. Select**, **Circle**, **10. Dynamic Mirror Entities**

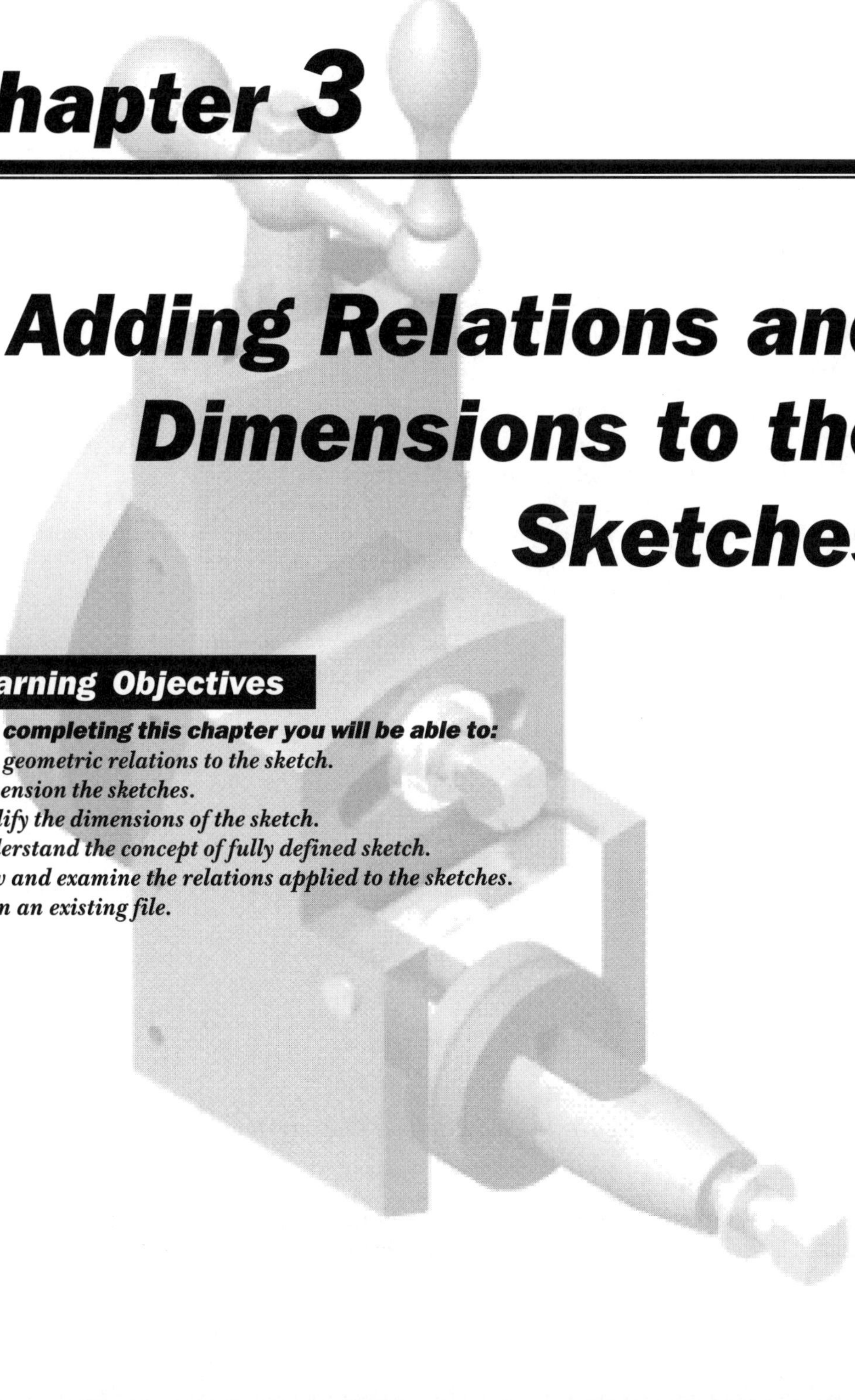

Chapter 3

Adding Relations and Dimensions to the Sketches

Learning Objectives

After completing this chapter you will be able to:

- *Add geometric relations to the sketch.*
- *Dimension the sketches.*
- *Modify the dimensions of the sketch.*
- *Understand the concept of fully defined sketch.*
- *View and examine the relations applied to the sketches.*
- *Open an existing file.*

ADDING GEOMETRIC RELATIONS TO THE SKETCHES

Geometric relations are the logical operations that are performed to add a relationship (such as tangent or perpendicular) between the sketched entities, planes, axes, edges, or vertices. The relations applied to the sketched entities are used to capture the design intent. Geometric relations constrain the degree of freedom of the sketched entities. There are two methods of applying the relations to the sketch.

1. Add Relations PropertyManager
2. Automatic Relations

Adding Relations Using the Add Relations PropertyManager

CommandManager:	Sketch > Add Relation
Menu:	Tools > Relations > Add
Toolbar:	Sketch > Add Relation

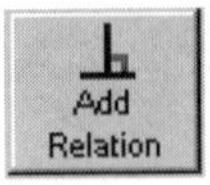

The **Add Relations PropertyManager** is widely used to apply relations to the sketch in the sketching environment of SolidWorks. It is invoked using the **Add Relation** button from the **Sketch CommandManager**. You can also invoke it by right-clicking in the drawing area and choosing the **Add Relation** option from the shortcut menu. The **Add Relations PropertyManager** is shown in Figure 3-1. The confirmation corner is also displayed at the top right corner of the drawing area. The options available in the **Add Relations PropertyManager** are discussed next.

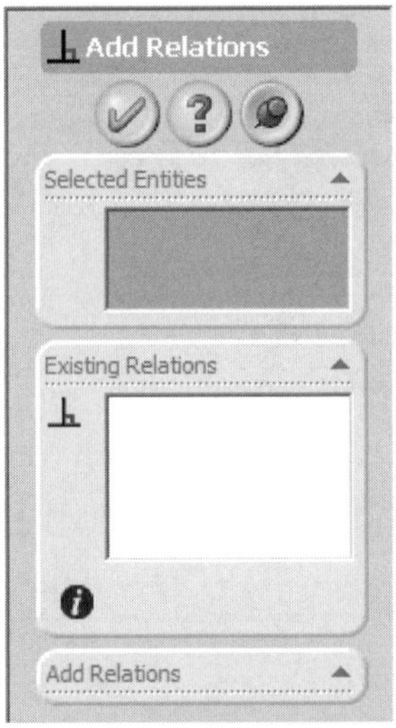

*Figure 3-1 The **Add Relations PropertyManager***

Selected Entities

The **Selected Entities** rollout displays the name of the entities that are selected to apply the relations. The entities that you select are displayed in green and are added in the area under the

Selected Entities rollout. You can remove the selected entity from the selection set by selecting the same entity again in the drawing area.

Tip. *You can also remove the selected entity from the selection set by selecting that entity in the **Selected Entities** rollout and right-clicking to invoke the shortcut menu. Choose the **Delete** option from this menu to remove the entity from the selection set. If you choose the **Clear Selections** option from the shortcut menu, all entities will be removed from the selection set.*

Existing Relations

The **Existing Relations** rollout displays the relations that are already applied to the selected sketch entities. It also shows the status of the sketch entities. You can delete the already existing relation from this rollout. Select the already existing relation from the selection area and right-click to display the shortcut menu. Choose the **Delete** option from this shortcut menu to delete the selected relation. If you choose the **Delete All** option, all relations displayed in the selection area of the **Existing Relations** rollout will be deleted.

Add Relations

The **Add Relations** rollout is used to apply the relations to the selected entity. The list of relations that can be applied to the selected entity or entities is shown in the **Add Relations** rollout. The most appropriate relation for the selected entities appears in bold letters.

Tip. *You can apply a relation to a single entity or between two or more entities. To apply a relation between two or more entities, at least one entity should be a sketched entity. The other entity or entities can either be sketch entities, edges, faces, vertices, origins, plane, or axes. The sketch curves from other sketches that form lines or arcs when projected on the sketch plane can also be included in the relation.*

All relations that can be applied to the sketches using the **Add Relations** rollout are discussed next.

Horizontal

The **Horizontal** relation forces one or more selected lines or centerlines to become horizontal. You can also select an external entity such as an edge, plane, axis, or sketch curve on an external sketch that projects as a line in the sketch to apply this relation. Using the **Horizontal** relation you can also force two or more points to become horizontal. A point can be a sketch point, a center point, an endpoint, a control point of a spline, or an external entity such as origin, vertex, axis, or point in an external sketch that projects as a point. To use this relation, invoke the **Add Relations PropertyManager**. Select the entity or entities to apply the **Horizontal** relation. Choose the **Horizontal** button from the **Add Relations** rollout provided in the **Add Relations PropertyManager**. You will notice that the name of the horizontal relation is displayed in the **Existing Relations** rollout.

Vertical

The **Vertical** relation forces one or more selected lines or centerlines to become vertical. Using the **Vertical** relation, you can also force two or more points to become vertical. To use this relation, invoke the **Add Relations PropertyManager** and select the entity or entities to apply the **Vertical** relation. Choose the **Vertical** button from the **Add Relations** rollout. You will notice that the name of the vertical relation is displayed in the **Existing Relations** rollout.

Collinear

The **Collinear** relation forces the selected lines to lie on the same infinite line. To use this relation, select the lines to apply the **Collinear** relation. Choose the **Collinear** button from the **Add Relations** rollout.

Coradial

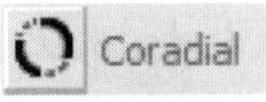

The **Coradial** relation forces the selected arcs or circles to share the same radius and the same center point. You can also select an external entity that projects as an arc or a circle in the sketch to apply this relation. To use this relation, invoke the **Add Relations PropertyManager**. Select two arcs or circles, or an arc and a circle to apply the **Coradial** relation. Choose the **Coradial** button from the **Add Relations** rollout.

Perpendicular

The **Perpendicular** relation forces the selected lines to become perpendicular to each other. To use this relation, invoke the **Add Relations PropertyManager**. Select two lines and choose the **Perpendicular** button from the **Add Relations** rollout. Figure 3-2 shows two lines before and after applying the perpendicular relation.

Parallel

The **Parallel** relation forces the selected lines to become parallel to each other. To use this relation, invoke the **Add Relations PropertyManager**. Select two lines and choose the **Parallel** button from the **Add Relations** rollout. Figure 3-3 shows two lines before and after applying this relation.

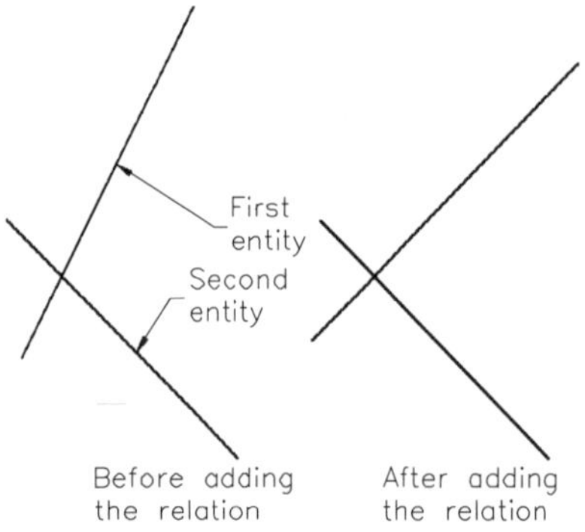

***Figure 3-2** Entities before and after applying the **Perpendicular** relation*

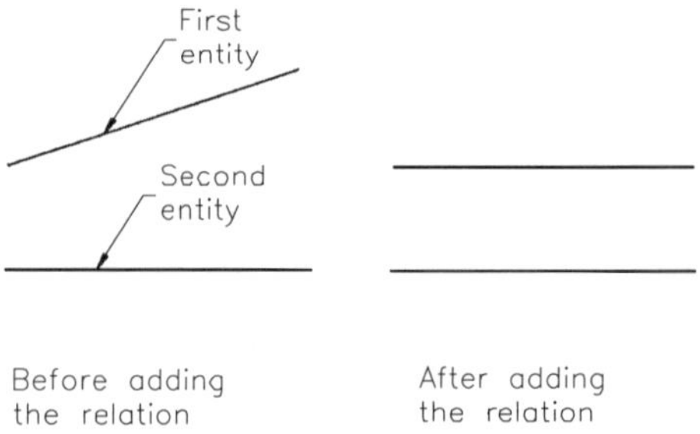

***Figure 3-3** Entities before and after applying the **Parallel** relation*

ParallelYZ

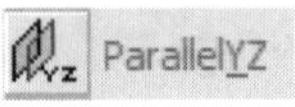

The **ParallelYZ** relation forces a line in the three-dimensional (3D) sketch to become parallel to the YZ plane with respect to the selected plane. To use this relation, invoke the **Add Relations PropertyManager**. Select the line in the 3D sketch and a plane and choose the **ParallelYZ** button from the **Add Relations** rollout.

Note

You will learn more about the 3D curves in later chapters.

ParallelZX

The **ParallelZX** relation forces a line in the 3D sketch to become parallel to the ZX plane with respect to the selected plane. To use this relation, invoke the **Add Relations PropertyManager**. Select the line in the 3D sketch and a plane and choose the **ParallelZX** button from the **Add Relations** rollout.

AlongZ

The **AlongZ** relation forces a line in the 3D sketch to become normal to the selected plane. To use this relation, invoke the **Add Relations PropertyManager**. Select the line in the 3D sketch and a plane and choose the **AlongZ** button from the **Add Relations** rollout.

Tangent

The **Tangent** relation forces the selected arc, circle, spline, or ellipse to become tangent to other arc, circle, spline, ellipse, line, or edge. To use this relation, invoke the **Add Relations PropertyManager**. Select two entities and choose the **Tangent** button from the **Add Relations** rollout. Figures 3-4 and 3-5 show the entities before and after applying this relation.

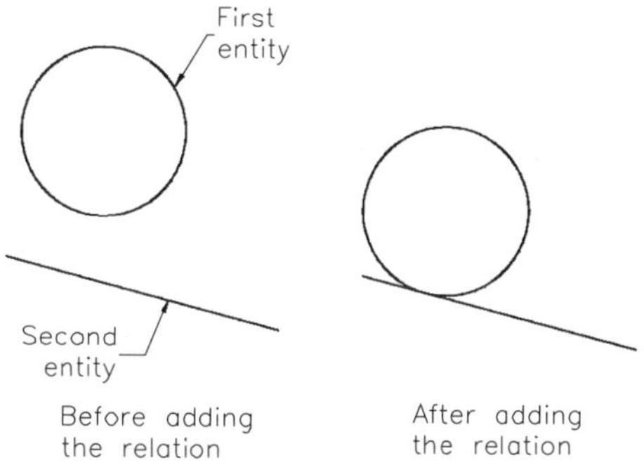

***Figure 3-4** Applying the **Tangent** relation to a line and a circle*

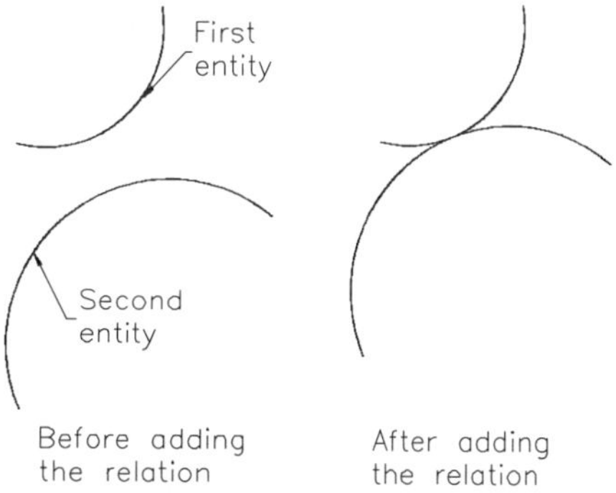

***Figure 3-5** Applying the **Tangent** relation to two arcs*

Concentric

The **Concentric** relation forces the selected arc or circle to share the same center point with other arc, circle, point, vertex, or circular edge. To use this relation, invoke the **Add Relations PropertyManager**. Select the required entity to apply the **Concentric** relation and then choose the **Concentric** button from the **Add Relations** rollout.

Midpoint

The **Midpoint** relation forces the selected point to move to the midpoint of a selected line. To use this relation, invoke the **Add Relations PropertyManager**. Select the point and the line to the midpoint of which the point will be moved. Choose the **Midpoint** button from the **Add Relations** rollout.

Intersection

The **Intersection** relation forces the selected point to move at the intersection of two selected lines. To use this relation, invoke the **Add Relations PropertyManager**. Select the required entity to apply the **Intersection** relation. Choose the **Intersection** button from the **Add Relations** rollout.

Coincident

The **Coincident** relation forces the selected point to be coincident with the selected line, arc, circle, or ellipse. To use this relation, invoke the **Add Relations PropertyManager**. Select the required entity to apply the **Coincident** relation. Choose the **Coincident** button from the **Add Relations** rollout.

Equal

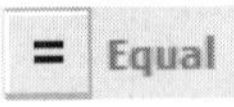

The **Equal** relation forces the selected lines to have equal length and the selected arcs, circles, or arc and circle to have equal radii. To use this relation, invoke the **Add Relations PropertyManager**. Select the required entity to apply the **Equal** relation and choose the **Equal** button.

Symmetric

The **Symmetric** relation forces two selected lines, arcs, points, and ellipses to remain equidistant from a centerline. This relation also forces the entities to have the same size and orientation. To use this relation, invoke the **Add Relations PropertyManager**. Select the required entity to apply the **Symmetric** relation and select a centerline. Choose the **Symmetric** button from the **Add Relations** rollout.

Fix

The **Fix** relation forces the selected entity to be fixed at the specified position. If you apply this relation to a line or an arc, its location will be fixed but you can change its size by dragging the endpoints. To use this relation, invoke the **Add Relations PropertyManager**. Select the required entity to apply the **Fix** relation and choose the **Fix** button.

Pierce

The **Pierce** relation forces a sketch point or an endpoint of an entity to be coincident with an entity of another sketch. To use this relation, invoke the **Add Relations PropertyManager**. Select the required entities to apply the **Pierce** relation and choose the **Pierce** button from the **Add Relations** rollout.

Merge Points

The **Merge Points** relation forces two sketch points or endpoints to merge in a single point. To use this relation, invoke the **Add Relations**

PropertyManager. Select the required entities to apply the **Merge Points** relation and choose the **Merge Points** button from the **Add Relations** rollout.

Tip. *You can also apply the relations using the **Properties PropertyManager**. This **PropertyManager** is automatically invoked if you select more than one entity from the drawing area. The possible relations for the selected geometry will be displayed in the **Add Relations** rollout. Choose the relation you want to apply to the selected geometry.*

*Another method of applying relations to the sketches is by selecting the entity or entities to which you have to apply the relation and then right-clicking to display the shortcut menu. The relation that can be applied to the selected entities will be displayed in the shortcut menu under the **Relations** section. Choose the relation from the shortcut menu.*

Automatic Relations

Automatic relations are applied automatically to the sketch while drawing. For example, you will notice that when you specify the start point of the line and move the cursor horizontally toward the right or left, the symbol is displayed below the line cursor. This is the symbol of the Horizontal relation that is applied to the line while drawing. If you move the cursor vertically downward or upward, the symbol for the Vertical relation will be displayed below the line cursor. If you move the cursor to the intersection of two or more sketched entities, the intersection symbol will appear below the cursor. Similarly, other relations are also automatically applied to the sketch while creating the sketch.

You can activate the automatic relations option if it is not available. Invoke the **System Options - Sketch** dialog box by choosing **Tools > Options** to invoke the **System Options - General** dialog box and then select the **Sketch** option from the area on the left. Select the **Automatic Relations** check box from the **System Options - Sketch** dialog box and choose the **OK** button. The relations that are applied automatically are listed below.

1. Horizontal
2. Vertical
3. Coincident
4. Midpoint
5. Intersection
6. Tangent
7. Perpendicular

Tip. *You will observe that while sketching, two types of inferencing lines are displayed. One is displayed in blue and the other in brown. The brown inferencing line indicates that the relation is applied automatically to the sketch. The blue inferencing line indicates that no automatic relation is applied.*

DIMENSIONING THE SKETCH

After drawing the sketches and adding the relations, dimensioning is the most important step in creating a design. As mentioned earlier, SolidWorks is a parametric software. This property of SolidWorks ensures that irrespective of the original size, the selected entity is driven by the dimension value you specify. Therefore, when you apply and modify a dimension of an entity, it is forced to change its size in accordance with the specified dimension value.

In SolidWorks, you can use the **Smart Dimension** tool to dimension any kind of entity. You can also select the individual type of dimensioning tool from the **Dimension/Relations CommandManager**. This **CommandManager** is not available by default. To invoke it, right-click on the **Sketch CommandManager** and choose **Customize Command Manager** from the shortcut menu; the **Customize CommandManager** flyout will be displayed. In this flyout, choose the **Dimension/Relations** option and then click anywhere on the screen. Again, right-click on the **Sketch CommandManager** and choose **Dimension/Relations** to display the tools available in this **CommandManager**. Figure 3-6 shows the default dimension types available in the **Dimension/Relations CommandManager**.

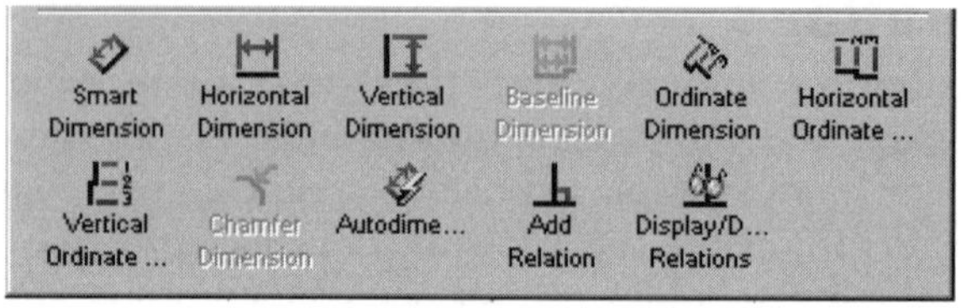

Figure 3-6 Dimension/Relations CommandManager

If you use the **Smart Dimension** tool, the type of dimension that will be applied will depend on the type of entity selected. For example, if you select a line, a horizontal, vertical, or aligned dimension will be applied. If you select a circle, a diametric dimension will be applied. Similarly, if you select an arc, a radial dimension will be applied. However, if you want to apply a particular type of dimension, choose its tool from the **Dimension/Relations CommandManager** and then apply the dimension.

In SolidWorks 2005, as soon as you place the dimension, the **Modify** dialog box will be displayed, as shown in Figure 3-7. You can enter a new value in this box to modify the dimension. You can modify the default dimension value using the spinner or by entering a new value in the edit box available in the **Modify** dialog box.

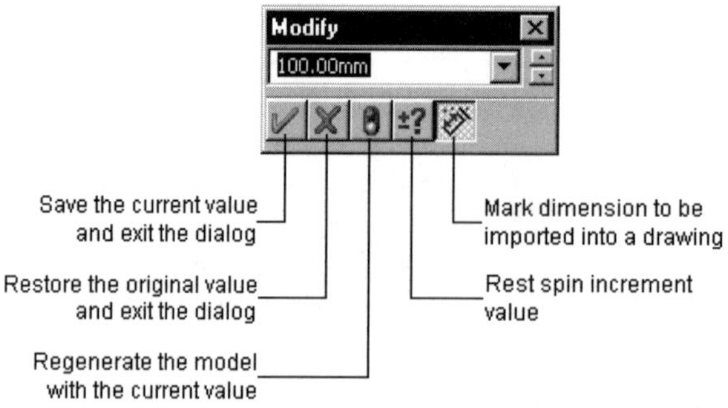

***Figure 3-7 Modify** dialog box*

Tip. *If the **Modify** dialog box is not displayed when you place the dimension, you need to set its preference manually. To do this, invoke the **System Options - General** dialog box and choose the **Input dimension value** check box.*

The buttons available in the **Modify** dialog box are discussed next.

The **Save the current value and exit the dialog** button is used to accept the current value and exit the dialog box. The **Restore the original value and exit the dialog** button is used to restore the last dimensional value applied to the sketch and exit the dialog box. The **Regenerate the model with the current value** button is used to preview the geometry of the sketch with the new modified dimensional value. The **Reset spin increment value** button is used to enter a new spin increment value. This is the value that is added to or subtracted from the current value when you click once on the spinner arrow. If you choose this button, the **Increment** dialog box will be displayed. The **Mark dimensions to be imported into a drawing** button is chosen to make sure that the selected dimension is generated as model annotation in the drawing views. If this button is not chosen, the dimensions will not be generated.

Tip. *You can also enter the arithmetic symbols directly into the edit box of the **Modify** toolbar to calculate the dimension. For example, if you have a dimension as a complex arithmetic function such as (220*12.5)-3+150, which is equal to 1247, you do not need to calculate this using the calculator. Just enter the statement in the edit box and press ENTER. SolidWorks will automatically solve the function to get the value of the dimension.*

The types of dimensions that can be applied to the sketches in the sketching environment of SolidWorks 2005 are discussed next.

Horizontal/Vertical Dimensioning

CommandManager:	Dimension/Relations > Horizontal/Vertical Dimension
Menu:	Tools > Dimension > Horizontal/Vertical
Toolbar:	Dimension/Relations > Horizontal/Vertical Dimension

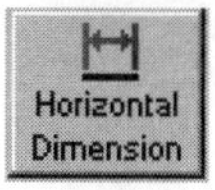

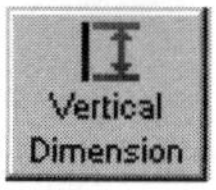

These dimensions are used to define horizontal or vertical dimensions of a selected line or between two points. The points can be the endpoints of lines or arcs, or the center points of circles, arcs, ellipses, or parabolas. You can dimension a vertical or a horizontal line by directly selecting it. Choose the **Horizontal/Vertical Dimension** button from the **Dimension/Relations CommandManager**. You can also right-click in the drawing area and choose the **More Dimensions > Horizontal/Vertical** option from the shortcut menu to activate these tools. When you move the cursor on the line, it will be highlighted and turn red. As soon as you select the line, it will turn green, and the dimension will be attached to the cursor. Move the cursor and place the dimension at an appropriate place using the left mouse button. As the priority of editing the dimension as it is placed is set by default, the **Modify** dialog box will be displayed with the default value in it. Enter the new value of the dimension in the **Modify** dialog box and press ENTER. Figure 3-8 shows horizontal and vertical dimensioning of lines.

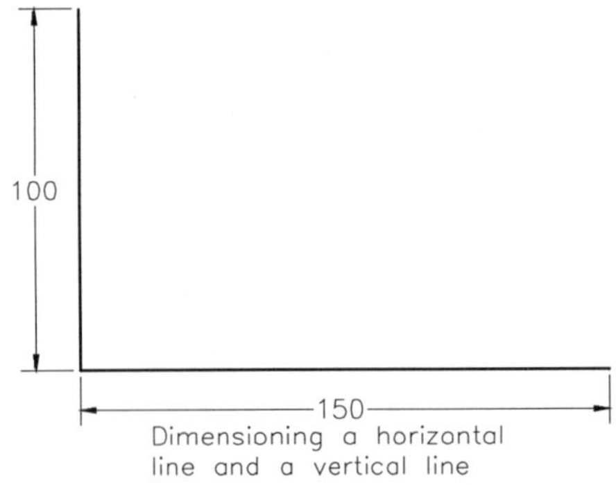

***Figure 3-8** Linear dimensioning of lines*

If the dimension is selected in the drawing area, the **Dimension PropertyManager** will be displayed on the left of the drawing area, as shown in Figure 3-9. The options available in the **Dimension PropertyManager** are discussed next.

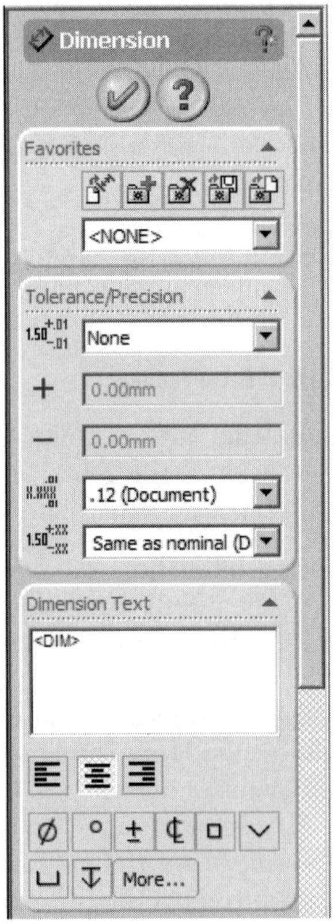

Figure 3-9** The **Dimension PropertyManager

> **Tip**. *There are some enhancements in the **Dimension PropertyManager**. But these enhancements are related to the **Drawing** mode of SolidWorks. Therefore you will learn about them in the later chapters.*

Favorite

The **Favorite** rollout shown in Figure 3-10 is used to create, save, delete, and retrieve the dimension style in the current document. You can also retrieve the dimensions styles saved in other documents using this rollout. The options available in this rollout are discussed next.

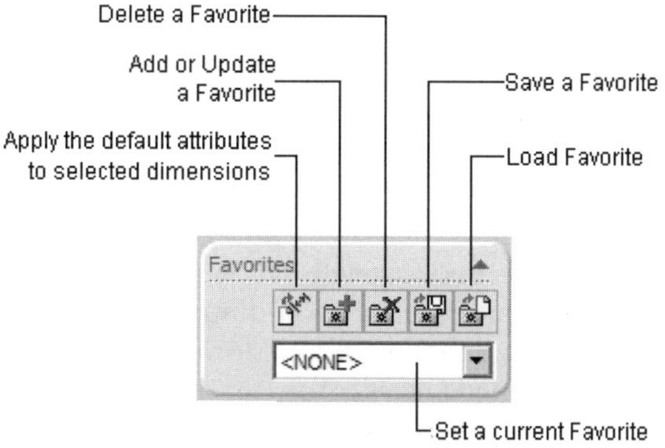

*Figure 3-10 The **Dimension Favorite** rollout*

Apply the default attributes to selected dimensions

The **Apply the default attributes to selected dimensions** button is used to apply the default attributes to the selected dimension or dimensions. The attributes include the tolerance, precision, arrow style, dimension text, and so on. This option is generally used when you have modified the settings applied to a dimension and then you want to restore the default settings on that dimension.

Add or Update a Favorite

The **Add or Update a Favorite** button is used to add a dimension style to the current document for a selected dimension. After invoking the **Dimension PropertyManager**, set the attributes using various options provided in this **PropertyManager**. Now choose the **Add or Update a Favorite** button; the **Add or Update a Favorite** dialog box will be displayed, as shown in Figure 3-11. Enter the name of the dimension style in the edit box and press

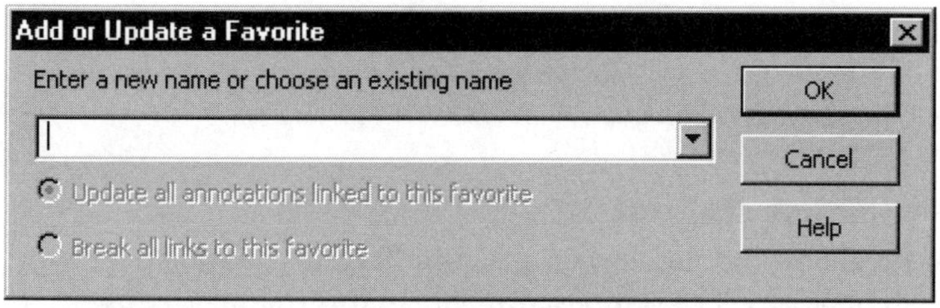

*Figure 3-11 The **Add or Update a Favorite** dialog box*

ENTER or choose the **OK** button from this dialog box. The dimension style will be added to the current document.

You can apply the new dimension style to the selected dimension by selecting the dimension style from the drop-down list in the **Favorite** rollout. You can also update the dimension style. To update a dimension style, select the dimension and set the options of the dimension style according to your need and then choose the **Add or Update a Favorite** button to invoke the **Add or Update a Favorite** dialog box. Select the dimension style to update from the drop-down list provided in the dialog box. The two radio buttons in this dialog box are enabled. Select the **Update all the annotations linked to this favorite** radio button and choose the **OK** button to update all dimensions linked with the selected favorite. If you select the **Break all links to this favorite** radio button and choose the **OK** button, then the link between the other dimensions having the same favorite and the selected favorite will be broken.

Delete a Favorite

The **Delete a Favorite** button is used to delete a dimension style. Select the dimension style or favorite from the **Set a current Favorite** drop-down list and choose the **Delete a Favorite** button. Note that even after you delete the dimension style, the properties of the dimensions will be the same as with the deleted favorite. You can set the properties of a dimension to the default settings using the **Apply the default attributes to the selected dimension** button.

Save a Favorite

The **Save a Favorite** button is used to save a dimension style so that it can be retrieved in some other document. Select the dimension style or favorite from the **Set a current Favorite** drop-down list and choose the **Save a Favorite** button. The **Save As** dialog box will be displayed. Browse the folder in which you want to save the favorite and enter the name of the favorite in the **File name** edit box. Choose the **Save** button from the **Save As** dialog box. The extension for the file in which the favorite is saved is *.sldfvt*.

Load Favorite

The **Load Favorite** button is used to open a saved favorite in the current document. The properties of that favorite will be applied to the selected dimension. To load a favorite, choose the **Load favorite** button to invoke the **Open** dialog box. Browse the folder in which the favorite is saved. Now, select the file with the extension *.sldfvt* and choose the **Open** button; the **Add or Update a Favorite** dialog box will be displayed. Choose the **OK** button from this dialog box.

Tip. *You can load more than one favorite by pressing the CTRL key and selecting the favorites from the **Open** dialog box. All favorites will be displayed in the **Set a current Favorite** drop-down list.*

Tolerance/Precision Rollout

The **Tolerance/Precision** rollout shown in Figure 3-12 is used to specify the tolerance and precision in the dimensions. The options available in this rollout are discussed next.

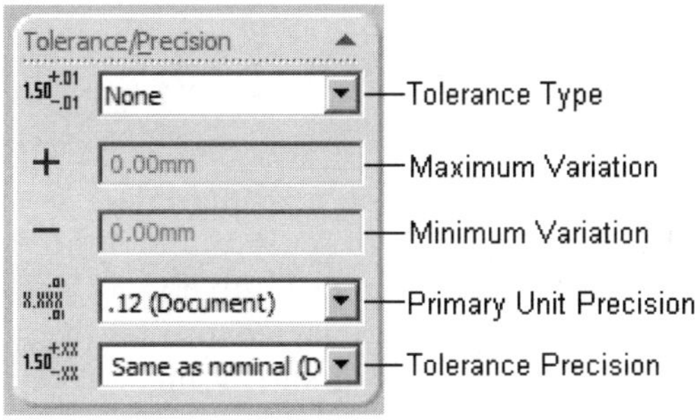

***Figure 3-12** The **Tolerance/Precision** rollout*

Tolerance Type

The **Tolerance Type** drop-down list is used to apply the tolerance to the dimension. By default, the **None** option is selected. Therefore, no tolerance is applied to the dimensions. The other tolerance methods available in this drop-down list are discussed next.

Basic

The basic dimension is the one that is enclosed in a rectangle. To display a basic dimension, select a dimension that you want to display as a basic dimension and then select the **Basic** option from the **Tolerance Display** drop-down list. You will notice that the dimension is enclosed in a rectangle, suggesting that it is a basic dimension, see Figure 3-13.

Bilateral

The bilateral tolerance provides the maximum and minimum variation in the value of the dimension that is acceptable in the design. To apply the bilateral tolerance, select the dimension and then the **Bilateral** option from the **Tolerance Type** drop-down list. The **Maximum Variation** and **Minimum Variation** edit boxes will be enabled. These edit boxes are used to apply the value of the maximum and minimum variation to a dimension. The **Show parentheses** check box will also be displayed. If you select the **Show parentheses** check box, the bilateral tolerance will be displayed with parentheses. The dimension with bilateral tolerance is shown in Figure 3-14.

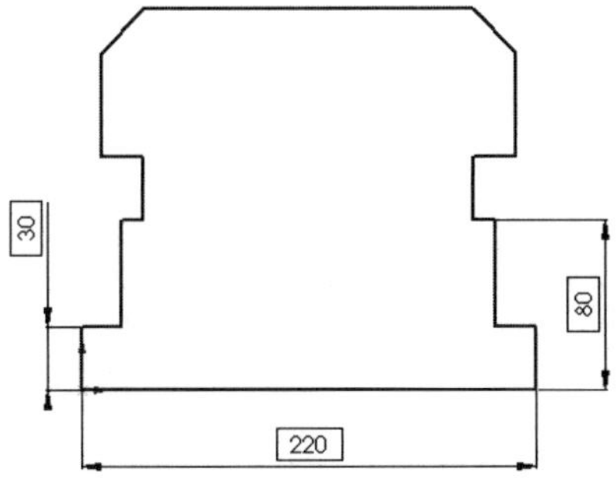

***Figure 3-13** Basic dimension*

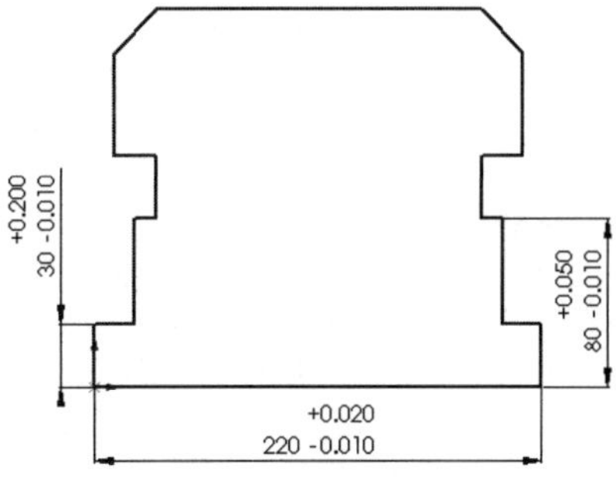

***Figure 3-14** Bilateral tolerance*

Note

The dimension standard used in the sketches in this book is the ISO standard.

Limit

In the limit dimension, the dimension is displayed as its maximum and minimum values that are allowed in the design. To apply this tolerance type, select the dimension to display as limit dimension and select the **Limit** option. The **Maximum Variation** and **Minimum Variation** edit boxes will be enabled to enter the value of the maximum and minimum variation. The dimension along with the limit tolerance is shown in Figure 3-15.

Symmetric

The symmetric tolerance is displayed with plus and minus signs. To use this tolerance, select the dimension and select the **Symmetric** option; the **Maximum Variation** edit box will be displayed to enter the value of the tolerance. You can select the **Show parentheses** check box to show the tolerance in parentheses. The dimension along with the symmetric tolerance is shown in Figure 3-16.

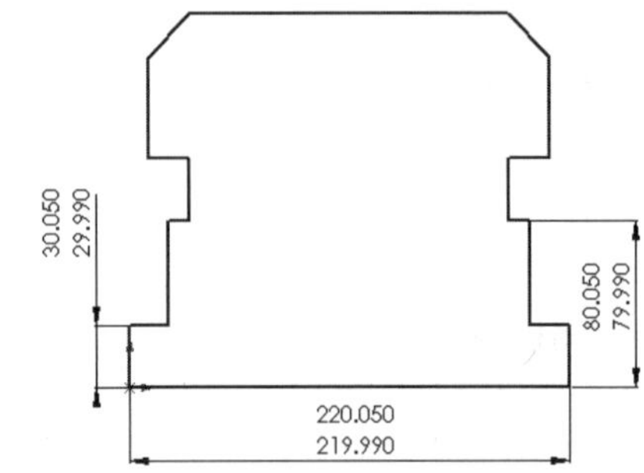

Figure 3-15 *Limit tolerance*

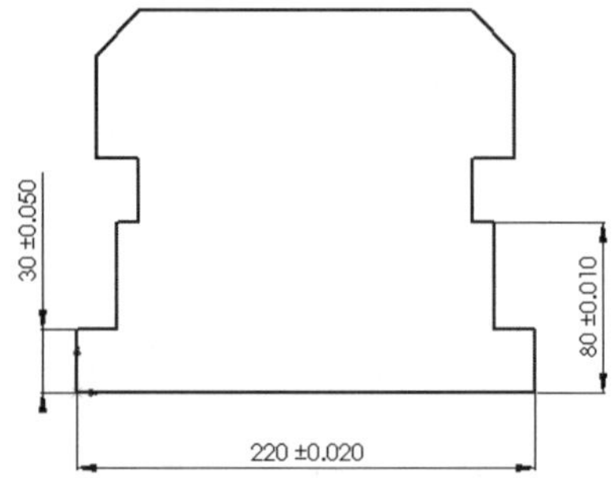

Figure 3-16 *Symmetric tolerance*

MIN

In this type of dimensional tolerance, the **min.** symbol is added to the dimension as a suffix. This implies that the dimensional value is the minimum value that is allowed in the design. To display this dimensional tolerance, select the dimension and the **MIN** option from the **Tolerance Type** drop-down list. The dimension along with the minimum tolerance is shown in Figure 3-17.

MAX

In this type of dimensional tolerance, the **max.** symbol is added to the dimension as suffix. This implies that the dimensional value is the maximum value that is allowed in the design. To display this dimensional tolerance, select the dimension and the **MAX** option from the **Tolerance Type** drop-down list. The dimension along with the maximum tolerance is shown in Figure 3-18.

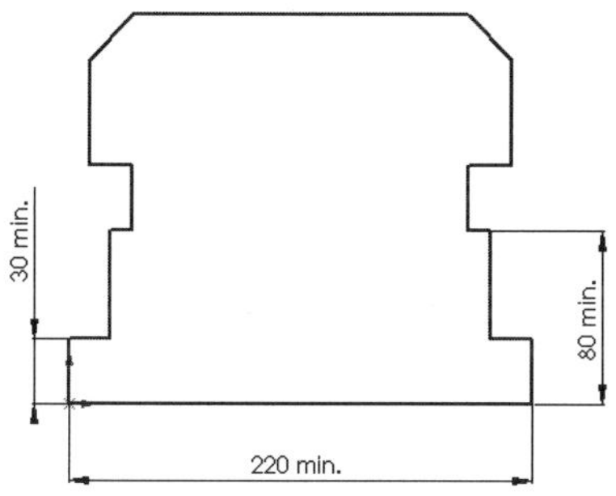

Figure 3-17 Minimum tolerance

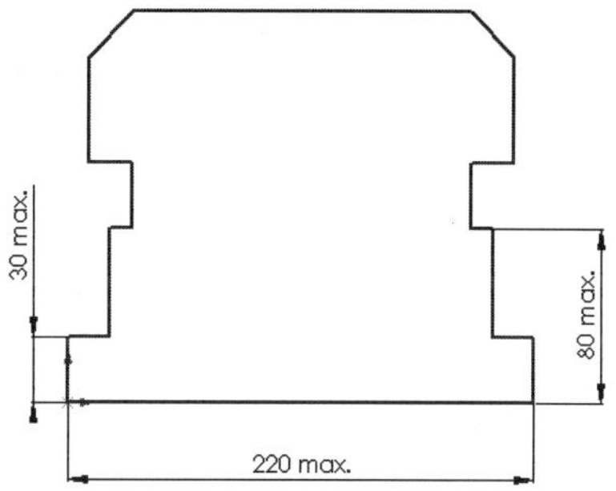

Figure 3-18 Maximum tolerance

Fit

The **Fit** option is used to apply the fit according to the **Hole Fit** and **Shaft Fit** systems. The **Tolerance/Precision** rollout with the **Fit** option selected is shown in Figure 3-19. Specify the type of fit from the **Classification** drop-down list. The **Classification** drop-down list is used to define the **User Defined** fit, **Clearance** fit, **Transitional** fit, and **Press** fit. To apply the fit using the hole fit system or the shaft fit system, select the dimension and the **Fit** option from the **Tolerance Type** drop-down list. The **Classification** drop-down list, the **Hole Fit** drop-down list, and the **Shaft Fit** edit drop-down list will be displayed below the **Tolerance Type** drop-down list. Choose the type of fit from the **Classification** drop-down list and select the standard of fit from the **Hole Fit** drop-down list or the **Shaft Fit** drop-down list. If you choose the **Clearance**, **Transitional**, or **Press** option from the **Classification** drop-down list and the standard of fit from the **Hole Fit** drop-down list, only the standards that match with the selected hole fit will be displayed in the **Shaft Fit** drop-down list and vice versa. If you choose the **User Defined** option from the **Classification** drop-down list, you can choose any standard from the **Hole Fit** and **Shaft Fit** drop-down lists. The **Stacked with line display** button provided under the **Shaft Fit** drop-down list is chosen to display stacked tolerance with a line. You can also display the tolerance as stacked without a line using the **Stacked without line display** button. If you choose the **Linear display** button, the tolerance will be displayed in linear form. The dimension along with the hole fit and shaft fit is shown in Figure 3-20.

Fit with tolerance

The **Fit with tolerance** option in the **Tolerance Type** drop-down list is used to display the tolerance along with the hole fit and shaft fit in a dimension. To apply the fit with the tolerance, select the dimension and the **Fit with tolerance** option from the **Tolerance Type** drop-down list. Select the type of fit from the **Classification** drop-down list. Now, select the fit standard from the **Hole Fit** drop-down list or the **Shaft Fit** drop-down list. The tolerance will be displayed with the fit standard only if you select only one fit system either from the hole drop-down list or from the shaft drop-down list. The tolerance will be displayed along with the fit standard in the drawing area. In this release of SolidWorks the tolerance is calculated automatically, depending on the type and standard

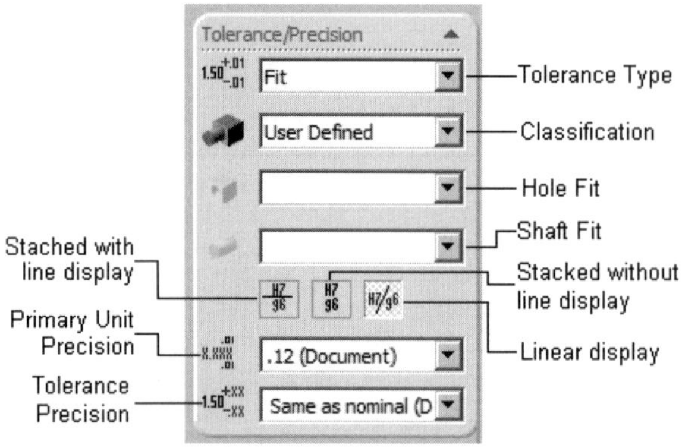

Figure 3-19 *The **Tolerance/Precision** rollout with the **Fit** option selected from the **Tolerance Type** drop-down list*

of fit selected. The **Show parentheses** check box can be selected to show the tolerance in parentheses. The dimension along with the fit and tolerance is shown in Figure 3-21.

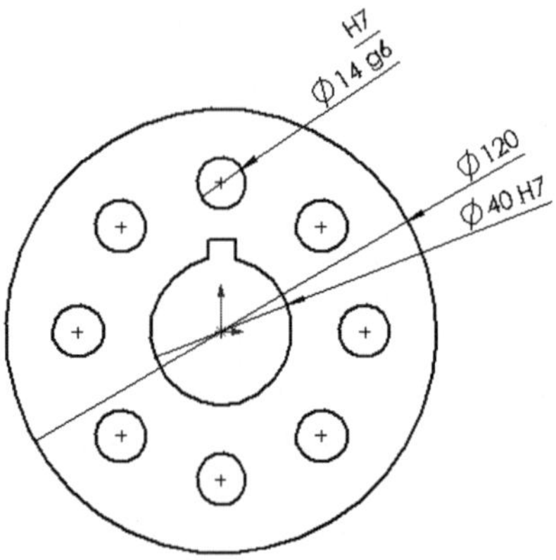

Figure 3-20 *Hole fit and shaft fit*

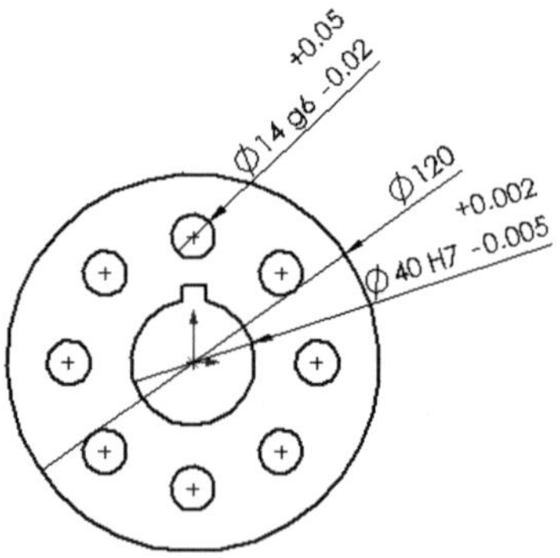

Figure 3-21 *Fit with tolerance*

Note

The diametrical dimension will be discussed later in this chapter.

Fit (tolerance only)

The **Fit (tolerance only)** option in the **Tolerance Type** drop-down list is used to display the tolerance in a dimension based on the hole fit or shaft fit.

Primary Unit Precision

The **Primary Unit Precision** drop-down list is used to specify the precision of the number of places after the decimal for dimensions. By default, the selected precision is two places after the decimal.

Tolerance Precision

The **Tolerance Precision** drop-down list is used to specify the precision of the number of places after the decimal for tolerance. By default, the selected precision is two places after the decimal.

Dimension Text

The **Dimension Text** rollout, shown in Figure 3-22, is used to add the text and symbols to the dimension. The text box provided in this rollout is used to add the text to the dimension. The **<DIM>** displayed in the text box symbolizes the dimensional value. You can add the text before or after the dimension value.

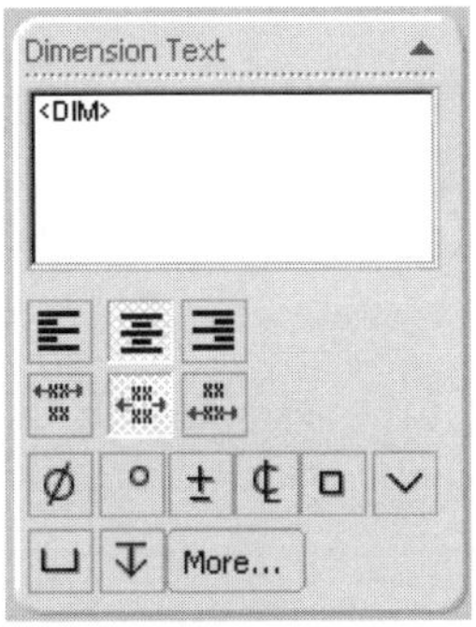

***Figure 3-22 Dimension Text** rollout*

This rollout also provides buttons to modify the text justification and add symbols such as **Diameter**, **Degree**, **Plus/Minus**, **Centerline**, and so on to the dimension text. You can add more symbols by choosing the **More Symbols** button from the **Dimension Text** rollout. When you choose this button, the **Symbols** dialog box will be displayed, as shown in Figure 3-23.

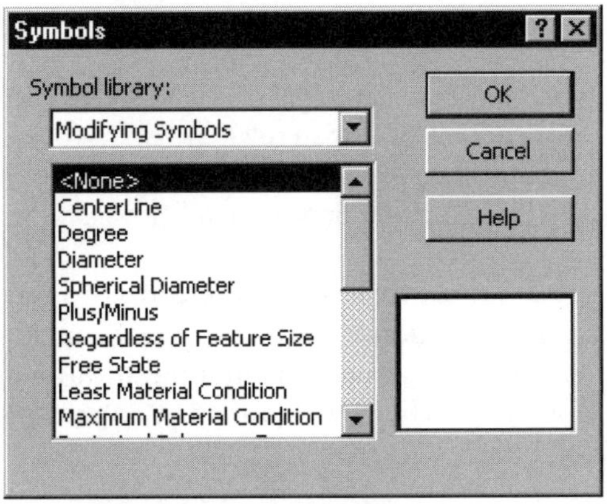

***Figure 3-23** The **Symbols** dialog box*

Display Options

The options available in the **Display Options** rollout are used to display the dimension value in parentheses, display dual dimensions, and convert the normal dimension into inspection dimension.

Arrows

The **Arrows** rollout shown in Figure 3-24 is used to specify the arrow style in the dimensions. The options available in this rollout are discussed next.

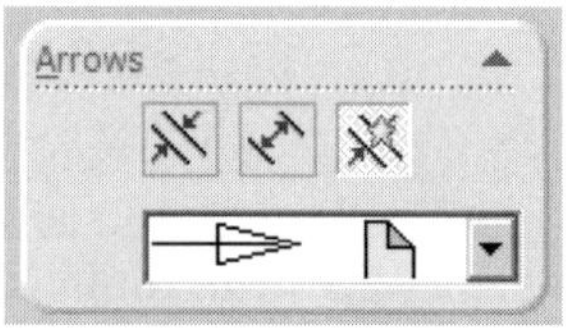

Figure 3-24 *The* ***Arrows*** *rollout*

Outside

The **Outside** button is used to display the arrows outside the dimension line. Select a dimension from the drawing area and choose the **Outside** button from the **Arrows** rollout.

Inside

The **Inside** button is used to display the arrows inside the dimension line. Select a dimension from the drawing area and choose the **Inside** button from the **Arrows** rollout.

Note

You can also click on the control point displayed on the arrowhead to reverse the direction of the arrowheads.

Smart

The **Smart** button is used to display the dimension inside or outside the dimension line, depending on the surrounding geometry. The **Smart** button is chosen by default in the **Arrows** rollout.

Style

The **Style** drop-down list is used to select the arrowhead style. The unfilled triangular arrow is selected by default. You can select any arrowhead style for a particular dimension or dimension style. To change the arrow style, select the dimension from the drawing area and then the arrowhead style from the **Style** drop-down list.

The **More Properties** button is used to invoke the **Dimension Properties** dialog box to modify the properties of the dimension. All options available in the **Dimension PropertyManager** are available in the **Dimension Properties** dialog box with some additional options. You can modify the dimension properties from the **Dimension PropertyManager** or from the **Dimension Properties** dialog box.

Horizontal/Vertical Dimensioning Between Points

As mentioned earlier, you can add a horizontal or a vertical dimension between two points. To add a horizontal or a vertical dimension between two points, choose the required button from the **Dimension/Relations CommandManager**. Select the first point, and then the second point. Specify a point to place the dimension; the **Modify** dialog box will be displayed. Enter a new dimension value in this dialog box and press ENTER. Figures 3-25 and 3-26 show the horizontal and vertical linear dimensioning between two points.

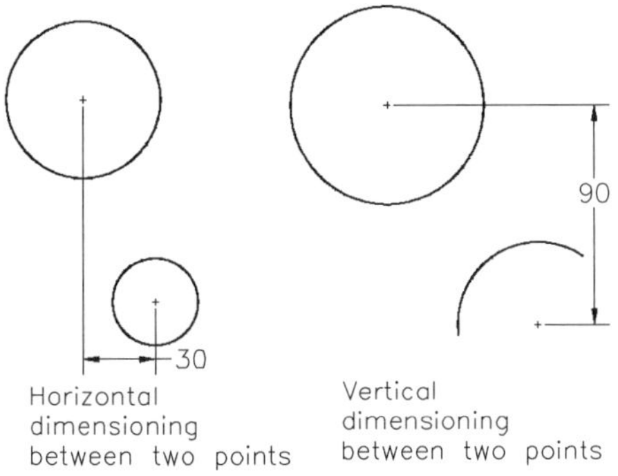

Figure 3-25 Applying horizontal and vertical dimensions to two points

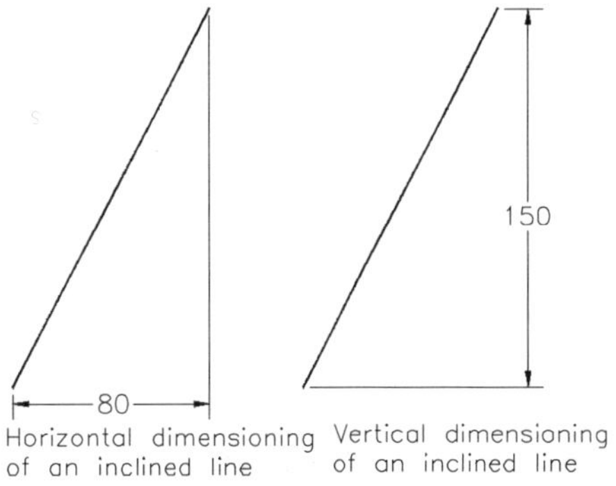

Figure 3-26 Applying horizontal and vertical dimensions to inclined lines

You can also apply horizontal or vertical (linear) dimensioning to a circle. However, note that these types of dimensions can only be applied to a circle by using the **Smart Dimension** tool. To apply this type of dimension to a circle, choose the **Smart Dimension** button and select the circle. The dimension will be attached to the cursor. If you want to create the vertical dimension, move the cursor to the right or left of the sketch. If you want to create the horizontal dimension, move the cursor to the top or bottom of the sketch. Using the left mouse button, place the dimension and enter a new value in the **Modify** dialog box. The linear dimensioning of the circle is shown in Figure 3-27.

Aligned Dimensioning

CommandManager:	Sketch > Smart Dimension
Menu:	Tools > Dimension > Smart
Toolbar:	Sketch > Smart Dimension

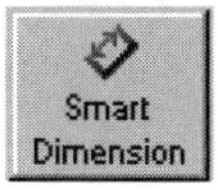

Aligned dimensions are used to dimension lines that are at an angle with respect to the X axis and the Y axis. These types of dimensions measure the actual distance of the inclined lines. You can directly select the inclined line to apply this dimension or select two points. The points that can be used to apply aligned dimensions include the endpoints of a line, arc, parabolic arc, or spline and the center points of arcs, circles, ellipse, or parabolic arc. To apply an aligned dimension to an inclined line, choose the **Smart Dimension** button from the **Sketch CommandManager** and select the line. Move the cursor at an angle such that the dimension line is parallel to the inclined line. Place the dimension at an appropriate place and enter a new value in the **Modify** dialog box.

To apply an aligned dimension between two points, choose the **Smart Dimension** button and select the first point to apply the dimension. Next, select the second point to apply the dimension. The dimension will be attached to the cursor. Move the cursor such that the dimension line is parallel to the imaginary line that joins the two points. Now, place the dimension at an appropriate location. Enter a new value in the **Modify** dialog box and press ENTER. Figure 3-28 shows aligned dimensioning of a line and an aligned dimension between two points.

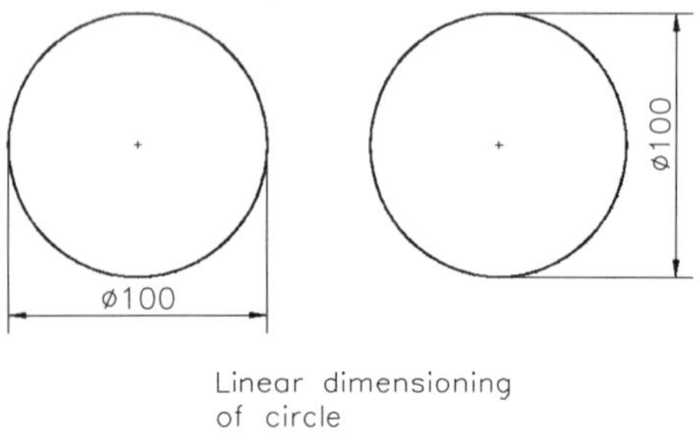

Figure 3-27 Linear dimensioning of a circle

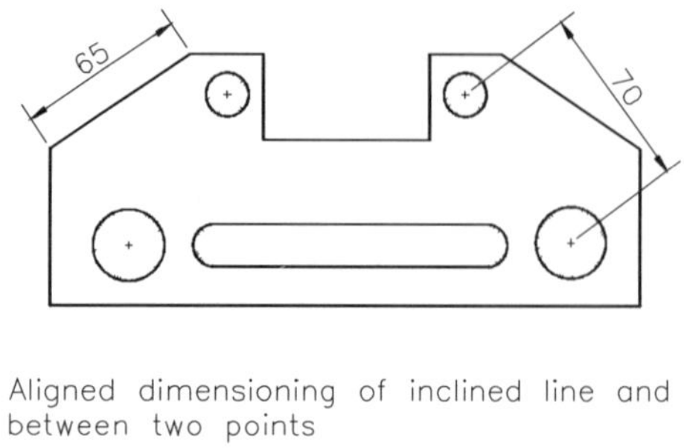

Figure 3-28 Aligned dimensioning

Angular Dimensioning

Angular dimensions are used to dimension angles. You can select two line segments or use three points to apply angular dimensions. You can also use angular dimensioning to dimension an arc. All these options of angular dimensioning are discussed next.

Angular Dimensioning Using Two Line Segments

CommandManager:	Sketch > Smart Dimension
Menu:	Tools > Dimension > Smart
Toolbar:	Sketch > Smart Dimension

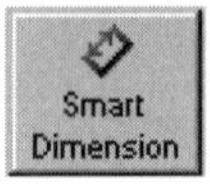

To apply angular dimensions to two lines, choose the **Smart Dimension** button from the **Sketch CommandManager** and select the first line segment; a dimension will be attached to the cursor. Now, select the second line segment; an angular dimension will be attached to the cursor. Place the angular dimension and enter the new value of angular dimension in the **Modify** dialog box. You need to be very careful while placing the angular dimension. This is because depending on the location of the dimension placement, the interior angle, exterior angle, major angle, or minor angle is displayed. Figures 3-29 through 3-32 illustrate various angular dimensions depending on the dimension placement point.

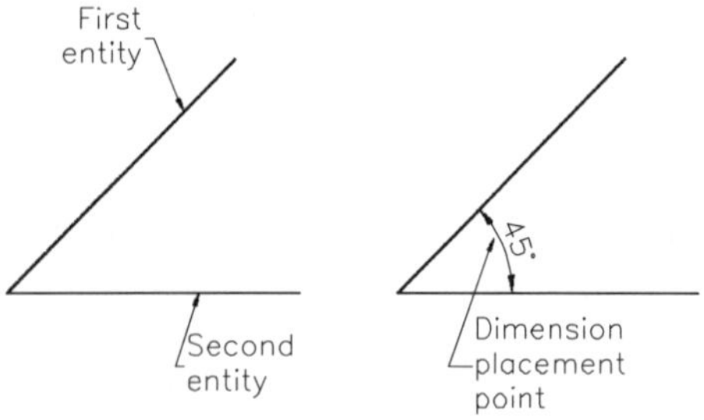

Figure 3-29 Angular dimension displayed according to the dimension placement point

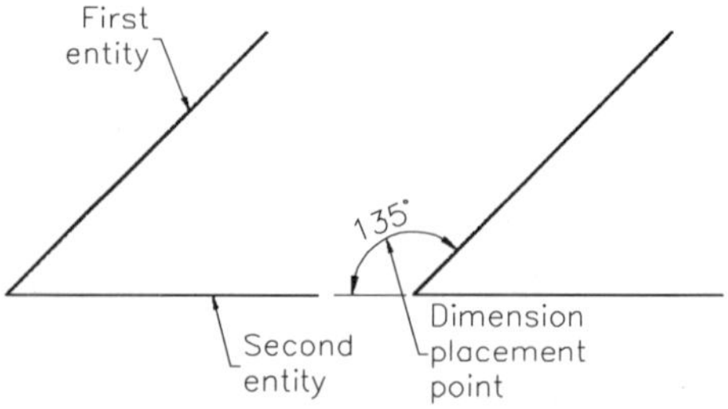

Figure 3-30 Angular dimension displayed according to the dimension placement point

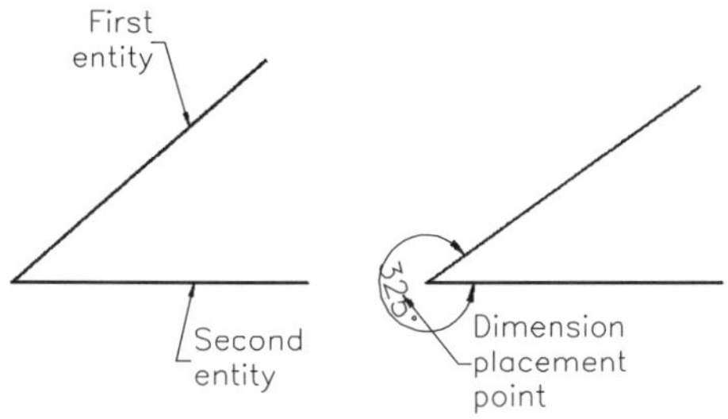

Figure 3-31 Angular dimension displayed according to the dimension placement point

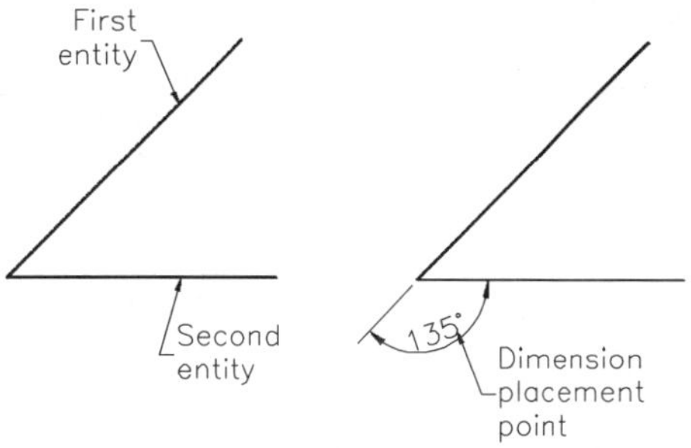

Figure 3-32 Angular dimension displayed according to the dimension placement point

Angular Dimensioning Using Three Points

While adding angular dimensions to three points, you need to be extremely careful in selecting the points. Remember that the first point that you select is the angle vertex point. To apply angular dimensions to three points, choose the **Smart Dimension** button from the **Sketch CommandManager**. Select the first point using the left mouse button. This is the angle vertex point. Select the second point; a linear dimension will be attached to the cursor. Next, select the third point; an angular dimension will attached to the cursor. Place the angular dimension and enter a new value of angular dimension in the **Modify** dialog box. Figure 3-33 shows the angular dimensioning using three points.

Angular Dimensioning of an Arc

You can use angular dimensions to dimension an arc. In the case of arcs, the three points that should be used are the endpoints and the center point of the arc. Figure 3-34 shows the angular dimensioning of an arc.

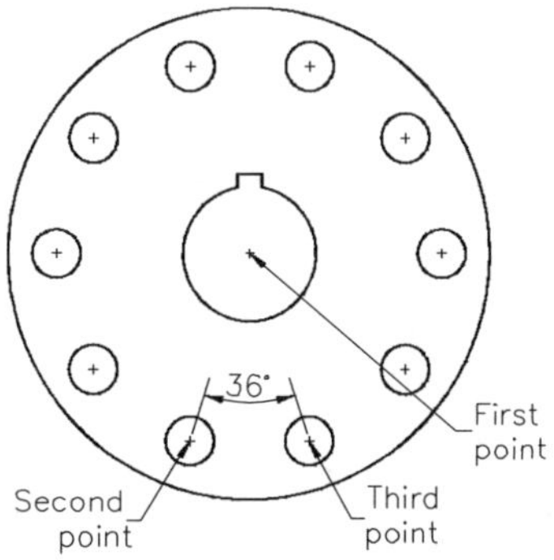

Figure 3-33 Angular dimension displayed according to the dimension placement point

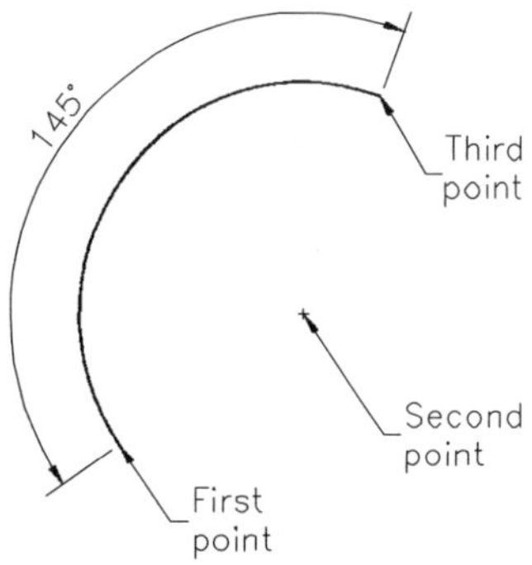

Figure 3-34 Angular dimension displayed according to the dimension placement point

Diameter Dimensioning

CommandManager:	Sketch > Smart Dimension
Menu:	Tools > Dimension > Smart
Toolbar:	Sketch > Smart Dimension

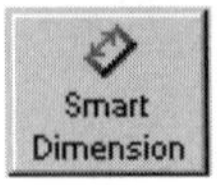

Diameter dimensions are applied to dimension a circle or an arc in terms of its diameter. To apply the diameter dimension, choose the **Smart Dimension** button from the **Sketch CommandManager**. Select a circle or an arc and place the dimension.

In SolidWorks, when you select a circle to dimension, the diameter dimension is applied to it by default. However, when you select an arc, the radius dimension is applied to it. To apply the diameter dimension to an arc, select the arc and place the radius dimension. Next right-click to display the shortcut menu and choose the **Properties** option from the shortcut menu; the **Dimension Properties** dialog box will be displayed. Select the **Diameter dimension** check box and choose **OK**. Figure 3-35 shows a circle and an arc with the diameter dimension.

Radius Dimensioning

CommandManager:	Sketch > Smart Dimension
Menu:	Tools > Dimension > Smart
Toolbar:	Sketch > Smart Dimension

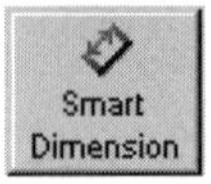

Radius dimensions are applied to dimension a circle or an arc in terms of its radius. As mentioned earlier, by default the dimension applied to a circle is in the diameter form and the dimension applied to an arc is a radius dimension. To apply radius dimension, choose the **Dimension** button from the **Sketch** toolbar and select an arc. A radius dimension will be attached to the cursor. Using the left mouse button place the dimension at an appropriate place. To convert the diameter dimension to the radius dimension clear the **Diameter dimension** check box from the **Dimension Properties** dialog box. Figure 3-36 illustrates the radial dimensioning of a circle and an arc.

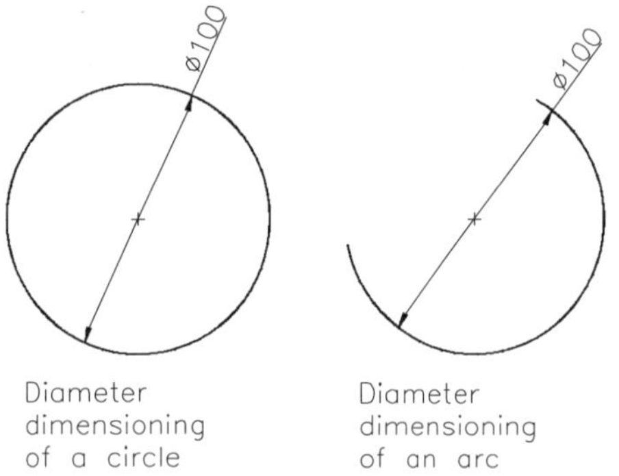

***Figure 3-35** Diameter dimensioning of a circle and an arc*

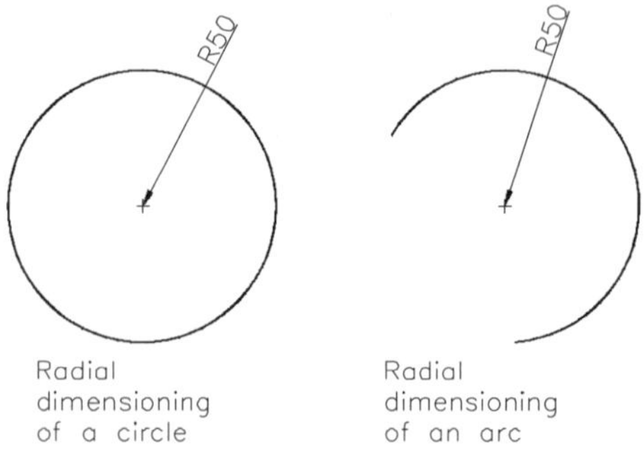

***Figure 3-36** Radial dimensioning of a circle and an arc*

Linear Diameter Dimensioning

CommandManager:	Sketch > Smart Dimension
Menu:	Tools > Dimension > Smart
Toolbar:	Sketch > Smart Dimension

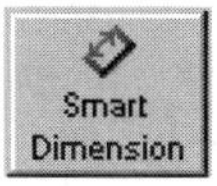

Linear diameter dimensioning is used to dimension the sketch of a revolved component. An example of a revolved component is shown in Figure 3-37. The sketch for a revolved component is drawn using simple sketcher entities as shown in Figure 3-38. If you dimension the sketch of the base feature of the given model using the linear dimensioning method, the same dimensions will be generated in drawing views. This may be confusing because in the shop floor drawing, you need the diameter dimension of a revolved model. To overcome this problem, it is recommended that you create a linear diameter dimension as shown in Figure 3-38. To create a linear diameter dimension, choose the **Smart Dimension** button from the **Sketch CommandManager**. Select the entity to be dimensioned and then select the centerline around which the sketch will be revolved. Move the cursor to the other side of the centerline; a linear diameter dimension will be displayed. Place the dimension and enter a new value in the **Modify** dialog box.

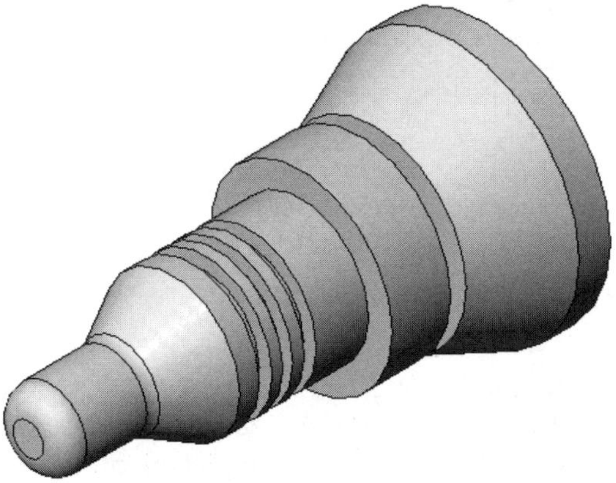

Figure 3-37 A revolved component

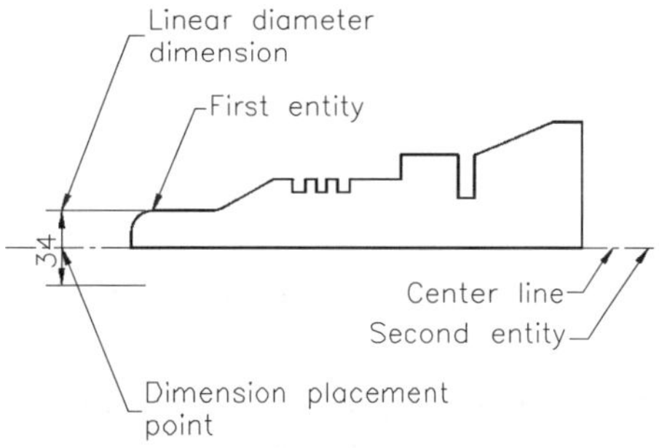

Figure 3-38 Sketch for the revolved feature and linear diameter dimension

Ordinate Dimensioning

The ordinate dimensions are used to dimension the sketch with respect to a specified datum. Depending on the requirement of the design, the datum can be an entity in the sketch or the origin. The ordinate dimensions are of two types, horizontal and vertical. The method of creating these types of ordinate dimensions is discussed next.

Horizontal Ordinate Dimensioning

CommandManager:	Dimension/Relations > Horizontal Ordinate Dimension
Menu:	Tools > Dimension > Horizontal Ordinate
Toolbar:	Dimension/Relations > Horizontal Ordinate Dimension

Horizontal ordinate dimensions are used to dimension the horizontal distances of the selected entities from the specified datum, see Figure 3-39. Note that when you apply the ordinate dimensions, the **Modify** dialog box is not displayed to modify the dimension values. After placing all ordinate dimensions, you need to exit the tool and then double-click on the dimensions to modify their values.

To apply a horizontal ordinate dimension, choose the **Horizontal Ordinate Dimension** button from the **Dimension/Relations CommandManager**. You will be prompted to select an edge or a vertex. Note that the first entity that you select is taken as the datum entity from where the remaining entities will be measured. Select the first entity and place the dimension above or below it. You will notice that the dimension shows a value of 0; refer to the dimension of the left vertical line in Figure 3-39.

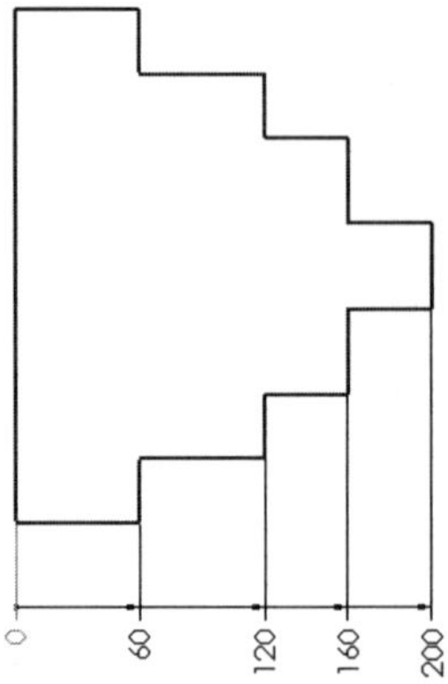

Figure 3-39 *Horizontal ordinate dimensions*

After placing the first dimension, you will again be prompted to select an edge or a vertex. Select the edge that you need to dimension using the first selected edge as datum. As soon as you select the edge, a horizontal dimension between the datum and this entity will be placed. Similarly, continue placing the dimensions to apply multiple horizontal ordinate dimensions, refer to Figure 3-39.

Vertical Ordinate Dimensioning

CommandManager:	Dimension/Relations > Vertical Ordinate Dimension
Menu:	Tools > Dimension > Vertical Ordinate
Toolbar:	Dimension/Relations > Vertical Ordinate Dimension

Vertical ordinate dimensions are used to dimension the vertical distances of the selected entities from the specified datum, see Figure 3-40. To add a vertical ordinate dimension, choose the **Vertical Ordinate Dimension** button from the **Dimension/ Relations CommandManager**. You will be prompted to select an edge or a vertex. As mentioned earlier, the first entity that you select is taken as the datum entity from where the remaining entities will be measured. Select the first entity and place the dimension on the right or left of it. You will notice that the dimension shows a value of 0; refer to the dimension of the left vertical line in Figure 3-40.

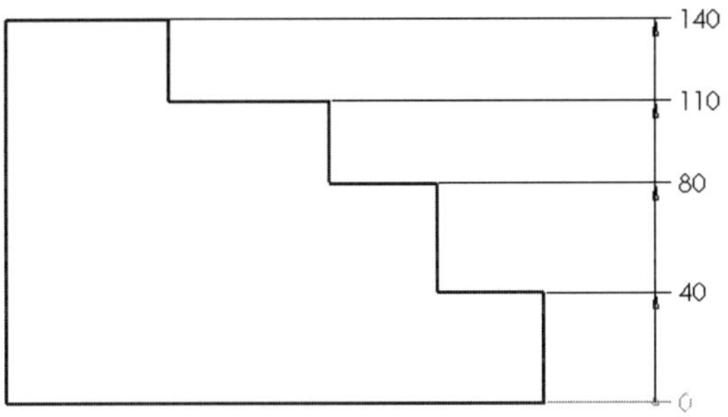

Figure 3-40 *Vertical ordinate dimensions*

Next, select the edge that you need to dimension using the first selected edge as datum. As soon as you select the edge, a vertical dimension will be placed between the datum and this entity. Similarly, continue placing the dimensions to create multiple horizontal ordinate dimensions, refer to Figure 3-40.

Additional Dimensioning Options

In SolidWorks, you are also provided with some dimensioning options other than those discussed earlier. The main additional dimensioning option is discussed next.

Dimensions Between Arcs or Circles

If you select two arcs, two circles, or an arc and a circle to add a dimension, the dimension placed depends on the point from where you select the circles or arcs, refer to Figures 3-41 through 3-43. These figures display the position of selection of circles and the resulting dimension applied.

Enhanced

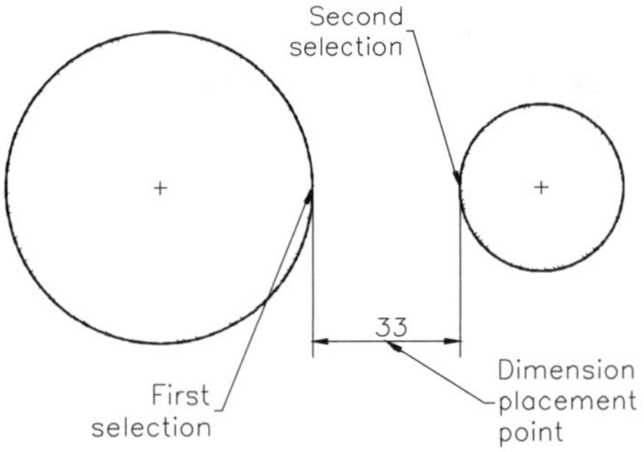

Figure 3-41 *Dimension with first arc condition as* ***Center*** *and second arc condition as* ***Center***

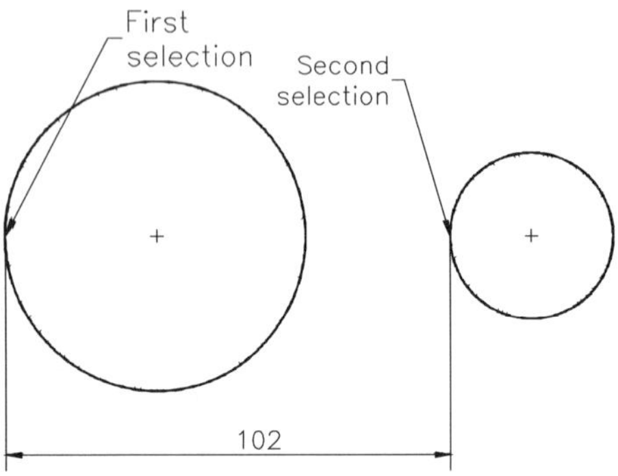

Figure 3-42 *Dimension with first and second arc conditions as* ***Min***

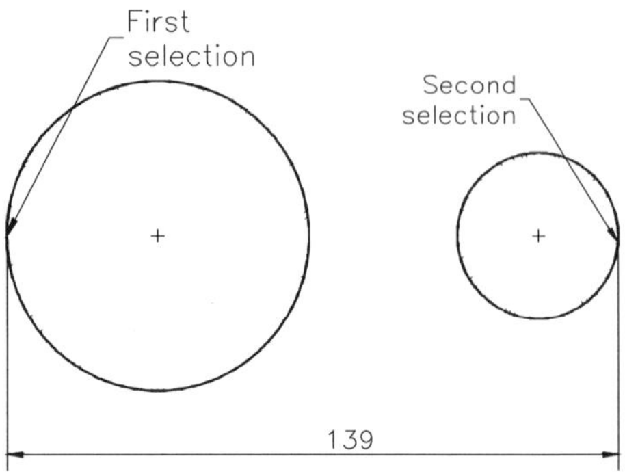

Figure 3-43 *Dimension with first and second arc conditions as* ***Max***

Tip. *Note that if you select the center points to add dimensions, you will not be able to place the dimensions tangential to the curves.*

After placing the dimension if you need to change the selection location, right-click to display the shortcut menu. Choose the ***Properties*** *option from the shortcut menu to display the* ***Dimension Properties*** *dialog box. Set the selection locations using the options available in the* ***First arc condition*** *and* ***Second arc condition*** *areas.*

If you select the ***Center*** *radio buttons from these areas, then a center-to-center dimension will be applied to the selected curve.*

CONCEPT OF FULLY DEFINED SKETCH

It is very necessary for you to understand the concept of fully defined sketches. While creating a model, you first have to draw the sketch of the base feature and then proceed further for creating other features. After creating the sketches, you have to add the required relations and dimensions to constrain the sketch with respect to the surrounding environment. After adding the required relations and dimensions, the sketch may exist in one of the six states discussed below.

1. Fully Defined
2. Overdefined
3. Underdefined
4. Dangling
5. No Solution Found
6. Invalid Solution Found

Fully Defined

A fully defined sketch is the one in which all entities of the sketch and their positions are fully defined by the relations or dimensions, or both. In the fully defined sketch, all degrees of freedom of a sketch are constrained using relations and dimensions and the sketched entities cannot move or change their size and location unexpectedly. If the sketch is not fully defined, it can change its size or position at any time during the design because all degrees of freedom are not constrained. All entities in a fully defined sketch are displayed in black.

Overdefined

An overdefined sketch is the one in which some of the dimensions, relations, or both are conflicting or the dimensions or relations have exceeded the required number. The overdefined sketch is displayed in red. In an overdefined sketch, you need to delete the extra and conflicting relations or dimensions. It is recommended that you do not proceed further for creating the feature with an overdefined sketch. The overdefined sketch is changed to fully defined or underdefined sketch by deleting the conflicting relations or dimensions. Deleting the overdefining relations or dimensions is discussed later in this chapter.

Underdefined

An underdefined sketch is the one in which some of the dimensions or relations are not defined and the degree of freedom of the sketch is not fully constrained. In these types of sketches, the entities may move or change their size unexpectedly. As a result, the sketched entities of the underdefined sketch are displayed in blue. When you add relations and dimensions, the color of the entities in the sketch changes to black, suggesting that the sketch is fully defined. If the entire sketch is displayed in black and only some of the entities are shown in blue, this means that the entities in blue require some dimension or relation.

Tip. *In SolidWorks, it is not necessary that you fully dimension or define the sketches before you use them to create the features of the model. However, it is recommended that you fully define the sketches before you proceed further for creating the feature.*

*If you want to always use fully defined sketches before proceeding further, you can set the option by choosing **Tools** > **Options** from the menu bar to display the **System Option - General** dialog box. Select the **Sketch** option from the area on the left. Select the **Use fully defined sketches** check box and choose **OK** from this dialog box.*

Note

From this chapter onwards, you will work with fully defined sketches. Therefore, follow the above procedure to use the fully defined sketches in future.

Dangling

In the dangling sketch, the dimensions or relations applied to an entity lose their reference because of deletion of the entity from which they were referenced. These entities are displayed in brown. You need to delete the dangling entities, dimensions, or relations that conflict.

No Solution Found

In the no solution found state, the sketch is not solved with the current constraints. Therefore,

Tip. *The status bar of the SolidWorks window is divided into four areas while working in the sketching environment. The* ***Sketch Definition*** *area of the status bar always displays the status of the sketch, dimension, and relation. If the sketch is underdefined, the status area will display the* ***Under Defined*** *message; if the sketch is overdefined, the message displayed in the status area will be* ***Over Defined****; if the sketch is fully defined, the message displayed in the status area will be* ***Fully Defined****.*

you need to delete the conflicting dimensions or relations and add other dimensions or relations. The sketched entity, dimension, or relation will be displayed in pink.

Invalid Solution Found

In the invalid solution found state, the sketch is solved but will result in invalid geometry such as a zero length line, zero radius arc, or self-intersecting spline. The sketch entities for this state are displayed in yellow.

Sketch Dimension or Relation Status

In SolidWorks, when you are applying the dimensions and relations to the sketches, sometimes you apply the dimensions or relations that are not compatible with the geometry of the sketched entities or that make the dimensioned entity overdefined. In addition to the fully defined state, the sketch dimensions or relations may have any of the following states.

1. Dangling
2. Satisfied
3. Overdefining
4. Not Solved
5. Driven

Dangling

A dangling dimension or relation is the one that cannot be resolved because the entity to which it was referenced is deleted. The dangling dimension appears in brown.

Satisfied

A satisfied dimension is the one that is completely defined and is displayed in black.

Overdefining

An overdefining dimension or relation overdefines one or more entities in the sketch. The overdefining dimension appears in red.

Not Solved

The not solved dimension or relation is not able to determine the position of the sketched entities. A not solved dimension appears in pink.

Driven

The driven dimension's value is driven by other dimensions in the sketch that solve the sketch. The driven dimension appears in gray.

DELETING THE OVERDEFINING DIMENSIONS

In SolidWorks, when you add a dimension that overdefines a sketch, the sketch and dimension turn red. Also, the **Make Dimension Driven?** dialog box is displayed as shown in Figure 3-44.

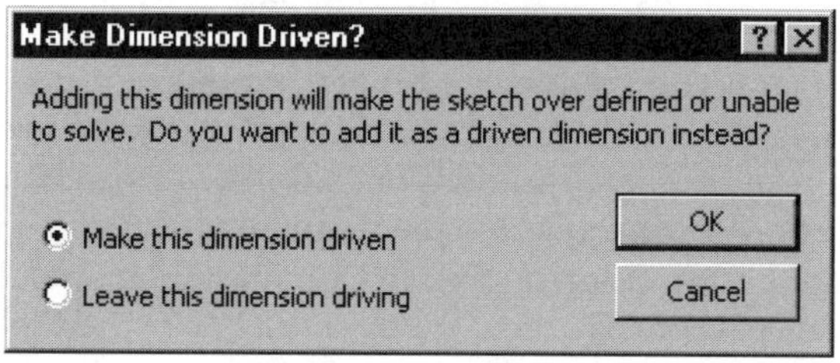

Figure 3-44 *The **Make Dimension Driven?** dialog box*

This dialog box informs you that adding this dimension will overdefine the sketch or the sketch will not be solved. You are also prompted to specify whether you want to add the dimension as driven dimension? If you keep the **Make this dimension driven** radio button selected and choose **OK**, then that dimension will become a driven dimension. The driven dimension is displayed in gray and you cannot modify it. Its value depends on the value of the driver dimension. If you change the value of the driver dimension, the value of the driven dimension will be automatically changed.

If you select the **Leave this dimension driving** radio button and choose **OK**, then some of the entities and dimensions in the sketch will be displayed in red. A callout will also be displayed attached to the status bar, which informs you that status is overdefined. You will learn more about this callout later in this chapter. You need to delete either the red sketched entity or the red dimension to make sure the sketch is no longer overdefined. When the sketch is not overdefined any more, the **SolidWorks** information dialog box will be displayed as shown in Figure 3-45. The information box informs you that the sketch is no longer overdefined. Choose **OK** from the **SolidWorks** information dialog box; the sketch will be displayed in black or blue, depending on the current state of the sketch.

Figure 3-45 *The **SolidWorks** information dialog box*

You can also prevent the sketch from being overdefined by choosing the **Cancel** button from the **Make Dimension Driven?** dialog box. If you choose **Cancel** from the **Make Dimension Driven?** dialog box, the **SolidWorks** information dialog box will be displayed, as shown in Figure 3-45, with a message that the sketch is no longer overdefined.

Displaying and Deleting Relations

CommandManager:	Sketch > Display/Delete Relations
Menu:	Tools > Relations > Display/Delete
Toolbar:	Sketch > Display/Delete Relations

If the sketch is overdefined after adding the dimensions and relations, you need to delete some of the overdefining, dangling, or not solved relations or dimensions. You can view and delete the relations applied to the sketch using the **Display/ Delete Relations PropertyManager**. To invoke this **PropertyManager**, choose the **Display/Delete Relations** button from the **Sketch CommandManager**. You can also right-click in the drawing area to display the shortcut menu and choose the **Display/Delete Relations** option. You can also invoke this **PropertyManager** by clicking on the status of the sketch on the status bar provided on the bottom of the SolidWorks window. Whenever a sketch is overdefined or do not find any solution, a callout is displayed that prompts you to click on the status bar to resolve the conflict. Figure 3-46 shows the **Display/Delete Relations PropertyManager**. The options available in the **Display/Delete Relations PropertyManager** are discussed next.

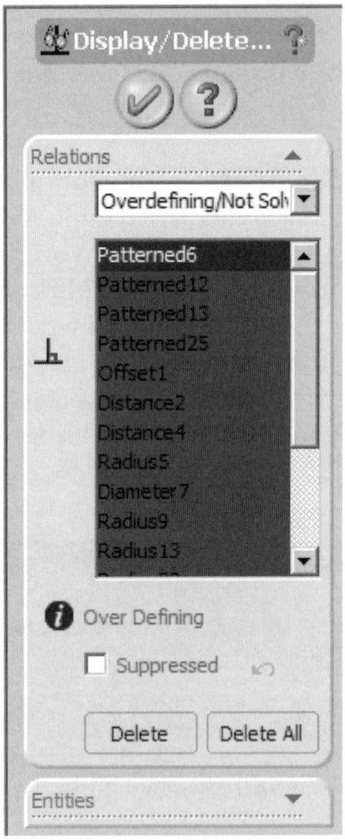

*Figure 3-46 The **Display/Delete Relations PropertyManager***

Relations

The **Relations** rollout is used to check, delete, and suppress the unwanted and conflicting relations. The status of the sketch or the selected entity is displayed in the **Information** area of this rollout. The options of the **Relations** rollout are discussed next.

Filter

The **Filter** drop-down list is used to select the filter to show the relations in the **Sketch Relation PropertyManager**. The options available in the **Filter** drop-down list are discussed next.

All in this sketch

The **All in this sketch** option is used to display all relations applied to the sketch. The first relation displayed in the list will be selected by default and will appear with blue background. The status of the selected relations is displayed in the **Information** area of the **Relations** rollout. The overdefined relations are highlighted in red. If you select a relation highlighted in red in the **Relations** area, the status of the selected relation will be displayed as **Over Defining**. The dangling relation is highlighted in brown. When you select the dangling relation, the status of the relation will be displayed as **Dangling** in the **Information** area. Similarly, the not solved relation will be highlighted in yellow and the driven relation in gray.

Dangling

The **Dangling** option is used to display only the dangling relations applied to the sketch.

Overdefining/Not Solved

The **Overdefining/Not Solved** option is used to display only the overdefining and not solved relations. The dangling relations are also not solved relations. Therefore, they will also be displayed.

External

The **External** option is used to display the relations that have a reference with an entity outside the sketch. This entity can be an edge, vertex, or origin within the same model or it can be an edge, vertex, or origin of a different model within an assembly.

Defined In Context

The **Defined In Context** option is used to display only the relations that are in the context of a design. They are the relations between the sketched entity in one part and an entity in another part. These relations are defined while working with the top-down assemblies.

Locked

The **Locked** option is used to display only the locked relations.

Broken

The **Broken** option is used to display only the broken relations.

Note

*The **Locked** and **Broken** relations are applied while creating a part within the assembly environment. You will learn more about creating parts within the assembly in the later chapters.*

Selected Entities

The **Selected Entities** option is used to display the relations of only the selected set of entities. When you select this option from the **Filter** drop-down list, the **Selected Entities** selection area will be displayed in the **Relations** rollout. When you select an entity to display the relations, the name of the selected entity will be displayed in the **Selected Entities** area and the relations applied to this entity will be displayed in the **Relations** area. To remove the selected entity from the selection set, select the entity to be removed and right-click to display the shortcut menu. Choose the **Delete** option from the shortcut menu. If you choose the **Clear Selections** option, all entities will be removed from the selection set.

Suppressed

The **Suppressed** check box is selected to suppress the selected relation. When you suppress a relation, it will be displayed in gray in the **Relations** area. The status of the suppress relation will be displayed as **Satisfied** or **Driven** in the information area. If you suppress overdefining dimensions, the **SolidWorks** information dialog box will be displayed with a message that **The sketch is no longer over defined.** Choose the **OK** button from this dialog box.

Delete

The **Delete** button is used to delete the relation selected in the **Relations** area.

Delete All

The **Delete All** button is used to delete all relations that are displayed in the **Relations** area.

Undo last relation change

The **Undo last relation change** button is used to undo the **Delete**, **Replace**, and **Suppressed** options used earlier. The replace option is discussed later in this chapter.

Entities

The **Entities** rollout is used to display the entities that are referred to in the selected relation. This rollout is also used to display the status of the selected relation and the external reference, if any. By default, the **Entities** rollout is closed. You can open this rollout by clicking on the blue arrow displayed on the right of this rollout. The **Entities** rollout is shown in Figure 3-47. The options available in the **Entities** rollout are discussed next.

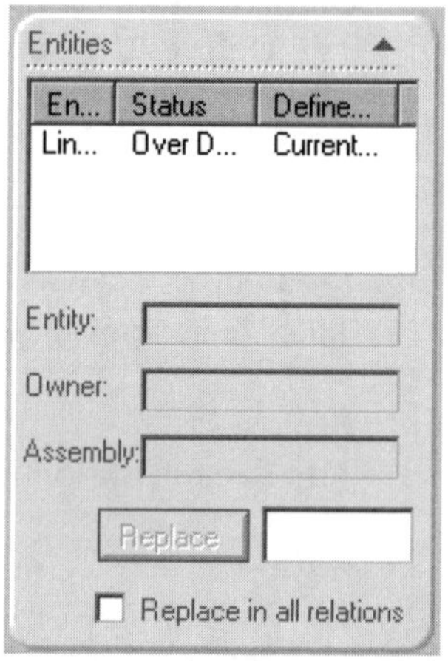

*Figure 3-47 The **Entities** rollout*

Entities used in the selected relation

The **Entities used in the selected relation** area is used to display the information about entities used in the selected relation. The information about the name of the entity, the status of the entity, and the place where the entity is defined is provided. This area is divided into three columns, which are discussed next.

Entity

The **Entity** column is used to display the entity or entities to which the selected relation is applied.

Status

The **Status** column is used to display the status of the selected relation. The status can be **Fully Defined**, **Dangling**, **Over Defined**, or **Not Solved**.

Defined In

The **Defined In** column is used to display the placement of the entity. The entity can be placed in any of the following places.

Current Sketch

The **Current Sketch** option is displayed in the **Defined In** column when the entity is placed in the same sketch.

Same Model

The **Same Model** option is displayed in the **Defined In** column when the entity is defined to be placed in the same model. This means that the entity is placed in the same model, but outside the sketch. This entity can be an edge, vertex, or origin of the same model.

External Model

The **External Model** option is displayed in the **Defined In** column when the entity is placed in some other model but within the assembly. This entity can be an edge, vertex, or origin of a different model, but in the same assembly.

Tip. *By default, the* ***Override Dims on Drag*** *option is not selected from* ***Tools > Sketch Settings*** *menu in the menu bar. As a result, if you drag a dimensioned sketched entity, the entity will not be modified automatically. However, if the sketch is not fully defined, the entities that are not properly dimensioned or constrained will move. If you select this option, you can change a dimensioned sketched entity by dragging it. However, if the sketch is not fully defined, the entities that are not properly dimensioned or constrained will move.*

By default, the ***Automatic Solve*** *option is selected from the* ***Tools > Sketch Settings*** *menu in the menu bar. This option helps you solve the relations and dimensions automatically when you drag or modify a sketched entity. If you clear this option, a message that* ***The sketch cannot be dragged because Auto Solve Mode is off. To drag the sketch, please turn the Auto Solve Mode on*** *appears. If you modify the dimension value using the* ***Modify*** *dialog box, the dimension will not update automatically, and you have to update the new dimension manually. To update and solve the dimension, you have to choose the* ***Rebuild*** *button from the* ***Standard*** *toolbar or press CTRL+B from the keyboard.*

Tip. *By default, the relations applied to the sketched entities are displayed when you draw it or apply a relation to it. Therefore, sometimes the sketch looks untidy as all relations concerned with the sketch are displayed. To turn off the display of these relations choose **View > Sketch Relations** from the menu bar. To again turn on the display of these relations again choose **View > Sketch Relations** from the menu*

Entity
The **Entity** display box is used to display the name of the entity and the name of the part in which the selected entity is placed. This entity is selected in the **Entity** column of the **Entities used in the selected relation** area. The selected entity is also highlighted in the drawing area.

Owner
The **Owner** display box is used to display the name of the model in which the entity is placed when the **External Model** option is displayed in the **Defined In** column.

Assembly
The **Assembly** display box is used to display the path of the assembly in which the entity is placed when the **External Model** option is displayed in the **Defined In** column.

Replace
The **Replace** button is used to replace the selected entity from the **Entity** column with some other entity from the drawing area. When you select the entity from the drawing area, the entity will be displayed in the display box provided below the **Replace** button. Choose the **Replace** button to replace the entity. If the sketch is overdefined, you will be given a warning message. Sometimes after replacing the entity, the status of the entity is changed to not solved or overdefining. You need to undo the last operation. Select the **Replace in all relations** check box to replace the entity in all relations.

Options
The check box available in this rollout is selected to display the **Display/Delete PropertyManager** when the sketch becomes overdefined or cannot be solved.

OPENING AN EXISTING FILE

Toolbar:	Standard > Open
Menu:	File > Open

The **Open** dialog box is used to open an existing SolidWorks part, assembly, or drawing document. You can also use this dialog box to import files from other applications saved in some standard file formats. Choose the **Open** button from the **Standard** toolbar, or press CTRL+O keys to invoke the **Open** dialog box, which is shown in Figure 3-48. The options available in this dialog box are discussed next.

Look in drop-down list
The **Look in** drop-down list is used to specify the drive or directory in which the file is saved. The location of the file and the folder that you browse is shown in this drop-down list.

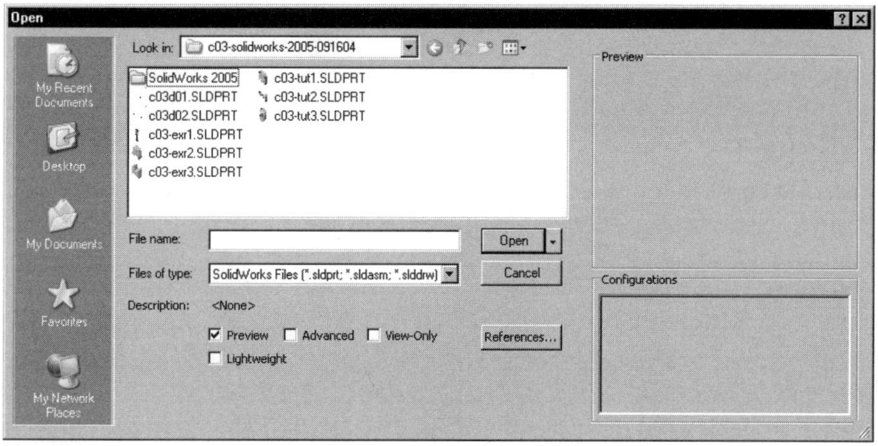

Figure 3-48 *The* ***Open*** *dialog box*

File name

The name of the file selected is shown in the **File name** edit box. You can also enter the name of the file to open in this edit box.

Files of type

The **Files of type** drop-down list is used to specify the type of file to open. Using this drop-down list, you can select a particular type of file type such as the part file, assembly file, drawing file, all SolidWorks files, and so on. You can also define the standard file format in this drop-down list to import the files saved in those file formats.

Open as read-only

The **Open as read-only** option is selected to open the document as a read-only file. This option is available in the flyout that will be displayed by choosing the down arrow available on the right of the **Open** button. If you modify the design in a read-only file, the changes will be saved in a new file. The original file will not be modified. This also allows another user to access the document while the document is open on your computer.

Preview

The **Preview** check box is selected to display the **Preview** area in this dialog box. In the **Preview** area, you can preview the selected part, assembly, or drawing document before opening.

Advanced

The **Advanced** check box is selected to display the configurations available in the selected file. The configurations available in the selected file are displayed in the **Configurations** area.

View-Only

The **View-Only** check box is selected to open a SolidWorks document in the view only format. When you open a view-only file, only the tools related to viewing the models are enabled. Rest of the tools are not available. This is the reason why you cannot do any modification in the view-only document. You can use only the zoom, pan, or dynamically rotate tools. If you want to edit the design, right-click in the drawing area and choose the **Edit** option from the shortcut menu to edit the design.

Lightweight

The **Lightweight** check box is selected to open the assembly document using the lightweight parts. You will learn more about the lightweight parts in later chapters.

TUTORIALS

Tutorial 1

In this tutorial, you will draw the sketch of the model shown in Figure 3-49. This is the same sketch that was drawn in Tutorial 1 of Chapter 1. You will draw the sketch using the mirror line and then add the required relations and dimensions. The sketch is shown in Figure 3-50. The solid model is given only for reference. **(Expected time: 30 min)**

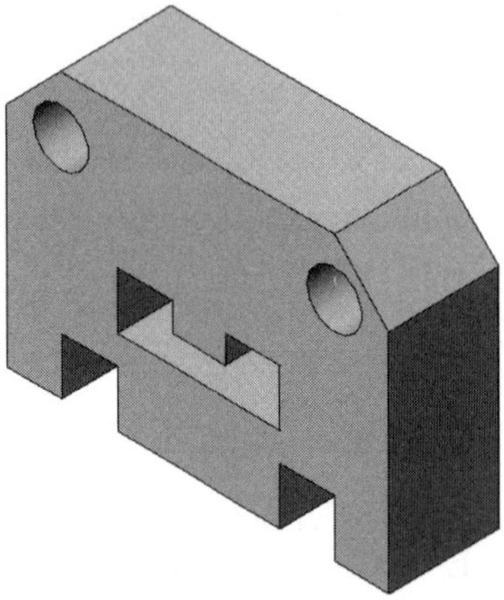

Figure 3-49 Solid Model for Tutorial 1

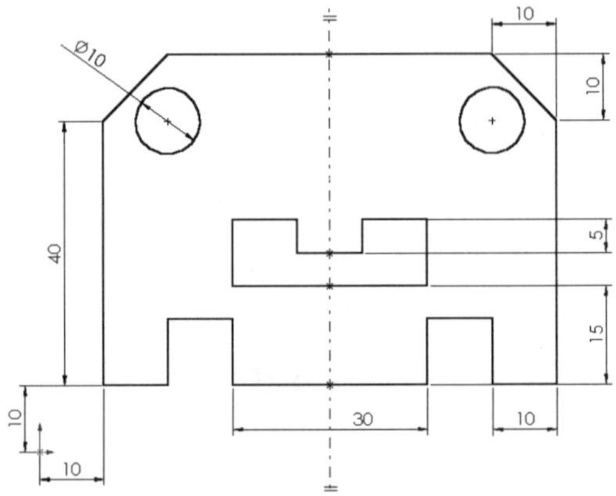

Figure 3-50 Sketch of the model

The steps that will be followed to complete this tutorial are listed below.

a. Start SolidWorks and then start a new part document.
b. Maximize the part document and then invoke the sketching environment.
c. Create a mirror line using the **Centerline** and **Mirror** tools.
d. Draw the sketch of the model on one side of the mirror line so that it is automatically drawn on the other side, refer to Figures 3-51 through 3-56.
e. Add the required relations to the sketch, refer to Figures 3-57 and 3-58.
f. Add the required dimensions to the sketch and fully define the sketch, refer to Figure 3-59.
g. Save the sketch and then close the document.

Starting SolidWorks and Starting a New Part Document

1. Start SolidWorks by double-clicking on the shortcut icon of SolidWorks 2005 available on the desktop of your computer.

The SolidWorks 2005 window is displayed and the **SolidWorks Resources Task Pane** is displayed on the right of the SolidWorks window.

2. Choose the **New Document** button from the **Standard** toolbar.

   The **New SolidWorks Document** dialog box is displayed.

3. The **Part** button is chosen by default. Choose the **OK** button from the **New SolidWorks Document** dialog box; a new SolidWorks part document is started.

4. Choose the **Sketch** button from the **Standard** toolbar and then select the **Front Plane** to invoke the sketching environment.

   In previous chapters, you used the grid and snap settings to create the sketches. However, from this chapter onwards, it is recommended that you do not use those settings. This way you can get familiar with drawing sketches at arbitrary locations and then using dimensions and relations to move them to their actual locations.

5. Invoke the **System Options - Relations/Snaps** dialog box and then clear the **Grid** check box from the **Sketch Snap** area.

6. Set the units for measuring linear dimensions to millimeters and the units for angular dimensions to degree using the **Document Properties - Units** dialog box. If you selected millimeters as units while installing SolidWorks, you can skip this point.

Drawing the Mirror Line

In this tutorial, you will draw the sketch of the given model with the help of the mirror line. As mentioned earlier, the sketches that are symmetrical along any axis are recommended to be drawn using the mirror line. The mirror line is drawn using the **Centerline** and **Mirror** tools. When you draw an entity on one side of the mirror line, the same entity is drawn automatically on the other side of it. Also, the symmetrical relation is applied to the entities on both sides of the mirror line. Therefore, if you modify an entity on one side of the mirror line, the same modification is reflected in the mirrored entity and vice versa. First, you need to draw a mirror line.

The origin of the sketching environment is placed at the center of the drawing area and you have to create the sketch in the first quadrant. Therefore, it is recommended that you modify the drawing area such that the area in the first quadrant is increased. This can be done using the **Pan** tool.

1. Choose the **Pan** tool button from the **View** toolbar. Press and hold the left mouse button down and drag the cursor toward the bottom left corner of the drawing area.

2. Choose the **Centerline** button from the **Sketch CommandManager** to display the **Insert Line PropertyManager**.

3. Move the line cursor to a location whose coordinates are close to 45 mm 70 mm 0 mm. You

do not need to move the cursor exactly to this location. You can move it to a point close to this location.

4. Specify the start point of the centerline and move the line cursor vertically downward to draw a line of length close to 80 mm.

   As soon as you specify the endpoint of the centerline, a rubber-band line is attached to the line cursor. Right-click to display the shortcut menu and choose the **Select** option from the shortcut menu to end line creation and exit the **Line** tool.

5. Choose the **Zoom to Fit** button from the **View** toolbar to fit the sketch on the screen.

6. Select the centerline and convert it into a mirror line using the **Dynamic Mirror Entities** tool.

Drawing the Sketch

You will draw the sketch on the right of the mirror line and the same sketch will be automatically drawn on the other side of the mirror line.

1. Choose the **Line** button from the **Sketch CommandManager**; the arrow cursor is replaced by the line cursor.

2. Move the line cursor to a location where the coordinates are close to 45 mm 10 mm 0 mm. At this point, the cursor snaps to the mirror line and a coincident symbol is displayed below the cursor.

3. Specify the start point of the line at this point and move the cursor horizontally toward the right. Specify the endpoint of the line when the length of the line above the line cursor shows a value close to 15. As soon as you press the left mouse button to specify the endpoint of the line, a line of the same length is drawn automatically on the other side of the mirror line. Figure 3-51 shows the mirrored entity created automatically on the other side of the mirror line. The display of relations in this figure is turned off by choosing **View > Sketch Relations** menu.

   Note that the mirrored entity that is automatically created on the left of the mirror line is merged with the line drawn on the right. Therefore, the entire line becomes a single entity. As mentioned earlier, the mirror image of the line is merged with the line you draw only if one of the endpoints of the line is coincident with the mirror line.

4. Move the cursor vertically upward. Specify the endpoint of the line when the length of the line on the line cursor displays a value close to 10. Figure 3-52 shows the sketch after drawing the vertical lines.

5. Move the line cursor horizontally toward the right. Specify the endpoint of the line when the length of the line on the line cursor displays a value close to 10.

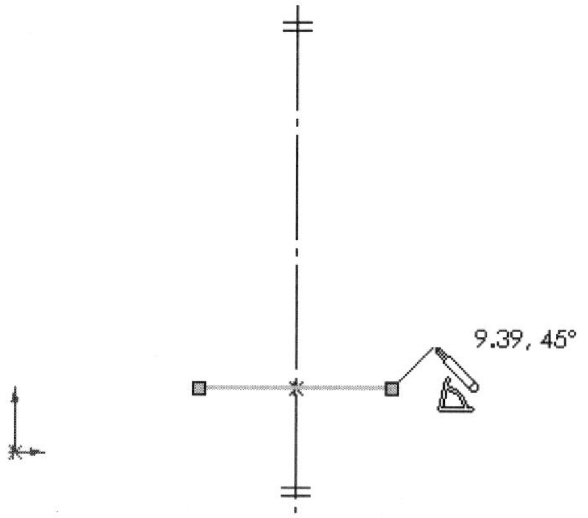

Figure 3-51 *Sketching using automatic mirroring*

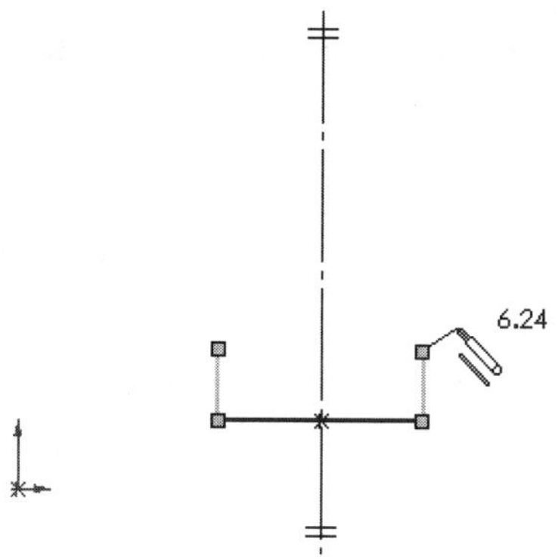

Figure 3-52 *Left line drawn by mirroring*

6. Move the line cursor vertically downward. Specify the endpoint when the length of the line on the line cursor displays a value close to 10.

7. Move the line cursor horizontally toward the right. Specify the endpoint when the length of the line on the line cursor displays a value close to 10.

8. Move the line cursor vertically upward. Specify the endpoint when the length of the line on the line cursor displays a value close to 40.

9. Move the line cursor at an angle close to 135-degree and specify the endpoint of the line. Figure 3-53 shows the sketch after drawing this inclined line.

10. Move the line cursor horizontally toward the left. Specify the endpoint when the cursor snaps to the mirror line and the mirror line is highlighted in red. Double-click anywhere in the drawing area to end line creation. The sketch after completing the outer loop is shown in Figure 3-54.

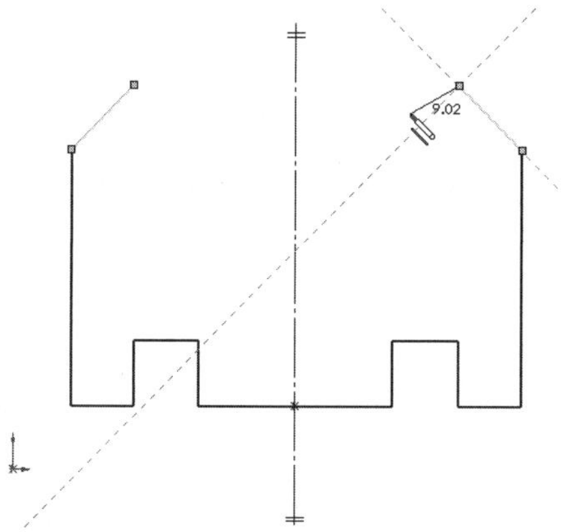

Figure 3-53 *Sketch after drawing the inclined line*

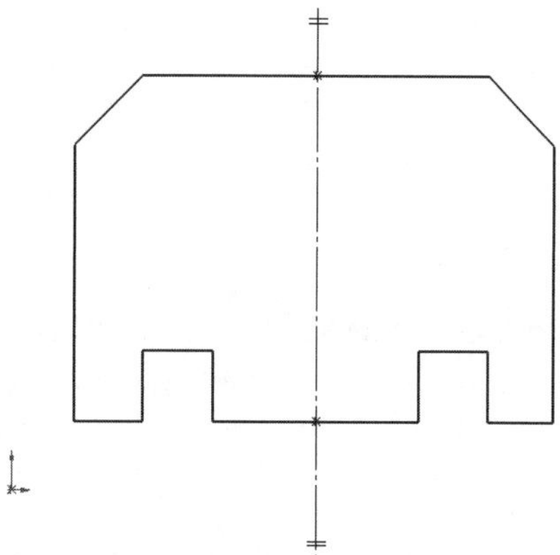

Figure 3-54 *Sketch after completing the outer profile of the sketch*

Next, you will draw the sketch of the inner cavity. To draw the sketch of the inner cavity, you will start with the lower horizontal line.

11. Move the line cursor to a location whose coordinates are close to 45 mm 25 mm 0 mm.
12. Specify the start point of the line at this point and move the cursor horizontally toward the right. Specify the endpoint when the length of the line above the line cursor shows a value close to 15.
13. Move the line cursor vertically upward. Specify the endpoint when the length of the line on the line cursor displays a value close to 10.
14. Move the line cursor horizontally toward the left. Specify the endpoint when the length of the line on the line cursor displays a value close to 10.
15. Move the line cursor vertically downward. Specify the endpoint when the length of the line on the line cursor displays a value close to 5.
16. Move the line cursor horizontally toward the left. Specify the endpoint when the line cursor snaps to the mirror line.
17. Double-click anywhere in the drawing area to end line creation. The sketch after completing the inner cavity is shown in Figure 3-55.
18. Choose the **Circle** button from the **Sketch CommandManager** to invoke the circle tool.
19. Move the circle cursor to the point where the inferencing line originating from the endpoints of the right inclined line intersect.
20. Specify the center of the circle at this point and move the circle cursor toward the left to define the radius of the circle. Press the left mouse button when the radius of the circle above the circle cursor shows a value close to 5.

    The circle is automatically mirrored on the other side of the mirror line. The sketch after drawing the circle is shown in Figure 3-56.
21. Clear the automatic mirroring option by choosing the **Dynamic Mirror Entities** button again.

Adding the Required Relations

After drawing the sketch, you need to add the relations using the **Add Relations PropertyManager**. The relations are applied to a sketch to constrain its degree of freedom, to reduce the number of dimensions in the sketch, and also to capture the design intent in the sketch.

1. Press the ESC key to remove the circles created previously from the selection set.

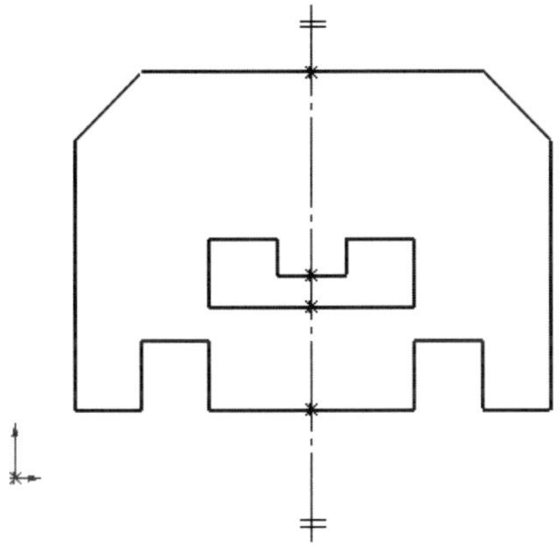

Figure 3-55 Sketch after drawing the inner cavity

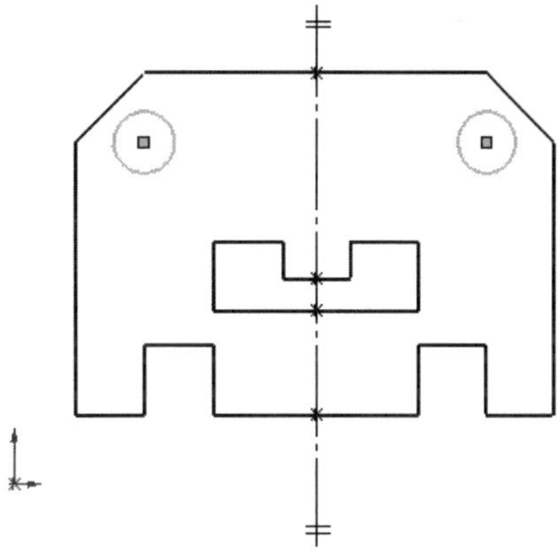

Figure 3-56 Sketch after drawing the circle

2. Choose the **Add Relation** button from the **Sketch CommandManager** to invoke the **Add Relations PropertyManager**. The confirmation corner is also displayed at the upper right corner of the drawing area.

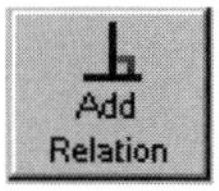

3. Select the center point of the circle on the right and then select the lower endpoint of the right inclined line. The names of the selected entities are displayed in the **Selected Entities** rollout of the **Add Relations PropertyManager**.

   The relations that can be applied to the two selected entities are displayed in the **Add Relations** rollout of the **Add Relations PropertyManager** as shown in Figure 3-57. The **Horizontal** option is highlighted, suggesting that the horizontal relation is the most appropriate relation for the selected entities.

Tip. *Sometimes unwanted inferencing lines are displayed when you are drawing a sketch. You can remove the unwanted inferencing lines by choosing* ***View > Redraw*** *from the menu bar. You can also redraw the drawing area by pressing CTRL+R keys.*

Note

The names of the entities displayed in the ***Selected Entities*** *rollout of the* ***Add Relations PropertyManager*** *may be different from those displayed on your computer screen.*

4. Choose the **Horizontal** button from the **Add Relations** rollout to apply the **Horizontal** relation to the selected entities.

5. Move the cursor to the drawing area and right-click to invoke the shortcut menu. Choose the **Clear Selections** option to remove the selected entities from the selection set.

6. Select the center point of the circle on the right and the upper endpoint of the right inclined line.

   The relations that can be applied to the selected entities are displayed and the **Vertical** option is highlighted.

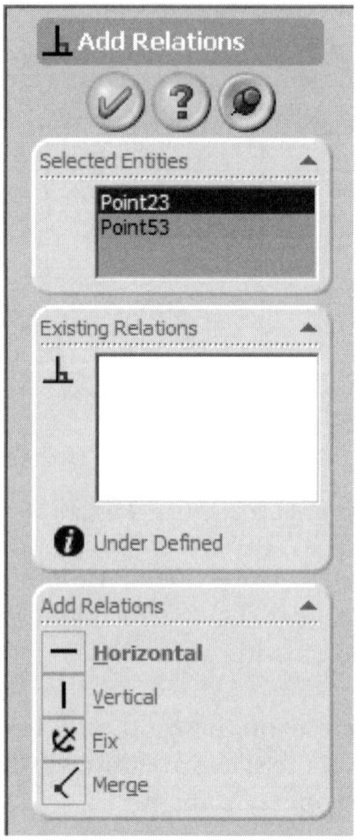

Figure 3-57 Add Relations PropertyManager

7. Choose the **Vertical** button from the **Add Relations PropertyManager**. Right-click in the drawing area and choose the **Clear Selections** option.

8. Select the entities shown in Figure 3-58. Choose the **Equal** button from the **Add Relations** rollout of the **Add Relations PropertyManager**.

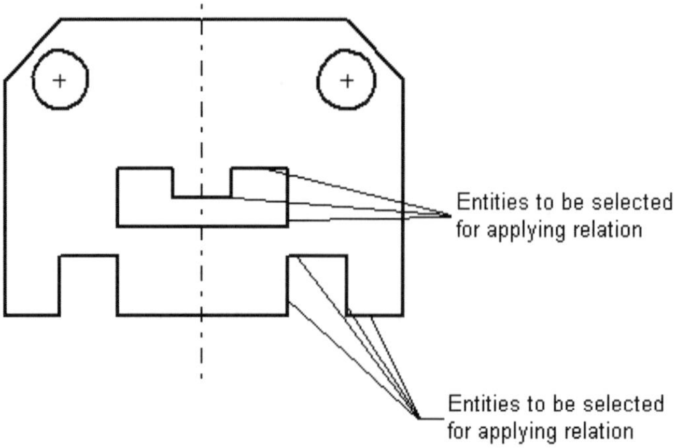

Figure 3-58 *Entities to be selected to apply the **Equal** relation*

9. Choose the **OK** button from the **Add Relations PropertyManager** or choose **OK** from the confirmation corner to close the **PropertyManager**. Click anywhere in the drawing area to clear the selected entities.

Applying the Dimensions to the Sketch

Next, you will apply the dimensions to the sketch and fully define the sketch. As mentioned earlier, the sketched entities are shown in blue, suggesting that the sketch is underdefined. It will be changed to black after applying the required dimensions to the sketch, suggesting that the sketch is fully defined.

1. Choose **Tools > Options** from the menu bar to display the **System Options - General** dialog box. Select the **Input dimension value** check box if cleared, and choose **OK** from the **System Options - General** dialog box. This check box is selected to invoke the **Modify** dialog box to enter a new dimension value and modify the sketch as you place the dimension.

2. Choose the **Smart Dimension** button from the **Sketch CommandManager**. You can also right-click in the drawing area and choose the **Smart Dimension** option to invoke the dimension option.

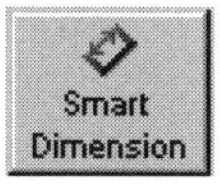

The cursor is replaced by the dimension cursor.

3. Move the dimension cursor to the lower right horizontal line; the line is highlighted in red.

4. Select the line; a linear dimension is attached to the cursor.

5. Move the cursor downward and click to place the dimension below the line, refer to Figure 3-59. As you place the dimension, the **Modify** dialog box is displayed.

6. Enter the dimension value of **10** in this dialog box and press ENTER. The dimension is placed and the length of line is also modified to **10**.

7. Move the dimension cursor to the lower-middle horizontal line, refer to Figure 3-59. Select the line when the line changes to red; a dimension is attached to the cursor.

8. Move the cursor downward and click to place the dimension. Enter a value of **30** in the **Modify** dialog box and press ENTER.

9. Move the cursor to the outer left vertical line and when the color of the line changes to red, select the line; a dimension is attached to the cursor.

10. Move the cursor to the left and then click to place the dimension. Enter a value of **40** in the **Modify** dialog box and press ENTER.

11. Select the right inclined line; a dimension is attached to the cursor. Move the cursor vertically upward to apply the horizontal dimension to the selected line. Click to place the dimension at an appropriate place, see Figure 3-59.

12. Enter a value of **10** in the **Modify** dialog box and press ENTER.

13. Again, select the right aligned line; a dimension is attached to the cursor. Move the cursor horizontally toward the right to apply the vertical dimension for the selected line. Click to place the dimension at an appropriate place, see Figure 3-59.

14. Enter a value of **10** in the **Modify** dialog box and press ENTER.

15. Move the cursor to the left circle and when the circle is highlighted in red, select it. A diameter dimension is attached to the cursor. Move the cursor outside the sketch.

16. Place the diameter dimension. Enter a value of **10** in the **Modify** dialog box and press ENTER.

17. Select the lower horizontal line of the inner cavity; a linear dimension is attached to the cursor. Select the lower right horizontal line of the outer loop.

    A vertical dimension between the lower horizontal line of the inner cavity and the lower right horizontal line of the outer sketch is attached to the cursor.

18. Move the cursor horizontally toward the right and place the dimension. Enter a value of **15** in the **Modify** dialog box and press ENTER.

19. Select the inner right vertical line of the cavity and place the dimension outside the sketch. Enter a value of **5** in the **Modify** dialog box and press ENTER.

20. Select the lower left horizontal line of the outer sketch and the origin.

21. Move the cursor horizontally toward the left and place the dimension. Enter a value of **10** in the **Modify** dialog box.

    Notice that some of the entities are displayed in black. This suggests that these entities are now fully defined. But you have to fully define the entire sketch. So you need to add some more dimensions.

22. Select the left vertical line of the outer sketch and the origin.

23. Move the cursor vertically downward and left-click to place the dimension. Enter a value of **10** in the **Modify** dialog box.

    Notice that all entities are displayed in black. This suggests that the sketch is fully defined. If the sketch is not fully defined, then you have to add a dimension between the outer right vertical line and the outer left vertical line. The value of the dimension should be maintained 70. The fully defined sketch, after applying all required relations and dimensions, is shown in Figure 3-59.

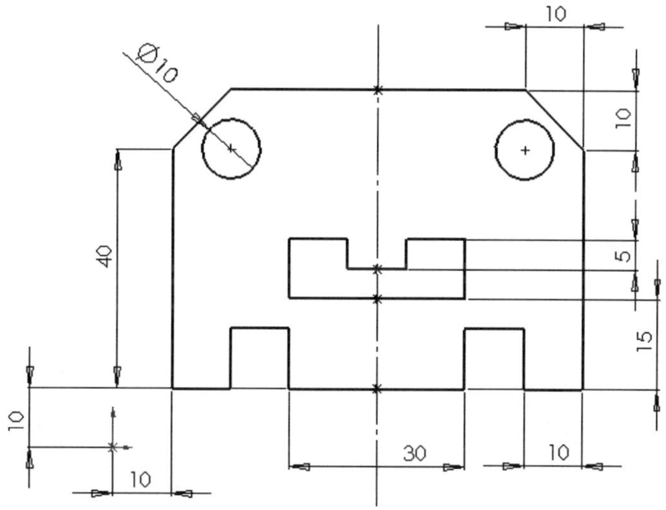

***Figure 3-59** Fully defined sketch after applying all the required relations and dimensions*

Saving the Sketch

1. Choose the **Save** button from the **Standard** toolbar to invoke the **Save As** dialog box.
2. Browse to the *My Documents\SolidWorks* folder. Choose the **Create New Folder** button from the **Save As** dialog box. Enter the name of the folder as *c03* and press ENTER.
3. Enter the name of the document as *c03tut1* in the **File name** edit box and choose the **Save** button. The document will be saved in the *\My Documents\SolidWorks\c03* folder.
4. Close the document by choosing **File > Close** from the menu bar.

Tutorial 2

In this tutorial, you will draw the sketch of the model shown in Figure 3-60. You will draw the sketch using the mirror line and then add the required relations and dimensions. The sketch is shown in Figure 3-61. The solid model is given only for reference. **(Expected time: 30 min)**

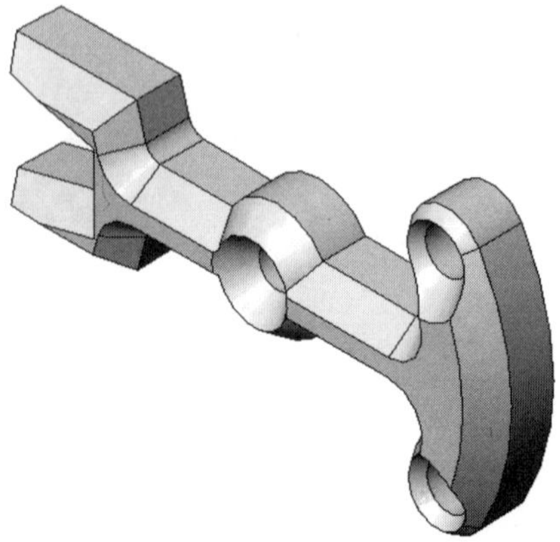

Figure 3-60 Solid Model for Tutorial 2

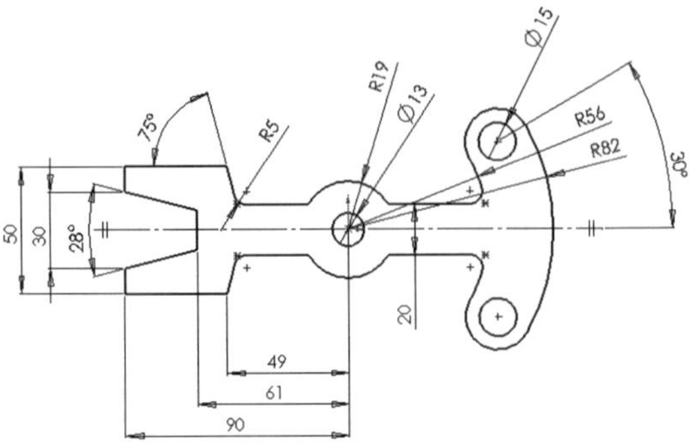

Figure 3-61 Sketch of the solid model

The steps that will be followed to complete this tutorial are listed below.

a. Start a new document file.
b. Maximize the part file document and then invoke the sketching environment.
c. Create a mirror line.
d. Draw the sketch on one side of the mirror line, refer to Figure 3-62.
e. Trim the arcs and circles and add the fillets, refer to Figures 3-63 through 3-65.
f. Add the required relations.
g. Add the required dimensions and fully define the sketch, refer to Figure 3-66.

Starting a New Document and Invoking the Sketching Environment

1. Choose the **New** button from the **Standard** toolbar to invoke the **New SolidWorks Document** dialog box.

2. The **Part** button is chosen by default. Choose the **OK** button; a new SolidWorks part document is started.

3. Choose the **Sketch** button from the **Standard** toolbar and select the **Front Plane** to invoke the sketching environment.

Drawing the Mirror Line

Similar to the last tutorial, you will draw the sketch of the given model with the help of a mirror line.

1. Increase the display of the drawing area using the **Zoom In/Out** tool and choose the **Centerline** button from the **Sketch CommandManager**.

2. Move the line cursor to a location whose coordinates are close to -102 mm 0 mm 0 mm. You do not need to move the cursor exactly to this location. You can move it to a point close to this location.

3. Specify the start point of the centerline at this point and move the line cursor horizontally

toward the right. Specify the endpoint of the centerline when the length of the line shows a value close to 204. Double-click anywhere in the drawing area to end line creation.

4. Choose the **Zoom to Fit** button from the **View** tollbar to fit the sketch in the drawing area.

5. Convert the centerline to mirror line and enable the automatic mirror option.

Drawing the Sketch

You will draw the sketch on the upper side of the mirror line and the same sketch will be automatically drawn on the other side of the mirror line.

1. Choose the **Centerpoint Arc** button from the **Sketch CommandManager**; the arrow cursor is replaced by the arc cursor.

2. Move the arc cursor close to the origin. Specify the center point of the arc when the cursor snaps the origin. Move the cursor horizontally toward the right. The cursor snaps to the mirror line. As you move the cursor, a reference circle is drawn. Specify the start point of the arc when the radius of the arc on the arc cursor shows a value close to 82.

3. Move the arc cursor in the counterclockwise direction. Specify the endpoint of the arc when the value of the angle above the arc cursor shows a value close to 30-degree. The mirror image of the sketched entity is automatically created on the other side of the mirror line.

4. Move the arc cursor to the origin. Specify the center point of the arc. Move it horizontally toward the right. The cursor snaps to the mirror line and a reference circle is drawn. Specify the start point of the arc when the radius of the arc on the arc cursor shows a value close to 56.

5. Move the arc cursor in the counterclockwise direction. Specify the endpoint of the arc when the value of the angle above the arc cursor shows a value close to 30-degree. The mirror image of the sketched entity is automatically created on the other side of the mirror line.

6. Choose the **Tangent Arc** button from the **Sketch CommandManager** .

7. Move the cursor to the upper endpoint of the left arc and when the red circle is displayed, specify the first point of the tangent arc. Move the cursor close to the upper endpoint of the right arc and when the red circle is displayed, specify the second point of the tangent arc.

8. Choose the **Circle** button from the **Sketch CommandManager**.

9. Move the cursor to the center point of the upper arc and when the red circle is displayed, specify the center point of the circle. Move the cursor horizontally toward the right. Press the left mouse button when the radius of the circle above the circle cursor shows a value close to 7.5.

10. Move the cursor to the origin and when the red circle is displayed, specify the center point

of the circle. Move the cursor horizontally toward the right. Press the left mouse button when the radius of the circle above the circle cursor shows a value close to 6.5.

The **SolidWorks** warning dialog box is displayed, which shows the warning message "**Unable to create the symmetric element**." This message is displayed because the circle is created on both sides of the mirror line. Therefore, a symmetric element is not created in this case.

11. Choose **OK** from the **SolidWorks** warning dialog box.
12. Move the cursor to the origin and when the red circle is displayed, specify the center point of the circle. Move the cursor horizontally toward the right. Press the left mouse button when the radius of the circle above the circle cursor displays a value close to 19. The **SolidWorks** warning dialog box is displayed.
13. Choose **OK** from the **SolidWorks** warning dialog box.
14. Choose the **Line** button from the **Sketch CommandManager**.
15. Move the line cursor to a location whose coordinates are close to 55 mm 10 mm 0 mm. The cursor snaps to the left arc.
16. Specify the start point of the line at this point and move the cursor horizontally toward the left. Specify the endpoint of the line when the line cursor snaps the outer circle. Double-click anywhere in the drawing area to end line creation.
17. Move the cursor to a location whose coordinates are close to -10 mm 10 mm 0 mm.
18. Specify the start point of the line on this location and move the cursor horizontally toward the left. Specify the endpoint when the length of the line above the line cursor shows a value close to 42.
19. Move the line cursor such that the line is drawn at an angle close to 105-degree. Specify the endpoint when the length of line is close to 15.

Note

*You may need to scroll down the **Line PropertyManager** to view the angle. Move the cursor to the scroll bar, press and hold the left mouse button down, and drag the cursor vertically downward to scroll down the **Line PropertyManager**.*

20. Move the line cursor horizontally toward the left. Specify the endpoint when the length of the line above the line cursor shows a value close to 34.
21. Move the line cursor vertically downward. Specify the endpoint when the length of the line above the line cursor shows a value close to 10.
22. Move the line cursor such that the line is drawn at an angle close to 346-degree. Specify the endpoint when the length of the line is around 27.

23. Move the line cursor vertically downward. Specify the endpoint when the line cursor snaps the mirror line. Double-click anywhere in the drawing area to end line creation.

The sketch after drawing the required arcs, circles, and lines is shown in Figure 3-62.

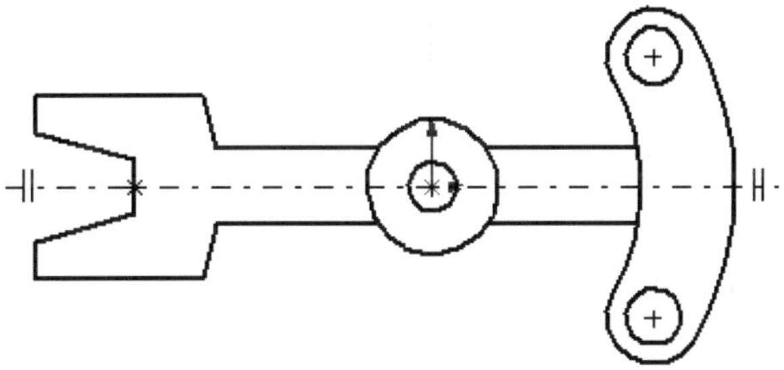

Figure 3-62 Sketch after drawing all entities

Trimming the unwanted entities

After drawing the sketch, you need to trim some of the unwanted sketched entities using the **Trim Entities** tool.

1. Choose the **Trim Entities** button from the **Sketch CommandManager** to display the **Trim PropertyManager**.

2. Choose the **Trim to closet** button from the **Options** rollout if it is not selected by default; the select cursor is replaced by the trim cursor.

3. Select the entities to be trimmed as shown in Figure 3-63; the entities are dynamically trimmed.

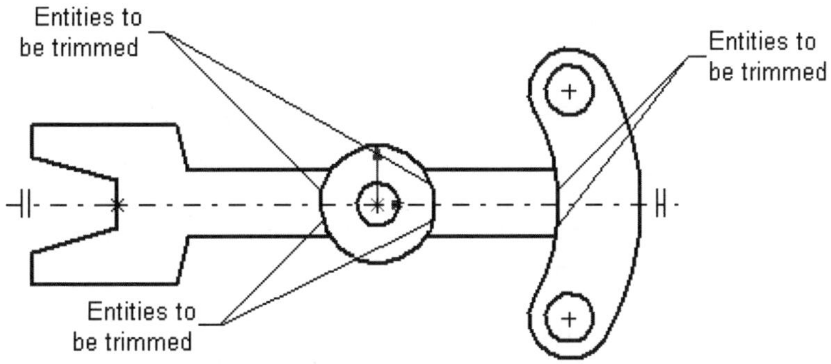

Figure 3-63 Entities to be trimmed

Filleting the Sketched Entities

Next, you need to fillet the sketched entities. The fillets are generally applied to avoid the stress concentration at sharp corners.

1. Choose the **Sketch Fillet** button from the **Sketch CommandManager**. Set the radius spinner to **5**.

2. Select the entities shown in Figure 3-64 to apply the fillet.

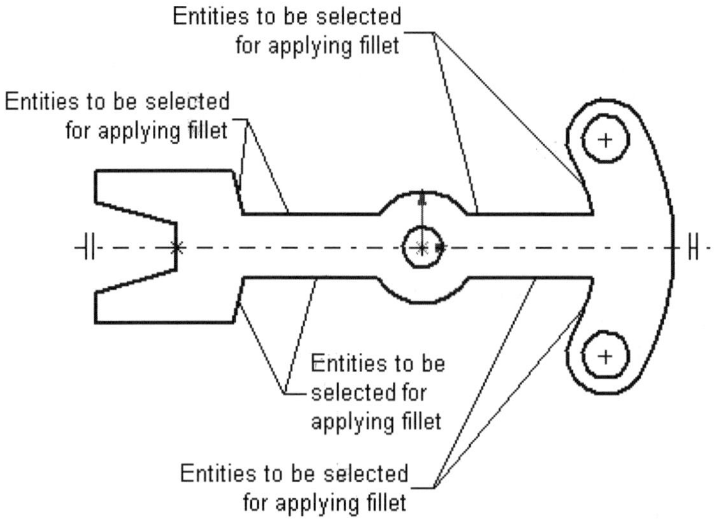

Figure 3-64 *Entities to be selected to apply fillet*

3. Choose the **OK** button from the **Sketch Fillet PropertyManager** to exit the **Fillet** tool. The sketch after applying the fillets to the sketch is shown in Figure 3-65.

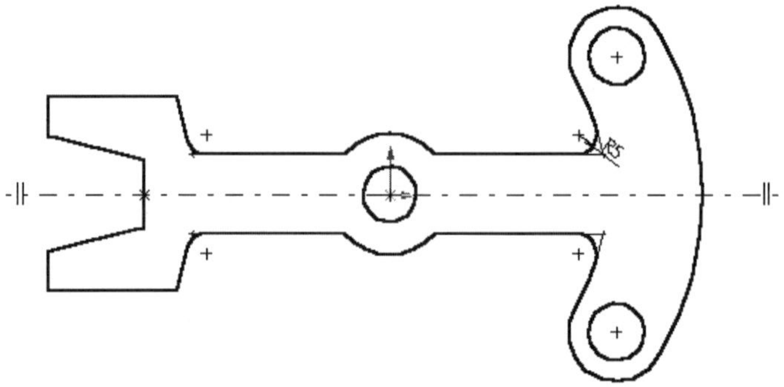

Figure 3-65 *Sketch after creating the fillets*

Adding Relations to the Sketch

Next, you need to add the required relations to the sketch.

1. Choose the **Add Relations** button from the **Sketch CommandManager** to invoke the **Add Relations PropertyManager**.

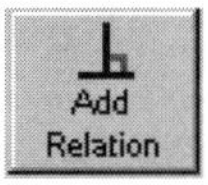

2. Select the lower left arc and the lower tangent arc; the **Tangent** button is highlighted in bold in the **Add Relations** rollout of the **Add Relations PropertyManager**. This suggests that the **Tangent** relation is the most appropriate relation for the selected entities.

3. Choose the **Tangent** button from the **Add Relations PropertyManager**.

4. Right-click in the drawing area and choose the **Clear Selections** option from the shortcut menu to clear the selections from the selection set. Select the right arc and the lower tangent arc; the **Tangent** button is highlighted.

5. Choose the **Tangent** button from the **Add Relations PropertyManager**. Clear the current selections from the selection set.

6. Select the upper right horizontal line and the upper left horizontal line that are coincident with the trimmed circle. Choose the **Collinear** button from the **Add Relations PropertyManager**.

7. Choose the **OK** button from the **Add Relations PropertyManager** or choose **OK** from the confirmation corner.

Adding Dimensions to the Sketch

Next, you will apply the required dimensions to the sketch and fully defined it.

1. Choose the **Smart Dimension** button from the **Sketch CommandManager** to invoke the dimension tool; the arrow cursor is replaced by the dimension cursor.

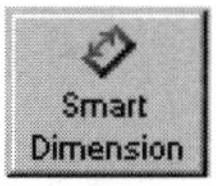

2. Select the right arc; a radius dimension is attached to the cursor. Move the cursor away from the sketch toward the right and place the dimension.

3. Enter a value of **82** in the **Modify** dialog box and press ENTER.

4. Select the left upper arc, refer to Figure 3-66; a radius dimension is attached to the cursor. Move the cursor away from the sketch toward the right and place the dimension.

5. Enter a value of **56** in the **Modify** dialog box and press ENTER.

6. Select the **Origin** and the center point of the upper circle; a dimension is attached to the cursor. Next, select the right endpoint of the mirror line; an angular dimension is attached to the cursor. Place the angular dimension outside the sketch.

7. Enter the angular dimension value of **30** in the **Modify** dialog box and press ENTER.

8. Select the upper right circle; a diameter dimension is attached to the cursor. Place the dimension outside the sketch.

9. Enter the diameter dimension value of **15** in the **Modify** dialog box and press ENTER.

10. Select the upper right horizontal line and the lower right horizontal line that coincide the trimmed circle, refer to Figure 3-66. A vertical dimension is attached to the cursor. Move the cursor vertically downward and click to place the dimension.

11. Enter a value of **20** in the **Modify** dialog box.

12. Select the smaller circle at the origin; a diameter dimension is attached to the cursor. Move the cursor upward and place the dimension outside the sketch.

13. Enter the diameter dimension value of **13** in the **Modify** dialog box.

14. Select the outer trimmed circle and place the radius dimension outside the sketch.

15. Enter the radial dimension value of **19** in the **Modify** dialog box.

16. Select the upper right inclined line; a dimension is attached to the cursor. Now, select the upper left horizontal line; an angular dimension is attached to the cursor. Place the dimension above the upper left horizontal line.

17. Enter the angular dimension value of **75** in the **Modify** dialog box.

18. Select the origin and the lower endpoint of the lower right inclined line. Move the cursor vertically downward and place the dimension.

19. Enter a value of **49** in the **Modify** dialog box.

20. Select the origin and the middle left vertical line, refer to Figure 3-66. Move the cursor vertically downward and place the dimension below the previous dimension.

21. Enter a value of **61** in the **Modify** dialog box.

22. Select the origin and the lower endpoint of the outer left vertical line. Move the cursor vertically downward and place the dimension below the last dimension.

23. Enter a value of **90** in the **Modify** dialog box.

24. Select the upper left inclined line and the lower left inclined line, refer to Figure 3-66; an angular dimension is attached to the cursor. Move the cursor horizontally toward the left and place the dimension.

25. Enter the angular dimension value of **28** in the **Modify** dialog box.

26. Select the upper left horizontal line and the lower left horizontal line; a linear dimension is attached to the cursor. Move the cursor horizontally toward the left and place the dimension.

27. Enter a value of **50** in the **Modify** dialog box.

28. Select the lower endpoint of the upper left vertical line and the upper endpoint of the lower left vertical line; a linear dimension is attached to the cursor. Move the cursor horizontally toward the left and left click to place the dimension.

29. Enter a value of **30** in the **Modify** dialog box.

    All sketched entities are displayed in black, suggesting that the sketch is fully defined. The fully defined sketch is shown in Figure 3-66.

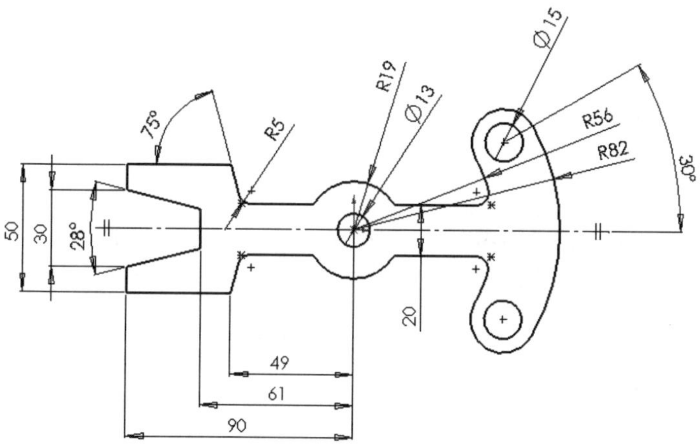

***Figure 3-66** Sketch after applying all relations and dimensions*

Saving the Sketch

1. Choose the **Save** button from the **Standard** toolbar to invoke the **Save As** dialog box.

2. Enter the name of the document as *c03tut2* in the **File name** edit box and choose the **Save** button.

3. Close the document by choosing **File > Close** from the menu bar.

Tutorial 3

In this tutorial, you will draw the sketch of a revolved model shown in Figure 3-67. The sketch is shown in Figure 3-68. The solid model is given only for your reference.

(Expected time: 30 min)

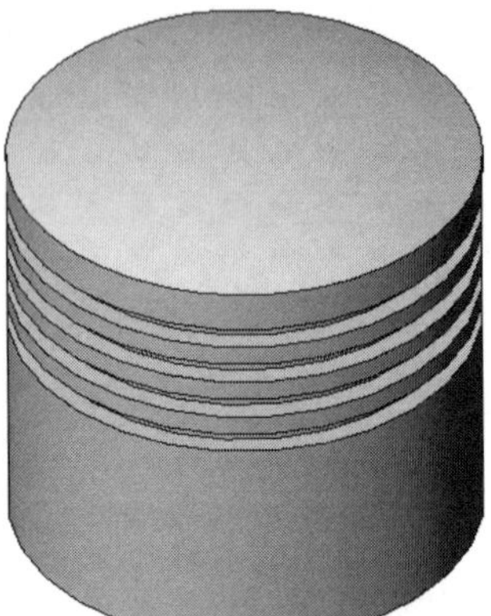

Figure 3-67 *Solid model of the piston*

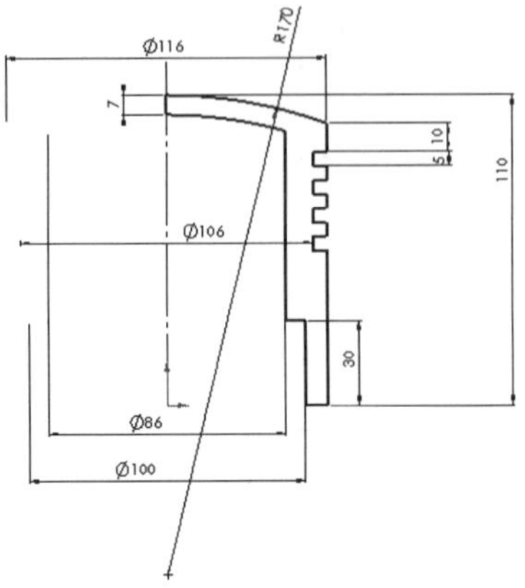

Figure 3-68 *The sketch of the base feature*

The steps that will be followed to complete this tutorial are listed below.

a. Start a new part document and then invoke the sketching environment.
b. Draw a centerline that will be used to add the linear diameter dimensions to the sketch of the piston.
c. Create the sketch using various sketching tools.
d. Use the **Offset Entities** tool to offset the required lines, refer to Figure 3-69.
e. Draw the arcs and trim the unwanted entities, refer Figure 3-70.
f. Add the required relations.
g. Add the required dimensions and fully define the sketch, refer Figure 3-71.

Starting a New Part Document

1. Choose the **New** button from the **Standard** toolbar and start a new part document using the **New SolidWorks Document** dialog box.

2. Choose the **Sketch** button from the **Standard** toolbar and select the **Front Plane** to invoke the sketching environment for drawing the sketch.

Drawing the Sketch

To draw the sketch of the revolved model, you will first need to draw the centerline around which the sketch of the base feature will be revolved.

1. Choose the **Centerline** button from the **Sketch CommandManager**.

2. Draw a vertical centerline starting from the origin and having a length of 120 mm.

   Next, you will draw the sketch of the piston.

3. Right-click in the drawing area and choose the **Line** option from the shortcut menu. Move the cursor to a location whose coordinates are close to 58 mm 0 mm 0 mm. Click at this point to specify the start point of the line.

4. Draw a vertical line of dimension close to 100. Right-click and choose the **End chain** option from the shortcut menu.

5. Move the cursor to the lower endpoint of the line drawn earlier. Specify the start point of the line when the red circle is displayed. Move the cursor horizontally toward the left and draw a horizontal line of dimension close to 8.

6. Move the line cursor vertically upward and draw a vertical line of dimension close to 30.

7. Move the cursor horizontally toward the left and draw a horizontal line of dimension close to 7.

8. Move the line cursor vertically upward and draw a vertical line of dimension close to 70.

9. Right-click and choose the **3 Point Arc** option from the shortcut menu. Move the cursor near the upper endpoint of the right vertical line.

10. Specify the first point of the arc when the red circle is displayed. Move the cursor horizontally toward the left; the reference arc is attached to the cursor. Specify the second point of the arc when the value of the length is close to 116.

11. Move the cursor toward the right. Specify the third point of the arc when the value of the radius is close to 170.

12. Right-click and choose the **Line** option from the shortcut menu. Move the line cursor to a location whose coordinates are close to 58 mm 90 mm 0 mm. Specify the start point of the line at this location and move the cursor horizontally toward the left to draw a line of dimension close to 5. Double-click anywhere in the drawing area to end line creation.

13. Choose the **Zoom To Fit** button from the **View** toolbar to fit the sketch to the drawing area.

Offsetting the Lines

Using the **Offset Entities** tool, you will offset the entities created earlier.

1. Select the line created previously using the select tool.

2. Choose the **Offset Entities** button from the **Sketch CommandManager** to invoke the **Offset Entities PropertyManager**. The confirmation corner is also displayed at the upper right corner of the drawing area.

3. Choose the **Keep Visible** button from the **Offset Entities PropertyManager** to pin the **PropertyManager**.

4. Set the value of **Offset Distance** spinner to **5**. Now, select the **Reverse** check box because you need to offset the entity in the reverse direction. The preview of the entity to be offset is modified in the drawing area.

5. Choose the **OK** button from the **Offset Entities PropertyManager**.

   You will notice that an entity is created at an offset distance of 5 from the original entity. Also, a dimension is applied between the newly created entity and the original entity with a value of **5**. This dimension is the offset distance between the two entities.

6. Select the newly created entity. Move the cursor vertically downward; the preview of the entity and the direction of offset creation are also displayed. Press the left mouse button to offset the selected line.

   Repeat this procedure of offsetting the entities until you get eight entities, including the original entity.

7. Set the value of the **Offset Distance** spinner to **7** and clear the **Select chain** check box. Then select the upper arc. The preview of the offset arc is shown in the drawing area.

8. Choose the **OK** button twice on the **Offset Entities PropertyManager**.

Completing the Remaining Sketch

Next, you will complete the remaining sketch using the line tool.

1. Right-click and choose the **Line** option from the shortcut menu. Move the line cursor to the left endpoint of the original line that was used to create offset lines. When the red circle is displayed, specify the start point. Specify the endpoint of the line at the endpoint of the last offset line.

2. Move the cursor to the intersection point of the upper arc and the centerline. When the cursor snaps the intersection, draw a vertical line that snaps the intersection point of the lower arc and the centerline.

   The sketch after drawing the entities using various sketch tools and the offset tool is shown in Figure 3-69.

Trimming the Unwanted Entities

Next, you will trim the unwanted entities using the trim tool.

1. Choose the **Trim Entities** button from the **Sketch CommandManager**.

2. Using the left mouse button trim the unwanted entities. The sketch after trimming the unwanted entities is shown in Figure 3-70.

Adding the Required Relations

Now, you will add the required relations to the sketched entities.

1. Right-click and choose the **Select** option from the shortcut menu. Press and hold the CTRL key down and select one of the endpoints of the lower horizontal line and then select the

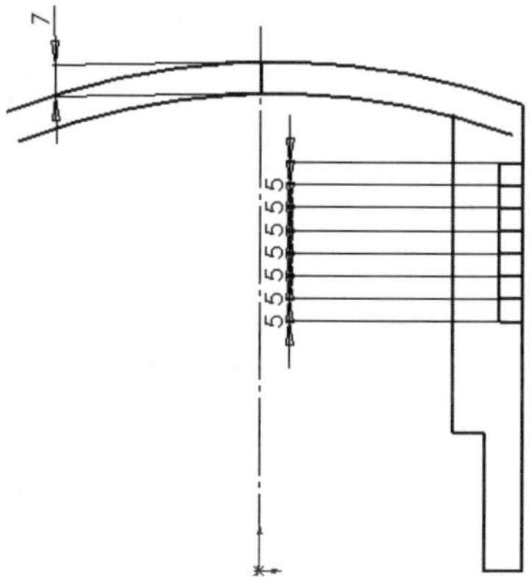

Figure 3-69 Sketch after creating various entities

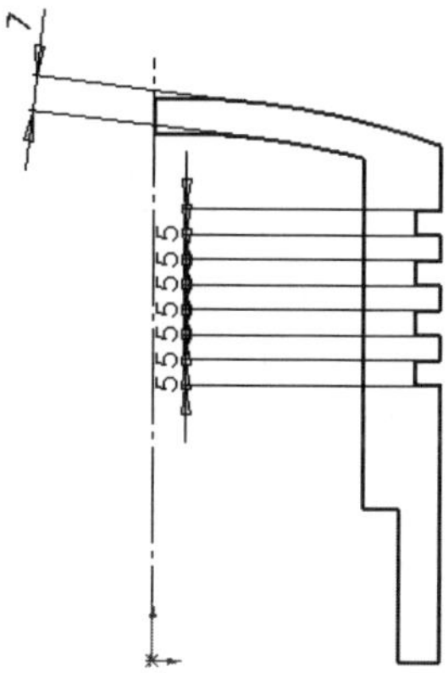

Figure 3-70 Sketch after trimming the unwanted entities

origin. Release the CTRL key after selection. Right-click and choose the **Horizontal** option from the shortcut menu to add the horizontal relation to the selected entities. Click anywhere in the drawing area to clear the selection set.

2. Press and hold the CTRL key down and select the small horizontal lines on the right of the sketch. Release the CTRL key after making the selection and right-click to display the shortcut menu. Choose the **Equal** option from the shortcut menu to apply the equal relation.

3. Similarly, apply the **Equal** relation to all small vertical lines.

   The **SolidWorks** warning message window is displayed. This warns you that the solution cannot be determined for the sketch. If the **Show property manager when the sketch becomes over defined or unsolved** check box is selected earlier in the **Display/Delete Relations Property Manager**, then this **PropertyManager** is displayed automatically and the warning message window is not displayed. You will notice that some of the entities of the sketch turn red. This indicates that the sketch is overdefined.

4. Choose **OK** from this message window or the **PropertyManager**.

   As discussed earlier, the overdefined sketch is not used to create any feature; therefore, you have to delete the conflicting relations. The sketch is overdefined after applying the previous relation. Therefore, if you delete the last applied relation using the **Undo** button from the **Standard** toolbar, then the sketch is not overdefined.

5. Choose the **Undo** button from the **Standard** toolbar. Click anywhere in the drawing area to clear the selections from the selection set.

   As evident from the color of the entities, which is blue, you still have to add some more relations or dimensions to fully define the sketch. The fully defined sketch is displayed in black.

6. Zoom out using the **Zoom In/Out** tool and apply the coincident relation between the center point of the upper arc and the centerline.

Adding the Dimensions to the Sketch

After drawing, editing, and applying the relations to the sketch, you will add the required dimensions to the sketch to fully define the sketch.

1. Select the dimension with a value of **7**, which is placed between the upper arcs and press the DELETE key from the keyboard.

   Delete this dimension because during the design and manufacturing practices the dimension between the tangents should be avoided.

2. Right-click and choose the **Smart Dimension** option from the shortcut menu to invoke this tool.

3. Select the outer upper right vertical line; a dimension is attached to the cursor. Now, select the centerline and move the cursor to the other side of the centerline. You will notice that the diameter dimension is displayed along the cursor.

4. Click to place the dimension above the sketch and enter a value of **116** in the **Modify** edit box and press ENTER.

5. Now, select the inner left vertical line and then select the centerline, refer to Figure 3-71. Move the cursor to the other side of the centerline and click to place the dimension below the sketch.

6. Enter a value of **86** in the **Modify** edit box and press ENTER.

   Add the remaining dimensions to fully define the sketch, refer to Figure 3-71.

7. Using the left mouse button, select one of the dimensions that are created after offsetting the entities and drag the cursor toward the right and using the left mouse button place the dimension at an appropriate place.

8. Arrange all dimensions using the above method. The fully defined sketch is shown in Figure 3-71. In this figure, the display of relations is turned off.

Saving the Sketch

1. Choose the **Save** button from the **Standard** toolbar and save the sketch with the name given below.

   \My Documents\SolidWorks\c03\c03tut3.sldprt.

2. Choose **File > Close** from the menu bar to close the document.

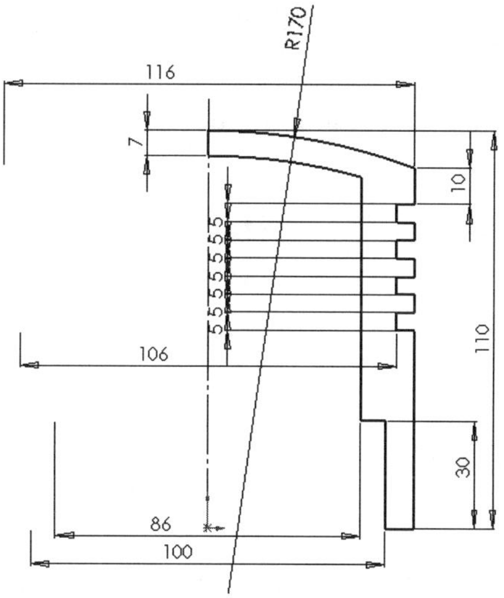

***Figure 3-71** Fully defined sketch after applying all the relations and dimensions*

SELF-EVALUATION TEST

Answer the following questions and then compare your answers with those given at the end of this chapter.

1. Some relations are automatically applied to the sketch while it is being drawn. (T/F)

2. When you choose the **Add Relations** button, the **Apply Relations PropertyManager** is displayed. (T/F)

3. You can modify the arrowhead style of a selected dimension. (T/F)

4. The dimension favorite created in one document cannot be retrieved in the other document. (T/F)

5. You can do modifications in the view-only file. (T/F)

6. The __________ **PropertyManager** is displayed using the **Add Relation** button from the **Sketch CommandManager**.

7. The __________ dimension is used to dimension a line that is at an angle with respect to the X axis or the Y axis.

8. A __________ defined sketch is the one in which all entities and their positions are described by the relations or dimensions, or both.

9. The __________ dimensions or relations are not able to determine the position of one or more sketched entities.

10. The __________ option is displayed in the **Defined In** column when the entity is defined as placed in the same sketch.

REVIEW QUESTIONS

Answer the following questions.

1. You can invoke the **Display/Delete Relations PropertyManager** using the **Display/Delete Relations** button from __________ **CommandManager**.

2. The linear diameter dimensions are applied to the sketches of __________ features.

3. You can modify the dimension to display minimum or maximum distance between two circles using the __________ dialog box.

4. The __________ sketch geometry is constrained by too many dimensions and/or relations. Therefore, you have to delete the extra and conflicting relations or dimensions.

5. The __________ relation forces two selected lines, arcs, points, or ellipses to remain equidistant from a centerline.

6. In SolidWorks, by default, the dimensioning between two arcs, two circles, or between an arc and a circle is done from

   (a) Center point to center point (b) Center point to tangent
   (c) Depends on the point where you select (d) None

7. Which relation forces the selected arc to share the same center point with another arc or a point?

   (a) **Concentric** (b) **Coradial**
   (c) **Merge points** (d) **Equal**

8. Which type of dimensions change the color of the entities to red?

   (a) Underdefined (b) Overdefined
   (c) Dangling (d) None

9. Which dialog box is displayed when you modify a dimension?

   (a) **Modify Dimensional Value** (b) **Insert a value**
   (c) **Modify** (d) None

10. Which dialog box is displayed when you add an extra dimension to a sketch or add an extra relation that overdefines the sketch?

   (a) **Over defining** (b) **Delete relation**
   (c) **Make Dimension Driven?** (c) **Add Geometric Relations**

EXERCISES

Exercise 1

Create the sketch of the model shown in Figure 3-72. Apply the required relations and dimensions to it and fully define it. The sketch is shown in Figure 3-73. The solid model is given only for reference. **(Expected time: 30 min)**

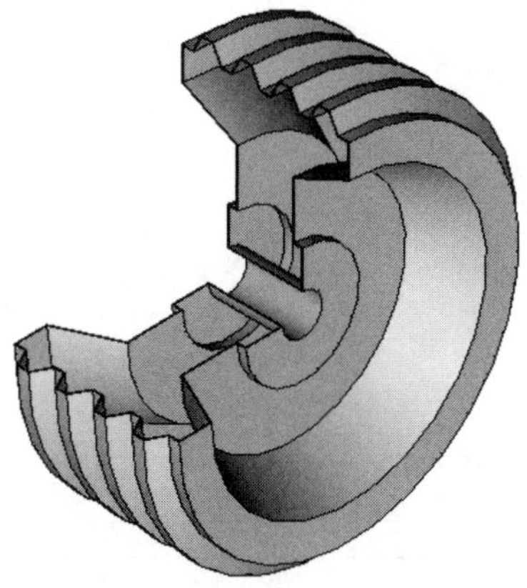

Figure 3-72 Solid model for Exercise 1

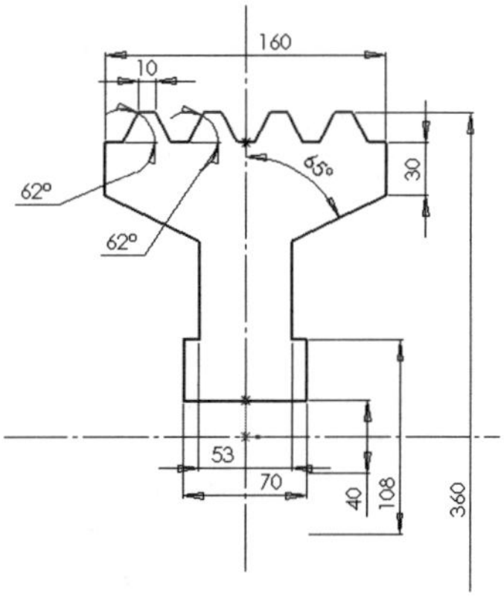

Figure 3-73 Sketch for Exercise 1

Exercise 2

Create the sketch of the model shown in Figure 3-74. Apply the required relations and dimensions to it and fully define it. The sketch is shown in Figure 3-75. The solid model is given only for reference. **(Expected time: 30 min)**

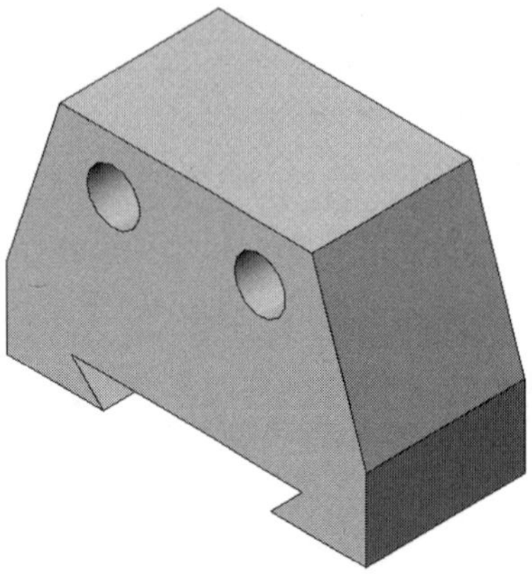

Figure 3-74 Solid model for Exercise 2

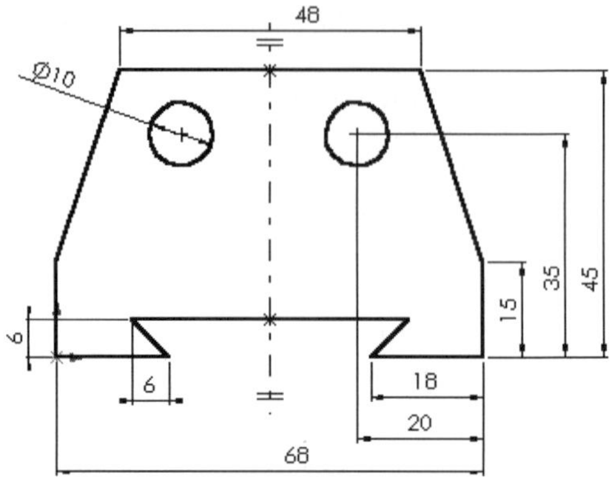

Figure 3-75 Sketch for Exercise 2

Exercise 3

Create the sketch of the model shown in Figure 3-76. Apply the required relations and dimensions to it and fully define it. The sketch is shown in Figure 3-77. The solid model is given only for reference. **(Expected time: 30 min)**

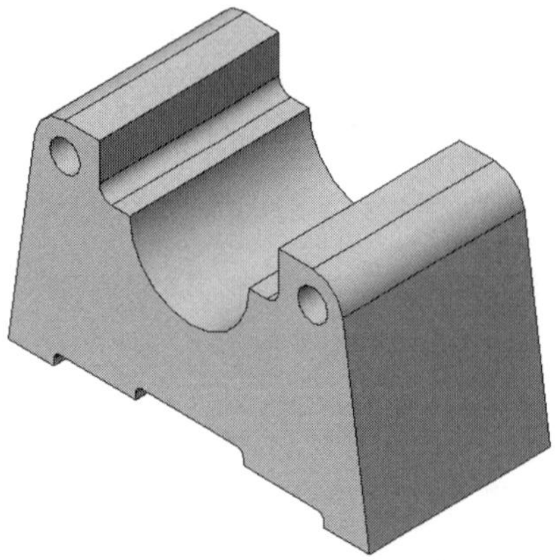

Figure 3-76 Solid model for Exercise 3

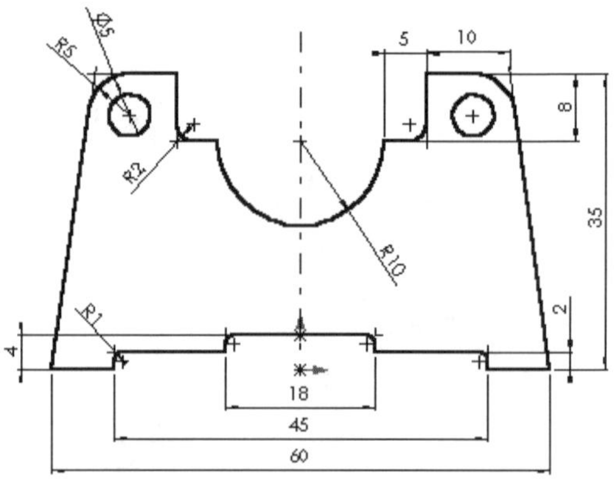

Figure 3-77 Sketch for Exercise 3

Answers to Self-Evaluation Test

1. T, **2.** F, **3.** T, **4.** F, **5.** T, **6. Add Relations**, **7.** aligned, **8.** fully defined, **9. Dangling**, **10. All in this sketch**

Chapter 4

Advance Dimensioning Techniques and Base Feature Options

Learning Objectives

After completing this chapter you will be able to:

- *Dimension the sketch using the Autodimension Sketch tool.*
- *Dimension the true length of an arc.*
- *Measure Distances and View Section Properties*
- *Create solid base extruded features.*
- *Create thin base extruded features.*
- *Create solid base revolved features.*
- *Create thin base revolved features.*
- *Dynamically rotate the view to display the model from all directions.*
- *Modify the orientation of the view.*
- *Change the display modes of the solid model.*
- *Apply materials and textures to the models.*

ADVANCE DIMENSIONING TECHNIQUES

In this chapter, you will learn about some of the advance dimensioning techniques used in dimensioning the sketches in SolidWorks. With this release of SolidWorks, you will be able to apply all possible dimensions to a sketch using a single option, which is known as **Autodimension Sketch**. The advanced dimensioning techniques are discussed next.

Autodimension the Sketches

CommandManager:	Dimension/Relations > Autodimension	*(Customize to Add)*
Menu:	Tools > Dimensions > Autodimension Sketch	
Toolbar:	Dimension/Relations > Autodimension	*(Customize to Add)*

The **Autodimension** tool is used to automatically apply the dimensions to the sketch. To apply autodimensions to a sketch, draw the sketch using standard sketching tools and then apply the required relations to the sketch. Now, choose the **Autodimension** button from the **Dimension/Relations CommandManager** to display the **Autodimension PropertyManager** as shown in Figure 4-1. The options available in the **Autodimension Sketch PropertyManager** are discussed next.

Entities to Dimension

The **Entities to Dimension** rollout is used to specify the entities to which the dimensions have to be applied. The **All entities in sketch** radio button is selected by default. As a result, all entities drawn in the current sketching environment are selected to apply the dimensions. The **Selected entities** radio button is selected if you have to dimension only the selected entities. When you select this radio button, the **Selected Entities to Dimension** selection area will be displayed in the **Entities to Dimension** rollout. Select the entities to dimension using the select cursor. The names of the selected entities will be displayed in the **Selected Entities to Dimension** selection area. If you select one or more entities before invoking the **Autodimension sketch PropertyManager**, the **Selected entities** radio button will be selected by default. Also, the names of the selected entities will be displayed in the **Selected Entities to Dimension** selection area.

Horizontal Dimensions

The **Horizontal Dimensions** rollout is used to specify the type of horizontal dimension, reference for the horizontal dimension, and the dimension placement. The options available in the **Horizontal Dimensions** rollout are discussed next.

Scheme

The **Scheme** drop-down list is used to specify the dimensioning scheme to be applied to the sketch. Various types of dimensioning schemes available in this drop-down list are discussed next.

Chain

The **Chain** option is used for the relative or incremental horizontal dimensioning of the sketch. When you invoke the **Autodimension PropertyManager** and select this scheme, a point or a vertical line will be selected as the reference entity. This reference entity is used as a datum for the generation of dimension. The name of the selected

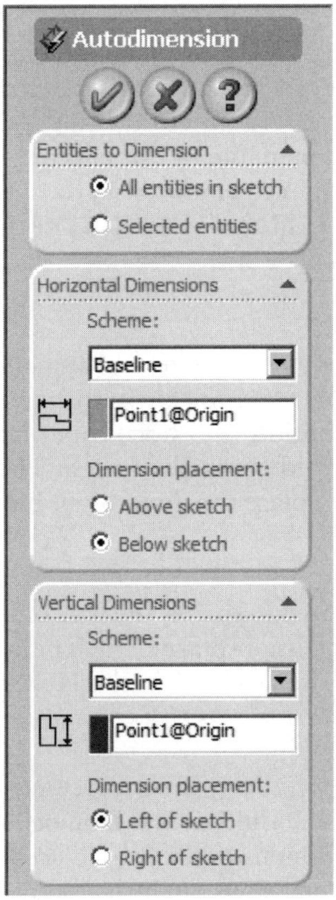

Figure 4-1** The **Autodimension PropertyManager

reference entity is displayed in the **Datums - Vertical Model Edge, Model Vertex, Vertical Line or Point** selection area and the reference entity is displayed in pink in the drawing area. You can also specify a user-defined reference entity.

Note

Chain dimensioning should be avoided if the tolerances relative to a common datum are required in the part.

Baseline

The **Baseline** option is used for absolute or datum vertical dimensioning of the sketch. In this dimensioning method, the dimensions are applied to the sketch with respect to the common datum. When you invoke the **Autodimension PropertyManager** and select this option, a point or a vertical line will be selected as the reference entity, which is used as a datum for the generation of dimension. The name of the selected reference entity will be displayed in the **Datums - Vertical Model Edge, Model Vertex, Vertical Line or Point** selection area and the reference entity will be displayed in pink in the drawing area. You can also specify a user-defined reference entity.

Ordinate

The **Ordinate** option is used for the ordinate dimensioning of the sketch. When you invoke the **Autodimension PropertyManager** and select this option, a point or a vertical line will be selected as the reference entity, which will be used as a datum for the generation of dimension. The name of the selected reference entity is displayed in the **Datums - Vertical Model Edge, Model Vertex, Vertical Line or Point** selection area and the reference entity will be displayed in pink in the drawing area. You can also specify a user-defined reference entity.

Dimension placement

The **Dimension placement** area is used to define the position where the generated dimensions will be placed. Two radio buttons are available in this area. The first is the **Above sketch** radio button. If you use this option, the horizontal dimensions generated using the **Autodimension** tool will be placed above the sketch. The **Below sketch** radio button is selected by default and is used to place the dimensions below the sketch.

Vertical Dimensions

The **Vertical Dimensions** rollout is used to specify the type of vertical dimension, reference for the vertical dimension, and the dimension placement. The options available in the **Scheme** area of this rollout are similar to those available in the **Horizontal Dimensions** rollout. The remaining option is discussed next.

Dimension placement

The **Dimension placement** area of the **Vertical Dimensions** rollout is used to define the position where the generated dimensions will be placed. The **Left of the sketch** radio button is selected to place the dimensions on the left of the sketch. The **Right of the sketch** radio button is selected to place the dimensions on the right of the sketch. The **Left of the sketch** radio button is selected by default.

After specifying all parameters in the **Autodimension Sketch PropertyManager**, choose the **OK** button or choose the **OK** icon from the **Confirmation Corner**. The dimension will be applied to the sketch with the selected dimension scheme. Figure 4-2 shows the autodimension created using the **Baseline** scheme taking the origin as the datum. Figure 4-3 shows the autodimension created using the **Ordinate** scheme taking the origin as the datum.

Dimensioning of the True Length of an Arc

In SolidWorks, you can also apply the dimension of the true length of an arc, which is one of the advantages of the sketching environment of SolidWorks. To apply the dimension of the true length, invoke the **Smart Dimension** tool and select the arc using the dimension cursor. A radial dimension will be attached to the cursor. Move the cursor to any of the endpoints of the arc. When the cursor snaps the endpoint, use the left mouse button to specify the first endpoint of the arc. A linear dimension will be attached to the cursor; move the cursor to the second endpoint of the arc and when the cursor snaps the endpoint, select it. A dimension will be attached to the cursor. Move the cursor to an appropriate place to place the dimension. The dimension of the true length of the arc is shown in Figure 4-4.

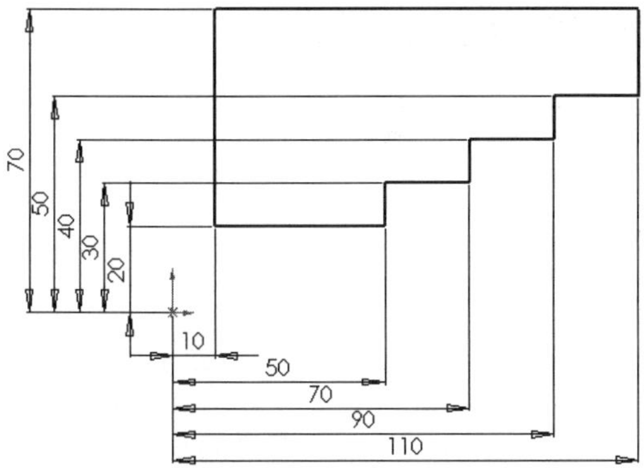

__Figure 4-2__ Baseline dimension created using the __Autodimension__ tool

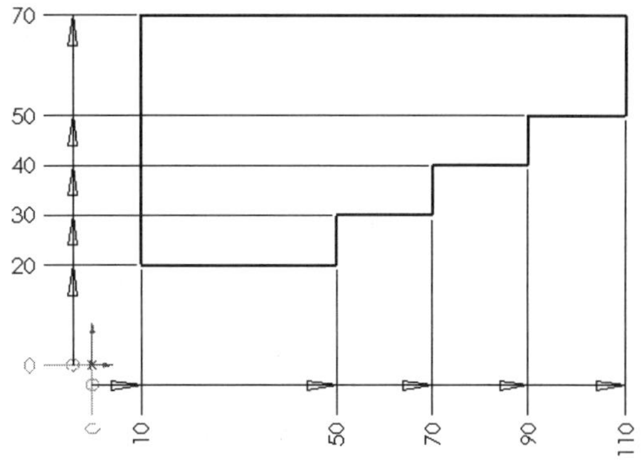

__Figure 4-3__ Ordinate dimension created using the __Autodimension__ tool

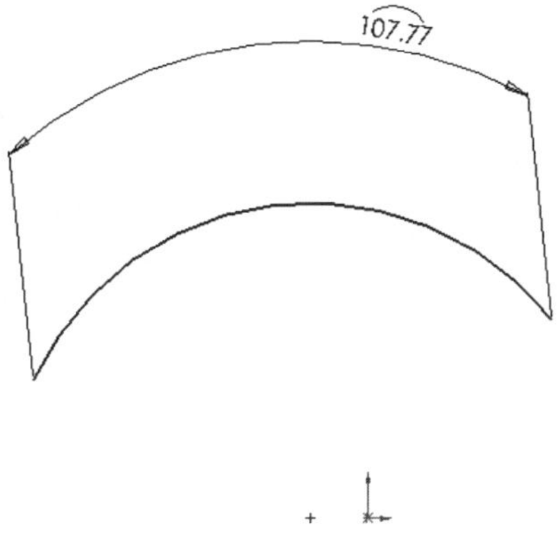

__Figure 4-4__ Dimensioning the true length of an arc

MEASURING DISTANCES AND VIEWING SECTION PROPERTIES

In SolidWorks, you can measure the distance of the entities and also view the section properties. These tools are discussed next.

Measuring Distances

CommandManager:	Tools > Measure
Menu:	Tools > Measure
Toolbar:	Tools > Measure

The **Measure** tool is used to measure the perimeter, angle, radius, and distance between lines, points, surfaces, and planes in sketches, 3D models, assemblies, or drawings. To use the measure tool, invoke the **Measure** toolbar by choosing the **Measure** button from the **Tools CommandManager**; the **Measure** toolbar will be displayed. The name of the document in which you are working will be displayed at the top of the **Measure** toolbar, refer to Figure 4-5. The current cursor will be replaced by the measure

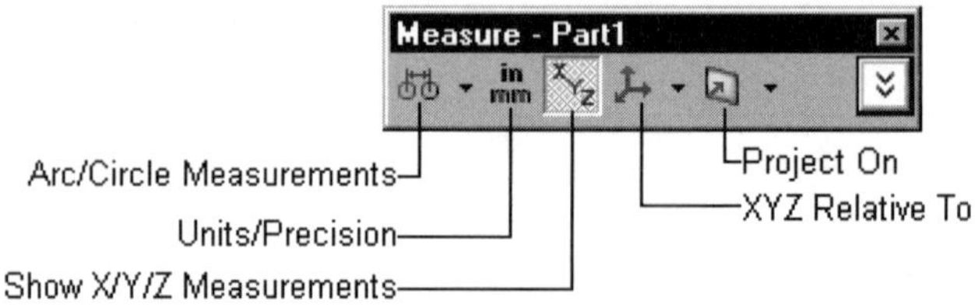

*Figure 4-5 The **Measure** toolbar*

cursor. Using the measure cursor, select the entity or entities to be measured. The result related to the selected element or elements will be displayed in the callout or callouts attached to the selected entity or entities. You can also view the result in the **Measure** toolbar by expanding it. To expand the toolbar, choose the button with down arrows on the right of the toolbar. The options in the **Measure** dialog box are discussed next.

Arc/Circle Measurements

The **Arc/Circle Measurements** button is used to specify the technique of measuring the distance between selected arcs or circles. When you choose the **Arc/Circle Measurements** button a flyout will appear. The **Center to Center** option will be selected by default in this flyout. With this option selected the center-to-center distance will be measured when you select two arcs or circles. If you choose the **Minimum Distance** option from this cascading menu, the minimum distance between the selected arcs or circles will be measured, which is the minimum tangential distance between two arcs or circles. However, if you choose the **Maximum Distance** option from this cascading menu, the maximum tangential distance between the selected arcs or circles will be measured.

Units/Precision

The **Units/Precision** button is used to set the type of units and their precision. To set the type and the precision of units, choose this button; the **Measure Units/Precision** dialog box will be displayed as shown in Figure 4-6. The **Use document settings** radio button is selected by default in this dialog box. The default units and precision of the document are used while measuring the entities. You can also set the type of units and their precision for measuring the entities. To do so, select the **Use custom settings** radio button from this dialog box. Other options available in this dialog box will be invoked. These options are discussed next.

Length unit Area

The **Length unit** area is used to set the units and options of linear measurements of the entities. The **Unit** drop-down list is provided at the top right corner of the **Length unit** area. In this drop-down list you can select any type of unit such as **Angstroms, Nanometers, Microns, Millimeters**, **Centimeters**, **Meters**, **Microinches, Miles, Inches**, **Feet**, and **Feet and Inches**. The **Decimal places** spinner is provided to control the decimal places. The other options in the **Length unit** area are discussed next.

Decimal

The **Decimal** radio button is available only when you select the units as **Microinches, Miles, Inches**, or **Feet and Inches** from the **Unit** drop-down list. This radio button is selected to display the dimension in the decimal form. You can also specify the decimal

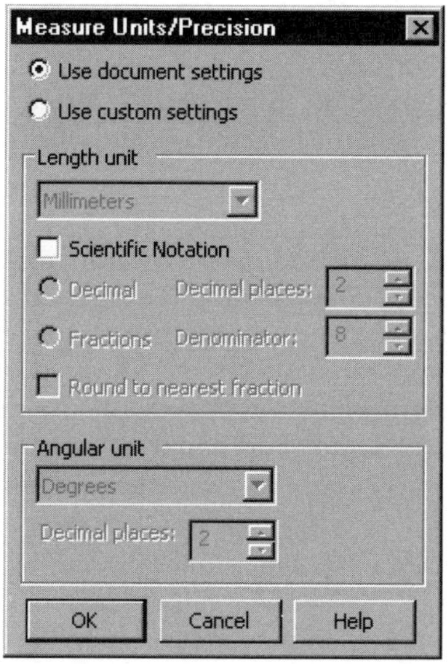

***Figure 4-6** The **Measurement Units/ Precision** dialog box*

places using the **Decimal places** spinner provided on the right of the **Decimal** radio button.

Fractions

The **Fractions** radio button is available only when you select the units as **Microinches, Miles, Inches**, or **Feet and Inches** from the **Units** drop-down list. This radio button is selected to display the dimension in the fraction form. You can also set the value of the denominator using the **Denominator** spin box provided on the right of the **Fractions** radio button.

Round to nearest fraction

The **Round to nearest fraction** check box is selected to display the value in fractions by rounding the value to the nearest fraction.

Scientific Notation

The **Scientific Notation** check box is selected to display the value in the scientific notation units.

Angular unit Area

The **Angular unit** area is used to set the units for angular measurement. This area is provided with a drop-down list to specify the angular measurement units such as **Degrees**, **Deg/Min**, **Deg/Min/Sec**, and **Radians**. The **Decimal places** spinner is provided to specify the decimal places.

Show XYZ Measurement

The **Show XYZ measurement** button is selected by default in the **Measure** toolbar. This button when chosen will show the dx, dy, and dz, values of the selected entities. If you clear this button, only the minimum distance between the selected entities will be displayed.

XYZ Relative To

The **XYZ Relative To** button is used to define the coordinate system along which the selected entities will be measured. By default the part origin is selected as the coordinate system. If you want to select any other coordinate system that you have created earlier, choose the **XYZ Relative To** button from the **Measure** toolbar. A flyout and all the coordinate systems that you have created along with the part origin will be displayed. Choose the coordinate system from this flyout. You will learn more about creating additional coordinate systems in the later chapters.

Projected On

The **Projected On** button is used to specify the location where the selected entity should be projected. You can project the selected entity on the screen or on a specific plane. The system will then calculate the measurement of the true projection. To specify the location choose the **Projected On** button from the **Measure** toolbar; a callout will be displayed and you can select the location where you need to project the selected entity.

Determining Section Properties of Closed Sketches

CommandManager:	Tools > Section Properties
Menu:	Tools > Section Properties
Toolbar:	Tools > Section Properties

The **Section Properties** tool enables you to determine the section properties of the sketch in the sketching environment or of the selected planar face in the **Part** mode and the **Assembly** mode. Remember that the section properties of only the closed sketches with nonintersecting closed loops can be determined. The section properties include the area, centroid relative to sketch origin, centroid relative to part origin, moment of inertia, polar moment of inertia, angle between the principle axes and sketch axes, and principle moment of inertia.

To calculate the section properties, draw the sketch and then choose the **Section Properties** button from the **Tools CommandManager**; the **Section Properties** dialog box will be displayed as shown in Figure 4-7.

When you invoke the **Section Properties** dialog box, a 3D triad will be placed at the centroid of the sketch. The section properties of the sketch are displayed in the **Section Properties** dialog box. The **Selected Items** display box is used to display the name of the selected planar face or sketch whose section properties are to be calculated. When you are in the **Part** mode, select the face to calculate the section properties and choose the **Recalculate** button to display the properties. If you want to calculate the section properties of some other face, clear the previously selected face from the selection set. Now, select the new face and choose the **Recalculate** button.

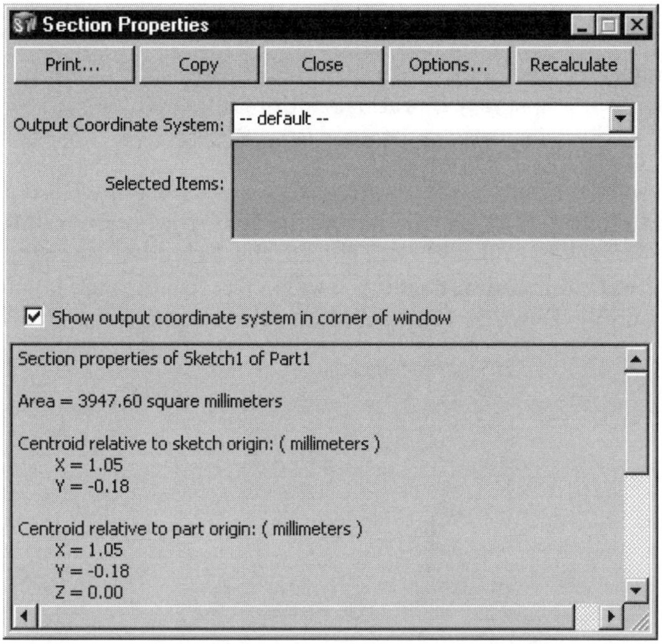

Figure 4-7 *Resized view of the* ***Section Properties*** *dialog box*

The **Print** button available in the **Section Properties** dialog box is used to print the section properties. The **Copy** button is chosen to copy the section properties to the Clip board from where you can copy them to a program such as MS Word. The rest of the options in this dialog box are similar to those discussed in the previous section.

CREATING BASE FEATURES BY EXTRUDING THE SKETCHES

The sketches that you have drawn until now can be converted into base features by extruding using the **Extruded Boss/Base** tool. This tool is available in the **Features CommandManager**. After drawing the sketch, choose the **Features** button from the **CommandManager** to display the **Features CommandManager**. From the **Features CommandManager**, choose the **Extruded Boss/Base** button. The sketching environment will be closed and the part modeling environment will be invoked. Also, the preview of the feature that will be created using the default options will be displayed in trimetric view. The trimetric view gives a better display of the solid feature.

On the basis of the options and the sketch selected for extruding, the resulting feature can be a solid feature or a thin feature. If the sketch is closed, it can be converted into a solid feature or a thin feature. However, if the sketch is open, it can be converted only into a thin feature. The solid and thin features are discussed next.

Creating Solid Extruded Features

CommandManager:	Features > Extruded Boss/Base
Menu:	Insert > Boss/Base > Extrude
Toolbar:	Features > Extruded Boss/Base

After you have completed drawing and dimensioning the closed sketch and converted it into a fully defined sketch, invoke the **Features CommandManager** by choosing the **Features** button and then choose the **Extruded Boss/Base** button from the **Features CommandManager**. You will notice that the view is automatically changed to trimetric view and the **Extrude PropertyManager** is displayed, as shown in Figure 4-8.

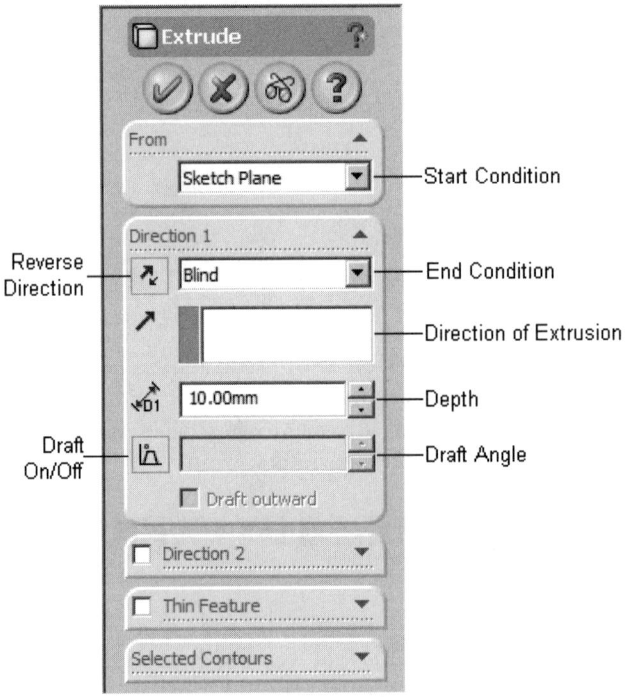

Figure 4-8** The **Extrude PropertyManager

You will notice that the preview of the base feature will be displayed in temporary graphics and an arrow will appear on the sketch. The arrow will appear in front of the sketch and is transparent. Figure 4-9 shows the preview of the sketch being extruded. Note that if the sketch consists of some closed loops inside the outer loop, they will be automatically subtracted from the outer loop while extruding. The options available in the **Extrude PropertyManager** are discussed next.

Direction 1

The **Direction 1** rollout is used to specify the end condition for extruding the sketch in one direction from the sketch plane. The options available in the **Direction 1** drop-down list are discussed next.

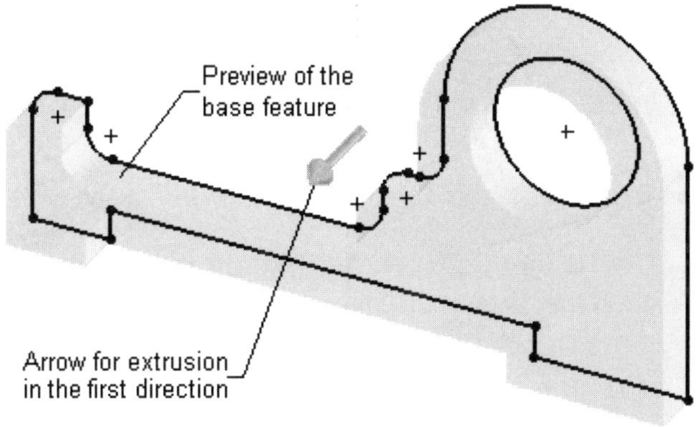

Figure 4-9 *Preview of the feature being extruded*

Tip. *You can also extrude an underdefined or an overdefined sketch. However, if you extrude an underdefined sketch, a - sign will be displayed on the left of the sketch in the* ***Feature Manager****. Similarly, if you extrude an overdefined sketch, you will find a + sign on the left of the sketch in the* ***Feature Manager****. To check these signs, click on the + sign available on the left of* ***Extrude 1*** *in the* ***Feature Manager****. The sketch will be displayed and you can see the + or the - sign.*

End Condition

The **End Condition** drop-down list provides the options to define the termination of the extruded feature. Note that because this is the first feature, some of the options available in this drop-down list will not be used at this stage. Also, some additional options will be available later in this drop-down list. The options available to define the termination of the base feature are discussed next.

Blind

The **Blind** option is selected by default and is used to define the termination of the extruded base feature by specifying the depth of extrusion. The depth of extrusion can be specified in the **Depth** spinner, which will be displayed in this rollout when you select the **Blind** option. You can reverse the extrusion direction by selecting the **Reverse Direction** button provided on the left of this drop-down list. Figure 4-10 shows the preview of the feature being created by extruding the sketch using the **Blind** option.

You can also extrude a sketch to a blind depth by dynamically dragging the feature using the mouse. Invoke the **Extrude PropertyManager** and move the mouse to the transparent arrow and when the color of the arrow changes to red, press the left mouse button. Now, move the cursor to specify the depth of extrusion; the **Blind** callout will also be displayed below the cursor. The value of the depth of extrusion will change dynamically in this callout as you move the cursor. Left-click to specify the termination of the extruded feature. The preview of the sketch being dragged is shown in Figure 4-11. The select cursor will be replaced by the mouse cursor. Use the right mouse button to

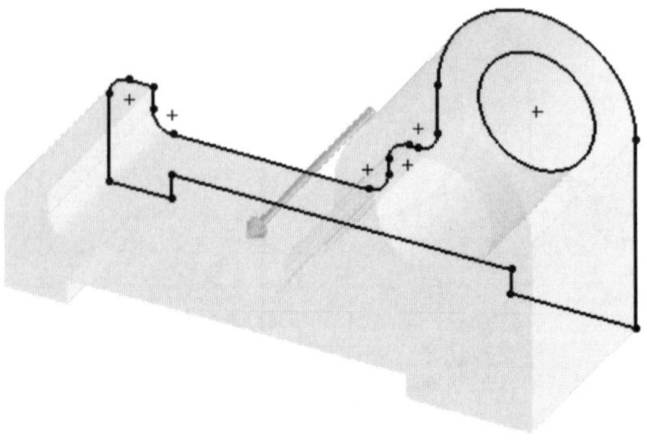

__Figure 4-10__ Preview of the feature being extruded using the __Blind__ option

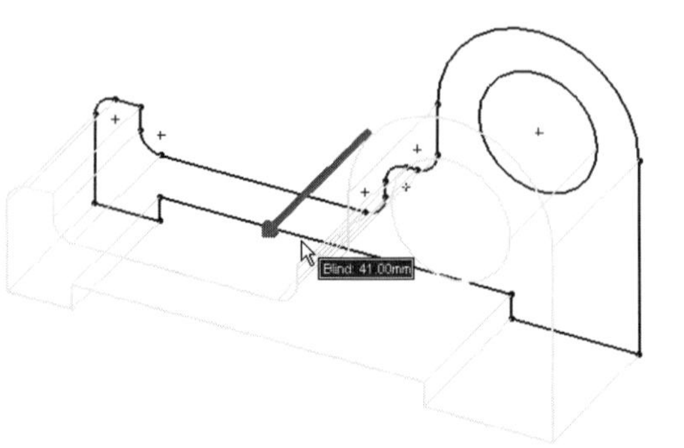

__Figure 4-11__ Preview of the feature being extruded by dynamically dragging

complete the feature creation or choose the **OK** button from the **Extrude PropertyManager**.

Mid Plane

The **Mid Plane** option is used to create the base feature by extruding the sketch equally in both directions of the plane on which the sketch is drawn. For example, if the total depth of the extruded feature is 30 mm, it will be extruded 15 mm toward the front of the plane and 15 mm toward the back. The depth of the feature can be defined in the **Depth** spinner, which is displayed below this drop-down list.

Figure 4-12 shows the preview of the feature being created by extruding the sketch using the **Mid Plane** option.

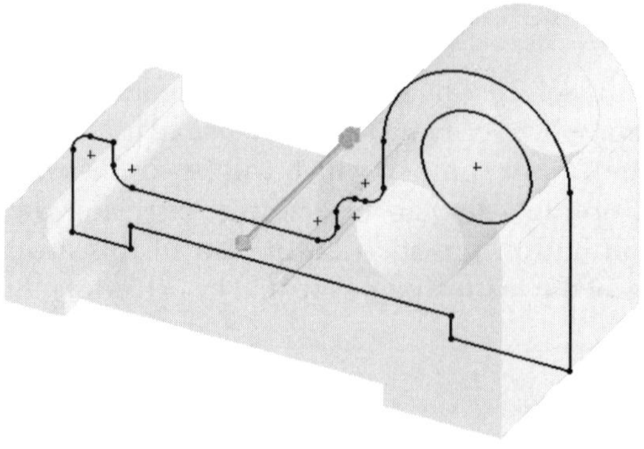

__Figure 4-12__ Preview of the feature being extruded using the __Mid Plane__ option

Draft On/Off

The **Draft On/Off** button is used to specify a draft angle while extruding the sketch. Applying the draft angle will taper the resulting feature. This button is not chosen by default. Therefore, the resulting base feature will not have any taper. However, if you want to add a draft angle to the feature, choose this button; the **Draft Angle** spinner and the **Draft outward** check box will be available. You can enter the draft angle for the feature in the **Draft Angle** spinner. By default, the feature will be tapered inward, as shown in Figure 4-13.

If you want to taper the feature outward, select the **Draft outward** check box, which is displayed below the **Draft Angle** spinner. The feature created with outward taper is shown in Figure 4-14.

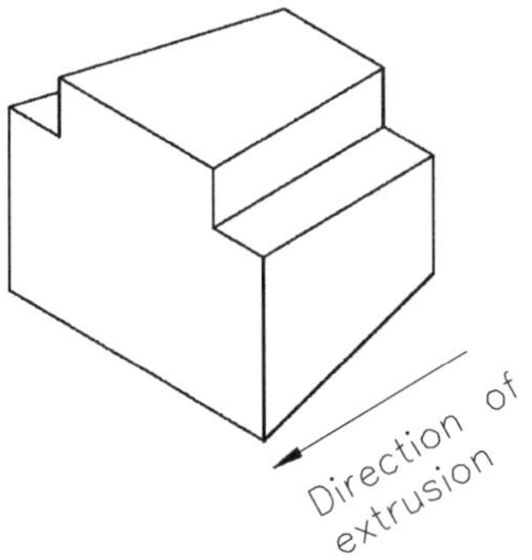

Figure 4-13 *Feature created with outward draft*

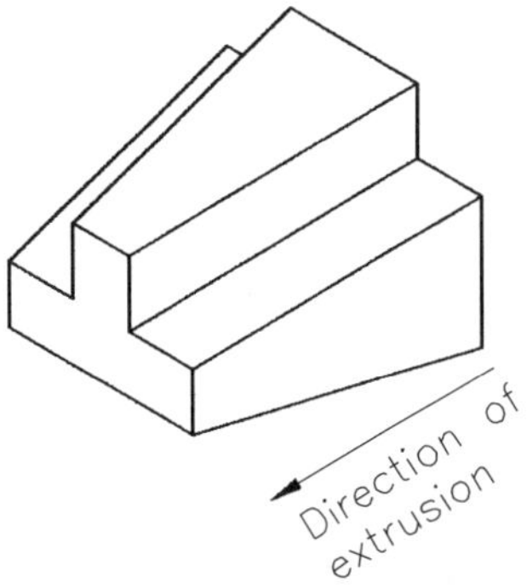

Figure 4-14 *Feature created with inward draft*

Note
*The **Direction of Extrusion** area will be discussed in later chapters.*

Tip. *You can also select the termination options using the shortcut menu that is displayed when you right-click in the drawing area.*

Direction 2

The **Direction 2** check box is selected to invoke the **Direction 2** rollout. This rollout is used to extrude the sketch with different values in the second direction of the sketching plane. This check box will not be available if you select the **Mid Plane** termination type.

Tip. *As soon as you select the **Direction 2** check box, another arrow will be displayed in the preview of the feature. You can use this arrow to define the depth of extrusion in the second direction.*

The options available in this rollout are similar to those in the **Direction 1** rollout. Note that unlike the **Mid Plane** termination option, the depth of extrusion and other parameters in both directions can be different. For example, you can extrude the sketch to a blind depth of 10 mm and an inward draft of 35-degree in front of the sketching plane and to a blind depth of 15 mm and an outward draft of 0-degree behind the sketching plane as shown in Figure 4-15.

After setting the values for both directions, choose the **OK** button or choose the **OK** icon from the confirmation corner. The feature will be created with the defined values.

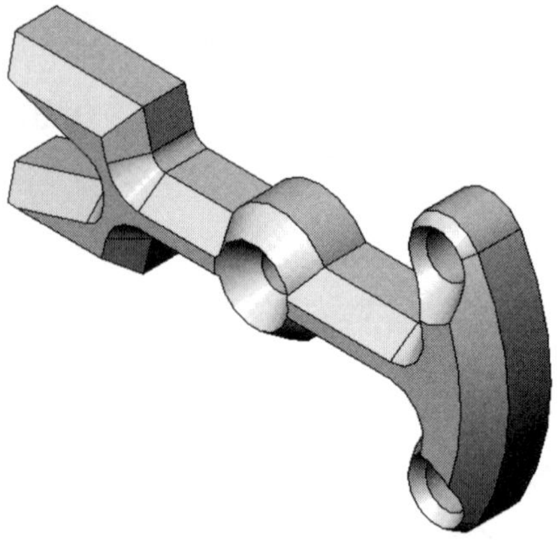

Figure 4-15 *Feature created in two directions with different values*

Note

*The **Selection Contours** rollout will be discussed in later chapters.*

Creating Thin Extruded Features

Thin extruded features can be created using a closed or an open sketch. If the sketch is closed, it will offset inside or outside to create a cavity inside the feature, as shown in Figure 4-16.

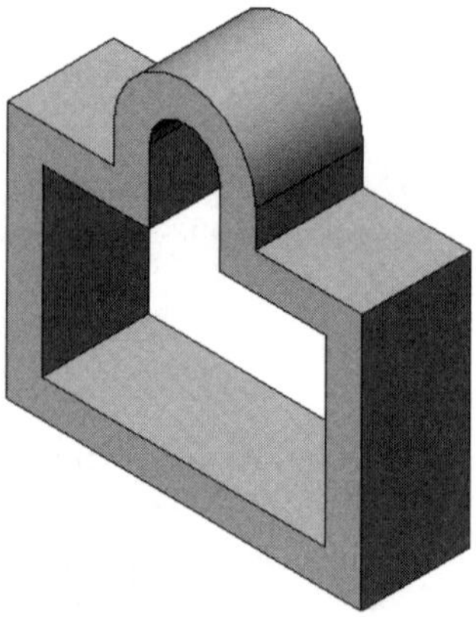

Figure 4-16 *Thin feature created using a closed loop*

If the sketch is open, as shown in Figure 4-17, the resulting feature will be as shown in Figure 4-18. To convert a closed sketch into a thin feature, choose the **Thin Feature** check box to invoke the **Thin Feature** rollout. The **Thin Feature** rollout, as shown in Figure 4-19, is used to create a thin feature. However, if the sketch to be extruded is open, the **Thin Feature** rollout will be invoked automatically when you invoke the **Extrude PropertyManager**.

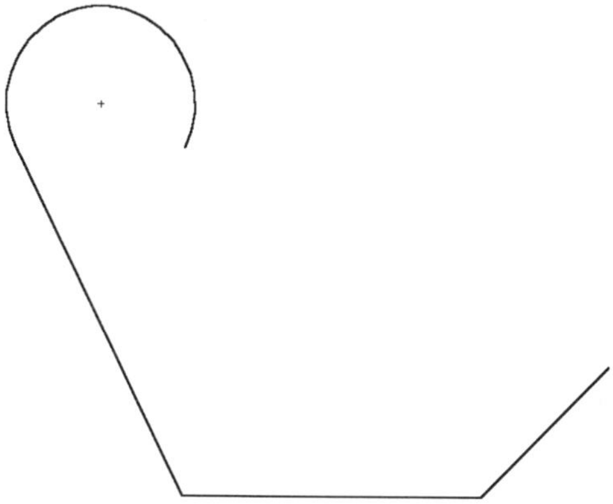

Figure 4-17 Open loop to be converted into thin feature

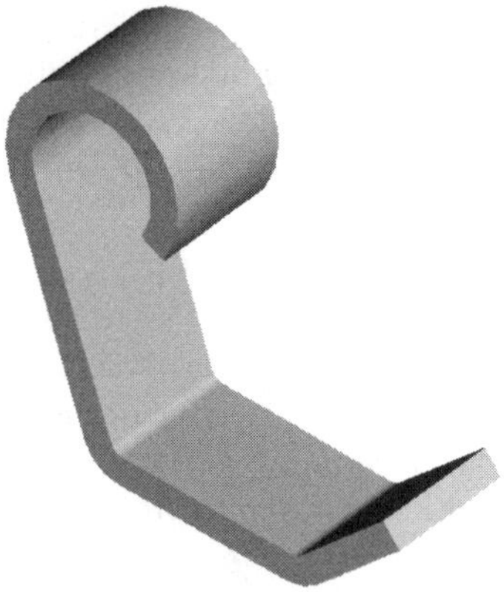

Figure 4-18 Resultant thin feature created with fillets at sharp corners

*Figure 4-19 The **Thin Feature** rollout*

The options available in the **Thin Feature** rollout of the **Extrude PropertyManager** are discussed next.

Type

The options provided in the **Type** drop-down list are used to select the method of defining the thickness of the thin feature. These options are discussed next.

One-Direction

The **One-Direction** option is used to add the thickness on one side of the sketch. The thickness can be specified in the **Thickness** spinner provided below this drop-down list. For the closed sketches, the direction can be inside or outside the sketch. Similarly, for open sketches, the direction can be below or above the sketch. You can reverse the direction of thickness using the **Reverse** check button available on the left of this drop-down list. This button will be available only when you select the **One-Direction** option from this drop-down list.

Mid-Plane

The **Mid-Plane** option is used to add the thickness equally on both sides of the sketch. The value of the thickness of the thin feature can be specified in the **Thickness** spinner provided below this drop-down list.

Two-Direction

The **Two-Direction** option is used to create a thin feature by adding different thicknesses on both sides of the sketch. The thickness values in direction 1 and direction 2 can be specified in the **Direction 1 Thickness** spinner and the **Direction 2 Thickness** spinner, respectively. These spinners will be automatically displayed below the **Type** drop-down list when you select the **Two-Direction** option from this drop-down list.

Cap ends

The **Cap ends** check box will be displayed only when you select a closed sketch to convert into a thin feature. This check box is selected to cap the two ends of the thin extruded feature. Both ends will be capped with a face of the thickness you specify. When you select this check box, the **Cap Thickness** spinner will be displayed below this check box. The thickness of the end caps can be specified using this spinner.

Auto-fillet corners

The **Auto-fillet corners** check box will be displayed only when you select an open sketch to convert into a thin feature. If you select this check box, all sharp vertices in the sketch will be automatically filleted while converting into a thin feature. As a result, the thin feature will have filleted edges. The radius of the fillet can be specified in the **Fillet Radius** spinner, which will be displayed below the **Auto-fillet corners** check box when you select this check box.

Figure 4-20 shows the thin feature created by extruding an open sketch in both directions. Note that a draft angle is applied to the feature while extruding in the front direction and the **Auto-fillet corners** option is selected while creating this thin feature.

Figure 4-20 *Thin feature created in both directions*

Note

Only the corners of the thin features that can accommodate the given radius will be filleted; other corners that cannot accommodate the given radius will not be filleted.

CREATING BASE FEATURES BY REVOLVING THE SKETCHES

CommandManager:	Features > Revolved Boss/Base
Menu:	Insert > Boss/Base > Revolve
Toolbar:	Features > Revolved Boss/Base

The sketches that you have drawn until now can also be converted into base features by revolving using the **Revolved Boss/Base** tool. This tool is available in the **Features CommandManager**. Using this tool the sketch is revolved about the revolution axis. The revolution axis could be an axis, an entity of the sketch, or an edge of another feature to create the revolved feature. Note that whether you use a centerline or an edge to revolve the sketch, the sketch should be drawn on one side of the centerline or the edge.

In SolidWorks, the right-hand thumb rule is used while determining the direction of revolution. The right-hand thumb rule states that if the thumb of your right hand points in the direction of the axis of revolution, the direction of curled fingers will determine the default direction of revolution, see Figure 4-21.

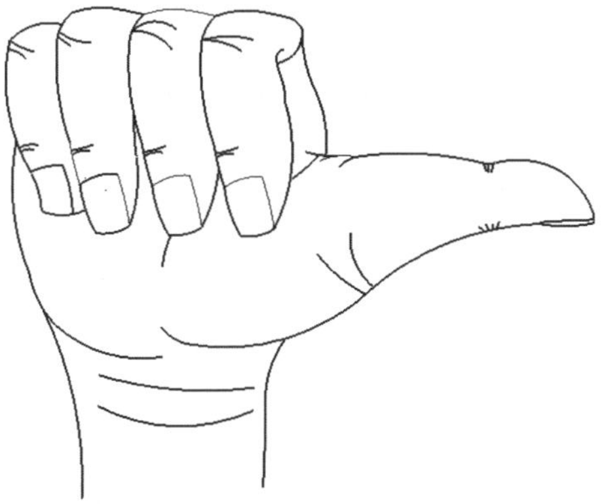

Figure 4-21 *Right-hand thumb rule*

For example, consider a case in which you draw a centerline from left to right (direction of thumb in Figure 4-21) in the drawing area. Now, if you use this centerline to create a revolved feature, the default direction of revolution will be in the direction of curled fingers.

Note

You can also reverse the direction of revolution using the button available in the ***Revolve PropertyManager****.*

After drawing the sketch, as you choose this tool, you will notice that the sketching environment is closed and the part modeling environment is invoked. Similar to extruding the sketches, the resulting feature can be a solid feature or a thin feature, depending on the sketch and the options selected to revolve. If the sketch is closed, it can be converted into a solid feature or a thin feature. However, if the sketch is open, it can be converted only into a thin feature. The solid and thin features are discussed next.

Creating Solid Revolved Features

After you have completed drawing and dimensioning the closed sketch and converted it into a fully defined sketch, choose the **Revolved Boss/Base** button from the **Features** toolbar. You will notice that the view is automatically changed to a 3D view, and the **Revolve PropertyManager** is displayed as shown in Figure 4-22, as well as the confirmation corner. Also, the preview of the base feature, as will be created using the default options, will be displayed in temporary shaded graphics. The direction arrow will also be displayed in gray. If you have not drawn the centerline, you will be prompted to select an axis of revolution. Select an edge or entity you need to define as an axis of revolution. The options available in the **Revolve Parameters** rollout of the **Revolve PropertyManager** are discussed next.

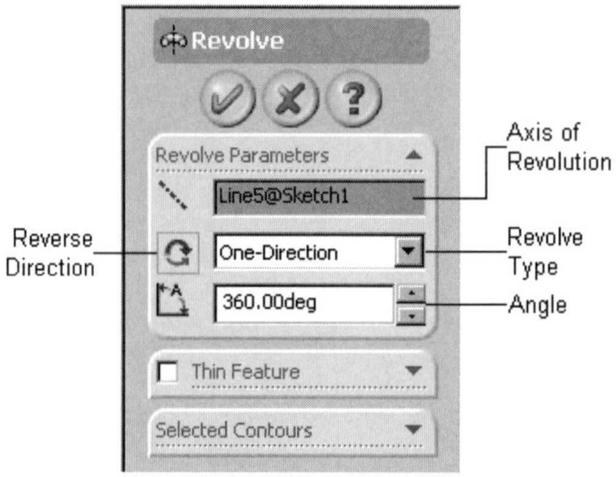

*Figure 4-22 The **Revolve PropertyManager***

Tip. *Even if you can revolve the sketch using an edge in the sketch, you need to draw a centerline. This is because you can create linear diameter dimensions for the revolved features only using a centerline.*

Revolve Type

The **Revolve Type** drop-down list provides the options to define the termination of the revolved feature. The options available in this drop-down list to terminate the revolved feature are discussed next.

One-Direction

The **One-Direction** option is used to revolve the sketch on one side of the plane on which it is drawn. The angle of revolution can be specified in the **Angle** spinner displayed below

Tip. *You can dynamically specify the angle in a revolved feature by dragging the direction arrows. You can also use the right mouse button to display the shortcut menu; the options available in the **PropertyManager** are also available in the shortcut menu.*

this drop-down list. The default value of the **Angle** spinner is **360deg**. Therefore, if you revolve the sketch using this value, a complete round feature will be created. You can also reverse the direction of revolution of the sketch by choosing the **Reverse Direction** button, which will be displayed when you select this option. Figure 4-23 shows the sketch of the piston and Figure 4-24 shows the resulting piston created by revolving the sketch through an angle of 360-degree. Note that the outer left vertical edge of the sketch that is vertically in line with the origin is used to revolve the sketch.

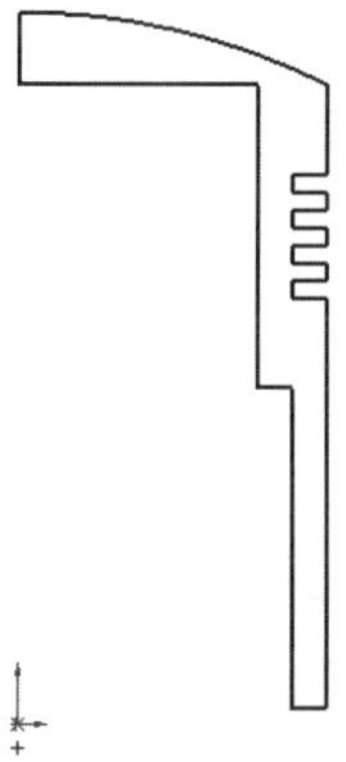

Figure 4-23 *Sketch of the piston to be revolved*

Figure 4-24 *Feature created by revolving the sketch through an angle of 360-degree*

Figure 4-25 shows a piston created by revolving the same sketch through an angle of 270-degree.

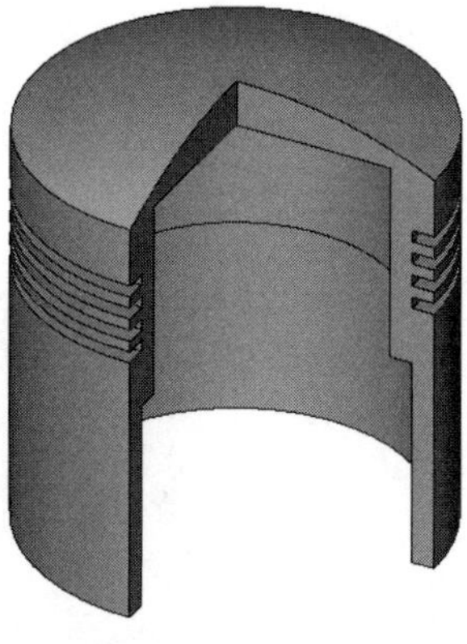

Figure 4-25 *Feature created by revolving the sketch through an angle of 270-degree*

Mid-Plane

The **Mid-Plane** option is used to revolve the sketch equally on both sides of the plane on which it is drawn. The angle of revolution can be specified in the **Angle** spinner. When you choose this option, the **Reverse Direction** button will be unavailable.

Two-Direction

The **Two-Direction** option is used to create a revolved feature by revolving the sketch using different values on both sides of the plane on which it is drawn. The angle values in direction 1 and direction 2 can be specified in the **Direction 1 Angle** spinner and the **Direction 2 Angle** spinner, respectively. These spinners will be displayed below the **Revolve Type** drop-down list automatically when you select the **Two-Direction** option from this drop-down list.

Creating Thin Revolved Features

Thin revolved features can be created using a closed or an open sketch. If the sketch is closed, it will offset inside or outside to create a cavity inside the feature as shown in Figure 4-26. In this figure, the sketch is revolved through an angle of 180-degree.

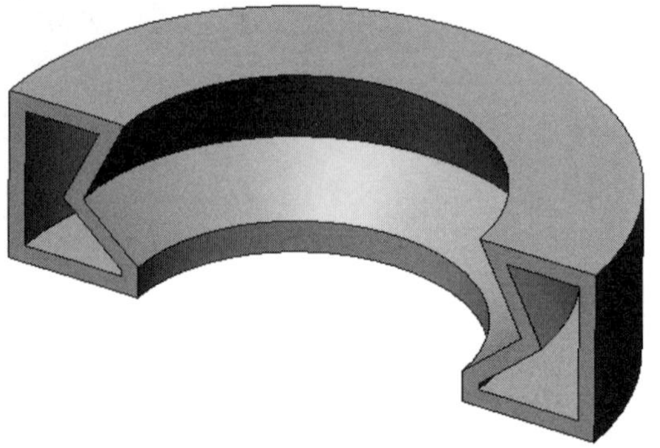

Figure 4-26 *Thin feature created by revolving the sketch through an angle of 180-degree*

If the sketch is open, as shown in Figure 4-27, the resulting feature will be as shown in Figure 4-28.

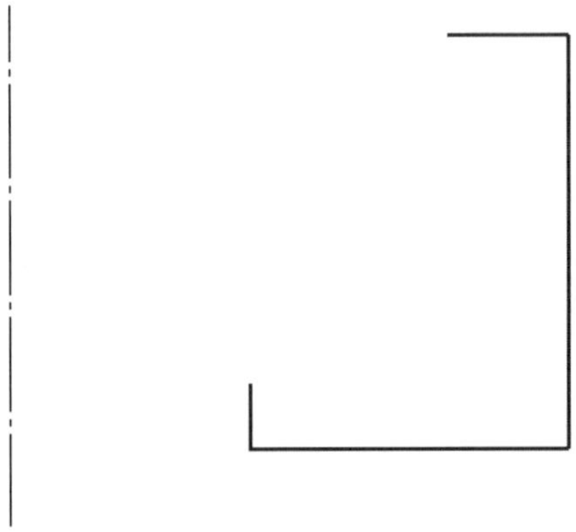

Figure 4-27 *The open sketch to be revolved and the centerline to revolve the sketch*

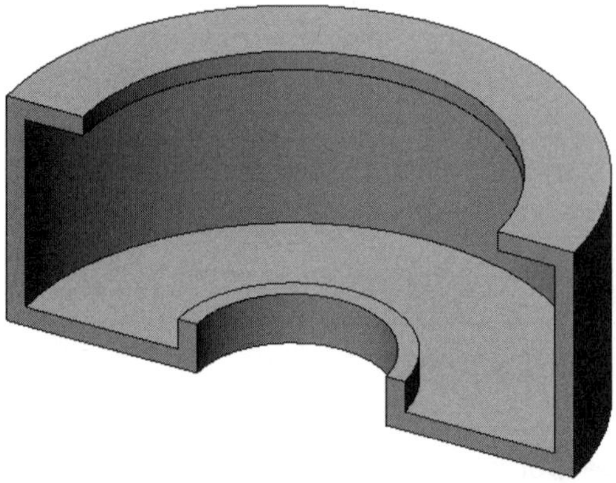

Figure 4-28 *Thin feature created by revolving the open sketch through an angle of 180-degree*

To convert a closed sketch into a thin feature, select the **Thin Feature** check box from the **Revolve PropertyManager** to invoke the **Thin Feature** rollout. However, if the sketch to be revolved is open and you invoke the **Revolved Boss/Base** tool, the **SolidWorks** information box will be displayed. This information box will inform you that the sketch is currently open and a nonthin revolved feature requires a closed sketch. You will be given an option of automatically closing the sketch. If you choose **Yes** from this dialog box, a line segment will be automatically drawn between the first and the last segment of the sketch and the **Revolve PropertyManager** will be displayed. However, if you choose **No** from this information box, the **Revolve PropertyManager** will be displayed and the **Thin Feature** rollout will be displayed automatically. The options available in the **Thin Feature** rollout of the **Revolve PropertyManager**, shown in Figure 4-29, are discussed next.

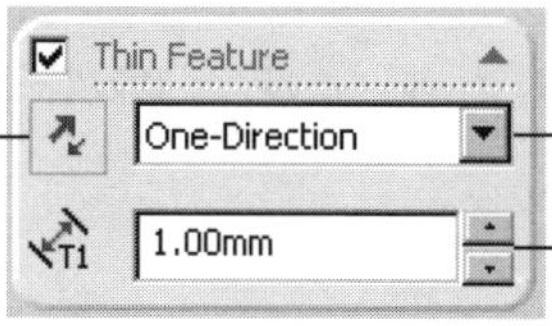

***Figure 4-29** The **Thin Feature** rollout*

Type

The options provided in the **Type** drop-down list are used to select the method of defining the thickness of the thin feature. These options are discussed next.

One-Direction

The **One-Direction** option is used to add the thickness on one side of the sketch. The thickness can be specified in the **Direction 1 Thickness** spinner provided below this drop-down list. For the closed sketches, the direction can be inside or outside the sketch. Similarly, for open sketches, the direction can be below or above the sketch. You can reverse the direction of thickness using the **Reverse Direction** button available on the right of this drop-down list. This button will be available only when you select the **One-Direction** option from this drop-down list.

Mid-Plane

The **Mid-Plane** option is used to add the thickness equally on both sides of the sketch. The value of the thickness of the thin feature can be specified in the **Direction 1 Thickness** spinner provided below this drop-down list.

Two-Direction

The **Two-Direction** option is used to create a thin feature by adding different thicknesses on both sides of the sketch. The thickness values in direction 1 and direction 2 can be specified in the **Direction 1 Thickness** spinner and the **Direction 2 Thickness** spinner, respectively.

Tip. *While defining the wall thickness of a thin revolved feature, remember that the wall thickness should be added such that the centerline does not intersect with the sketch. If the centerline intersects with the sketch, the sketch will not be revolved.*

These spinners will be displayed below the **Type** drop-down list automatically when you select the **Two-Direction** option from this drop-down list.

DYNAMICALLY ROTATING THE VIEW OF THE MODEL

In SolidWorks, you can dynamically rotate the view in the 3D space so that the solid models in the current document can be viewed from all directions. This allows you to visually maneuver around the model so that all features in the model can be clearly viewed. This tool can be invoked even when you are inside some other tool. For example, you can invoke this tool when the **Extrude Feature** dialog box is displayed. You can freely rotate the model in the 3D space or rotate it around a selected vertex, edge, or face. Both methods of rotating the model are discussed next.

Rotating the View Freely in 3D Space

Toolbar:	View > Rotate View
Menu:	View > Modify > Rotate

To rotate the view freely in the 3D space, choose the **Rotate View** button from the **View** toolbar. You can also invoke this tool by choosing the **Rotate View** option from the shortcut menu that will be displayed when you right-click in the drawing area. When you are inside some other tool, use the right mouse button in the drawing area and choose the **Zoom/Pan/Rotate** > **Rotate View** option from the shortcut menu to invoke the rotate view tool. When you invoke this tool, the cursor will be replaced by the rotate view cursor. Now, press the left mouse button and drag the cursor to rotate the view. Figure 4-30 shows a model being viewed from different directions by rotating the view.

Figure 4-30 *Rotating the view to display the model from different directions*

Rotating the View Around a Selected Vertex, Edge, or Face

To rotate the view around a selected vertex, edge, or face, invoke this tool and move the rotate view cursor close to the vertex, edge, or the face around which you want to rotate the view. When it is highlighted, select it using the left mouse button. Next, drag the cursor to rotate the view around the selected vertex, edge, or face.

Tip. *To resume rotating the view freely after you have completed rotating it around a selected vertex, edge, or face, double-click anywhere in the drawing area. Now when you drag the cursor, you will notice that the view is rotated freely in the 3D space.*

If a three-button mouse is configured to your computer, you can press and drag the middle mouse button to rotate the model freely in the 3D space. Note that in this case the rotate view cursor will not be displayed.

MODIFYING THE VIEW ORIENTATION

As mentioned earlier, when you invoke the **Extruded Boss/Base** tool or the **Revolved Boss/Base** tool, the view will be automatically changed to a trimetric view and the preview of the model will be displayed. In SolidWorks you can manually change the view orientation using some predefined standard views or user-defined views. To invoke these standard views, choose the **Standard Views** button from the **View** toolbar; a flyout will be displayed, as shown in Figure 4-31. You can select the required standard view from this flyout.

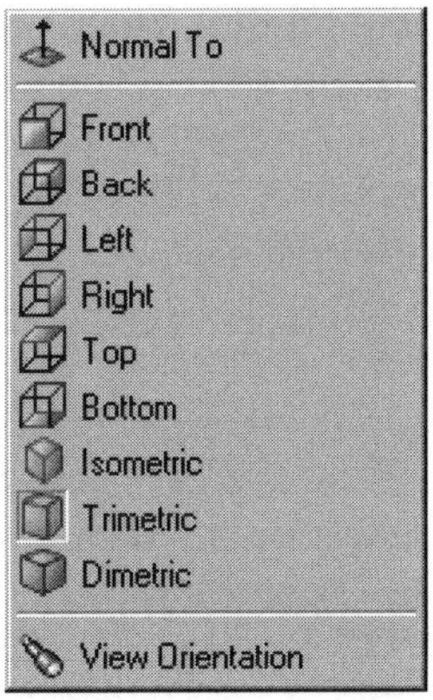

***Figure 4-31 Standard View** flyout*

Apart from the standard views, this flyout also provides the **Normal To** option. This option is chosen to reorient the view normal to a selected face or plane. To do this select the face along which you need to reorient the model and choose the **Normal To** option from this flyout. If you have not selected a face before invoking this option, the **Normal To PropertyManager** will be displayed and you will be prompted to select a reference along which the view will be reoriented.

You can also invoke these standard views using the **Orientation** dialog box. This dialog box is invoked by choosing the **View Orientation** option from the shortcut menu. This dialog box can also be invoked by pressing the SPACEBAR on the keyboard. Note that when you invoke this dialog box by pressing the SPACEBAR, the dialog box will be displayed at the location where the cursor is placed currently. The **Orientation** dialog box is shown in Figure 4-32.

You can invoke the view from this dialog box by double-clicking on it. You will notice that in

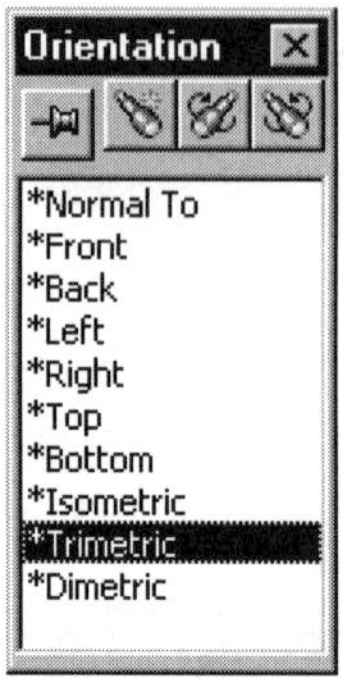

***Figure 4-32 Orientation** dialog box*

addition to the standard views, two more views are displayed. The buttons available on top of this dialog box are discussed next.

Push-Pin

You will notice that the **Orientation** dialog box is automatically closed when you select a view, select a point somewhere on the screen, or invoke a tool. If you want that this dialog box should be retained on the screen, you can pin it at a location by choosing the **Push-Pin** button. This is the first button on top of this dialog box. Move the dialog box to the desired location and then choose this button. The dialog box will be pinned to that location and will not close when you perform any operation.

New View

The **New View** button is chosen to create a user-defined view and save it in the list of views in the **Orientation** dialog box. Using various drawing display tools and the **Rotate View** tool, modify the current view and then choose this button. When you choose this button, the **Named View** dialog box will be displayed. Enter the name of the view in the **View Name** edit box and then choose the **OK** button. You will notice that a user-defined view is created and it is saved in the list available in the **Orientation** dialog box.

Update Standard Views

The **Update Standard Views** button is chosen to modify the orientation of the standard views. For example, if you want that the view that is displayed when you invoke the **Back** option from this dialog box should be the front view, change the current view to the back view by double-clicking on it in the **Orientation** dialog box. Now, select the **Front** option from the list of views available in the **Orientation** dialog box and then choose the **Update Standard Views** button. The **SolidWorks** warning box will be displayed and you will be informed that if you change the standard view, all other named views in the model will also be changed. If you make the change using the **Yes** button, the current view that was originally the back view will become the front view. Also, all other views will be modified automatically.

Reset Standard Views

The **Reset Standard Views** button is chosen to reset the standard settings of all standard views in the current drawing. When you choose this button, the **SolidWorks** warning box will be displayed and you will be prompted to confirm whether you want to reset all standard views to their original settings or not. If you choose **Yes,** all standard views will be reset to their default settings.

RESTORING PREVIOUS VIEW

In SolidWorks, while working on a model or while viewing it from different directions, you need to temporarily change the view of the model. Once you have finished editing or viewing the model in that view, you can restore the previous view using the **Previous View** button in the **View** toolbar. This tool saves ten previous views of the model.

DISPLAY MODES OF THE MODEL

SolidWorks provides you with various predefined modes to display the model. You can select any of these display modes from the **View** toolbar. These modes are discussed next.

Wireframe

When you choose the **Wireframe** button, all hidden lines will be displayed along with the visible lines in the model. Sometimes it becomes difficult to recognize the visible lines and the hidden lines if you set this display mode for complex models.

Hidden Lines Visible

When you choose the **Hidden Lines Visible** button, the model will be displayed in the wireframe and the hidden lines in the model will be displayed as dashed lines.

Hidden Lines Removed

When you choose the **Hidden Lines Removed** button, the hidden lines in the model will not be displayed. Only the edges of the faces visible in the current view of the model will be displayed.

Shaded With Edges

The **Shaded With Edges** mode is the default mode in which the model is displayed. In this display mode, the model is shaded and the edges of the visible faces of the model are displayed.

Shaded

This display mode is similar to the **Shaded With Edges** mode, with the only difference that the edges of the visible faces will not be displayed.

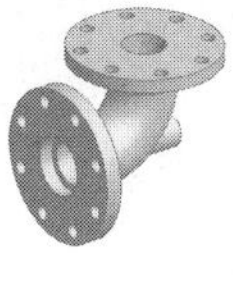

Tip. *Sometimes when you rotate the view of a large assembly or a model with large number of features with **Shaded** or the **Hidden Lines Removed** shading modes, the regeneration of the model takes a lot of time. This can be avoided by choosing the **Draft Quality HLR/HLV** button in combination with the other shading modes. Choosing this button speeds up the regeneration time and you can easily rotate the view. This is a toggle mode and is turned on when you choose this button. This button can be chosen in combination with any of the other display modes.*

Shadows In Shaded Mode

The **Shadows In Shaded Mode** button is used to display the shadow of the model. A light appears from the top of the model to display the shadow in the current view. With this option activated, the performance of the system is affected during the dynamic orientation. Remember that the position of the shadow is not changed when you rotate the model in the 3D space. To change the placement of the shadow, first remove the shadow in the shaded model using the **Shadow in Shaded Mode** button and rotate the model. After rotating the model, use the same button to activate the shadow in the shaded mode. Figure 4-33 shows a T-section with shadow in the shaded mode.

Perspective

The **Perspective** button is not available in the **View** toolbar by default. You need to customize this toolbar to add this button. Using this button, you can display the perspective view of a model. Figure 4-34 shows the T-section with shadow in perspective view.

Figure 4-33 *Shadow in active shaded mode*

Figure 4-34 *T-section displayed in perspective view with shadow*

You can also modify the settings of the perspective view. Choose **View > Modify > Perspective** from the menu bar to invoke the **Perspective View PropertyManager**. Using the **Object Sizes Away** spinner of this **PropertyManager**, you can modify the observer's position.

Tip. *You can also save a perspective view as a named view. To do this, invoke the perspective view and define the orientation by rotating the view. Now, press the SPACEBAR to invoke the* ***Orientation*** *dialog box. In this dialog box, choose the* ***New View*** *button and specify the name of the view in the* ***Named View*** *dialog box. The named view will be saved in the* ***Orientation*** *dialog box and you can invoke it whenever required.*

ASSIGNING MATERIALS AND TEXTURES TO THE MODELS

You can assign materials and textures to the models. When you apply a material to a model the physical properties such as density, young's modulus, and so on will be assigned to the model. When you apply a texture to a model or on a face of the model, the image of that texture will be applied to the model or its selected face. Physical properties are not applied to the model when you apply the texture. The method of assigning materials and textures is discussed next.

Assigning Materials to a Model

Toolbar: Standard > Edit Material *(Customize to Add)*
Menu: Edit > Appearance > Material

Whenever you assign a material to a model, all physical properties of the selected material are also assigned to the model. As a result, when you calculate the mass properties of the model, they will be based on the physical properties of the material. To assign a material to a model, choose the **Edit Material** button from the **Standard** toolbar; the **Materials Editor PropertyManager** will be displayed, as shown in Figure 4-35.

Figure 4-35** Partial display of the **Materials Editor PropertyManager

A number of material families are available in the **Materials** rollout. Click on the + sign located on the left of the material family to display all materials under that family. Select the material from that family.

You can change the scale and angle of the textures in the selected material using the **Scale** and **Angle** spinners available in the **Visual Properties** rollout.

Assigning Textures to a Model

Toolbar:	Standard > Edit Texture
Menu:	Edit > Appearance > Texture

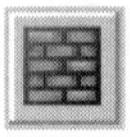

In SolidWorks 2005 you can assign a texture to a model, feature, or a selected face. This can be done using the **Texture PropertyManager**, which will be displayed when you invoke the **Edit Texture** tool. Figure 4-36 shows this **PropertyManager**.

Figure 4-36** Partial display of the **Texture PropertyManager

The **Selection** rollout has some buttons available on the left. These buttons are the filters for making a selection to assign the textures. For example, if you want to assign the texture on a face of the model, choose the **Select Faces** button and clear all remaining buttons. This allows you to select only a specified face. You can select the required face from the model. The selected face or model is displayed in the **Selected Entities** area of the **Selection** rollout.

Various texture families that you can select are available in the **Texture Tree** area of the **Texture**

Figure 4-37 *Texture applied to the top face of a model*

Selection rollout. Click on the + sign located on the left of a texture family to select a texture of that family. Figure 4-37 shows a model with Gravel type of stone applied to its top face.

The preview of the selected texture is displayed on the model or the face and also in the **Texture Preview** area of the **Texture Properties** rollout. You can use the **Scale** and **Angle** spinners available in this rollout to change the scale and angle of the selected texture.

You can remove the texture applied to a face or a model by choosing the **Remove Textures** button available below the **Selected Entities** area of the **Selection** rollout.

The drop-down list available below the **Texture Tree** area of the **Texture Selection** rollout lists the last ten textures used. You can also select a texture from this drop-down list.

TUTORIALS

Tutorial 1

In this tutorial, you will open the sketch drawn in Tutorial 2 of Chapter 3. You will then convert this sketch into an extruded model by extruding it in two directions as shown in Figure 4-38. The parameters for extruding the sketch are given next.

Direction 1
Depth = 10 mm
Draft angle = 35-degree

Direction 2
Depth = 15 mm
Draft angle = 0-degree

After creating the model, you will rotate the view using the **Rotate View** tool and then modify

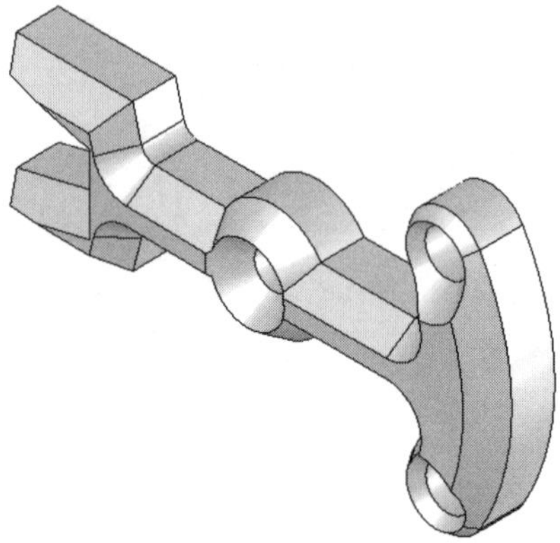

Figure 4-38 Model for Tutorial 1

the standard views such that the front view of the model becomes the top view. You will then save the model with the current settings. **(Expected time: 30 min)**

The steps that will be followed to complete this tutorial are given next.

a. Open the document of Tutorial 2 of Chapter 3, refer to Figure 4-39.
b. Save this document in the *c04* folder with a new name.
c. Invoke the **Extruded Boss/Base** tool and convert the sketch into a model, refer to Figures 4-40 and 4-41.
d. Rotate the view using the **Rotate View** tool to view the model from all directions, refer to Figure 4-42.
e. Invoke the **Orientation** dialog box and then modify the standard view, refer to Figure 4-43.

Opening the Document of Tutorial 2 of Chapter 3

As the required document is saved in the *\My Documents\SolidWorks\c03* folder, you need to select this folder and then open the *c03tut2.sldprt* document.

1. Start SolidWorks 2005 by double-clicking on its shortcut icon on the desktop of your computer.

2. Choose the **Open a Document** option from the **SolidWorks Resources Task Pane** to display the **Open** dialog box.

3. Browse and select the *\My Documents\SolidWorks\c03* folder. All documents that were created in Chapter 3 are displayed in this folder.

4. Select the *c03tut2.sldprt* document and then choose the **Open** button. Close the **SolidWorks Resources Task Pane**.

   Because the sketch was saved in the sketching environment in Chapter 3, it is opened in the sketching environment.

Saving the Document in the c04 Folder

It is recommended that when you open a document of some other chapter, you should save it in the folder of the current chapter with some other name before you proceed with modifying the document. This is because if you save the document in the folder of the current chapter, the original document of the other chapter will not be modified.

1. Choose **File > Save As** from the menu bar to display the **Save As** dialog box.

   Because the *c03* folder was selected last to open the document, it will be the current folder.

2. Choose the **Up One Level** button available on the right of the **Save in** drop-down list to move to the *\SolidWorks* folder. Create a new folder with the name *c04* using the **Create New Folder** button. Make the *c04* folder current by double-clicking on it.

3. Enter the new name of the document as *c04tut1* in the **File name** edit box and then choose the **Save** button to save the document.

   The document is saved with the new name and the new document is now opened in the drawing area, as shown in Figure 4-39.

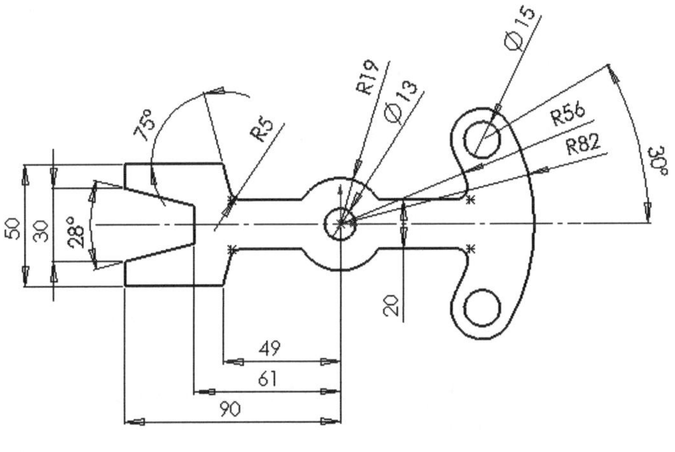

Figure 4-39 *Sketch that will be opened in the drawing area*

Extruding the Sketch

Next, you need to invoke the **Extruded Boss/Base** tool and extrude the sketch using the parameters given in the tutorial description.

1. Choose the **Features** button from the **CommandManager** to display the **Features CommandManager**. From this **CommandManager**, choose the **Extruded Boss/Base** button. The sketch is automatically oriented in the trimetric view and the **Extrude PropertyManager** is displayed, as shown in Figure 4-40.

   Because the sketch to be converted into a feature is closed, only the **Direction 1** rollout is

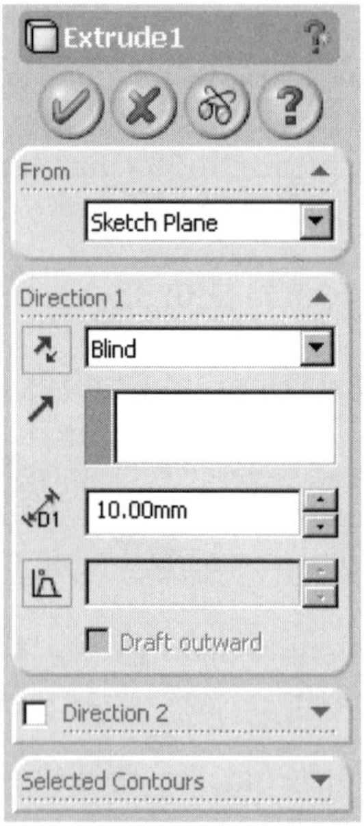

Figure 4-40 Extrude PropertyManager

displayed in the **Extrude PropertyManager**. The preview of the feature in the temporary shaded graphics with the default values is shown in the drawing area.

2. Choose the **Draft On/Off** button from the **Direction 1** rollout and then set the value of the **Draft Angle** spinner to **35**.

   These are the settings in direction 1. Next, you need to specify the settings for direction 2.

3. Select the **Direction 2** check box to invoke the **Direction 2** rollout.

   You will notice that the default values in this rollout are the same as you specified in the **Direction 1** rollout. Because the **Draft On/Off** button is chosen when you invoke the **Direction 2** rollout, you need to turn this button off. This is because you do not require the draft angle in the second direction.

4. Choose the **Draft On/Off** button from the **Direction 2** rollout to turn this option off. Set the value of the **Depth** spinner to **15** as the depth in the second direction is 15 mm. This completes all settings for the model in both directions.

5. Choose the **OK** button to create the feature or choose **OK** from the confirmation corner.

   It is recommended that you change the view to isometric view after creating the feature so that you can view the feature properly.

6. Choose the **Standard Views** button from the **View** toolbar to invoke the flyout and then choose **Isometric** from this flyout. Turn off the display of the origin in the model by

choosing **View > Origin** from the menu bar. The isometric view of the resulting solid model is shown in Figure 4-41.

Rotating the View

As mentioned earlier, you can rotate the view so that you can view the model from all directions.

1. Choose the **Rotate View** button from the **View** toolbar; the arrow cursor is replaced by the rotate view cursor.

2. Press and hold the left mouse button down and drag the cursor in the drawing area to rotate the view, as shown in Figure 4-42.

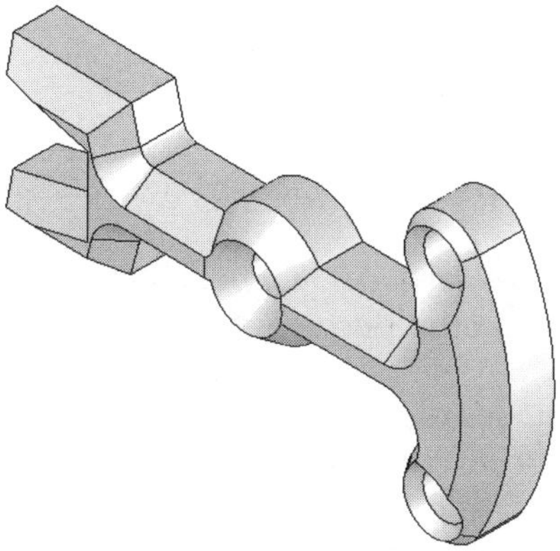

Figure 4-41 *Isometric view of the solid model*

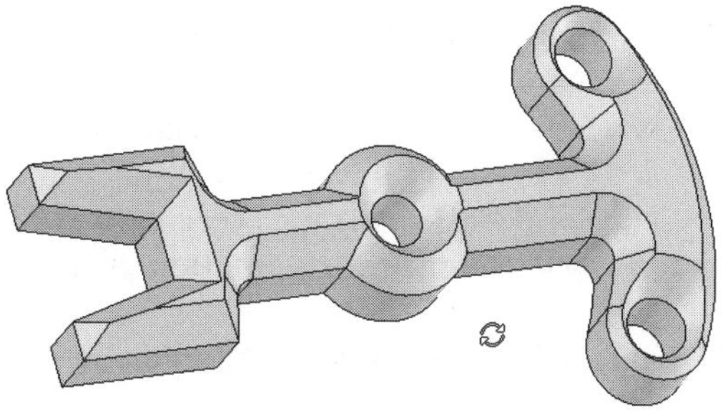

Figure 4-42 *Rotating the view to display the model from different directions*

You will notice that the model is being displayed from different directions. Remember that when you rotate the view, the model is not being rotated. The camera that is used to view the model is being rotated around the model.

3. After viewing the model from all directions, choose **Isometric** from the flyout that is displayed when you choose the **Standard Views** button from the **View** toolbar.

Modifying the Standard Views

As mentioned in the tutorial description, you need to modify the standard views such that the front view of the model becomes the top view. This is done using the **Orientation** dialog box.

1. Press the SPACEBAR on the keyboard to invoke the **Orientation** dialog box.

The orientation dialog box is automatically closed as soon as you perform any other operation. Therefore, you need to pin this dialog box so that it is not closed automatically.

2. Hold the **Orientation** dialog box by selecting it on the blue bar on the top of this dialog box and then drag it to the top right corner of the drawing area.

3. Choose the **Push Pin** button to pin this dialog box at the top right corner of the drawing area. Pinning the dialog box ensures that the dialog box is not automatically closed when you perform any other operation.

4. Double-click on the **Front** option in the list box of the **Orientation** dialog box.

   The current view is automatically changed to the front view and the model is now reoriented and displayed from the front.

5. Select the **Top** option from the list box by selecting it once.

   Make sure you do not double-click on this option. This is because if you double-click on this option, the model is reoriented and displayed from the top.

6. Now, choose the **Update Standard Views** button to update the standard views.

   The **SolidWorks** warning box is displayed and you are warned that modifying the standard views will change the orientation of any named view in this document.

7. Choose **Yes** from this warning box to modify the standard views.

8. Now, double-click on the **Isometric** option provided in the list box of the **Orientation** dialog box. You will notice that the isometric view is different now, see Figure 4-43.

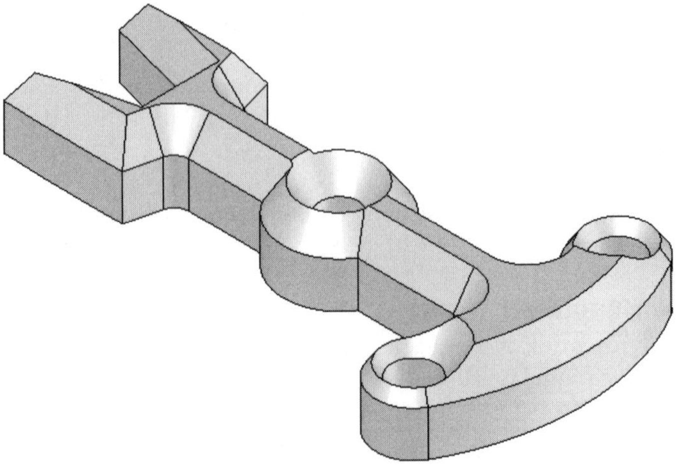

***Figure 4-43** Model displayed from modified isometric view*

9. Choose the **Push Pin** button in the **Orientation** dialog box again and select a point anywhere in the drawing area to close the dialog box.

Saving the Model

As the name of the document was specified at the beginning, you just have to choose the save button now to save the document.

1. Choose the **Save** button from the **Standard** toolbar to save the document.

   The model is saved with the name *\My Documents\SolidWorks\c04\c04tut1.sldprt*.

2. Choose **File > Close** from the menu bar to close the document.

Tutorial 2

In this tutorial, you will open the sketch drawn in Exercise 1 of Chapter 3. You will then create a thin feature by revolving the sketch through an angle of 270-degree, as shown in Figure 4-44. You will offset the sketch outward while creating the thin feature.

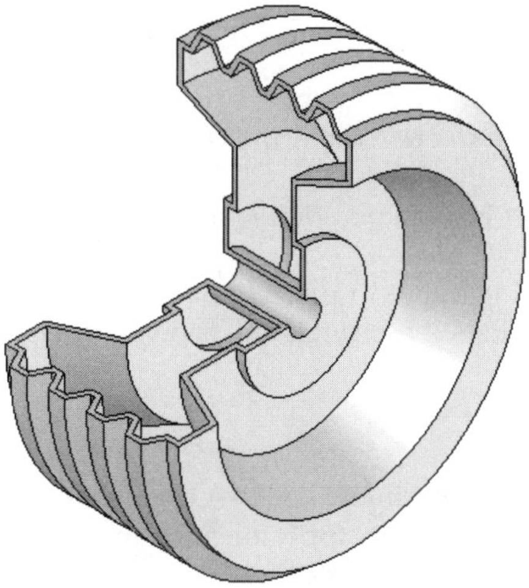

Figure 4-44 *Revolved model for Tutorial 2*

After creating the model, you will turn the option on to display the shadows and also apply Copper material to the model. **(Expected time: 30 min)**

The steps that will be followed to complete this tutorial are given next.

a. Open the sketch of Exercise 1 of Chapter 3, refer to Figure 4-45.
b. Save it in the folder of the current chapter.
c. Invoke the **Revolved Boss/Base** tool and revolve the sketch through an angle of 270-degree, refer to Figure 4-47.
d. Change the current view to isometric view and then display the model in shadow, refer to Figure 4-48.

e. Assign copper material to the model, refer to Figure 4-49.

Opening the Document of Exercise 1 of Chapter 3

As the required document is saved in the *\My Documents\SolidWorks\c03* folder, you need to select this folder and then open the *c03exr1.sldprt* document.

1. Choose the **Open** button from the **Standard** toolbar to display the **Open** dialog box. The *c04* folder is current in this dialog box.

2. Browse and select the *\My Documents\SolidWorks\c03* folder. All documents that were created in Chapter 3 are displayed in this folder.

3. Select the *c03exr1.sldprt* document and then choose the **Open** button. The document is opened in the sketching environment.

Saving the Document in the c04 Folder

As mentioned earlier, it is recommended that you save the document with a new name in the folder of the current chapter so that the original document is not modified.

1. Choose **File > Save As** from the menu bar to display the **Save As** dialog box. Because the *c03* folder was selected last to open the document, it is the current folder.

2. Browse and select the *c04* folder and double-click on it to make it current.

3. Enter the name of the document in the **File name** edit box as *c04tut2*. Choose the **Save** button to save the document. The sketch that is displayed in the drawing area is shown in Figure 4-45.

Revolving the Sketch

The sketch consists of two centerlines. The first centerline was used to mirror the sketched entities and the second was drawn to apply linear diameter dimensions. You need to revolve the sketch around the second centerline.

1. Choose the **Features** button from the **CommandManager** to display the **Features CommandManager**. From this **CommandManager**, choose the **Revolved Boss/ Base** button. The sketch is automatically oriented in the trimetric view and the **Revolve PropertyManager** is displayed.

   Because the sketch has two centerlines, SolidWorks cannot determine which one to use as an axis of revolution. This is the reason why you are prompted to select the axis of revolution.

2. Select the horizontal centerline that was used to create linear diameter dimensions as the axis of revolution. The preview of a complete revolved feature in temporary shaded graphics is displayed in the drawing area. Because the preview of the model is not displayed properly in the current view, you need to zoom the drawing.

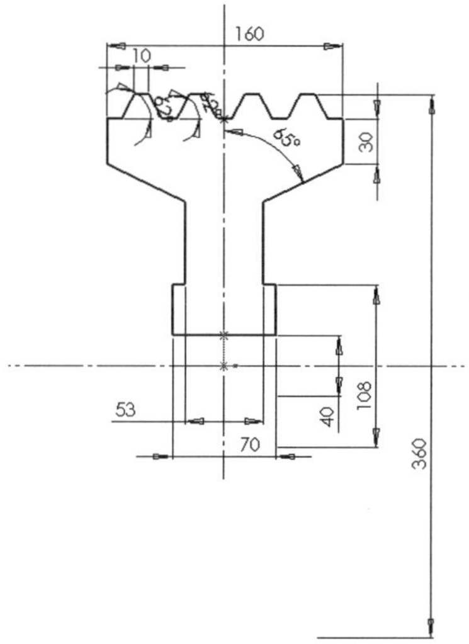

Figure 4-45 *Sketch for the revolved model*

3. Choose the **Zoom to Fit** button from the **View** toolbar or press the F key on the keyboard.

4. Set the value of the **Angle** spinner in the **Revolve PropertyManager** to **270**. The preview of the revolved model is also modified accordingly. If you enter the value in the **Angle** spinner, you need to click anywhere on the screen to make sure the preview is modified.

   Note that if the horizontal centerline was drawn from left to right, then the direction of revolution has to be reversed to get the required model; refer to the right-hand thumb rule. You can reverse the direction of revolution using the **Reverse Direction** button available on the left of the **Revolve Type** drop-down list.

5. Select the **Thin Feature** check box to invoke the **Thin Feature** rollout, as shown in Figure 4-46. Set the value of the **Direction 1 Thickness** spinner to **5**. You will notice that the preview of the thin feature is shown outside the original sketch.

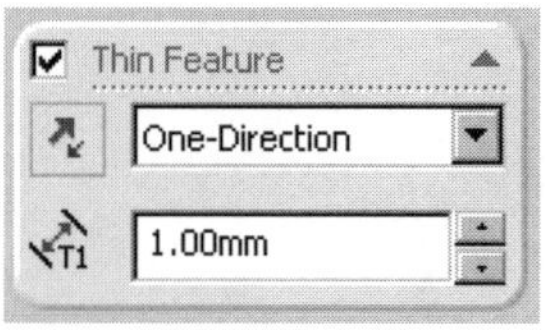

Figure 4-46 Thin Feature *rollout*

6. Choose the **OK** button to create the revolved feature. You will notice that the revolved feature is created.

7. Choose the **Isometric** option from the flyout that is displayed on choosing the **Standard Views** button from the **View** toolbar.

You will notice that the sketch that was used to create the thin feature is still displayed. You need to turn off the display of this sketch.

8. Click on the + sign located on the left of **Revolve-Thin1** in the **Feature Manager Design Tree**. The tree view expands and the sketch is displayed. Right-click on the sketch and choose **Hide** from the shortcut menu to hide the sketch. The revolved feature is shown in Figure 4-47.

Rotating the View

Next, you need to rotate the view so that you can view the model from all directions. As mentioned earlier, the view can be rotated using the **Rotate View** tool.

1. Choose the **Rotate View** button from the **View** toolbar; the arrow cursor is replaced by the rotate view cursor.

2. Press the left mouse button and drag the cursor in the drawing area to rotate the view.

3. Press the SPACEBAR to invoke the **Orientation** dialog box. In this dialog box, double-click on the **Isometric** option.

Displaying the Shadow

As mentioned in the tutorial description, you need to display the shadow of the model if it is not displayed by default. You can turn the display of the shadow on using the **View** toolbar.

1. Choose the **Shadows In Shaded Mode** button from the **View** toolbar; the shadow of the model is displayed below the model, see Figure 4-48.

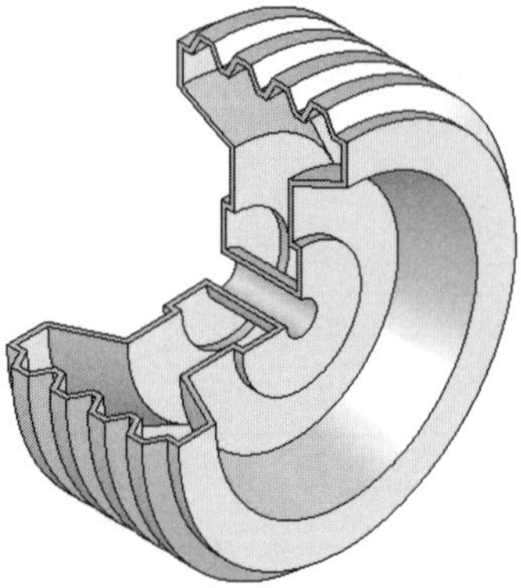

Figure 4-47 *Model created by revolving the sketch*

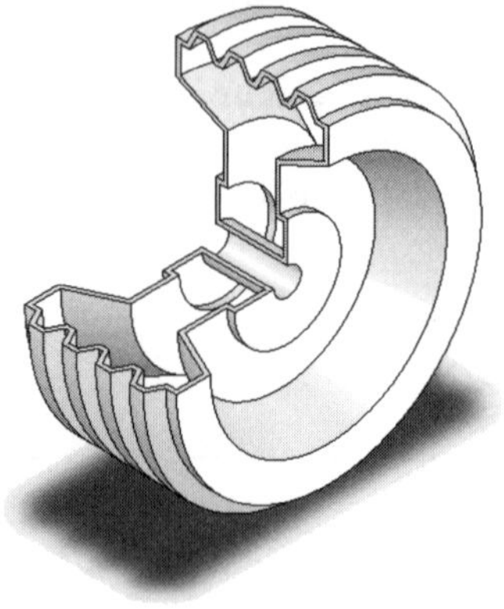

Figure 4-48 *Model with the display of shadow turned on*

Assigning Materials to a Model

As mentioned earlier, you can assign a material to a model using the **Materials Editor PropertyManager**, which is displayed by choosing the **Edit Material** button. Note that you can also invoke this **PropertyManager** using the **Material** option available in the **Feature Manager Design Tree**.

1. Right-click on the **Material <not specified>** option in the **Feature Manager Design Tree** and choose the **Edit Material** option; the **Materials Editor FeatureManager** is displayed.

2. Click on the + sign located on the left of the **Copper and its Alloys** option from the list box available in the **Materials** rollout. The tree view expands and the materials available under this family are displayed in the tree view.

3. Select the **Copper** option and choose **OK** from the **Materials Editor PropertyManager**. The model after assigning the material is shown in Figure 4-49.

Figure 4-49 *Model after assigning Copper material*

Saving the Sketch

As the name of the document was specified at the beginning, you just need to choose the save button now to save the document.

1. Choose the **Save** button from the **Standard** toolbar to save the model. If the SolidWorks warning box is displayed, choose **Yes** from it to rebuild the model before saving.

   The model is saved with the name *\My Documents\SolidWorks\c04\c04tut2.sldprt*.

2. Choose **File > Close** from the menu bar to close the document.

Tutorial 3

In this tutorial, you will create the model shown in Figure 4-50. The dimensions of the model are shown in Figure 4-51. The extrusion depth of the model is 20 mm. After creating the model, rotate the view and then change the view back to isometric view before saving the model.

(Expected time: 45 min)

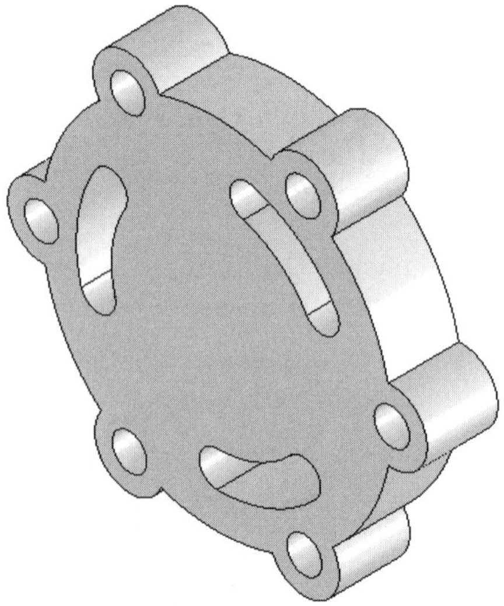

Figure 4-50 *Model for Tutorial 3*

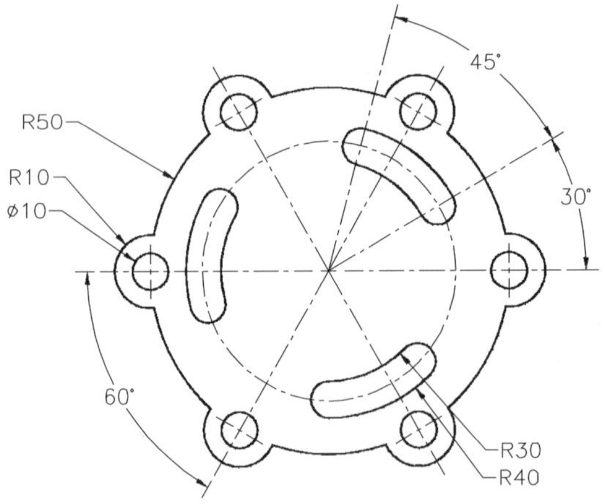

Figure 4-51 *Dimensions of the model for Tutorial 3*

The steps that will be used to complete this tutorial are given next.

a. Start a new SolidWorks part document and then invoke the sketching environment.

b. Create the outer loop and then create the sketch of three inner cavities. Finally, draw the six circles inside the outer loop for the holes, refer to Figures 4-52 through 4-56.
c. Invoke the **Extruded Boss/Base** tool and extrude the sketch through a distance of 20 mm, refer to Figure 4-57.
d. Rotate the view using the **Rotate View** tool.
e. Change the current view to isometric view and then save the document.

Starting a New Part Document

1. Choose the **New** button from the **Standard** toolbar and start a new part document using the **New SolidWorks Document** dialog box.

2. Choose the **Sketch** button from the **Standard** toolbar and select the **Front Plane** to invoke the sketching environment.

Drawing the Outer Loop

When the sketch consists of more than one closed loop, it is recommended that you draw the outer loop first and add relations and dimensions to it so that it is fully defined. Next, draw the inner loops one by one and add relations and dimensions to them. Therefore, you need to first draw the outer loop and add relations and dimensions to it.

1. Draw a circle in the first quadrant and then dimension it so that it is forced to a diameter of 100 mm.

2. Locate the center of the circle at a distance of 70 mm along the X and Y directions from the origin by adding dimensions in both directions. Choose the **Zoom to Fit** button to fit the display on the screen.

3. Draw a horizontal centerline from the center of the circle and then draw a circle with the center point at the intersection of the centerline and the bigger circle.

4. Trim the part of the sketch so that the sketch looks similar to that shown in Figure 4-52.

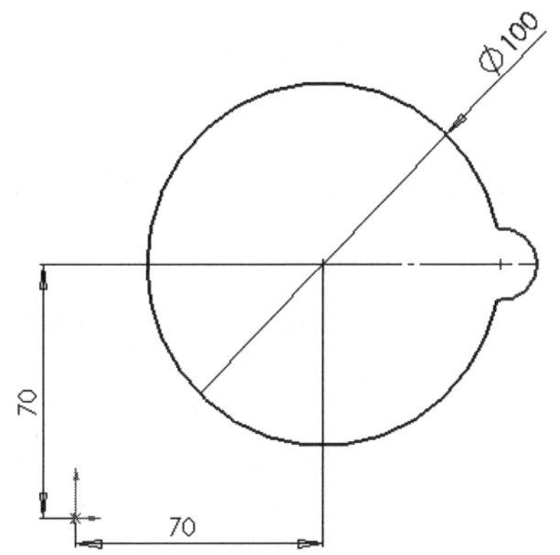

Figure 4-52 *Sketch after trimming the unwanted portion of circles*

5. Dimension the smaller arc so that it is forced to a radius of 10 mm. Add the **Coincident** relation to the center point of the smaller arc and the circumference of the bigger arc.

   You will notice that as you add the dimensions and relations to the smaller arc, it turns black. This suggests that the sketch drawn so far is fully defined.

   Next, you need to create a circular pattern of the smaller arc. The total number of instances in the pattern is 6 and the total angle is 360-degree.

6. Select the smaller arc and then choose the **Circular Sketch Step and Repeat** button from the **Sketch CommandManager**; the **Circular Sketch Step and Repeat** dialog box is displayed.

7. Drag the center of the circular pattern to the center of the bigger arc.

8. Set the value of the **Number** spinner in the **Step** area to **6**. Accept all other default values and choose the **OK** button to create the pattern.

   You will notice that all instances of the pattern are black in color. This is because you have already applied the dimensions and relations to the original instance and so the other instances are also fully defined.

9. Trim the unwanted portion of the bigger arc. This completes the outer loop. The sketch at this stage should look similar to that shown in Figure 4-53.

Drawing the Sketch of the Inner Cavities

Now, you need to draw the sketch of the inner cavities. Draw the sketch of one of the cavities and then add the required relations and dimensions to it. Next, you need to create a circular pattern of this cavity. The number of instances in the circular pattern is 3.

1. Using the **Centerpoint Arc** tool, draw an arc with the center at the center point of the bigger arc of 100 mm diameter.

2. Dimension this arc such that it is forced to a radius of 30 mm. Also, add the angular dimensions to the two endpoints of the arc, refer to Figure 4-54. The arc turns black, suggesting that it is fully defined.

3. Offset the last drawn arc outward through a distance of 10 mm using the **Offset Entities** tool.

   The new arc created using the **Offset Entities** tool is also black. Also, a dimension with the value 10 mm is placed between the two arcs.

4. Close the two ends of the arc using the **Tangent Arc** tool. This completes the sketch of one of the inner cavities. All entities in the sketch at this stage should be displayed in black, as shown in Figure 4-54.

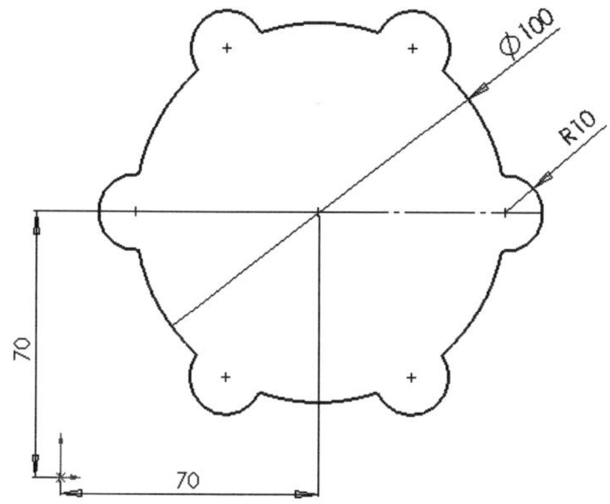

Figure 4-53 Outer loop of the sketch

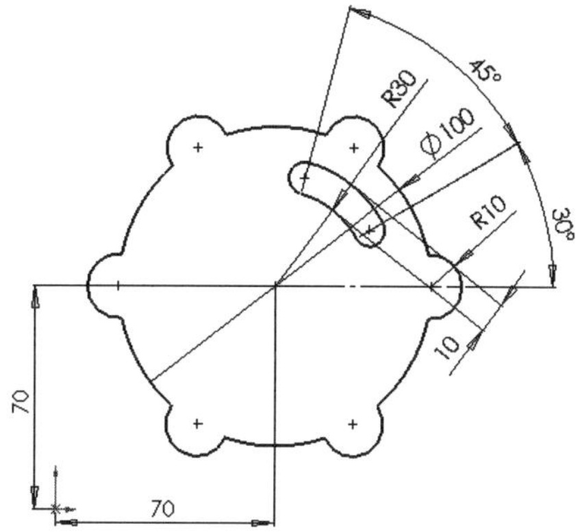

Figure 4-54 *Sketch after drawing the sketch of the inner cavity*

Next, you need to create a circular pattern of the inner cavity.

5. Select all entities in the sketch of the inner cavity and then invoke the **Circular Sketch Step and Repeat** dialog box.

6. Drag the center of the circular pattern to the center of the outer arc in the sketch.

7. Set the value of the **Number** spinner in the **Step** area to **3** and then choose the **OK** button to create the circular pattern. This completes the sketch of the inner cavities, see Figure 4-55.

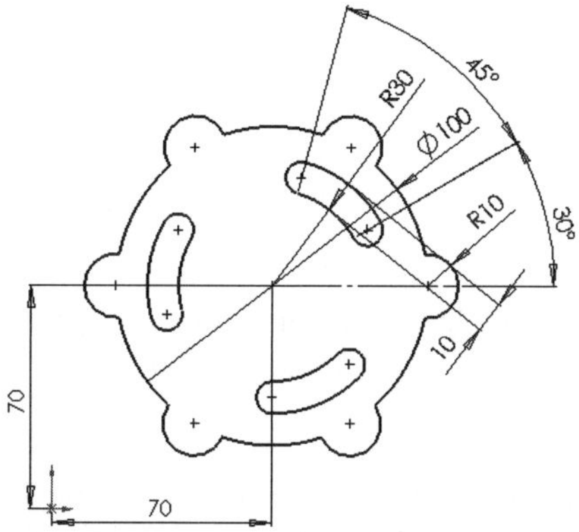

Figure 4-55 *Sketch after creating the circular pattern of the inner cavity*

Drawing the Sketch for the Holes

Next, you need to draw the sketch for the holes. Draw one of the circles and then add dimension to it. Then you need to create a circular pattern of the circle.

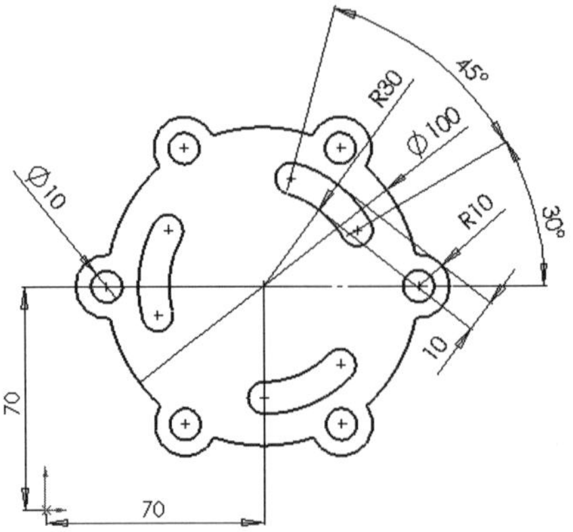

Figure 4-56 Final sketch of the model

1. Taking the center point of one of the smaller arcs of the outer loop, draw a circle. Add a diameter dimension to it so that the value of the diameter is forced to 10 mm.

   The circle turns black when you dimension it.

2. Select the circle and invoke the **Circular Sketch Step and Repeat** dialog box.

3. Drag the center of the circular pattern to the center of the outer arc of 100 mm diameter.

4. Set the value of the **Number** spinner in the **Step** area to **6**. Choose **OK** to create the pattern. All instances in the pattern are displayed in black.

   This completes the sketch of the model. The final sketch of the model is shown in Figure 4-56.

Extruding the Sketch

The next step after creating the sketch is to extrude it. The sketch is extruded using the **Extruded Boss/Base** tool.

1. Choose the down arrow available on the right of the **Features** button in the **CommandManager** to display a flyout. Choose the **Extruded Boss/Base** option from this flyout.

   The current view is changed to trimetric view and the **Extrude PropertyManager** is displayed. Also, the preview of the model, as will be created using the default values, is displayed in the drawing area.

2. Set the value of the **Depth** spinner to **20** and then choose the **OK** button to extrude the sketch.

   Because the option to display shadow was turned on in the previous tutorial, the shadow is displayed in this tutorial also.

3. Press the SPACEBAR and then double-click on the **Isometric** option in the **Orientation** dialog box to change the current view to isometric view. The completed model for Tutorial 3 is shown in Figure 4-57.

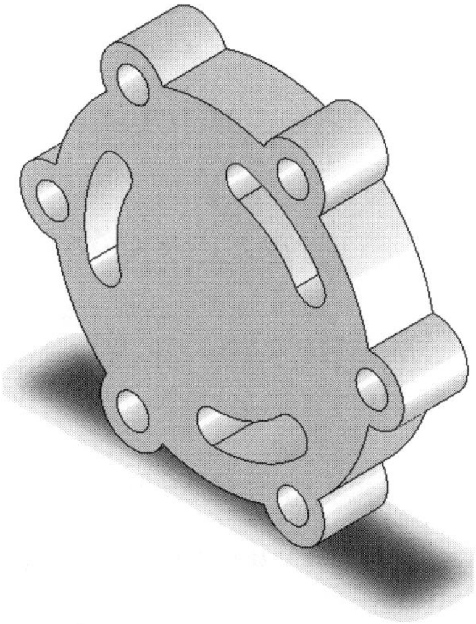

Figure 4-57 *Completed model for Tutorial 3*

Rotating the View

Before you start rotating the view of the model, it is recommended that you turn off the display of the shadow.

1. Choose the **Shadows In Shaded Mode** button from the **View** toolbar to turn off the display of the shadow.

2. Now, choose the **Rotate View** button from the **View** toolbar; the arrow cursor is replaced by the rotate view cursor.

3. Press the left mouse button and drag the cursor in the drawing area to rotate the view.

4. Change the current view to isometric view using the **Orientation** dialog box.

Saving the Sketch

1. Choose the **Save** button from the **Standard** toolbar and save the model with the name given below.

   \My Documents\SolidWorks\c04\c04tut3.sldprt.

2. Choose **File > Close** from the menu bar to close the document.

SELF-EVALUATION TEST

Answer the following questions and then compare your answers with those given at the end of this chapter.

1. In SolidWorks, a sketch is revolved using the **Extrude PropertyManager**. (T/F)

2. You can also specify the depth of extrusion dynamically in the preview of the extruded feature. (T/F)

3. You can invoke the drawing display tools such as **Zoom to Fit** while the preview of a model is displayed on the screen. (T/F)

4. When you rotate the view with the current display mode set to **Hidden Lines Removed**, the hidden lines in the model are automatically displayed while the view is being rotated. (T/F)

5. __________ tool is used to display the perspective view of a model.

6. The **Cap ends** check box is displayed in the **Extrude PropertyManager** only when the sketch for the thin base feature is __________.

7. The __________ check box is used to create a feature with different values in both directions of the sketching plane.

8. The __________ check box is used to apply automatic fillets while creating a thin feature.

9. The __________ button is used to display the shadow in the shaded mode.

10. To resume rotating the view freely after you have completed rotating it around a selected vertex, edge, or face, __________ any where in the drawing area.

REVIEW QUESTIONS

Answer the following questions.

1. You can also invoke the **Rotate View** tool by choosing the **Rotate View** option from the __________ that is displayed when you right-click in the drawing area.

2. When you choose the **Wireframe** button, all __________ lines will be displayed along with the visible lines in the model.

3. You can also modify the parallel view to perspective view by choosing __________ from the menu bar.

4. When you invoke the **Extruded Boss/Base** tool or the **Revolved Boss/Base** tool, the view is automatically changed to a __________.

5. The thin revolved features can be created using a _________ or an _________ sketch.

6. Which of the following buttons is chosen to modify the orientation of the standard views?

   (a) **Update Standard Views** (b) **Reset Standard Views**
   (c) None (d) Both

7. Which of the following buttons is not available in the **View** toolbar by default?

   (a) **Hidden Lines Removed** (b) **Hidden Lines Visible**
   (c) **Shaded** (d) **Perspective**

8. Which of the following parameters is not displayed in the preview of the model?

   (a) Depth (b) Draft angle
   (c) None (d) Both

9. If the sketch is open, it can be converted into

   (a) Thin feature (b) Solid feature
   (c) None (d) Both

10. In SolidWorks, which tool is used to apply automatic dimensions to the sketch?

    (a) **Autodimension** (b) **Smart Dimension**
    (c) None (d) Both

EXERCISES

Exercise 1

Create the model shown in Figure 4-58. The sketch of the model is shown in Figure 4-59. Create the sketch and dimension it using the autodimension option. The extrusion depth of the model is 15 mm. After creating the model, rotate the view. **(Expected time: 30 min)**

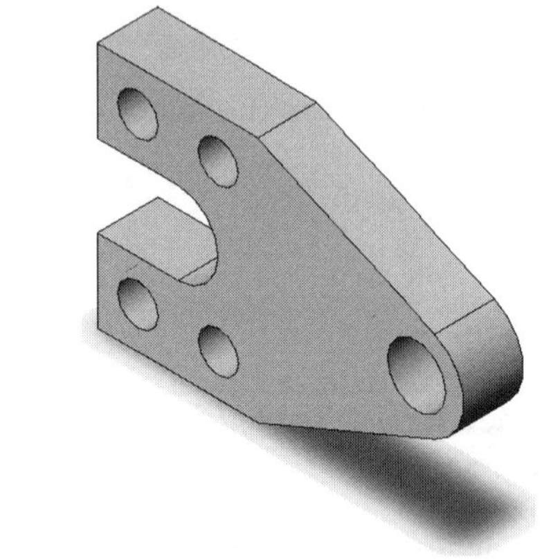

Figure 4-58 Model for Exercise 1

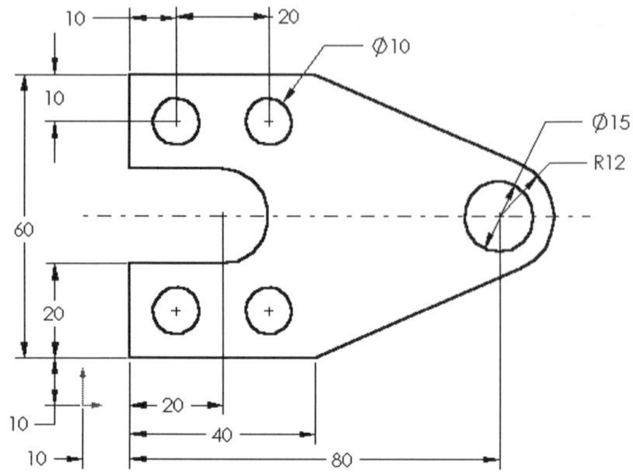

Figure 4-59 Sketch of the model for Exercise 1

Exercise 2

Create the model shown in Figure 4-60. The sketch of the model is shown in Figure 4-61. Create the sketch and dimension it using the autodimension tool. The extrusion depth of the model is 25 mm. Modify the standard view such that the current front view of the model is displayed when you invoke the top view. **(Expected time: 30 min)**

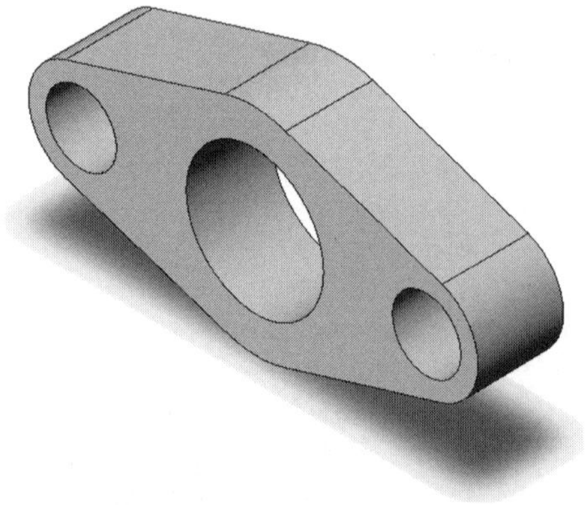

Figure 4-60 Model for Exercise 2

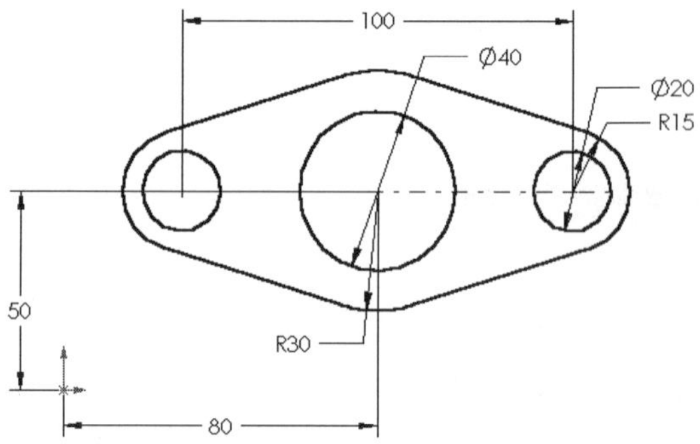

Figure 4-61 Sketch of the model for Exercise 2

Answers to Self-Evaluation Test

1. F, **2.** T, **3.** T, **4.** F, **5. Perspective**, **6.** closed, **7. Direction 2**, **8. Auto-fillet corners**, **9. Shadows In Shaded Mode**, **10.** double-click

Chapter 5

Creating Reference Geometries

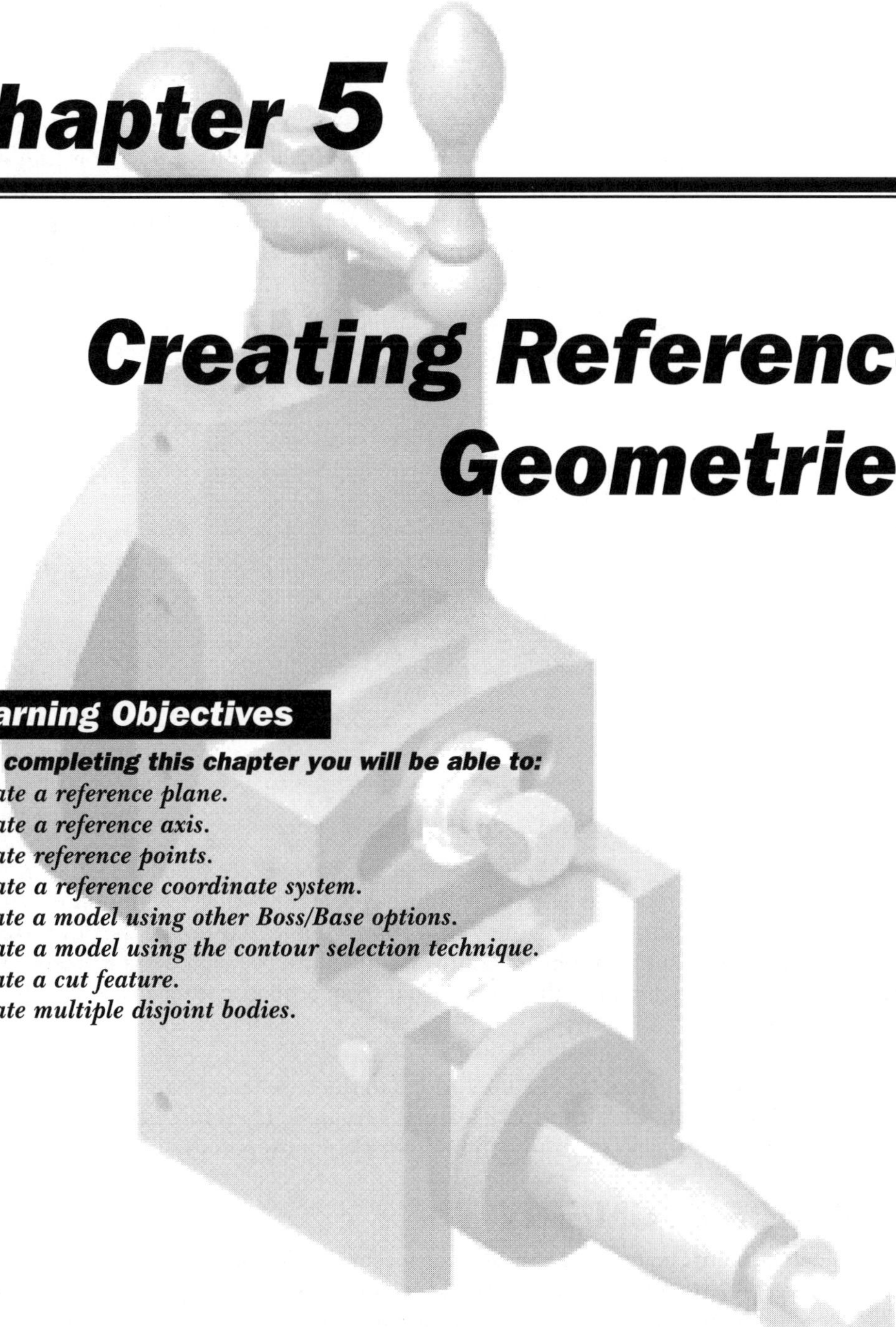

Learning Objectives

After completing this chapter you will be able to:

- *Create a reference plane.*
- *Create a reference axis.*
- *Create reference points.*
- *Create a reference coordinate system.*
- *Create a model using other Boss/Base options.*
- *Create a model using the contour selection technique.*
- *Create a cut feature.*
- *Create multiple disjoint bodies.*

IMPORTANCE OF THE SKETCHING PLANES

In previous chapters, you created basic models by extruding or revolving the sketches. All these models were created on a single sketching plane, the **Front Plane**. But most of the mechanical designs consist of multiple sketched features, referenced geometries, and placed features. These features are integrated together to complete a model. Most of these features lie on different planes. When you start a new SolidWorks document and try to invoke the sketching plane, you are prompted to select the plane on which you want to draw the sketch. On the basis of the design requirements, you can select any plane to create the base feature. To create additional sketched features, you need to select an existing plane or a planar surface, or you need to create a plane that will be used as a sketching plane. For example, consider the model shown in Figure 5-1.

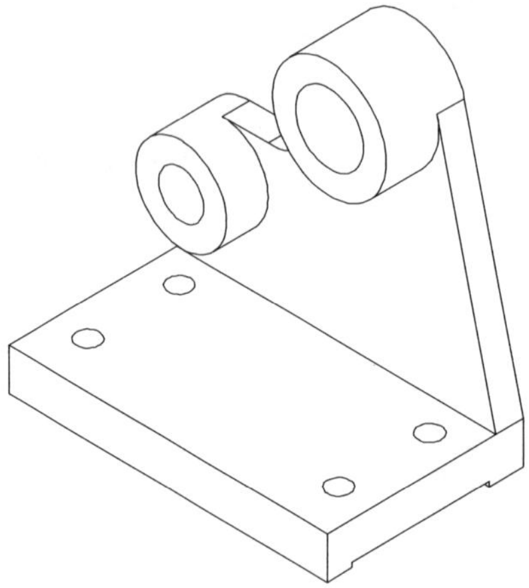

Figure 5-1 *A multifeatured model*

The base feature of this model is shown in Figure 5-2. The sketch for the base feature is drawn on the **Top Plane**. After creating the base feature, you need to create the other sketched features, placed features, and referenced features, see Figure 5-3. The boss features and cut features are the sketched features that require sketching planes where you can draw the sketch of the features.

It is evident from Figure 5-3 that the features added to the base feature are not created on the same plane on which the sketch for the base feature is created. Therefore, to draw the sketches of other sketched features, you need to define other sketching planes.

REFERENCE GEOMETRY

The reference geometry features are those that are available only to assist you in creating models. The reference geometries in SolidWorks include planes, axes, points, and coordinate systems. These reference geometries act as a reference for drawing the sketches for sketched features, defining the sketch plane, assembling the components, references for various placed features and sketched features, and so on. These features have no mass or volume. Because

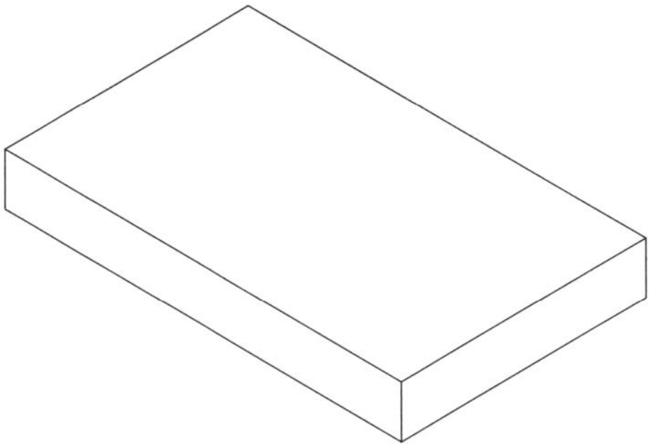

Figure 5-2 *Base feature for the model*

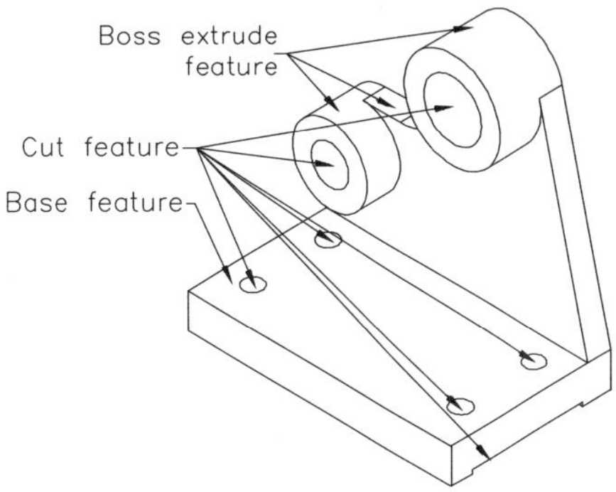

Figure 5-3 *Model after adding other features*

the reference geometries are widely used in creating complex models, you must have a good understanding of these geometries.

Reference Planes

Generally, all engineering components or designs are multifeatured models. Also, as discussed earlier, all features of a model are not created on the same plane on which the base feature is created. Therefore, you have to select one of the default planes or create a new plane that will be used as the sketching plane for the second feature. It is clear from the above discussion that you can use the default planes as the sketching plane or you can create a plane that can be used as a sketching plane. The default planes and the creation of a new plane are discussed next.

Default Planes

When you start a new SolidWorks part document, SolidWorks provides you with three default planes: the **Front Plane**, the **Top Plane**, and the **Right Plane**.

The orientation of the component depends on the sketch of the base feature. Therefore, it is recommended that you carefully select the sketching plane for drawing the sketch for the base feature. The sketching plane for drawing the sketch of the base feature can be one of the three datum planes provided by default.

If you create the base feature and choose the **Sketch** tool to invoke the sketching environment, the three default planes will be automatically displayed in the drawing area. However, if you create an additional feature, these planes will not be displayed when you invoke the **Sketch** tool. You need to manually select the required plane.

To select a plane for an additional feature, choose the **Sketch** button; the **Edit Sketch PropertyManager** will be displayed. Whenever a **PropertyManager** is displayed, the **Feature Manager Design Tree** shifts to the drawing area. Click on the + sign located on the left of the part document name in the **Feature Manager Design Tree**, which is displayed in the drawing area. The tree view will expand, displaying the names of the planes. Select the required plane from the tree view.

Tip. *You can turn the display of the default planes on in the drawing area using the following procedure.*

Press and hold the CTRL key down and select one by one the ***Front Plane****, the* ***Top Plane****, and the* ***Right Plane*** *from the* ***FeatureManager Design Tree****. Right-click to display the shortcut menu and choose the* ***Show*** *option to display the planes in the drawing area. Choose the* ***Isometric*** *option from the* ***Standard Views*** *flyout. By default, the reference planes are transparent.*

To display the shaded planes, choose ***Tools > Options*** *from the menu bar to invoke the* ***System Options - General*** *dialog box. Select the* ***Display/Selection*** *option from the left of this dialog box; the name of the dialog box will be changed to* ***System Options - Display/Selection****. Select the* ***Display shaded planes*** *check box from this dialog box and choose the* ***OK*** *button.*

After displaying the planes in shaded, invoke the ***Rotate View*** *tool and drag the rotate view cursor to rotate the shaded planes. You will observe that one side of the plane is displayed in reddish color and the other side is displayed in brownish color. The reddish side of the plane symbolizes the positive side and the brownish side symbolizes the negative side. This means that when you create an extruded feature, the depth of extrusion will be assigned to the positive side of the plane by default. When you create a cut feature, the depth of the cut feature will be assigned to the negative direction by default.*

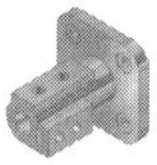

Note

You can also select a plane before choosing the ***Sketch*** *button to invoke the sketching environment. In this case, the* ***Edit Sketch PropertyManager*** *is not displayed and you are not prompted to select a plane for sketching. The selected plane is automatically taken as the sketching plane.*

Creating New Planes

CommandManager:	Reference Geometry > Plane
Menu:	Insert > Reference Geometry > Plane
Toolbar:	Reference Geometry > Plane

Default planes or the reference planes are used to draw sketches for the sketched features. These planes are also used to create a placed feature such as holes, reference an entity or a feature, and so on. You can also select a planar face of a feature that will be used as a sketching plane. Generally, it is recommended that you use the planar faces of the features as the sketching planes. However, sometimes you have to create a sketch on a plane that is at some offset distance from a plane or a planar face. In this case, you have to create a new reference plane at an offset distance from a plane or a planar face.

Consider another case where you have to define a sketching plane tangent to a cylindrical face of a shaft. You have to create a plane tangent to the cylindrical face of the shaft and this plane will be used as a sketching plane. In SolidWorks, there are six methods of creating planes. To create planes, choose the **Plane** button from the **Reference Geometry CommandManager**; the **Plane PropertyManager** will be displayed, as shown in Figure 5-4. The confirmation corner will also be displayed at the top right corner of the drawing area.

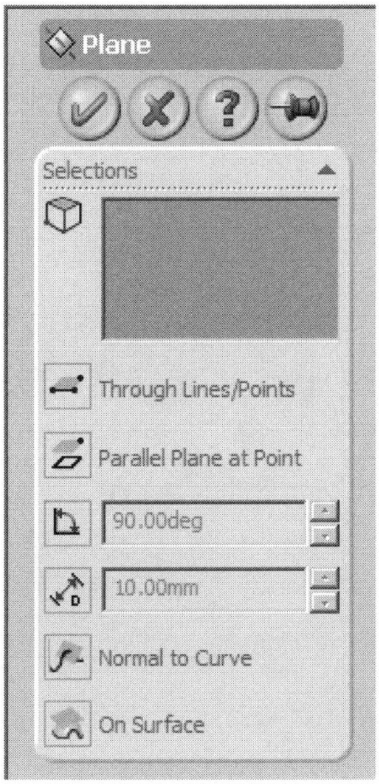

*Figure 5-4 The **Plane PropertyManager***

The options available in the **Plane PropertyManager** to create new planes are discussed next.

Creating a Plane Using Through Lines/Points

The **Through Lines/Points** option is used to create a plane that passes through an edge and a point, an axis and a point, or a sketch line and a point. Using this option, you can also create a plane that passes through three points. The selected point can be a sketched point or a vertex. To create a plane using this option, invoke the **Plane PropertyManager** and choose the **Through Lines/Points** button. Select the required entities from the drawing area. The name of the selected entities will be displayed in the **Reference Entities** selection area. Choose the **OK** button from the **Plane PropertyManager**. Figure 5-5 shows an edge and a vertex selected to create a plane. The resulting plane is displayed in Figure 5-6. The creation of a new plane by selecting three points is displayed in Figures 5-7 and 5-8.

Creating a Plane Parallel to an Existing Plane or Planar Face and Passing Through a Point

The **Parallel Plane at Point** option is used to create a plane that is parallel to another plane or a planar face and passes through a point. To create a plane using this option, invoke the **Plane PropertyManager** and then choose the **Parallel Plane at Point** button. Now, select a

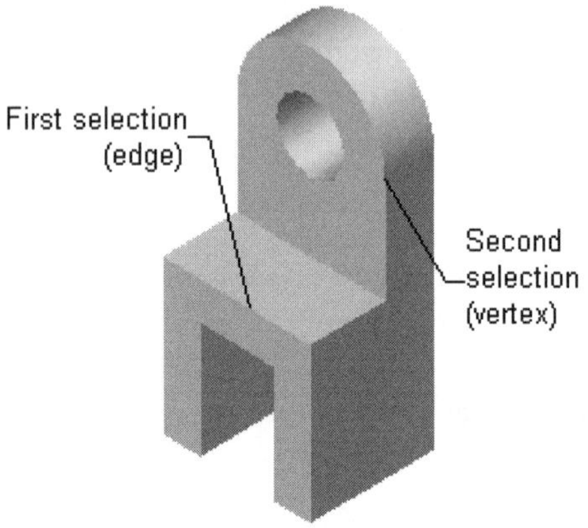

Figure 5-5 Selecting the edge and vertex

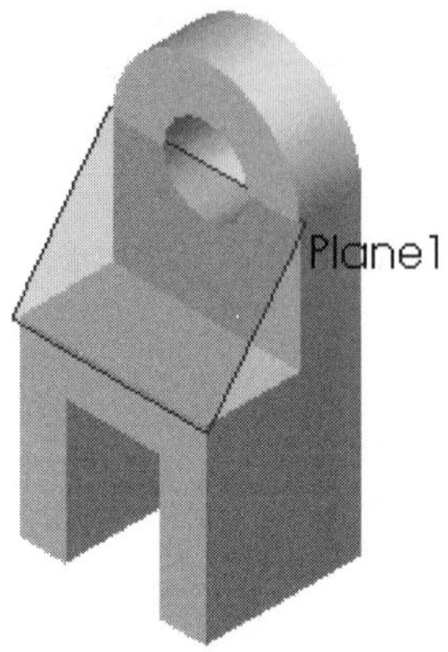

Figure 5-6 Resulting plane

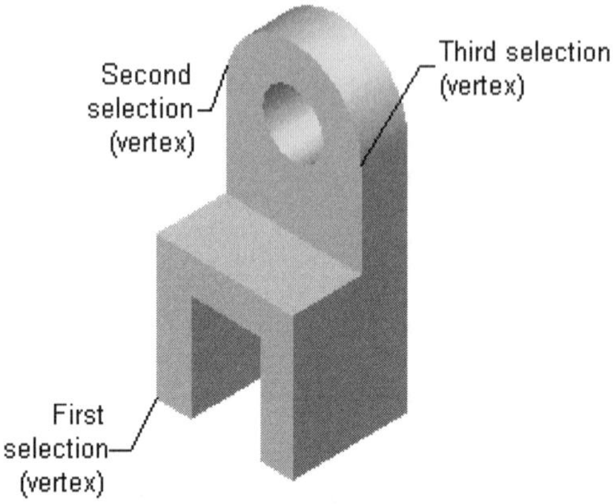

Figure 5-7 Selecting the vertices

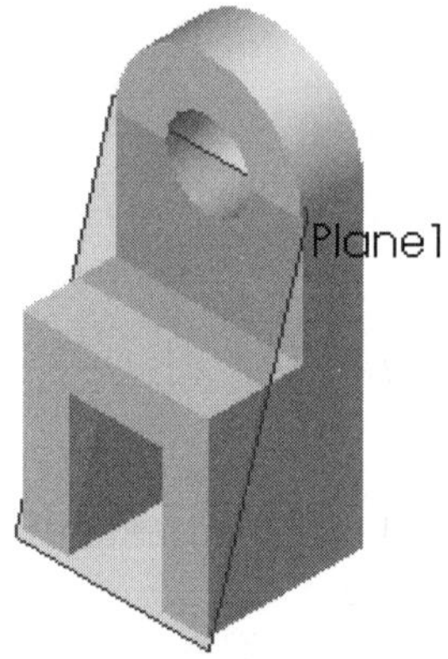

Figure 5-8 Resulting plane

plane or a planar face to which the new plane will be parallel. Next, select a sketched point, an endpoint of an edge, or a midpoint of an edge. The new plane will pass through this point. Choose the **OK** option. Figure 5-9 shows a planar face and the point selected to create the parallel plane. Figure 5-10 shows the resulting plane.

Creating a Plane at an Angle to an Existing Plane or a Planar Face

The **At Angle** option is used to create a plane that is at an angle to the selected plane or a planar face and also passes through an edge, axis, or sketched line. To create a plane at an angle, choose the **At Angle** button from the **Plane PropertyManager**; the **Angle** spinner will be invoked. The **Reverse direction** check box and the **Number of Plane to Create** spinner will appear below the **Distance** spinner in the **Plane PropertyManager**, as shown in Figure 5-11. Now, select an edge, an axis, or a sketched line through which the plane will pass. Next, you have to select a planar face or a plane to define the angle. After selecting the plane or the planar face, set the angle value using the **Angle** spinner. You can reverse the direction of plane creation by selecting the **Reverse direction** check box. You can also create multiple planes by increasing the value of the **Number of Planes to Create** spinner. Each plane will be incremented by the angle value defined in the **Angle** spinner. Figure 5-12 shows

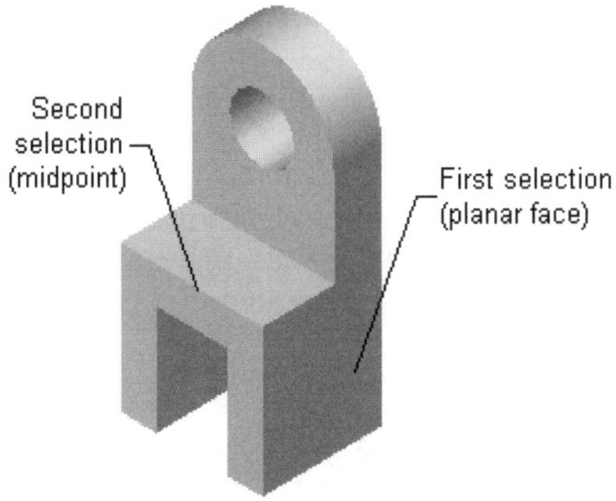

Figure 5-9 *Selecting a planar face and a vertex*

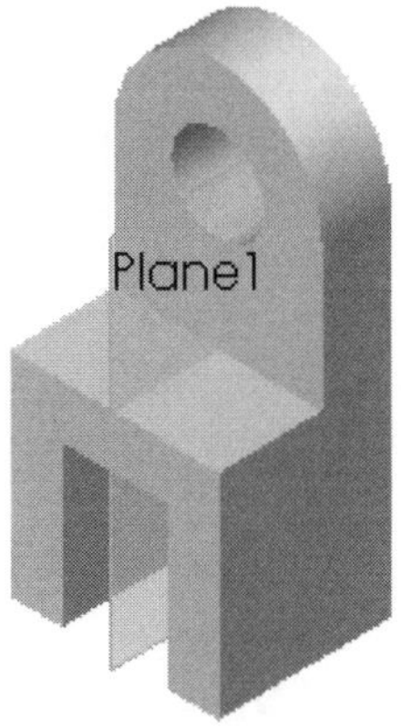

Figure 5-10 *Resulting plane*

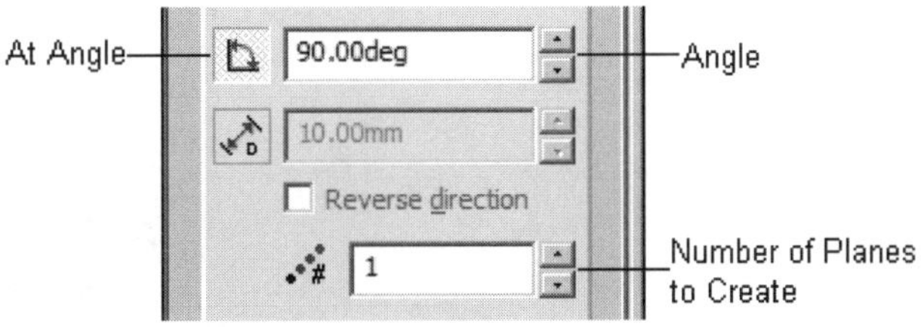

Figure 5-11 ***Plane PropertyManager*** *with the* ***At Angle*** *option selected*

a planar face and edge selected. Figure 5-13 shows the resulting plane created at an angle of 45-degree to the selected plane.

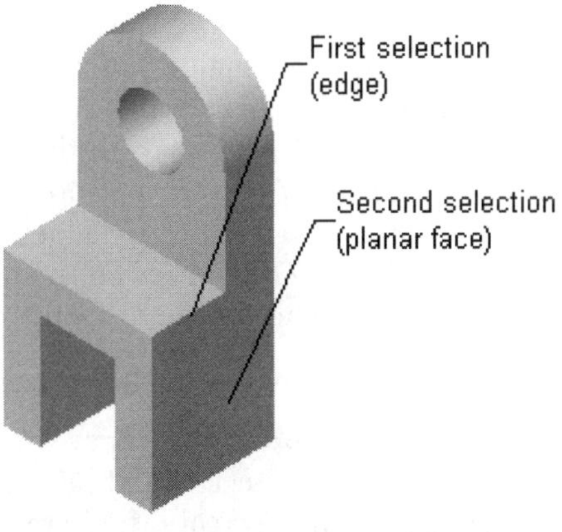

Figure 5-12 *Selecting the edge and the planar face*

Figure 5-13 *Resulting plane*

Creating a Plane at Some Offset From an Existing Plane or a Planar Face

The **Offset Distance** option is used to create a plane that is at some offset distance from a selected plane or planar face. To create a plane using this option, choose the **Offset Distance** button from the **Plane PropertyManager**. When you invoke this option, the **Distance** spinner will be invoked. Also, the **Reverse direction** check box and the **Number of Planes to Create** spinner will be displayed below the **Distance** spinner in the **Plane PropertyManager**. Select a plane or a planar face and set the value of the distance in the **Distance** spinner and choose the **OK** button from the **Planar PropertyManager**. You can reverse the direction of plane creation by selecting the **Reverse direction** check box. You can also create multiple planes by increasing the value of the **Number of Planes to Create** spinner. Each plane will be placed at the offset distance specified in the **Distance** spinner. Figure 5-14 shows a plane selected to create a parallel plane and Figure 5-15 shows the resulting plane created at the required offset.

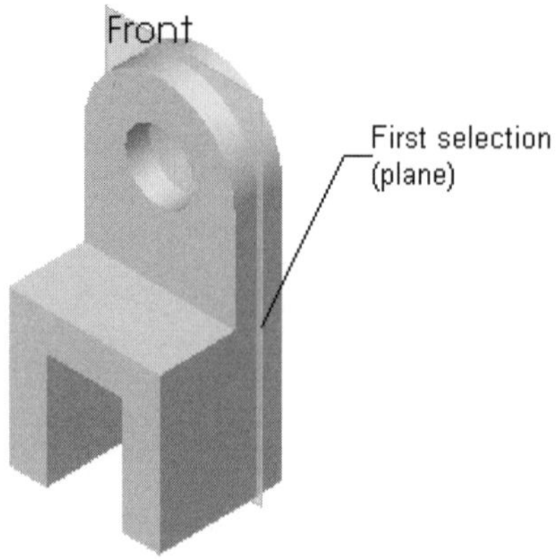

Figure 5-14 *Selecting the plane*

Figure 5-15 *Resulting plane*

Creating a Plane Normal to Curve

This option is used to create a plane normal to a selected curve. The curve can be a sketched entity or an edge of a feature. To create a plane normal to a curve, choose the **Normal to Curve** button from the **Plane PropertyManager** and select a curve normal to which the new plane will be created. As soon as you select the curve, the preview of the plane will be displayed in the drawing area. This plane is displayed by default at the endpoint closest to the point from where you selected the curve. If you have selected the curve close to its midpoint, the plane will be displayed at the midpoint of the selected curve. If you want to place it at the other endpoint or at any other point on the curve, select that point. The preview of the plane will change accordingly. After selecting the plane and the point, choose the **OK** button to create the plane.

When you choose the **Normal to Curve** button from the **Plane PropertyManager**, the **Set origin on curve** check box will be displayed below this button. This check box is selected to set the origin of the plane on the curve. Figure 5-16 shows a curve to create the plane and Figure 5-17 shows the resulting plane created normal to the selected curve.

Creating a Plane On Surface

The **On Surface** option is used to create a plane passing through a point on the selected plane or planar surface. To create a plane on a surface, choose the **On Surface** button from the **Plane PropertyManager** and select the surface on which you want to create the plane. Next, select the sketched point. The preview of the plane will be displayed in the drawing

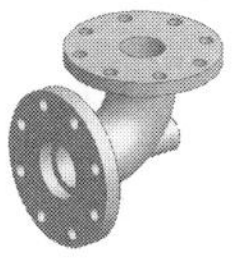

Tip. *You can also create the planes by dynamically dragging an existing plane. For creating a plane by dragging, you do not need to invoke the **Plane PropertyManager**. Using the left mouse button select the plane from the **Feature Manager Design Tree** or the **drawing** area. Press and hold the CTRL key down and drag the cursor. You will notice that the value of the distance in the **Distance** spinner of the **Plane PropertyManager** is modified and the preview of the plane is displayed in the drawing area. After dragging the plane to a required location release the left mouse button. Right-click and choose the **OK** option or choose the **OK** button from the **Plane PropertyManager**.*

*You can also create a plane at an angle by dragging. To create a plane at an angle by dragging, select an edge or an axis and an existing plane. Now, hold the CTRL key and drag the mouse. Enter the angle value in the **Angle** spinner.*

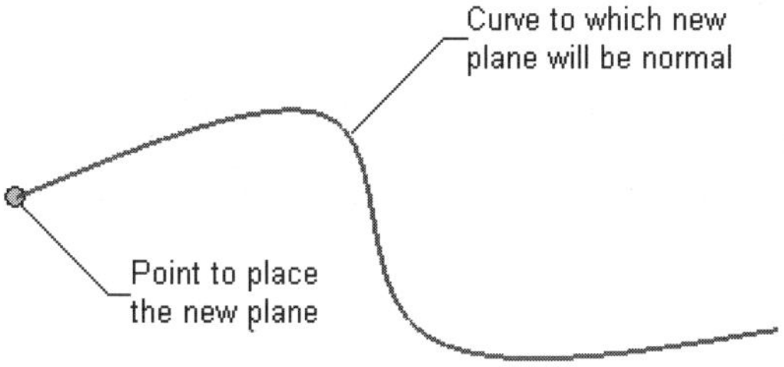

Figure 5-16 *Edge to be selected*

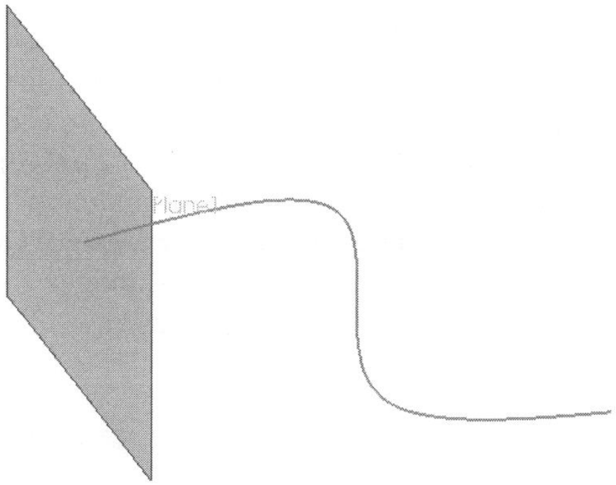

Figure 5-17 *Resulting plane*

area and you have to right-click to choose the **OK** option. If the sketch is created on a plane at an offset distance from the selected surface, the **Project to nearest location on surface** and **Project onto surface along sketch normal** radio buttons, and the **Other Solutions** button will be displayed on the **Plane PropertyManager**. Select any of the radio buttons according to the requirement. You can also view the other solutions of the plane creation using the **Other Solutions** button from the **Plane PropertyManager**. Figure 5-18 shows the selection of references for the plane creation and Figure 5-19 shows the resulting plane created.

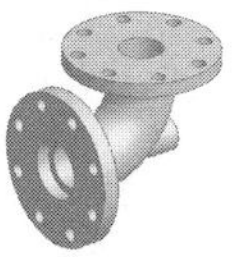

Tip. *If you select an edge of a model or an existing curve and invoke the sketching environment, a reference plane will be automatically created normal to that edge or curve and is selected as the sketching plane.*

Creating Reference Axis

CommandManager:	Reference Geometry > Axis
Menu:	Insert > Reference Geometry > Axis
Toolbar:	Reference Geometry > Axis

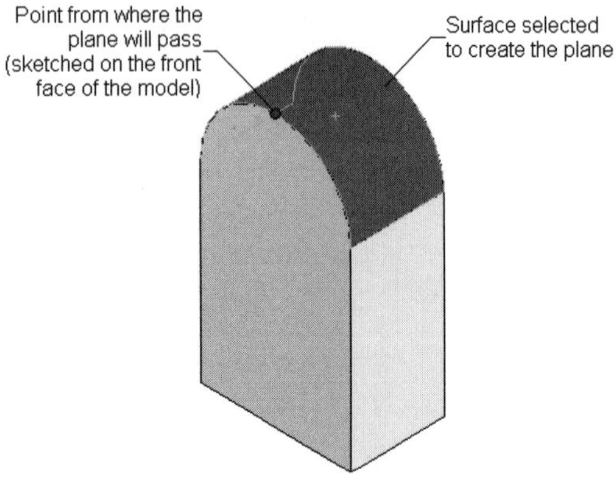

Figure 5-18 *References to be selected*

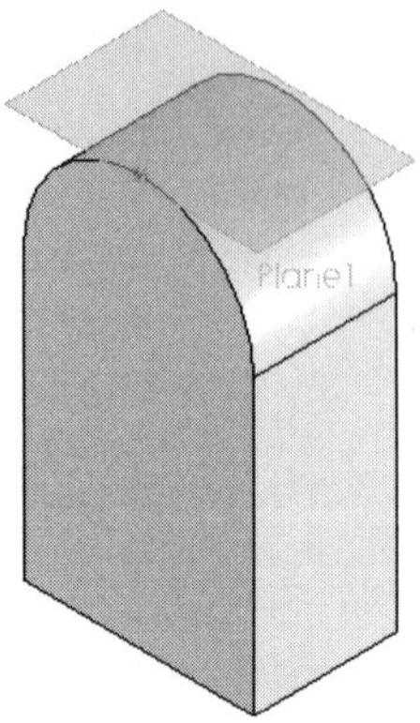

Figure 5-19 *Resulting plane*

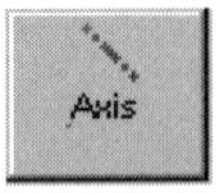

The **Reference Axis** option is used to create a reference axis or construction axis. These axes are the parametric lines passing through a model, feature, or reference entity. The reference axes are used to create reference planes, coordinate systems, circular patterns, and for applying mates in the assembly. These are also used as reference while sketching, or creating features. The reference axes are displayed in the model as well as in the **Feature Manager Design Tree**. When you choose the **Axis** button from the **Reference Geometry CommandManager** to create an axis, the **Axis PropertyManager** will be displayed, as shown in Figure 5-20.

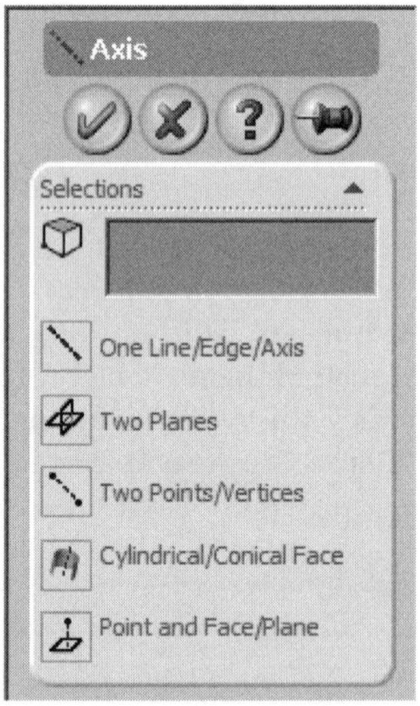

Figure 5-20 *The **Axis PropertyManager***

Tip. *If the axis is not displayed in the drawing area even after you have created it, choose **View** > **Axes** from the menu bar.*

The options available in this dialog box are discussed next.

Creating a Reference Axis Using One Line/Edge/Axis

The **One Line/Edge/Axis** option is used to create a reference axis by selecting a sketched line or construction line, an edge, or a temporary axis. To use this option, invoke the **Axis PropertyManager** and choose the **One Line/Edge/Axis** button. Select a sketched line, edge, or a temporary axis. The name of the selected entity will be displayed in the **Reference Entities** selection area and the preview of the reference axis will be displayed in the drawing area. Choose the **OK** button to create the reference axis. Figure 5-21 shows a construction line selected as a reference for creating an axis and Figure 5-22 shows an axis created using this option.

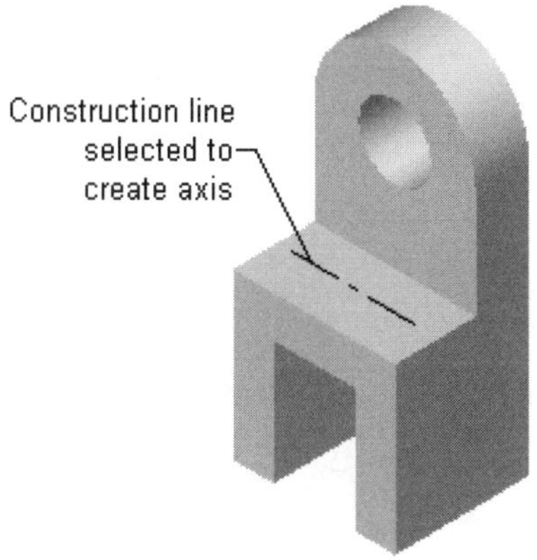

Figure 5-21 *Line to be selected*

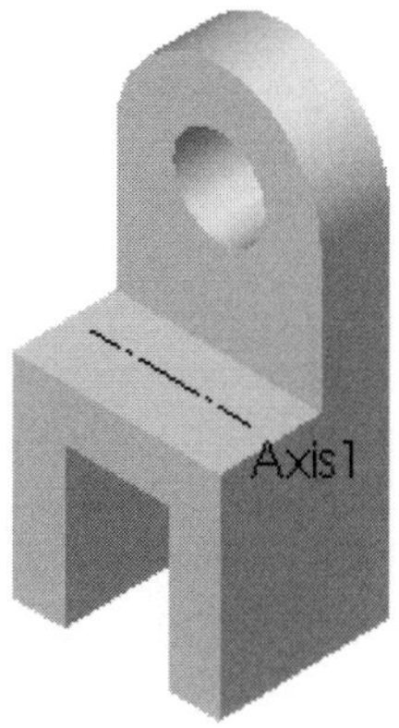

Figure 5-22 *Resulting reference axis*

Creating a Reference Axis Using Two Planes

Using the **Two Planes** option, you can create a reference axis at the intersection of two planes. To create a reference axis using this option, invoke the **Axis PropertyManager** and choose the **Two Planes** button. Now, select two planes, two planar faces, or a plane and a planar face you want to use to create the axis. The preview of the axis will be displayed in the drawing area. Choose the **OK** button from the **Axis PropertyManager**. Figure 5-23 shows two planes selected and Figure 5-24 shows the resulting reference axis created using this option.

Creating a Reference Axis Using Two Points or Vertices

Using the **Two Points/Vertices** option, you can create a reference axis that passes through two points or two vertices. To create a reference axis using this option, invoke the **Axis PropertyManager** and choose the **Two Points/Vertices** button. Now, select two points or two vertices through which you want the reference axis to pass. The preview of the reference axis will be displayed in the drawing area. Choose the **OK** button from the **Axis PropertyManager**. Figure 5-25 shows two vertices to be selected and Figure 5-26 shows the resulting reference axis created using this option.

Creating a Reference Axis Using a Cylindrical or a Conical Face

Using the **Cylindrical/Conical Face** option, you can create a reference axis that passes through

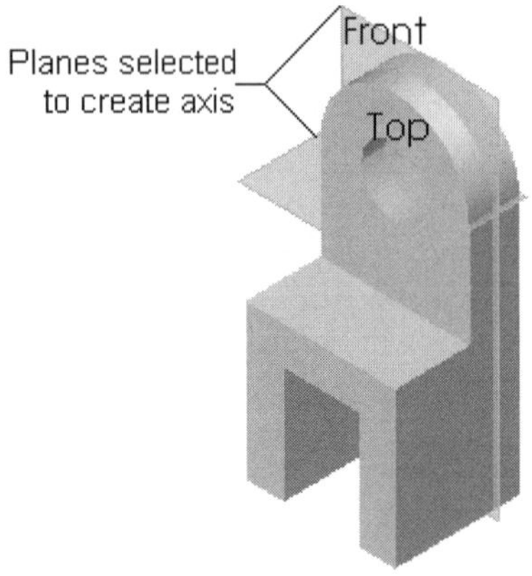

Figure 5-23 Planes to be selected

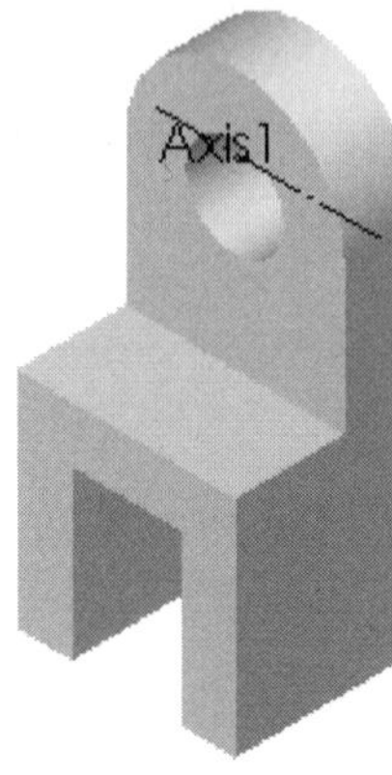

Figure 5-24 Resulting reference axis

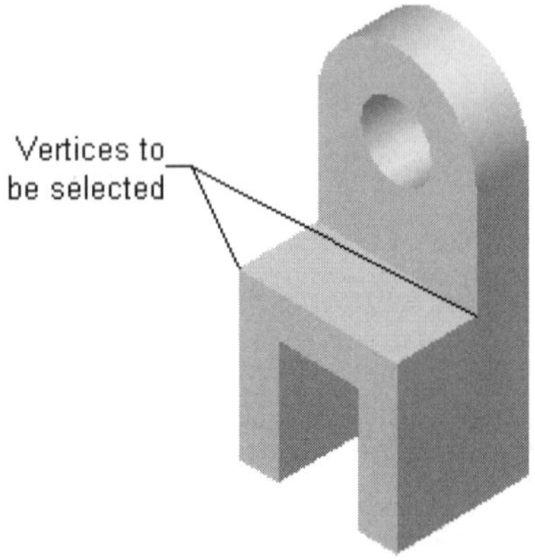

Figure 5-25 Vertices to be selected

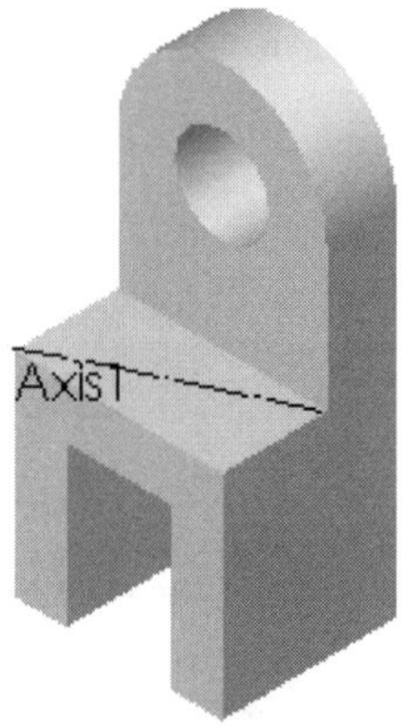

Figure 5-26 Resulting reference axis

the center of a cylindrical or a conical face. To create a reference axis using this option, invoke the **Axis PropertyManager** and choose the **Cylindrical/Conical Face** button. Now, select the cylindrical or the conical face through the center of which the axis needs to pass. The preview of the reference axis will be displayed in the drawing area. Choose the **OK** button from the **Axis PropertyManager**. Figure 5-27 shows a cylindrical face selected and Figure 5-28 shows the resulting reference axis created using this option.

Creating a Reference Axis on a Face/Plane Passing Through a Point

Using the **Point and Face/Plane** option you can create a reference axis that passes through a point and is normal to the selected face/plane. If the selected face is a nonplanar face, the selected point should be created on the face. To create a reference axis using this option, invoke the **Axis PropertyManager**. Select the **Point and Face/Plane** button from this **PropertyManager**. Now, select a point, vertex, or midpoint and then select a face or a plane. The preview of the axis will be displayed in the drawing area. Choose the **OK** button from the **Axis PropertyManager**. The newly created axis will be normal to the selected face or plane. Figure 5-29 shows the point and face selected and Figure 5-30 shows the resulting axis created using this option.

Figure 5-27 *Cylindrical face to be selected*

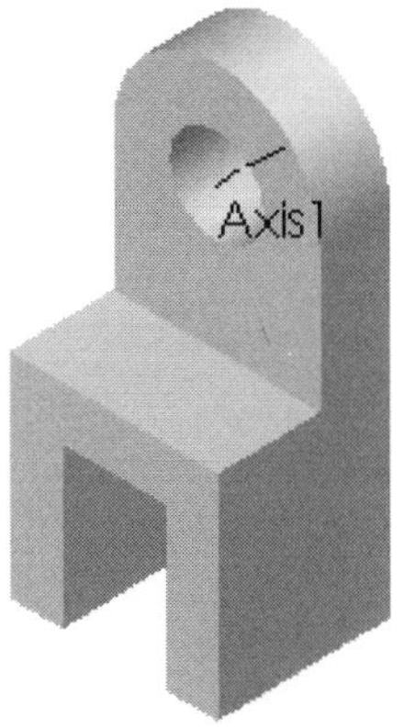

Figure 5-28 *Resulting reference axis*

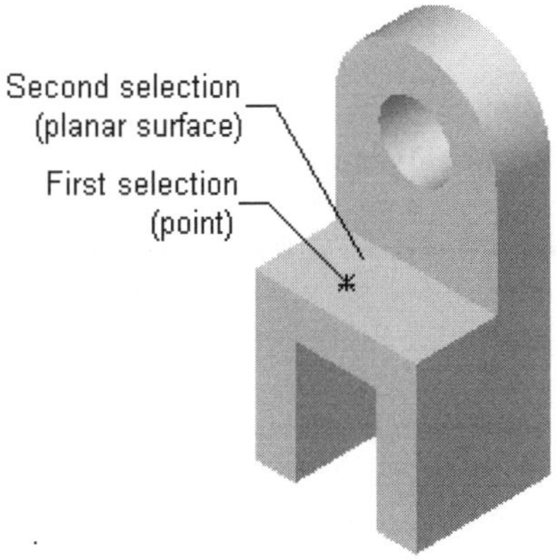

Figure 5-29 *Point and face to be selected*

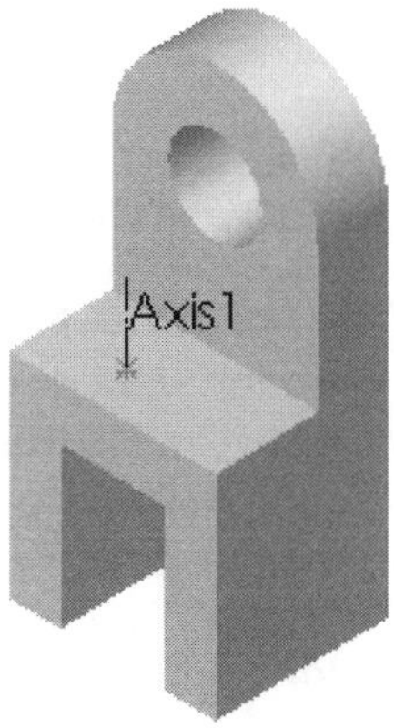

Figure 5-30 *Resulting reference axis*

Note

*When you create a circular feature, a temporary axis is automatically created. You can display the temporary axis by choosing **View > Temporary Axis** from the menu bar.*

Creating Reference Points

CommandManager:	Reference Geometry > Point
Menu:	Insert > Reference Geometry > Point
Toolbar:	Reference Geometry > Point

Reference points are also created to assist you in designing. They work as an aid for creating another reference geometry or feature. To create a reference point, choose the **Point** button from the **Reference Geometry CommandManager**; the **Point PropertyManager** will be displayed, as shown in Figure 5-31.

This **PropertyManager** allows you to use five methods of creating reference points. These methods are discussed next.

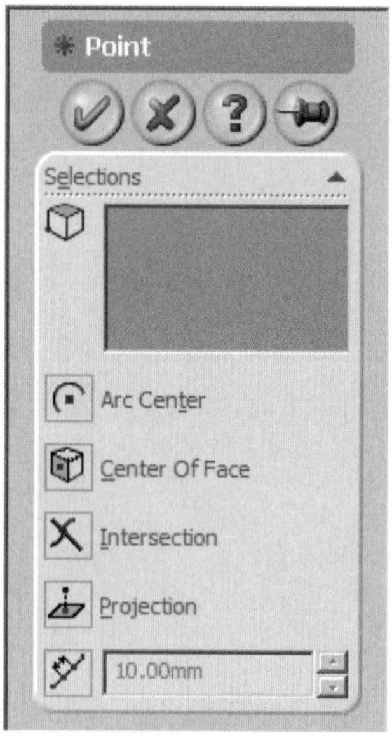

Figure 5-31 *The* ***Point PropertyManager***

Creating a Reference Point at the Center of an Arc or a Curved Edge

The **Arc Center** option is used to create a reference point at the center of a sketched arc or a curved edge. When you invoke the **Point PropertyManager** and choose the **Arc Center** button, you will be prompted to select an arc or a circular edge to define the reference point. As soon as you select a sketched arc or circle, or a curved edge, the preview of the reference point will be displayed at its center. Choose **OK** to confirm the creation of the reference point. Figure 5-32 shows a curved edge selected to create the reference point and the preview of the resulting reference point.

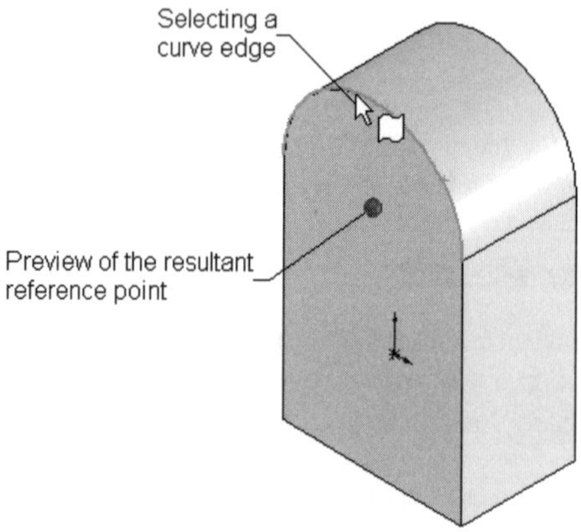

Figure 5-32 *Reference point at the center of a curved edge*

Creating a Reference Point at the Center of a Face

The **Center Of Face** option allows you to create a reference point at the center of a face. When you choose this button, you will be prompted to select a face to define the reference point. You can select a planar or a curved face. The resulting point is automatically placed at the point of the center of gravity of the selected face.

Creating a Reference Point at the Intersection of Two Edges, Sketched Segments, or Reference Axes

The **Intersection** option allows you to create a reference point at the intersection of two edges, sketched segments, or reference axes.

Creating a Reference Point by Projecting an Existing Point

The **Projection** option allows you to create a reference point by projecting a point from some other plane to a specified plane. The point that you can project can be a sketched point, endpoints, or center points of sketched entities, or a reference point.

Creating Single or Multiple Reference Points Along the Distance of a Sketched Curve or an Edge

The **Along curve distance or multiple reference point** option allows you to create single or multiple reference points along the distance of a selected curve. When you choose this button, the **Selections** rollout of the **Point PropertyManager** will expand and provide additional options. These options are discussed next.

Enter the distance/percentage value according to distance

This spinner is used to specify the distance or the percentage value between the individual reference points along the selected curve.

Distance

This radio button is selected to define the gap between the individual points in terms of distance value. If this radio button is selected, the value entered in the **Enter the distance/percentage value according to distance** spinner will be in terms of linear units of the current document.

Percentage

This radio button is selected to define the gap between the individual points in terms of the percentage of the length of the selected curve. If this radio button is selected, the total length of the selected curve will be taken as 100%. Now, the value entered in the **Enter the distance/percentage value according to distance** spinner will be in terms of the percentage of selected curve.

Evenly Distribute

This radio button is selected to evenly distribute the specified number of reference points through the length of the selected curve. If this radio button is selected, the **Enter the distance/percentage value according to distance** spinner will not be available.

Enter number of reference points to be created along the selected entity

This spinner is used to specify the number of reference points to be created. The specified number of reference points will be placed with the gap defined in the **Enter the distance/percentage value according to distance** spinner along the selected curve.

Creating Reference Coordinate System

CommandManager:	Reference Geometry > Coordinate System
Menu:	Insert > Reference Geometry > Coordinate System
Toolbar:	Reference Geometry > Coordinate System

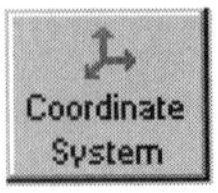

In SolidWorks, you may need to define some reference coordinate systems other than the default coordinate system for creating features, analyzing the geometry, analyzing the assemblies, and so on. To create a user-defined coordinate system, choose the **Coordinate System** button from the **Reference Geometry CommandManager**; the **Coordinate System PropertyManager** will be displayed, as shown in Figure 5-33. Also, a coordinate system will be displayed at the origin of the current document.

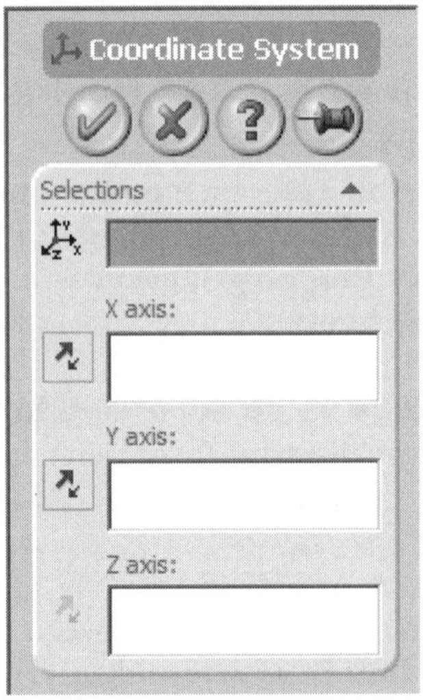

Figure 5-33** The **Coordinate System PropertyManager

To create a new coordinate system, you need to select a point that will be selected as the origin for the new coordinate and then define the directions of the X and Y, Y and Z, or Z and X axes. This is the reason that when you invoke the **Coordinate System PropertyManager**, the **Origin** selection area will be highlighted. After selecting the origin, you can define the direction of any two axes. The direction of the third axis will be automatically determined. You can select edges, points, or reference axes to define these directions.

You can reverse the directions of the axes by choosing the buttons available on the right of their respective selection areas.

OTHER BOSS/BASE OPTIONS

Enhanced

Some of the boss/base extrusion options were discussed in the previous chapter. In this chapter, the remaining boss/base extrusion options are discussed.

From

The **From** option available in the **Extrude PropertyManager** is one major enhancement in this release of SolidWorks. Using this option, you can specify the parameters at the start of the extrude feature. The **From** rollout of the **Extrude PropertyManager** is shown in Figure 5-34. The options available in this rollout are discussed next.

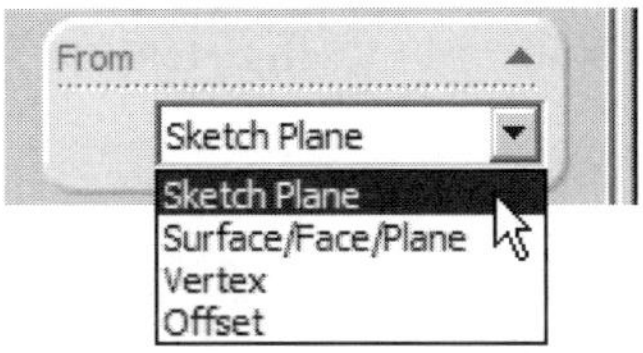

Figure 5-34 *The **From** rollout of the **Extrude PropertyManager***

Sketch Plane

When you invoke the **Extrude PropertyManager**, the **Sketch Plane** option will be chosen by default in the **From** rollout. With this option chosen, the extrude feature starts from the sketching plane on which the sketch is drawn. This option is mostly used while creating the extrude features.

Surface/Face/Plane

The **Surface/Face/Plane** option is used to start the extrude feature from a selected surface, face, or a plane, instead of the plane on which the sketch is drawn. To do this, invoke the **Extrude PropertyManager** and choose the **Surface/Face/Plane** option from the **From** rollout; the **Select A Surface/Face/Plane** selection area will be displayed in the **From** rollout. Select a surface, face, or plane from which you need start the extrude feature. Figure 5-35 shows the sketch to be extruded and the face selected as reference for starting the extrude feature. Figure 5-36 shows the resulting feature extruded from the selected face up to a specified depth.

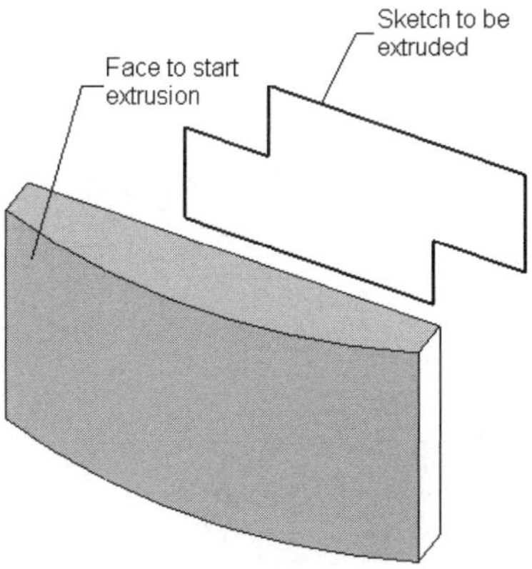

Figure 5-35 *Sketch to be extruded and a reference face to be selected*

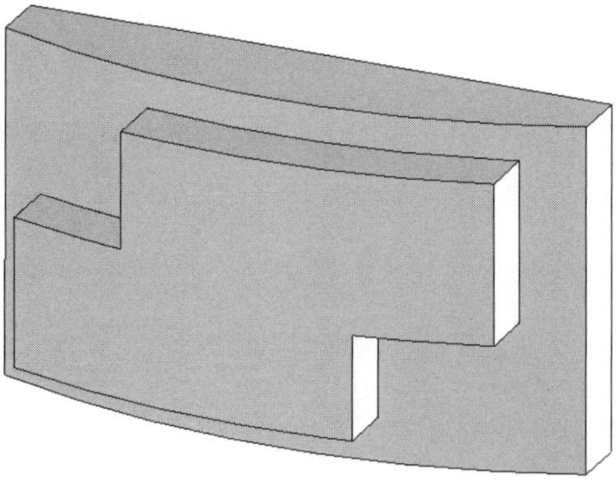

Figure 5-36 *Resulting extruded feature*

Vertex

The **Vertex** option available in the **From** rollout is used to specify a vertex as the reference for starting the extrude feature. To do this, invoke the **Extrude PropertyManager** and choose the **Vertex** option from the **From** rollout; the **Select A Vertex** selection area will be displayed. Select a vertex of an existing feature or an endpoint of an existing sketch. Figure 5-37 shows the sketch to be extruded and the vertex to be selected as reference to start the extrude feature. Figure 5-38 shows the resulting extruded feature from the selected vertex to the defined depth.

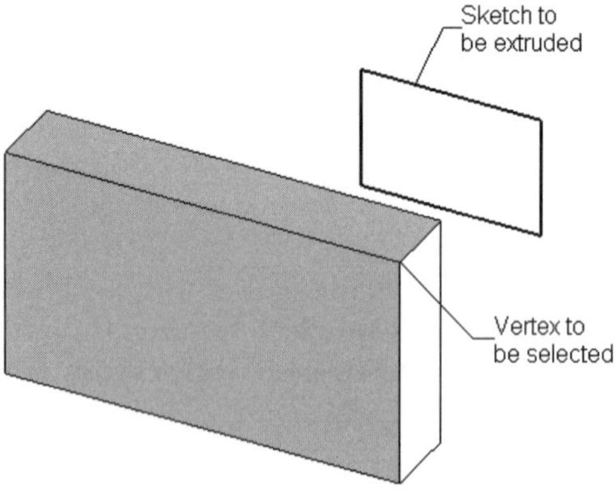

Figure 5-37 *Sketch to be extruded and a reference vertex to be selected*

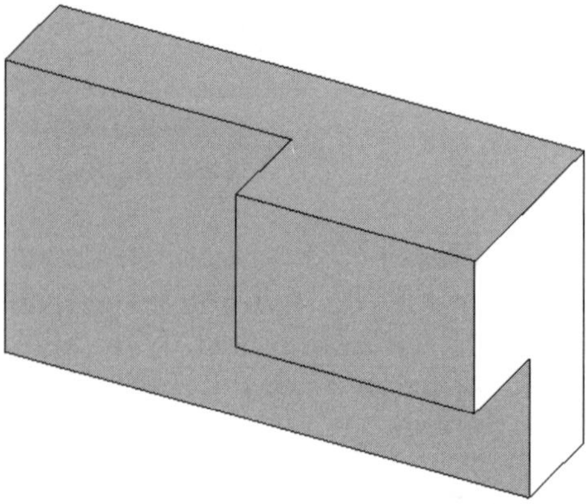

Figure 5-38 *Resulting extruded feature*

Offset

The **Offset** option available in the **From** rollout is used to start the extrude feature at an offset from the plane on which the sketch is drawn. To do this, choose the **Offset** option from the **From** rollout; the **Enter Offset Value** spinner will be displayed. Set the value of the offset in this spinner. Figure 5-39 shows the preview of the sketch drawn on the **Front Plane** being extruded using the **Offset** option. Figure 5-40 shows the resulting extruded feature from an offset distance from the sketching plane to a defined depth.

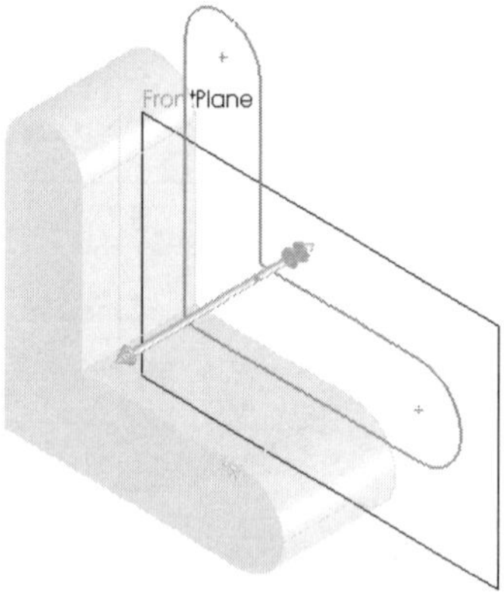

Figure 5-39 *Sketch being extruded using the **Offset** option*

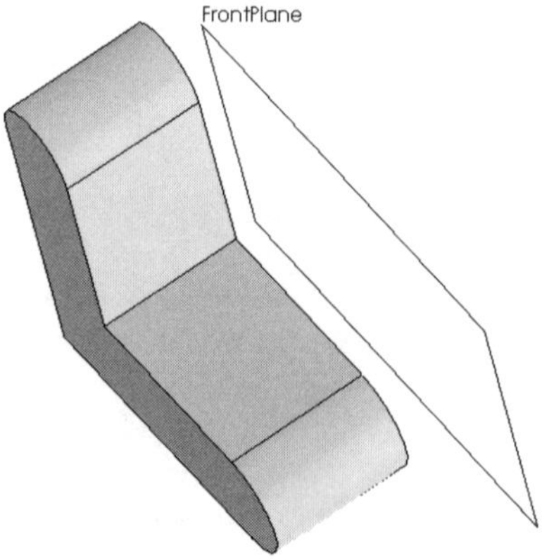

Figure 5-40 *Resulting extruded feature*

End Condition

The options available in the **End Condition** drop-down list are discussed next.

Through All

The **Through All** option is available in the **End condition** drop-down list only after you create a base feature. After creating a base feature, define a new sketching plane and create the sketch using the standard sketching tools. Now, choose the **Extruded Boss/Base** button from the **Features CommandManager** to invoke the **Extrude PropertyManager**. The preview of the extruded feature that will be created using the default settings will be displayed in temporary graphics in the drawing area. Select the **Through All** option from the **End Condition** drop-down list; the preview of the extruded feature will extend from the sketching plane through all existing geometric entities. You can also reverse the direction of extrusion using the **Reverse Direction** button available on the left of the **End Condition** drop-down list.

While creating an extruded feature using the **Through All** option, the sketch extrudes through all existing geometries. You will observe that the **Merge result** check box is displayed in the **Extrude PropertyManager**. This check box is selected by default. Therefore, the newly created extruded feature will merge with the base feature. If you clear this check box, this extruded feature will not merge with the existing base feature, resulting in the creation of another body. The creation of a new body can be confirmed by observing the **Solid Bodies** folder in the **Feature Manager Design Tree**. The value of the number of disjoint bodies in the model is displayed in parentheses on the right of the **Solid Bodies** folder. You can click on the + sign on the left of the **Solid Bodies** folder to expand the folder. To collapse the folder back, click on the - sign.

Figure 5-41 displays a sketch created on the sketching plane at an offset distance from the right planar face of the model. Figure 5-42 displays the feature created by extruding the sketch using the **Through All** option.

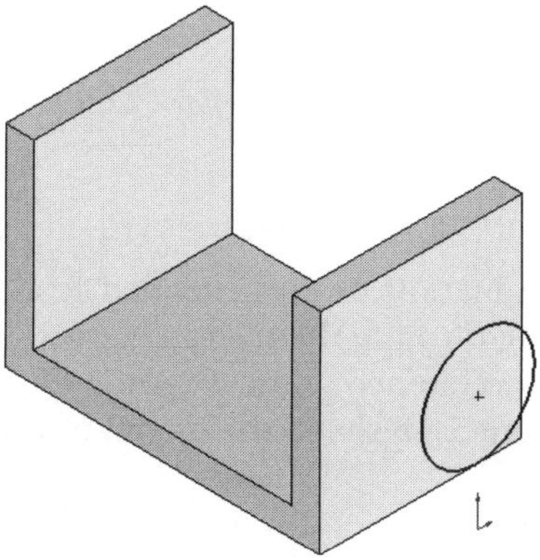

Figure 5-41 *A sketch drawn at an offset distance from the right planar surface*

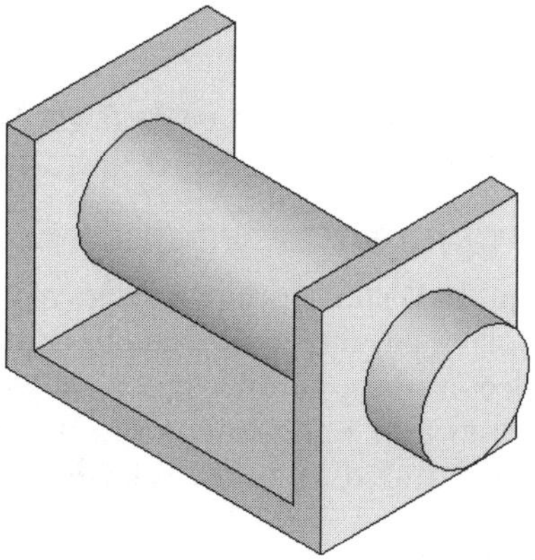

Figure 5-42 *Sketch extruded using the **Through All** option*

Tip. *The feature created from multiple disjoint closed contours results in creation of disjoint bodies.*

Note

It is recommended that while creating additional features after the base feature, you should always select the ***Merge results*** *check box in the* ***Feature PropertyManager****.*

Up To Next

The **Up To Next** option is used to extrude the sketch from the sketching plane to the next surface that intersects the feature. After creating a base feature, create a sketch by selecting or creating a sketching plane. Invoke the **Extrude PropertyManager**; the preview of the base feature will be displayed with the default options. Select the **Up To Next** option from the **End Condition** drop-down list. You can also reverse the direction of feature creation using the **Reverse Direction** button. The preview of the feature will be modified and the sketch will be displayed as extruded from the sketching plane to the next surface that intersects the feature geometry. Figure 5-43 shows the sketch that will be extruded using the **Up To Next** option and Figure 5-44 shows the resulting feature.

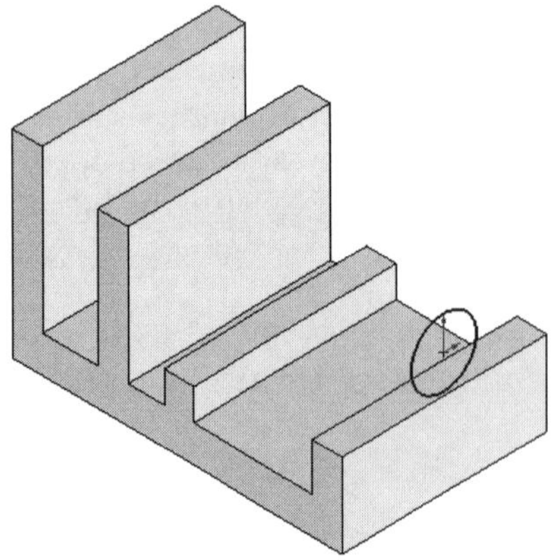

Figure 5-43 *A sketch drawn on the* ***Right Plane*** *as the sketching plane*

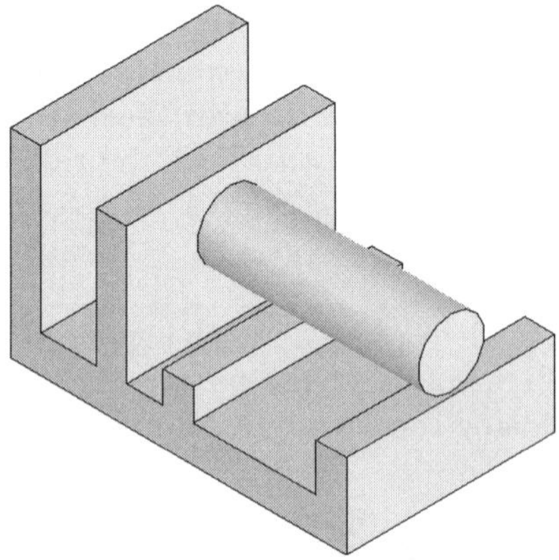

Figure 5-44 *Sketch extruded using the* ***Up To Next*** *option*

Up To Vertex

The **Up To Vertex** option is used to define the termination of the extruded feature at a virtual plane parallel to the sketching plane and passing through the selected vertex. You can also select a point on an edge, a sketched point, or a reference point. Figure 5-45 shows a sketch drawn on a plane at an offset distance and Figure 5-46 shows the model in which the sketch is extruded up to the selected vertex.

Up To Surface

The **Up To Surface** option is used to define the termination of the extruded feature using a selected surface or a face. To create an extruded feature using this option, draw a sketch using the normal sketching options and then invoke the **Extrude PropertyManager**. Select

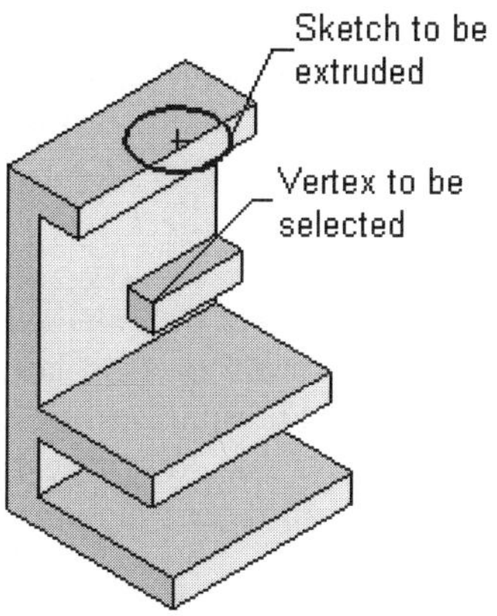

***Figure 5-45** Sketch drawn on a plane created at an offset distance and a vertex to be selected*

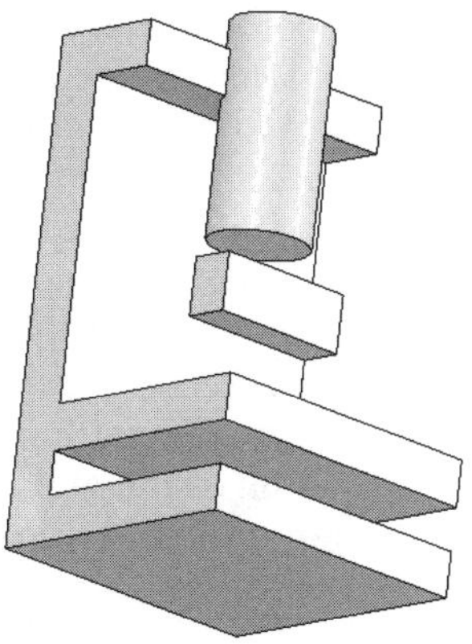

***Figure 5-46** Sketch extruded using the **Up To Vertex** option.*

the **Up To Surface** option from the **End Condition** drop-down list; the **Face/Plane** selection area will be displayed and you will be prompted to select a face or a surface. Select a surface up to which you want to extrude the feature. Figure 5-47 shows the sketch drawn at an offset distance and the surface to be selected. Figure 5-48 shows the resulting feature extruded up to the selected surface.

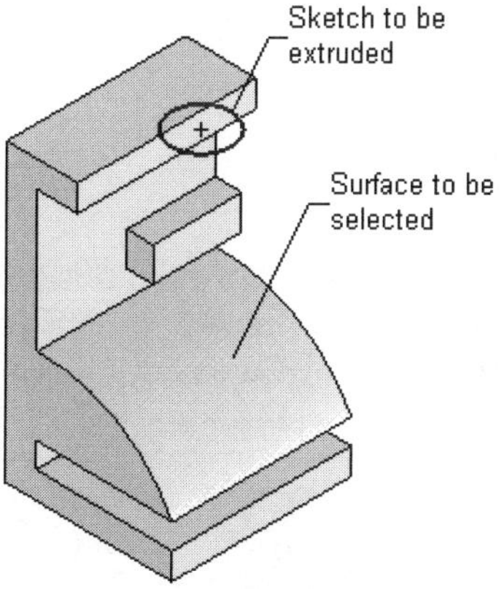

***Figure 5-47** Sketch drawn on a plane created at an offset distance and a surface to be selected*

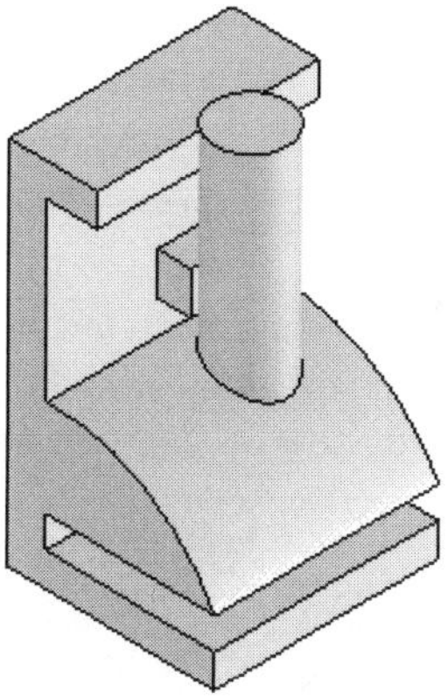

***Figure 5-48** Sketch extruded using the **Up To Surface** option.*

Offset From Surface

The **Offset From Surface** option is used to define the termination of the extruded feature on a virtual surface created at an offset distance from the selected surface. To create an extruded feature using the **Offset From Surface** option, create a sketch and invoke the **Extrude**

PropertyManager. Select the **Offset From Surface** option from the **End Condition** drop-down list; the **Face/Plane** selection area will be displayed along with the **Offset Distance** spinner. You will be prompted to select a face or a surface. Select the surface and set the offset distance in the **Offset Distance** spinner. You can reverse the direction of offset by selecting the **Reverse offset** check box from the **Direction 1** rollout. If the **Translate surface** check box is cleared, the virtual surface created for the termination of the extruded feature will have a concentric relation with the selected surface. Therefore, it reflects the true offset of the selected surface. If the **Translate surface** check box is selected, the virtual surface will be translated to the distance provided as the offset distance from the reference surface. Therefore, a virtual surface is created to define the termination of the extruded feature and it does not reflect the true offset of the selected surface. This concept will be more clear if you refer to Figure 5-49.

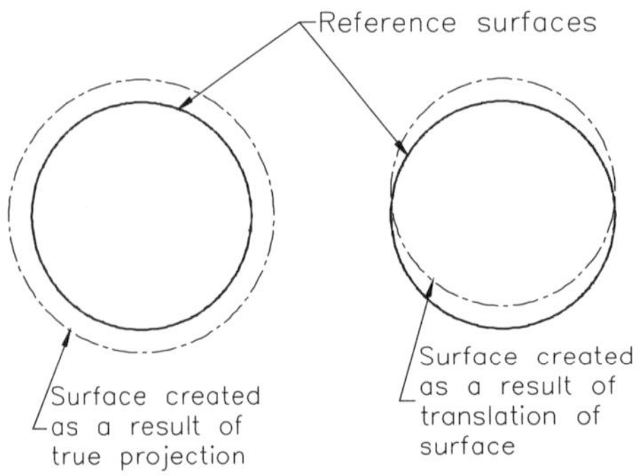

Figure 5-49 *Surfaces with true projection and translation*

Figure 5-50 shows the front view of the sketch extruded with termination at an offset distance from the selected cylindrical surface with the **Translate surface** check box cleared. Figure 5-51 shows the front view of the extruded feature with the **Translate surface** check box selected.

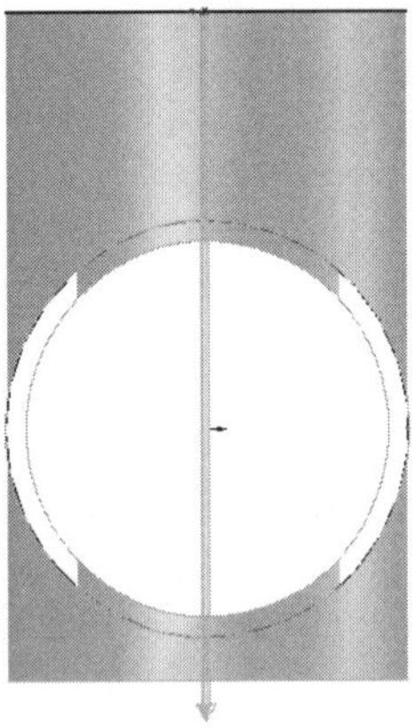

Figure 5-50 *Sketch extruded using the **Offset From Surface** option with the **Translate surface** check box cleared*

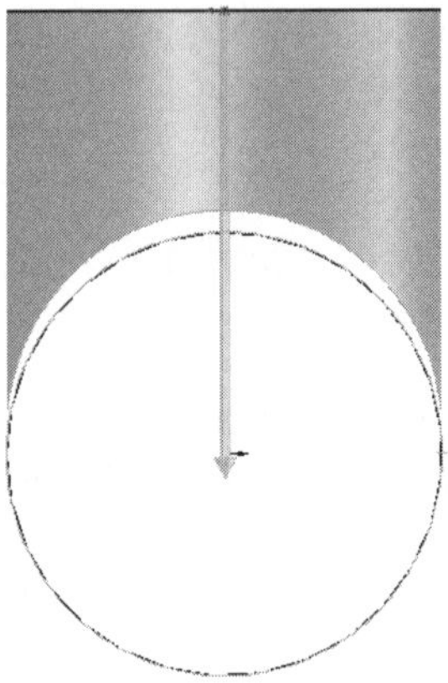

Figure 5-51 *Sketch extruded using the **Offset From Surface** option with the **Translate surface** check box selected*

Up To Body

The **Up To Body** option is used to define the termination of the extruded feature to another body. To create an extruded feature using the **Up To Body** option, invoke the **Extrude PropertyManager** and select the **Up To Body** option from the **End Condition** drop-down list; the **Solid/Surface Body** selection area will be displayed. Select the body to terminate the feature and choose the **OK** button. Figure 5-52 shows the sketch for the extruded feature and a body up to which the sketch will be extruded. Figure 5-53 shows the resulting feature.

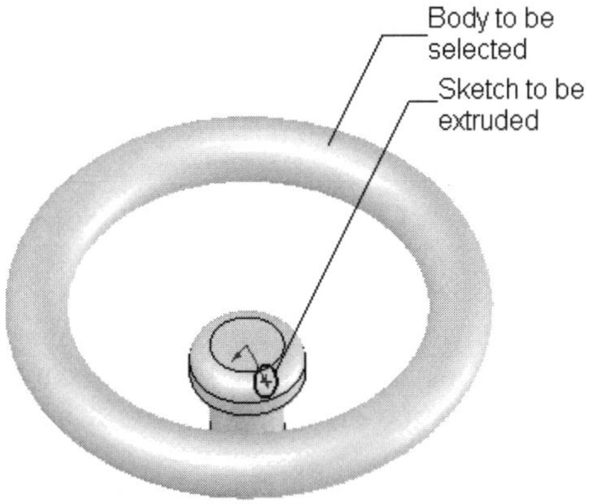

Figure 5-52 *Sketch to be extruded and the body to be selected for the extrude feature*

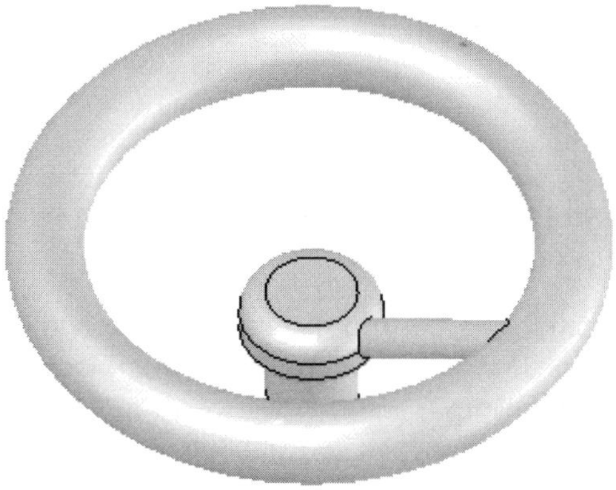

Figure 5-53 *Sketch extruded using the* ***Up To Body*** *option*

Direction of Extrusion

In SolidWorks 2005, you can define the direction for extruding the sketches. As mentioned in the previous chapter, the direction of extrusion is generally normal to the sketching plane. You can also define the direction of extrusion using a sketched line, an edge, or a reference axis. Note that the entity you want to use for defining the direction of extrusion should not be drawn on the sketch plane parallel to the plane on which the sketch to be extruded is drawn.

To define the direction of extrusion, click on the **Direction of Extrusion** selection box in the **Direction 1** area of the **Extrude PropertyManager**. Next, select an edge, a sketched line segment, or an axis. Figure 5-54 shows a sketch drawn on the top face of a rectangular block and a line sketched on the left face of the block to define the direction of extrusion. Figure 5-55 shows the resulting extruded feature.

MODELING USING THE CONTOUR SELECTION METHOD

Modeling using the contour selection method allows you to use partial sketches for creating the features. Using this method, you can draw sketches of the entire model in a single sketching environment and then manipulate the sketches by sharing them between various features. To understand this concept, consider the multifeatured solid model shown in Figure 5-56.

For a multifeatured model as shown in this figure, ideally you first need to draw the sketch for the base feature and convert that sketch into the base feature. Next, you need to draw the

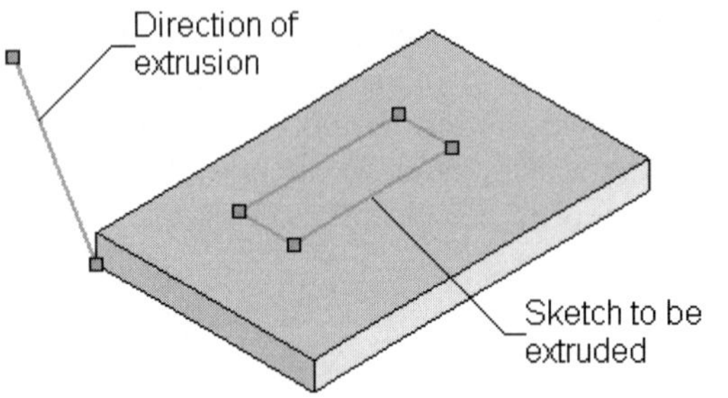

Figure 5-54 *Sketch to be extruded and the direction of extrusion*

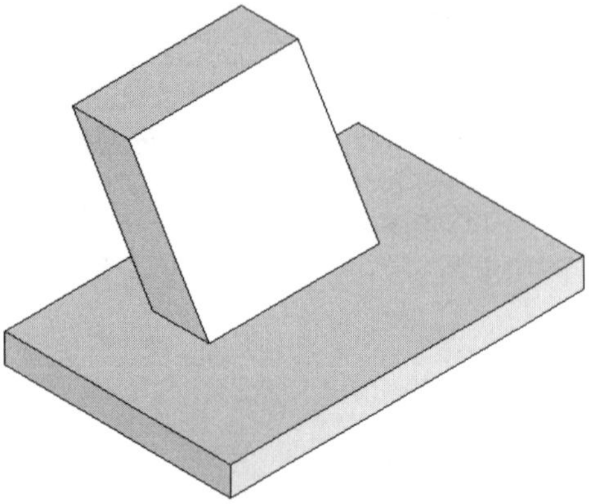

Figure 5-55 *Resulting extruded feature*

sketch for the second sketched feature and so on. In other words, you have to draw the sketch for each sketched feature. But using the contour selection method, you can share the contour created using the sketch for creating features. Figure 5-57 shows the sketch to be drawn for modeling using the contour selection method.

Figure 5-56 *Multifeatured solid model*

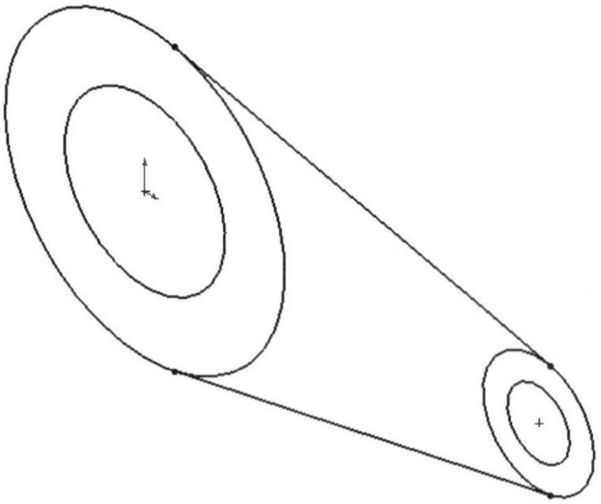

Figure 5-57 *Sketch for creating the model*

After drawing the entire sketch right-click in the drawing area to invoke the shortcut menu. Make sure you are still in the sketching environment. Choose the **Contour Selection Tool** option from the shortcut menu. If this option is not displayed by default, choose the down arrows in the shortcut menu to expand it. This option will be displayed now. When you choose this option, the select cursor will be replaced by the contour selection cursor and contour selection confirmation corner will be displayed. Using the contour selection cursor, click between the two circles on the left. The area between the two circles will be selected, as shown in Figure 5-58. Invoke the **Extrude PropertyManager** and extrude the selected contour using the **Mid Plane** option as shown in Figure 5-59.

Now, right-click in the drawing area and again choose the **Contour Selection Tool** option

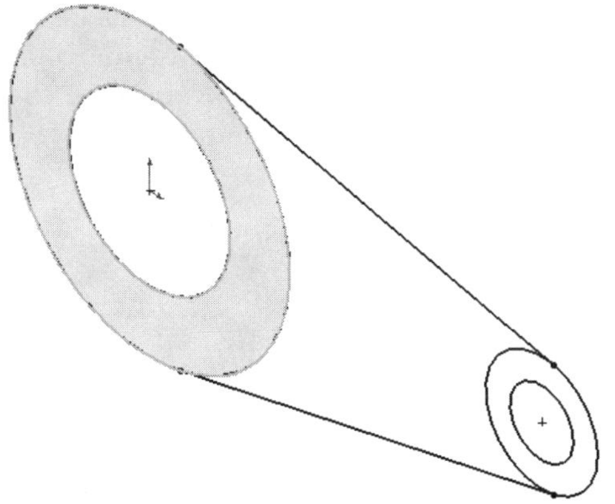

Figure 5-58 *Contour selected for creating the extruded feature*

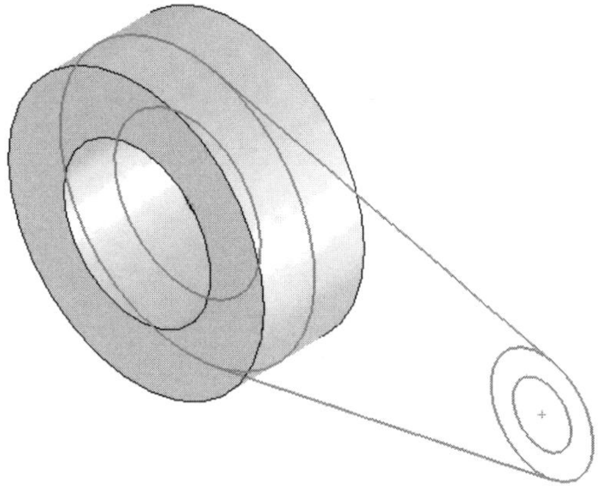

Figure 5-59 *Isometric view of the feature created by extruding the selected contour*

Tip. *When you move the contour selection cursor in the sketch, the areas where the contour selection is possible are dynamically highlighted.*

from the shortcut menu. Using the contour selection cursor, select any entity in the sketch and then select the middle contour of the sketch as shown in Figure 5-60. Invoke the **Extrude PropertyManager** and extrude the selected contour using the **Mid Plane** option. Again, invoke the contour selection tool and select an entity in the sketch. Similarly, again invoke the contour selection tool and select any entity of the sketch. Next, specify a point between the two circles on the right and extrude the same using the **Mid Plane** option.

After creating the model using this option, you will notice that the sketches are displayed in the model. This is the reason why you need to hide them. Click on the **+** sign located on the left of any of the extruded features to expand the tree view. Select the sketch icon and right-click to invoke the shortcut menu. Choose the **Hide** option. Figure 5-61 shows the model after creating all features using contour selection method and after hiding the sketch.

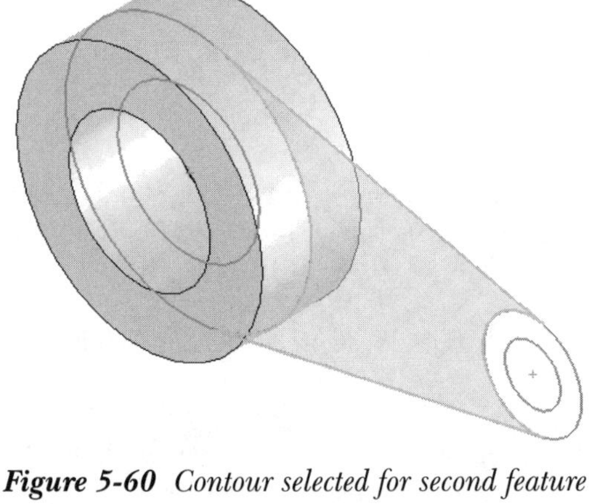

Figure 5-60 *Contour selected for second feature*

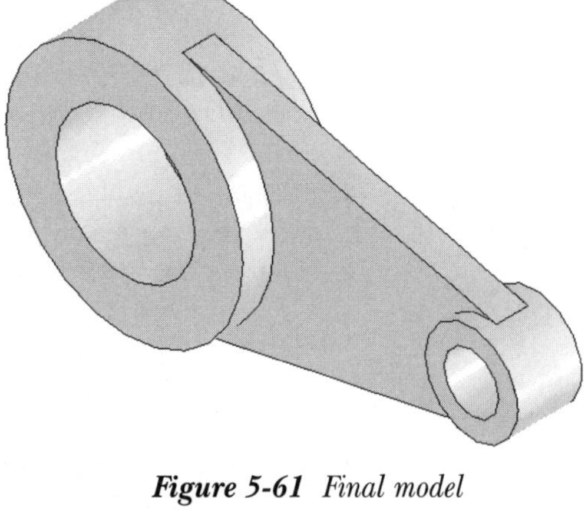

Figure 5-61 *Final model*

In SolidWorks 2005, you can also select the model edges as a part of the contour. For example, consider Figure 5-62. This figure shows a line drawn on the top face of a rectangular block. You can use the edges of the top face that form a contour with the line as a sketch to be extruded, see Figure 5-63.

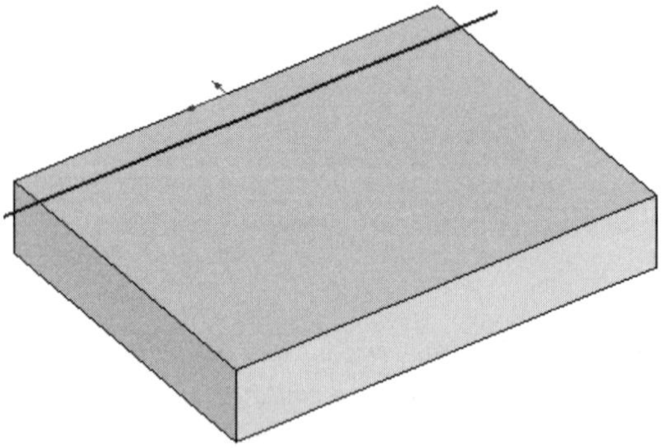

Figure 5-62 *Line drawn on the top face of a rectangular model*

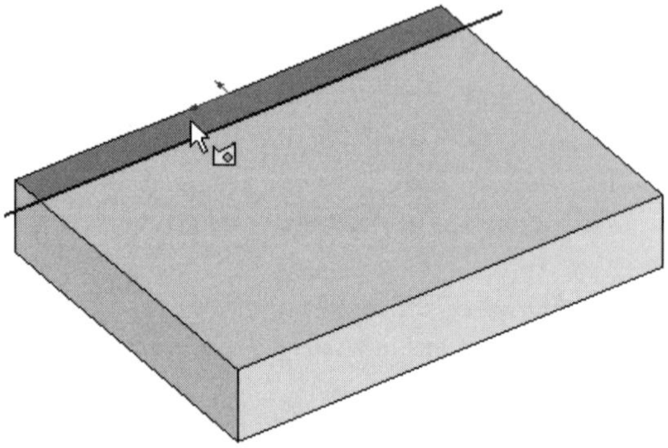

Figure 5-63 *Selecting the contour formed by the line and the model edges*

Figure 5-64 shows the resulting extruded feature.

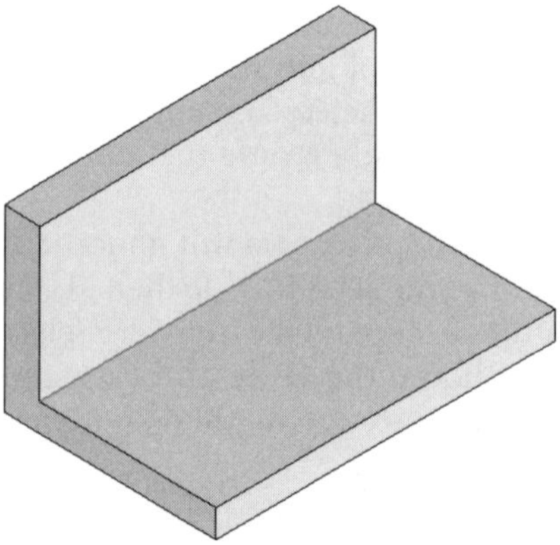

Figure 5-64 *Extruded feature created using the model edges as a part of the contour*

Tip. *When you select the contour using the contour selection tool and invoke the* ***Extrude PropertyManager****, you can observe the name of the selected contour in the selection area of the* ***Contour Selection*** *rollout.*

You can select the contours for all sketched features such as revolve, cut, sweep, loft, and so on.

You can also select a single sketched entity from a sketch using the contour selection tool instead of selecting the contour for creating the sketched features.

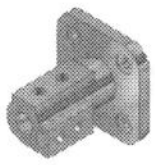

Note

*If you click on the +sign to expand the extruded feature in the **FeatureManager Design Tree**, you will notice that instead of showing the icon of a simple sketch, it will show the icon of contour selected sketch.*

CREATING CUT FEATURES

The cut is a material removal process. You can define a cut feature by extruding a sketch, revolving a sketch, sweeping a section along a path, lofting sections, or using a surface. You will learn more about sweep, loft, and surface in later chapters. The cut feature can be created only if a base feature exists. The extruded and revolved cut features are discussed next.

Creating Extruded Cuts

CommandManager:	Features > Extruded Cut
Menu:	Insert > Cut > Extrude
Toolbar:	Features > Extruded Cut

To create an extruded cut feature, create a sketch for the cut feature and then choose the **Extruded Cut** button from the **Features CommandManager**. The **Cut-Extrude PropertyManager** will be displayed, as shown in Figure 5-65. Also, the preview of the cut feature with default options will be displayed in the drawing area. Figure 5-66 shows the preview of the cut feature when you invoke the **Cut-Extrude**

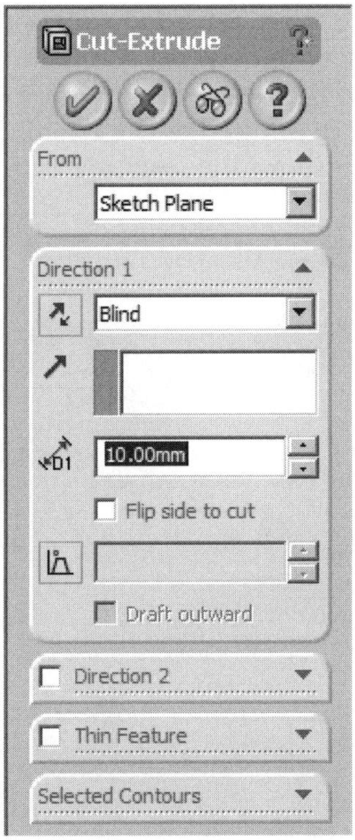

*Figure 5-65 The **Cut-Extrude PropertyManager***

PropertyManager after creating a sketch. Remember that when you are creating a cut feature, the current view is not automatically changed to a 3D view. You need to change it manually. The material to be removed is displayed in temporary graphics. Figure 5-67 shows the model after creating the cut feature.

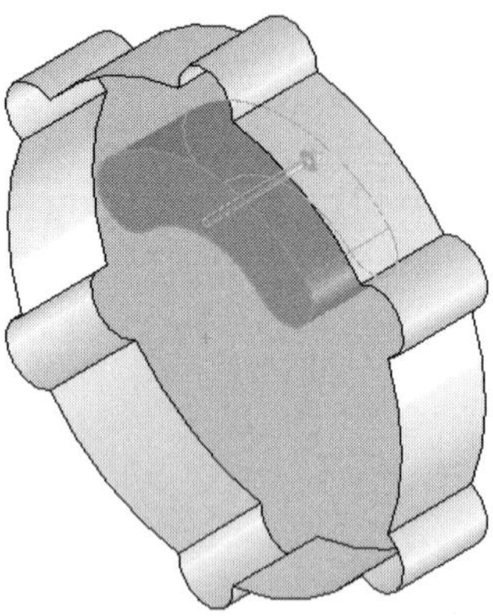

Figure 5-66 *The preview of the cut feature*

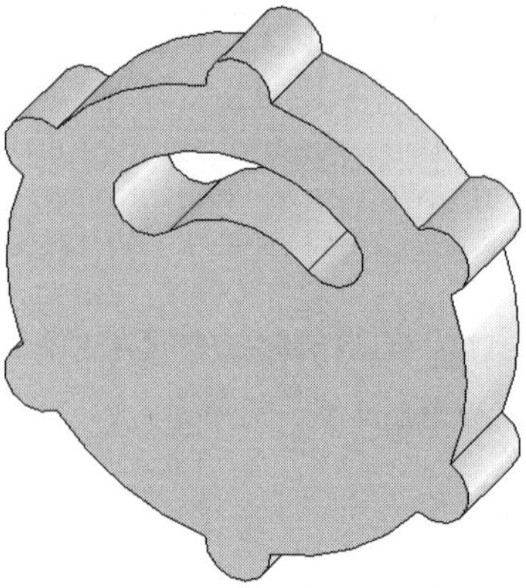

Figure 5-67 *Cut feature added to the model*

The options available in the **Cut-Extrude PropertyManager** are discussed next.

From

From this release of SolidWorks, you are provided with options of specifying parameters at the start of the extruded cut. These options are available in the **From** rollout and are the same as those discussed earlier for the **Boss/Base** option. Figure 5-68 shows the sketch to be extruded using with the curved face selected as the reference face for the end condition for starting the extrusion. Figure 5-69 shows the resulting extruded cut feature.

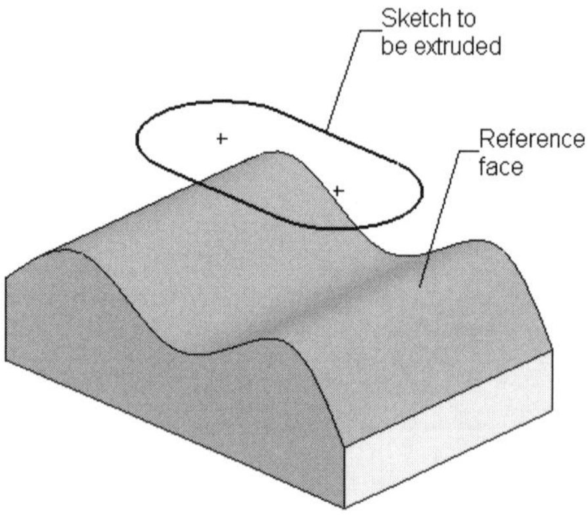

Figure 5-68 *Sketch to be extruded and the reference face*

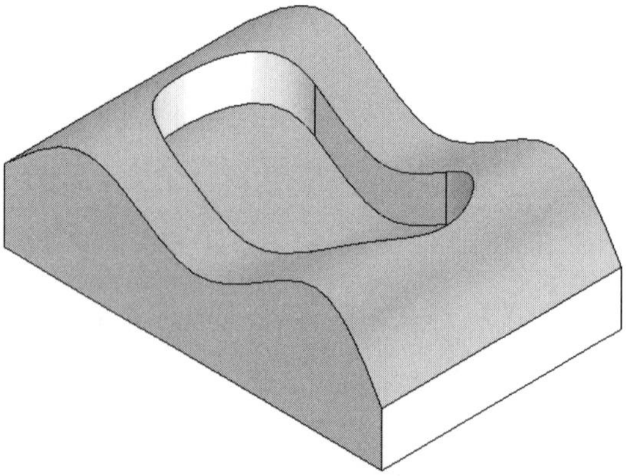

Figure 5-69 *Resulting extruded cut feature*

Direction 1

The **Direction 1** rollout is used to define the termination of the extrude in the first direction. The options available in the **Direction 1** rollout are discussed next.

End Condition

The **End Condition** drop-down list available in the **Direction 1** rollout is used to specify the type of termination option available. The feature termination options available in this drop-down list are **Blind**, **Through All**, **Up To Next**, **Up To Vertex**, **Up To Surface**, and **Mid Plane**. These options are the same as those discussed for boss/base options. By default, the **Blind** option is selected in the **End Condition** drop-down list. Therefore, the **Distance** spinner is displayed to specify the depth. If you choose the **Through All** or the **Through Next** options, the spinner will not be displayed. The type of spinner or the selection area that is displayed depends on the option selected from the **End Condition** drop-down list. The **Reverse Direction** button is used to reverse the direction of feature creation. If you choose the **Mid Plane** option from the **End Condition** drop-down list, the **Reverse Direction** button will not be available.

Flip side to cut

The **Flip side to cut** check box is selected to define the side of material removal. By default, the **Flip side to cut** check box is cleared. Therefore, the material will be removed from inside the profile of the sketch drawn for the cut feature. If you select this check box, the material will be removed from outside the profile of the sketch. Figure 5-70 shows a cut feature with the **Flip side to cut** check box cleared and Figure 5-71 shows a cut feature with the **Flip side to cut** check box selected.

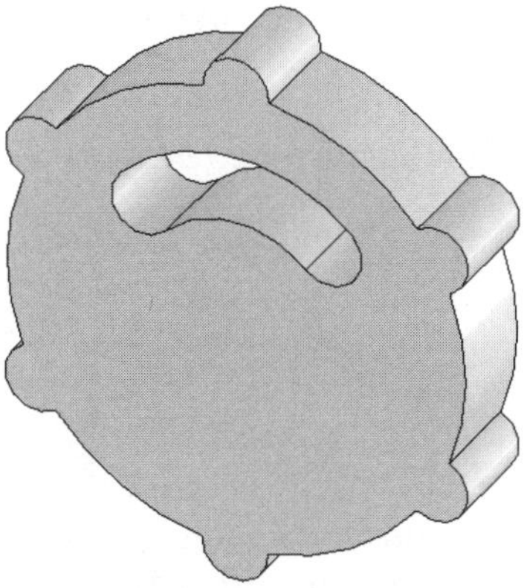

Figure 5-70 *Cut feature with the* ***Flip side to cut*** *check box cleared*

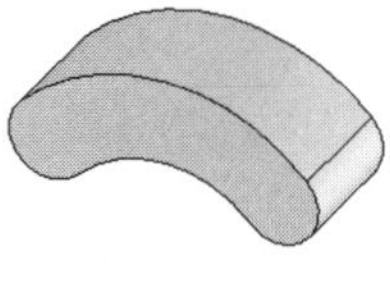

Figure 5-71 *Cut feature with the* ***Flip side to cut*** *check box selected*

Tip. *You can also reverse the direction of material removal by clicking on the arrow available on the sketch while creating the cut feature. This arrow is only available if you have selected or cleared the* ***Flip side to cut*** *check box once.*

Draft On/Off

The **Draft On/Off** button is used to apply the draft angle to the extruded cut feature.

The **Draft Angle** spinner available on the right of the **Draft On/Off** button is used to set the value of the draft angle. By default, the **Draft outward** check box is cleared. Therefore, the draft is created inward with respect to the direction of feature creation. If you select this check box, the draft added to the cut feature will be created outward with respect to the direction of feature creation. Figure 5-72 shows the draft added to the cut feature with the **Draft outward** check box cleared and Figure 5-73 shows the draft added to the cut feature with the **Draft outward** check box selected.

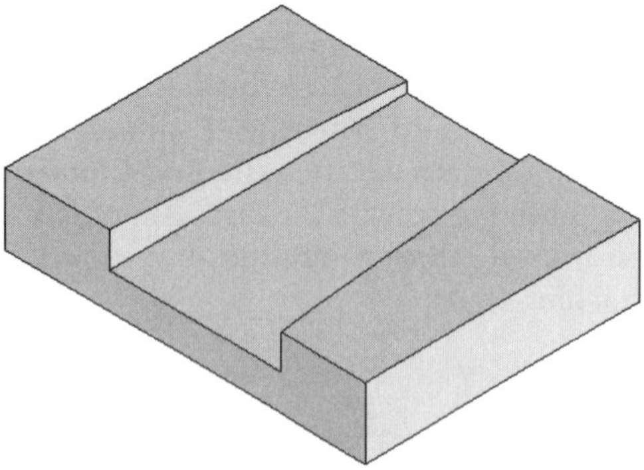

Figure 5-72 *Cut feature with the* ***Draft outward*** *check box cleared*

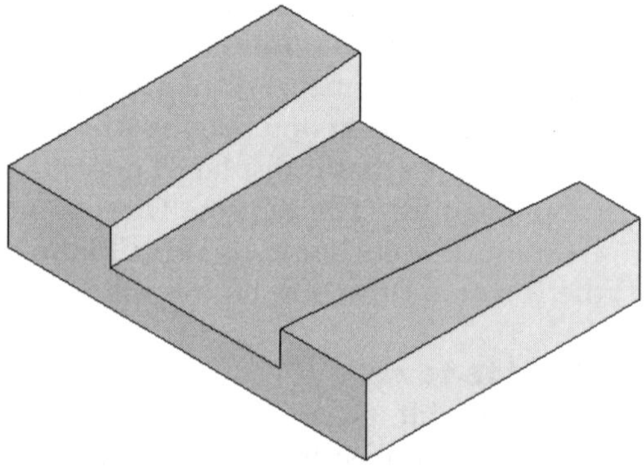

Figure 5-73 *Cut feature with the* ***Draft outward*** *check box selected*

The **Direction 2** rollout is used to specify the termination of feature creation in the second direction. The options available in the **Direction 2** rollout are the same as those discussed for the **Direction 1** rollout.

The **Selected Contour** rollout is used to select specific contours from the current sketch.

Tip. *The sketch used for the cut feature can be a closed loop or an open sketch. Note that if the sketch is an open sketch, the sketch should completely divide the model into two or more parts.*

Thin Feature

The **Thin Feature** rollout is used to create a thin cut feature. When you create a cut feature, you have to apply the thickness to the sketch in addition to the end condition. This rollout is used to specify the parameters to create the thin feature. To create a thin cut feature invoke the cut tool after creating the sketch and specify the end conditions in the **Direction 1** and **Direction 2** rollouts. Now, select the check box available in the **Thin Feature** rollout to activate the **Thin Feature** rollout. The options available in this rollout are the same as those discussed for the thin boss feature.

Handling Multiple Bodies in Cut Feature

While creating the cut feature, sometimes because of geometric conditions, feature termination, or end conditions, the cut feature results in the creation of multiple bodies. Figure 5-74 shows a sketch created on the top planar surface of the base feature to create a cut feature.

Figure 5-75 shows the cut feature created with the end condition as **Through All**. Using this type of sketch and end condition, if you choose **OK** from the **Extrude-Cut PropertyManager**, the **Bodies to Keep** dialog box will be displayed. As multiple bodies are created while applying the cut feature, this dialog box is used to define which body do you want to keep.

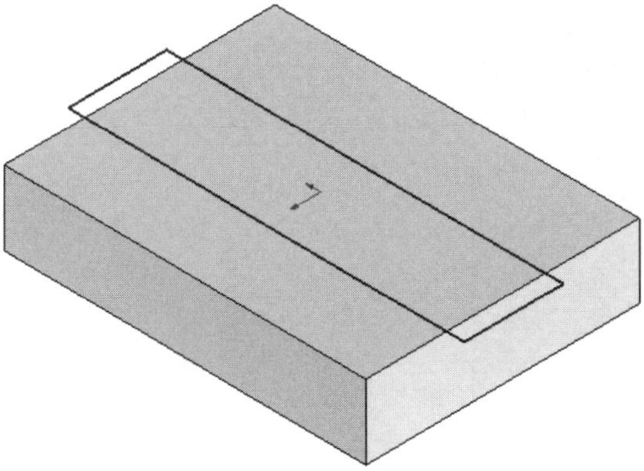

Figure 5-74 *Sketch created for the cut feature.*

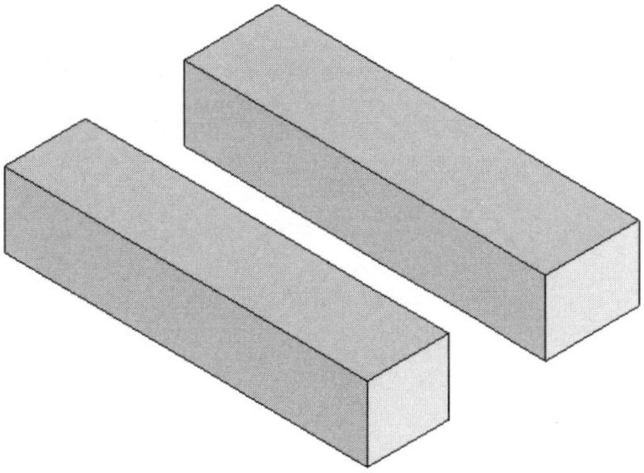

Figure 5-75 *Multiple bodies created using the cut feature*

The **Bodies to Keep** dialog box is displayed in Figure 5-76. By default, the **All Bodies** radio button is selected. Therefore, if you choose **OK** from this dialog box, all bodies created after the cut feature will remain in the model. If you want the cut feature to consume any of the bodies, select the **Selected bodies** radio button to expand the dialog box, as shown in Figure 5-77.

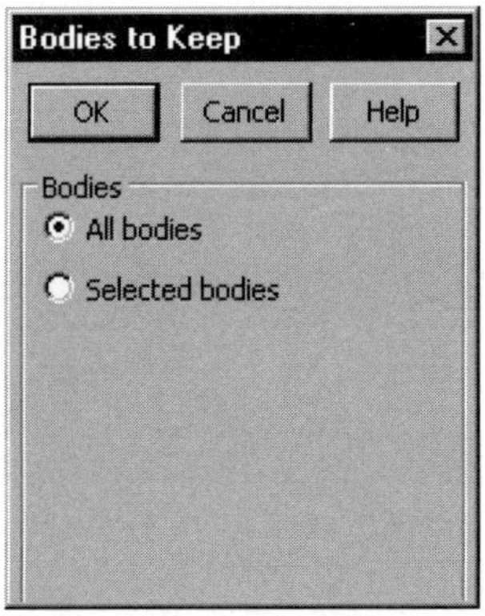

Figure 5-76 *The **Bodies to Keep** dialog box*

You can select the check box provided on the left of the name of the body to keep that body. The selected body is displayed in green in temporary graphics. Select the bodies to keep and choose the **OK** button from the **Bodies to Keep** dialog box. Figure 5-78 shows a sketch created for the cut feature. Figure 5-79 shows the cut feature created using the thin option and the **All bodies** option selected from the **Bodies to Keep** dialog box.

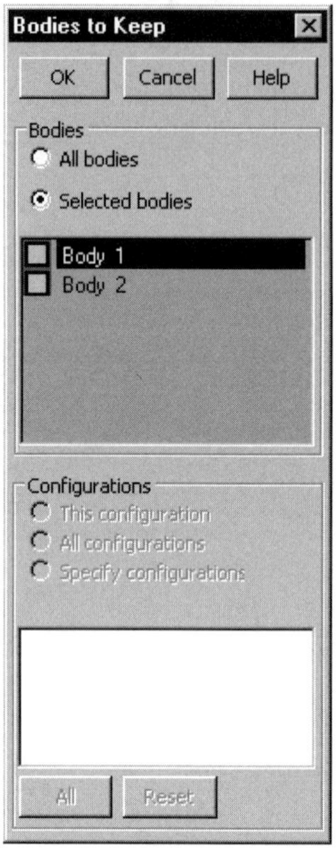

Figure 5-77 *The* ***Bodies to Keep*** *dialog box with the* ***Selected bodies*** *option selected*

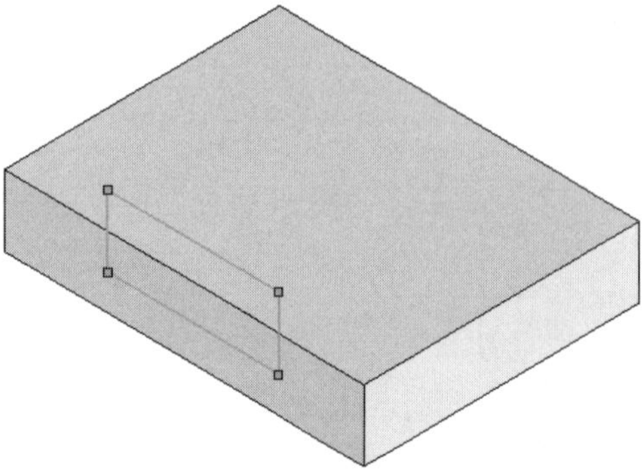

Figure 5-78 *Sketch to create a cut feature using the thin option*

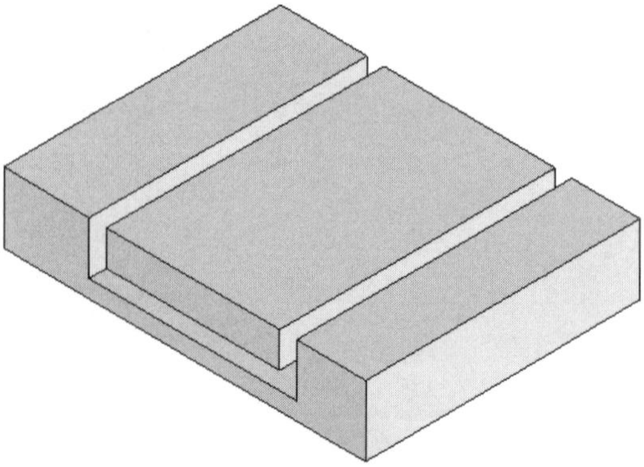

Figure 5-79 *A thin cut feature created with all resulting bodies retained*

Note

You will learn more about configurations in the later chapters.

Creating Revolved Cuts

CommandManager:	Features > Revolved Cut
Menu:	Insert > Cut > Revolve
Toolbar:	Features > Revolved Cut

Revolved cuts are used to remove the material by revolving a sketch around a selected axis. Similar to revolved boss/base features, you can define the revolution axis using a centerline or using an edge in the sketch. When you invoke the **Revolved Cut** tool, the **Cut-Revolve PropertyManager** will be displayed, as shown in Figure 5-80.

*Figure 5-80 The **Cut-Revolve PropertyManager***

Various options available in this **PropertyManager** are similar to those discussed earlier. Figure 5-81 shows a sketch for a revolved cut feature and Figure 5-82 shows the resulting cut feature. Note that in Figure 5-82, a texture is applied to the cut feature.

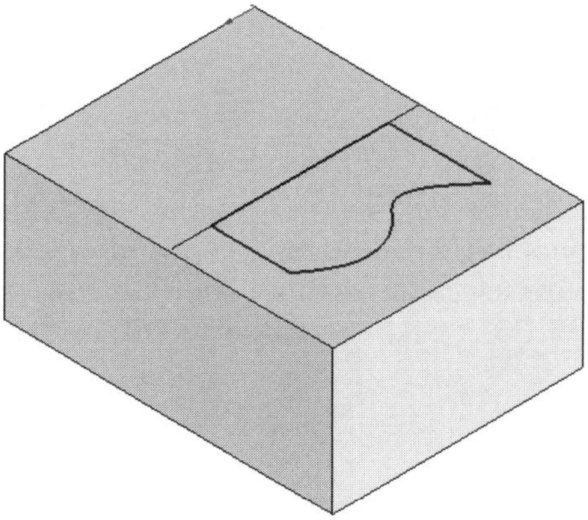

Figure 5-81 *Sketch for the revolved cut feature*

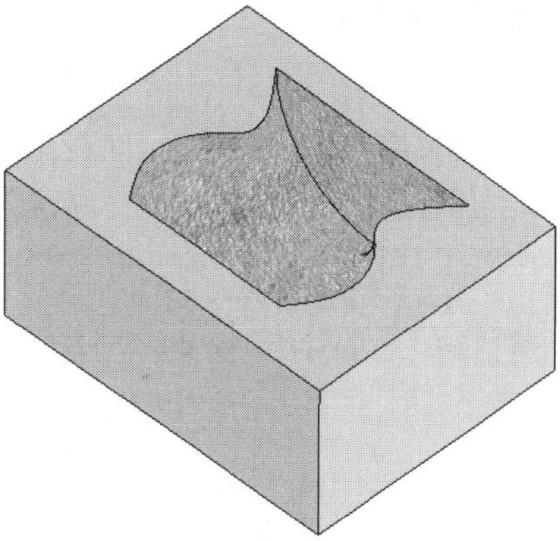

Figure 5-82 *Resulting cut feature with a texture*

Note

You can also select the tool to create the feature first and then select the plane to create the sketch. This way, after you exit the sketching environment, the feature tool will be activated automatically and you can define the parameters in their respective rollouts. For example, you can choose the ***Revolved Cut*** *tool without creating any sketch. In this case, the* ***Revolve PropertyManager*** *will be displayed and you will be prompted to select a plane or a planar face to create a sketch or to select a sketch. As soon as you exit the sketching environment after creating the sketch, the* ***Cut-Revolve PropertyManager*** *will be automatically displayed.*

CONCEPT OF THE FEATURE SCOPE

As discussed earlier, you can create different disjoint bodies in a single part file in SolidWorks. After creating two or more disjoint bodies, when you create another feature, the **Feature Scope** rollout will be displayed in the **PropertyManager**. This rollout is used to define the bodies that will be affected by the creation of the feature. The feature scope option is used with the Extrude boss and cut, Revolve boss and cut, Sweep boss and cut, Loft boss and cut, Boss cut and thicken, Surface cut, and Cavity features.

In the **Feature Scope** rollout, the **Selected bodies** radio button and the **Auto-select** check box are selected by default. With the **Auto-select** check box selected, all disjoint bodies are selected and are affected by the feature creation. If you clear the **Auto-select** check box, a selection area will be invoked. You can select the bodies that you want to be affected by the feature creation. The name of the selected body is displayed in the selection area. If you select the **All bodies** radio button, all bodies available in the part file will be selected and affected by the creation of the feature.

Tip. *You can invoke the previous command by pressing the ENTER key.*

TUTORIALS

Tutorial 1

In this tutorial, you will create the model shown in Figure 5-83 by drawing the sketch of the front view of the model and then select the contours to extrude them. As a result, you will learn the procedure of modeling using the contours selection method. The dimensions of the model are shown in Figure 5-84. **(Expected time: 30 min)**

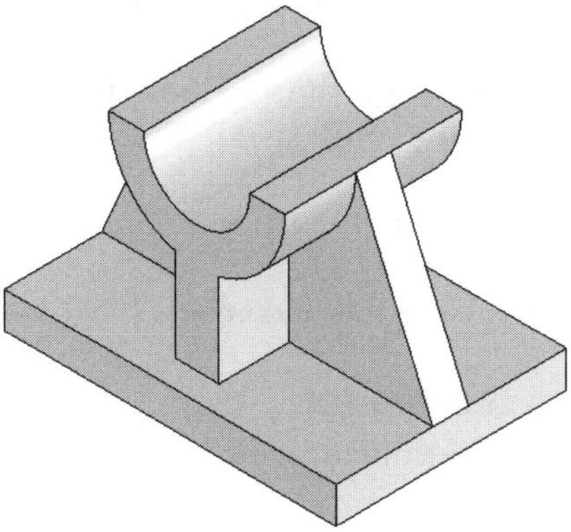

Figure 5-83 *Solid model for Tutorial 1*

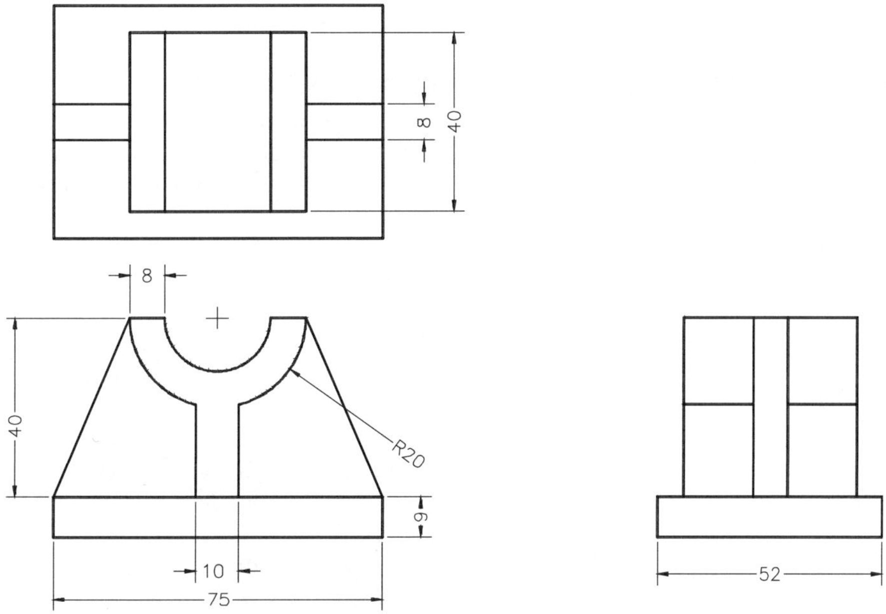

Figure 5-84 *Dimensions and views for Tutorial 1*

It is clear from the figures that the given model is a multifeatured model. It consists of various extruded features. You first need to draw the sketch for each feature and then convert that sketch into a feature. In conventional methods, you create a separate sketch for each sketched feature. But in this tutorial, you use the contour selection method to draw a sketch and select the contours and share the same sketch for creating all the features.

The steps to be followed to complete this tutorial are listed next.

a. Create the sketch on the default plane and apply the required relations and dimensions, refer to Figure 5-85.
b. Invoke the extrude option and extrude the selected contour, refer to Figures 5-86 and 5-87.
c. Select the second set of contours and extrude them to the required distance, refer to Figures 5-88 and 5-89.
d. Select the third set of contours and extrude them to the required distance, refer to Figures 5-90 and 5-91.
e. Save the document and then close the document.

Creating the Sketch of the Model

1. Start a new SolidWorks part document using the **New SolidWorks Document** dialog box.

2. Draw the sketch of the front view of the model on the **Front Plane**. Apply the required relations and dimensions to the sketch as shown in Figure 5-85. Make sure that you do not exit the sketching environment.

Selecting and Extruding the Contours of the Sketch

In this tutorial, you need to use the contour selection method to create the model. Therefore, you first need to select one of the contours from the given sketch and extrude it. For a better representation of the sketch, you also orient the sketch to isometric view.

1. Choose the **Isometric** option from the flyout that is displayed when you choose the **Standard Views** button from the **View** toolbar.

2. Right-click in the drawing area to invoke the shortcut menu and choose the **Contour Select Tool** option. The select cursor is replaced by the contour selection cursor and selection confirmation corner is displayed.

3. Move the cursor to the lower rectangle of the sketch; the area of the rectangle is highlighted. This indicates that this rectangle is a closed profile.

4. Click at this location when the rectangular area is highlighted. Figure 5-86 shows the lower rectangle selected using the contour selection tool.

5. Choose the **Extruded Boss/Base** button from the **Features CommandManager**; the **Extrude PropertyManager** is invoked and the preview of the base feature is displayed in the drawing area in temporary graphics.

   The name of the selected contour is displayed in the selection area of the **Selected Contours** rollout.

6. Right-click in the drawing area and choose the **Mid Plane** option from the shortcut menu. The preview of the feature is modified dynamically when you choose the **Mid Plane** option.

7. Set the value of the **Depth** spinner to **52** and choose the **OK** button from the **Extrude**

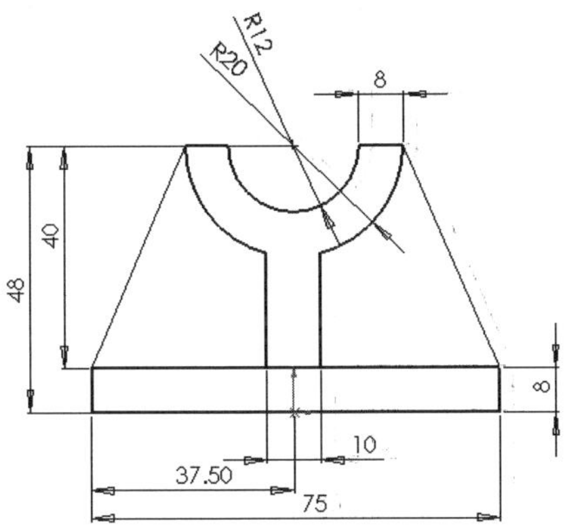

Figure 5-85 *Fully defined sketch for creating the model*

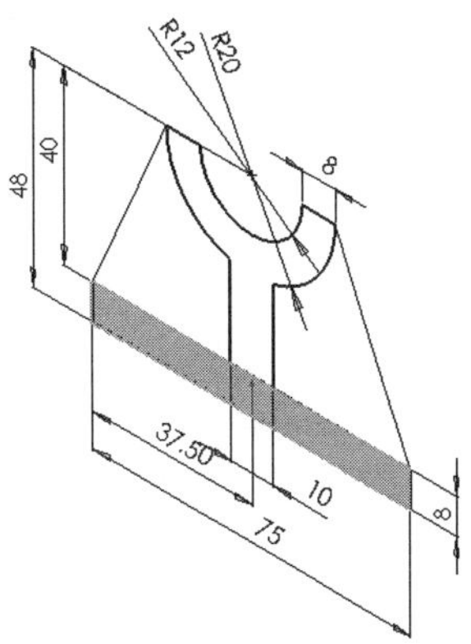

Figure 5-86 *Lower rectangle selected as a contour*

PropertyManager. The base feature of the model by extruding the selected contour is shown in Figure 5-87.

8. Right-click in the drawing area and choose the **Contour Select Tool** option from the shortcut menu; the select cursor is replaced by the contour selection cursor.

9. Using the contour selection cursor, select any entity of the sketch to invoke the selection mode of the sketch.

10. Using the left mouse button, select the middle contour of the sketch, as shown in Figure 5-88. The selected region is highlighted.

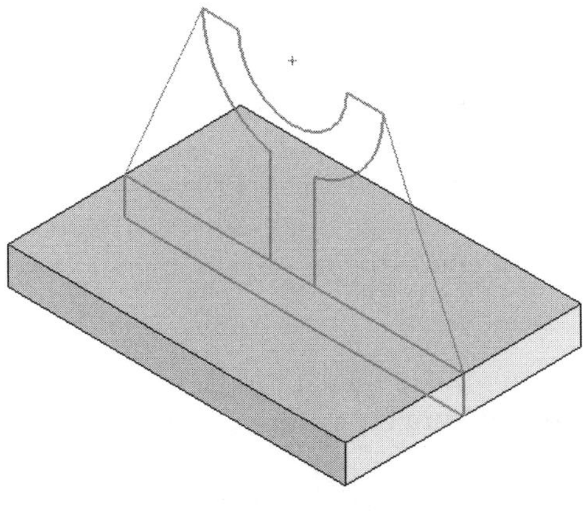

Figure 5-87 *Base feature of the model*

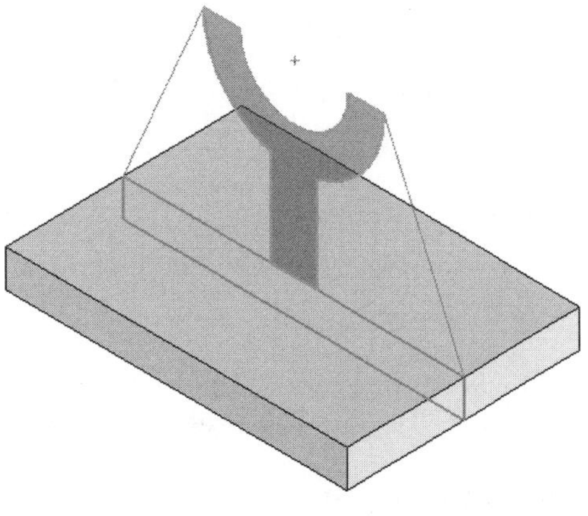

Figure 5-88 *Middle contour is selected using the contour selection tool*

11. Invoke the **Extrude PropertyManager**. Right-click in the drawing area and choose the **Mid Plane** option from the shortcut menu.

12. Set the value of the **Depth** spinner to **40** and choose the **OK** button from the **Extrude PropertyManager**. The feature created by selecting the middle contour is shown in Figure 5-89.

13. Again invoke the **Contour Select Tool** and select a sketched entity. Next, select the right contour of the sketch. Press and hold the CTRL key down and select the left contour of the sketch, see Figure 5-90.

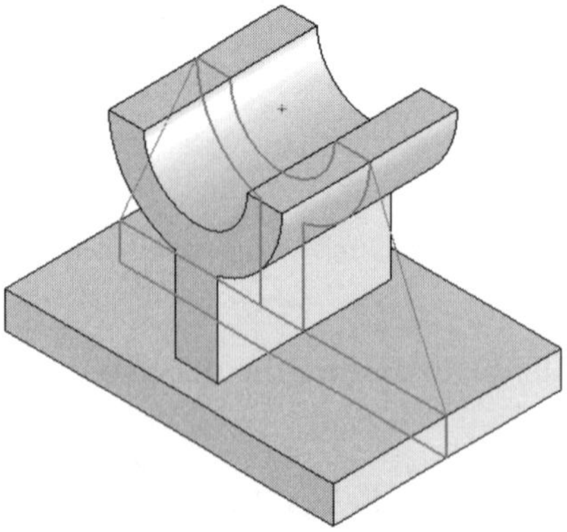

Figure 5-89 *Second feature created by extruding the middle contour*

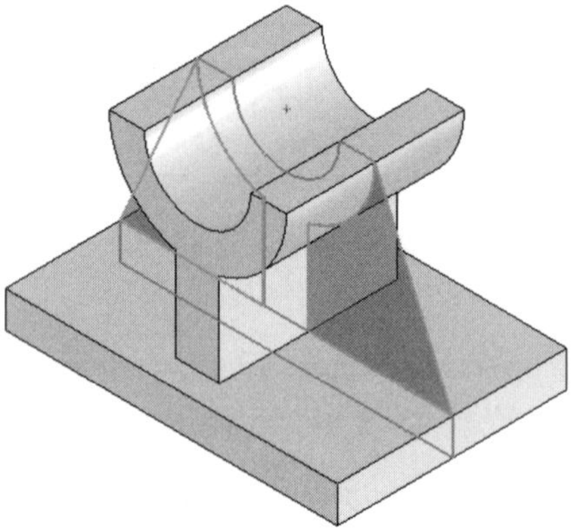

Figure 5-90 *The right and the left contours selected*

14. Invoke the **Extrude PropertyManager**. Right-click and choose the **Mid Plane** option from the shortcut menu.

15. Set the value of the **Depth** spinner to **8** and choose the **OK** button from the **Extrude PropertyManager**.

    The model is completed, but the sketch is displayed in the model. Therefore, you need to hide the sketch.

16. Move the cursor to any of the sketched entities and when the entity turns red, select the entity. The selected sketched entity is displayed in green. Now, right-click and choose the **Hide** option from the shortcut menu.

    The isometric view of the final model with the visibility of sketch turned off is shown in Figure 5-91. The **FeatureManager Design Tree** of the model is shown in Figure 5-92.

Saving the Model

1. Choose the **Save** button from the **Standard** toolbar and save the model with the name given below.

   \My Documents\SolidWorks\c05\c05tut1.sldprt

2. Choose **File > Close** from the menu bar to close the document.

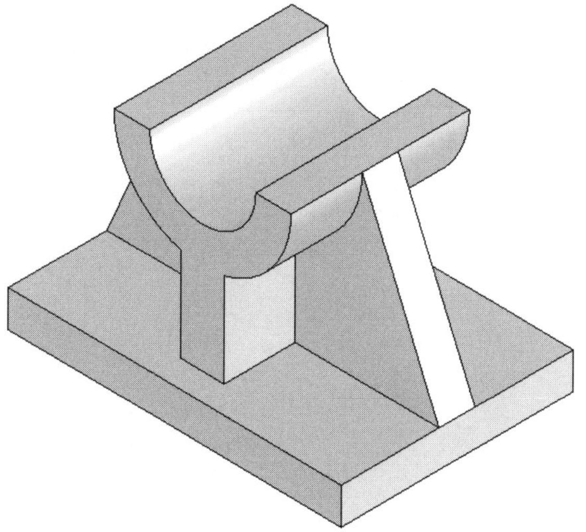

Figure 5-91 *Final solid model*

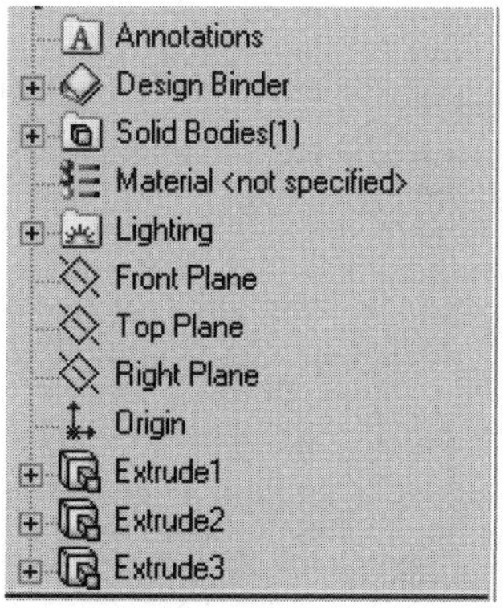

Figure 5-92 *The* ***FeatureManager Design Tree***

Tutorial 2

In this tutorial, you will create a model whose dimensions are shown in Figure 5-93. You will use a combination of the conventional modeling method and the contour selection method to create this model. The solid model is shown in Figure 5-94. **(Expected time: 30 min)**

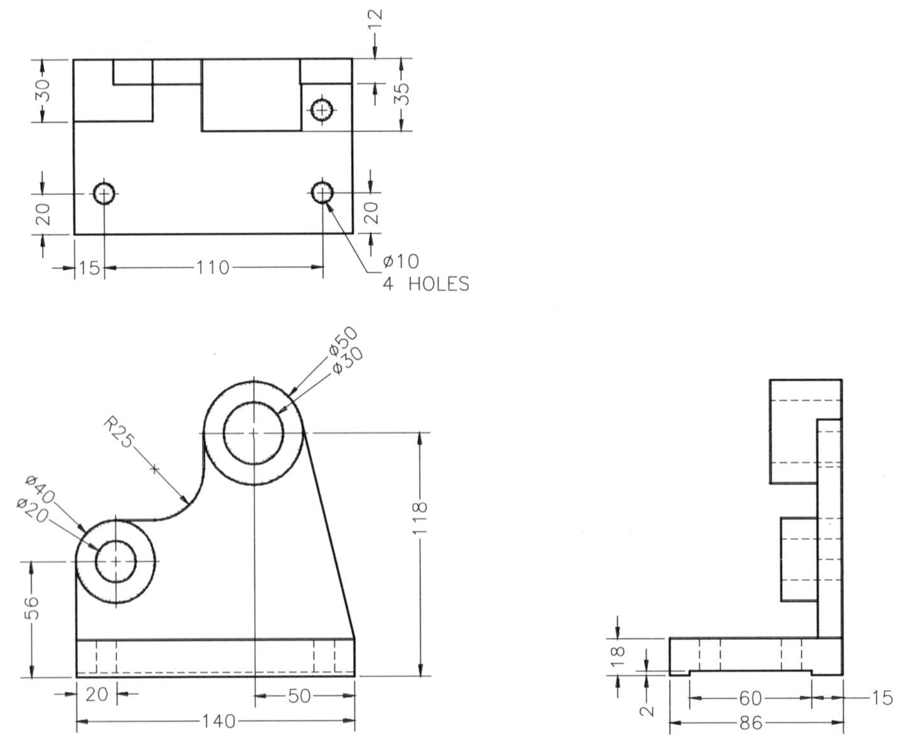

Figure 5-93 Drawing views of the solid model

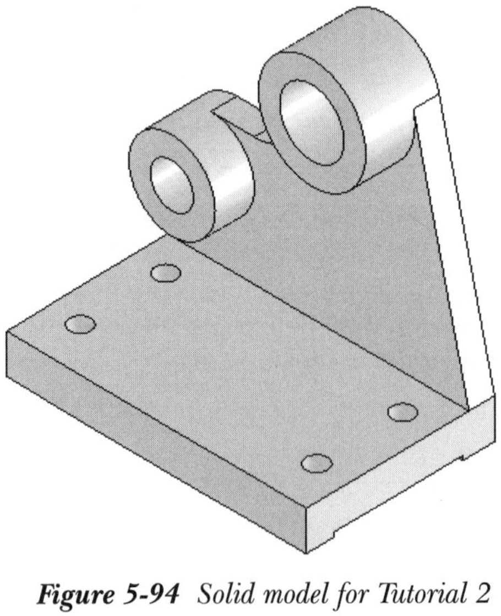

Figure 5-94 Solid model for Tutorial 2

The steps that will be used to complete the model are listed next.

a. Draw the sketch of the front view of the model, refer to Figure 5-95.
b. Extrude the selected contours, refer to Figures 5-96 through 5-98.
c. Add recess feature to the model by drawing the sketch on the right planar face, refer to Figures 5-99 and 5-100.
d. Create four holes using the cut feature on the top face of the base feature, refer to Figures 5-101 and 5-102.
e. Save and close the document.

Drawing the Sketch for Contour Selection Modeling

1. Start a new SolidWorks part document. Draw the sketch of the front view of the model on the **Front Plane** using the standard sketching tools.

2. Apply the required relations and dimensions to fully define the sketch, see Figure 5-95.

   Orient the view to isometric view because it will help you in the selection of contours.

3. Change the current view to isometric view.

4. Right-click and choose the **Contour Select Tool** option from the shortcut menu; the select cursor is replaced by the contour selection cursor.

5. Using the contour selection cursor select the area enclosed by the lower rectangle as shown in Figure 5-96.

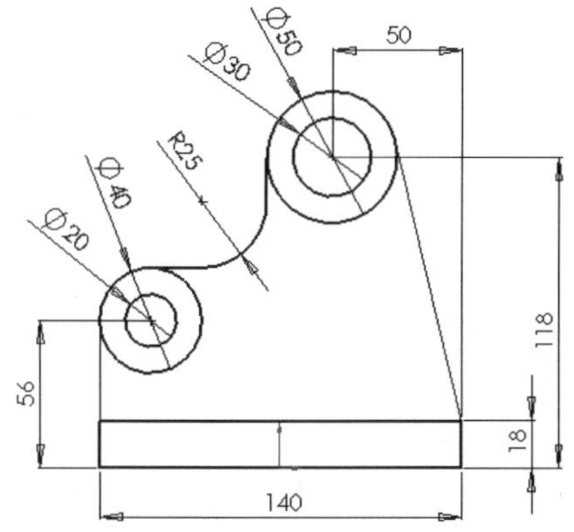

Figure 5-95 *Fully defined sketch*

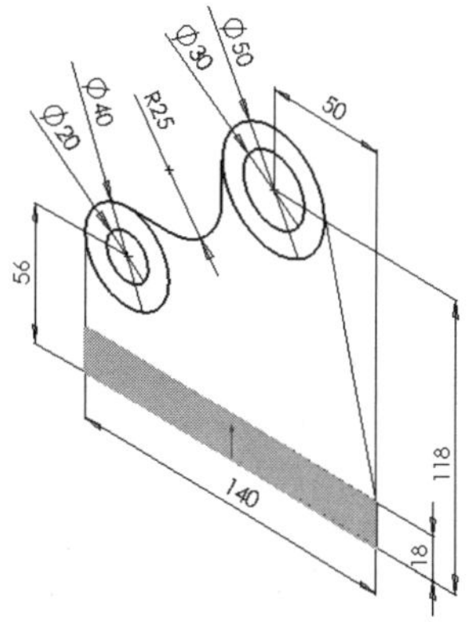

Figure 5-96 *Lower rectangle selected as contour*

6. Choose the **Extruded Boss/Base** button from the **Features CommandManager** to invoke the **Extrude PropertyManager**.

7. Set the value of the **Depth** spinner to **86** and choose the **OK** button from the **Extrude PropertyManager**. The base feature created after extruding the selected contour is shown in Figure 5-97.

8. Using the contour selection tool and the extrude tool create the other features and then hide the sketch. The model created after extruding all the contours and hiding the sketch is shown in Figure 5-98.

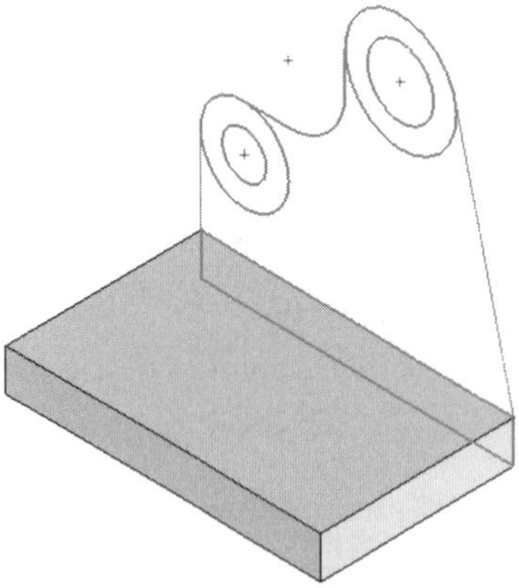

Figure 5-97 Base feature created after extruding the selected contour

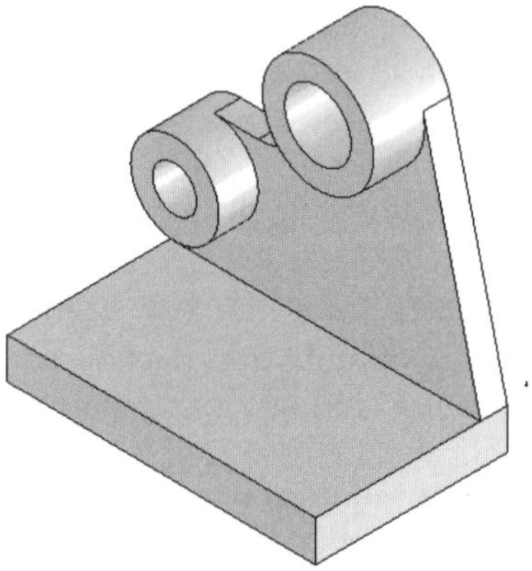

Figure 5-98 Model created after extruding all contours

Creating the Recess at the Base of the Model

After creating the extruded features of the model, you need to create the recess provided at the base of the model. The recess is created by a cut extrude feature created using a sketch drawn on the right planar face of the model.

1. Select the right planar face of the base feature as the sketching plane; the selected face is displayed in green.

2. Right-click in the drawing area and choose the **Insert Sketch** option from the shortcut menu to invoke the sketching environment.

   Now, you need to orient the view such that the selected face is normal to your eye view.

3. Choose the **Normal To** option from the flyout that is displayed when you choose the **Standard Views** button to orient the selected plane normal to the view.

4. Using the standard sketching tools, draw the sketch for the recess and apply the required relations and dimensions to the sketch. The fully defined sketch for the cut feature is shown in Figure 5-99.

5. Choose the **Extruded Cut** button from the **Features CommandManager** to invoke the **Cut-Extrude PropertyManager**. The preview of the cut feature is displayed in the drawing area in temporary graphics.

6. Right-click in the drawing area and choose **Through All** from the shortcut menu.

7. Choose **OK** from the **Cut-Extrude PropertyManager** to complete feature creation. The isometric view of the model after creating the cut feature is shown in Figure 5-100.

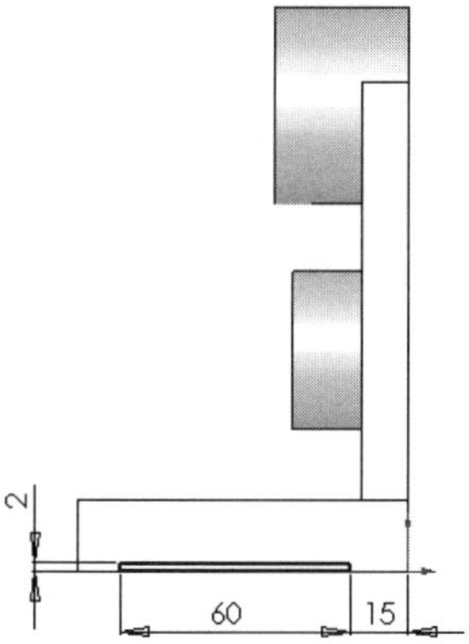

***Figure 5-99** Sketch for the cut feature*

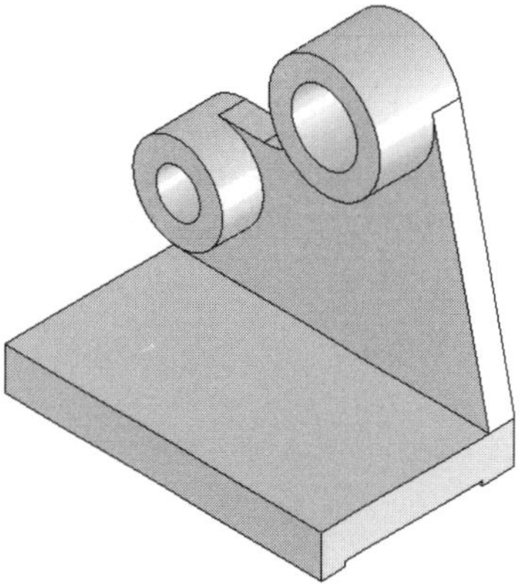

***Figure 5-100** Cut feature added to the model*

Creating the Holes

Next, you need to create the holes at the base of the model. These holes will be created as the extruded cut features. You need to draw the sketch of the hole feature on the top planar face of the base feature of the model. To draw the sketch of the holes, you need to first draw a circle and then using the **Linear Sketch Step and Repeat** option create the remaining circles.

1. Select the top planar face of the base feature as the sketching plane. Right-click and choose the **Insert Sketch** option from the shortcut menu.

2. Orient the current view normal to the viewing direction. Using the standard sketching tools, draw the sketch for the holes and apply the required relations and dimensions to fully define the sketch. You may need to apply the horizontal relations between the center points of the top circles to fully define the sketch. The fully defined sketch is shown in Figure 5-101.

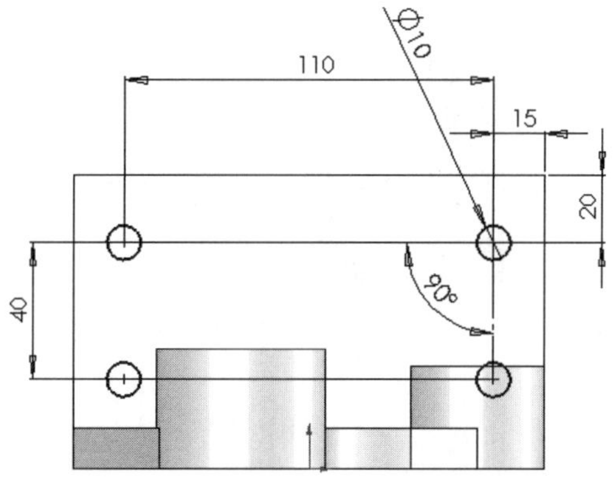

***Figure 5-101** Holes sketched for the cut feature*

3. Change the current view to isometric view and then choose the **Extruded Cut** button from the **Features CommandManager** to invoke the **Cut-Extrude PropertyManager**.

4. Right-click in the drawing area and choose the **Through All** option from the shortcut menu. Choose the **OK** button from the **Cut-Extrude PropertyManager**. The isometric view of the final model after hiding the sketch is shown in Figure 5-102. The **FeatureManager Design Tree** of the model is shown in Figure 5-103.

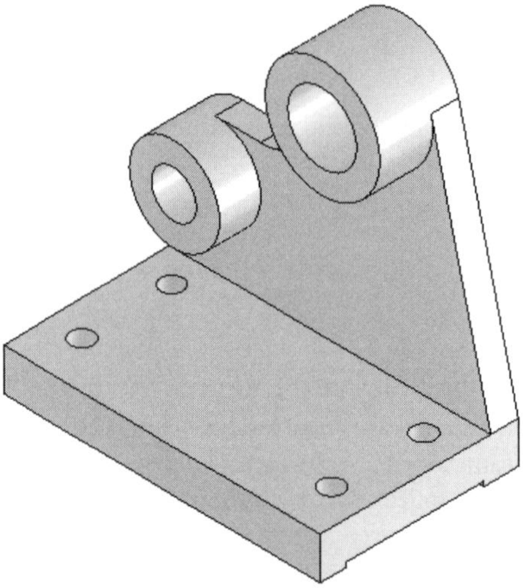

Figure 5-102 *Final solid model*

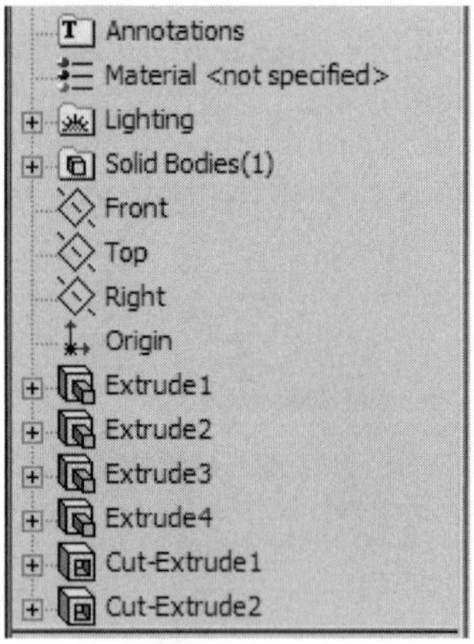

Figure 5-103 *The* ***FeatureManager Design Tree***

Saving the Model

1. Choose the **Save** button from the **Standard** toolbar and save the model with the name given below.

   \My Documents\SolidWorks\c05\c05tut2.sldprt

2. Choose **File > Close** from the menu bar to close the document.

Tutorial 3

In this tutorial, you will create a model whose dimensions are shown in Figure 5-104. The solid model is shown in Figure 5-105. **(Expected Time: 30 min)**

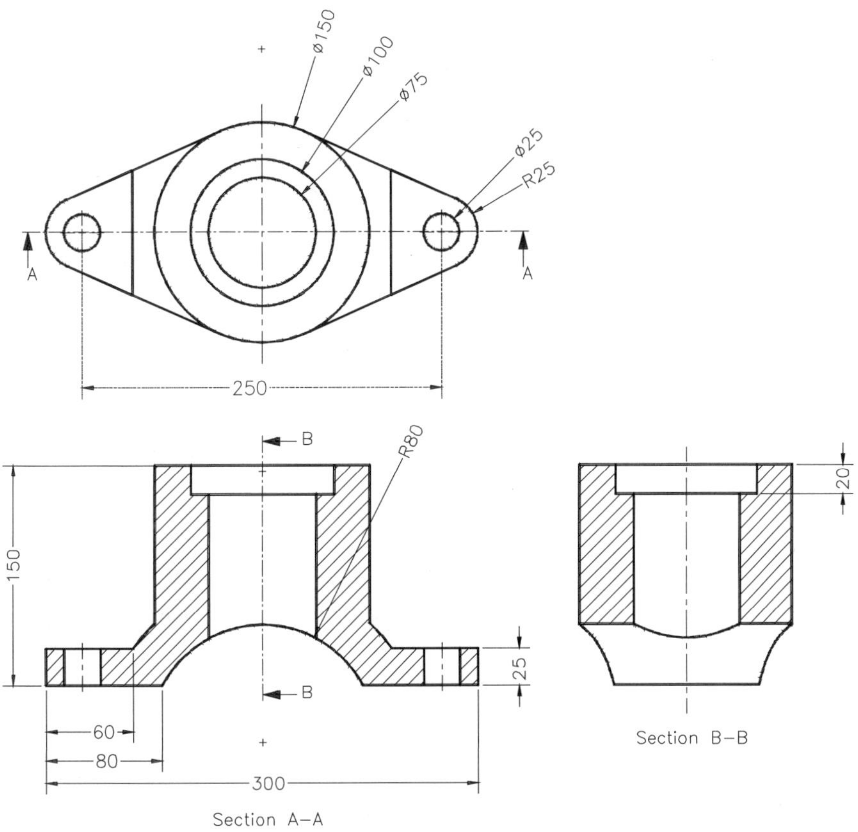

Figure 5-104 *Top view, sectioned front view, and sectioned right view of the model*

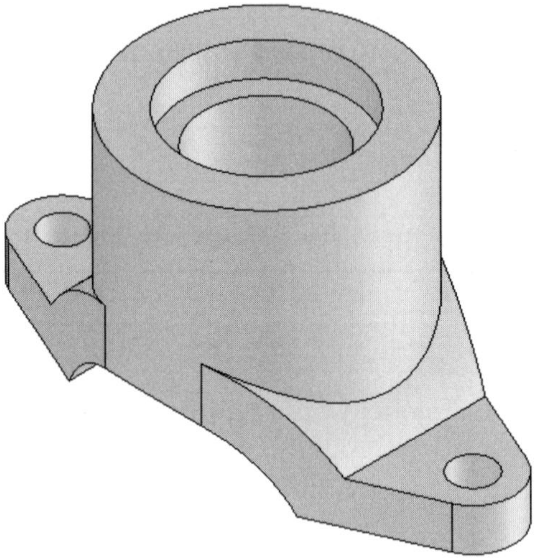

Figure 5-105 Solid model for Tutorial 3

The steps that will be followed to complete this tutorial are listed next.

a. Create the base feature by extruding the sketch drawn on the **Front Plane**, refer to Figures 5-106 and 5-107.
b. Extrude the sketch created on the **Top Plane** to create a cut feature, refer to Figures 5-108 through 5-110.
c. Create a plane at an offset distance of 150 from the **Top Plane**.
d. Draw a sketch on the newly created plane and extrude it to the selected surface, refer to Figures 5-111 and 5-112.
e. Create a contour bore using the cut revolve option, refer to Figures 5-113 and 5-114.
f. Create the holes using the cut feature, refer to Figures 5-115 and 5-116.
g. Save the document and close it.

Creating the Base Feature

It is evident from the model that the base of the model comprises a complex geometry. You need to create the base feature and then apply the cut feature to the base of the model to get the desired shape. You need to create the base feature on the **Front Plane** as the sketching plane. After drawing the sketch, you need to extrude the sketch using the mid plane option to complete feature creation.

1. Start a new SolidWorks part document and invoke the sketching environment to draw the sketch for the base feature.

2. Using the standard sketching tools, draw the sketch of the base feature and then apply the required relations and dimensions to the sketch as shown in Figure 5-106.

3. Invoke the **Extrude PropertyManager**. Right-click in the drawing area and choose the **Mid Plane** option from the shortcut menu.

4. Set the value of the **Depth** spinner to **150** and choose the **OK** button from the **Extrude PropertyManager**. The isometric view of the base feature of the model is shown in Figure 5-107.

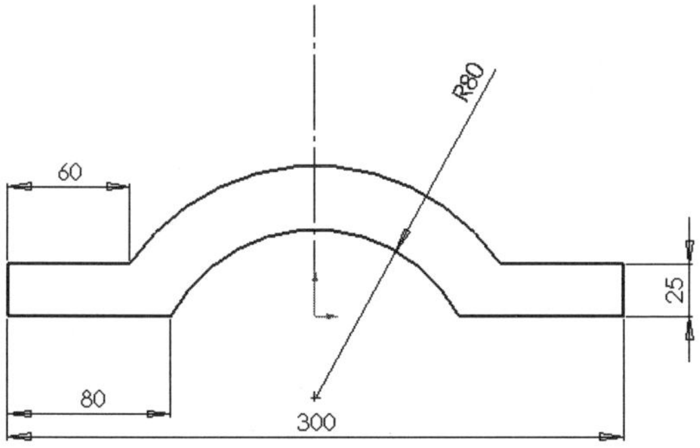

Figure 5-106 Sketch of the base feature

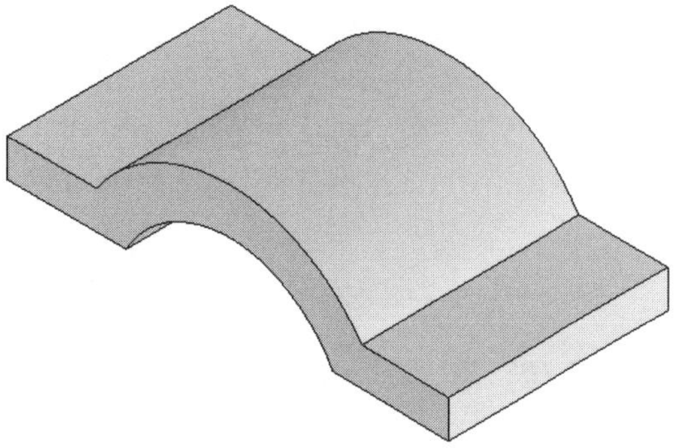

Figure 5-107 Base feature of the solid model

Creating the Cut Feature

Now, you need to create a cut feature to get the required shape of the base feature. The sketch for this cut feature is created using a reference plane defined tangent to the curved face of the previous feature.

1. Choose the **Plane** button from the **Reference Geometry CommandManager** to display the **Plane PropertyManager**.

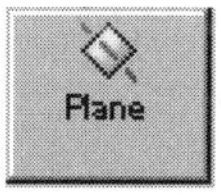

2. Select the upper curved face of the existing feature. The **On Surface** button is automatically chosen. Now, move the cursor close to the midpoint of the curved edge of the upper curved face; the point is highlighted in red, see Figure 5-108. Select this point; the preview of a plane tangent to the curved face and passing through the midpoint of the curved edge is displayed. Choose **OK** to create the reference plane.

3. Draw the sketch for the cut feature using the standard sketching tools and then apply the required relations and dimensions to the sketch as shown in Figure 5-109.

4. Choose the **Extruded Cut** button from the **Features CommandManager** to invoke the **Cut-Extrude PropertyManager**. Change the current view to isometric view.

You will also observe that the direction of material removal is not the required direction. Therefore, you need to flip the direction of material removal.

5. Select the **Flip side to cut** check box. You will observe that the direction of material removal is also changed in the preview.

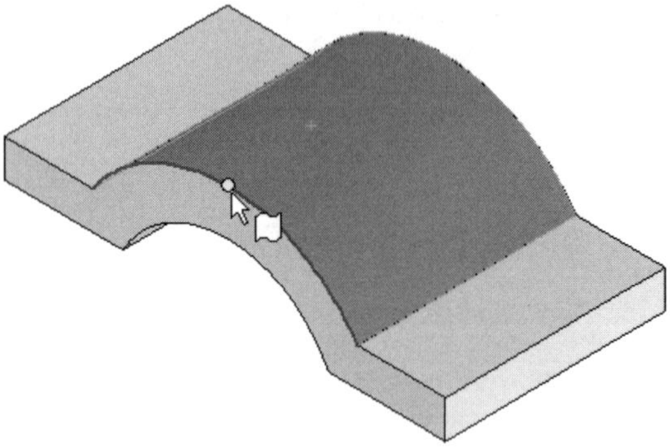

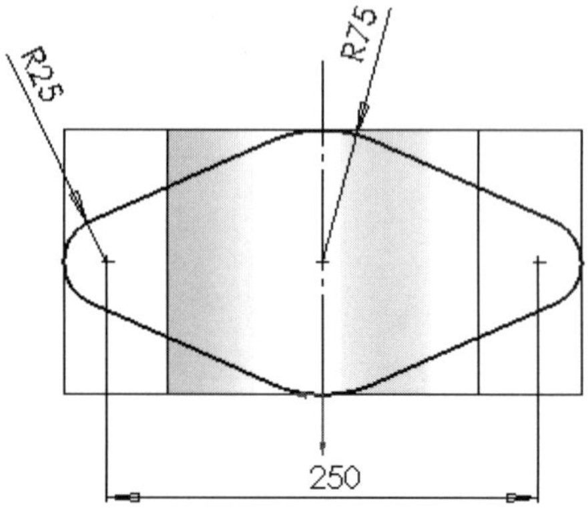

Figure 5-108 Selecting the point to define the tangent plane

Figure 5-109 Fully dimensioned sketch for the cut feature

6. Right-click in the drawing area and choose the **Through All** option from the shortcut menu and choose the **OK** button from the **Cut-Extrude PropertyManager**.

   Because the reference plane is displayed in the drawing area, you need to hide it.

7. Right-click on **Plane1** in the drawing area and choose **Hide** from the shortcut menu; the display of the reference plane is turned off. The model after creating the base feature is shown in Figure 5-110.

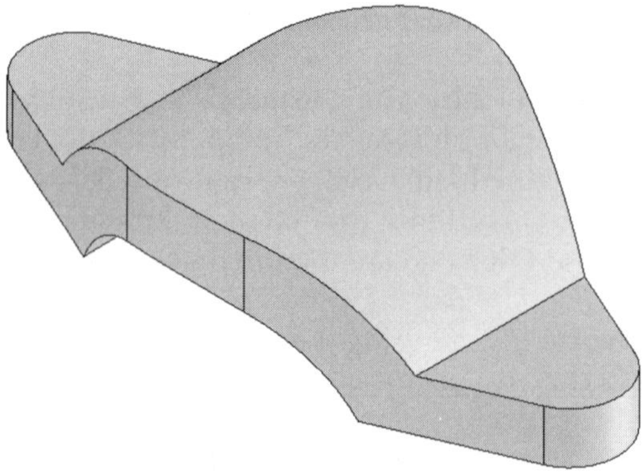

Figure 5-110 Cut feature added to the base feature

Creating a Plane at an Offset Distance for the Extruded Feature

After creating the base of the model you need to create a plane at an offset distance of 150 mm from the **Top Plane**. This newly created plane is used as a sketching plane for the next feature.

1. Choose the **Plane** button from the **Reference Geometry CommandManager** to invoke the **Plane PropertyManager**.

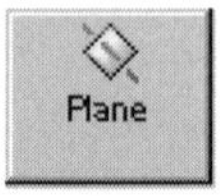

2. Choose the **Offset Distance** button from the **Plane PropertyManager**. The **Distance** spinner, the **Reverse direction** check box, and the **Number of Planes to Create** spinner are displayed in the **Plane PropertyManager**.

3. Click on the + sign located on the left of the **Feature Manager Design Tree**, which is now displayed in the drawing area. The tree view expands and the three default planes are now visible in the tree view.

4. Select the **Top Plane** from the **FeatureManager Design Tree**; the preview of the newly created plane at a default offset is displayed in the drawing area.

5. Set the value of the **Distance** spinner to **150** and choose the **OK** button from the **Plane PropertyManager**. The required plane is created.

Creating the Extruded Feature

After creating the plane at an offset distance from the **Top Plane**, you need to draw the sketch for the next feature.

1. Select the reference plane you just created if it is not selected, and invoke the sketching environment. Set the current view normal to the eye view.

2. Draw the sketch of the circle and apply the required relations to the sketch, as shown in Figure 5-111.

3. Change the current view to isometric view and invoke the **Extrude PropertyManager**. You will observe from the preview that the direction of feature creation is opposite to the required direction. Therefore, you need to change the direction of feature creation.

4. Choose the **Reverse Direction** button provided on the left of the **End Condition** drop-down list to reverse the direction of feature creation. The preview of the feature is changed when you reverse the direction of feature creation.

5. Right-click in the drawing area and choose the **Up To Surface** option from the shortcut menu. You are prompted to select a face or a surface to complete the specification of Direction 1 and the **Face/Plane** selection area is displayed under the **End Condition** drop-down list in the **Direction 1** rollout.

6. Select the upper curved surface of the model using the left mouse button. You will observe that the preview shows the feature extruded up to the selected surface.

7. Choose the **OK** button from the **Extrude PropertyManager**.

   Because the plane is displayed in the drawing area, you need to turn off its display.

8. Select **Plane2** from the **FeatureManager Design Tree** or from the drawing area and invoke the shortcut menu. Choose the **Hide** option from the shortcut menu. The model created after creating the extruded feature is shown in Figure 5-112.

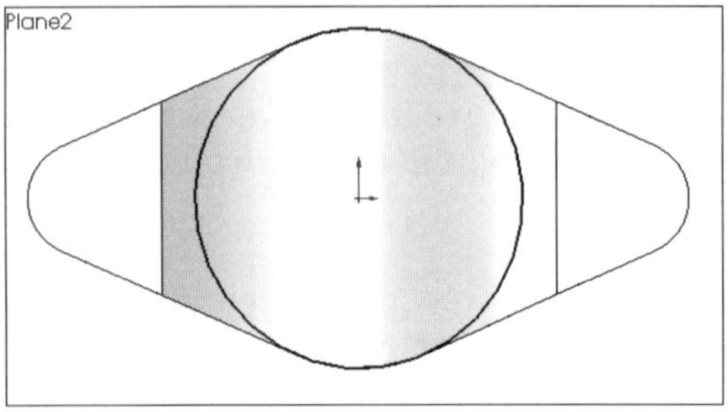

Figure 5-111 *Sketch created on the newly created plane*

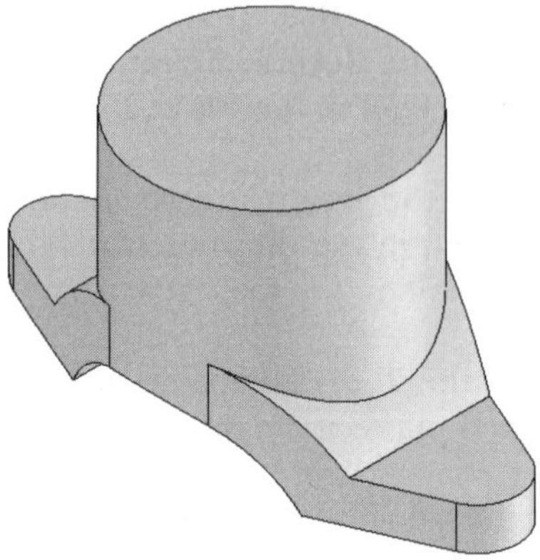

Figure 5-112 *Sketch extruded up to a selected surface*

Creating the Counterbore Hole

Next, you need to create the counterbore hole. It will be created as a revolved cut feature using a sketch drawn on the **Front Plane**.

1. Invoke the sketching environment by selecting the **Front Plane** as the sketching plane and orient the plane normal to the eye view.

2. Draw the sketch of the counterbore hole using the standard sketching tools. Add the required relations and then add linear diameter dimensions, as shown in Figure 5-113.

3. Set the current view to isometric view and then choose the **Revolved Cut** button from the **Features CommandManager**; the **Cut-Revolve PropertyManager** is displayed.

The preview of the cut feature is displayed in the drawing area in temporary graphics. The value of the angle in the **Angle** spinner is set to **360** by default. Therefore, you do not need to set the value of the **Angle** spinner.

4. Choose the **OK** button from the **Cut-Revolve PropertyManager**. Figure 5-114 shows the model after creating the revolve cut feature.

Creating the Holes

After creating all features, you need to create the holes using the extruded cut feature to complete the model. The sketch for the cut feature is to be drawn using the top planar surface of the base feature as the sketching plane.

1. Select the top planar surface of the base feature and invoke the sketching environment. Orient the model so that the selected face of the model is oriented normal to the eye view.

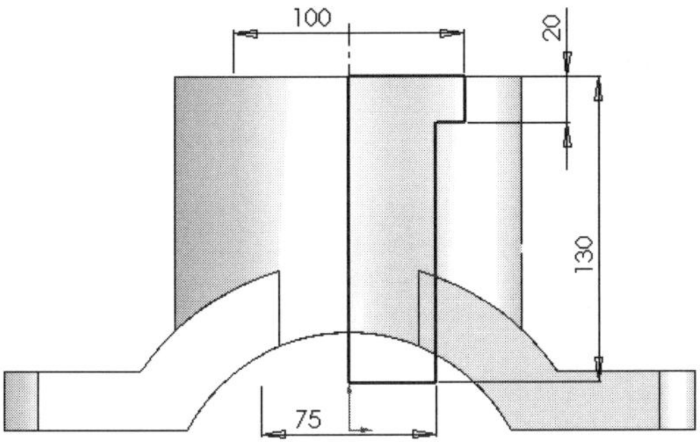

Figure 5-113 *Fully defined sketch for the counterbore*

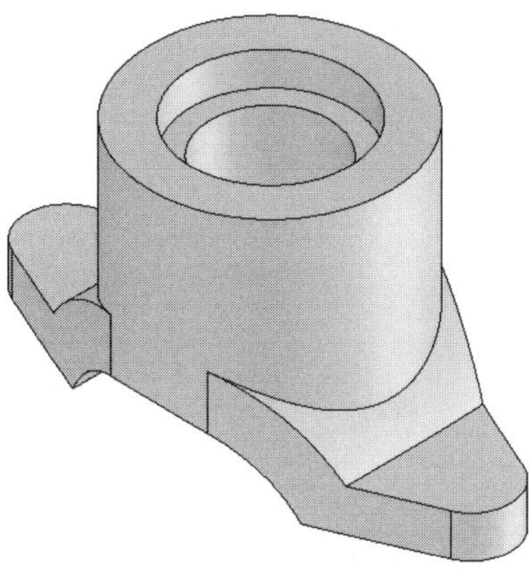

Figure 5-114 *Cut feature added to the model*

2. Draw the sketch using the standard sketching tools and apply the required relations and dimensions as shown in Figure 5-115.

3. Change the current view to isometric view. Choose the **Extruded Cut** button from the **Features CommandManager** to invoke the **Cut-Extrude PropertyManager**.

4. Right-click and choose the **Through All** option from the shortcut menu and choose the **OK** button from the **Cut-Extrude PropertyManager**. The final model is shown in Figure 5-116.

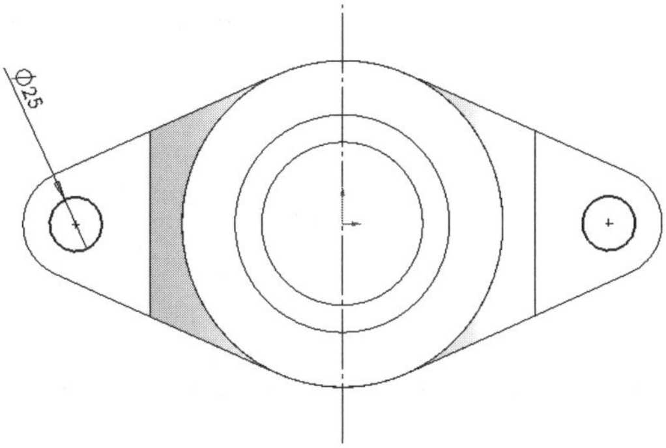

Figure 5-115 *Fully defined sketch for the cut feature*

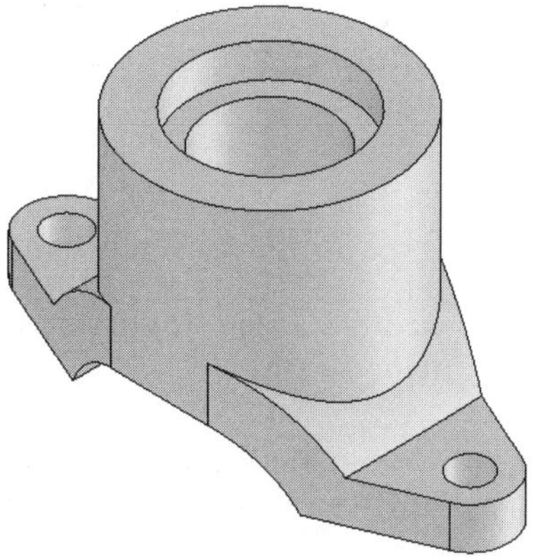

Figure 5-116 *Final model*

The **FeatureManager Design Tree** of the model is shown in Figure 5-117.

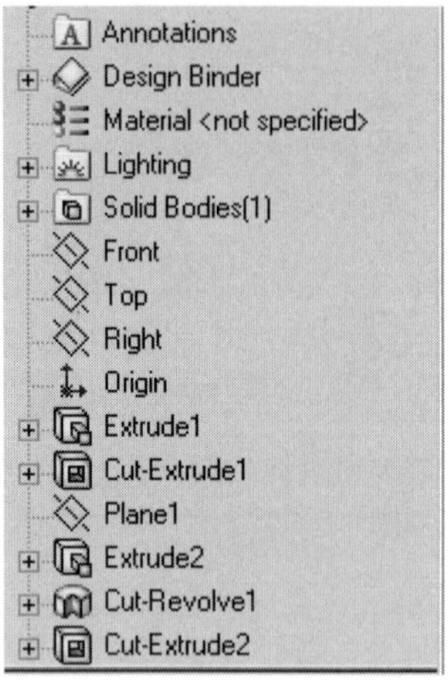

*Figure 5-117 The **FeatureManager Design Tree***

Saving the Model

1. Choose the **Save** button from the **Standard** toolbar and save the model with the name given next.

   \My Documents\SolidWorks\c05\c05tut3.sldprt

2. Choose **File** > **Close** from the menu bar to close the document.

SELF-EVALUATION TEST

Answer the following questions and then compare your answers with those given at the end of this chapter.

1. When you draw a sketch for the first time in the sketching environment, the sketch is drawn on the default plane, which is the **Front Plane**. (T/F)

2. When you start a new SolidWorks part document, SolidWorks provides you with two default planes. (T/F)

3. You can choose the **Plane** button from the **Reference Geometry CommandManager** to invoke the **Plane PropertyManager**. (T/F)

4. You cannot create a plane at an offset distance by dynamically dragging. (T/F)

5. When you create a circular feature, a temporary axis is automatically created. (T/F)

6. The __________ option is used to extrude a sketch from the sketching plane to the next surface that intersects the feature.

7. The __________ option available in the **End Condition** drop-down list is used to define the termination of the extruded feature up to another body.

8. The __________ check box is used to merge the newly created body with the parent body.

9. Using the __________ option, you can create a reference axis that passes through the center point of a cylindrical or a conical surface.

10. Sometimes multiple bodies are created while applying the cut feature. Therefore, the __________ dialog box is displayed, which allows you to define which body to keep.

REVIEW QUESTIONS

Answer the following questions.

1. If the __________ check box is cleared, the virtual surface created for the termination of the extruded feature will have a concentric relation with the selected surface.

2. The __________ option is selected from the shortcut menu to select the contours.

3. The __________ option is available in the **End condition** drop-down list only after you create the base feature.

4. The __________ check box is used to define the side of material removal.

5. The __________ check box is used to create an outward draft in a cut feature.

6. Which check box is selected while creating any other feature in single-body modeling?

   (a) **Combine results** (b) **Fix bodies**
   (c) **Merge results** (d) **Union results**

7. Which button is used to add a draft angle to a cut feature?

   (a) **Add Draft** (b) **Create Draft**
   (c) **Draft On/Off** (c) None of these

8. Which **PropertyManager** is invoked to create a cut feature by extruding a sketch?

   (a) **Extruded Cut** (b) **Cut-Extrude**
   (c) **Extrude-Cut** (d) **Cut**

9. Which of the following options is used to define the termination of feature creation at an offset distance to a selected surface?

   (a) **Distance To Surface** (b) **Normal From Surface**
   (c) **Distance From Surface** (d) **Offset From Surface**

10. Which of the following options is used to define the termination of feature creation to the selected surface?

    (a) **To Surface** (b) **Selected Surface**
    (c) **Up To Surface** (d) None of these

EXERCISES

Exercise 1

Create a model whose dimensions are shown in Figure 5-118. The solid model is shown in Figure 5-119. **(Expected time: 30 min)**

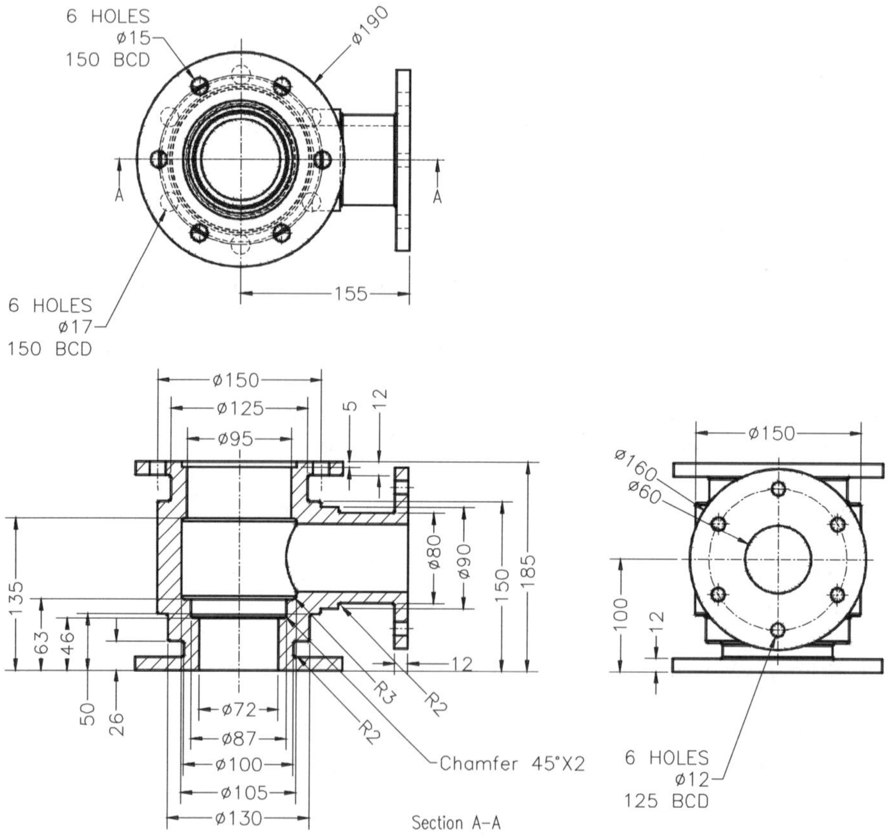

***Figure 5-118** Dimensions of the model for Exercise 1*

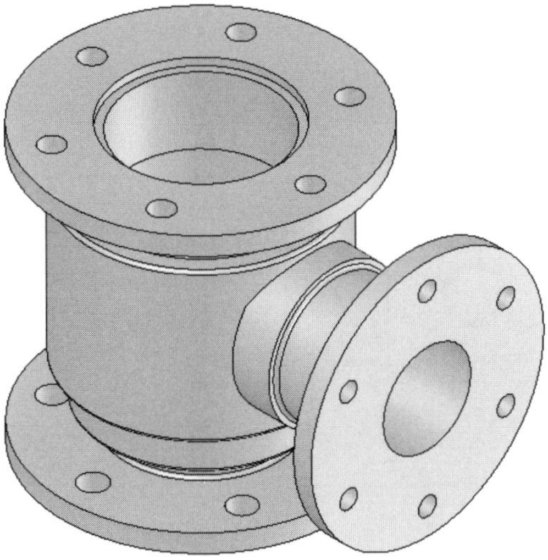

Figure 5-119 Model for Exercise 1

Exercise 2

Create the model shown in Figure 5-120 . The dimensions of the model are given in Figure 5-121. **(Expected time: 30 min)**

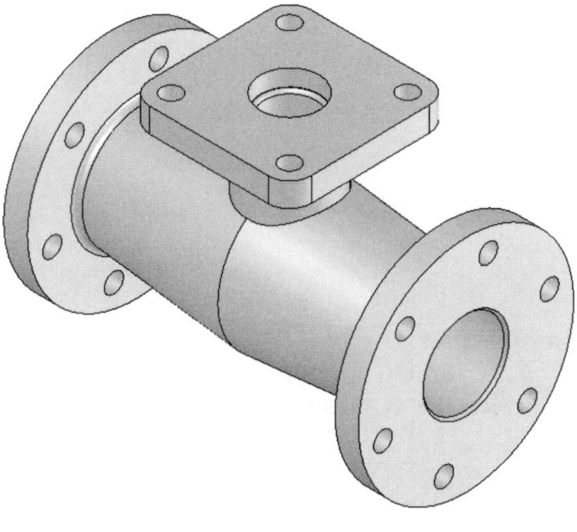

Figure 5-120 Model for Exercise 2

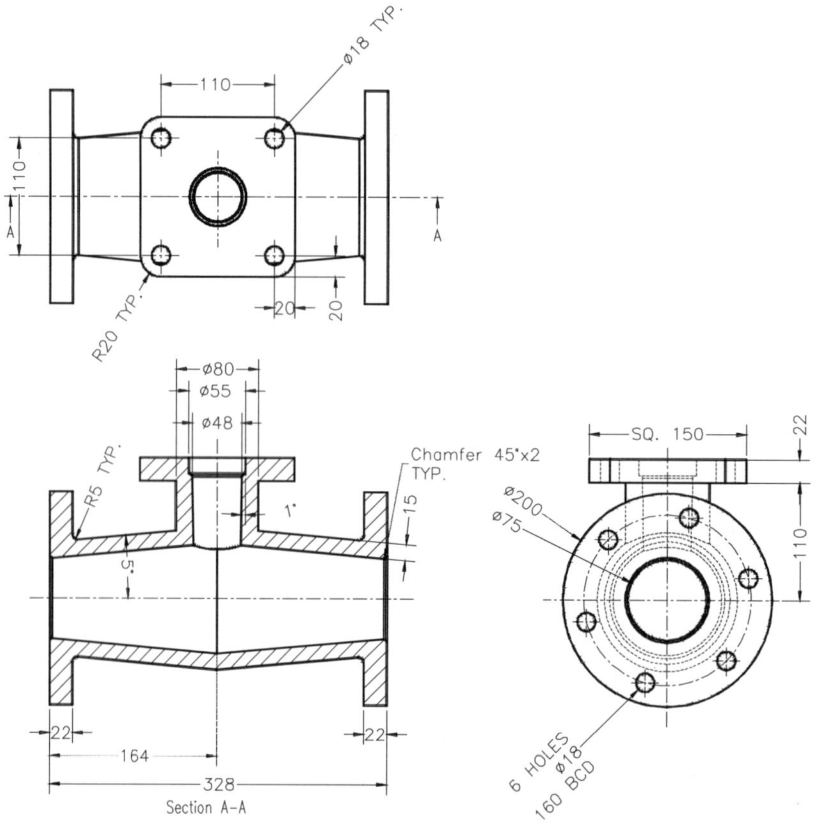

Figure 5-121 Dimensions for Exercise 2

Answers to Self-Evaluation Test

1. F, **2.** F, **3.** T, **4.** F, **5.** T, **6. Up To Next**, **7. Up To Body**, **8. Merge results**, **9. Cylindrical/Conical Surface**, **10. Bodies to Keep**

Chapter 6

Advanced Modeling Tools-I

Learning Objectives

After completing this chapter you will be able to:

- *Create holes using the Simple Hole option.*
- *Create standard holes using Hole Wizard.*
- *Apply simple and advanced fillets.*
- *Understand various selection methods.*
- *Chamfer the edges and vertices of the model.*
- *Create the Shell feature.*
- *Create the Wrap feature.*

ADVANCED MODELING TOOLS

This chapter discusses various advanced modeling tools available in SolidWorks that assist you in creating a better and accurate design by capturing the design intent in the model. For example, in the previous chapters you learned how to create a hole using the cut option. But in this chapter you will create holes using the **Simple Hole** option and the **Hole Wizard** option. Using the hole wizard, you can create standard holes classified on the basis of the industrial standard, screw type, and size. Hole wizard of SolidWorks is one of the largest standard industrial virtual hole generation machines available in any CAD package. You will also learn about some other advanced modeling tools such as fillet, chamfer, shell, and warp in this chapter.

Creating Simple Hole

CommandManager:	Features > Simple Hole	*(Customize to Add)*
Menu:	Insert > Features > Hole > Simple	
Toolbar:	Features > Simple Hole	*(Customize to Add)*

The **Simple Hole** option is used to create a simple hole feature. In the previous chapter you learned how to create holes by sketching a circle and then using the cut option. But using the simple hole option you do not need to draw a sketch of the hole. This is the reason why the holes created with this option act as a placed feature. To create a hole using this option, first you need to select the plane on which you want to place the hole feature. If you invoke this tool before selecting a plane, the **Hole PropertyManager** will be displayed, which will prompt you to select a placement plane. Now, choose the **Simple Hole** button from the **Features CommandManager** after customizing it; the **Hole PropertyManager** will be displayed as shown in Figure 6-1.

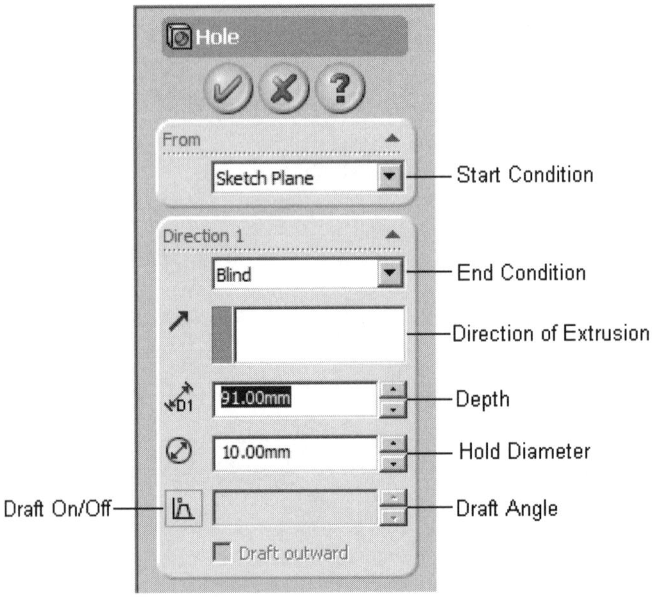

Figure 6-1 *The* ***Hole PropertyManager***

When you invoke the **Hole PropertyManager**, the preview of the hole feature will be displayed in the drawing area in temporary graphics with the default values. The preview of the hole feature with default values is shown in Figure 6-2.

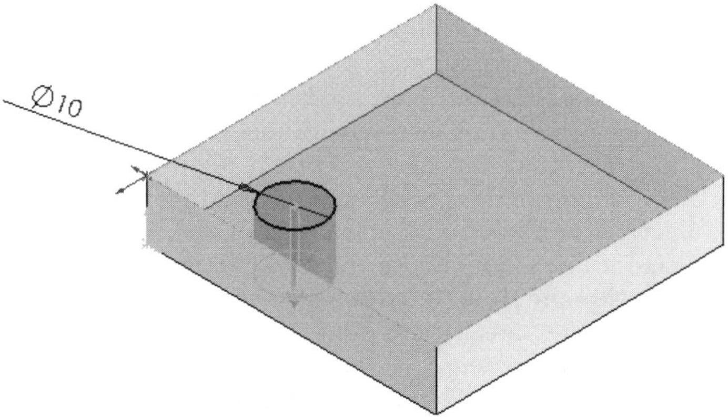

Figure 6-2 *Preview of the hole being created using the* ***Simple Hole*** *tool*

> **Tip**. *It is recommended that before creating the hole feature, you create a point that will define the placement of the hole feature. While creating the hole feature, you can dynamically move the hole feature by selecting the center point of the sketch and dragging the cursor to the point sketched earlier. The cursor will snap to that point and coincident relation will be applied between the sketched point and the center point of the hole.*

Using the **End Condition** drop-down list, specify the termination type and set the value of the hole diameter in the **Hole Diameter** spinner. You can also specify the direction of extrusion using the **Direction of Extrusion** selection area. You can also specify a draft angle in the hole feature using the **Draft On/Off** button and set the value of the draft angle using the **Draft Angle** spinner. The preview of the draft angle is also displayed in the drawing area in temporary graphics. After setting all parameters choose the **OK** button from the **Hole PropertyManager**.

The hole feature created using this option is placed on the selected plane but the placement of the hole is not yet defined. Therefore, select the hole feature from the **FeatureManager Design Tree** and right-click to display the shortcut menu. Choose the **Edit Sketch** option from the shortcut menu; the sketching environment will be invoked and you can apply the relations and dimensions to define the placement of the hole feature on the selected face and exit the sketching environment.

Creating Standard Holes Using the Hole Wizard

CommandManager:	Features > Hole Wizard
Menu:	Insert > Features > Hole > Wizard
Toolbar:	Features > Hole Wizard

The **Hole Wizard** tool is used to add standard holes to the model. The holes applied using the hole wizard include the standard counterbore, countersink, drilled, tapped, and pipe tap holes. You can also create a user-defined counterbored drilled hole, counter-drilled drilled hole, counterbored hole, counterdrilled hole, countersunk hole, countersunk drilled hole, simple hole, simple drilled hole, tapered hole, and tapered drilled hole. You can control all parameters of the holes including the termination options and you can also modify the holes according to your requirement after placing them. Thus, it results in the placement of standard parametric holes using this option. There are two methods of placing a hole feature using the **Hole Wizard** tool. The first method is the preselection method, in which you have to select a placement face or a placement plane before invoking this tool. The placement face can be a planar face or a curved face. Then choose the **Hole Wizard** button from the **Features CommandManager** to invoke the **Hole Definition** dialog box. The preview of the hole feature will be displayed in the graphics area. If the preview is not visible properly, move the **Hole Definition** dialog box to see the preview. If you change the settings and parameters or the type of hole, the preview of the hole will also get modified dynamically. The options available in the **Hole Definition** dialog box are discussed next.

Counterbore Tab

The **Counterbore** tab, shown in Figure 6-3, available in the **Hole Definition** dialog box is used to define the parameters of a counterbore hole. You can also add the frequently used

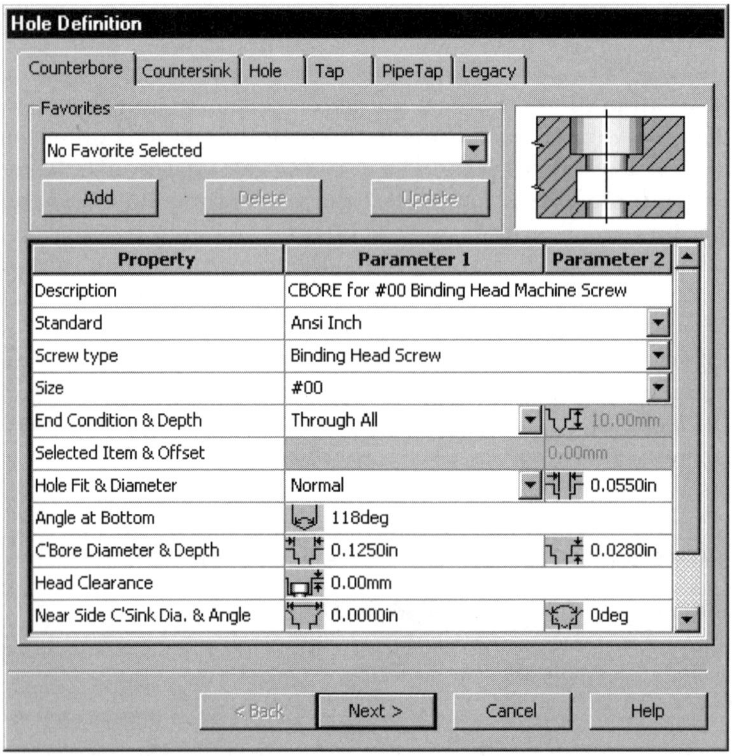

***Figure 6-3** The **Counterbore** tab of the **Hole Definition** dialog box*

counterbore holes in the favorites list option in this tab. The options available in the **Counterbore** tab are discussed next.

Property

The **Property** column of the **Hole Definition** dialog box is used to define various properties of the standard hole, which include the description, standard, screw type, size, and so on. The parameters available for all properties are divided into two columns: **Parameter 1** and **Parameter 2**. These two columns are used to define the values and other related information to create the hole. The options available in the **Property** column are discussed next.

Description

The **Description** option is used to display the name and the description of the standard hole.

Standard

The **Standard** option is used to specify the industrial dimensioning and hole standard. You can select the standard from the **Standard** drop-down list available in the **Parameter 1** column. By default, the **Ansi Inch** standard is selected. Various other dimensioning standards are available in this drop-down list such as **Ansi Metric**, **BSI**, **DIN**, **ISO**, **JIS**, **DME Mould Bases**, **Hasco Metric Mould Bases**, **PCS Mould Bases**, **Progressive Mould Bases**, and **Superior Mould Bases**.

Screw type

The **Screw type** option is used to define the type of fastener to be inserted in the hole. The standard holes created using the **Hole Wizard** tool depend on the fastener to be inserted in that hole and the size of the fastener. Therefore, creation of a hole based on the fastener to be used in that hole is a good practice and a unique feature of SolidWorks. You can select the screw type from the **Screw Type** drop-down list available in the **Parameter 1** column. The types of screws available in the drop-down list depend on the standard selected from the **Standard** drop-down list.

Size

The **Size** option is used to define the size of the fastener that will be inserted in the hole that is created using the **Hole Wizard** tool. The size of the fastener is selected from the **Size** drop-down list available in the **Parameter 1** column. The sizes of the fasteners available in the **Size** drop-down list depend on the standard selected from the **Standard** drop-down list.

End Condition and Depth

The **End Condition and Depth** option is used to define the end condition of feature termination. To define the end condition for feature termination, select the end condition option from the **End Condition** drop-down list available in the **Parameter 1** area. By default, the **Blind** option is selected in this drop-down list. Therefore, if the end condition is **Blind** you have to specify the depth of feature termination. This is the reason that the **Depth** area available on the right of the **End Condition** drop-down list in the **Parameter 2** column is replaced by the edit box when clicked once. You can specify the depth of feature termination in this edit box.

Note

A preview area is provided on the top right corner of the ***Hole Definition*** *area. You can observe the preview of the hole being created. The preview of the hole feature will be changed dynamically as you change the parameters of the hole feature.*

Selected Item and Offset

The **Selected Item and Offset** option is used to define the surface for feature termination. The **Selected Item** selection area in the **Parameter 1** column will be active only when the **Up To Vertex**, **Up To Surface**, or **Offset From Surface** option is selected from the **End Condition** drop-down list. The **Offset** area available on the right of the **Select Item** selection area will be replaced by the **Offset** edit box and you can enter the value of offset in this edit box. This option will be available only when you select the **Offset From Surface** option from the **End Condition** drop-down list.

Hole Fit and Diameter

The **Hole Fit and Diameter** option is used to specify the types of fits of **Fit** to be applied to the hole to be created. The fit types are available in the **Hole Fit** drop-down list in the **Parameter 1** column, they are **Close**, **Normal**, and **Loose**. The **Diameter** area on the right of the **Hole Fit** drop-down list in the **Parameter 2** area will be replaced by the **Diameter** edit box by clicking once in this area. The value of the diameter in this area is the default value according to the standard. The changes in value depend on the type of hole fit selected from the **Hole Fit** drop-down list. You can also modify this value according to the requirement.

Angle at Bottom

The **Angle at Bottom** option is used to specify the angle at the bottom of the hole feature. The **Angle at Bottom** area available in the **Parameter 1** column will be replaced by the **Angle at Bottom** edit box by clicking once. You can modify the default value for the angle at the bottom by entering a new value in this edit box.

C'Bore Diameter and Depth

The **C'Bore Diameter and Depth** option is used to specify the diameter and the depth of the counterbore. The **C'Bore Diameter** area will be available in the **Parameter 1** column and replaced by the **C'Bore Diameter** edit box by clicking once to modify the default value. The **Depth** area available on the right of the **C'Bore Diameter** area will be replaced by the **Depth** edit box by clicking once and you can modify the default depth of the counterbore.

Head Clearance

The **Head Clearance** option available in the **Property** column is used to specify the clearance distance between the head of the fastener and the placement plane of the hole feature. The **Head Clearance** area available in the **Parameter 1** column will be replaced by the **Head Clearance** edit box by clicking once and you can modify the default head clearance value.

Near Side C'Sink Dia. and Angle

The **Near Side C'Sink Dia. and Angle** option is used to specify the diameter and the angle for the countersink on the upper face, which is the placement plane of the hole feature. The diameter is specified in the **Near Side C'Sink Dia.** area of the **Parameter 1** column and the angle is specified in the **Angle** area of the **Parameter 2** column.

Under Hd C'Sink Dia. and Angle

The **Under Hd C'Sink Dia. and Angle** option is used to specify the diameter and the angle for the countersink to be applied at the end of the counterbore head. The diameter is specified in the **Under Hd C'Sink Dia.** area in the **Parameter 1** column and the angle is specified in the **Angle** area in the **Parameter 2** column.

Far Side C'SinkDia. and Angle

The **Far Side C'Sink Dia. and Angle** option is used to specify the diameter and the angle for the countersink to be applied at the end of the hole feature. This option will be available only if the **Through All** or the **Up To Next** option is selected from the **End Condition** drop-down list. The diameter is specified in the **Far Side C'Sink Dia.** area in the **Parameter 1** column and the angle is specified in the **Angle** area in the **Parameter 2** column.

Favorites

The **Favorites** area available in the **Hole Definition** dialog box is used to add the frequently used holes to the favorite list. If you add a hole to the favorite list, you will not have to perform the same settings to add similar types of holes every time. To add a particular hole to the favorite list, choose the **Add** button available in the **Favorite** area of the **Hole Definition** dialog box. The **New Favorite Name** dialog box will be displayed with the default name of the hole in the **Favorite Name** edit box as shown in Figure 6-4.

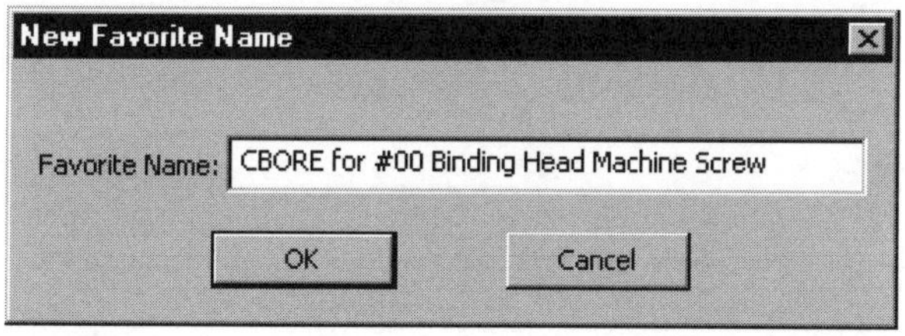

***Figure 6-4** The **New Favorite Name** dialog box*

You can enter a new name in the **Favorite Name** edit box and choose the **OK** button to add the hole to the favorite list. If you have to apply a hole feature from the favorite list, you just need to select the name of the hole feature from the **Favorites** drop-down list. The current parameters of the hole will be changed to the parameters of the hole feature that is selected from the **Favorites** drop-down list. Figure 6-5 shows the counterbore holes created using the **Hole Wizard** tool.

Countersink Tab

The **Countersink** tab, as shown in Figure 6-6, available in the **Hole Definition** dialog box is

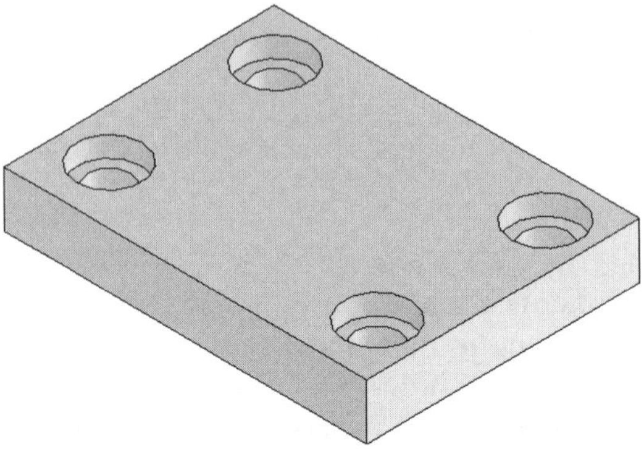

Figure 6-5 *A model with counterbore holes*

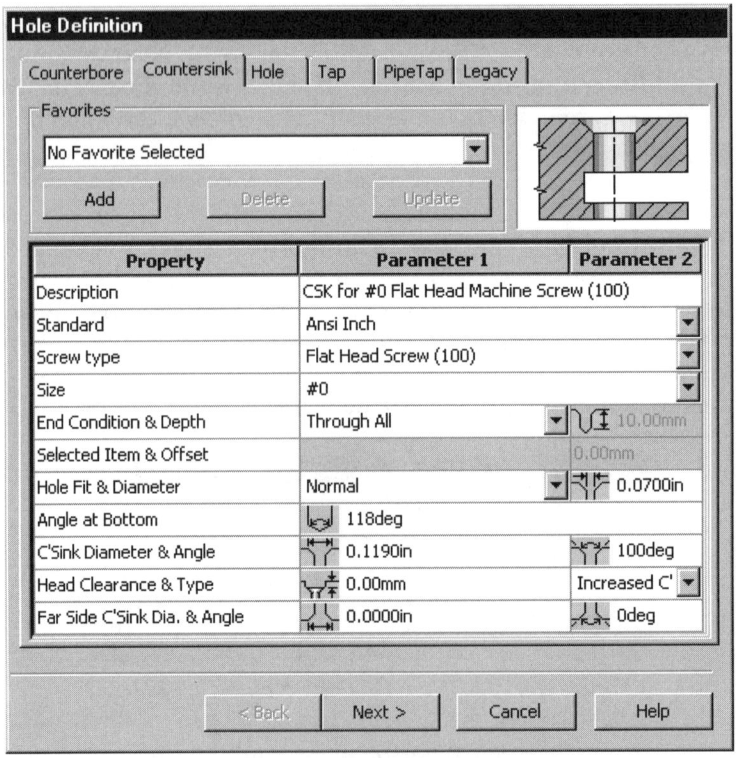

Figure 6-6 *The* ***Countersink*** *tab of the* ***Hole Definition*** *dialog box*

used to define the parameters of the countersink hole. Most of the options available in the **Property** column are the same as those in the **Property** column of the **Counterbore** tab. However, there are some additional options to define the angle and diameter of the countersink. These options of the **Property** column of the **Countersink** tab are discussed next.

Property

The **Property** column of the **Hole Definition** dialog box is used to define the parameters of the countersink hole. As mentioned earlier, most of the options are the same as those discussed in the **Property** column of the **Counterbore** tab. The options that have been discussed earlier are not included here. Some options that are used to create only a counterbore and are not used in countersink are not available in this tab. Only the additional options available in the **Property** column of the **Countersink** tab are discussed next.

C'Sink Diameter and Angle

The **C'Sink Diameter and Angle** option is used to define the diameter and the angle of the countersink. The **C'Sink Diameter** area in the **Parameter 1** column will be changed to the **C'Sink Diameter** edit box when clicked once. The value of the countersink diameter is specified in this edit box. The **Angle** area available in the **Parameter 2** area will be replaced by the **Angle** edit box when clicked once. The value of the angle of the countersink is specified in this edit box.

Head Clearance and Type

The **Head Clearance and Type** option is used to define the head clearance distance between the placement surface of the hole feature and the top face of the fastener and to define the type of the counter to be added. The **Head Clearance** area available in the **Parameter 1** column will be replaced by the **Head Clearance** edit box when clicked once. The value of the clearance distance is specified in this edit box. The **Type** drop-down list available on the right of the **Head Clearance** area is used to specify whether the clearance distance is compensated by extending the countersink or by adding a counterbore.

Figure 6-7 shows the countersink holes created using the **Hole Wizard**.

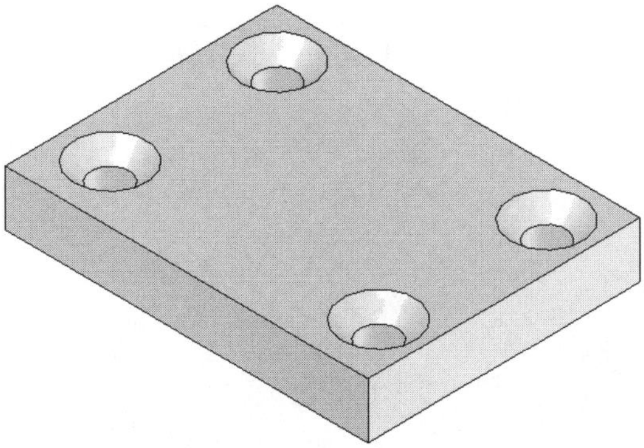

Figure 6-7 *A model with countersink holes*

Hole Tab

The **Hole** tab, shown in Figure 6-8, available in the **Hole Definition** dialog box is used to define the parameters to add a standard drilled hole to the model. Using this tab you can

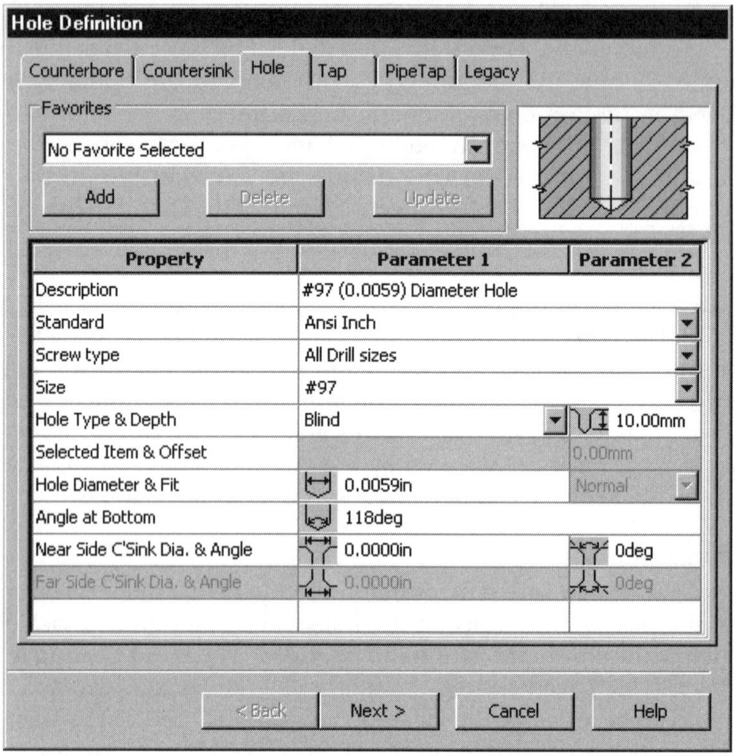

Figure 6-8 *The* ***Hole*** *tab of the* ***Hole Definition*** *dialog box*

add simple holes that can be drilled using the standard drills available in the market. The options available in the **Property** column of this tab are used to specify the parameters for creating a standard drilled hole.

Figure 6-9 shows the drilled holes created using the **Hole Wizard** tool.

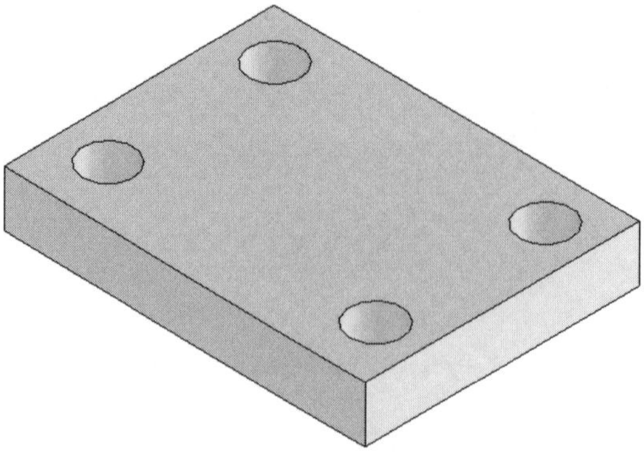

Figure 6-9 *A model with drilled holes*

Tap Tab

The **Tap** tab, shown in Figure 6-10, available in the **Hole Definition** dialog box is used to define the parameters to add a tapped hole to the model. Tapped hole is basically a hole in which internal threading (also known as tapping) is shown. Using this tab, you can add the standard tapped holes in your design. Almost all options available in the **Property** column of this tab are the same as those discussed earlier. The remaining options are discussed next.

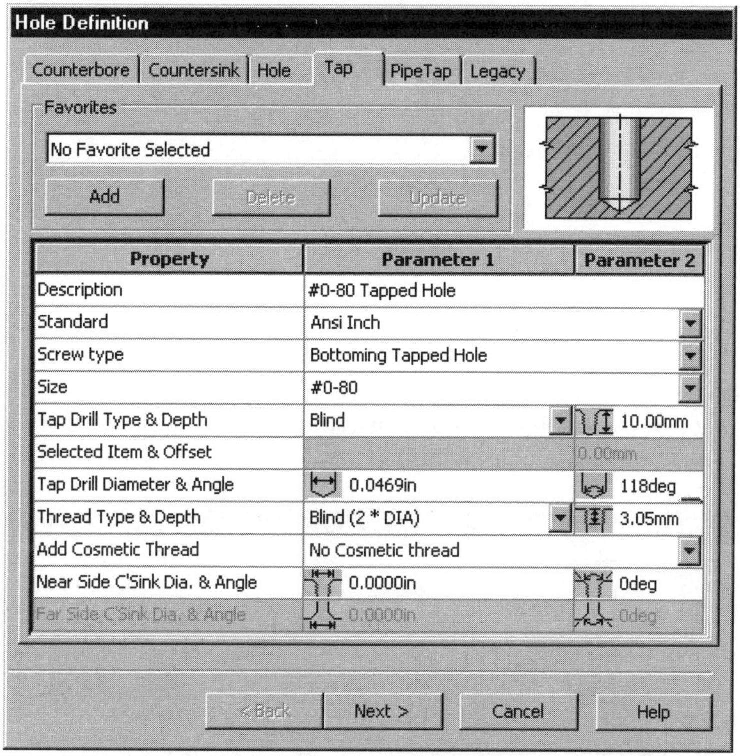

*Figure 6-10 The **Tap** tab of the **Hole Definition** dialog box*

Thread Type and Depth

The **Thread Type and Depth** option is used to define the type of depth and the value of the depth of the thread. The **Thread Type** drop-down list available in the **Parameter 2** column is used to specify the type of thread depth. The options available in the **Thread Type** drop-down list depend on the feature termination option selected from the **Tap Drill Type** drop-down list. The depth of the thread is specified in the **Depth** area in the **Parameter 2** column.

Add Cosmetic Thread

The enhancement in the **Hole Wizard** tool is the introduction of graphic threads. The **Add Cosmetic Thread** option is used to add the cosmetic threads while creating a tapped hole. The **Add Cosmetic Thread** drop-down list is available in the **Parameter 1** column. If you select the **No Cosmetic thread** option from this drop-down list, the cosmetic thread will not be added to the tapped hole. However, if you select the **Add Cosmetic thread**

with thread callout option from this drop-down list, the graphic cosmetic thread and the thread callout will be added to the tapped hole feature. The **Add Cosmetic thread without thread callout** option is used to add only the graphic cosmetic thread to the hole feature. Figure 6-11 shows the tapped holes created using the **Hole Wizard**.

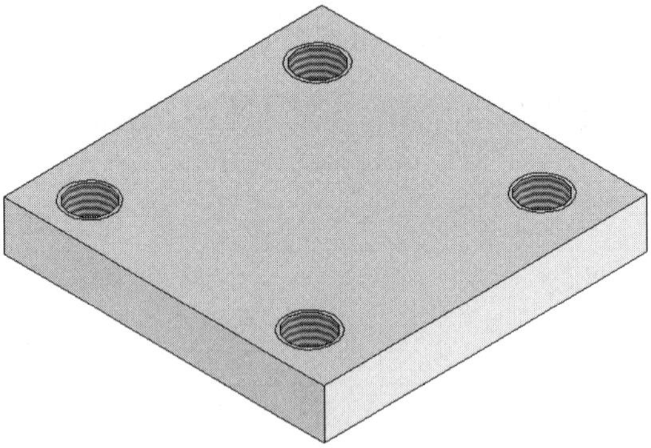

Figure 6-11 *Model with tapped holes*

Pipe Tap Tab

The **Pipe Tap** tab, shown in Figure 6-12, is used to specify the parameters for the tapered pipe tap hole. The options available in this tab are the same as those discussed in earlier tabs.

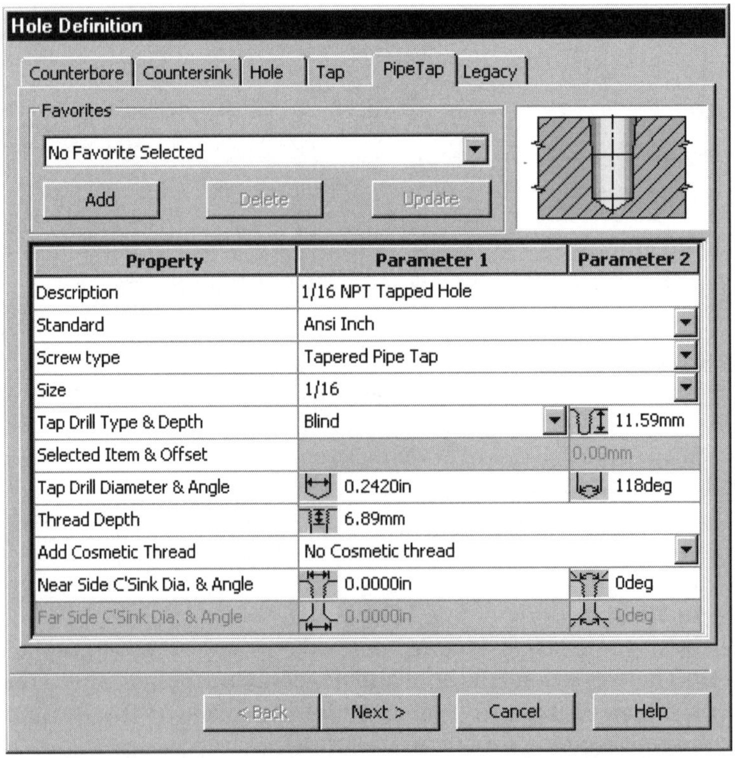

Figure 6-12 *The **Pipe Tap** tab of the **Hole Definition** dialog box*

Legacy Tab

The **Legacy** tab of the **Hole Definition** dialog box, shown in Figure 6-13, is used to add a user-defined hole feature. The options available in this tab are discussed next.

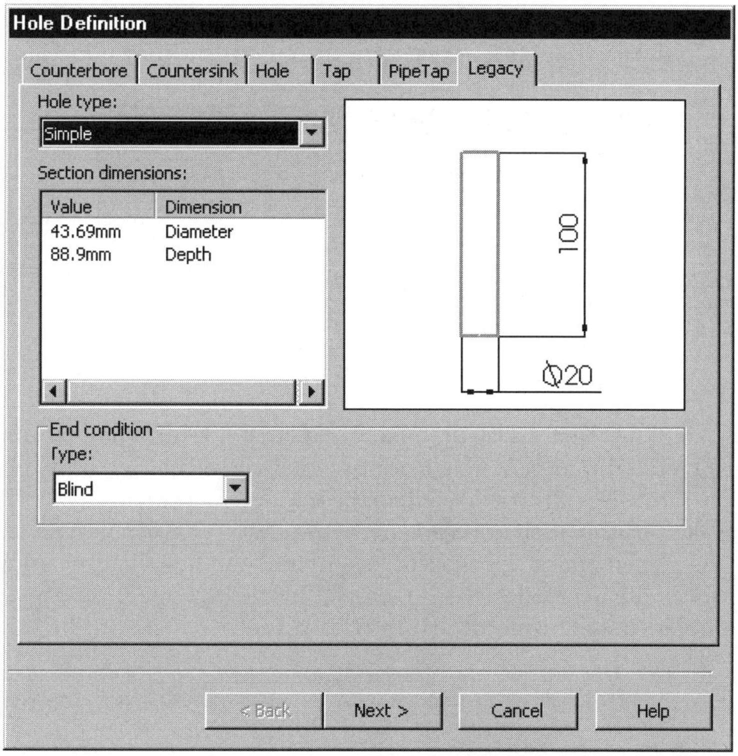

*Figure 6-13 The **Legacy** tab of the **Hole Definition** dialog box*

Hole type

The **Hole type** drop-down list is used to specify the type of hole you want to create. The types of holes available in this drop-down list are **C-Bored Drilled**, **C-Drilled Drilled**, **Counterbored**, **Counterdrilled**, **Countersunk**, **C-Sunk Drilled**, **Simple, Drilled**, **Tapered**, and **Tapered Drilled**. The preview of the selected hole type is displayed in the preview area provided on the right of the **Hole Definition** dialog box.

Section dimensions

The **Section dimensions** area is used to display various dimensions of the selected hole and their default values. You can also modify the values by replacing the **Value** area by the **Value** edit box.

End condition

The **Type** drop-down list available in the **End condition** area is used to specify the end condition of feature termination.

As discussed earlier, the preview of the hole feature is dynamically updated in the drawing area while you are setting the parameters of the hole feature. This is because you have already selected the placement plane for the placement of the hole feature. If you do not select a placement plane for the placement of the hole feature, the preview of the hole feature will not be displayed in the drawing area. After setting all parameters of the hole feature, choose the **Next** button to display the **Hole Placement** dialog box, as shown in Figure 6-14.

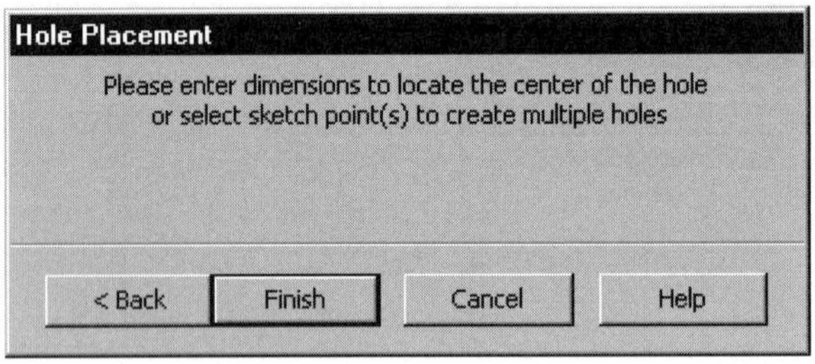

Figure 6-14 *The* ***Hole Placement*** *dialog box*

The select cursor will be replaced by the placement cursor. Using the placement cursor you can place more holes in the current hole features. As discussed earlier, if the placement plane is selected earlier, the hole is already placed on the selected placement plane. If the placement plane is not selected earlier then you can specify a point to place the hole feature. You can also constrain the placement of the placement point of the hole feature using the relations and dimensions. Choose the **Finish** button to complete feature creation. Figure 6-15 shows a base plate with holes created using the **Hole Wizard** tool.

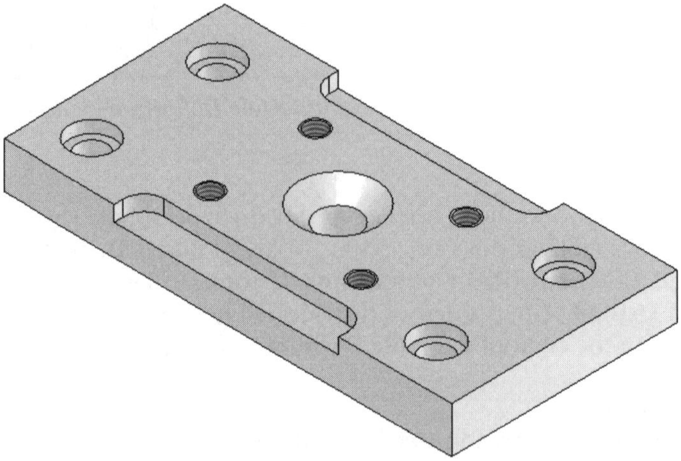

Figure 6-15 *Base Plate with holes created using the* ***Hole Wizard*** *option*

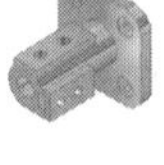

Note

If you create a pattern feature of a tapped hole feature, the graphic threads will not be displayed in the other instances of the pattern, except the parent instance. Therefore, to add graphic in other pattern instances use the ***Texture PropertyManager****. You will learn more about patterns in later chapters.*

Tip. *The hole feature created using the* ***Hole Wizard*** *tool consists of two sketches. The first sketch is the sketch of the placement point and the second sketch is the sketch of the profile of the hole feature. If you preselect the placement plane before invoking the* ***Hole Definition*** *dialog box, the resulting placement sketch will be a 2D sketch. Instead of preselecting the placement plane if you select the placement point after invoking the* ***Placement Point*** *dialog box, the resulting placement sketch will be a 3D sketch. You will learn more about 3D sketches in the later chapters.*

In the modern modeling practice, the creation of threads is avoided in models because it results in the creation of complex geometry. The views generated from those models that contain complex geometry, which is difficult to understand. Therefore, it is a better practice to avoid the creation of threads in the model and add the cosmetic threads. Using the cosmetic threads you will get the thread convention in the drawing views, which is recommended, instead of creating complete thread.

If a cosmetic thread is added in a tapped hole, the cosmetic thread will also be displayed along with the placement and hole profile sketches. You can edit the cosmetic threads by selecting them from the ***FeatureManager Design Tree*** *and right-clicking to display the shortcut menu. Choose the* ***Edit Definition*** *option to display the* ***Cosmetic Thread*** *dialog box. The* ***Cosmetic Thread*** *dialog box and the cosmetic threads are discussed in later chapters.*

You can also view the convention of the thread if the cosmetic thread is added to a tapped hole feature. Orient the model to the top view to observe the thread convention from the top view. Orient the model in the front view, back view, or any side view to observe the thread convention from the side views.

Creating Fillets

CommandManager:	Features > Fillet
Menu:	Insert > Features > Fillet/Round
Toolbar:	Features > Fillet

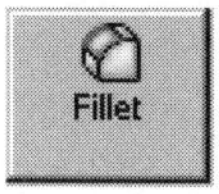

In SolidWorks, you can add fillets as a feature in the model using the **Fillet** tool. As discussed earlier you can also add fillets within a sketch. But adding fillets in a sketch is not a good practice from the design point of view. This is because you have to keep the sketch as simple as possible. Using the fillet tool you can round an internal or external face or edge of a model. You can also use the advanced fillet options to add advanced fillets to the model. You can preselect the face, edge, or feature to which the fillet has to be applied. You can also select the entity to be filleted after invoking the fillet tool. Choose the **Fillet** button from the **Features CommandManager** to invoke the **Fillet PropertyManager**. The **Fillet PropertyManager** is shown in Figure 6-16. The preview of the fillet feature will also be displayed in the drawing area if the entities to be filleted are selected and the **Full Preview** radio button will be selected by default. If preselection is not done, you will be prompted to select the edges, faces, features, or loops to add the fillet feature. Using

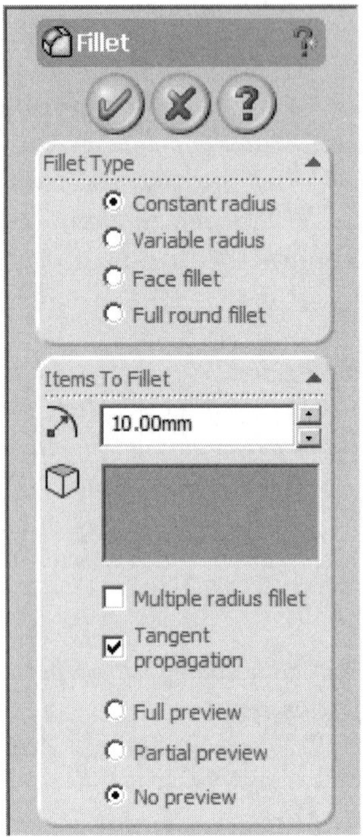

Figure 6-16 The **Fillet PropertyManager**

the select cursor, select the entity to be filleted. A fillet callout will also be displayed along the preview of the fillet. Figure 6-17 shows the preview of the fillet feature with fillet callout.

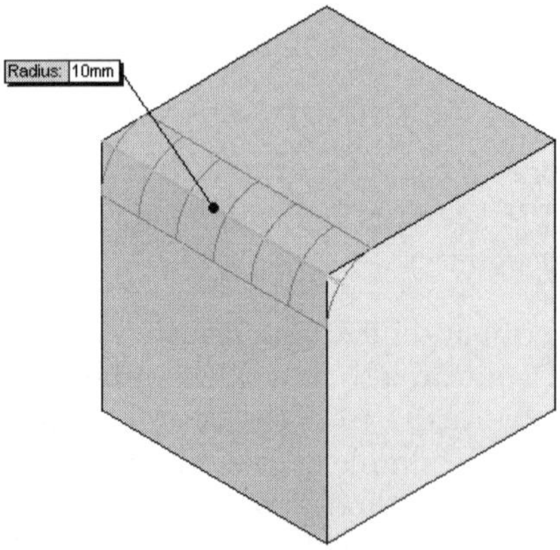

Figure 6-17 Preview of the fillet feature

Using the fillet tool you can create various types of fillets, which given next.

1. Constant radius fillet
2. Variable radius fillet

3. Face Fillet
4. Full round fillet

The above-mentioned fillets are discussed next.

Constant Radius Fillet

The constant radius fillet option creates a fillet of constant radius along the selected entity. This is the default option selected in the **Fillet Type** rollout. You can set the value of the fillet radius using the **Fillet** spinner provided in the **Items To Fillet** rollout or clicking in the value area of the fillet callout. The value area of the callout will be replaced by the radius edit box. Enter the value of the radius and press the ENTER key on the keyboard. The preview of the fillet will be changed dynamically when the value of the radius of the fillet is changed. The names of the selected entities are displayed in the **Edges, Faces, Features, and Loops** selection area. The entities that you can select to add the fillet feature are faces, edges, features, and loops. Now, choose the **OK** button from the **Fillet PropertyManager**. Figures 6-18 through 6-23 show the selection of different entities and the resulting fillet creation from the selected entities.

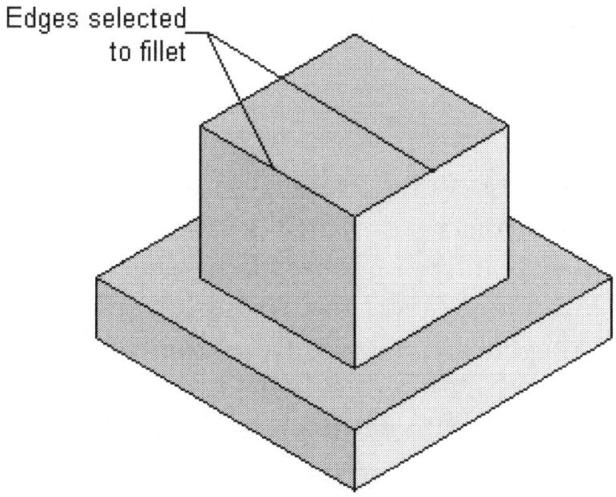

Figure 6-18 *Selecting the edges*

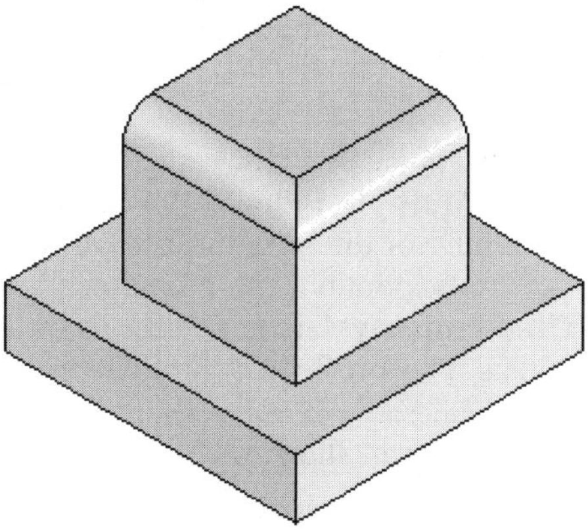

Figure 6-19 *Resulting fillet*

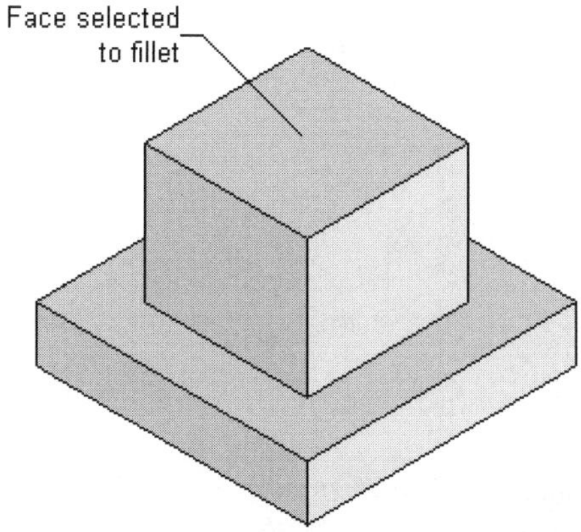

Figure 6-20 *Selecting the face*

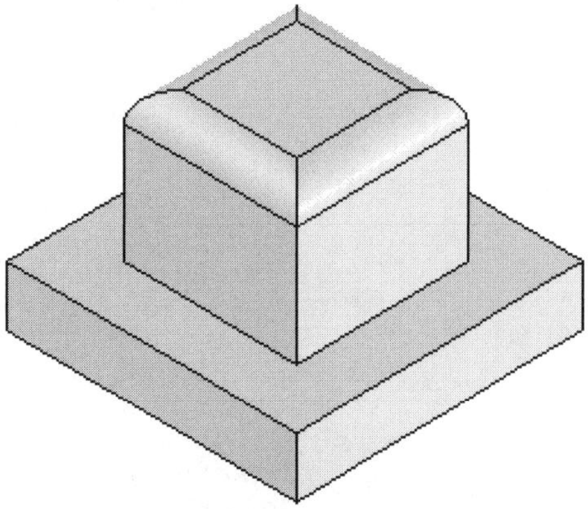

Figure 6-21 *Resulting fillet*

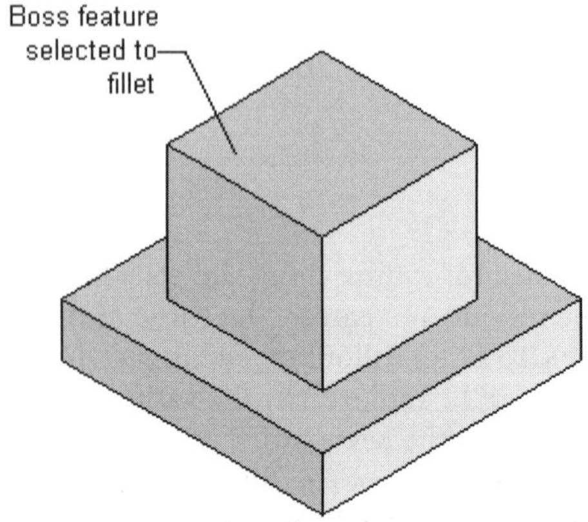

Figure 6-22 Selecting the feature

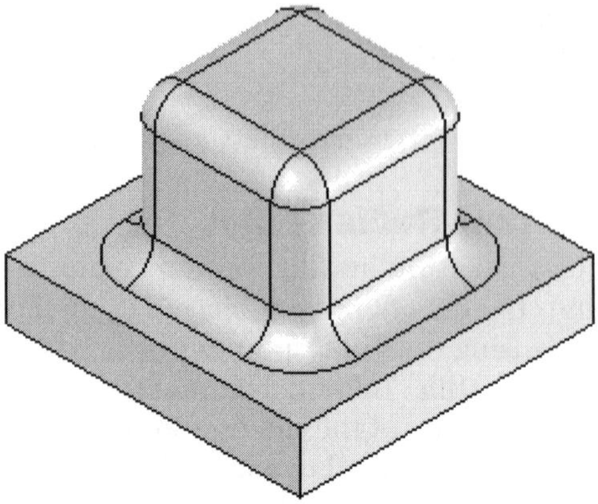

Figure 6-23 Resulting fillet

Tip. *You can preselect a feature for creating a fillet feature or select the feature after invoking the* ***Fillet*** *tool. For postselection of the feature to be filleted, expand the* ***FeatureManager Design Tree****, which is displayed in the drawing area, and select the feature. You can also select the feature from the drawing area. The preview of the filleted feature will be displayed in the drawing area.*

Multiple Radius Fillet

Using the **Multiple radius fillet** option provided in the **Fillet PropertyManager** you can specify a fillet of different radii to all selected edges. For creating a fillet feature using the multiple radius option, preselect the edges, faces, or features or select them after invoking the **Fillet PropertyManager**. After invoking the **Fillet** tool, select the **Multiple radius fillet** check box. The preview of the fillet feature with the default value will be displayed in the drawing area. You will notice that you are provided with different callouts for each selected entity. Figure 6-24 shows the preview of the fillet feature with **Multiple radius fillet** check box selected.

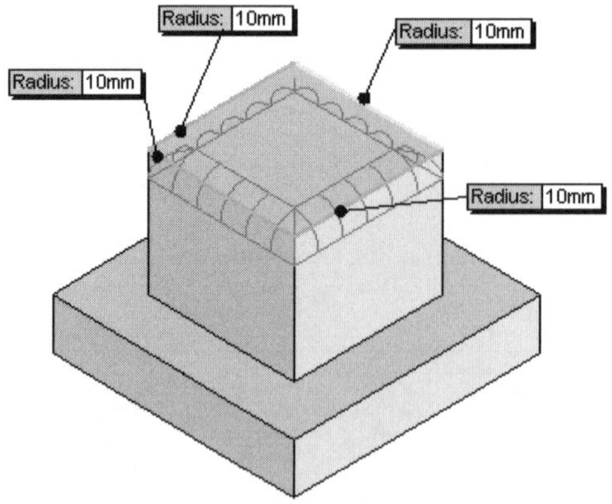

Figure 6-24 *Preview of the fillet feature with the* ***Multiple radius fillet*** *check box selected*

The names of the selected entities are displayed in the **Edges, Faces, Features and Loops** display box. The boundaries of the currently selected entity in the selection box are highlighted in green with a thick line. You can set the value of each selected entity by using the **Radius** spinner or specify the value of fillet radius in the radius callout as shown in Figure 6-25. As you modify the value of the radius, the preview of the fillet feature will be modified dynamically in the drawing area. Figure 6-26 shows the fillet created using the multiple radius fillet.

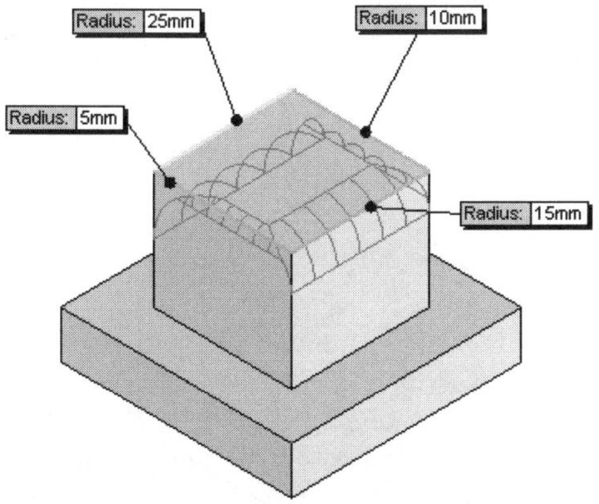

Figure 6-25 *Different radii specified in each radius callout*

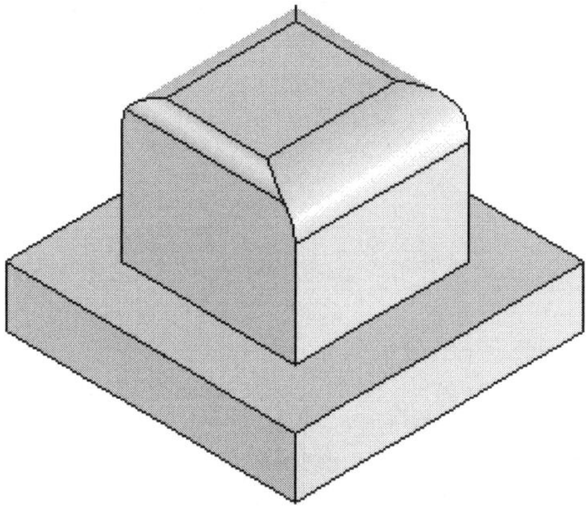

Figure 6-26 *Resulting fillet*

Fillet With and Without Tangent Propagation

In SolidWorks, you can add a fillet feature to a model with or without the tangent propagation. When you invoke the **Fillet PropertyManager**, you can observe that by default the **Tangent propagation** check box is selected. Therefore, if you select an edge, face, feature, or loop to fillet, it will automatically select other entities that are tangential to the selected entity. Thus, it will apply the fillet feature to all entities that are tangential to the selected one. If you clear the **Tangential propagation** check box, the fillet will be applied only to the selected entity. Figure 6-27 shows the entity to be selected to add a fillet feature. Figures 6-28 and 6-29, show the fillet feature created with the **Tangent propagation** check box cleared and selected respectively.

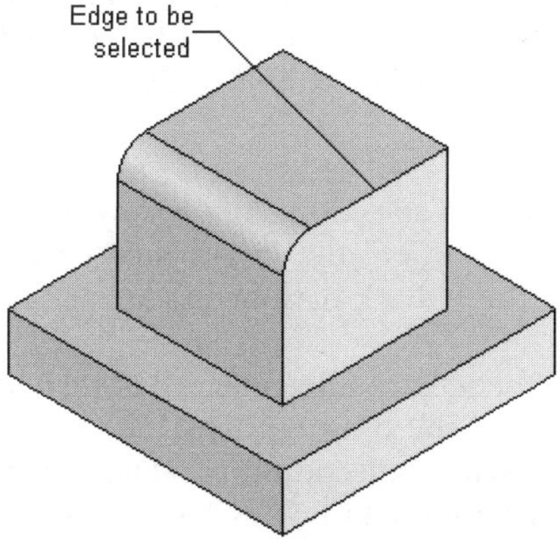

Figure 6-27 *Edge to be selected to apply the fillet feature*

Tip. *The **Full preview** radio button available in the **Fillet PropertyManager** is used to preview the fillet feature before actually creating the feature. If you select the **Partial preview** radio button, you can view only the partial preview of the fillet feature. If you select a face to add a fillet feature and select the **Partial preview** radio button, you cannot preview the fillet feature created to all edges adjacent to the selected face. You can preview only the fillet on the single edge of the selected face. Press the A key on the keyboard to cycle the preview of the fillet feature on other edges of the selected face. If you select the **No preview** radio button, you cannot see the preview of the fillet feature.*

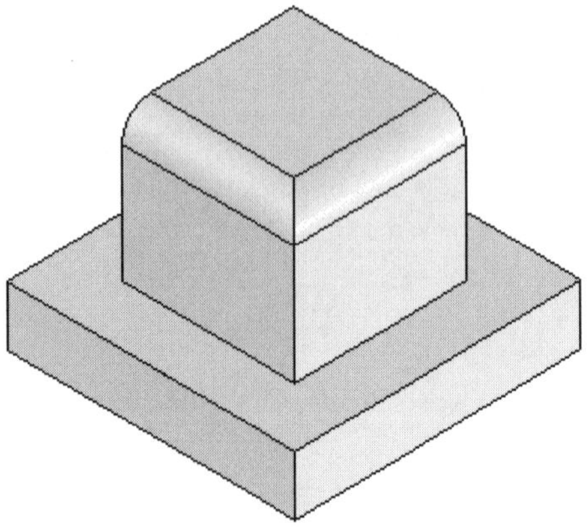

***Figure 6-28** Fillet feature created with the **Tangent propagation** check box cleared*

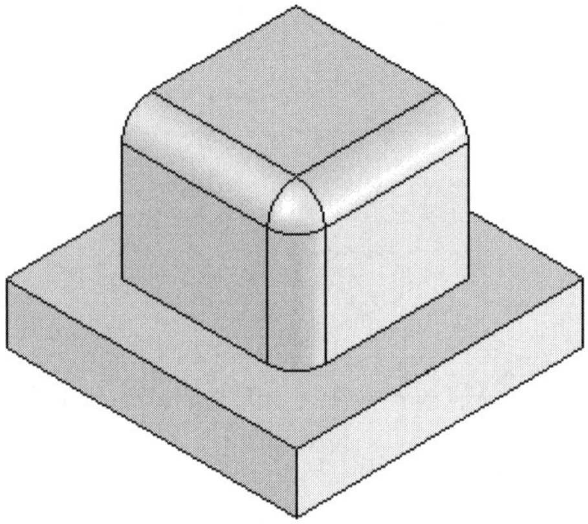

***Figure 6-29** Fillet feature created with the **Tangent propagation** check box selected*

Setback Fillets

The setback fillet is created where three or more edges are merged into a vertex. This type of fillet is used to smoothly blend transition surfaces generated from the edges to the fillet vertex. This smooth transition is created between all selected edges and the vertex selected for the setback type of fillet. To create a setback fillet, invoke the **Fillet PropertyManager** and select three or more edges to apply the fillet. Note that the edges should share the same vertex. The preview of the fillet will be displayed in the drawing area. Now, click on the black arrow in the **Setback Parameters** rollout to expand the rollout. The **Setback Parameters** rollout will be displayed, as shown in Figure 6-30. This rollout will be used to specify the setback parameters.

Click once in the **Setback Vertices** selection area to invoke the setback vertex selection command. Now, select the vertex where the edges meet. Figure 6-31 shows the selected edges and the vertex to which the setback parameters are assigned.

When you select the vertex for the setback fillet, you will observe that callouts with unassigned setback distances are displayed in the drawing area. The name of the selected vertex will be displayed in the **Setback Vertices** selection area. The names of the selected edges will be displayed in the **Setback Distances** edit box. Select the name of the edge in the **Setback Distances** edit box to assign a setback distance to that edge; a magenta arrow will be displayed

Tip. *You can also drag and drop the fillet features created on one edge to the other edge. Using the left mouse button select the fillet feature from the* ***FeatureManager Design Tree*** *or from the drawing area and hold the left mouse button down and drag the cursor and release the left mouse button to drop the feature on the required edge or face. You can also copy the fillet feature and paste it on the selected entity.*

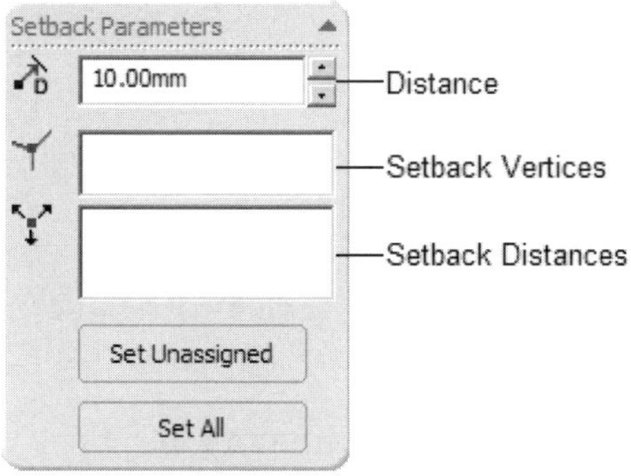

Figure 6-30 *The* ***Setback Parameters*** *rollout*

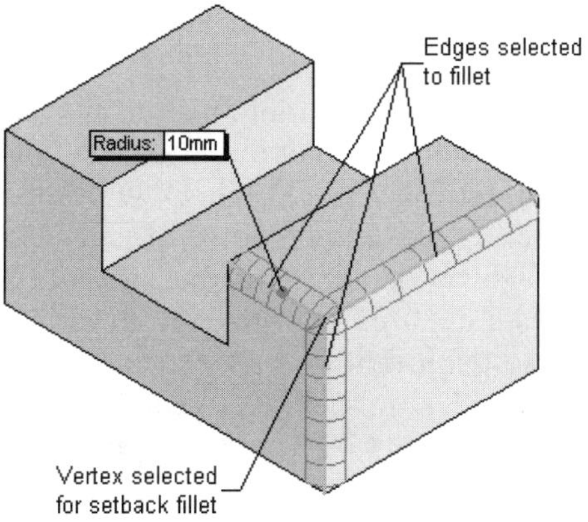

Figure 6-31 *Edges and vertex to be selected to apply the setback fillet feature*

along that edge. Using the **Distance** spinner provided in the **Setback Parameters** rollout, assign a setback distance. Similarly, assign the setback distance to all edges. You can also assign the setback distance directly by specifying the value in the setback callouts displayed in the drawing area. As discussed earlier, the preview of the fillet will be updated automatically when you assign any value. The **Set Unassigned** button available in the **Setback Parameters** rollout is used to assign the setback distance displayed in the **Distance** spinner to all unassigned edges. The **Set All** button is used to assign the setback distance displayed in the **Distance** spinner to all edges. Figure 6-32 shows the preview of the setback fillet and Figure 6-33 shows a setback fillet on the right of the model and a normal fillet on the other side of the model.

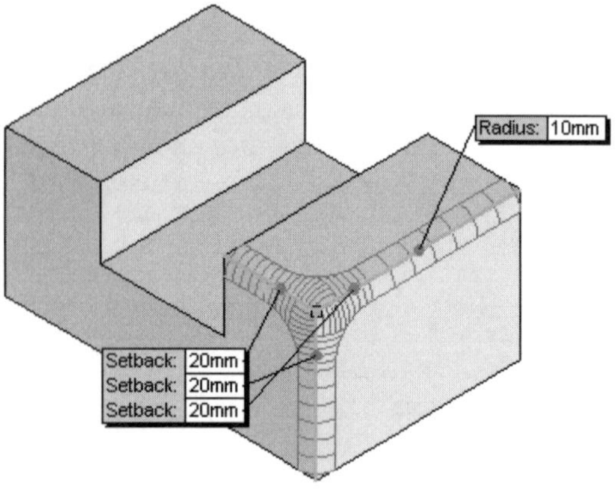

Figure 6-32 Preview of the setback fillet

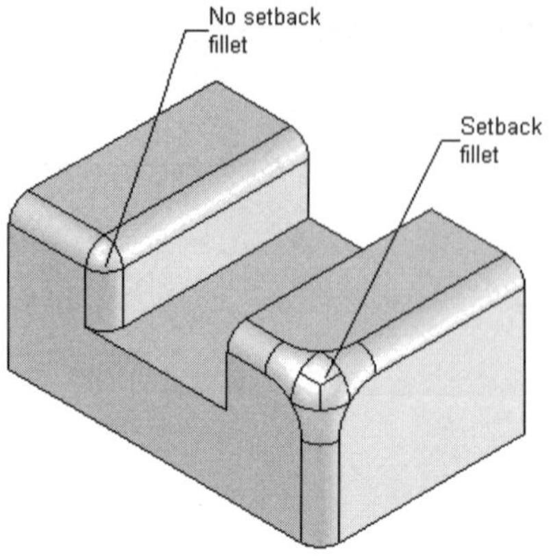

Figure 6-33 Simple and setback fillet features

Other Fillet Options

You are also provided with various other fillet options using which you can create an accurate and aesthetic design. The other fillet options available in the **Fillet PropertyManager** are **Keep features**, **Round corners**, **Controlling the Overflow type**, and so on. These options are discussed next.

Keep feature

If you have boss or cut features in a model and the fillet created is large enough to consume those features, it is recommended that you select the **Keep features** check box available in the **Fillet Options** rollout. This check box is selected by default, but you should confirm before creating any fillet feature. If you clear this check box, the fillet feature will consume the features that will obstruct its path. Note that the features that are consumed by the fillet feature are not deleted from the model. They disappear from the model because of some geometric inconsistency. When you rollback, suppress, or delete the fillet, the consumed features will reappear. You will learn more about rollback and suppress in later chapters. Figure 6-34 shows the model and the edge to be selected for fillet.

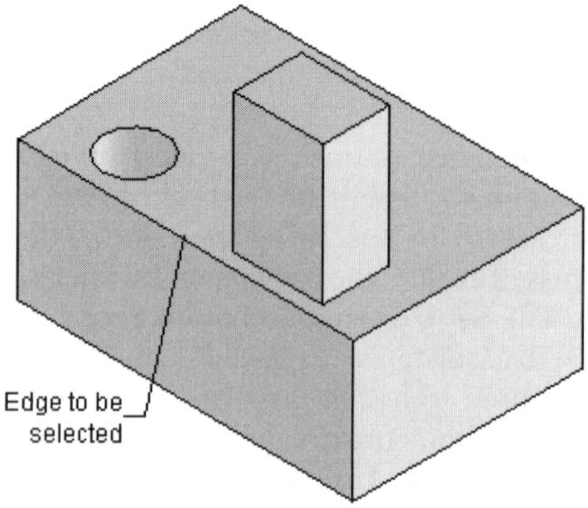

Figure 6-34 Edge to be selected to apply the fillet feature

Figure 6-35 shows the fillet feature with the **Keep features** check box cleared and Figure 6-36 shows the fillet feature with the **Keep features** check box selected.

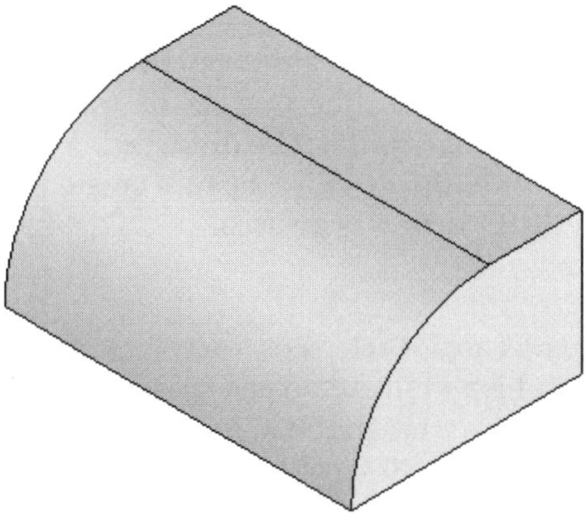

Figure 6-35 *Fillet feature created with the* ***Keep features*** *check box cleared*

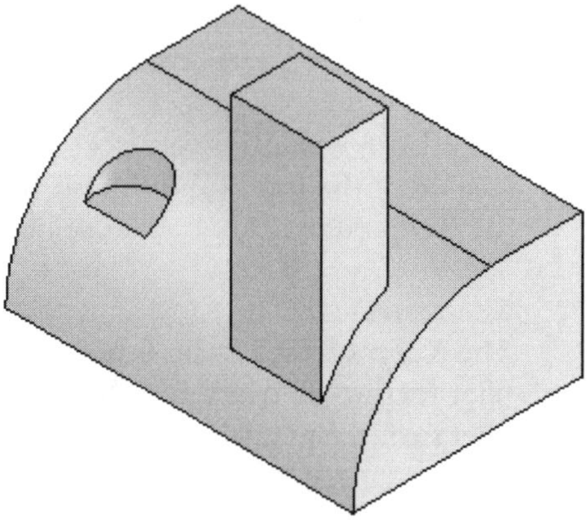

Figure 6-36 *Fillet feature created with the* ***Keep features*** *check box selected*

Round corners

The **Round corners** option is used to round the edges at the corners of the fillet feature. To create a fillet feature with round corners, select the **Round corners** check box from the **Fillet Options** rollout after specifying all parameters of the fillet feature. Figure 6-37 shows a fillet feature created with the **Round corners** check box cleared and Figure 6-38 shows a fillet feature created with the **Round corners** check box selected.

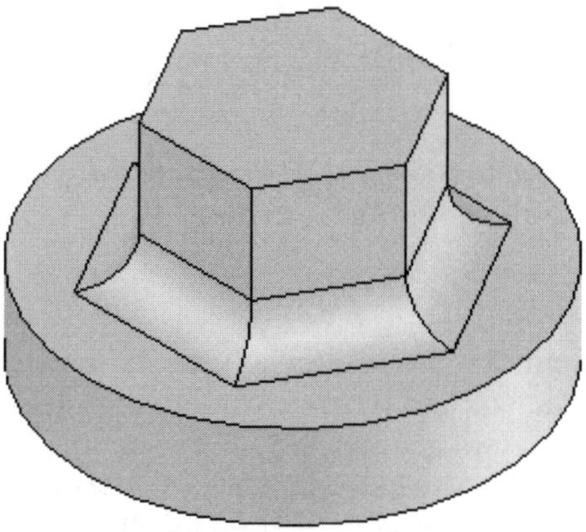

Figure 6-37 *Fillet feature created with the* ***Round corners*** *check box cleared*

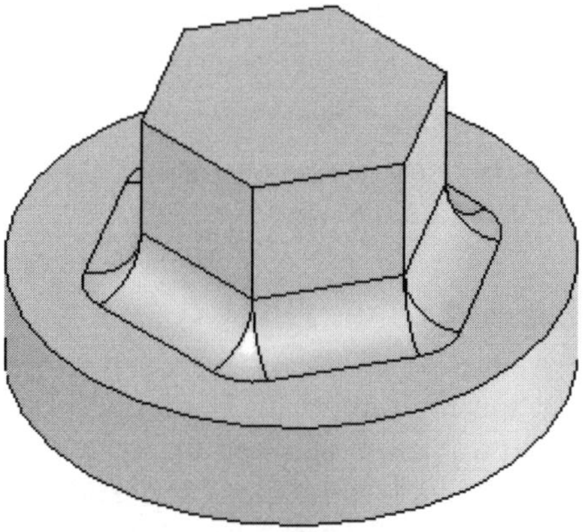

Figure 6-38 *Fillet feature created with the* ***Round corners*** *check box selected*

Overflow type

The **Overflow type** area is used to specify the physical condition that the fillet feature should adopt when it extends beyond an area. By default, SolidWorks automatically adopts the best possible flow type to accommodate the fillet, depending on the geometric conditions. This is because the **Default** option is selected by default in the **Overflow type** area. The other two options available in this area are discussed next.

Keep edge

The **Keep edge** radio button is selected when the fillet feature extends beyond a specified area. Therefore, to accommodate the fillet feature this option will divide the fillet into multiple surfaces and the adjacent edges will not be disturbed. A dip is created at the top of the fillet feature. Figure 6-39 shows a fillet feature created with the **Keep edge** radio button selected from the **Overflow type** area.

Keep surface

The **Keep surface** radio button available in this area is selected to accommodate the fillet feature by trimming the fillet feature. This will maintain the smooth rounded fillet surface but it will disturb the adjacent edges. As this option maintains the smooth fillet surface, it extends the adjacent surface. Figure 6-40 shows a fillet feature created with the **Keep surface** radio button selected from the **Overflow type** area.

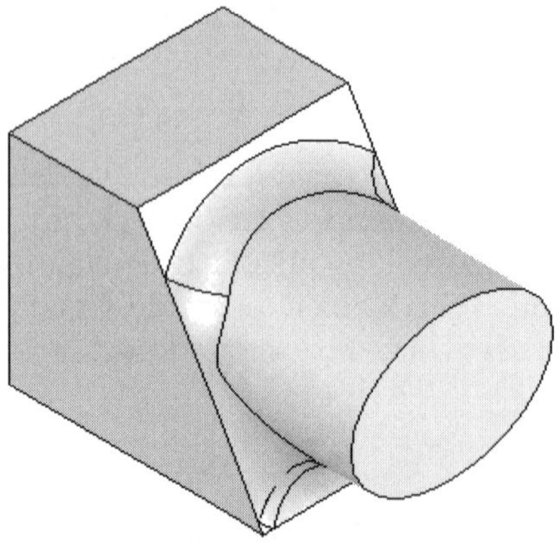

Figure 6-39 *Fillet feature created with the overflow type as* ***Keep edge***

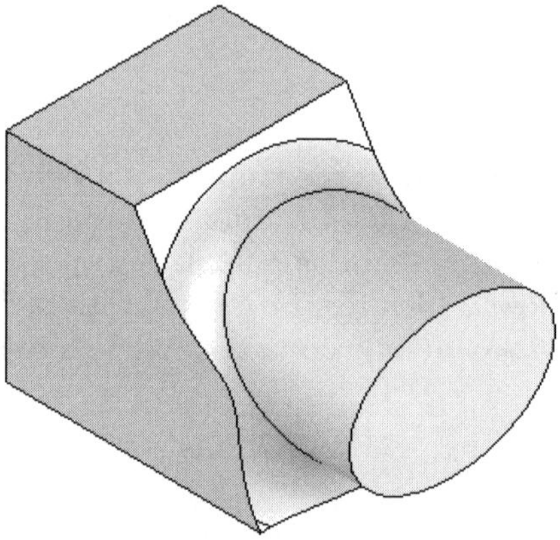

Figure 6-40 *Fillet feature created with the overflow type as* ***Keep surface***

Variable Radius Fillet

The variable radius fillet is created by specifying different radii along the length of the selected edge at specified intervals. Depending on the options, you can create a smooth transition or a straight transition between the vertices to which the radii are applied. To create a variable radius fillet invoke the **Fillet PropertyManager**. Select the **Variable radius** radio button from the **Fillet Type** rollout; the **Variable Radius Parameters** rollout will be automatically displayed in the **Fillet PropertyManager** as shown in Figure 6-41.

You will be prompted to select the edge to fillet. Using the left mouse button select the edge or edges you want to fillet. The name of the selected edge will be displayed in the **Edges, Faces, Features and Loops** selection area. By default, the radius is applied at the start point and the endpoint. The variable radius callouts are displayed at the ends of the selected edge as shown in Figure 6-42.

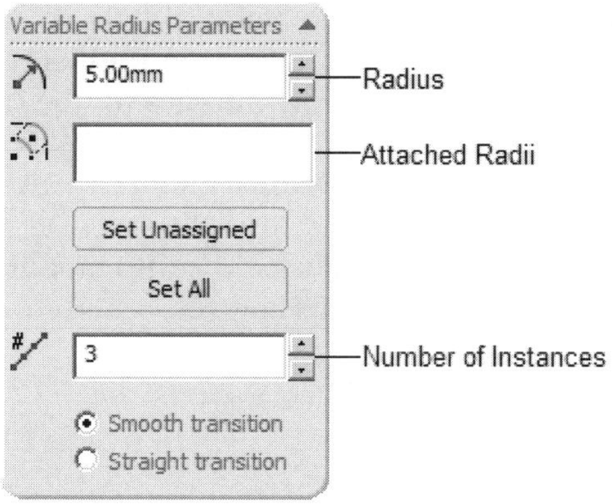

Figure 6-41 *The* ***Variable Radius Parameters*** *rollout*

The names of the vertices on which the callouts are added are displayed in the **Attached Radii** display box. You will find three red points on the selected edge because the value of the control point in the **Number of Instances** spinner is set to **3**. You can create a number of control points using the **Number of Instances** spinner. These control points are also called movable points because you can change the position of these control points. The additional radii are specified on these points on the selected edge.

Using the left mouse button select the three control points available on the selected edge. As you select the control point the **Radius and Position** callouts will be displayed for each control point as shown in Figure 6-43. The name of the selected point will also be displayed in the **Attached Radii** display box.

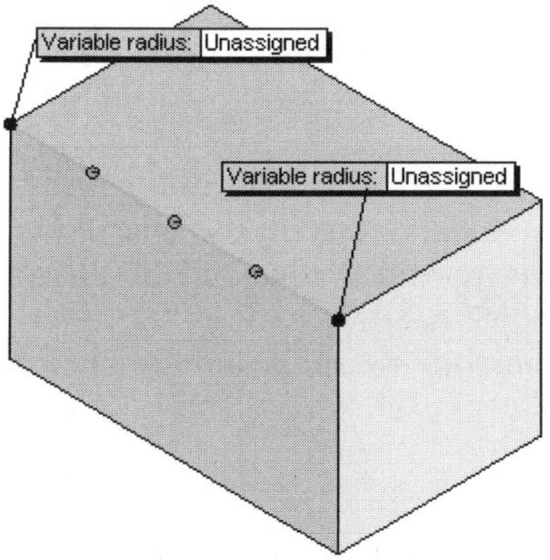

Figure 6-42 *Variable radius callouts displayed at the vertices of the selected edge*

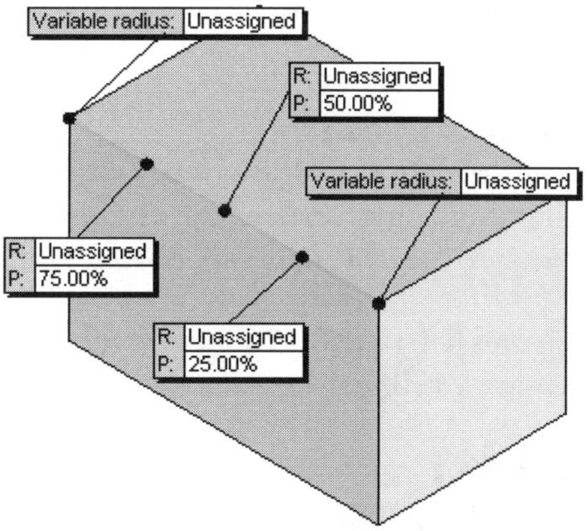

Figure 6-43 ***Radius and Position*** *callouts are displayed after selecting the control points*

You will observe that the position of the three points is described in terms of percentage. You can modify the position of the points by modifying the value of percentage in the **Position** area of the **Radius and Position** callout. By following this procedure you can also modify the placement of other points. You will observe that radius value is not assigned to any of the callouts. Therefore, you need to specify the value of the radius in the callouts. Using the left mouse button select the name of the vertex or point; the name of the selected item will be highlighted in yellow in its respective callout. Using the **Radius** spinner set the value of the radius for the selected item. You can also specify the value of the radius in the radius area of the callout. Set the value of each unassigned radius. You can also use the **Set Unassigned** button to assign the value displayed in the **Radius** spinner to all unassigned radii. The **Set All** button is used to assign the same value that is displayed in the **Radius** spinner to all radii. Figure 6-44 shows the preview of the fillet feature with modified positions of the control points with radius values specified to all points and vertices. Figure 6-45 shows the resulting fillet feature.

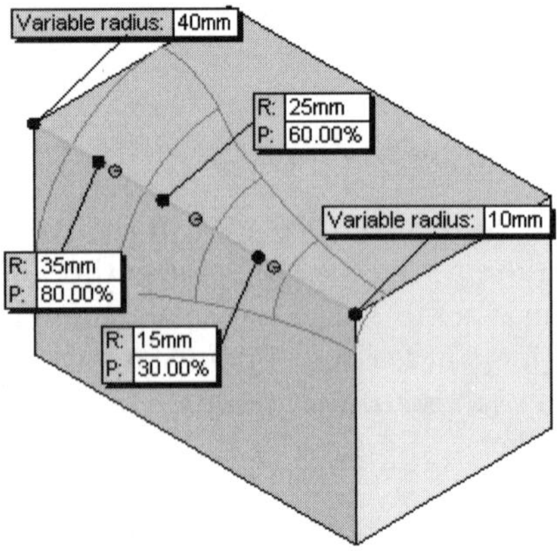

Figure 6-44 Preview of the variable radius fillet

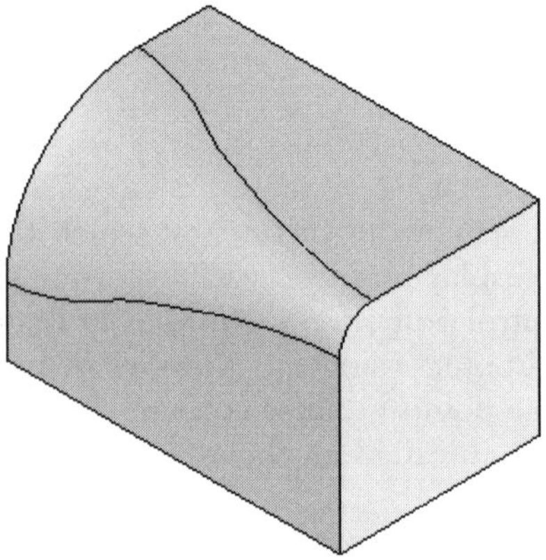

Figure 6-45 Resulting fillet

Smooth transition

The **Smooth transition** radio button when selected creates a smooth transition by smoothly blending the points and vertices on which you have defined the radius on the selected edge.

Straight transition

The **Straight transition** radio button when selected creates a linear transition by blending the points and vertices on which you have defined the radius on the selected edge. In this case the edge tangency is not maintained between one fillet radius and the adjacent face.

Figures 6-46 and 6-47 show the fillets created with smooth transition and straight transition options, respectively.

Face Fillet

Using the **Face Fillet** you can add a fillet between two sets of faces. It blends the first set of faces with the second set of faces. It adds or removes the material according to the geometric conditions. It can also completely or partially remove the faces to accommodate the fillet feature. To create a face fillet feature, invoke the **Fillet PropertyManager** and select the **Face fillet** radio button from the **Fillet Type** rollout; the **Item To Fillet** rollout will be modified and provide the **Face Set 1** and **Face Set 2** selection areas. The **Fillet PropertyManager** is shown in Figure 6-48 with the **Face fillet** radio button selected.

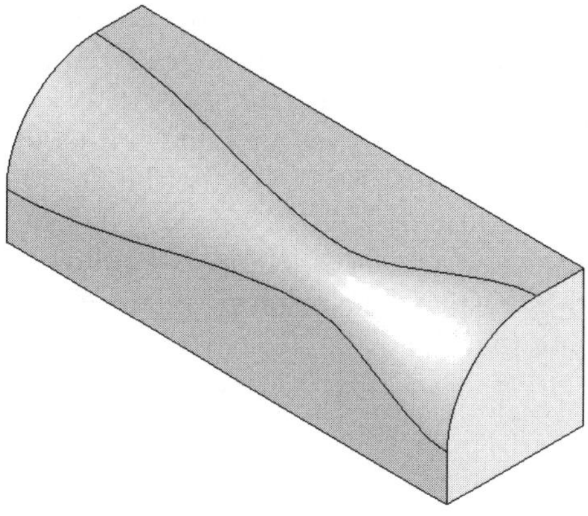

Figure 6-46 *Variable radius fillet with smooth transition*

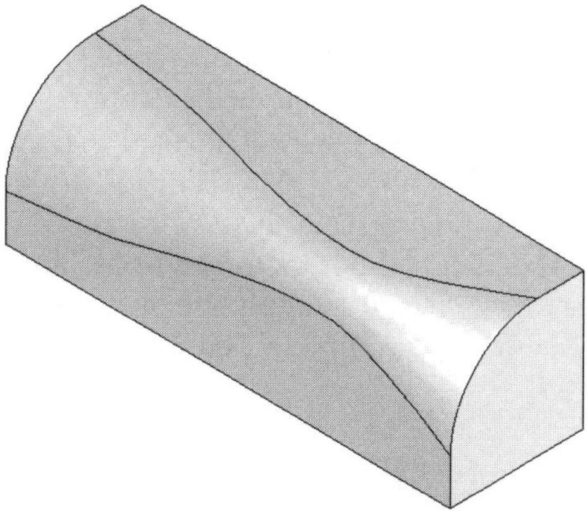

Figure 6-47 *Variable radius fillet with straight transition*

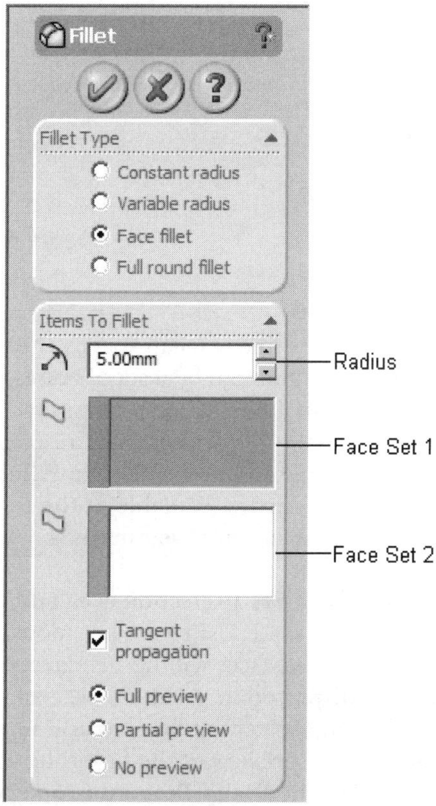

Figure 6-48 *The **Fillet PropertyManager** with the **Face fillet** radio button selected*

You will be prompted to select the faces to fillet for face set 1 and face set 2. Using the left mouse button select the first set of faces. You can select even more than one face in a set. The name of the selected faces will be displayed in the **Face Set 1** display box and the selected faces will be displayed in green. The **Face Set 1** callout with radius will be displayed in the drawing area. Click in the **Face Set 2** selection area to invoke the selection tool and select the second set of faces. The second set of selected faces will be displayed in magenta and the **Face Set 2** callout will be displayed in the drawing area. With this release of SolidWorks the preview of the face fillet will be displayed in the drawing area. Now, set the value of the radius in the **Radius** spinner. The **Tangent propagation** check box is used to create the face fillet tangent to the adjacent faces. This check box is selected by default. If you clear this check box, the fillet will not be forced to be tangent to the adjacent faces. Figure 6-49 shows the faces to be selected to apply the face fillet. Figure 6-50 displays the resulting fillet with three faces of the slot completely eliminated after applying the fillet.

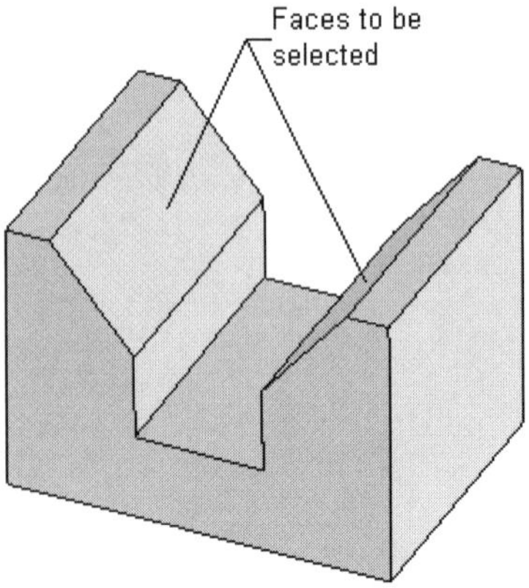

Figure 6-49 *Faces to be selected*

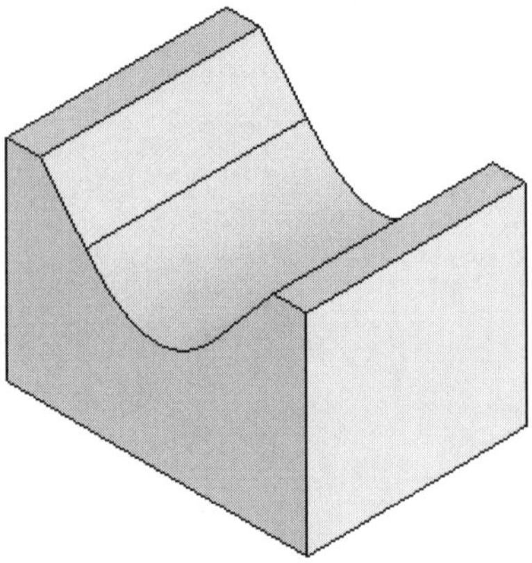

Figure 6-50 *Resulting fillet*

Face Fillet Using the Hold Line

Using the face fillet created with the hold line you can specify the radius and the shape of the fillet by determining a hold line. A hold line can be a set of edges, or a split line projected on a face. You will learn more about split lines in later chapters. The radius of the fillet is determined by the distance between the hold line and the edges or faces selected to be filleted. To create a face fillet using the hold line invoke the **Fillet PropertyManager**. Using the left mouse button click on the black arrow available at the top right corner of the **Fillet Options** rollout to display the rollout as shown in Figure 6-51.

The selection mode is active in the **Face Set 1** selection area and you are prompted to select the faces to fillet for face set 1 and face set 2. Using the left mouse button select the faces for the face set 1. The names of the selected faces will be displayed in the **Face Set 1** selection area and the selected faces will be displayed in green. Now, click in the **Face Set 2** selection area to activate the selection mode and select the faces to add in the face set 2. The selected faces will be displayed in magenta. The preview of the face fillet with the default settings will be displayed in the drawing area. By default, the **Tangent propagation** check box is selected. Therefore, you do not need to select the tangent faces in both face sets. This is because it will automatically select the faces tangent to the selected face. Now, using the left mouse button click on the **Hold line** display box and select the hold line or lines. The preview of the face fillet will be modified automatically. Now, choose the **OK** button from the **Fillet**

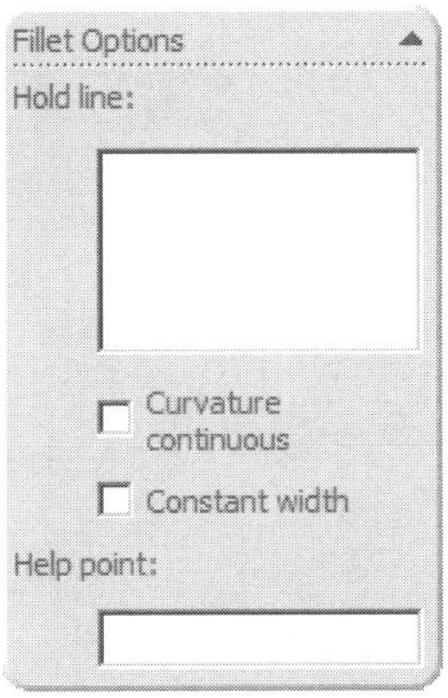

*Figure 6-51 The **Fillet Options** rollout*

PropertyManager. Figure 6-52 shows an example in which the faces and the hold line are selected. Figure 6-53 shows the resulting face fillet using the hold line.

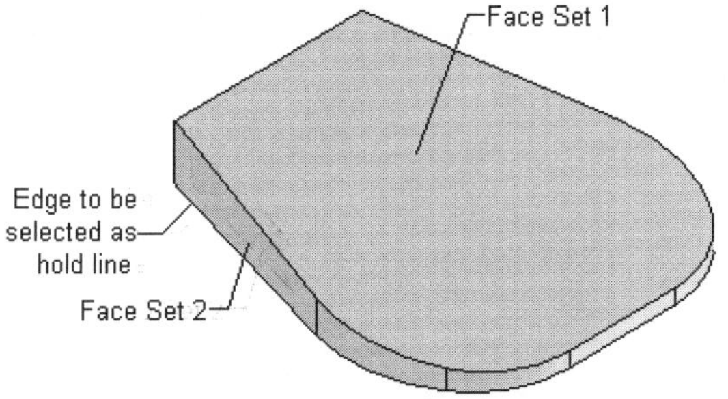

Figure 6-52 Faces and hold line to be selected

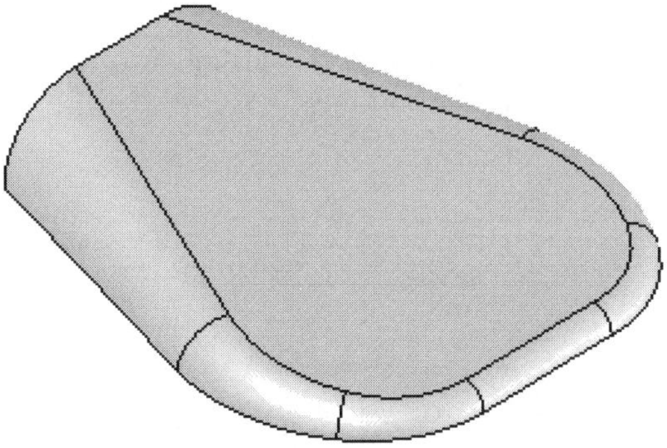

Figure 6-53 Resulting fillet

Curvature Continuous in the Face Fillet With Hold Line

The **Curvature continuous** check box is selected to apply the face fillet feature with continuous curvature throughout the fillet feature. Note that fillet with continuous curvature is possible only by creating a face fillet feature with hold line. You have to specify the hold lines on both sets of faces. Figure 6-54 shows a model in which a face fillet using the hold line is created on both the pillars. On the right pillar the face fillet is created with the **Curvature continuous** check box cleared and on the left pillar the face fillet is created with the **Curvature continuous** check box selected.

Constant Width

In SolidWorks 2005, the **Fillet** tool is enhanced with the addition of the **Constant width** option. Consider a case in which you have applied a face fillet to faces that are at an angle other than 90-degree. You will notice that addition material is added to the fillet on the side which forms an acute angle with the other face, refer to Figure 6-55. If you select the **Constant**

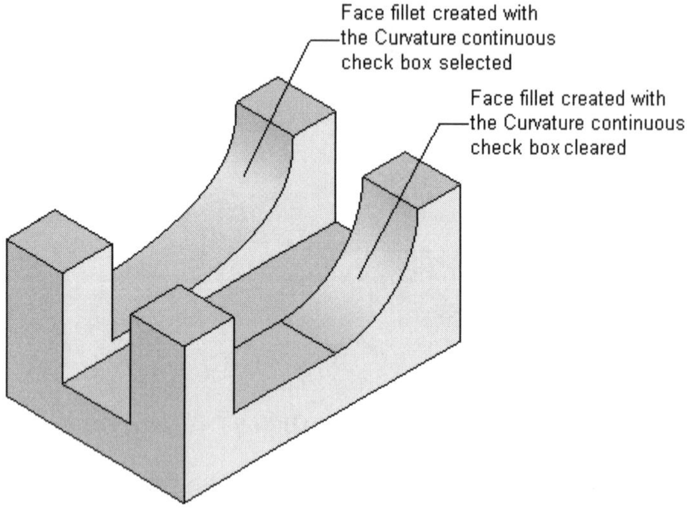

Figure 6-54 *Face fillet created with the* ***Curvature continuous*** *check box selected and cleared*

width check box from the **Fillet Options** rollout, a fillet of constant width will be applied between the selected face, refer to Figure 6-56.

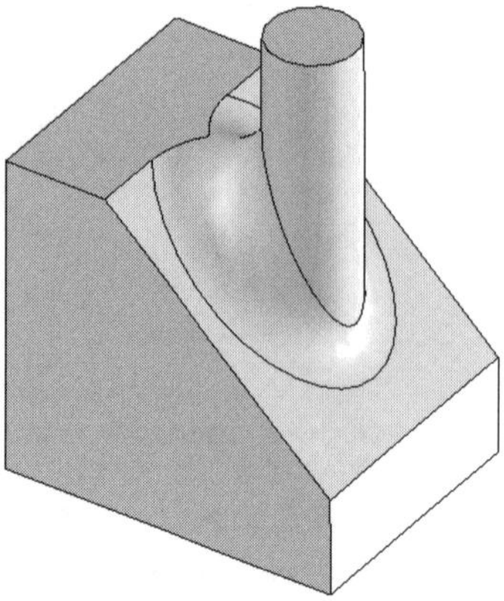

Figure 6-55 *Face fillet with the* ***Constant width*** *check box cleared*

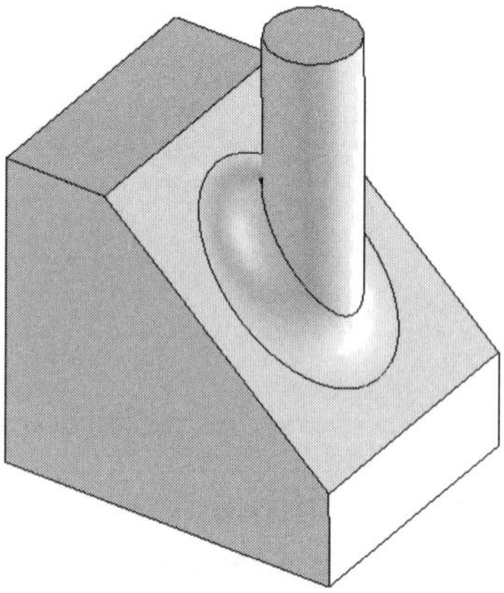

Figure 6-56 *Face fillet with the* ***Constant width*** *check box selected*

Full Round Fillet

The full round fillet is used to add a semicircular fillet feature. To create a full round fillet invoke the **Fillet PropertyManager** and select the **Full round fillet** radio button; the **Item To Fillet** rollout will be modified as shown in Figure 6-57. The selection mode is active in the **Side Face Set 1** selection area and you are prompted to select faces for the center and side face sets. Select the first face for the side face set 1. The selected face will be displayed in green. Now, click in the **Center Face** selection area and select the center face; the selected face will be displayed in blue. Next, click in the **Side Face Set 2** selection area and select the face for the side face set 2. The preview of the full round fillet will be displayed in the drawing

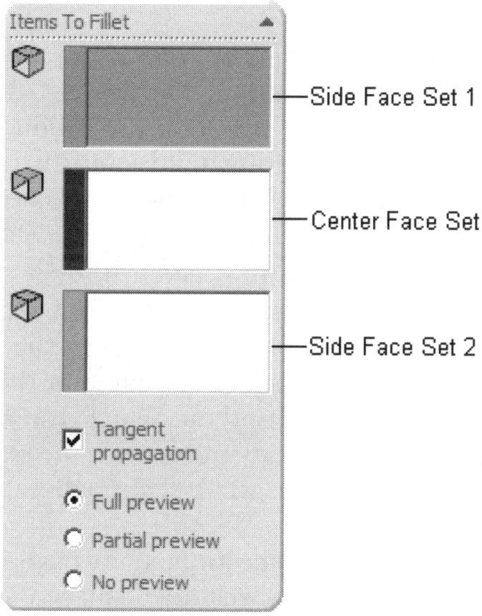

Figure 6-57 *The **Item to Fillet** rollout when the **Full round fillet** radio button is selected from the **Fillet Type** rollout*

area. Choose the **OK** button from the **Fillet PropertyManager**. Figure 6-58 shows the two faces being selected; the third face to be selected is the left face parallel to the first selected face. Figure 6-59 shows the resulting full round fillet.

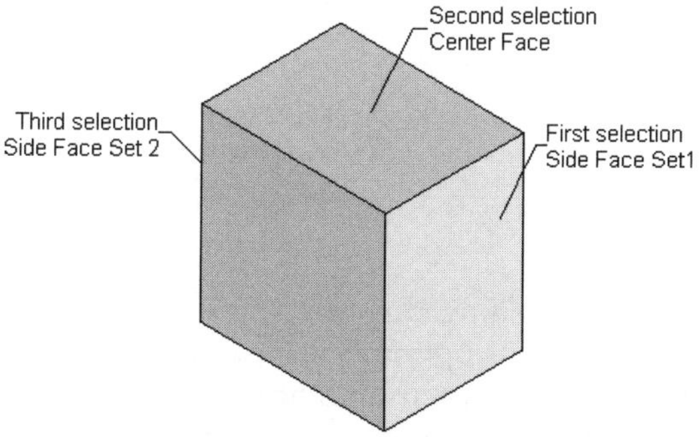

Figure 6-58 *Faces selected to create the full round fillet.*

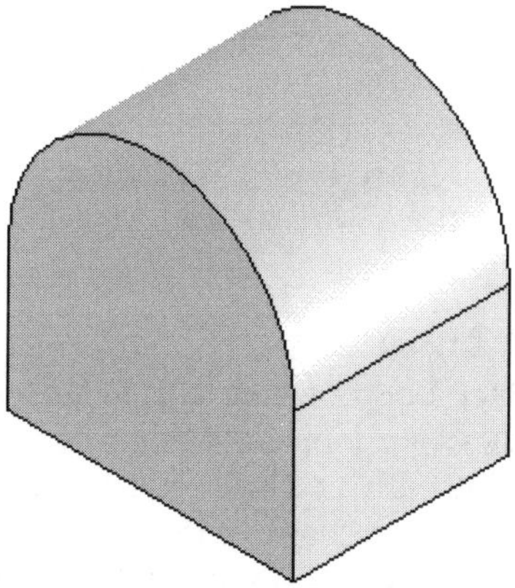

Figure 6-59 *Resulting full round fillet*

Selection Methods

As you have learned about basic and advanced modeling tools, it is necessary for you to learn about some selection methods using which you can increase your productivity and speed of modeling. The selection methods that can increase your speed of creating fillets and chamfers are discussed next.

Select Other

One of the major enhancements in selection options is the addition of the **Select Other** list box in the **Select Other** option. The **Select Other** option is the most common tool to cycle through the entities for selection. This option is used when the selection is difficult in a multifeatured complex model. Before invoking any other tool, select any entity and right-click to invoke the shortcut menu. Choose the **Select Other** option from the shortcut menu to display the **Select Other** list box as shown in Figure 6-60, and the select cursor will be replaced by the cursor. The entities that are surrounding the selected entities will be displayed in this list box. When you move the cursor on the name of the entity, the entity will be highlighted in the drawing area. To select an entity using this list box you just need to select the name of the entity in the list box; the resulting entity will be selected in the model and the **Select Other** list box will disappear. The last item in this list box is the name of the body from which the entity is selected. If you select the last item in the list box then whole body will be selected.

If you select a face and invoke the **Select Other** list box, the display of the selected face will be turned off and you can easily select the faces that are displayed under the selected face. The name of the hidden face will be displayed in the **Hidden Faces** area of the **Select Other** dialog box, refer to Figure 6-61. To turn off the display of any other face, press the SHIFT key and right-click on the face whose display you need to turn off.

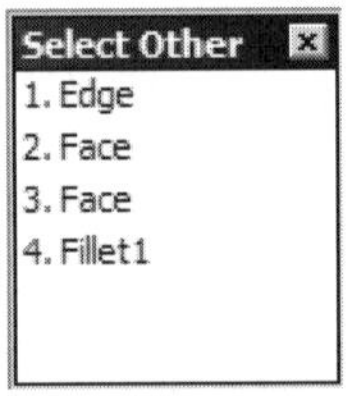

***Figure 6-60** The **Select Other** selection box*

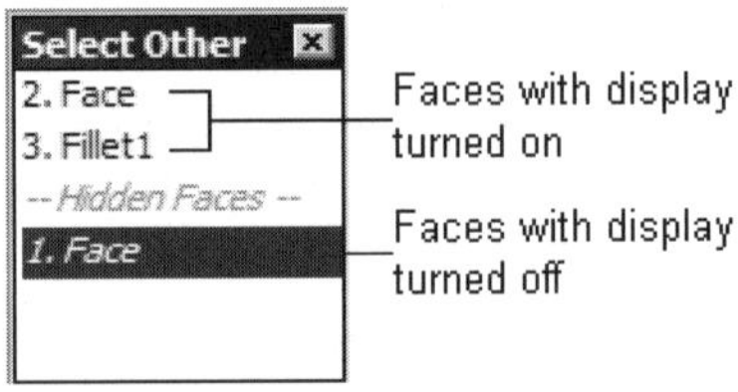

***Figure 6-61** The **Select Other** selection box after selecting a face*

Select Loop

The **Select Loop** option is used to select the loops and using this tool you can also cycle through various loops before confirming the selection. This option is very useful when you are working with a complex model and you have to select a loop from that model. Using the select cursor, select any of the edges of the loop and right-click to invoke the shortcut menu. Choose the **Select Loop** option from the short cut menu. The loop that is possible by selecting that edge will be highlighted in green and an arrow will be displayed in yellow. Move the cursor on that arrow and when the arrow is displayed in red, use the left mouse button to cycle through the loops. Repeat this until you select the required loop. Figure 6-62 shows a loop selected using the **Select Loop** option. Figure 6-63 shows the second loop selected when the left mouse button is used on the arrow to cycle through the loops.

Select Partial Loop

Using this technique you can select the partial loop created by joining two edges. To select a partial loop, select two edges as shown in Figure 6-64. Invoke the shortcut menu and select the **Select Partial Loop** option from the shortcut menu. Generally, the selection point on

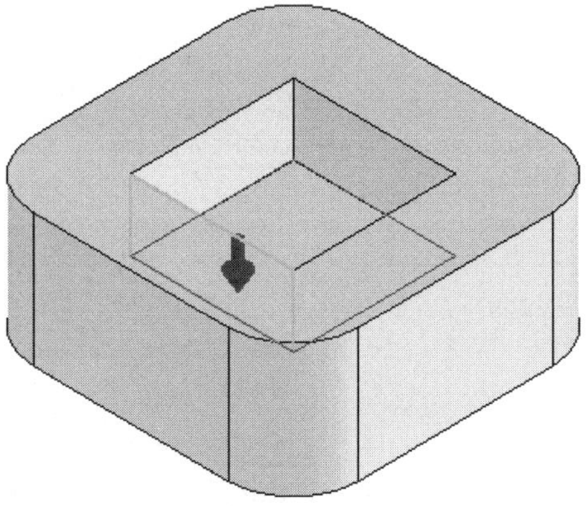

***Figure 6-62** Loop selected using the **Select Loop** tool*

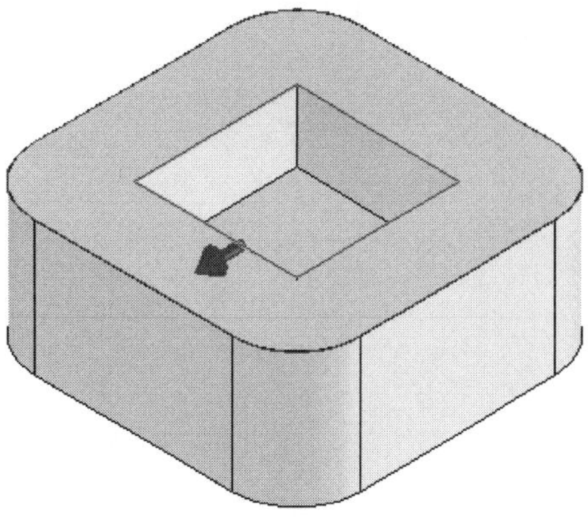

***Figure 6-63** Second loop is selected while cycling through the loops*

second edge defines the major or minor partial loop selection. Figure 6-65 shows the resulting partial loop selected.

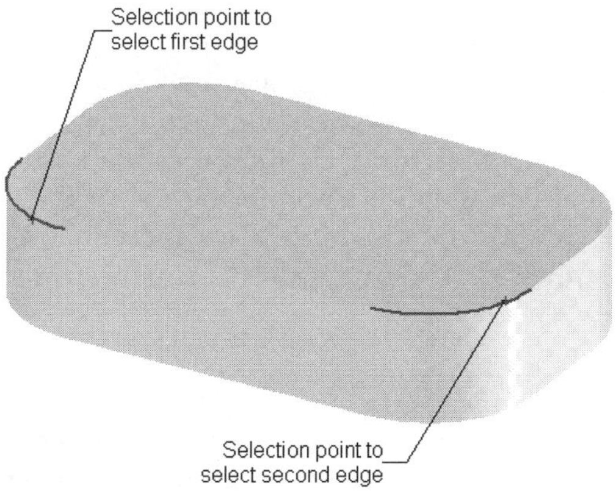

***Figure 6-67** Selecting edges to define the partial loop*

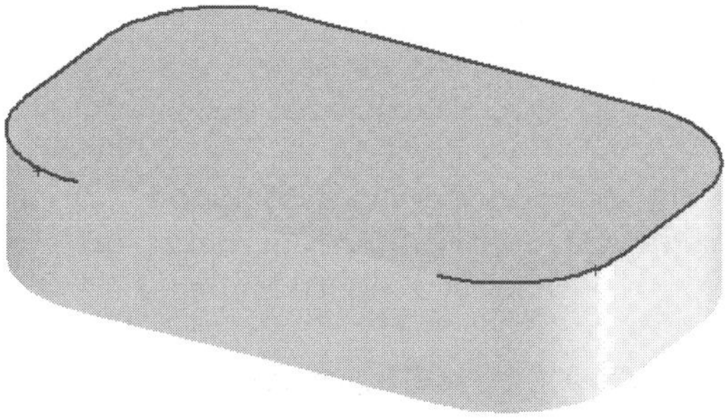

***Figure 6-68** Resulting partial loop that is selected*

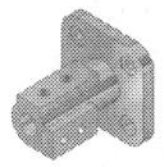

Note

For better visualization of the selected partial loop the display of the model edges is turned off in Figures 6-64 and 6-65.

Tip. *The **Select Midpoint** option available in the shortcut menu is generally used in the sketching environment or while creating 3D sketches. You will learn more about 3D sketches in later chapters.*

Select Tangency

The **Select Tangency** option is used to automatically select the edges or faces that are tangent to the selected face. This option will be available in the shortcut menu only when a face or an `edge is tangent to the selected face or edge. For using this option select any face or edge

> **Tip**. *For creating a fillet or a chamfer feature if you select a face, the fillet or chamfer will be applied to all edges of that face. Consider a case in which you have a blind cut feature on the top face of a block. You want to fillet only the upper edges of the cut feature. You can use the* ***Select Loop*** *option to execute feature creation. If you select the top face of the model, then all edges of the top of the block and the edges of the cut will be filleted. Therefore, after selecting the top face of the block press and hold the CTRL key down and select any one upper edge of the cut. Now, apply the fillet feature to this combination of selection. You will observe that only the upper loop of edges of the cut will be filleted instead of the whole upper face. In the same way you can also fillet only the edges of the block when you are also having a slot on the top face of the model.*

using the select tool and right-click to invoke the shortcut menu and choose the **Select Tangency** option from the shortcut menu.

Creating Chamfer

CommandManager:	Features > Chamfer
Menu:	Insert > Features > Chamfer
Toolbar:	Features > Chamfer

Chamfering is defined as a process in which the sharp edges are beveled in order to reduce the area of stress concentration. This process also eliminates the sharp edges and corners that are not desirable. In SolidWorks, chamfer is created using the **Chamfer** tool. This tool is invoked by choosing the **Chamfer** button available in the **Features CommandManager**. When you invoke the **Chamfer** tool, the **Chamfer PropertyManager** will be displayed as shown in Figure 6-66. Various types of chamfers created using the **Chamfer PropertyManager** are discussed next.

Creating Edge Chamfer

The chamfers that are applied to the edges are known as edge chamfer. To create an edge chamfer invoke the **Chamfer PropertyManager** and then select the edges to chamfer. When you select an edge to chamfer, the preview of the chamfer feature with a distance and angle callout will be displayed in the drawing area. The name of the selected edge will be displayed in the **Edges and Faces or Vertex** selection area. Also, the selected entity will be highlighted in green and a yellow arrow is also displayed in the preview. The **Tangent propagation** check box is selected by default. Therefore, the edges tangent to the selected edge are selected automatically. By default the **Partial preview** button is selected by default; you can select the **Full preview** button to display the full preview of the chamfer feature. Figure 6-67 shows the edge to be selected for chamfering and Figure 6-68 shows the full preview of the chamfer feature.

By default, the **Angle distance** radio button is selected. Therefore, the distance and angle callout is displayed in the drawing area. You can set the value of the distance and angle using the **Distance** and **Angle** spinners or you can insert the value of distance and angle directly in the **Distance** and **Angle** callouts. The **Flip direction** check box is used to specify the direction of distance measurement. If you select the **Flip direction** check box, the arrow will also be

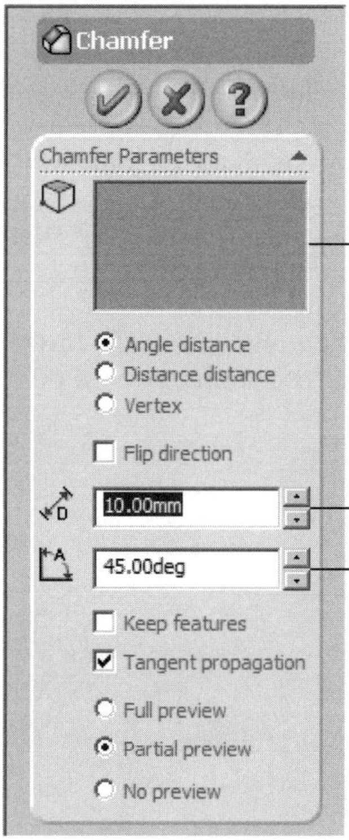

Figure 6-66 *The* ***Chamfer PropertyManager***

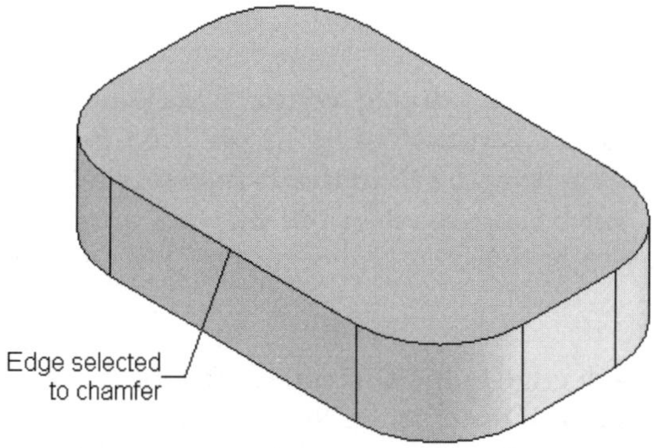

Figure 6-67 *Edge selected to chamfer*

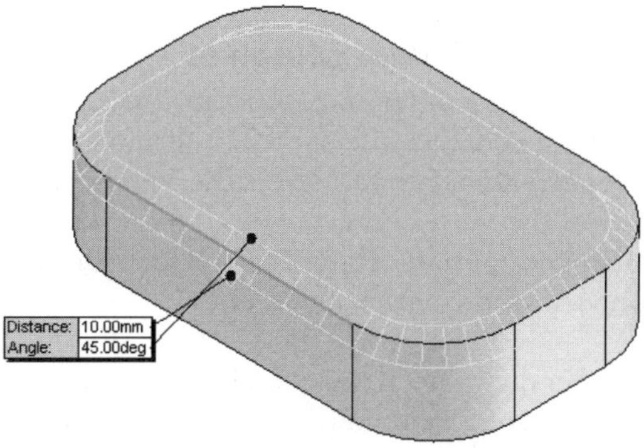

Figure 6-68 *Preview of the chamfer feature*

Tip. *You can also select the face for applying the chamfer feature. If you select the face for applying the chamfer, the chamfer will be applied to all edges of the selected face.*

flipped in the preview in the drawing area. You can also flip the direction by clicking on the arrow in the drawing area.

If you select the **Distance distance** radio button from the **Chamfer Parameters** rollout, the **Flip direction** check box will be replaced by the **Equal distance** check box. The **Angle** and **Distance** callouts will be replaced by **Distance 1** and **Distance 2** callouts. By default, the **Equal distance** check box is cleared. Set the value of the chamfer distance in the **Distance 1** spinner or specify the value in the callout. Now, set the value of distance 2 in the respective spinner or callout. If you need to specify the name distance for creating the chamfer, select the **Equal distance** check box. The **Distance 2** will spinner will disappear from the **Chamfer Parameters** rollout.

After specifying all parameters choose the **OK** button from the **Chamfer PropertyManager**. Figure 6-69 shows the chamfer created on a model.

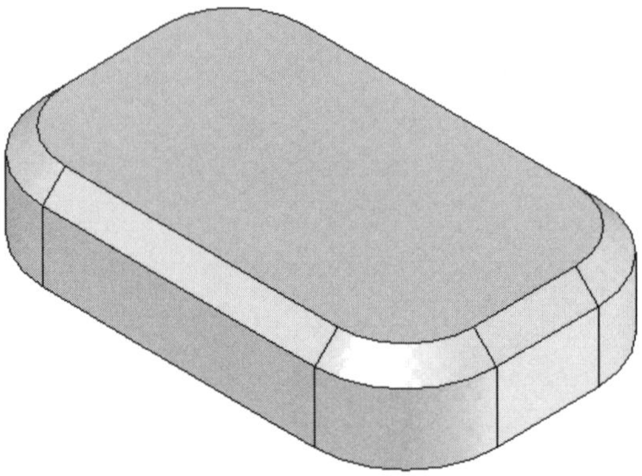

Figure 6-69 *Chamfer created on a base plate*

Creating Vertex Chamfer

Using the chamfer tool you can also add a chamfer to a selected vertex. It will chop the selected vertex to the specified distance. To create the vertex chamfer invoke the **Chamfer PropertyManager** and select the **Vertex** radio button from the **Chamfer Parameters** rollout. Select the vertex; the preview of the chamfer will be displayed in the drawing area with **Distance** callouts. Figure 6-70 shows the vertex to be selected and Figure 6-71 shows the preview of the vertex chamfer.

Set the value of the chamfer distance along each edge in the **Distance 1**, **Distance 2**, and **Distance 3** spinners. You can also specify the value of the chamfer distance in the distance callouts. If you want to specify equal distance to all edges, select the **Equal distance** check box. After specifying all parameters choose the **OK** button from the **Chamfer PropertyManager**. Figure 6-72 shows the vertex chamfer feature applied to the base feature.

Chamfer With and Without Keep Feature Option

If you have boss or cut features in a model and the chamfer created is large enough to consume those features, it is recommended that you select the **Keep features** check box. If this check box is cleared, the chamfer feature will consume the features that will obstruct its path. Note that the features that are consumed by the chamfer feature are not deleted

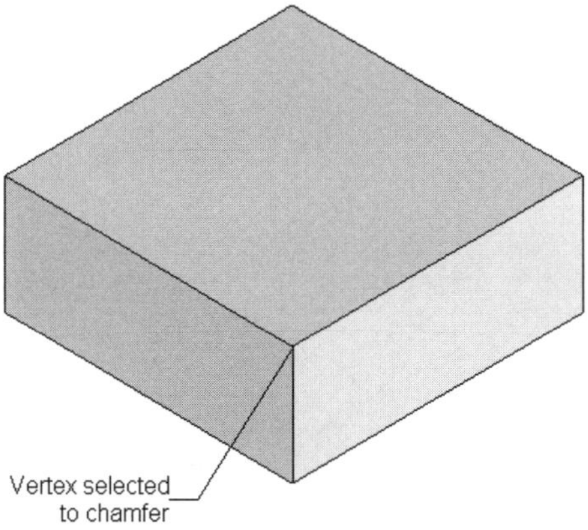

Figure 6-70 Vertex to be selected

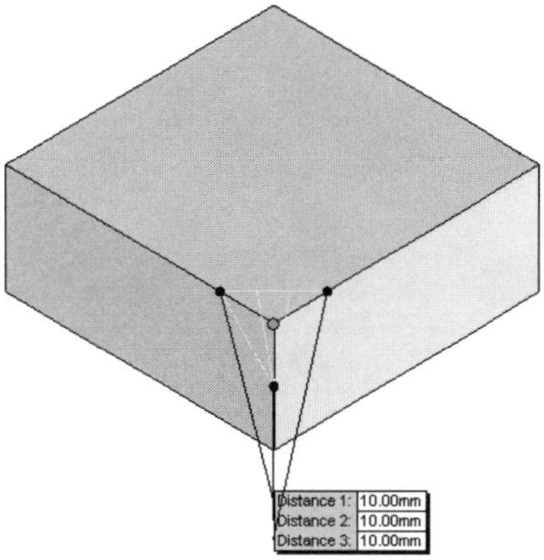

Figure 6-71 Preview of the vertex chamfer

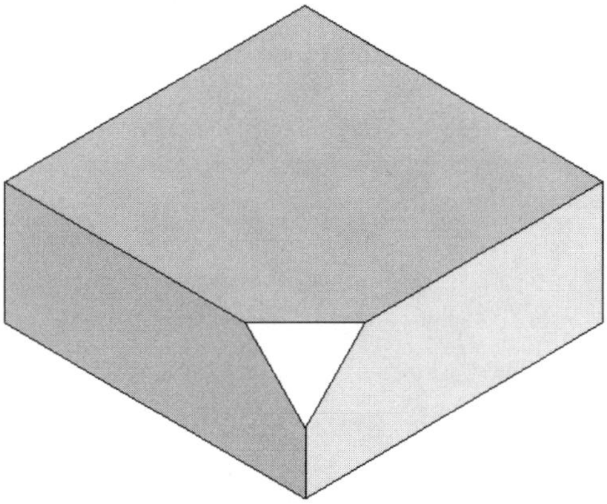

Figure 6-72 Vertex chamfer created on a base feature

from the model. They are removed from the model because of some geometric inconsistency. When you rollback or delete the chamfer, the consumed features will reappear. Figure 6-73 shows the chamfer feature with the **Keep features** check box cleared and Figure 6-74 shows the chamfer feature with the **Keep features** check box selected.

Creating the Shell Feature

CommandManager:	Features > Shell
Menu:	Insert > Features > Shell
Toolbar:	Features > Shell

Shelling is defined as the process in which the material is scooped out from a model and the resulting model is hollowed from inside. The resulting model will be a hollow model with walls of specified thickness and cavity inside. The selected face or faces of the model are also removed in this operation. If you do not select any face to remove, a closed hollow model will be created. You can also specify multiple thicknesses to the walls. You can create a shell feature using the **Shell** tool. Use the **Shell**

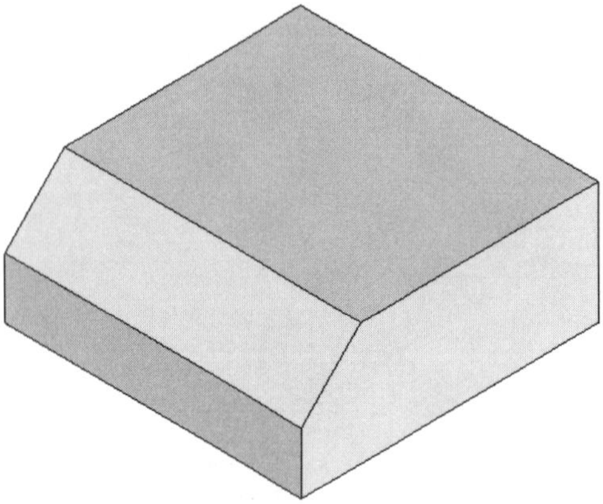

Figure 6-73 *Chamfer feature with the **Keep features** check box cleared*

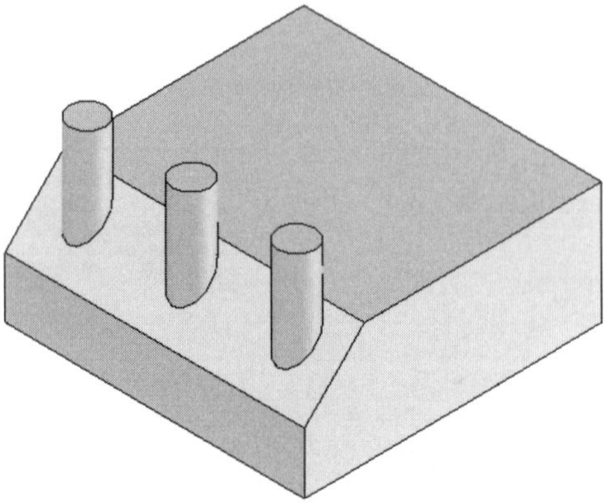

Figure 6-74 *Chamfer feature with the **Keep features** check box selected*

button from the **Features CommandManager** to invoke The **Shell1 PropertyManager**, as shown in Figure 6-75.

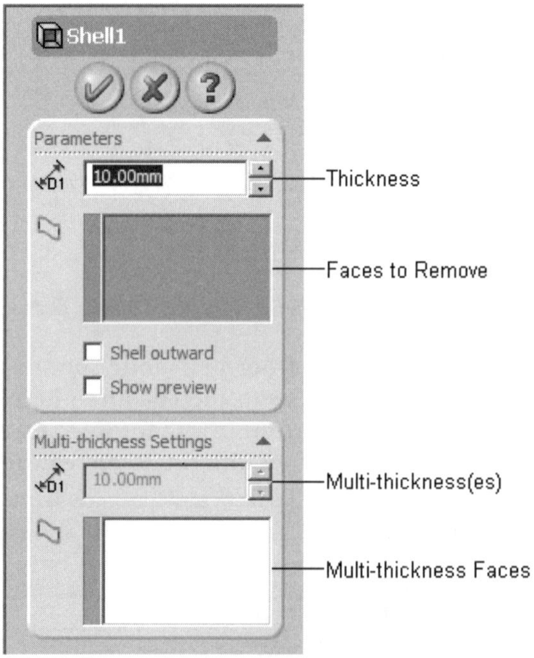

Figure 6-75 *The **Shell1 PropertyManager***

You will be prompted to select the faces to remove. Select the face or faces of the model that you want to remove. The selected faces will be highlighted in green and their names will be displayed in the **Faces to Remove** selection area. Set the value of the wall thickness in the **Thickness** spinner and choose the **OK** button from the **Shell1 PropertyManager**. Figure 6-76 shows the face selected to remove and Figure 6-77 shows the resulting shell feature created.

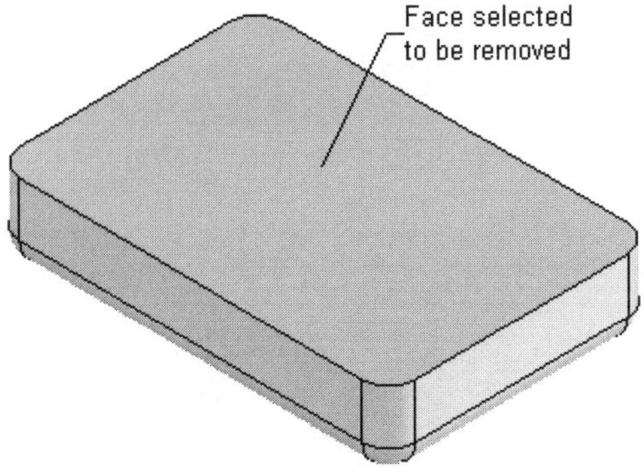

Figure 6-76 *Face selected to be removed*

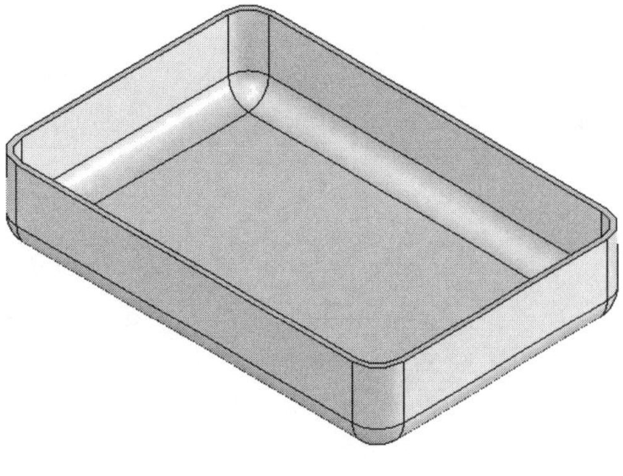

Figure 6-77 *Resulting shell feature*

If none of the faces are selected to be removed, the resulting model will be hollowed from inside with no face removed. Figure 6-78 shows a model in the **Hidden Line Visible** mode with a shell feature in which faces are not selected to be removed.

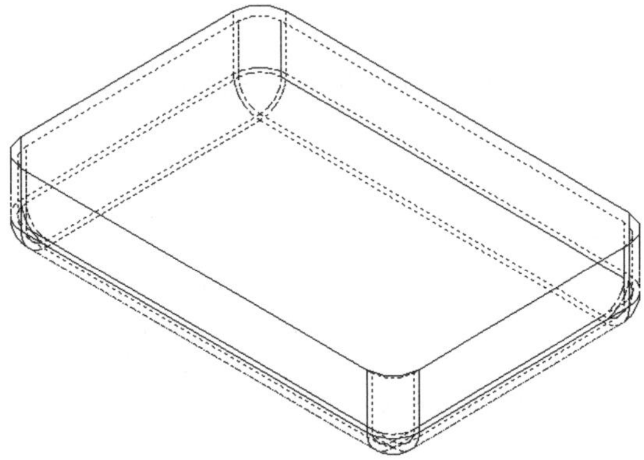

Figure 6-78 *Shell feature with faces not selected to remove*

Tip. *If the thickness of the shell feature is more than the radius of the fillet feature, the fillet will not be included in the shell feature. Therefore, it results in sharp edges after adding the fillet. Same is with the chamfer feature. The faces selected to be removed in the shell feature can be a planar face or a curved face. But creating a shell by removing a curved face depends on the geometry of the curved face to adopt the specified shell thickness and other geometric conditions.*

The ***Shell outward*** *check box is selected to create the shell feature on the outer side of the model.*

The in-built artificial intelligence of shelling command in SolidWorks enables the shell feature to decide how much quantity of material to be removed depending on the geometric conditions. Figure 6-79 shows the shell feature whose wall thickness is small for uniform shelling of the entire model. Figure 6-80 shows the shell feature whose wall thickness is large

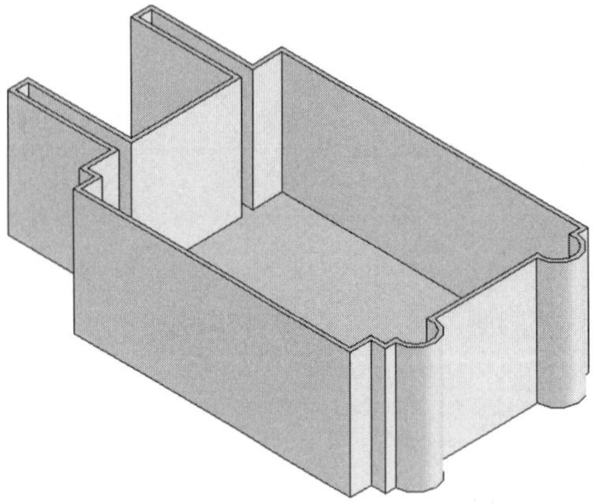

Figure 6-79 Shell feature with smaller shell thickness

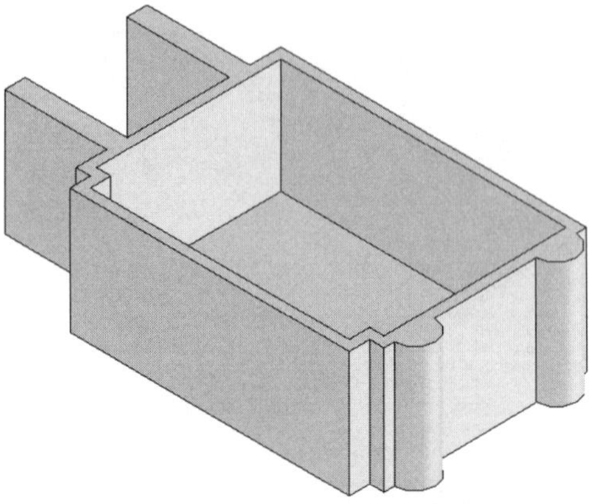

Figure 6-80 Shell feature with larger shell thickness

because of which it cannot accommodate the uniform shelling of the entire model. Therefore, the shell feature will not remove the material from that area where the material removal is not possible.

With this release of SolidWorks, you can also display the preview of the shell feature by selecting the **Show preview** check box from the **Parameters** rollout of the **Shell1 PropertyManager**.

Creating Multi-thickness Shell

Using this option available in the shell tool you can specify different thickness values to the selected faces. To use this option, invoke the **Shell1 PropertyManager** and select the faces to be removed and then specify the uniform thickness in the **Thickness** spinner of the **Parameters** rollout. Click once in the **Multi-thickness(es) Faces** selection area to activate the selection mode.

Select the faces to which you want to specify the special thickness. Set the value of the thickness using the **Multi-thickness(es)** spinner and choose the **OK** button. Figure 6-81 shows the faces to be selected to specify the multithickness shell and Figure 6-82 shows the resulting shell feature.

Error Diagnostics

This option activates automatically if the shell feature creation fails. With this option you can figure out the possible reasons for the failure of the shell feature creation.

While creating the shell feature, specify all parameters in the **Shell1 PropertyManager** and choose the **OK** button. If the shell feature creation fails because of geometric inconsistency, the **What's Wrong** dialog box will be displayed, which will informs you about the possible errors because of which the feature creation failed. You will learn more about the **Rebuild Errors** dialog box later in this chapter. The **Error Diagnostics** rollout shown in Figure 6-83 is displayed in the **Shell 1 PropertyManager**.

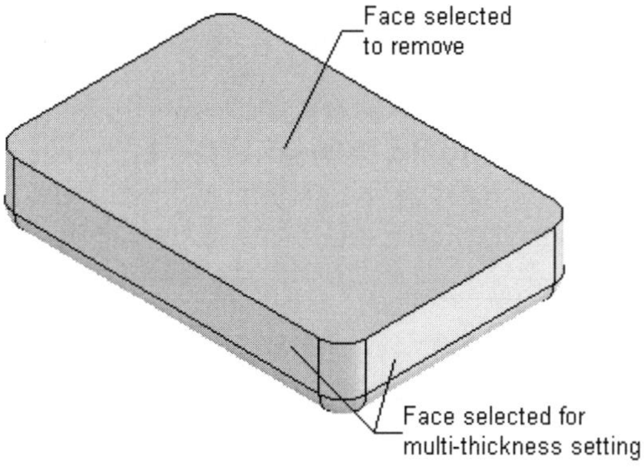

***Figure 6-81** Faces to be selected to apply multithickness shell*

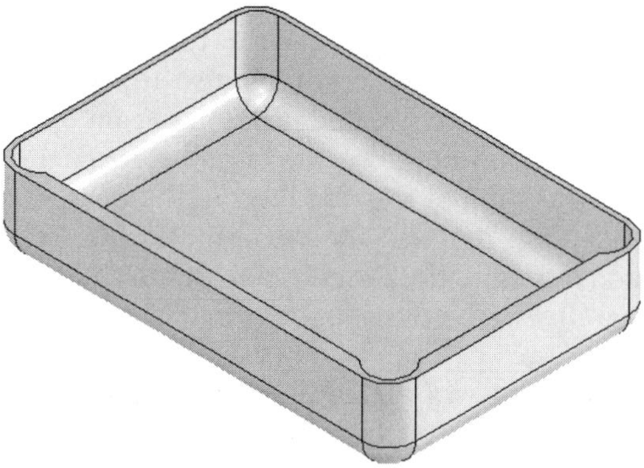

***Figure 6-82** Resulting multithickness shell*

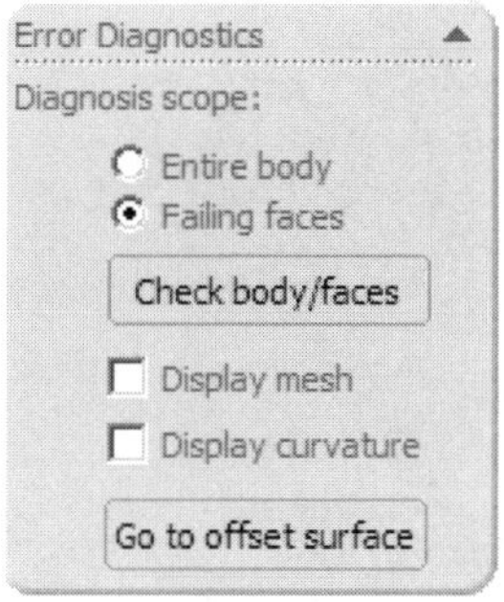

***Figure 6-83** The **Error Diagnostics** rollout*

The radio buttons available in the **Diagnosis scope** area of the **Error Diagnostics** rollout are used to define the diagnosis to be done on the entire body or only on the faces that failed to be shelled. The **Check body/faces** button is used to run the diagnostic tool. When you choose this button, the areas of the model that are responsible for the feature creation failure will be highlighted using callouts. The **Display mesh** check box is used to display the surface curvature mesh. The **Display curvature** check box is selected to display the surface curvature. The **Go to offset surface** button is used to display the **Offset Surface PropertyManager** and the offset surfaces.

Creating the Wrap Feature

CommandManager:	Features > Wrap *(Customize to Add)*
Menu:	Insert > Features > Wrap
Toolbar:	Features > Wrap *(Customize to Add)*

The **Wrap** tool is used to emboss, deboss, or scribe a closed multiloop sketch on a selected planar or curved face that is tangent to the plane on which the selected sketch is created. You can also create this type of geometry by extruding the sketch from a selected surface using the boss or cut option. There are two main differences between the emboss and deboss created using the **Wrap** tool and those created using the boss or cut tool. The first difference is the method of projection. The projection of the geometry using the **Wrap** tool follows the rule of true length, which means the actual length of the geometry remains the same after projecting it on the surface. The projection of the geometry created using the **From** option of the boss or cut tool follows the rule of true projection. Therefore, the original size of the geometry is distorted. The second difference is the direction of the side faces of the geometry. The side faces of the geometry created using the **Wrap** tool are always normal to the reference surface. The side faces of the geometry created using the **From** option of the cut or extrude tool are normal to the sketching plane or parallel to the direction vector.

To create a wrap feature, create a closed multiloop or a single loop sketch and then choose the **Wrap** button from the **Features CommandManager**. The **Message PropertyManager** will be displayed and you will be prompted to select a plane or face on which you need to create a closed counter or select an existing sketch. Select an existing sketch to display the **Wrap1 PropertyManager**, as shown in Figure 6-84.

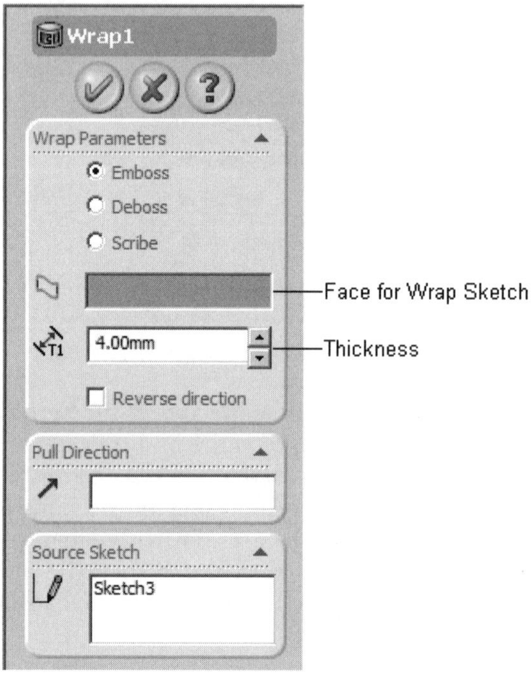

Figure 6-84 *The **Wrap1 PropertyManager***

You will notice that the **Emboss** radio button is selected by default. This radio button is selected to create an embossed wrap feature. Next, select the face on which you need to wrap the sketch. As soon as you select the face, the preview of the wrap feature that will be created with the default settings will be displayed in the drawing area. Set the value of the thickness using the **Thickness** spinner. Choose the **OK** button from the **Wrap1 PropertyManager**. Figure 6-85 shows the sketch and the face to be selected to create the wrap feature and Figure 6-86 shows the resulting embossed wrap feature.

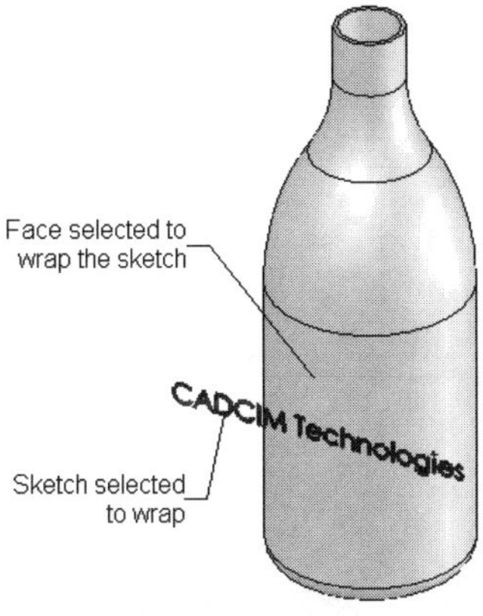

Figure 6-85 *Sketch and face selected to create the wrap feature*

Figure 6-86 *Resulting embossed wrap feature*

Using the **Deboss** radio button available in the **Wrap Parameters** rollout you can engrave the sketch on a selected planar or curved face. Figure 6-87 shows a wrap feature created using the **Deboss** option. The **Scribe** radio button is selected to project the selected sketch on a planar or a curved face. The projected sketch will split the face on which it is projected. Using the **Pull Direction** rollout you can specify the normal direction in which you need to create the wrap feature. Figure 6-88 shows the wrap feature created using the **Scribe** option.

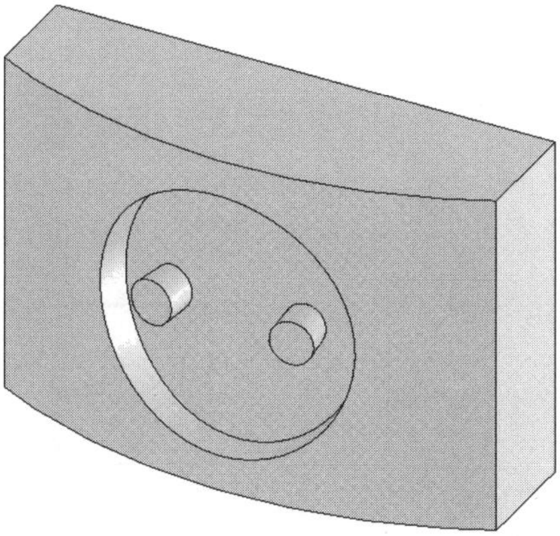

Figure 6-87 *Wrap feature created using the* ***Deboss*** *option*

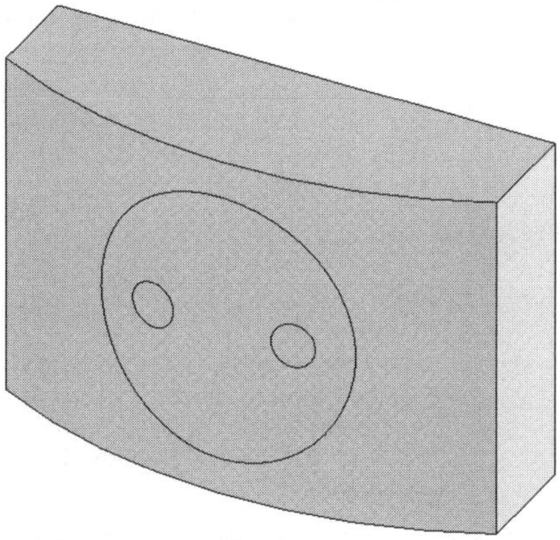

Figure 6-88 *Wrap feature created using the* ***Scribe*** *option*

TUTORIALS

Tutorial 1

In this tutorial, you will create the model of the Plummer Block Casting shown in Figure 6-89. The dimensions of the model are shown in Figure 6-90. **(Expected time: 30 min)**

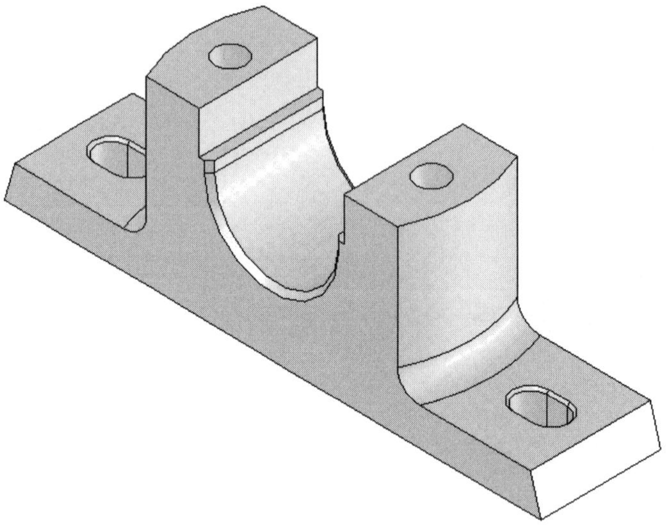

Figure 6-89 Solid model of the Plummer Block Casting

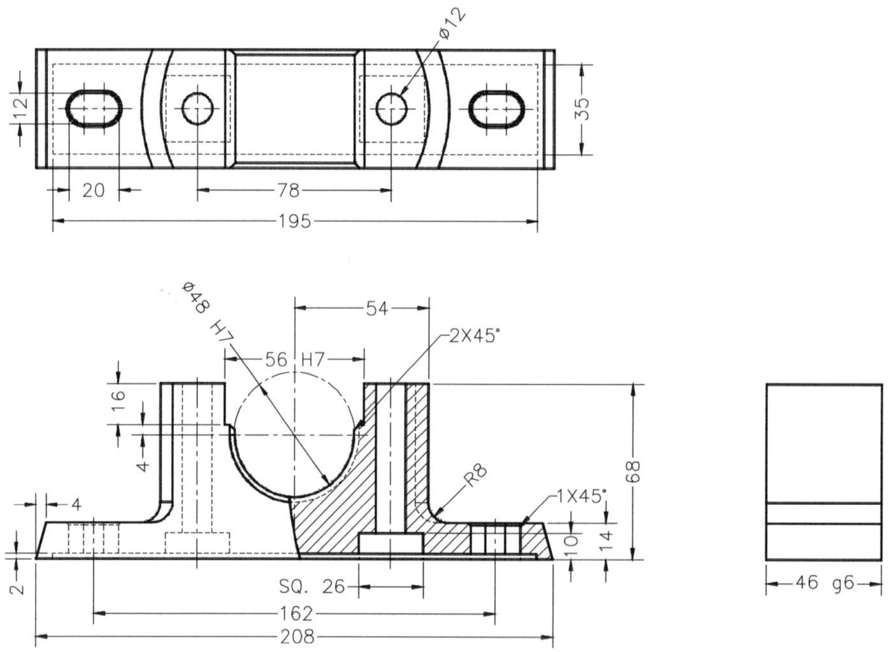

Figure 6-90 Dimensions of the Plummer Block Casting

The steps to be followed to complete this tutorial are listed next.

a. Create the base feature of the model on the **Front Plane**, refer to Figures 6-91 and 6-92.
b. Create the second feature, which is a cut feature, on the top planar face of the base feature, refer to Figures 6-93 through 6-95.
c. Create the rectangular recess at the bottom of the base as a cut feature, refer to Figure 6-96.

d. Create the square cuts that will act as the recess for the head of the square head bolts, refer to Figure 6-96.
e. Create a hole feature using the **Hole PropertyManager** and modify the placement of the hole feature, refer to Figure 6-97.
f. Create the cut feature on the second top planar face of the base feature, refer to Figure 6-98.
g. Add the fillet feature to the model, refer to Figures 6-99 and 6-100.
h. Add chamfers to the model, refer to Figures 6-101 and 6-102.

Creating the Base Feature

1. Start a new SolidWorks part document using the **New SolidWorks Document** dialog box.

2. Invoke the **Extruded Boss/Base** tool and select the **Front Plane** as the sketching plane.

3. Draw the sketch of the front view of the model using the automatic mirroring option to capture the design intent of the model.

4. Add the required relations and dimensions to the sketch as shown in Figure 6-91.

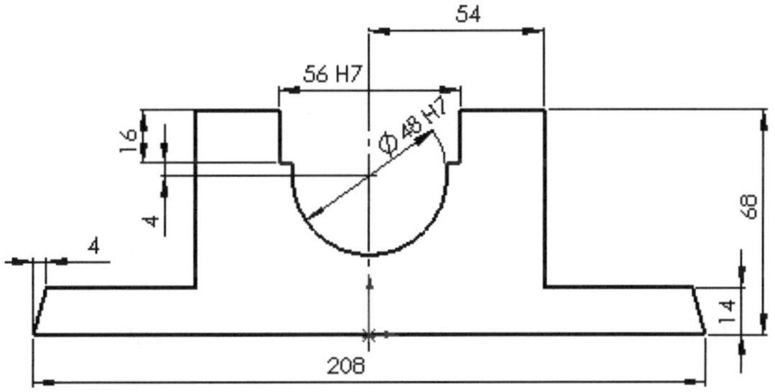

***Figure 6-91** Fully defined sketch of the base feature*

Note

*As shown in Figure 6-91, tolerances are applied to some dimensions. After adding the dimensions to the sketch, select the dimension and add the tolerance to the dimension using the **Dimension PropertyManager**, as discussed in the earlier chapters.*

*Using the **Dimension Properties** dialog box, change the radial dimension of the arc to the diameter dimension.*

You need to extrude the sketch to a distance of 46 mm using the **Mid Plane** option. You need to extrude the sketch using the **Mid Plane** option because it is recommended that parts that are to be assembled should have default planes in the center of the model.

5. Exit the sketching environment to display the **Extrude PropertyManager**.

6. Right-click in the drawing area and select the **Mid Plane** option from the shortcut menu to extrude the sketch symmetrically to both sides of the sketching plane.

7. Set the value of the **Depth** spinner to **46** and choose the **OK** button from the **Extrude PropertyManager**.

8. Change the view orientation to isometric. The resulting base feature created after extruding the sketch to a given depth is shown in Figure 6-92.

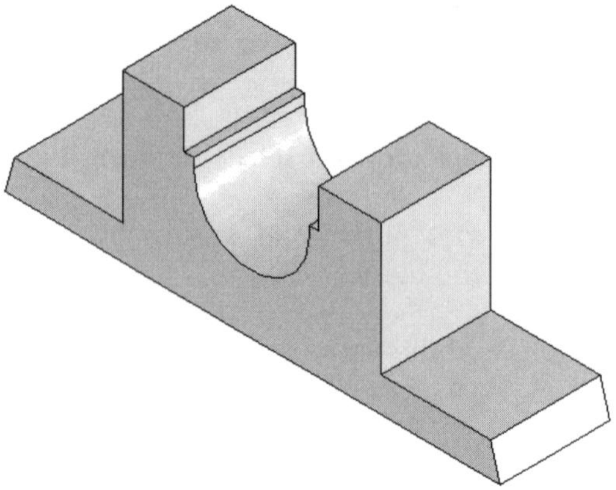

Figure 6-92 *Isometric view of the base feature of the model*

Tip. *Refer to Figure 6-90; you will observe that the dimension that reflects the depth of the extruded feature has a tolerance applied to it. Therefore, double-click on the extruded feature in the* ***FeatureManager Design Tree*** *or in the drawing area. All dimensions applied to the model are displayed. Select the dimension that reflects the depth of the extruded feature and apply the tolerance using the* ***Dimension PropertyManager****.*

Creating the Second Feature

The second feature of the model is a cut feature. The sketch of the cut-extrude feature is drawn on the top planar face of the base feature. This sketch is extruded up to the specified plane to create the resulting cut feature.

1. Invoke the **Extruded Cut** tool and select the top planar face of the base feature as the sketching plane.

2. Orient the model normal to the sketching plane and using the standard sketching tools draw the sketch for the cut feature and apply the required relations to the sketch.

   The sketch of the cut-extrude feature is shown in Figure 6-93.

3. Exit the sketching environment to display the **Cut-Extrude PropertyManager**.

   The preview of the cut feature is displayed in the drawing area with the default values of

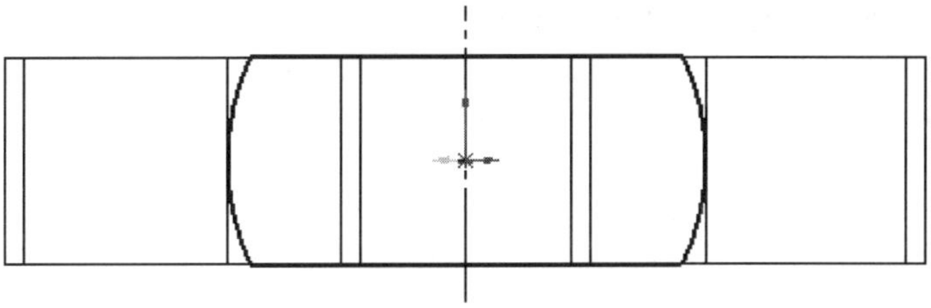

Figure 6-93 *Sketch of the second feature*

the blind option. You need to extrude the cut feature up to the selected surface. Therefore, orient the model in the isometric view because the feature termination surface can be easily selected in the isometric view.

4. Change the view orientation to isometric.

   You will observe that the preview of the cut feature is inside the model, but you need to remove the outer part of the sketch profile. Therefore, you need to change the side of the cut feature.

5. Select the **Flip side to cut** check box; the preview of the cut feature is modified dynamically.

6. Right-click in the drawing area and select the **Up To Surface** option from the shortcut menu.

   You are prompted to select a face or a surface to complete the specification of Direction 1.

7. Select the surface for feature termination shown in Figure 6-94. Right-click and choose **OK** from the shortcut menu to exit the tool.

   The model after creating the cut feature is shown in Figure 6-95.

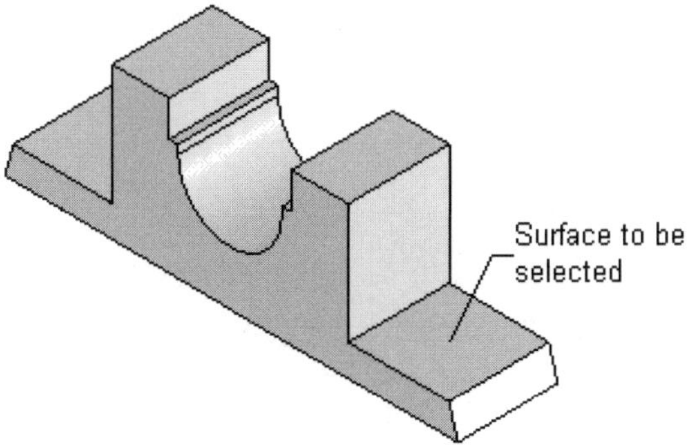

Figure 6-94 *Surface to be selected for the cut feature*

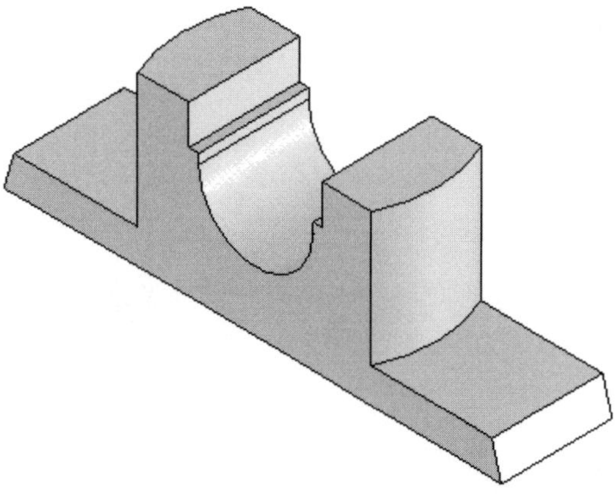

Figure 6-95 *Model after creating the cut feature*

Creating the Rectangular Recess

The third feature is the rectangular recess. This feature is created using a rectangular cut feature that will be created on the bottom face of the model.

1. Orient the model using the **Rotate View** tool and select the bottom face of the base feature as the sketching plane.

2. Invoke the **Extruded Cut** tool to invoke the sketching environment.

3. Using the standard sketching tools, draw a rectangle of 195 mm x 35 mm as the sketch of the rectangular recess. Apply the required relations and dimensions to the sketch.

4. Exit the sketching environment.

5. Set the value of the **Depth** spinner to **2** and choose **OK** from the **Cut-Extrude PropertyManager**.

Creating the Recess for the Head of Square-headed Bolt

It is evident from Figure 6-90 that the bolt to be inserted in the part will be a square-headed bolt. Therefore, you need to create the recess for the head of the square-headed bolt. A square of 26 mm length is used to create this recess.

1. Rotate the model and select the upper face of the recess created in the previous section as the sketching plane. Invoke the **Extruded Cut** tool and orient the model normal to the sketching plane.

2. Using the standard sketching tools draw the sketch. The sketch includes two squares of 26 mm length and the distance between the centers of the squares is 78 mm, refer to Figure 6-90. Apply the required relations and dimensions.

3. Exit the sketching environment and extrude the sketch to a depth of 10 mm. Exit the tool. The rotated model after adding this cut feature is shown in Figure 6-96.

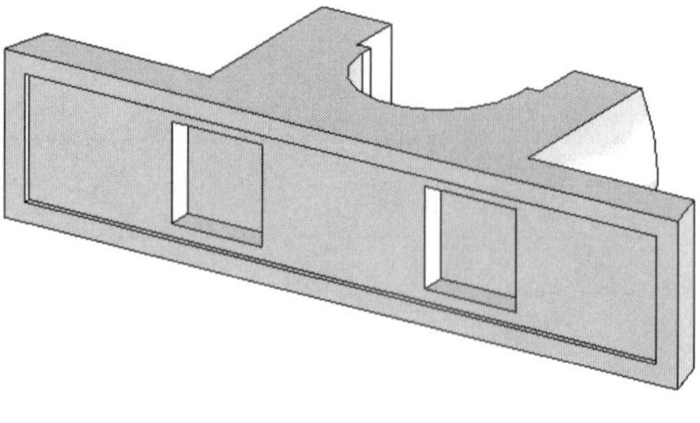

Figure 6-96 *Model after creating recess for square-headed bolts*

Creating the Hole Feature

After creating the recess for the head of the square-headed bolt, you need to create a hole using the **Hole** tool. To create the hole, you first need to place sketched points on the placement plane of the hole feature. After placing the sketched points, select the hole placement plane and invoke the **Hole** tool.

1. Invoke the sketching environment using the top right planar surface of the base feature as the sketching plane.

2. Create two sketched points on the right and the left top planar faces of the base feature.

   Next, you need to define the placement of these sketched point. It is evident from Figure 6-90 that the center point of the hole is coincident with the center point of the square recess feature. Therefore, you need to coincide the sketched points with the center of the squares. First you have to change the model view display from **Shaded** to **Hidden Lines Visible** so that the square recess feature is visible.

3. Orient the view normal to the viewing direction and choose the **Hidden Lines Visible** button from the **View** toolbar.

4. Invoke the **Centerline** tool from the **Sketch CommandManager** and move the cursor to the lower left corner of the left square. When a red dot is displayed along with the cursor, specify the start point of the line. Move the cursor to the upper right corner of the square and when the red dot is displayed, specify the endpoint of the line.

5. Right-click in the drawing area and choose the **Select** option from the shortcut menu. This will end line creation and also exit the **Line** tool. A centerline is drawn as a diagonal of the square.

6. Press and hold the CTRL key down and select the centerline created earlier. Now, select the left sketched point.

   The **Properties PropertyManager** is displayed. Using this add the relation between the left sketched point and the centerline. The centerline is drawn as the diagonal of the square and it is a general geometric fundamental that the midpoint of the diagonal of a square or a rectangle lies at its center. Therefore, you add a relation such that the sketched point is placed at the midpoint of the diagonal.

7. Choose the **Midpoint** button from the **Add Relations** rollout of the **Properties PropertyManager** and then choose the **OK** button from the **Properties PropertyManager**.

8. Similarly, draw the diagonal centerline for the right square. Add the midpoint relation between this centerline and the right sketched point.

9. Exit the sketching environment and change the model display mode to shaded. Select the right top planar surface of the base feature as the placement plane for the hole feature.

10. Choose **Insert > Features >Hole > Simple** from the menu bar to invoke the **Hole PropertyManager**. The preview of the hole feature with default settings is displayed in the drawing area.

11. Right-click in the drawing area and choose the **Through All** option from the shortcut menu to define the feature termination.

12. Set the value of the **Hole Diameter** spinner to **12**.

13. Select the center point of the hole and drag the cursor to the right sketched point. Release the left mouse button when the cursor snaps to the sketched point. Now, choose the **OK** button from the **Hole PropertyManager**

14. Using the same procedure create the second hole feature on the left of the model. The model after creating both hole features is displayed in Figure 6-97.

15. Similarly, create the cut feature on the second top planar face of the base feature, refer to Figure 6-90 for the dimensions of the sketch.

    Note that the dimension of 162 mm between the cut on the left and the cut on the right is given from the center of the sketches. To get this dimension, you need to create vertical centerlines in both sketches. Using the dimensions and relations, you need to make sure that the centerlines pass through the center of the sketches. Now, add the dimension between these two vertical centerlines. The model after creating the cut feature is shown in Figure 6-98.

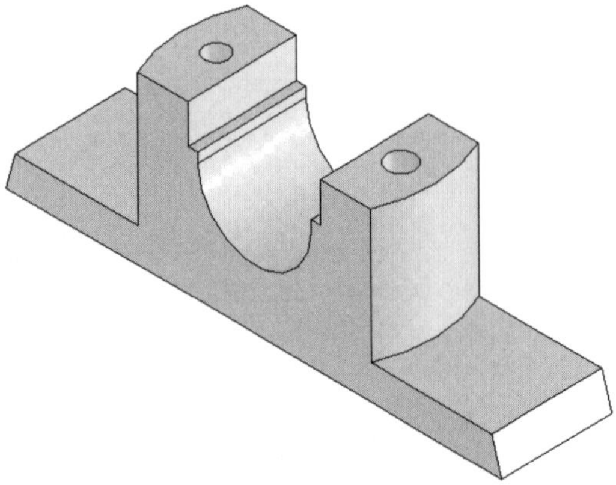

Figure 6-97 *Model after adding the hole features*

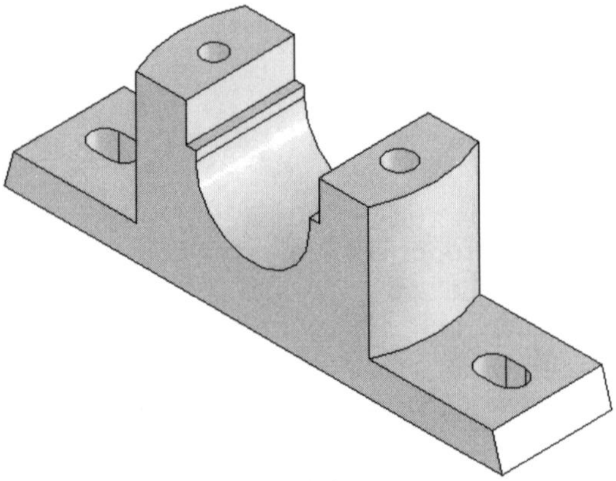

Figure 6-98 *Model after adding the cut feature*

Adding a Fillet to the Model

After creating all other features, you now need to add the fillet feature to the model.

1. Choose the **Fillet** button from the **Features CommandManager** to invoke the **Fillet PropertyManager**.

After invoking the **Fillet PropertyManager** you are prompted to selected edges, faces, features, or loops to fillet. As evident from the model you need to select only the edges to apply the fillet feature.

2. Select the edges shown in Figure 6-99.

As soon as you select the edges the preview of the fillet feature with the default values is shown in the drawing area. A radius callout is also displayed along the selected edge. Now, you need to modify the default radius value of the fillet feature.

Note

*If the preview of the fillet feature is not displayed in the drawing area, select the **Full Preview** radio button available in the **Items To Fillet** rollout of the **Fillet PropertyManager**.*

3. Set the value of the **Radius** spinner to **8** and choose the **OK** button from the **Fillet PropertyManager**.

The model after adding the fillet is shown in Figure 6-100.

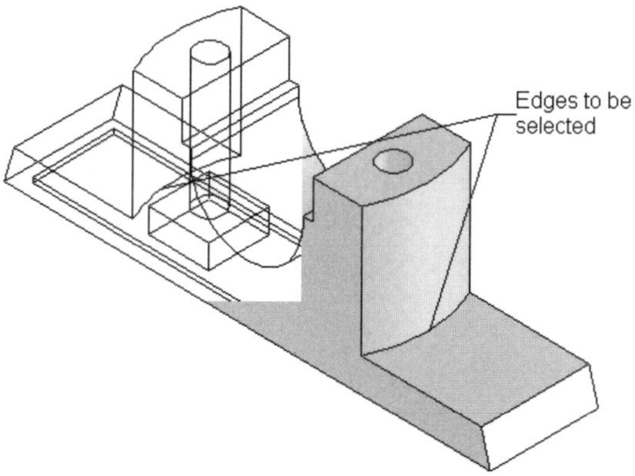

Figure 6-99 *Edges to be selected for the fillet feature*

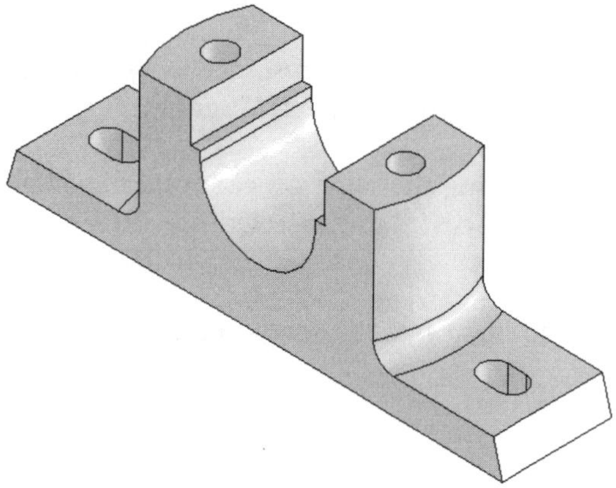

Figure 6-100 *Model after adding the fillet*

Adding a Chamfer to the Model

The next feature that you need to add to this model is chamfer.

1. Choose the **Chamfer** button from the **Features CommandManager** to invoke the **Chamfer PropertyManager**.

2. Select the edges of the cut features as shown in Figure 6-101. The edges that are tangent to the selected edges are selected automatically because the selection mode of the chamfer feature uses tangent propagation by default.

As soon as you select the edges the preview of the chamfer feature is displayed in the drawing area with the default values. The angle and distance callouts are also displayed. Now, you need to set the required value of the chamfer.

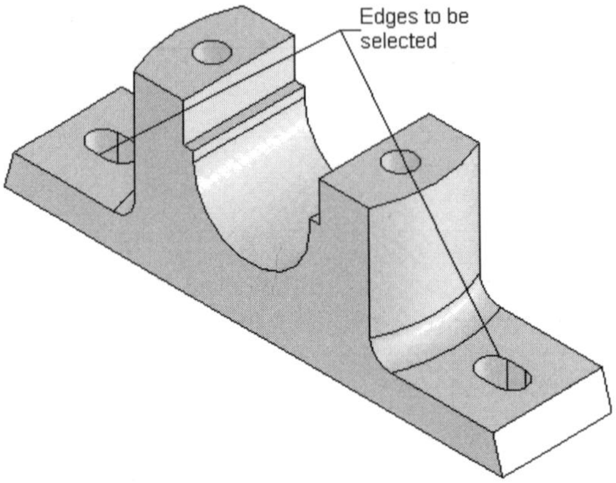

Figure 6-101 *Edges to be selected*

The required values of the chamfer parameters are 1 mm and 45-degree. The **Angle distance** radio button is selected by default in the **Chamfer Parameters** rollout. The value of the angle in the **Angle** spinner is set as 45-degree. Therefore, you do not need to modify this value. You need to set only the value of the distance in the **Distance** spinner.

3. Set the value of the distance in the **Distance** spinner to **1** and choose the **OK** button from the **Chamfer PropertyManager**.

4. Following the same procedure add a chamfer to the semicircular edges on the front and back faces of the base feature. Refer to Figure 6-90 for the parameters of the chamfer feature.

The final solid model is shown in Figure 6-102.

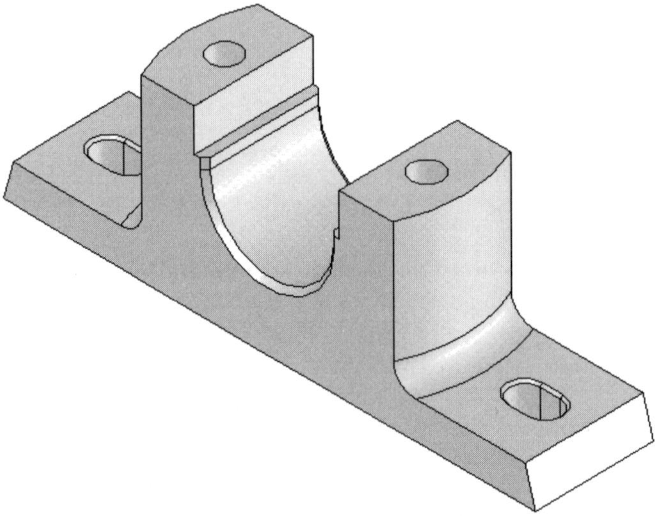

Figure 6-102 *Final solid model*

Saving the Model

Next, you need to save the document.

1. Choose the **Save** button from the **Standard** toolbar and save the document with the name given below.

   \My Documents\SolidWorks\c06\c06tut1.sldprt.

2. Choose **File > Close** from the menu bar to close the document.

Tutorial 2

In this tutorial, you will create the model shown in Figure 6-103. For better understanding of the model the section view of the model is shown in Figure 6-104. The dimensions of the model are shown in Figure 6-105. **(Expected time: 30 min)**

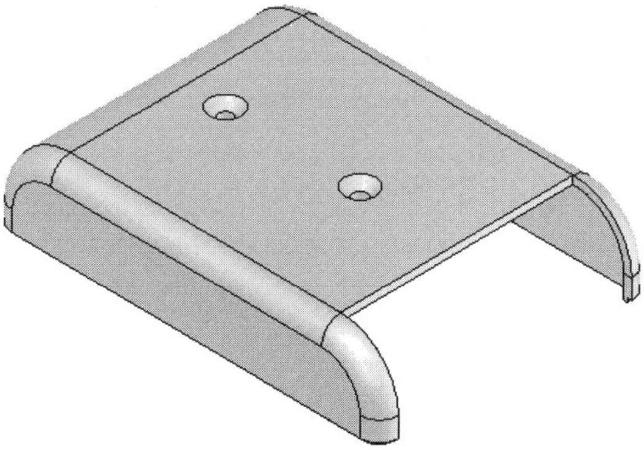

Figure 6-103 *Solid model for Tutorial 2*

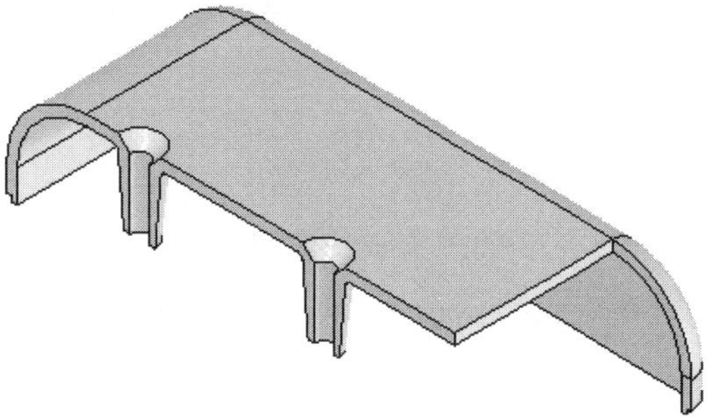

Figure 6-104 *Section view of the model*

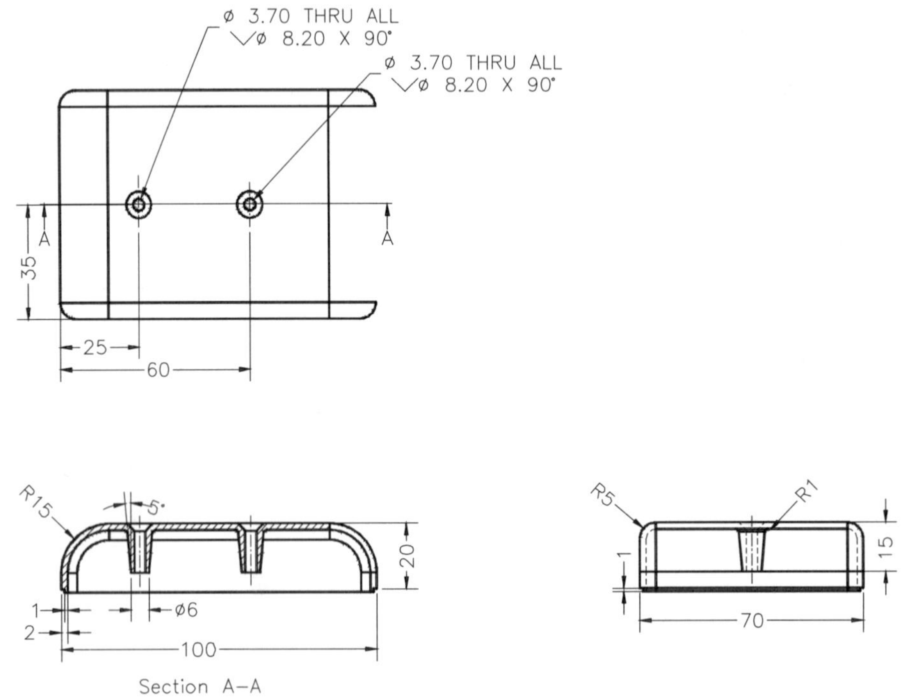

Figure 6-105 Top view, front section view, and right-side view with dimensions

The steps to be followed to complete this tutorial are listed next.

a. Create the base feature of the model by extruding a rectangle of 100 mm x 70 mm to a distance of 20 mm.
b. Add a fillet to the base feature, refer to Figures 6-106 through 6-109.
c. Create the shell feature to create a thin-walled part and remove some of the faces, refer to Figures 6-110 and 6-111.
d. Create a reference plane at an offset distance from the top planar face of the base feature and extrude the sketch created on the new plane, refer to Figure 6-112.
e. Using the **Hole Wizard** tool add the countersink hole to the model, refer to Figure 6-113.
f. Add a fillet to the extruded feature, refer to Figures 6-114 and 6-115.
g. Create the lip of the component by extruding the sketch, refer to Figure 6-116.

Creating the Base Feature

The base feature of the model is created by extruding a rectangle of 100 mm x 70 mm to a distance of 20 mm. It is evident from the model that the sketch of the base feature is created on the **Top Plane**. Therefore, you have to select the **Top Plane** as the sketching plane.

1. Start a new SolidWorks part document and invoke the **Extruded Boss/Base** tool. Select the **Top Plane** from the **FeatureManager Design Tree**.

2. Orient the sketch plane normal to the viewing direction.

3. Using the **Rectangle** tool, draw a rectangle and force it to a dimension of 100 mm x 70 mm. Add the other required dimensions to fully define the sketch.

4. Exit the sketching environment and extrude the rectangle to a depth of 20 mm.

Creating the Fillet Features

After creating the base feature, you need to add fillets to the model. In this model, you need to add three fillet features. Two fillet features will be added at this stage of the design process and the remaining one will be added at a later stage of the design process.

1. Choose the **Fillet** button from the **Features CommandManager** to invoke the **Fillet PropertyManager**.

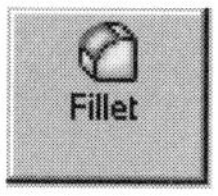

2. Select the edges of the model as shown in Figure 6-106.

As soon as you select the edges of the model, the preview of the fillet with the default values and the radius callout are displayed in the drawing area.

3. Set the value of the **Radius** spinner to **15** and choose the **OK** button from the **Fillet PropertyManager**.

Figure 6-107 shows the model after adding the first fillet feature.

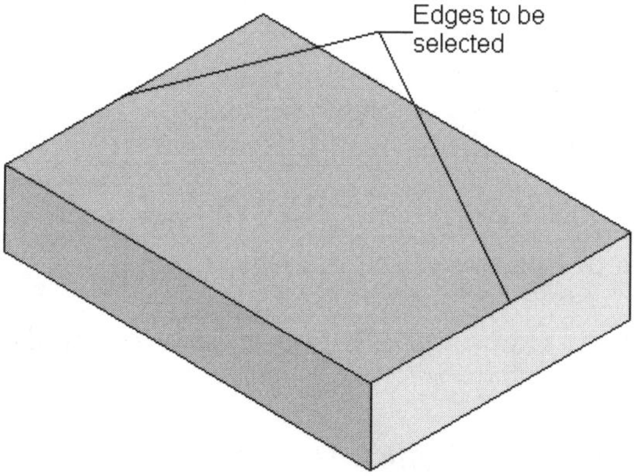

Figure 6-106 *Edges to be selected*

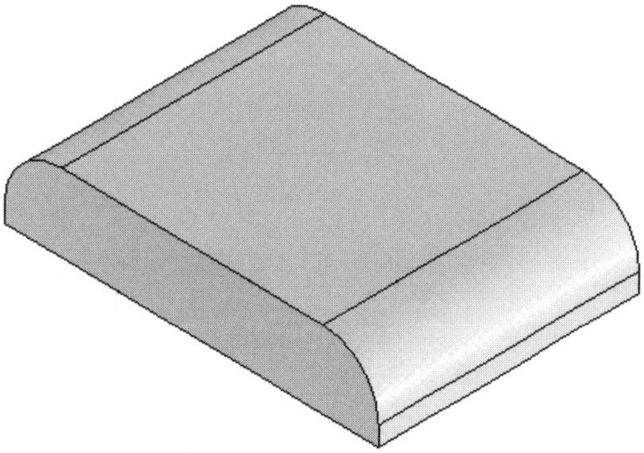

Figure 6-107 *Fillet added to the base feature*

Now, add the second fillet feature to the model.

4. Invoke the **Fillet PropertyManager** and set the value of the **Radius** spinner to **5**.

5. Select the edges of the model as shown in Figure 6-108. Right-click and choose **OK** from

the shortcut menu to complete feature creation. The model after adding the second fillet feature is shown in Figure 6-109.

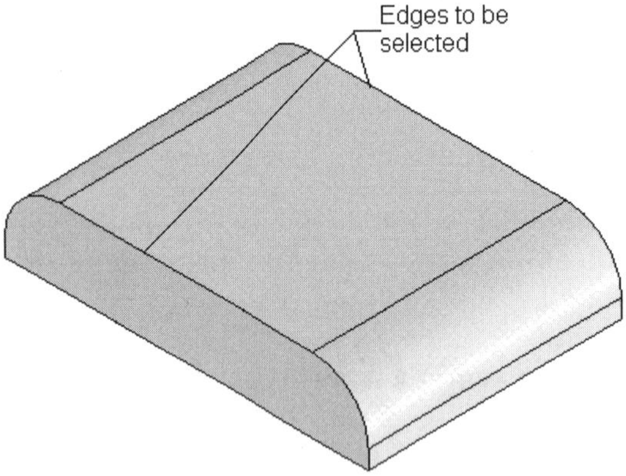

Figure 6-108 Edges to be selected

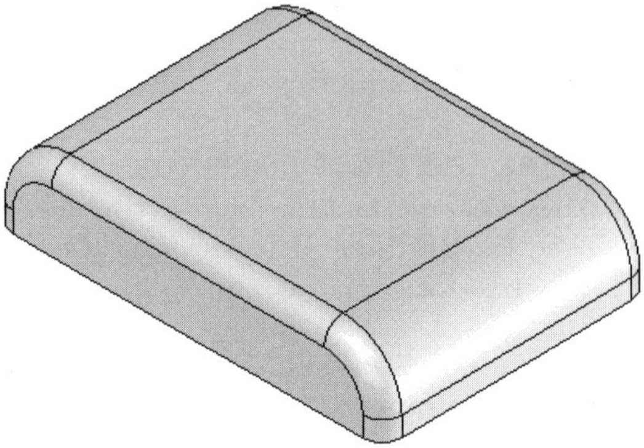

Figure 6-109 Second fillet added to the model

Tip. *The edges tangent to the selected edges are filleted automatically because by default, the **Tangent propagation** check box is selected in the **Fillet PropertyManager**.*

Creating the Shell Feature

It is evident from Figures 6-103 and 6-104 that a shell feature is required to create a thin-walled structure. As discussed earlier, the shell feature is used to scoop out the material from the model, leaving behind a thin-walled hollow part.

1. Choose the **Shell** button from the **Features CommandManager** to invoke the **Shell1 PropertyManager**; you are prompted to select the faces to remove.

2. Rotate the model and select the faces to remove, as shown in Figure 6-110. The names of the selected faces are displayed in the **Faces to Remove** selection area.

3. Set the value of the **Thickness** spinner to **2** and choose the **OK** button from the **Shell1 PropertyManager**.

The model after creating the shell feature is shown in Figure 6-111.

Creating the Extruded Feature

The next feature that you are going to create is an extruded feature. But, before creating this feature, you need to create a reference plane at an offset distance from the top planar face of the base feature.

1. Invoke the **Plane PropertyManager** and create a plane at an offset distance of 15 mm from the top planar face of the base feature. You have to select the **Reverse direction** check box from the **Plane PropertyManager** to create the plane.

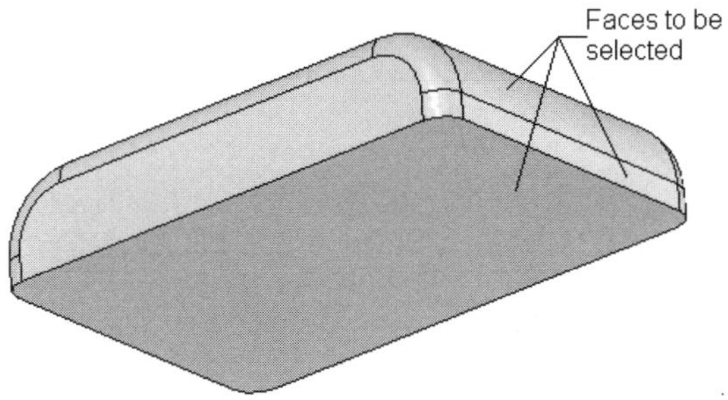

Figure 6-110 Faces to be selected to remove

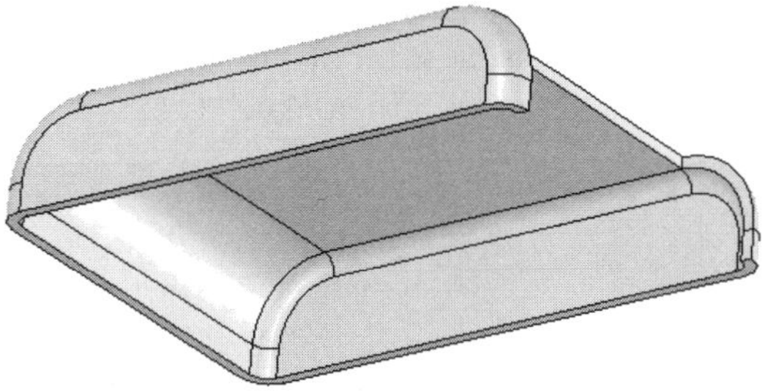

Figure 6-111 Model after creating the shell feature

2. Invoke the **Extruded Boss/Base** tool and select the newly created plane as the sketching plane and create the sketch using the standard sketching tools. The sketch consists of two circles of 6 mm diameter. For other dimensions refer to Figure 6-105.

3. Exit the sketching environment and extrude the sketch using the **Up To Next** option and add an outward draft of 5-degree. Hide the reference plane.

   Figure 6-112 shows the rotated model after creating the extruded feature with draft.

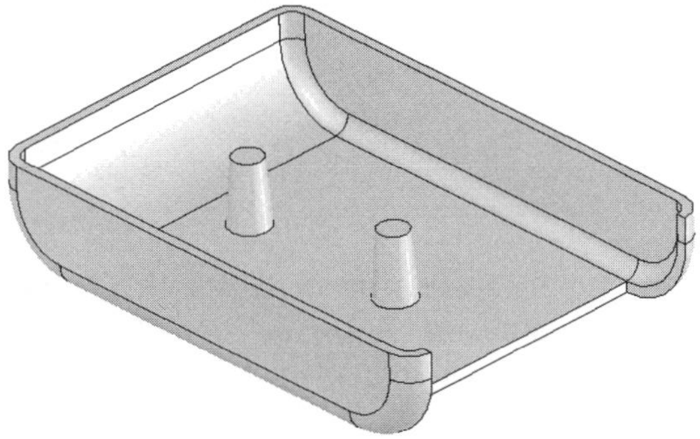

Figure 6-112 Model after creating the extruded feature

Adding the Countersink Hole Using the Hole Wizard Tool

The next feature that you are going to create is a **M3.5 Flat Head Machine Screw** countersink hole, refer to Figure 6-105. In SolidWorks you are provided with one of the largest standard hole-generating techniques known as **Hole Wizard**. Using the **Hole Wizard** tool, you can create the standard holes to the model that can accommodate standard fasteners.

1. Rotate the model and select the top planar face of the base feature and then choose the **Hole Wizard** button from the **Features CommandManager**.

The **Hole Definition** dialog box is displayed and the preview of the hole feature on the selected face with the default values is shown in the drawing area.

Tip. *If you select the placement plane for the placement of the hole feature before invoking the* ***Hole Definition*** *dialog box, the placement sketch will be a 2D sketch. If you select the placement plane after invoking the* ***Hole Definition*** *dialog box, the hole placement sketch will be a 3D sketch. You will learn more about 3D sketches in the later chapters.*

2. Choose the **Countersink** tab of the **Hole Definition** dialog box.

Now, set the parameters to define the standard hole. The preview of the standard hole is modified dynamically as you set the parameters for the hole feature.

3. Select the **Ansi Metric** option from the **Standard** drop-down list.

4. Select the **Flat Head Screw - ANSI B18.6.7M** option from the **Screw type** drop-down list.

5. Select the **M3.5** option from the **Size** drop-down list and choose the **Through All** option from the **End Condition & Depth** drop-down list.

6. Choose the **Next** button from the **Hole Definition** dialog box; the **Hole Placement** dialog box is displayed and the sketching environment is invoked. The select cursor is replaced by the point cursor.

7. Using the point cursor, specify another point anywhere on the top planar face of the base feature to create another hole.

Because both placement points are not properly placed, you need to add the required relations and dimensions to define the proper location of these placement points. Before doing that, you need to change the model display from **Shaded With Edges** to **Hidden Lines Visible** for a better display.

8. Choose the **Hidden Lines Visible** button from the **View** toolbar to display the model with hidden lines visible.

9. Press the ESC key to invoke the select cursor.

10. Right-click in the drawing area and choose the **Add Relation** option from the shortcut menu to display the **Add Relations PropertyManager**.

11. Select the left placement point and then select the upper left hidden circle. Choose the **Concentric** button from the **Add Relations** rollout.

12. Right-click in the drawing area and choose **Clear Selections** from the shortcut menu. Select the right placement point and the upper right hidden circle. Now, choose the **OK** option from the confirmation corner.

13. Choose the **Finish** button from the **Hole Placement** dialog box and choose the **Shaded With Edges** button from the **View** toolbar.

    The isometric view of the model after adding the hole feature is shown in Figure 6-113.

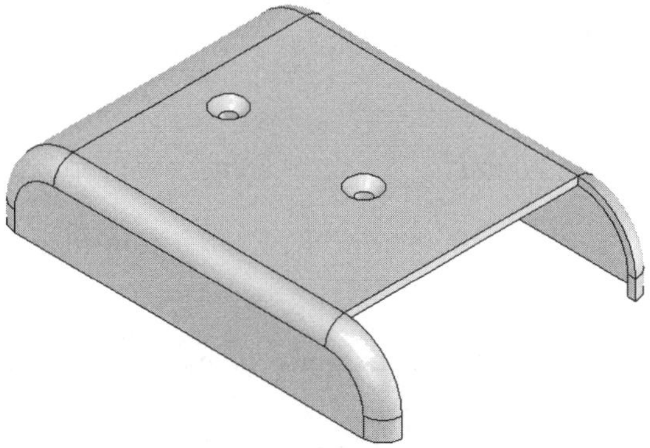

Figure 6-113 *Model after adding the hole feature using the* ***Hole Wizard*** *tool*

Adding a Fillet to the Model

Now, you need to add a fillet to the edges of the extruded feature with draft that was created earlier.

1. Rotate the model and invoke the **Fillet PropertyManager**. Select the edges of the model, as shown in Figure 6-114.

2. Set the value of the **Radius** spinner to **1** and choose the **OK** button to end feature creation.

   The model after adding the fillet is shown in Figure 6-115.

Adding a Lip to the Model

The last feature you need to add to the model is a lip. It is created by extruding an open sketch using the thin option.

1. Invoke the **Extrude Boss/Base** tool and select the bottom face of the base feature as the sketching plane and invoke the sketching environment.

2. Using the select tool, select any one of the inner edges of the model on the current sketching plane and right-click to display the shortcut menu. Now, choose the **Select Tangency** option from the shortcut menu.

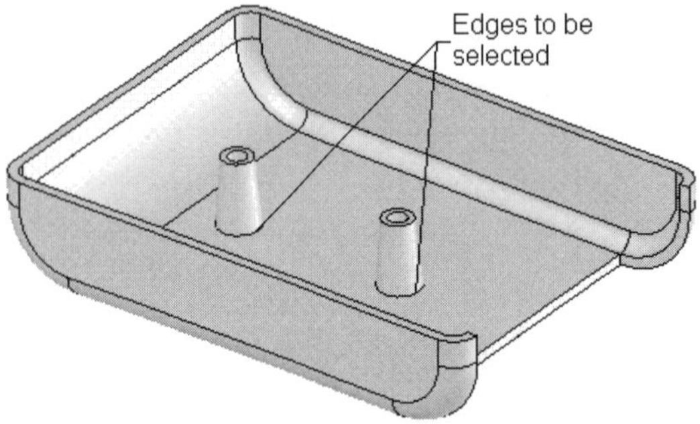

Figure 6-114 Edges to be selected

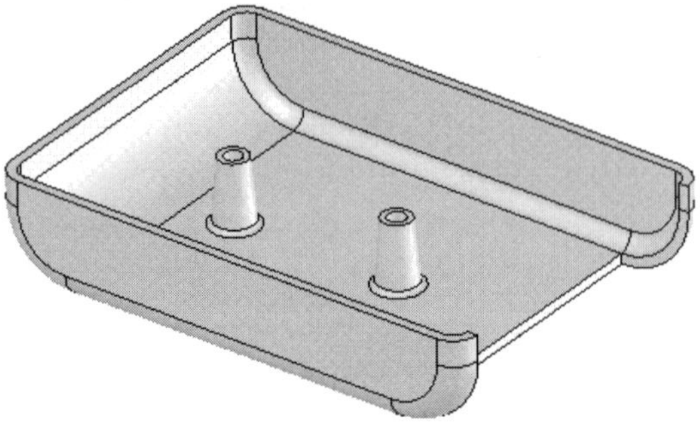

Figure 6-115 Model after adding the fillet

3. Choose the **Convert Entities** button from the **Sketch CommandManager**; the selected edges are converted into sketched entities.

4. Exit the sketching environment; the **Extrude PropertyManager** is displayed and the **Thin Feature** rollout is invoked automatically because you are extruding an open sketch.

5. Set the value of the **Distance** spinner in the **Direction 1** rollout to **1** and set the value of the **Thickness** spinner in the **Thin Feature** rollout to **1**.

6. Choose the **OK** button from the **Extrude PropertyManager**. The rotated view of the final model is shown in Figure 6-116.

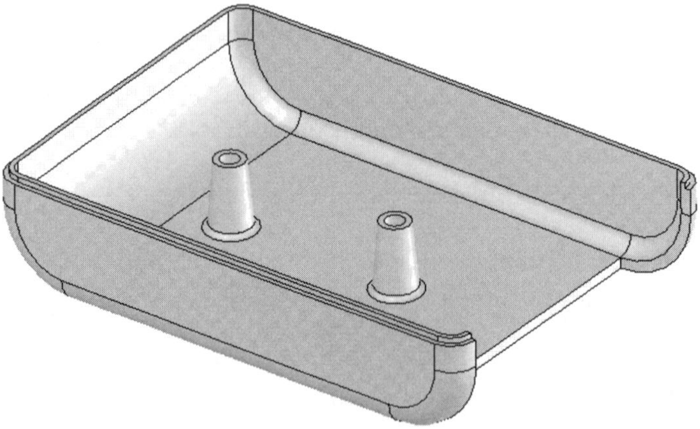

Figure 6-116 Final model rotated to display the maximum features

Saving the Model

1. Choose the **Save** button from the **Standard** toolbar and save the part document with the name given below.

   \My Documents\SolidWorks\c06\c06tut2.sldprt

2. Choose **File > Close** from the menu bar to close the document.

SELF-EVALUATION TEST

Answer the following questions and then compare your answers with those given at the end of this chapter.

1. Using the **Hole PropertyManager** you can create counterbore, countersink, and tapped holes. (T/F)
2. The hole features created using the **Hole Wizard** tool and the **Hole PropertyManager** are not parametric. (T/F)
3. You cannot define a user-defined hole using the **Hole Wizard** tool. (T/F)
4. You cannot preselect the edges or faces for creating a fillet feature. (T/F)
5. In SolidWorks you can create a multithickness shell feature. (T/F)
6. The __________ check box is selected to create the shell feature on the outer side of the model.
7. The __________ is created by specifying different radii along the length of the selected edge at specified intervals.
8. The names of the faces to be removed in the shell features are displayed in the __________ selection area.
9. If you want to specify different distances while creating the chamfer, clear the __________ check box.
10. The __________ check box is selected to apply the face fillet feature with continuous curvature throughout the fillet feature.

REVIEW QUESTIONS

Answer the following questions.

1. The __________ option is used to add standard holes to the model.

2. After specifying all parameters to a hole feature using the **Hole Definition** dialog box, the __________ dialog box is displayed to specify the placement of the hole feature.

3. The __________ radio button is selected to create a smooth transition by smoothly blending the points and vertices on which you have defined the radius on the selected edge.

4. The __________ tab available in the **Hole Definition** dialog box is used to define the parameters to add a standard drilled hole.

5. By default, the __________ radio button is selected in the **Chamfer PropertyManager**.

6. If you preselect the placement surface for creating a hole feature using the **Hole Wizard** tool, the resulting placement sketch will be a

   (a) 2D sketch (b) Planar sketch
   (c) Bezier spline (d) 3D sketch

7. Using which options you do not use radius while creating a fillet feature?

   (a) Face fillet with hold line (b) Constant radius fillet
   (c) Variable radius fillet (d) Full round fillet

8. Which radio button from the **Variable Radius Parameters** rollout is used to create smooth transition while creating a variable radius fillet?

   (a) **Straight transition** (b) **Parametric transition**
   (c) **Smooth transition** (d) **Surface transition**

9. If you do not select any face to be removed while creating a shell feature, what will be the resulting model?

   (a) Remains complete solid model (b) Thin walled hollow model
   (c) Automatically removed one face (d) None of these

10. Which dialog box is displayed when you choose the **Hole Wizard** button from the **Features CommandManager**?

   (a) **Hole** (b) **Hole Definition**
   (c) **Hole Wizard** (d) **Hole Parameters**

EXERCISES

Exercise 1

Create the model shown in Figure 6-117. The dimensions of the model are shown in Figure 6-118. **(Expected time: 30 min)**

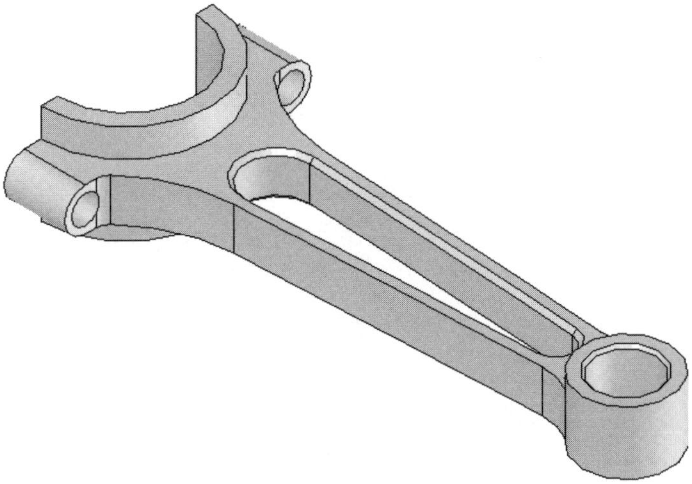

Figure 6-117 *Solid model for Exercise 1*

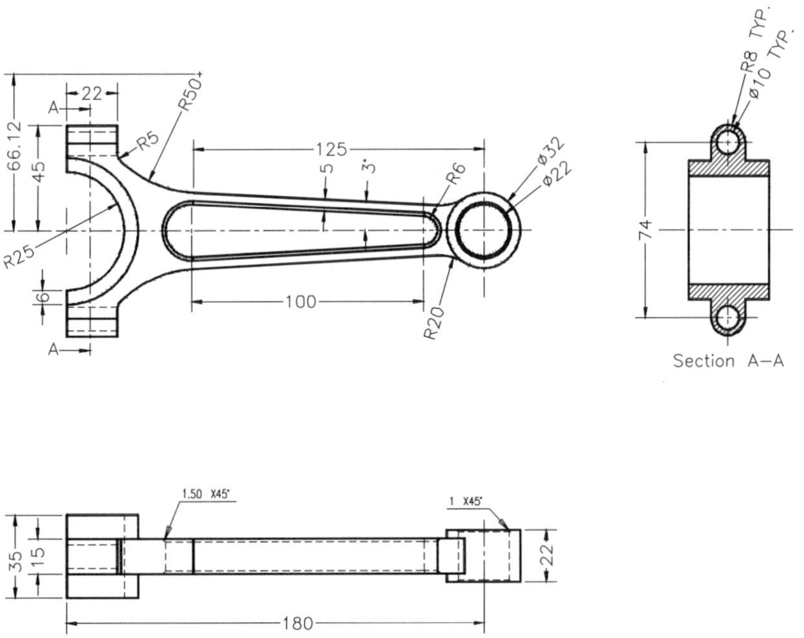

Figure 6-118 *Views and dimensions of the model for Exercise 1*

Exercise 2

Create the model shown in Figure 6-119. The dimensions of the model are shown in Figure 6-120. **(Expected time: 30 min)**

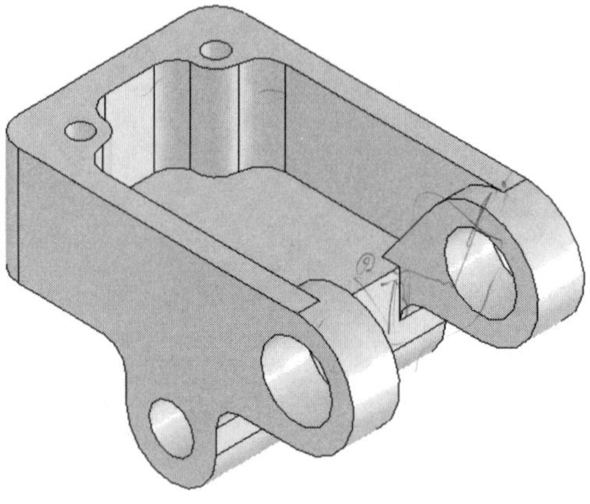

Figure 6-119 *Solid model for Exercise 2*

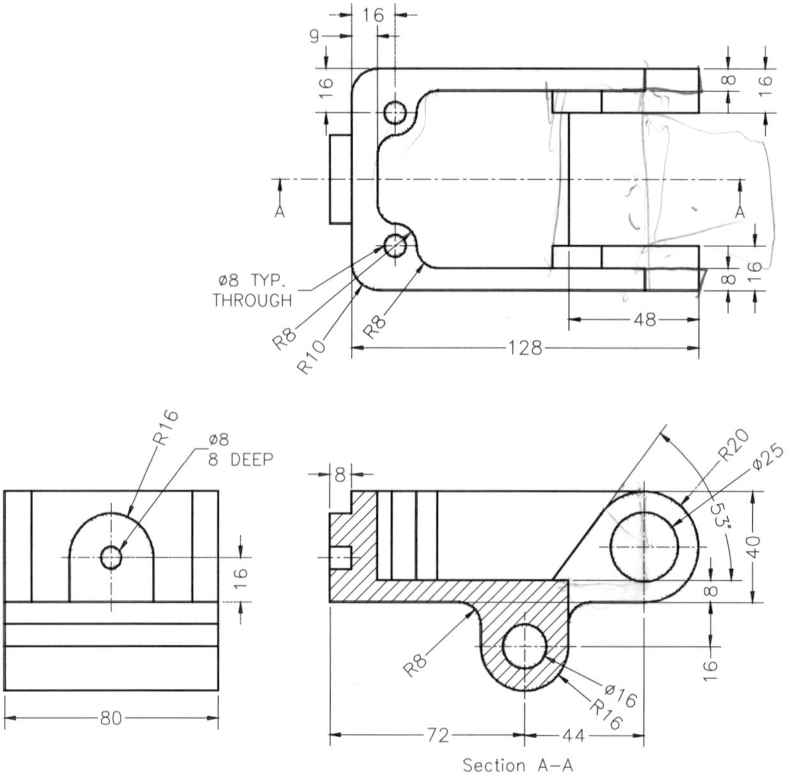

Figure 6-120 *Views and dimension of the model for Exercise 2*

Answers to Self-Evaluation Test

1. T, **2.** F, **3.** F, **4.** F, **5.** T, **6. Shell outward**, **7.** Variable radius fillet, **8. Faces to Remove**, **9. Equal distance**, **10. Curvature continuous**

Chapter 7

Advanced Modeling Tools-II

Learning Objectives

After completing this chapter you will be able to:

- *Mirror features, faces, and bodies.*
- *Create a Linear Pattern.*
- *Create a Circular Pattern.*
- *Create a Sketch-driven Pattern.*
- *Create a Curve-driven Pattern.*
- *Create a Table-driven Pattern.*
- *Create a Rib Feature.*
- *Create a Dome Feature.*
- *Create a Shape Feature.*
- *Create a Indent Feature.*
- *Display the section view of the model.*

ADVANCED MODELING TOOLS

Some of the advanced modeling options were discussed in Chapter 6, Advanced Modeling Tools-I. In this chapter you will learn about more advanced modeling tools, using which you can capture the design intent of the model. The rest of the advanced modeling tools are discussed in later chapters.

Creating a Mirror Feature

CommandManager:	Features > Mirror
Menu:	Insert > Pattern/Mirror > Mirror
Toolbar:	Features > Mirror

The **Mirror** tool is used to copy or mirror the selected feature, face, or body about a specified mirror plane, which can be a reference plane or a planar face. To use this tool, choose the **Mirror** button from the **Features CommandManager**, or choose **Insert > Pattern/Mirror > Mirror** to invoke the **Mirror PropertyManager**, as displayed in Figure 7-1. The confirmation corner is also displayed in the drawing area.

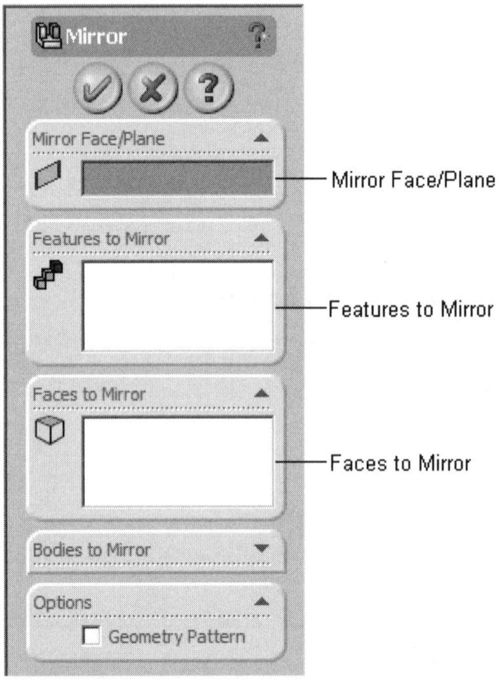

Figure 7-1** The **Mirror PropertyManager

The options that are used to mirror features, faces, and bodies are discussed next.

Mirroring Features

You can mirror the selected feature along the specified mirror plane or face by using this feature. To do so, invoke the **Mirror PropertyManager**. You are prompted to select a plane or a planar face about which to mirror, followed by the features to mirror. Select a plane or a planar face that will act as mirror plane or mirror face. After selecting the mirror plane or face, the selection mode of the **Features to Mirror** selection area is invoked and you are prompted to select features to mirror. Select the feature or features from the drawing area or from the **FeatureManager Design Tree** which is displayed in the drawing area. When you select the features to be mirrored, the preview of the mirrored image is displayed in the drawing area. After selecting all the required features, choose the **OK** button from the **Mirror PropertyManager**. Figure 7-2 shows the mirror plane and features to be mirrored and Figure 7-3 shows the resulting mirrored features.

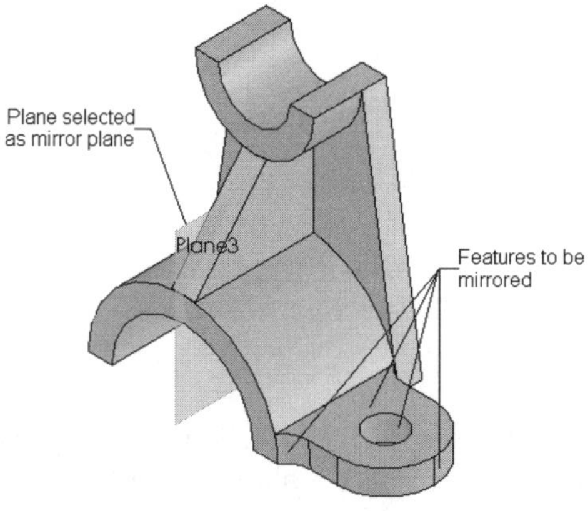

Figure 7-2 *Mirror plane the features to be mirrored*

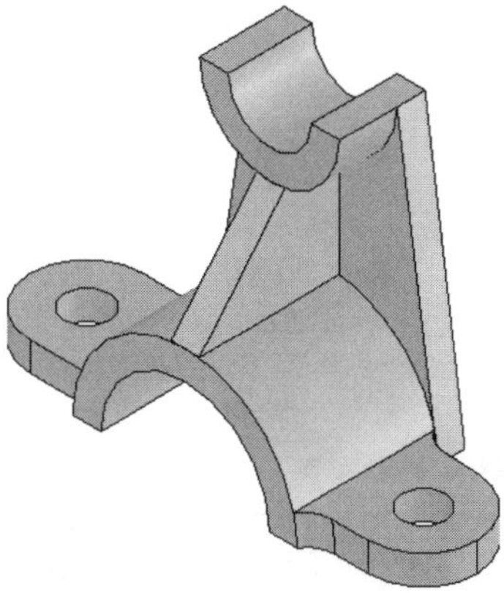

Figure 7-3 *The resulting mirror feature*

Tip. *You can also preselect the mirror plane or mirror face and the features to be mirrored before invoking the **Mirror PropertyManager**.*

Mirroring with and without Geometric Pattern

When you create a mirror feature, you are provided with an option known as the geometric pattern. This option is available in the **Options** rollout, as shown in Figure 7-4.

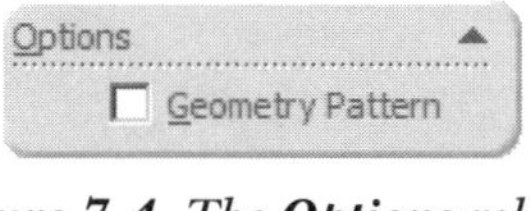

Figure 7-4 *The **Options** rollout*

By default, the **Geometry Pattern** check box is cleared. Therefore, if you mirror a feature that is related to some other entity, the same relationship will be applied to the mirrored feature. Consider a case in which an extruded cut is created using the **Offset From Surface** option. If you mirror the cut feature along a plane, the same relationship will be applied to the mirrored cut feature. The mirrored cut feature will be created with the same end condition of feature termination. Figure 7-5 shows a hole feature created on the right and mirrored along **Plane 1**, with the **Geometry Pattern** check box cleared.

If you select the **Geometry Pattern** check box, the mirror feature created will not depend on the relational references. It will just create a replica of the selected geometry. Figure 7-6 shows the mirror feature created with the **Geometry Pattern** check box selected.

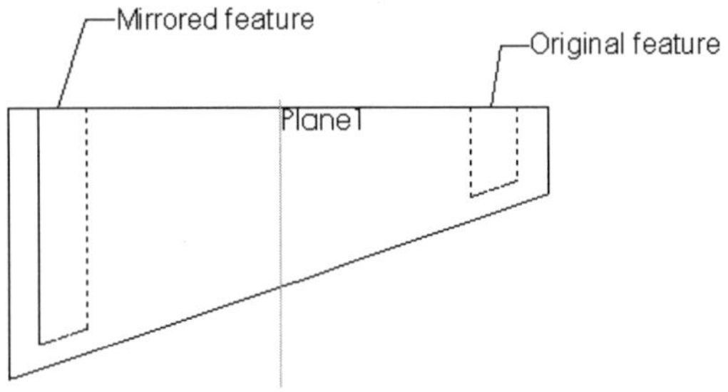

Figure 7-5 Mirror feature created with the ***Geometry Pattern*** *check box cleared*

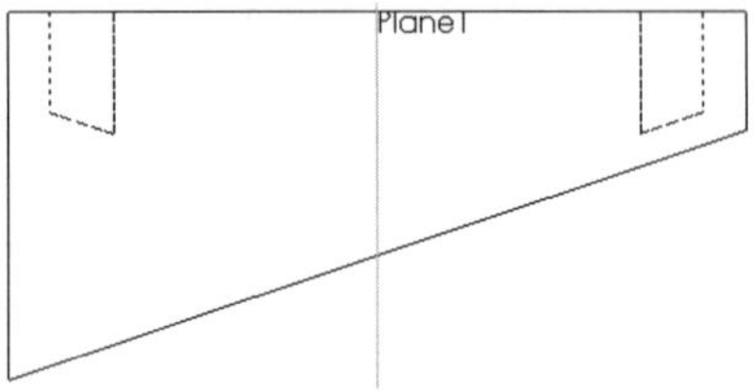

Figure 7-6 Mirror feature created with the ***Geometry Pattern*** *check box selected*

Mirroring Faces

Using this option you can mirror faces along a mirror plane or mirror face. To use this option, invoke the **Mirror PropertyManager**. You are prompted to select a plane or a planar face about which to mirror. Select the planar face or a plane about which the selected faces will be mirrored. Now, click once in the **Faces to Mirror** selection area to invoke the selection mode, and select the faces to be mirrored. The selected faces must form a closed body. Otherwise, feature creation is not possible. Use the **OK** button from the **Mirror PropertyManager** to end feature creation. Figure 7-7 shows the faces and the mirror plane to be selected. Figure 7-8 shows the resulting mirror feature created.

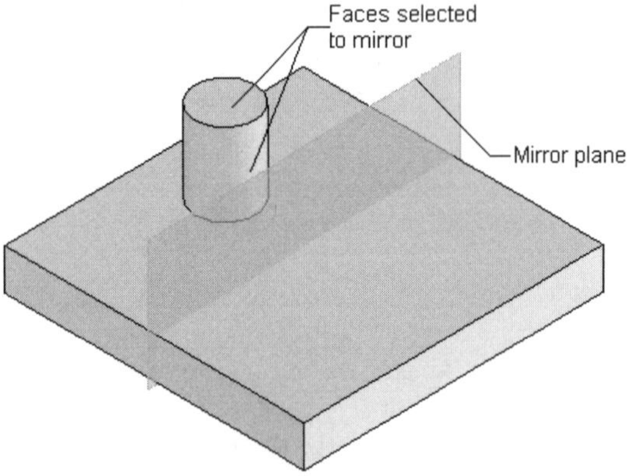

Figure 7-7 Mirror plane and faces selected to mirror

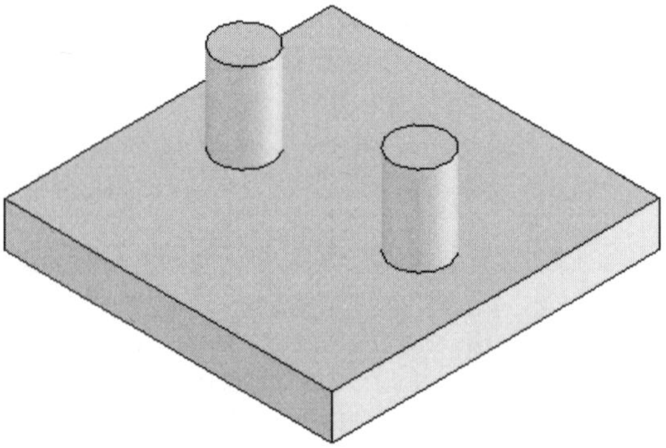

Figure 7-8 The resulting mirror created

Note

The following are some of the factors that should be considered, while creating a mirror feature by mirroring the faces along the selected plane or planar face:

1. If the replica of the faces is not coincident to the parent part body, SolidWorks will give an error while creating the mirror feature.

2. If the replica of the faces exists on faces other than the original face, SolidWorks will give an error while creating the mirror feature.

3. If the selected faces form a complex geometry, SolidWorks will give an error while creating the mirror feature.

4. If the mirrored faces exist on more than one face, SolidWorks will give an error while creating the mirror feature.

5. The selected faces should form a closed body. If the selected faces do not form a closed body, SolidWorks will give an error while creating the mirror feature.

Mirroring Bodies

As discussed in the earlier chapters, the SolidWorks supports the multibodies environment. Therefore, using the **Mirror** tool, you can also mirror the disjoint bodies. To mirror a body along a plane, invoke the **Mirror PropertyManager** and select a plane or a planar face that will act as a mirror plane. Invoke the **Bodies to Mirror** rollout and select the body from the drawing area. Alternatively, you can expand the **FeatureManager Design Tree** flyout and select the body to mirror from the **Solid Bodies** folder. The name of the selected body is displayed in the **Solid/Surface Bodies to Mirror** selection area. The preview of the mirrored body is displayed in the drawing area. Choose the **OK** button from the **Mirror PropertyManager**. Figure 7-9 shows the plane and the body selected to be mirrored. Figure 7-10 shows the resulting mirrored feature.

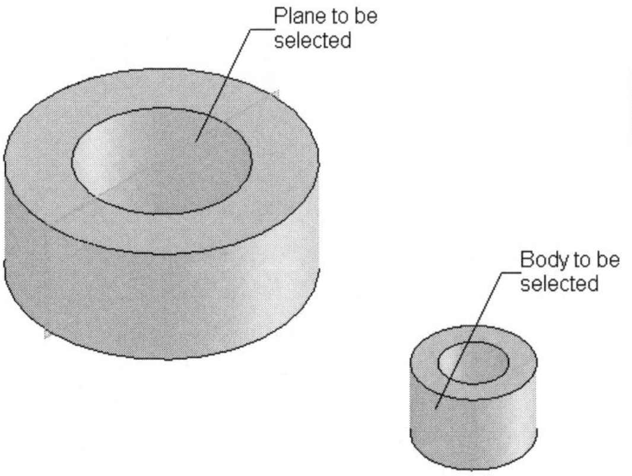

Figure 7-9 *Selecting the mirror plane and body to be mirrored*

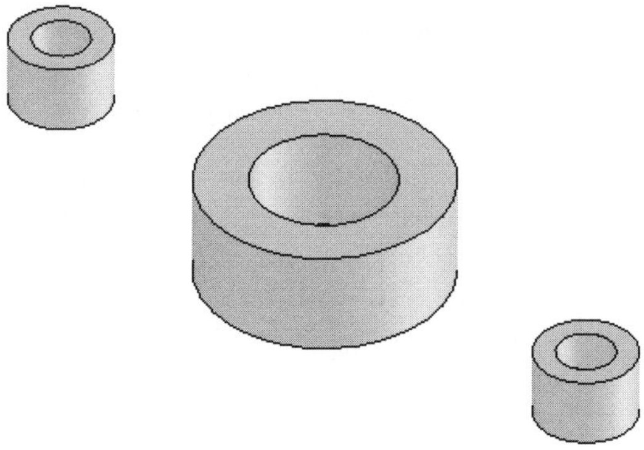

Figure 7-10 *Resulting mirrored feature*

The options available in the **Options** rollout, while mirroring the bodies are discussed next. The **Options** rollout with these options, is displayed in Figure 7-11.

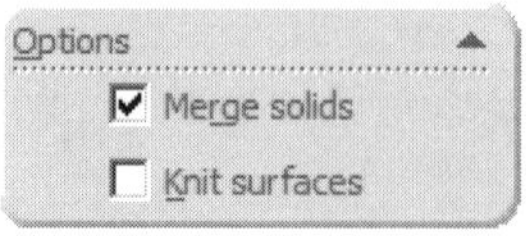

***Figure 7-11** The **Options** rollout*

Merge solids

The **Merge solids** option is used to merge the mirrored body with the parent body. Consider a case, in which you mirror a body along a selected plane or a planar face of the same body and the resulting mirrored body is joined to the parent body. In this case, if you select the **Merge solids** check box, the resulting mirrored body will merge with the parent body to become a single body. If the **Merge solids** check box is cleared, the resulting body will be joined with the parent body, but it will not merge with the parent body, therefore, resulting in two separate bodies.

Knit surfaces

If you mirror a surface body, then the **Knit Surface** check box is selected to knit the mirrored and the parent body together.

Tip. *As discussed earlier, the design intent is captured in the model using the mirror option. Therefore, if you modify the parent feature, face, or body the same will be reflected on the mirrored feature, face, or body.*

*If you want to mirror all the features of the model using the **Features to Mirror** option, you need to select all the features. But for using **Bodies to Mirror** option, you need to select the body from the **Solid Bodies** folder. By selecting the body, all the features are added to the mirror image.*

Creating a Linear Pattern Feature

CommandManager:	Features > Linear Pattern
Menu:	Insert > Pattern/Mirror > Linear Pattern
Toolbar:	Features > Linear Pattern

As discussed in the previous chapters, you can arrange the sketched entities in a particular arrangement or pattern. In the same manner, you can also arrange the features, faces, and bodies in a particular pattern. In SolidWorks, you are provided with various types of patterns such as linear patterns, circular patterns, sketch-driven patterns, curve-driven patterns, and table-driven patterns.

In this section, you will learn how to create linear patterns. The other type of patterns are discussed later in this chapter.

To create a linear pattern, choose the **Linear Pattern** button from the **Features CommandManager**, or choose **Insert > Pattern/Mirror > Linear Pattern** from the menu bar. The **Linear Pattern PropertyManager** is invoked, and the confirmation corner is also displayed. A partial view of the **Linear Pattern PropertyManager** is displayed in Figure 7-12.

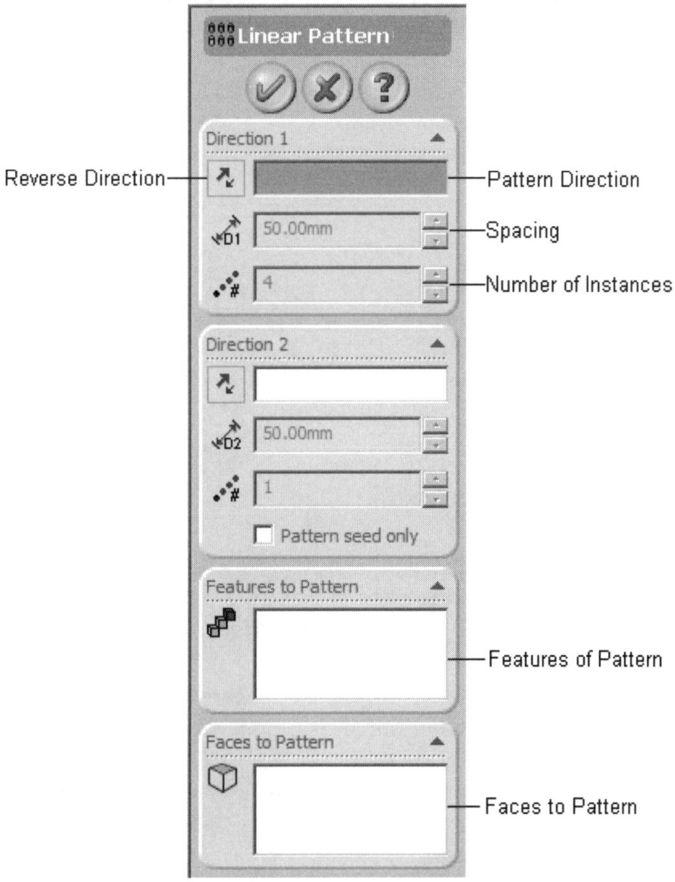

Figure 7-12 *Partial view of the* ***Linear Pattern PropertyManager***

The various options available in the **Linear Pattern PropertyManager** are discussed next.

Linear Pattern in One Direction

When you invoke the **Linear Pattern PropertyManager**, the **Direction 1** rollout and the **Features** rollout are invoked by default. Also, you are prompted to select an edge or axis for direction reference and face of the feature for pattern features. Therefore, first you need to select an edge or axis as a direction reference. Select an edge or axis as the direction reference. The name of the selected reference will be displayed in the **Pattern Direction** selection area of the **Direction 1** rollout. The selected reference is displayed in green and the **Direction 1** callout is attached to it. The **Direction 1** callout has two edit boxes to define the number of instances and spacing. You are also provided with the **Reverse Direction** arrow along with the selected reference. Now, select a face of the feature to be patterned. The name of the selected feature will be displayed in the **Features to pattern** area of the **Features to Pattern** rollout. The preview of the pattern is displayed in the drawing area with the default values. Set the value of the center to center spacing between the pattern instances in the **Spacing** spinner. Set the value of the number of instances to be patterned in the **Number of Instances** spinner. You can also set these values in the **Direction 1** callout. Using the **Reverse Direction** button from the **PropertyManager** or the **Reverse Direction** arrow from the drawing area, you can reverse the direction of pattern feature creation. Figure 7-13 shows the feature and edge to be selected for directional reference and Figure 7-14 shows the model after pattern creation.

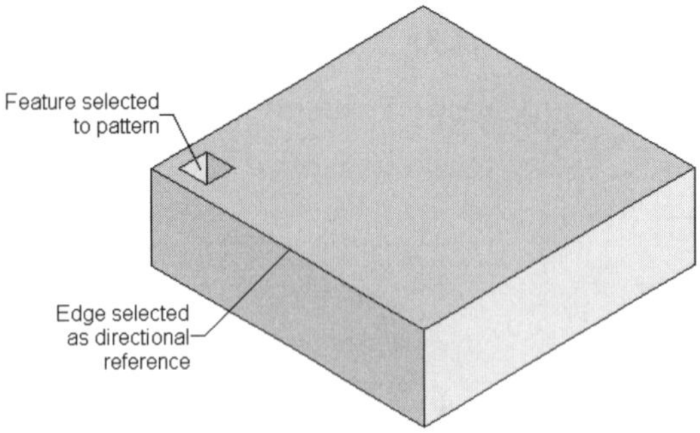

Figure 7-13 *Feature and the edge to be selected*

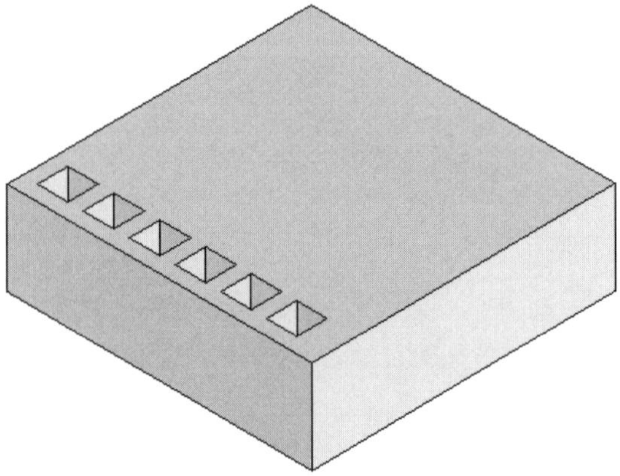

Figure 7-14 *Linear pattern created using the **Direction 1** option*

Tip. *When you select the feature to be patterned, its dimensions are also displayed in the drawing area. You can also select the dimensions as the directional reference.*

As discussed earlier, you can mirror the faces and bodies. In addition you can also pattern them. It should be noted that the selected faces must form a closed body, otherwise the patterning of faces will give an error.

Linear Pattern in Two Directions

As discussed earlier, you can create a linear pattern of features, faces, and bodies by defining a single direction using the **Direction 1** rollout. You can also define the parameters in the **Direction 2** rollout to define the pattern in the second direction. The **Direction 2** rollout is shown in Figure 7-15. When you define the pattern in the second direction, the entire row created by specifying the parameters in the first direction is patterned in the second direction. To create a pattern by specifying the parameters in both the directions, select the feature to be patterned, and invoke the **Linear Pattern PropertyManager**. Select the first directional reference. Next, specify the parameters in the **Direction 1** rollout. Now, select the second directional reference. If the **Direction 2** rollout is not invoked by default in the **Linear Pattern PropertyManager**, use the blue arrow in the **Direction 2** callout to invoke it. The options available in the **Direction 2** rollout are the same as those discussed in the **Direction 1** rollout. Figure 7-16 shows the directional references and the feature to be selected. Figure 7-17 shows the linear pattern created using the **Direction 1** and **Direction 2** rollouts.

By default, all rows of the instances created in the first direction are patterned in the second

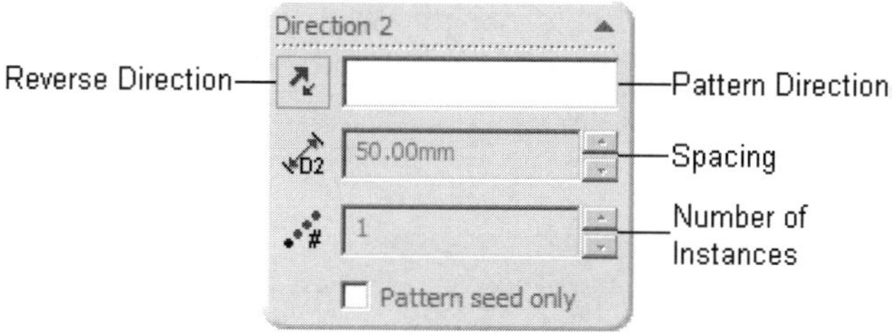

*Figure 7-15 The **Direction 2** rollout*

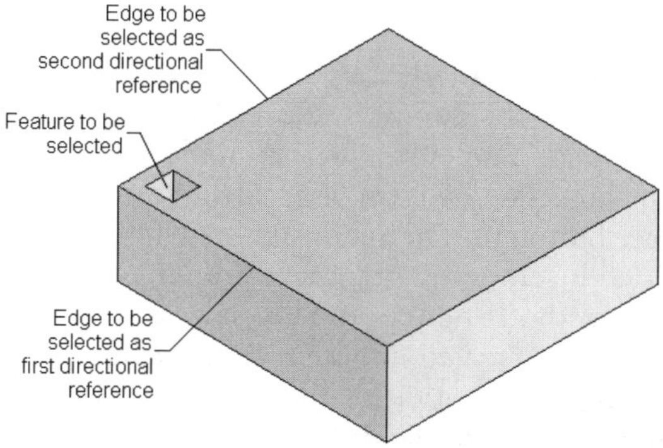

Figure 7-16 References and feature to be selected

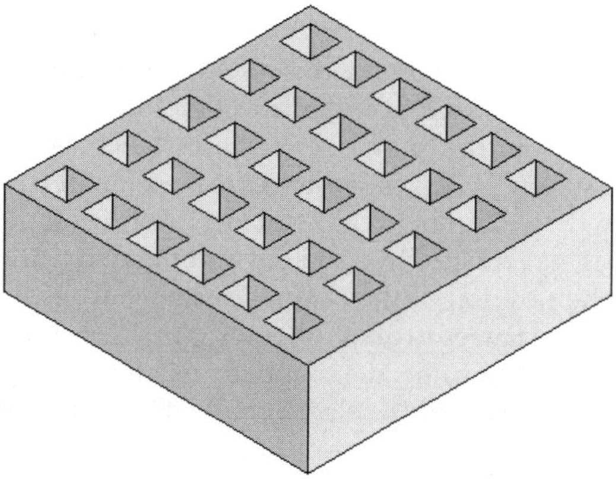

*Figure 7-17 Linear pattern created using the **Direction 1** and **Direction 2** rollouts*

direction also. This is because the **Pattern seed only** check box in the **Direction 2** rollout is cleared. You can select this check box to pattern only the original selected feature (also called seed feature) in the second direction. Figure 7-18 shows the pattern created with the **Pattern seed only** check box selected.

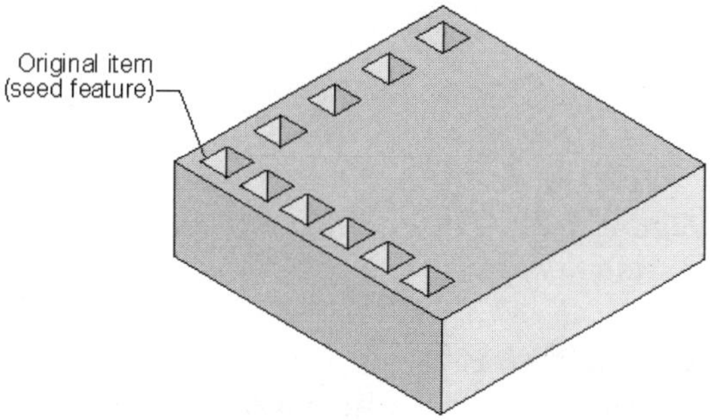

*Figure 7-18 Linear pattern created with the **Pattern seed only** check box selected*

Instances to Skip

Using the **Instances to Skip** option you can skip some of the instances from the pattern. These instances are not actually deleted. These only disappear from the pattern feature and you can resume them at any time of your design cycle. To skip the pattern instances, invoke the **Instances to Skip** rollout from the **Linear Pattern PropertyManager**. The **Instances to Skip** rollout is displayed in Figure 7-19.

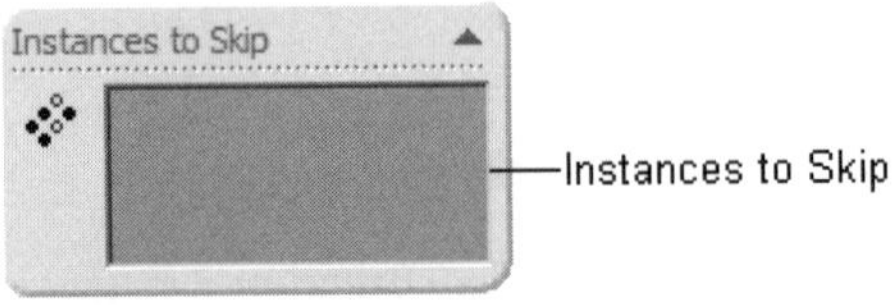

Figure 7-19 *The **Instances to Skip** rollout*

As soon as you invoke this rollout, pink dots are displayed at the center of all pattern instances except the parent instance. Therefore, you cannot skip the parent instance. Now, move the cursor to the pink dot of the instance to be skipped. The cursor will be replaced by the instance to skip cursor and the position of that instance, in the form of a matrix, is displayed in the tooltip below this cursor. Use the left mouse button to skip that instance. The pink dot will be replaced by a red dot and the preview of that instance will disappear from the pattern. The position of the skipped instance is displayed in the **Instances to Skip** selection area of the **Instances to Skip** rollout. Figure 7-20 shows a pattern created with some instances skipped. You can resume the skipped instances by deleting the position of the instance from the **Instances to Skip** selection area or selecting the red dot from the drawing area.

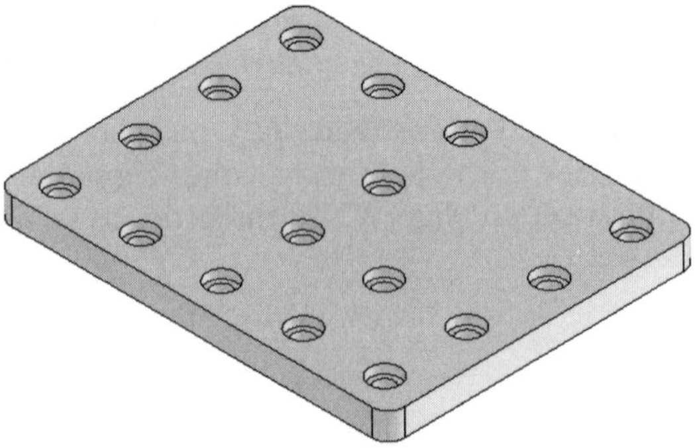

Figure 7-20 *Linear pattern created with some instances skipped*

Pattern Using a Varying Sketch

The **Vary Sketch** option is used in a pattern where the shape and size of each pattern instance is controlled by the relations and dimensions of the sketch of that feature. In this type of pattern, the dimension of the sketch of the feature to be patterned is selected as the directional reference, which drives the shape and size of the sketch of the feature to be patterned. In Figure 7-21 a cut feature is created on the base feature. Figure 7-22 shows the linear pattern created using the varying sketch option. For creating this type of pattern, the sketch of the feature to be patterned should be in relation with the geometry along which it will vary. The dimensions of the sketch should allow it to change the shape and size easily. You

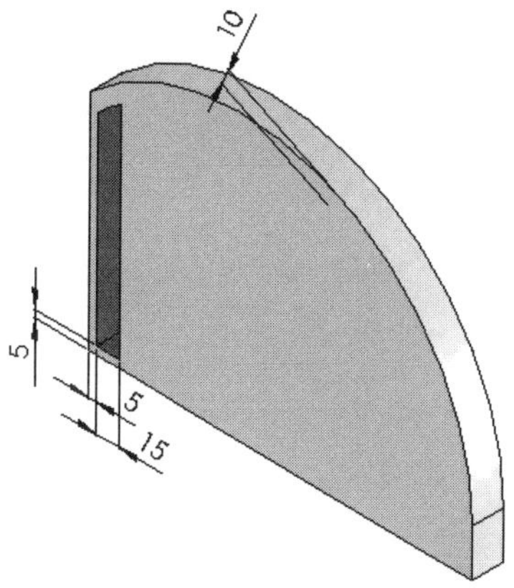

Figure 7-21 *Cut feature created on the base feature*

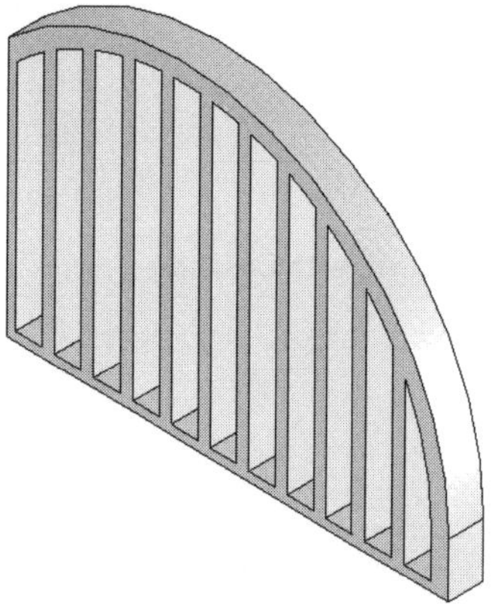

Figure 7-22 *Linear pattern created with the* ***Vary sketch*** *check box selected*

should also provide a linear dimension that will drive the entire sketch and will also be the directional reference. From the **FeatureManager Design Tree** select the feature to be patterned and then invoke the **Linear Pattern PropertyManager**. Now, select the dimension to specify the directional reference and set the value of spacing and the number of instances. Invoke the **Options** rollout and select the **Vary sketch** check box. As you select it, the preview of the pattern disappears from the drawing area. Choose the **OK** button from the **Linear Pattern PropertyManager** to end the feature creation.

The **Geometry pattern** option available in the **Options** rollout of the **Linear Pattern PropertyManager** is the same as that discussed earlier in the **Mirror PropertyManager**.

Creating a Circular Pattern Feature

CommandManager:	Features > Circular Pattern
Menu:	Insert > Pattern/Mirror > Circular Pattern
Toolbar:	Features > Circular Pattern

As discussed in the previous chapters, you can arrange the sketched entities in a circular pattern using the **Circular Sketch Step and Repeat** option. In this section, you will learn how to create the circular pattern of a feature, face, or body. You can do so, by using the **Circular Pattern** tool. To invoke this tool, choose the **Circular Pattern** button from the **Features CommandManager** or choose **Insert > Pattern/Mirror > Circular Pattern** from the menu bar. The **Circular Pattern PropertyManager** is invoked and the confirmation corner is displayed, as shown in Figure 7-23.

After invoking the **Circular Pattern PropertyManager**, you are prompted to select an edge or axis for direction reference, and a face of the feature to be patterned. If the base feature of the model is a circular feature, the temporary axis is created automatically at the center of the model. Choose **View > Temporary Axes** from the menu bar to display the temporary axis. Otherwise, before invoking the **Circular Pattern** tool, you need to create an axis. Now, select the axis that will act as the pattern axis and the feature to pattern. The preview of the pattern feature with the default values is displayed in the drawing area. The **Direction 1** callout is also

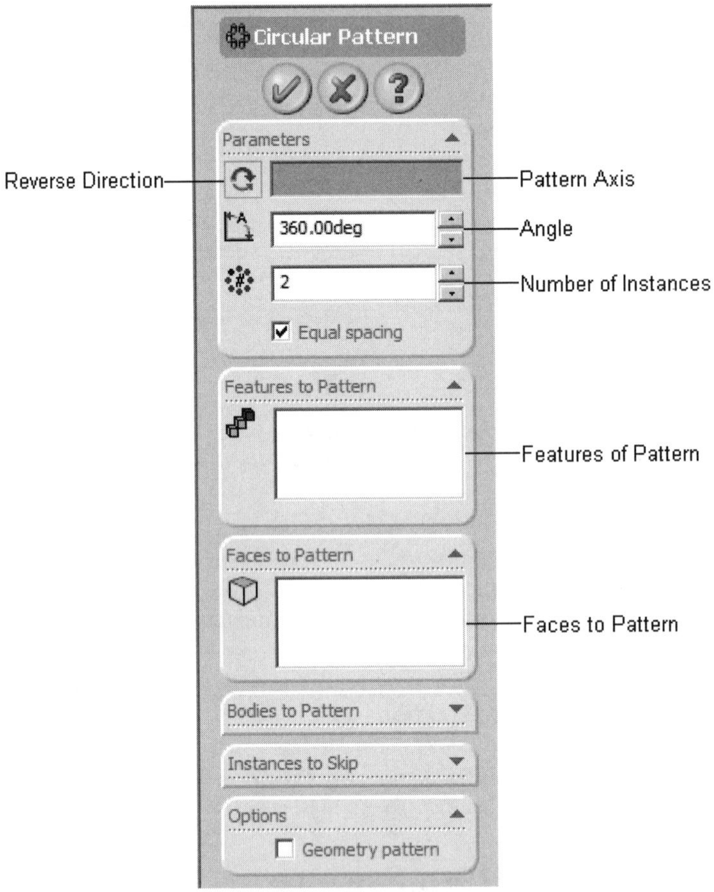

Figure 7-23** The **Circular Pattern PropertyManager

displayed with the **Reverse Direction** arrow in the drawing area. By default, the **Equal Spacing** check box is cleared. Therefore, you need to set the value of the incremental angle between the instances in the **Total Angle** spinner. Set the value of number of instances to pattern in the **Number of Instances** spinner. If you select the **Equal Spacing** check box, you need to enter the value of the total angle, along which all the instances of the pattern will be accommodated. The angular spacing between the instances will be automatically calculated.

The **Reverse Direction** button available on the left of the **Pattern Axis s**election area is used to change the direction of rotation. By default, the direction of the pattern creation is clockwise. If you choose this button then the resulting pattern will be created in the counterclockwise direction. You can also change the direction of the pattern creation using the **Reverse Direction** arrow from the drawing area. Figure 7-24 shows the feature and the temporary axis being selected. Figure 7-25 shows the resulting pattern feature.

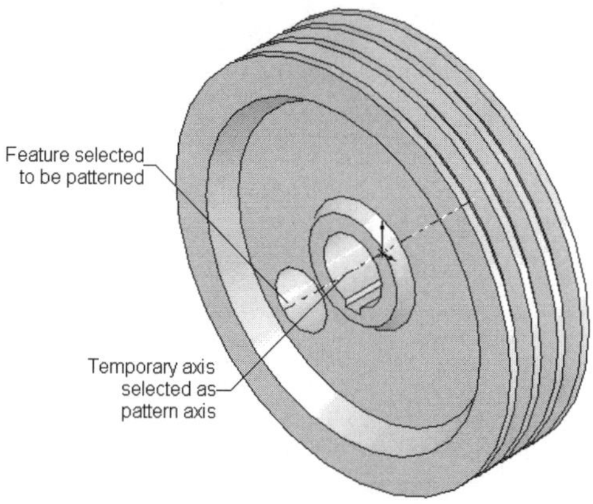

Figure 7-24 The reference to be selected for creating a circular pattern

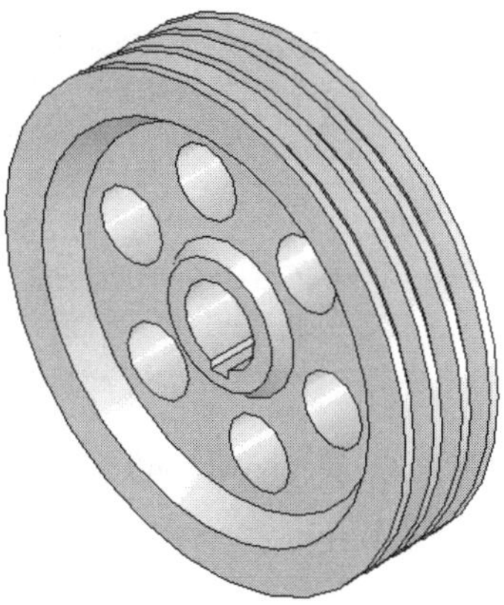

Figure 7-25 The resulting circular pattern

Circular Pattern Using a Dimensional Reference

You can also create a circular pattern by selecting an angular dimension. To create a pattern using this option you need to create an angular dimension in the sketch of the feature that is to be patterned. Invoke the **Circular Pattern PropertyManager** and select the feature to be patterned. The dimensions of the feature to be patterned will be displayed in the drawing area, as shown in Figure 7-26. Now, select the angular dimension and set the value of the total angle and spacing in the **Circular Pattern PropertyManager**. Figure 7-27 shows the pattern created by selecting the angular dimension as the angular reference. The other options available in the **Circular Pattern PropertyManager** are the same as those discussed earlier for the **Linear Pattern PropertyManager**.

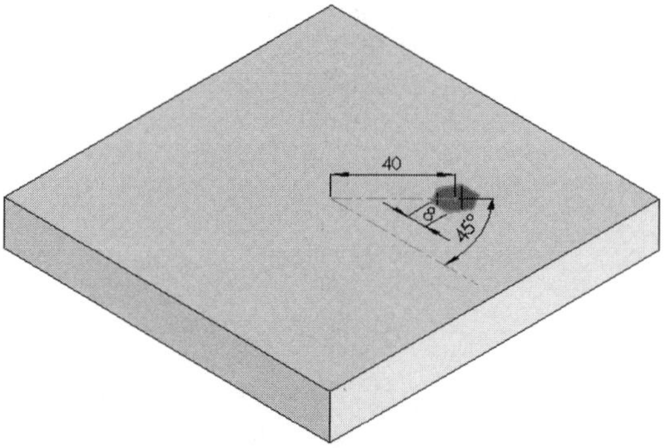

Figure 7-26 Dimensions displayed after selecting the feature to be patterned

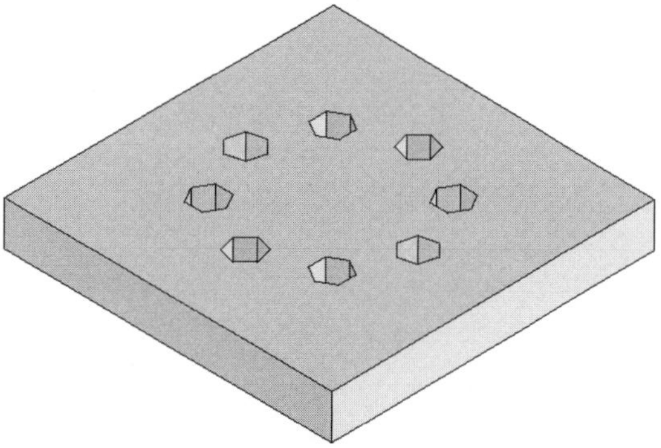

Figure 7-27 Circular pattern created by selecting the angular dimension as the angular reference

Tip. *Instead of setting the value of the angle and the number of instances in the respective spinners, you can also set the values in the callout. Entering values in the callout is a better option and a timesaving practice.*

It is always a good practice to create patterns of features instead of creating complex sketches or repeatedly creating the same feature again and again. It also helps in capturing the design intent of the model. The patterns created in the ***Part*** *mode are very useful in assembly modeling. You will learn more about it in later chapters.*

Creating a Sketch-driven Pattern

CommandManager:	Features > Sketch Driven Pattern	*(Customize to Add)*
Menu:	Insert > Pattern/Mirror > Sketch Driven Pattern	
Toolbar:	Features > Sketch Driven Pattern	*(Customize to Add)*

A sketch-driven pattern is created when the features, faces, or bodies are to be arranged in a non-uniform manner, which is neither rectangular nor circular. For creating a sketch-driven pattern, first you have to create an arrangement of the sketch points in a single sketch. This arrangement of sketch points will drive the instances in the pattern feature. After creating the feature to pattern and placing the points in the sketch, invoke the **Sketch Driven Pattern PropertyManager**, as shown in Figure 7-28.

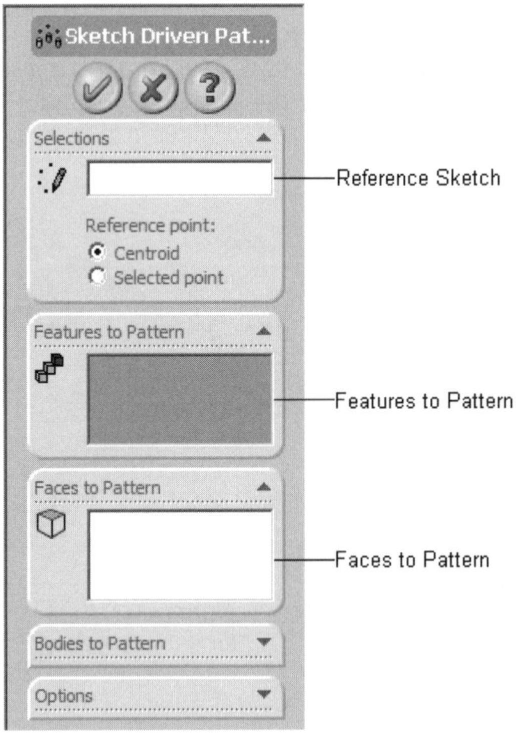

Figure 7-28 The ***Sketch Driven Pattern PropertyManager***

You are prompted to select a sketch for pattern layout, and the face of the feature to pattern. Select the feature or features to be patterned. Now, click in the **Reference Sketch** display box in the **Selections** rollout and select any one of the sketched point from the drawing area. You can also select the sketch from the **FeatureManager Design Tree** flyout. Choose the **OK** button from the **Sketch Driven Pattern PropertyManager**.

Figure 7-29 shows the feature and the sketch point to be selected and Figure 7-30 shows the resulting pattern feature.

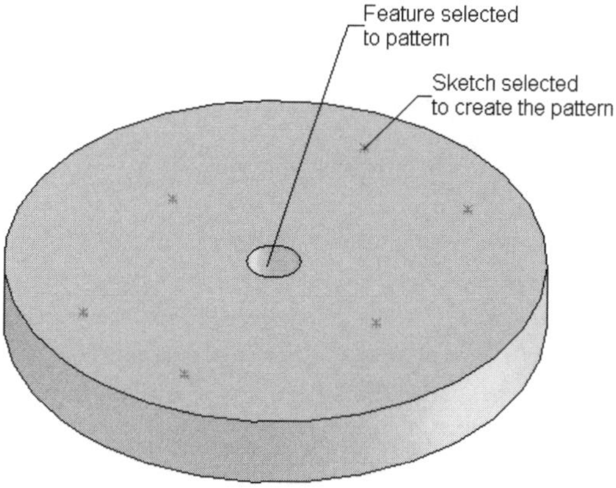

Figure 7-29 The feature and the sketch point to be selected

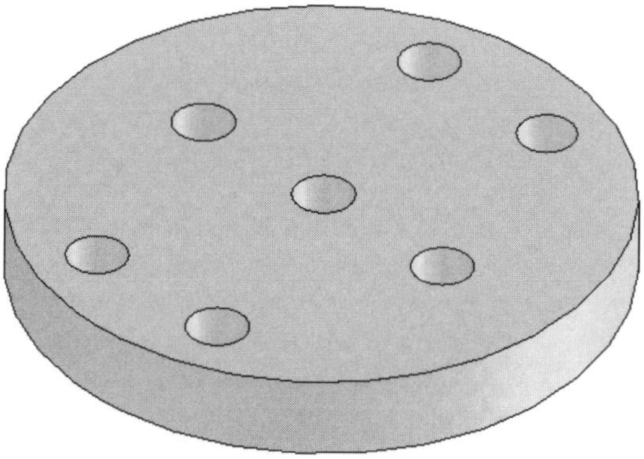

Figure 7-30 The resulting pattern feature

The options available in the **Sketch Driven Pattern PropertyManager** are discussed next.

Sketch Driven Pattern Using a Centroid

When you invoke the **Sketch Driven Pattern PropertyManager**, the **Centroid** radio button is selected by default in the **Reference Point** area of the **Selections** rollout. Therefore, the pattern created is with reference to the centroid.

Sketch Driven Pattern Using a Selected Point

If you select the **Selected point** radio button from the **Reference Point** area of the **Selections** rollout, the pattern created is with reference to the selected point. When you select this radio button, a **Reference Vertex** selection area is displayed. Select a vertex; the pattern will be created in reference to that vertex.

Creating a Curve-driven Pattern

CommandManager:	Features > Curve Driven Pattern	*(Customize to Add)*
Menu:	Insert > Pattern/Mirror > Curve Driven Pattern	
Toolbar:	Features > Curve Driven Pattern	*(Customize to Add)*

The **Curve Driven Pattern** option is used to pattern the features, faces, or bodies along a selected reference curve. The reference curve can be a sketched entity or an edge, or an open profile, or a closed loop. To create a pattern using this option, choose the **Curve Driven Pattern** button from the **Features CommandManager** or choose **Insert > Pattern/Mirror > Curve Driven Pattern** from the menu bar. When you choose this button, the **Curve Driven Pattern PropertyManager** is displayed, as shown in Figure 7-31. Figure 7-32 shows the feature and the curve that will be used to create the pattern. Figure 7-33 shows the resulting curve-driven pattern.

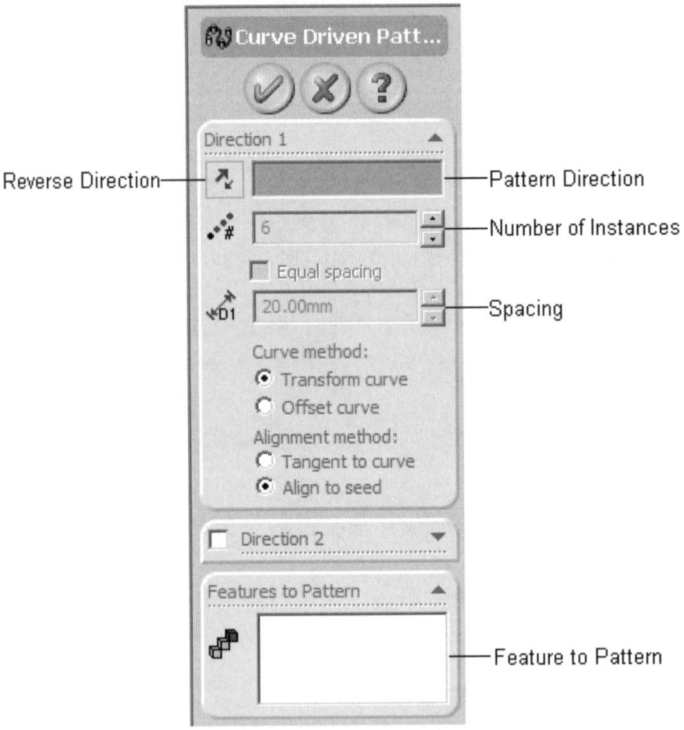

Figure 7-31 *Partial view of the* ***Curve Driven Pattern PropertyManager***

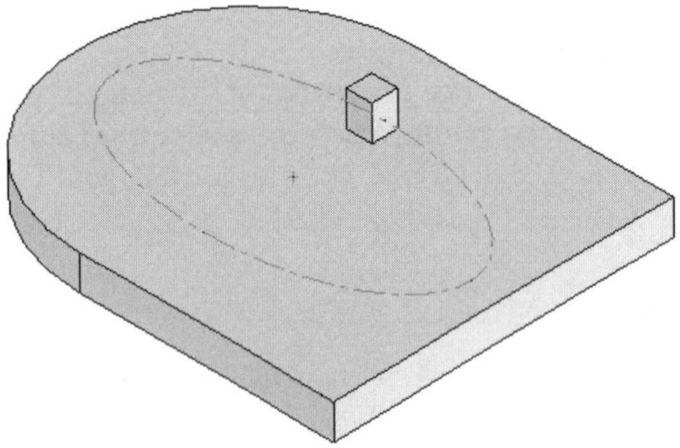

Figure 7-32 *The feature and curve to be used to create the curve-driven pattern*

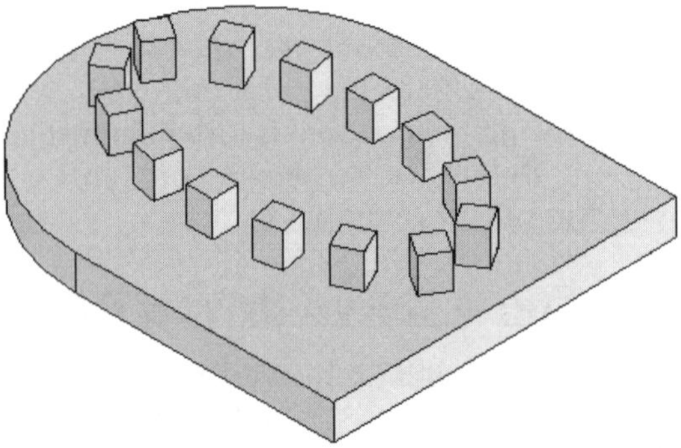

Figure 7-33 *The resulting pattern feature*

After invoking the **Curve Driven Pattern PropertyManager**, you are prompted to select an edge, curve, or sketch segment for pattern layout and select a face of feature to be patterned. Select the reference curve along which the feature, face, or body to be patterned. When you select the reference curve, its name is displayed in the **Pattern Direction** selection area and the **Direction 1** callout is also displayed. As discussed earlier, the **Direction 1** callout is divided in two areas. Select the feature to pattern; the preview of the pattern is displayed in the drawing area. Set the various parameters available in the **Direction 1** rollout and choose the **OK** button from the **Curve Driven Pattern PropertyManager**.

The various options available in the **Direction 1** rollout are discussed next.

Equal Spacing

The **Equal Spacing** check box is used to accommodate all the instances of the pattern along the selected curve. By default, this check box is cleared. Therefore, you have to specify the distance between the instances and the total number of instances to be created along the selected curve. When you select this check box, the **Spacing** spinner is not available and you have to specify only the total number of instances. The distance between the instances is calculated automatically.

Curve method and Alignment method

The **Curve method** area of the **Direction 1** rollout is used to specify the type of curve method to be followed while patterning along it. The two options available in this area are in the form of radio buttons. The options are the **Transform curve** method and the **Offset curve** method. The **Alignment method** area of the **Direction 1** rollout is used to specify the type of alignment method to be applied. The two alignment methods are the **Tangent to curve** method and the **Align to seed** method. Figure 7-34 shows the curve-driven pattern created with selected **Transform curve** and the **Tangent to curve** radio buttons. Figure 7-35 shows the curve-driven pattern created with the **Transform curve** and **Align to seed** radio buttons selected.

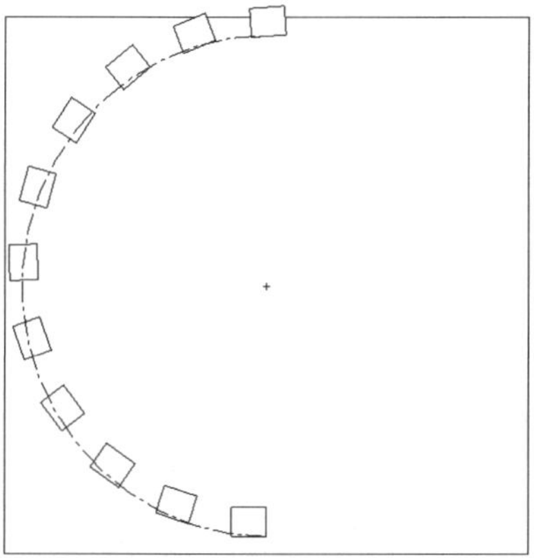

Figure 7-34 *Pattern created with **Transform curve** and **Tangent to curve** radio buttons selected*

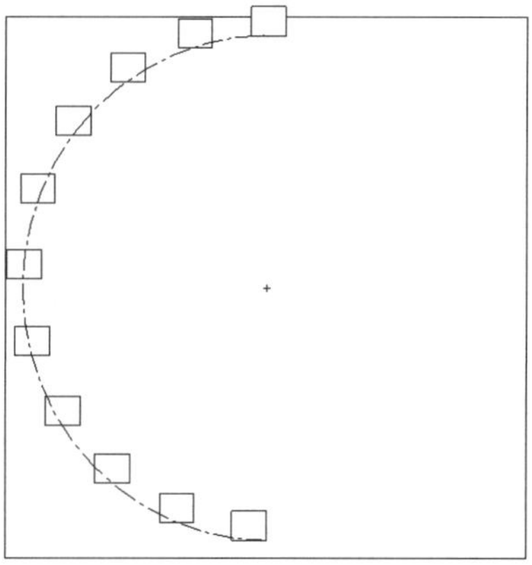

Figure 7-35 *Pattern created with **Transform curve** and **Align to seed** radio buttons selected*

Figure 7-36 shows the curve-driven pattern created with the **Offset curve** and the **Tangent to curve** radio buttons selected. Figure 7-37 shows the curve-driven pattern created with the **Offset curve** and **Align to seed** radio buttons selected.

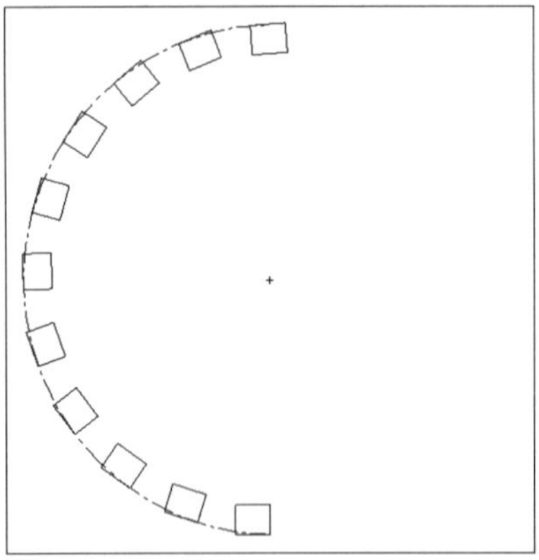

Figure 7-36 *Pattern created with* ***Offset curve*** *and* ***Tangent to curve*** *radio buttons selected*

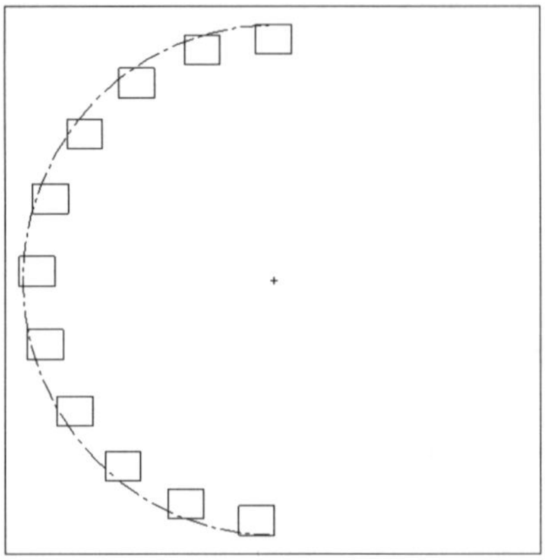

Figure 7-37 *Pattern created with* ***Offset curve*** *and* ***Align to seed*** *radio buttons selected*

Other options available in the **Curve Driven PropertyManager** are the same as those discussed earlier for the mirror and other pattern features. By selecting the check box available in the **Direction 2** rollout, you can also specify the parameters in the second direction. Figure 7-38 shows the curve-driven pattern feature created with the pattern defined in first and second direction.

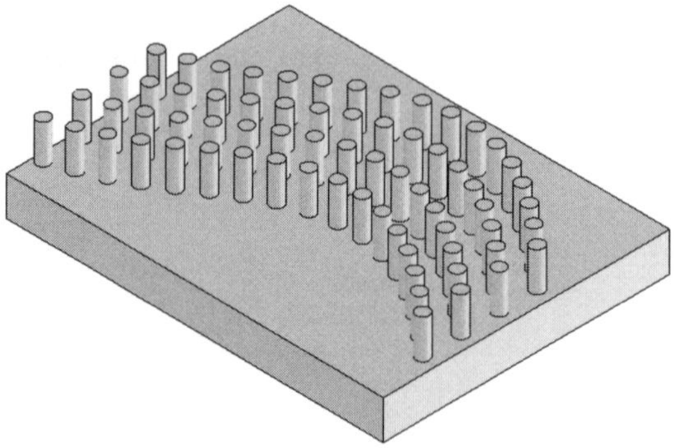

Figure 7-38 *A curve-driven pattern created by specifying parameters in both the directions*

Tip. *In SolidWorks you can pattern a pattern feature. You can also pattern a mirror feature. The mirror of the pattern feature is also possible in SolidWorks.*

Creating a Table-driven Pattern

CommandManager:	Features > Table Driven Pattern	*(Customize to Add)*
Menu:	Insert > Pattern/Mirror > Table Driven Pattern	
Toolbar:	Features > Table Driven Pattern	*(Customize to Add)*

The **Table Driven Pattern** is created by specifying the X and Y coordinates with reference to a coordinate system. The instances of the selected features, faces, or bodies are created at the points specified using the X and Y coordinates. For creating this pattern, you first need to create a coordinate system using the **Coordinate System** button from the **Reference Geometry** toolbar. The coordinate system defines the direction along which the selected feature will be patterned. Choose the **Table Driven Pattern** button from the **Features CommandManager** or choose **Insert > Pattern/ Mirror > Table Driven Pattern** from the menu bar. The **Table Driven Pattern** dialog box is displayed, as shown in Figure 7-39.

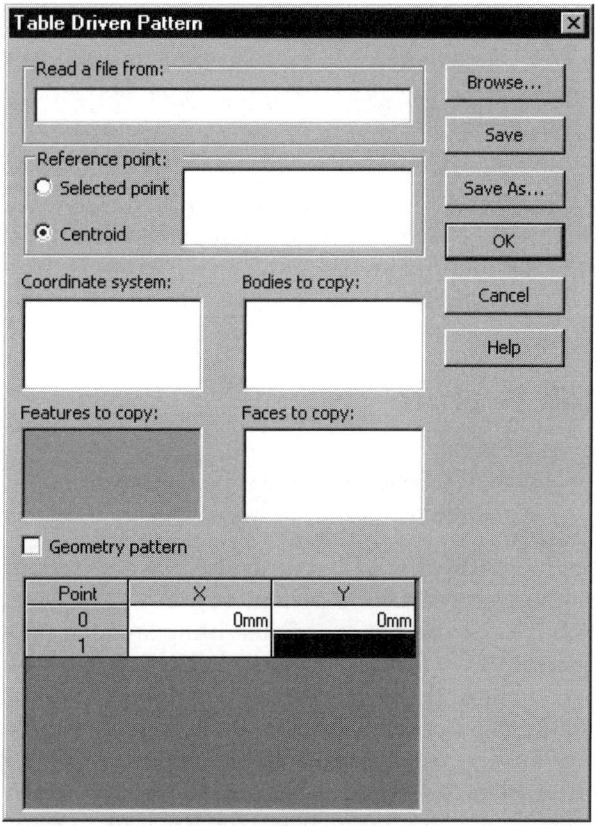

Figure 7-39 *The* ***Table Driven Pattern*** *dialog box*

Select the feature to be patterned and the coordinate system, from the drawing area or from the **FeatureManager Design Tree**. Enter the coordinates for creating the instances in the **Coordinate points** area of the **Table Driven Pattern** dialog box. As you enter the coordinates for the instances, the preview of the pattern is displayed in the drawing area. After entering all the coordinate points, choose the **OK** button from the **Table Driven Pattern** dialog box. Figure 7-40 shows the feature and the coordinate system to be selected. Figure 7-41 shows the table-driven pattern created after entering the coordinate values in the **Table Driven Pattern** dialog box.

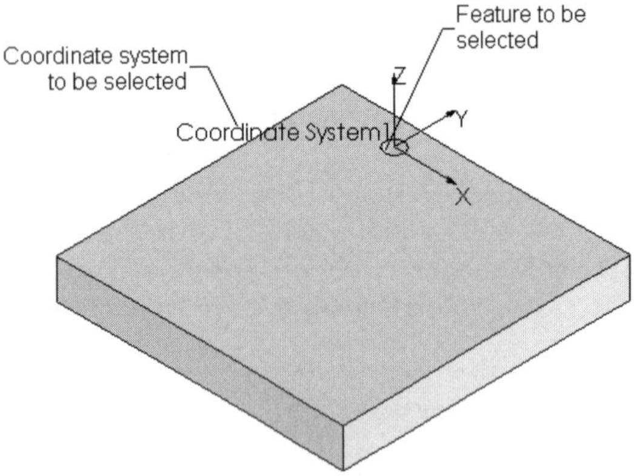

Figure 7-40 Feature and the coordinate system to be selected for creating a table-driven pattern

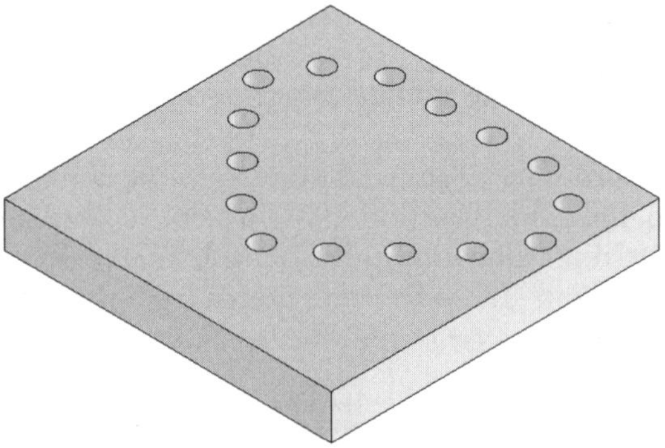

Figure 7-41 The resulting pattern after specifying the coordinate points

You can also save the table-driven pattern file and retrieve the same coordinates by simply browsing the saved file using the **Browse** button from the **Table Driven Pattern** dialog box. You can also simply write the coordinates in a text file and browse the same file while creating a table-driven pattern. The other options available in this dialog box are the same as those discussed earlier.

Creating the Rib Feature

CommandManager:	Features > Rib
Menu:	Insert > Features > Rib
Toolbar:	Features > Rib

Ribs are defined as the thin walled structures that are used to increase the strength of the entire structure of the component, so that it does not fail under an increased load. In SolidWorks, the ribs are created using an open sketch as well as a closed sketch. To create a rib feature, invoke the **Rib PropertyManager** and select the plane on which you need to draw the sketch for creating the rib feature. Draw the sketch and exit the sketching environment. Specify the rib parameters in the **Rib PropertyManager** and view the detailed preview using the **Detailed Preview** button. The **Rib** tool is invoked by choosing the **Rib** button from the **Features CommandManager** or by choosing **Insert > Features > Rib** from the menu bar. After invoking the **Rib** tool, draw the sketch and exit the sketching environment; the **Rib PropertyManager** is displayed, as shown in Figure 7-42.

The preview of the rib feature with the direction arrow and the confirmation corner is displayed in the drawing area. Figure 7-43 shows the sketch drawn for the rib feature and Figure 7-44 shows the resulting rib feature.

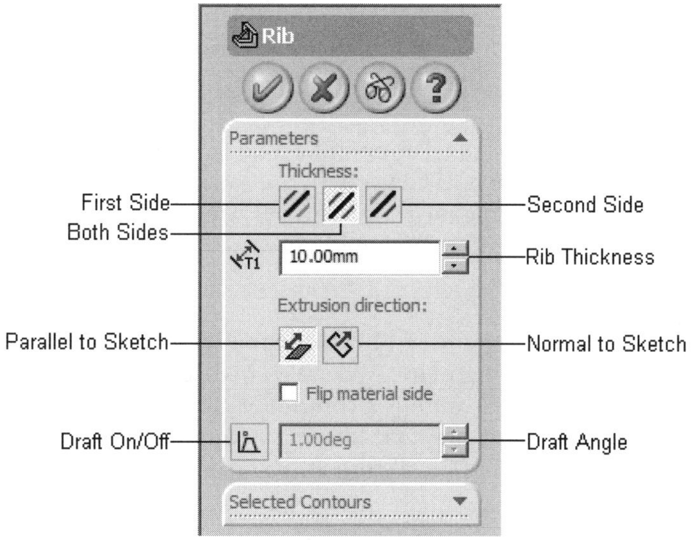

*Figure 7-42 The **Rib PropertyManager***

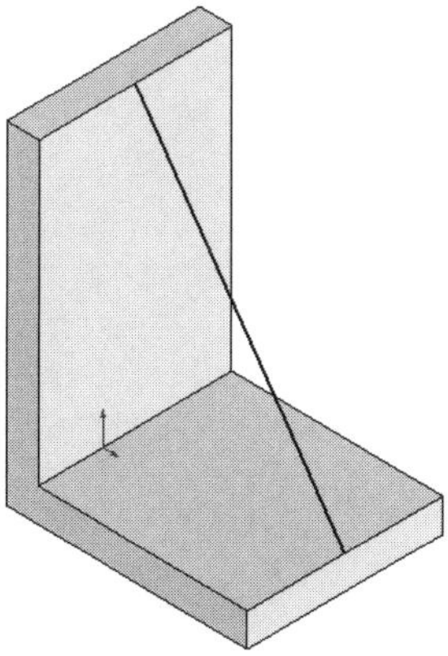

Figure 7-43 Sketch for the rib feature

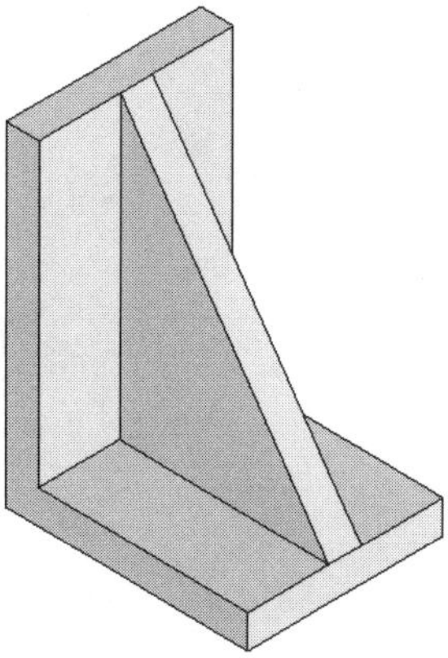

Figure 7-44 The resulting rib feature

Tip. *You can also create the rib feature by first drawing the sketch and then invoking the **Rib** tool within the sketching environment. If you exit the sketching environment after creating the sketch for a rib feature, invoke the **Rib** tool and select the sketch from the drawing area.*

The various options available in the **Rib PropertyManager** are discussed next.

Thickness

The **Thickness** area of the **Parameters** rollout is used to specify the side of the rib thickness and the thickness of the rib feature. The buttons available in this area are used to control the side on which you want to add the rib thickness. By default, the **Both Sides** button is chosen. Therefore, the rib is created on both sides of the sketch. You can choose the **First Side** or

Second Side button to create ribs on either side of the sketch. The **Rib Thickness** spinner in this area of the **Parameters** rollout is used to specify the rib thickness.

Extrusion direction

The **Extrusion direction** area of the **Parameters** rollout is used to specify the method of extruding the closed or open sketch. When you invoke the **Rib PropertyManager**, by default the option that is suitable for creating the rib feature will be active, depending on the geometric conditions. The options available in this area are discussed next.

Parallel to Sketch

The **Parallel to Sketch** option is used to extrude the sketch in a direction that is parallel to both the sketch and sketching plane. When you invoke the **Rib PropertyManager** and the sketch created for the rib feature is a continuous single entity open sketch, then this option is selected by default. Figure 7-45 shows an open sketch suitable for creating a rib using the **Parallel to Sketch** option. Figure 7-46 shows the rib feature created using the sketch.

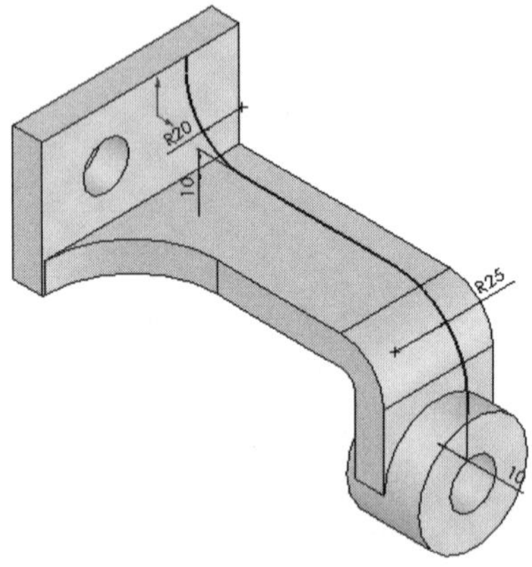

Figure 7-45 *An open sketch for the rib feature*

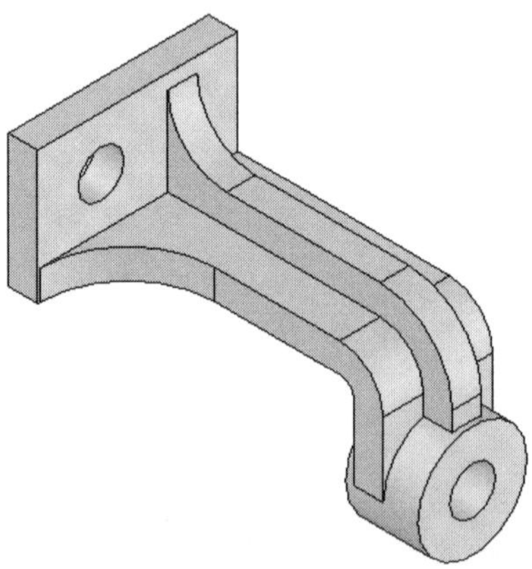

Figure 7-46 *The resulting rib feature*

Normal to Sketch

The **Normal to Sketch** option is used to create a rib feature when the sketch of the rib feature is a closed loop sketch, or it consists of multiple sketched entities. The sketch with multiple entities can be closed loops or open profiles. If you draw a sketch with a closed loop or with multiple sketched entities and invoke the **Rib** tool, the **Normal to Sketch** button will be selected by default. You can also choose the **Normal to Sketch** button from the **Extrusion Direction** area to use this option. Figure 7-47 shows a multiple entity sketch for the rib feature. Figure 7-48 shows the resulting rib feature.

When this option is selected, the **Type** area is displayed under the **Draft Angle** spinner. The options available in the **Type** area are discussed next.

Type

The **Type** area is available only when you choose the **Normal to Sketch** button from the **Extrusion direction** area of the **Parameters** rollout. The **Type** area is provided

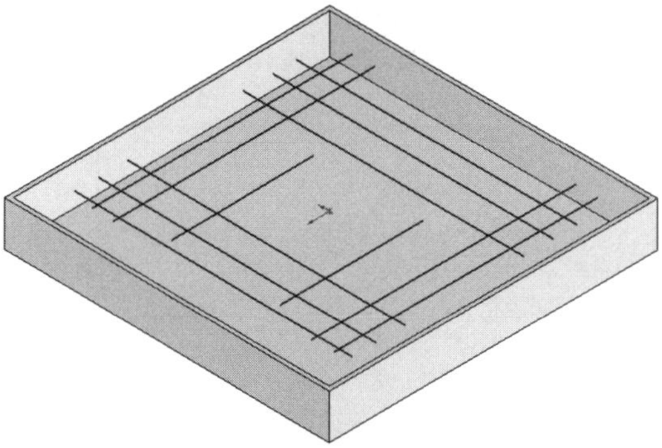

Figure 7-47 Multiple entities for the rib feature

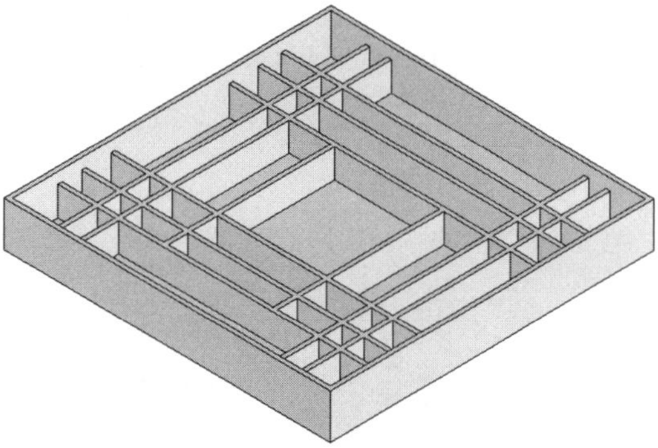

Figure 7-48 The resulting rib feature

Tip. *You will observe that the endpoints of the sketched lines drawn in Figure 7-47 do not merge with the model edges. However, the rib created using this sketch merges with the model edges. This is because while creating the sketch for the rib feature, you do not have to create a complete sketch. The ends of the rib feature automatically extend to the next surface.*

with two radio buttons, **Linear** and **Natural**. These radio buttons are used if the endpoints of the open sketch for the rib are not coincident with the faces of the existing feature. If the **Linear Radio** button is selected, the rib is created by extending the sketch normal to the sketched entity direction. The sketch will be extended up to a point where it meets a boundary.

On the other hand, if the **Natural** radio button is selected, the rib feature is created by extending the sketch along the direction of the sketched entities. For example, consider the sketch shown in Figure 7-49 which shows a multiple entity sketch created for the rib feature.

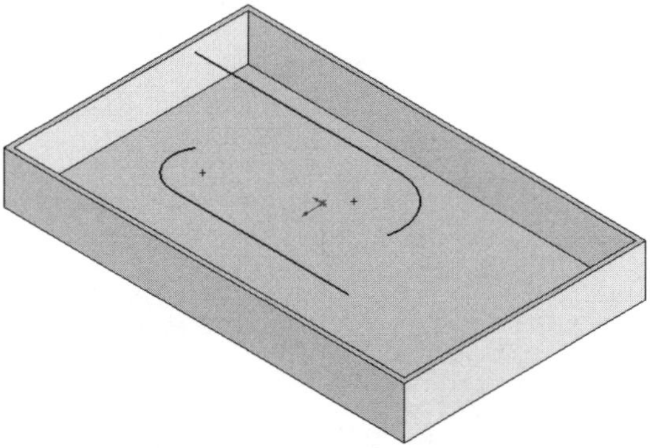

Figure 7-49 Sketch for the rib feature

Figure 7-50 shows a rib feature created by extending the sketch normal to the arc and the line. This is because the **Linear** radio button is selected. Similarly, in Figure 7-51, the feature is created by extending the sketch along the line and arc using the **Natural** radio button. This is the reason a circular feature is created at the end where the sketch has the arc.

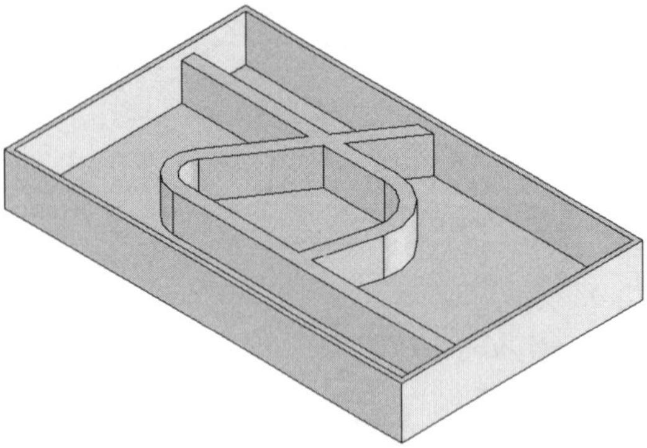

Figure 7-50 *Rib feature created with* ***Linear*** *radio button selected from the* ***Type*** *area of the* ***Rib PropertyManager***

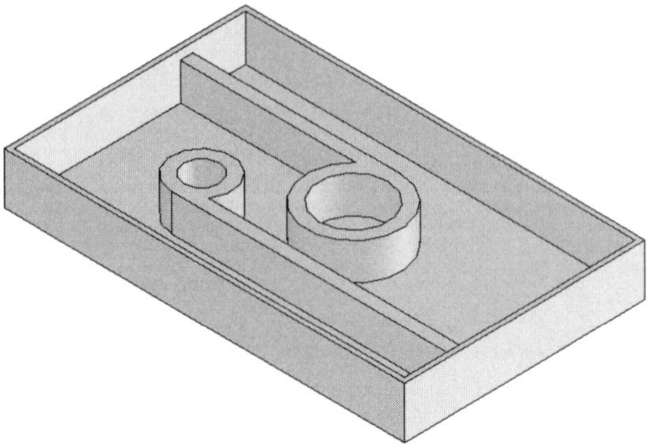

Figure 7-51 *Rib feature created with* ***Natural*** *radio button selected from the* ***Type*** *area of the* ***Rib PropertyManager***

Flip material side

The **Flip material side** check box is selected to reverse the direction of material addition, while creating the rib feature. You can also reverse the direction of material addition using the **Flip material side** arrow available in the drawing area.

Draft On/Off

The **Draft On/Off** button is used to add the taper to the faces of the rib feature. When you choose the **Draft On/Off** button, the **Draft Angle** spinner is invoked. If you are creating a rib feature using multiple sketched entities, you can add only a simple draft to it. Figure 7-52 shows the draft angle added to the rib feature. By default, the draft is added inwards to the rib feature.

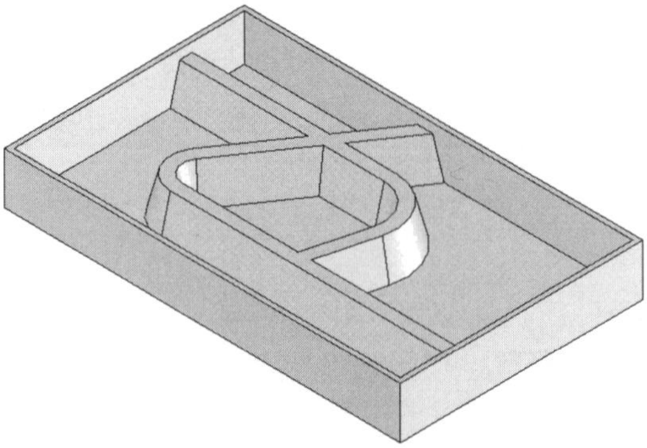

Figure 7-52 *Draft angle added to the rib feature*

Using **Draft outward** check box you can add the draft outwards. If the rib feature to be created consists of a single continuous sketch and if you choose the **Draft On/Off** button, the **Next Reference** button is displayed below the **Draft Angle** spinner. A reference arrow is also displayed in the drawing area. Using the **Next Reference** button you can cycle through the reference along which you want to add the draft angle.

Figure 7-53 shows the preview of the rib feature and Figure 7-54 shows the resulting rib feature with the **Draft outward** check box selected.

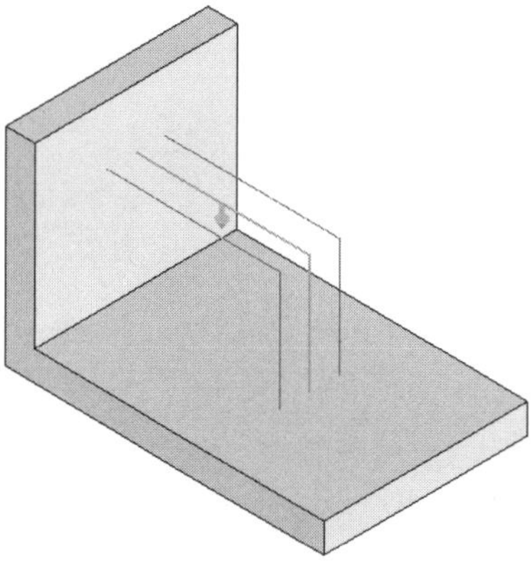

Figure 7-53 *Preview of the rib feature*

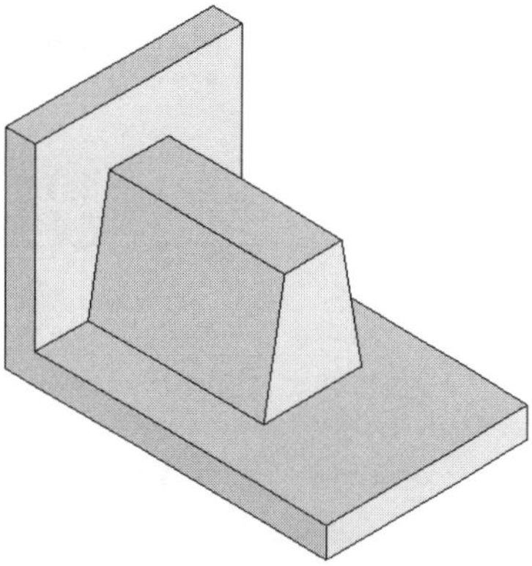

Figure 7-54 *The resulting rib feature with the* ***Draft outward*** *check box selected*

Figure 7-55 shows the preview of the rib feature and Figure 7-56 shows the resulting rib feature with the **Draft outward** check box cleared.

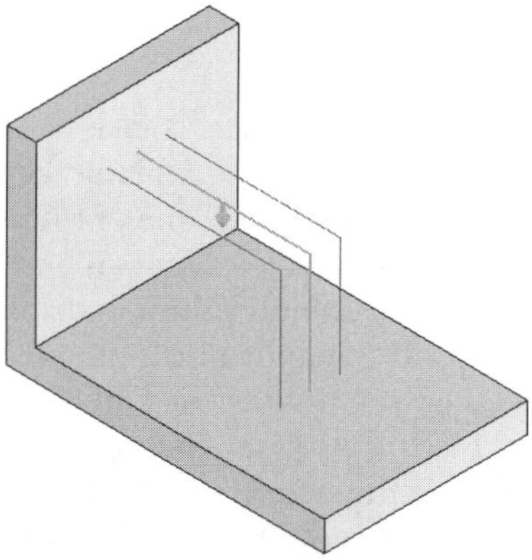

Figure 7-55 *Sketch selected to create a rib feature with an inward draft*

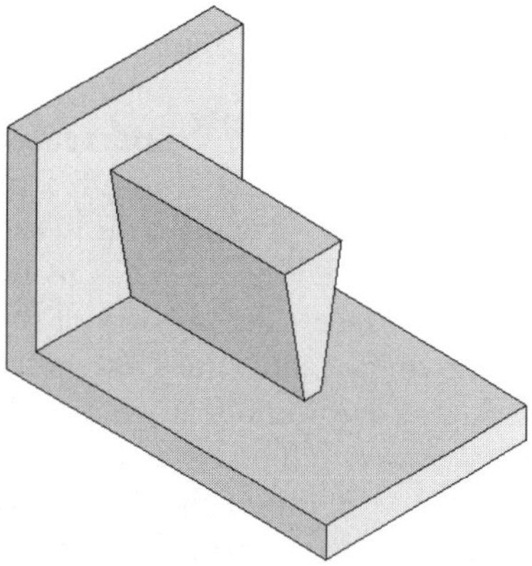

Figure 7-56 *The resulting rib feature*

Creating the Dome Feature

CommandManager:	Features > Dome	*(Customize to Add)*
Menu:	Insert > Features > Dome	
Toolbar:	Features > Dome	(Customize to Add)

Using the **Dome** tool you can create a dome feature on the selected face. Depending upon the feature creation direction, a dome can be of a convex or concave shape. To create a dome feature choose the **Dome** button from the **Features CommandManager**, or choose **Insert >Features > Dome** from the menu bar. When you choose this button, the **Dome PropertyManager** is displayed, as shown in Figure 7-57.

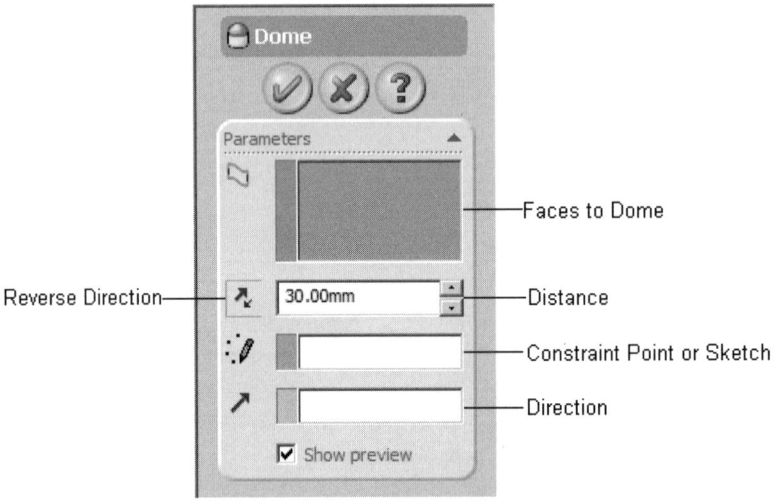

*Figure 7-57 The **Dome PropertyManager***

After invoking the **Dome PropertyManager**, you are prompted to select a face or faces on which you want to add the dome feature. The face to be selected can be a planar or curved face. Select the face on which you need to create the dome feature; the name of the selected face will be displayed in the **Faces to Dome** selection area. The selected face can be a planar face or a curved face. The preview of the dome feature, with the default values, is displayed in the drawing area. Using the **Distance** spinner, you can set the height of the dome feature. The preview of the dome feature modifies dynamically when you modify the height using the **Distance** spinner. The height of the dome feature is calculated from the centroid of the selected face to the top of the dome feature. After specifying the height of the dome feature, choose the **OK** button from the **Dome PropertyManager**. Figure 7-58 shows the planar face to be selected for the dome feature. Figure 7-59 shows the resulting dome feature.

If the selected planar face belongs to a circular or an elliptical feature, the **Elliptical dome** check box is displayed in the **Dome PropertyManager**. You can create an elliptical dome by selecting this check box. Figure 7-60 shows a circular dome created by selecting a circular planar face. Figure 7-61 shows an elliptical dome created by first selecting the **Elliptical dome** check box and then a circular planar face.

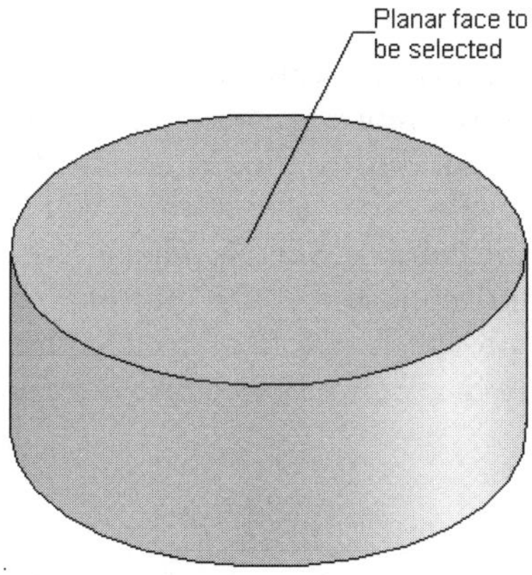

Figure 7-58 Planar face to be selected

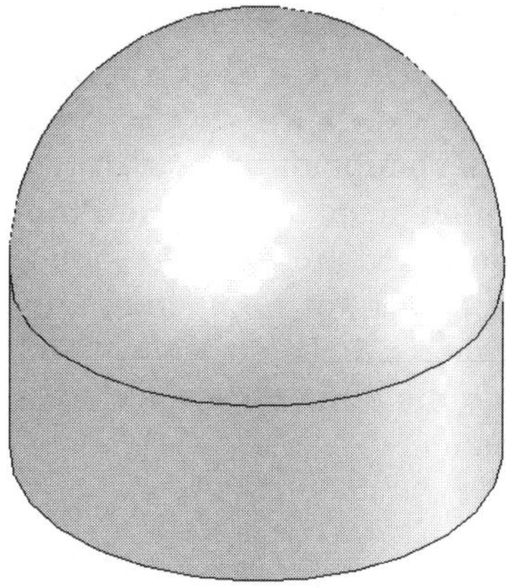

Figure 7-59 Resulting dome feature

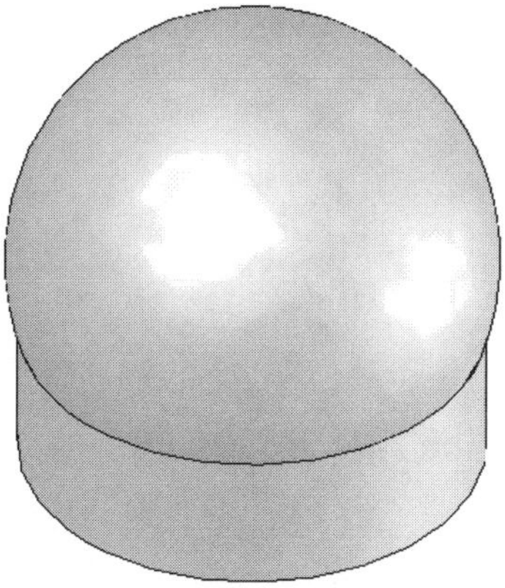

Figure 7-60 Circular dome

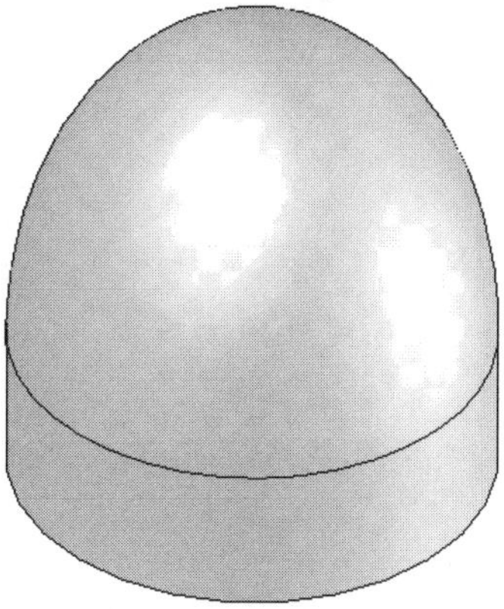

Figure 7-61 Elliptical dome

The **Reverse direction** button, available in the **Dome PropertyManager**, is used to create a concave dome. A concave dome is one in which the material is removed by creating a cavity in the form of a dome. For removing the material using the dome feature, select the **Reverse direction** button. Figure 7-62 shows the dome feature created with the **Reverse direction** button selected.

Other options available in the **Dome PropertyManager** are discussed next.

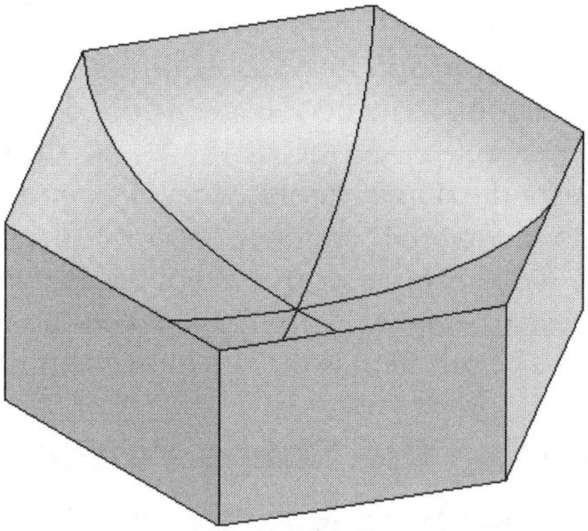

Figure 7-62 *A dome feature created with the **Reverse direction** button selected*

Constraint Point or Sketch

The **Constraint Point or Sketch** option is provided in the **Dome PropertyManager** to constraint the dome creation to the selected reference. The selected reference can be a point, a sketched point, or the endpoint of an entity. To create a dome feature using the **Constraint Point or Sketch** option, invoke the **Dome PropertyManager** and select the face on which you need to add the dome feature. Now, click once in the **Constraint Point or Sketch** selection area and select the constraint point from the drawing area and choose the **OK** button from the **Dome PropertyManager**. Figure 7-63 shows the face on which you need to add the dome feature and the point to be selected as the constraint point. Figure 7-64 shows the resulting dome feature.

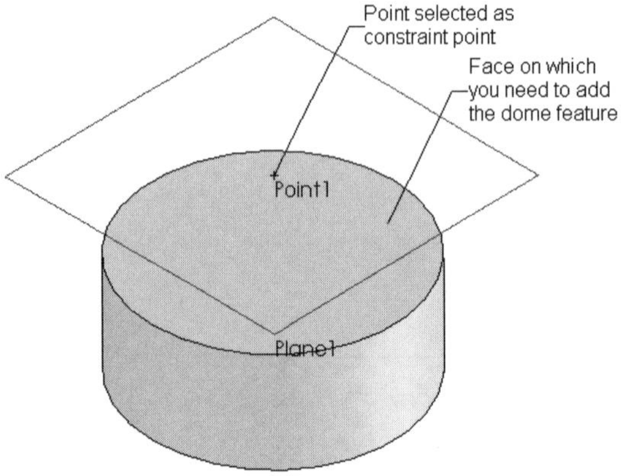

Figure 7-63 *Face and the point to be selected*

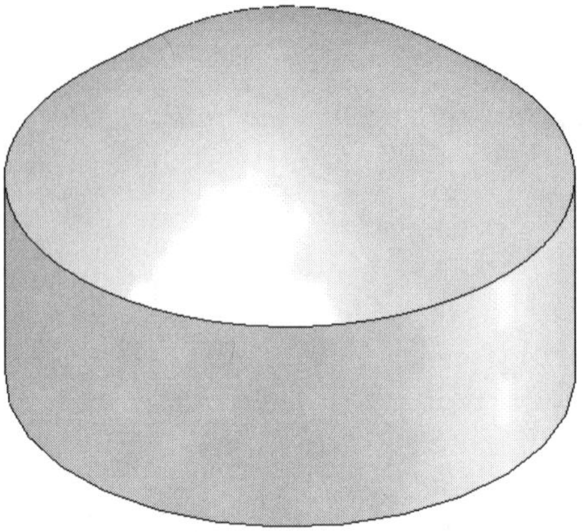

Figure 7-64 *Resulting dome feature*

Note

The dome feature will be created depending upon the height of the dome and the position of the constraint point.

Direction

The **Direction** option available in the **Dome PropertyManager** is used to specify the direction vector, in which you need to create the dome feature. For creating the dome feature by defining the direction vector, invoke the **Dome PropertyManager**. Select the face on which you need to create the dome feature. Now, click once in the **Direction** selection area and select an edge as the directional reference from the drawing area. The selected edge will be displayed in cyan. Set the value of the height of the dome and choose the **OK** button from the **Dome PropertyManager**. Figure 7-65 shows the faces on which you need to add the dome and the edge to be selected as the direction vector. Figure 7-66 shows the resulting dome.

Creating the Shape Feature

CommandManager:	Features > Shape	*(Customize to Add)*
Menu:	Insert > Features > Shape	
Toolbar:	Features > Shape	(Customize to Add)

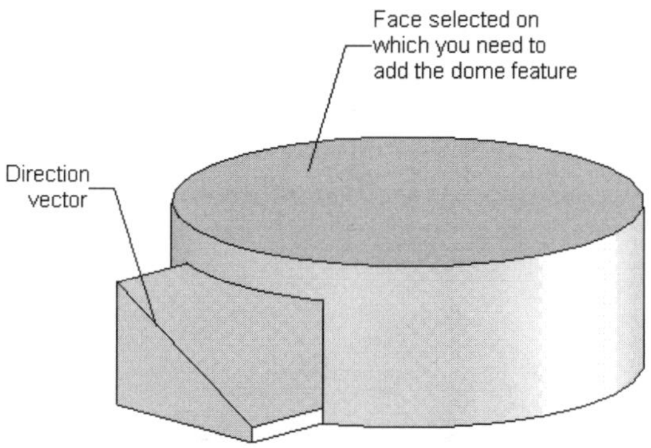

Figure 7-65 Face and direction to be selected

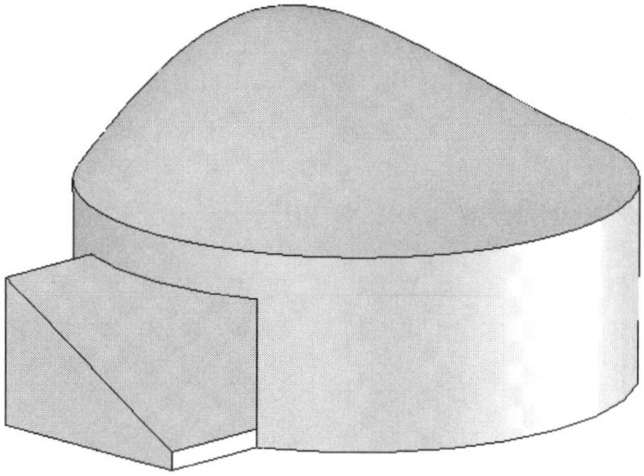

Figure 7-66 Resulting dome feature

SolidWorks provides you with a free form styling tool named the **Shape** tool, using which you can create free form shapes by manipulating the faces of the models. This free form styling tool is named as the **Shape** tool. On using the **Shape** tool, the selected face acts like a rubber membrane on which pressure is applied. As a result the face inflates or deflates like a balloon. You can also adjust the shape of the membrane by changing its stretch and bend characteristics. To create a shape feature, choose **Insert > Features > Shape** from the menu bar, or select the **Shape** button from the **Features CommandManager** after customizing it. When you invoke this tool, the **Shape Feature** dialog box is displayed, as shown in Figure 7-67.

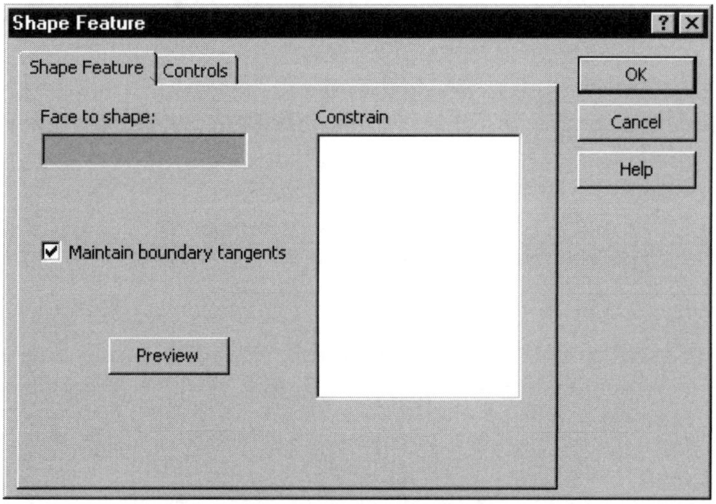

*Figure 7-67 The **Shape Feature** dialog box*

After invoking the **Shape Feature** dialog box, select the face that you need to manipulate. The selected face will be highlighted in green and its name will be displayed in the **Face to shape** selection area. You can also constrain the selected face along a curve, sketch, or an edge. To select the constraining curve, click once in the **Constrain to** selection area and select the constraining reference from the drawing area. The **Maintain boundary tangents** check box is selected by default and is used to maintain the tangency between the shape feature and the boundary of the selected face. Choose the **Preview** button to display the preview of the shape feature. The preview of the shape feature in the form of surface mesh will be displayed in the drawing area. Figure 7-68 shows the face to be manipulated and the sketch that will constrain the shape feature. Figure 7-69 shows the preview in the form of a mesh.

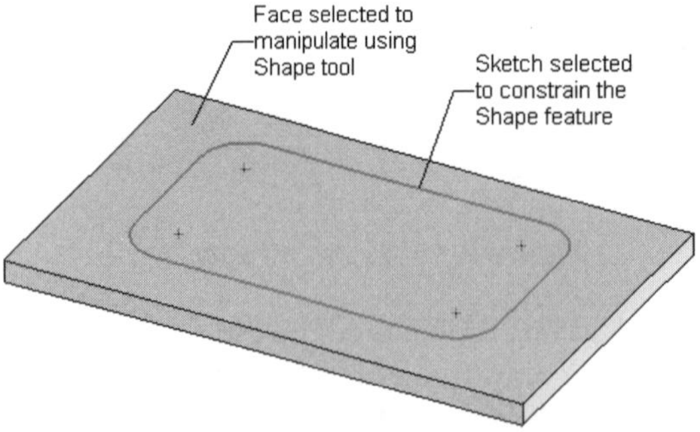

Figure 7-68 *Face to be manipulated and the sketch to constrain the shape feature*

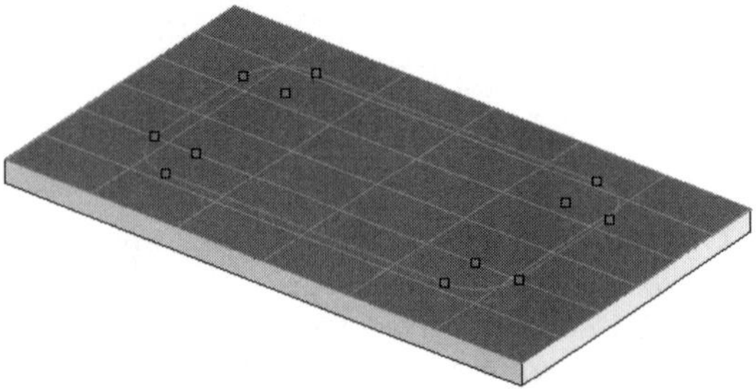

Figure 7-69 *Preview in the form of a mesh*

Tip. *The constraint that will constrain the shape feature can be a point, sketch point, endpoint, vertex, sketch, edge, or a curve. The constrain can be created on the same face or another face at an offset, or angle.*

Next, you need to set the controls to manipulate the shape feature. Choose the **Control** tab to set its parameters. The options available in the **Control** tab are discussed next.

Gains

The **Gains** area of the **Control** tab is used to set the pressure and value of the curve influence, while creating the shape feature. These options are discussed next.

Pressure

The **Pressure** slider available in the **Gains** area is used to specify the pressure on the selected face. By default the slider is set at **0**. The positive pressure inflates and the negative pressure deflates it.

Curve Influence

The **Curve Influence** slider is used to specify the percentage of the influence of the constraining curve on the shape feature creation.

Characteristics

The **Characteristics** area is provided with the options, such as stretch and bend, that are used to adjust the smoothness of the shape feature.

Advanced controls

The **Advanced controls** area is used to set the resolution of the mesh. Using the **Resolution** slider available in the **Advanced controls** area, you can set the value of the resolution. The higher the resolution, the better will be the shape of the feature and more the time it will take to rebuild.

Figure 7-70 shows the model with top face manipulated using the shape feature.

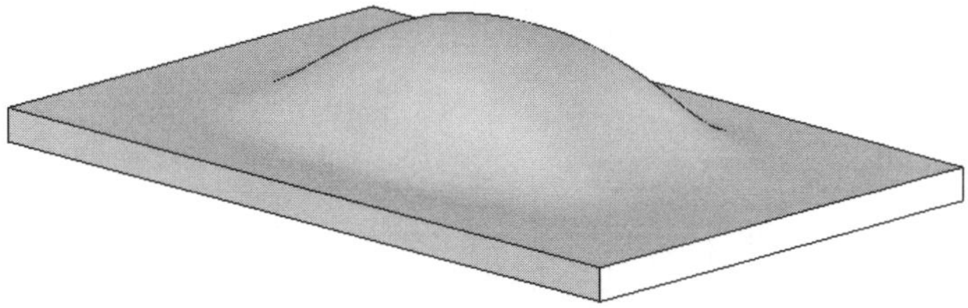

***Figure 7-70** Shape feature added on the top face of the model*

Creating Indents

CommandManager:	Features > Indent	*(Customize to Add)*
Menu:	Insert > Features > Indent	
Toolbar:	Features > Indent	(Customize to Add)

The **Indent** tool is introduced in this release of SolidWorks. This tool is extremely useful for the packaging industry, industrial designers, and product designers. You will learn about its benefits, while going through the examples in this section. The **Indent** tool is used to add or scoop out material defined by the geometry a body, from a target body. The body that is used to add or scoop out material is known as the tool body. Depending upon the geometry, the tool is sometimes is used to add as well as scoop out material from the target body.

To use this tool, choose the **Indent** button from the **Features CommandManager,** the **Indent PropertyManager** is displayed, as shown in Figure 7-71. The options available in this dialog box are discussed next.

Selections

The options available in the **Selections** rollout are used to select the target body, tool body, and set some parameters. These option are discussed next.

Target Body

After invoking the **Indent PropertyManager**, you need to select the body that will be deformed using this tool. This selected body is known as the target body. The name of the selected body is displayed in the **Target Body** selection area.

Tool Body Region

After selecting the target body, you need to select a body that will be used as the tool for deforming the target body. This body is known as tool body. Select the body that needs to be defined as the tool body. The selected portion of the body is highlighted and its name

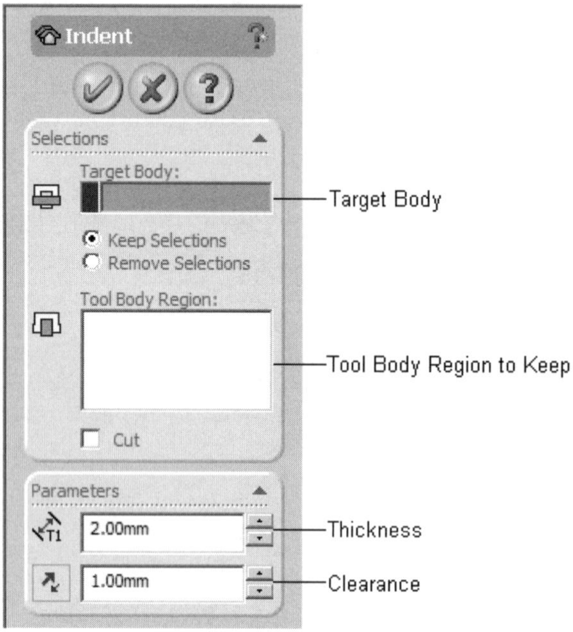

*Figure 7-71 The **Indent PropertyManager***

is displayed in the **Tool Body Region** selection area. You will notice that by default, the **Keep Selections** radio button is selected. Therefore, the selected portion of the tool body will be retained and the other portion will be removed. If you select the **Remove Selections** radio button, the selected portion of the tool body will be removed and the remaining portion will be retained.

Cut

The **Cut** check box is selected to create a cut in the target body. This cut is defined by the geometry of the tool body.

Parameters

The **Parameters** rollout is used to define the value of clearance and thickness in the **Clearance** and **Thickness** spinners, respectively. Refer to Figure 7-72 to know more about clearance and thickness. On selecting the **Cut** check box from the **Selections** rollout, the **Thickness** spinner is frozen.

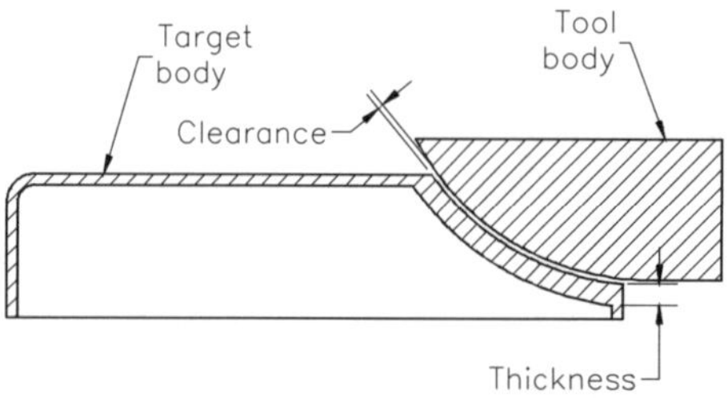

*Figure 7-72 The **Indent PropertyManager***

Figure 7-73 shows two different bodies, with the smaller rectangular body placed inside the main body. Figure 7-74 shows the target body and the portion of the tool body to be selected. Figure 7-75 shows the resulting body, after removing the material using the **Indent** tool. In this figure, the display of the tool body is turned off. To turn off the display of a body, select it and invoke the shortcut menu. Choose the **Hide** option from the **Body** area of the shortcut menu. You will learn more about toggling the display of the bodies in the later chapters. Figure 7-76 shows the resulting body after rotating the view of the model.

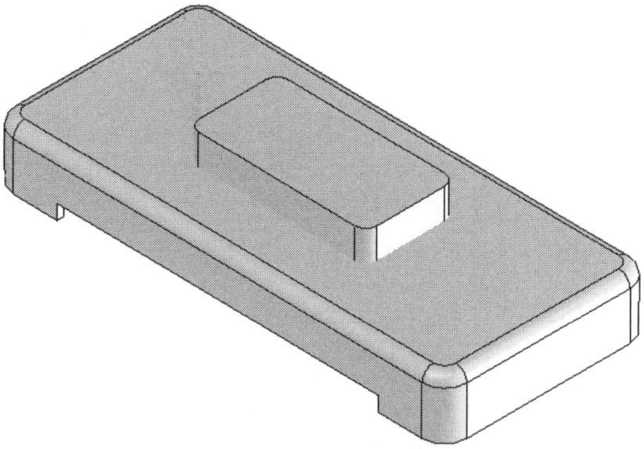

Figure 7-73 *Two separate bodies*

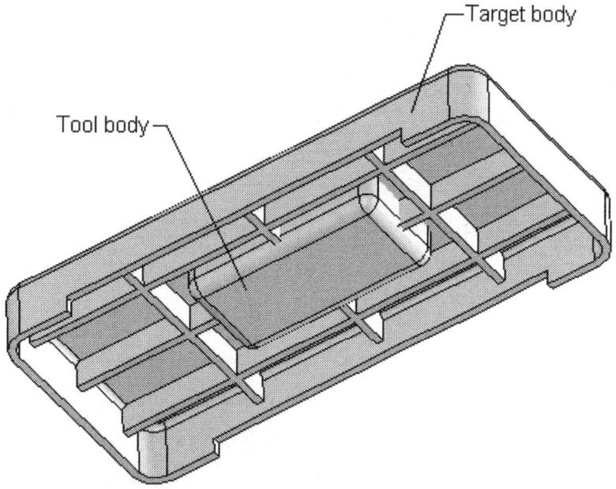

Figure 7-74 *Bodies to be selected*

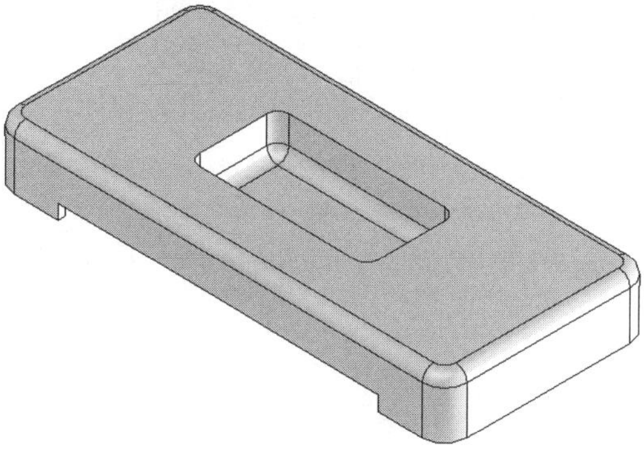

Figure 7-75 *Resulting body*

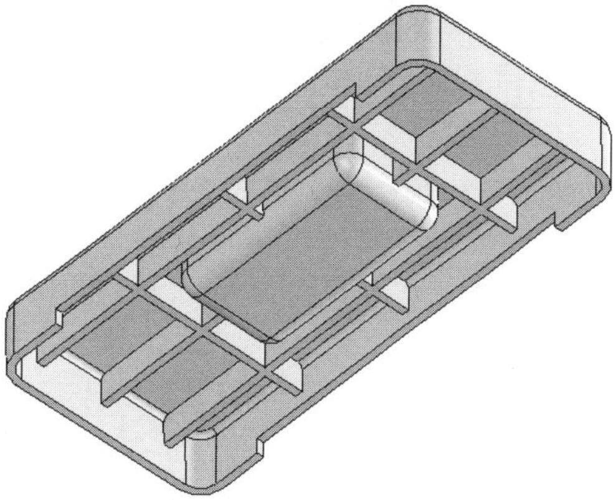

Figure 7-76 *Resulting body after rotating its view*

Figure 7-77 shows the target body and the tool body to be selected. Figure 7-78 shows the resulting body created using the **Indent** tool with the **Cut** check box selected. The display of the tool body is turned off in this figure.

Figure 7-79 shows the eggs created by revolving the sketch, and then patterned using the **Linear Pattern** tool. Figure 7-80 shows the base plate created for the egg tray as a separate body and placed with the eggs. In this example, the base plate is selected as the target body and all eggs are selected from the bottom portion as tool bodies. Figure 7-81 shows the egg tray created using the **Indent** tool. The display of all eggs is turned off in this figure.

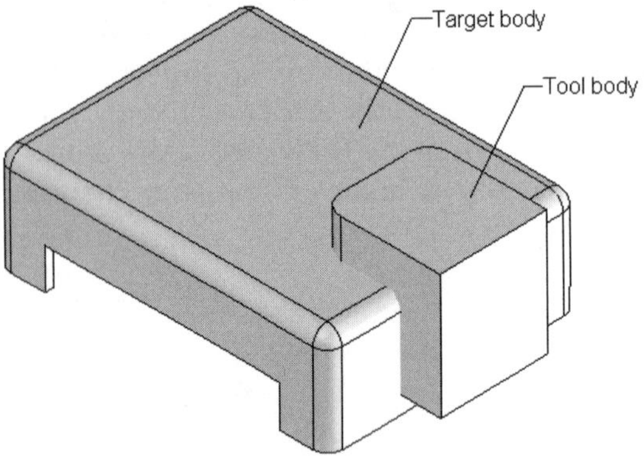

Figure 7-77 *Bodies to be selected*

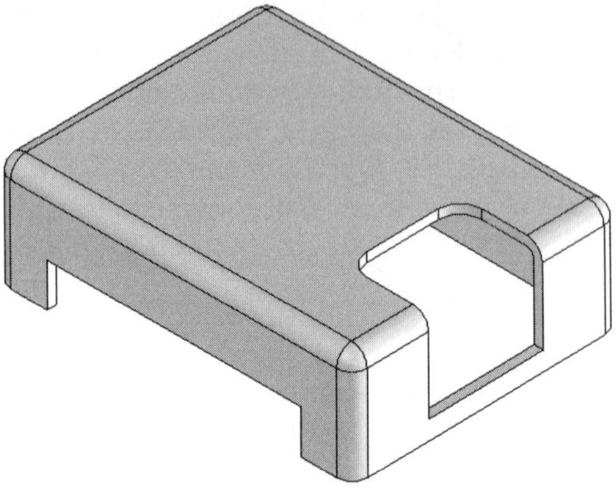

Figure 7-78 *Resulting body created using the* ***Indent*** *tool with the* ***Cut*** *check box selected*

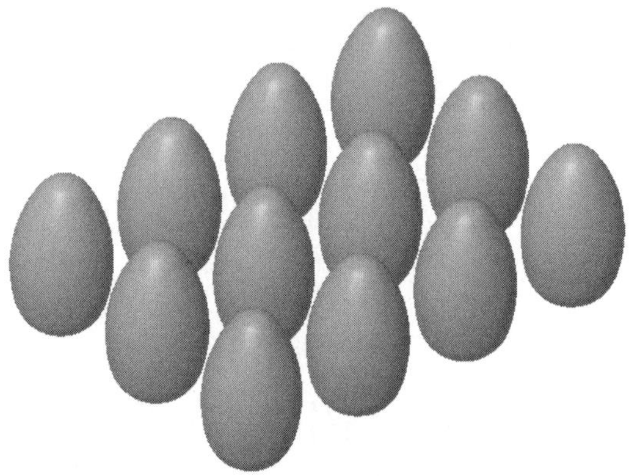

Figure 7-79 *Eggs patterned using the* ***Linear Pattern*** *tool*

Figure 7-80 *Base plate for creating the egg tray*

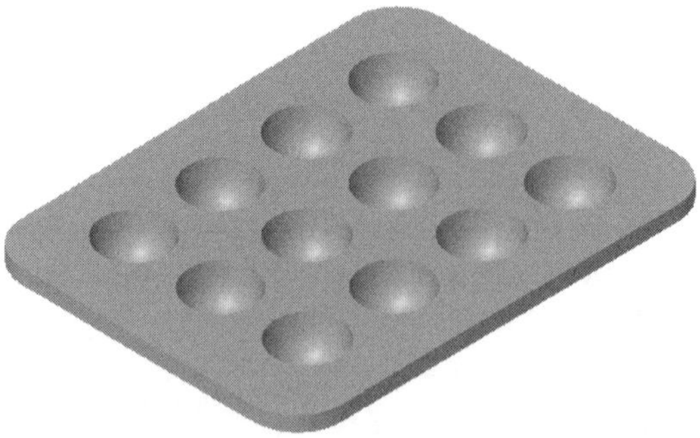

Figure 7-81 *Resulting egg tray*

Displaying the Section View of the Model

CommandManager:	View > Section View	*(Customize to Add)*
Menu:	View > Display > Section View	
Toolbar:	View > Section View	

The **Section View** tool is used to display the section view of the model by cutting it by using a plane or face. To display the section view of a model, choose the **Section View** button from the **View CommandManager**, or choose **View > Display > Section View** from the menu bar. You need to invoke the **View CommandManager**, if it not is not available by default. Move the cursor on any one of the buttons available in **CommandManager** and right-click to invoke the shortcut menu. Choose the **Customize CommandManager** option from it. The **Customizing CommandManager** menu is displayed. Select the **View** check box to display the **View CommandManager** and click once anywhere in the drawing area. On invoking this tool the **Section View PropertyManager** is displayed, as shown in Figure 7-82.

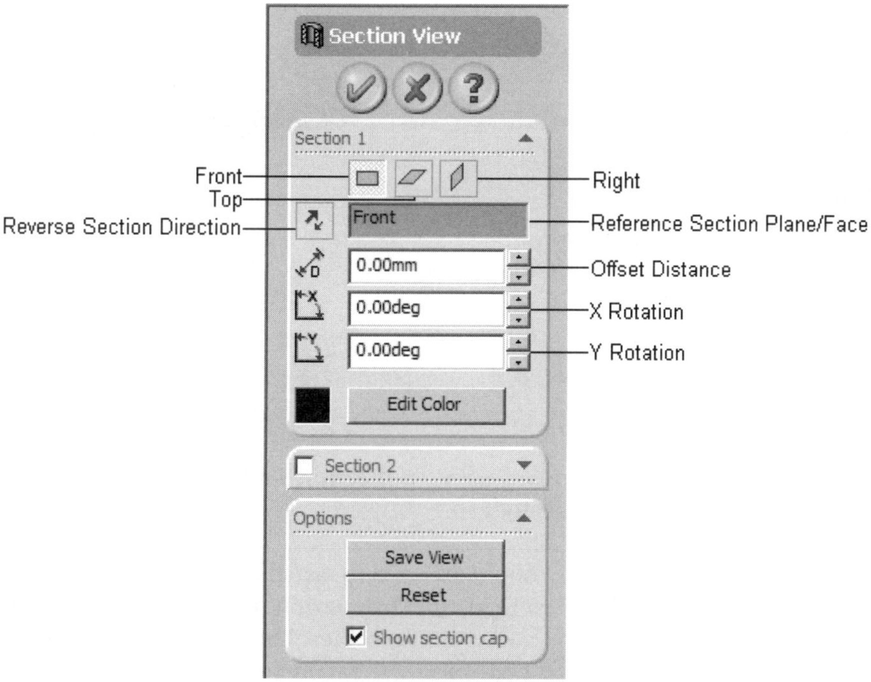

Figure 7-82** The **Section View PropertyManager

By default, the **Front Plane** is automatically selected in the **Section View PropertyManager** and the section view of the model, created using the **Front Plane** as the section plane, is displayed in the drawing area. A drag handle is provided in the drawing area to dynamically adjust the offset distance of the section plane, as shown in Figure 7-83.

If you need to select the **Right Plane** or **Top Plane** as the section plane, choose the respective buttons available in the **Section 1** rollout. You can also select a face or a user-defined plane as the section plane. You can also specify the offset distance using the **Offset Distance** spinner. You will observe that as you modify the offset distance, the preview of the section view is automatically modified. You can also rotate the section plane along the X axis and the Y axis using the **X Rotation** and **Y Rotation** spinners respectively.

The **Edit Color** button is used to modify the color of the preview of the section cap. This color is displayed only when the **Section View PropertyManager** is active. When you exit the **PropertyManager**, the color is not displayed in the section view.

To create a half section view, you need to invoke the **Section 2** rollout and specify the section plane in it. You can also specify the offset distance and the rotation of the plane, as discussed earlier for **Section 1**. After setting all the parameters, choose the **OK** button from the **Section View PropertyManager**. Figure 7-84 shows the preview of the model after defining the section plane.

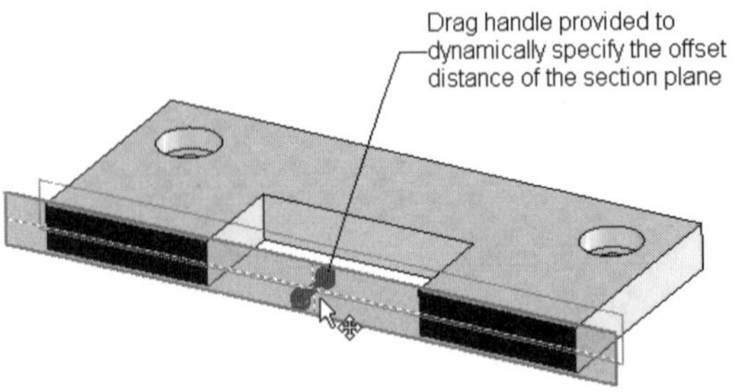

Figure 7-83 *Drag handle for dynamic offset*

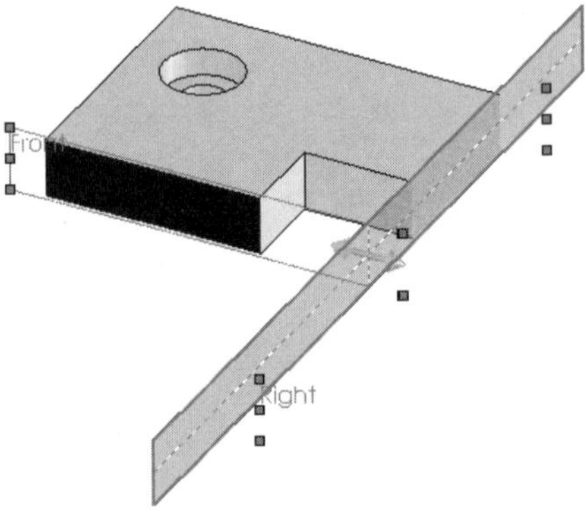

Figure 7-84 *Section preview after selecting the second section plane*

You can also define the third section plane using the **Section 3** rollout. This rollout is displayed only after you invoke the **Section 2** rollout.

The options available in the **Options** rollout are used to save the section view as a named view. You can retrieve the saved section view from the **Orientation** dialog box at any stage of your design cycle. To save a named view, choose the **Save View** button from the **Options** rollout. The **Named View** dialog box is displayed; specify the name of the view and choose the **OK** button from this dialog box. Using the **Reset** button you can reset the section view setting to the default settings. The **Show section cap** check box is selected by default. If you clear this check box, the preview of the section view is not capped. Figure 7-85 shows the section view of a model.

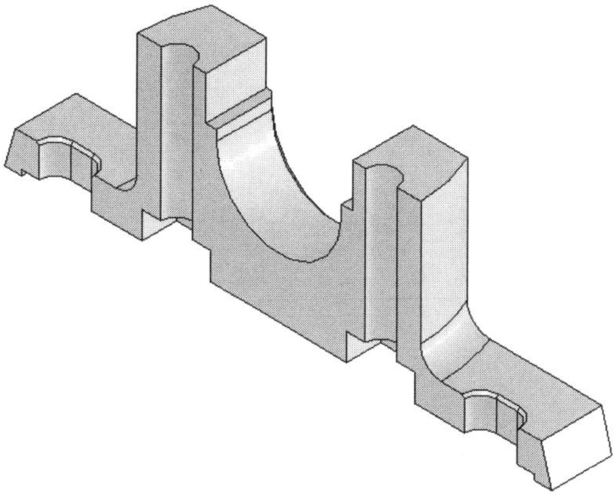

Figure 7-85 *Section view of a model*

Tip. *You can also rotate the section plane dynamically by moving the cursor on the edge of the section plane. The cursor will be replaced by the rotate cursor. Drag the cursor to dynamically rotate the section plane.*

To switch back to the full view mode, you need to choose the ***Section View*** *button from the* ***View CommandManager****. You can also select any face of the sectioned model and then right-click to invoke the shortcut menu. Choose the* ***Section View*** *option from the shortcut menu.*

To modify the section view, select any face of the sectioned model and invoke the shortcut menu. Choose the ***Section View Properties*** *option from it. The* ***Section View PropertyManager*** *is displayed and you can modify the section view.*

TUTORIALS

Tutorial 1

In this tutorial you will create the model shown in Figure 7-86. The dimensions of the model are shown in Figure 7-87. **(Expected time: 30 min)**

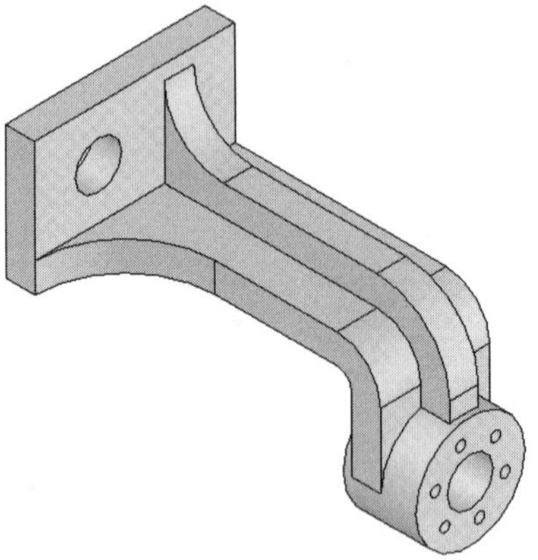

Figure 7-86 *Solid model for Tutorial 1*

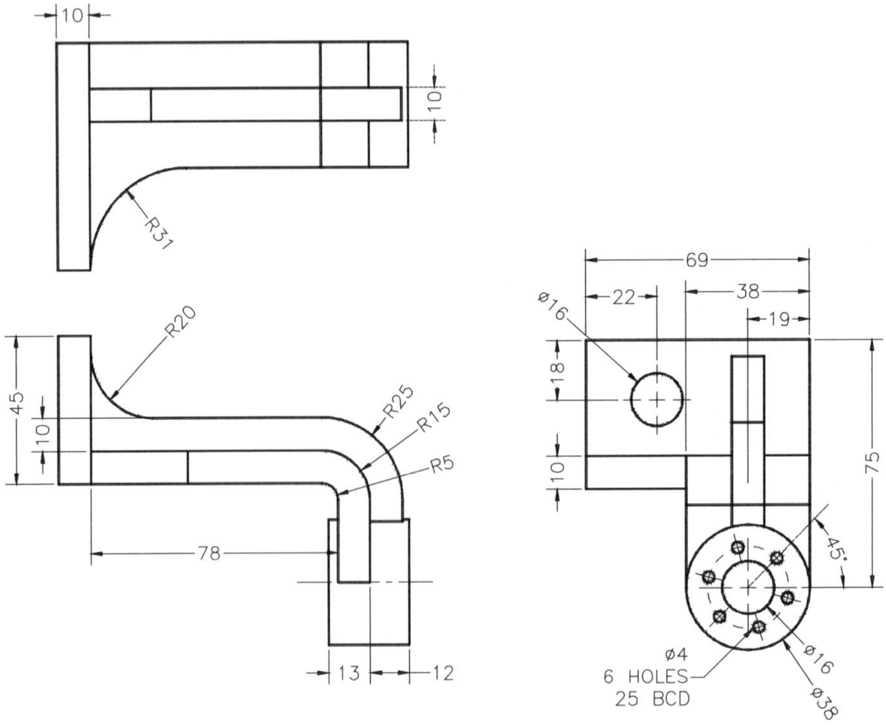

Figure 7-87 *Views and dimensions of the model for Tutorial 1*

The steps to be followed to complete this tutorial are discussed next.

a. Create the base feature of the model by extruding a rectangle of 69 mm x 45 mm, created on the **Right Plane** to a depth of 10 mm, refer to Figure 7-88.
b. Create the second feature, which is created by extruding the sketch created on the back face of the base feature, refer to Figure 7-89 (a).

c. The third feature of the model is a circular feature, refer to Figure 7-89 (b).
d. Create a hole feature placed concentric to the circular feature.
e. Create the hole on the specified BCD and pattern the hole feature using the circular pattern option.
f. Create the hole feature on the base feature, refer to Figure 7-90.
g. Create a fillet feature to add the required fillets, refer to Figures 7-91 and 7-92.
h. Create the rib feature, refer to Figures 7-93 and 7-94.

Creating the Base Feature

1. Start SolidWorks and then start a new part document from the **New SolidWorks Document** dialog box.

   It is evident from the model that the sketch of its base feature is drawn on the **Right Plane**. Therefore, you need to select the **Right Plane** from the **FeatureManager Design Tree** to create the base feature.

2. Select the **Right Plane** from the **FeatureManager Design Tree** and choose the **Extruded Boss/Base** button from the **Features CommandManager**. The sketching environment is invoked and the **Right Plane** is oriented normal to the view.

3. Draw the sketch of the base feature of the model, which consists of a rectangle of dimensions 69 mm x 45 mm.

4. Add the required relations and dimensions to the sketch. Exit the sketching environment.

5. Set the value of the **Depth** spinner to 10 mm and exit the **Extrude PropertyManager**. The base feature of the model is shown in Figure 7-88.

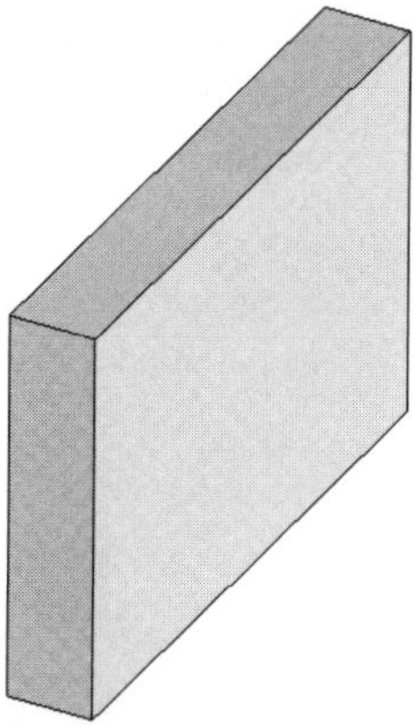

Figure 7-88 *Base feature of the model*

Creating the Second Feature of the Model

The second feature of the model is also an extruded feature. Draw the sketch for the second feature on its back face, and extrude this sketch to the given depth.

1. Select the back face of the base feature as the sketching plane and invoke the **Extruded Boss/Base** tool.

2. Draw the sketch of the second feature and add the required relations and dimensions.

3. Exit the sketching environment and select the **Reverse Direction** button from the **PropertyManager**.

4. Set the value of the **Depth** spinner to **38** and end the feature creation.

   The model, after creating the second feature, is shown in Figure 7-89 (a).

Note
You can also create the first and second features using the contour selection method.

Creating the Third Feature

The third feature of this model is a circular extruded feature, which is created by extruding a circular sketch on both sides of the sketching plane. The sketch for this feature is drawn on the right planar face of the second feature.

1. Select the right planar face of the second feature as the sketching plane and invoke the **Extruded Boss/Base** tool.

2. Draw the sketch using the **Circle** tool. Add the required relations and dimensions it.

3. Exit the sketching environment and set the value of the **Depth** spinner available in the **Direction 1** rollout to **12**. Since you have to extrude the sketch in both the directions with variable values, therefore, you need to invoke the **Direction 2** rollout. Set the value of the **Depth** spinner available in the **Direction 2** rollout to **13** and end the feature creation.

   Figure 7-89 (b) shows the model, after adding the third feature.

Figure 7-89 (a) *Second feature added to the model*

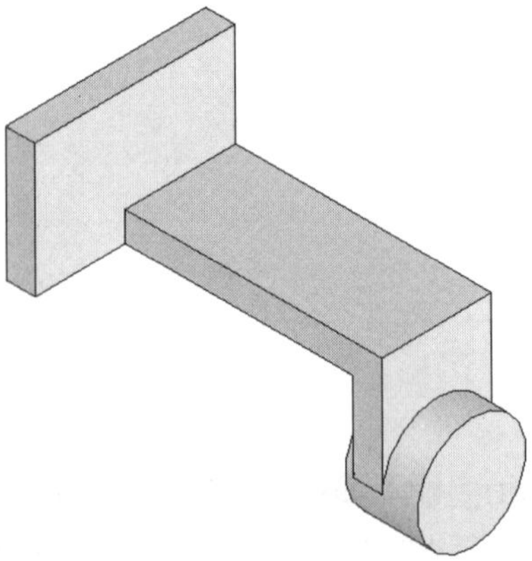

Figure 7-89 (b) *Third feature added to the model*

Creating the Fourth Feature

The fourth feature of this model is a hole feature. You will create a hole using the **Simple Hole** option on the right face of the third feature. To create a hole using this option, you need to place a point on which the hole will be placed.

1. Select the right face of the third feature as the sketching plane for placing the sketch point and invoke the sketching environment. The center point of the hole to be created, using the **Hole PropertyManager**, will be placed coincident to this sketched point.

2. Place a point on the right face of the circular feature.

3. Add the **Concentric** relation between the sketched point and the circular edge of the third feature and exit the sketching environment.

4. Now, select the face on which the point is placed. Choose the **Simple Hole** button from the **Features CommandManager**, or choose **Insert > Features > Hole > Simple** from the menu bar to invoke the **Hole PropertyManager**.

5. Choose the **Through All** option from the **End Condition** drop-down list and set the value of the **Hole Diameter** spinner to **16**.

6. Select the center point of the hole feature and drag the cursor to the sketched point drawn earlier. Release the left mouse button when the cursor snaps the sketched point.

7. Choose the **OK** button from the **Hole PropertyManager**.

Creating the Fifth Feature

1. Using the procedure given in the previous section, create the fifth feature, which is also a hole feature placed on the same placement plane. The hole feature is created using the **Through All** option, with the diameter of the hole as **4**. Next, define the placement of the feature by adding the required relations and dimensions.

Patterning the Hole Feature

After creating the fifth feature, which is a hole feature, you will pattern it using the **Circular Pattern** tool.

1. In case all the buttons of the **CommandManager** are not visible, select the black arrow; a flyout appears. Choose the **Circular Pattern** button from the **CommandManager** flyout.

The **Circular Pattern PropertyManager** is invoked and you are prompted to select an edge or axis for direction reference; select a face of the feature for pattern features.

To create a circular pattern, you need an edge or axis that will be used as the central axis. As discussed in the earlier chapters, when you create a circular feature, a temporary axis is automatically created passing through its center. In this tutorial, the feature to be patterned is created on a circular feature. Therefore, you will display the temporary axis of the circular feature which will be used as the central axis.

2. Choose **View > Temporary Axes** from the menu bar to display the temporary axis.

   Temporary axes are displayed in the model.

3. Select the temporary axis that passes through the center of the first hole feature.

   The **Direction 1** callout is displayed and is attached to the selected axis.

4. Click in the **Features to Pattern** selection area to invoke the selection mode.

5. Select the smaller hole from the drawing area, or expand the **FeatureManager Design Tree**, which is displayed in the drawing area, and select the **Hole2** feature.

   The preview of the pattern feature is displayed with the default settings.

6. Select the **Equal spacing** check box, if cleared and set the value of the **Number of Instances** spinner to **6**. Choose the **OK** button from the **PropertyManager**.

7. Choose **View** > **Temporary Axes** to hide the temporary axes.

Creating the Seventh Feature

1. The seventh feature of this model is a hole feature. You will create this hole feature using the procedure given to create the fourth feature. This hole feature will be placed on the right planar face of the base feature. Therefore, after selecting the right planar face of the base feature, place the hole feature. Next, define the placement of the hole feature by adding the required relations and dimensions. Figure 7-90 shows the model after adding the hole features.

Figure 7-90 *Model after adding all the hole features*

Creating the Fillet Feature

Next, you need to create the fillet feature. It is evident from the model that the fillets to be added to the model are of different radii. In SolidWorks you can specify different radii to individual selected edges, faces, or loops in a single fillet feature.

1. Choose the **Fillet** button from the **Features CommandManager** to invoke the **Fillet PropertyManager**.

   You are prompted to select edges, faces, features, or loops to fillet.

2. Select the **Multiple radius fillet** check box from the **Items To Fillet** rollout of the **Fillet PropertyManager**.

3. Select the edges to fillet, as shown in Figure 7-91. As the **Multiple radius fillet** check box is selected, therefore, each selected edge has a separate **Radius** callout.

4. Modify the value of the radii, as required in the respective **Radius** callouts.

5. Choose the **OK** button from the **Fillet PropertyManager**.

   The isometric view of the model, after adding the fillet feature, is shown in Figure 7-92.

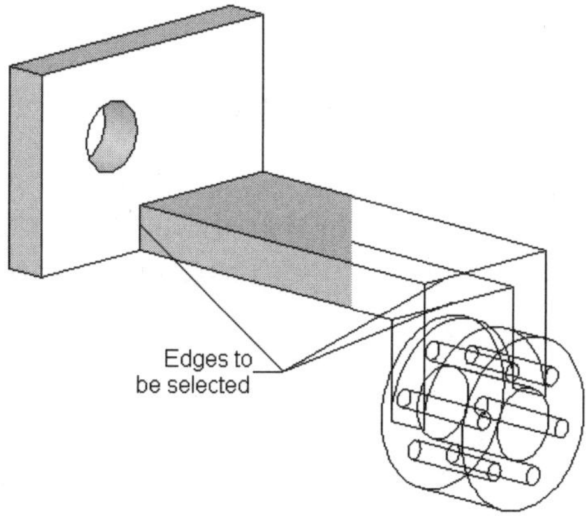

Figure 7-91 *Edges to be selected*

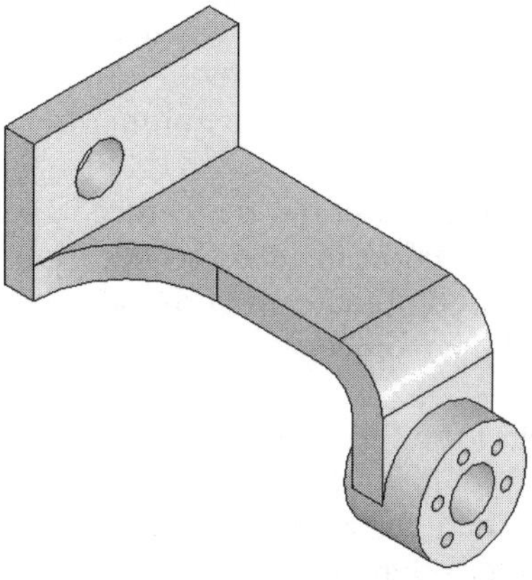

Figure 7-92 *Model after adding the fillet feature*

Creating the Rib Feature

The next feature that you need to create is a rib feature. The sketch of the rib feature needs to be drawn on a sketching plane at an offset distance from the back planar face of the model. Therefore, first you need to create a reference plane at an offset distance from the back planar face of the model.

1. Invoke the **Plane PropertyManager**.

2. Rotate the model and select its back planar face. Choose the **Reverse direction** check box under the **Distance** spinner and set the value of the **Distance** spinner to **19**.

3. Choose the **OK** button from the **Plane PropertyManager** to end the feature creation.

   A new plane is created at an offset distance from the back planar face of the model.

4. Choose the **Rib** button from the **Features CommandManager**.

5. Create the sketch for the rib feature on the newly created plane and add the required relations and dimensions to the sketch, as shown in Figure 7-93.

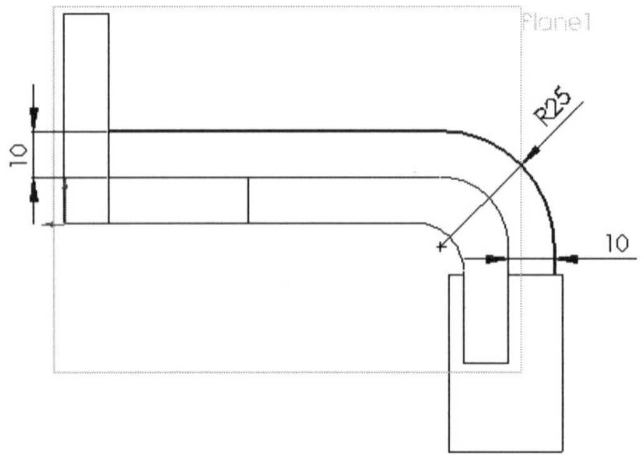

***Figure 7-93** Sketch for the rib feature*

6. Exit the sketching environment. The **Rib PropertyManager** is displayed.

   The preview of the rib feature is displayed in the drawing area and you will observe that the direction of material addition is displayed by an arrow in the drawing area. The direction of material addition is opposite to the required direction. Therefore, you need to flip its direction.

7. Select the **Flip material side** check box to flip the direction of material addition.

   The default value of the rib thickness is 10, which is the required value, and so you do not need to change it.

8. Choose **OK** from the **Rib PropertyManager**. Also, hide the newly created plane.

   The last feature of the model is the fillet feature. Add the fillet feature on the left edge of the rib using the **Fillet** tool. Figure 7-94 shows the isometric view of the final model.

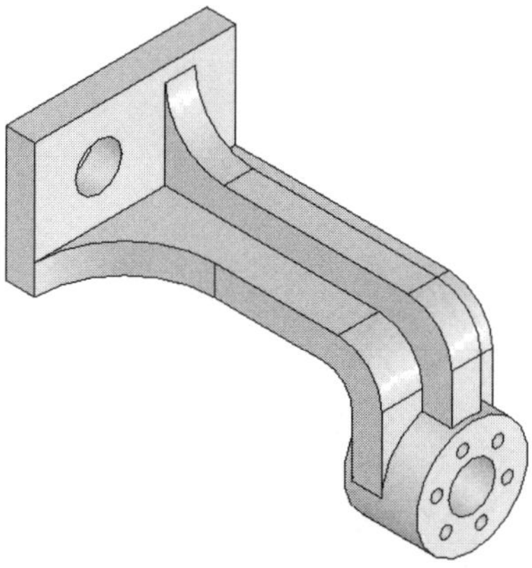

Figure 7-94 *The final solid model*

Saving the Model

1. Create a *c07* folder in the *SolidWorks* folder and choose the **Save** button from the **Standard** toolbar. Save the model with the name given below.

   \My Documents\SolidWorks\c07\c07tut1.sldprt

2. Choose **File > Close** from the menu bar to close the document.

Tutorial 2

In this tutorial you will create the model shown in Figure 7-95. The dimensions of the model are shown in Figure 7-96. **(Expected time: 30 min)**

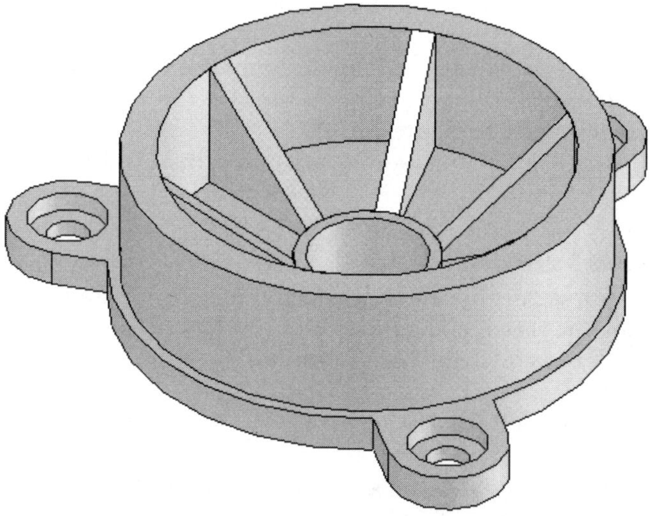

Figure 7-95 *Model for Tutorial 2*

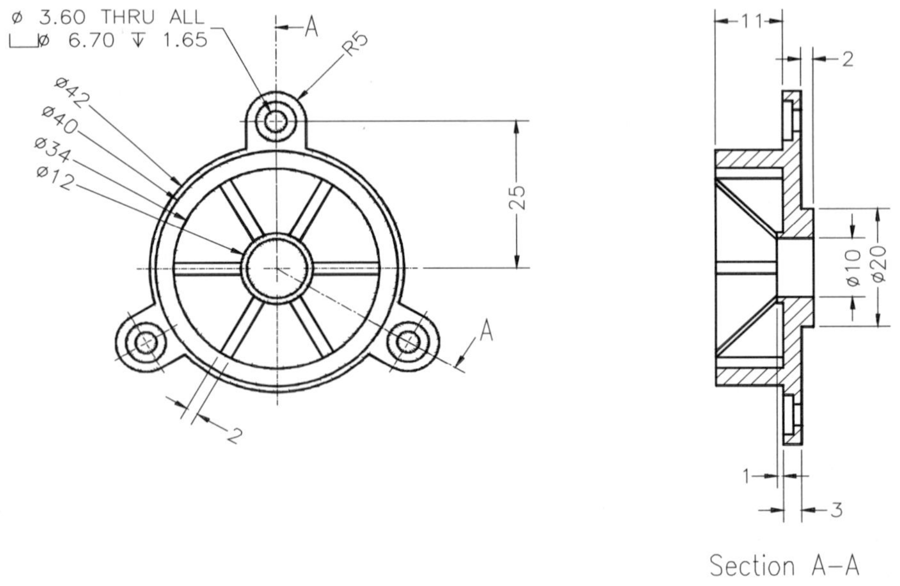

Figure 7-96 *Front view and aligned section view with dimensions*

The steps to be followed to complete this tutorial are discussed next.

a. Create the base feature of the model by revolving the sketch along a centerline, refer to Figures 7-97 and 7-98.
b. Create the second feature, which is on the outer periphery of the base feature, by extruding the sketch from the sketch plane to a selected surface, refer to Figures 7-99 and 7-100.
c. Place a counterbore hole feature on the top face of the second feature using the hole wizard.
d. Pattern the second and third features along the temporary axis using the **Circular Pattern** tool, refer to Figure 7-101.
e. Create the rib feature, refer to Figure 7-102.
f. Pattern the rib feature along a temporary axis using the **Circular Pattern** tool, refer to Figure 7-93.

Creating the Base Feature

Start a new SolidWorks part document. First you need to create the base feature of the model by revolving the sketch along the axis of revolution. The axis of revolution will be a centerline and the sketch for the base feature will be drawn on the **Right Plane**.

1. Invoke the **Revolved Boss/Base** tool and select the **Right Plane** as the sketching plane.

2. Create the sketch for the base feature and add the required relations and dimensions to the sketch, as shown in Figure 7-97.

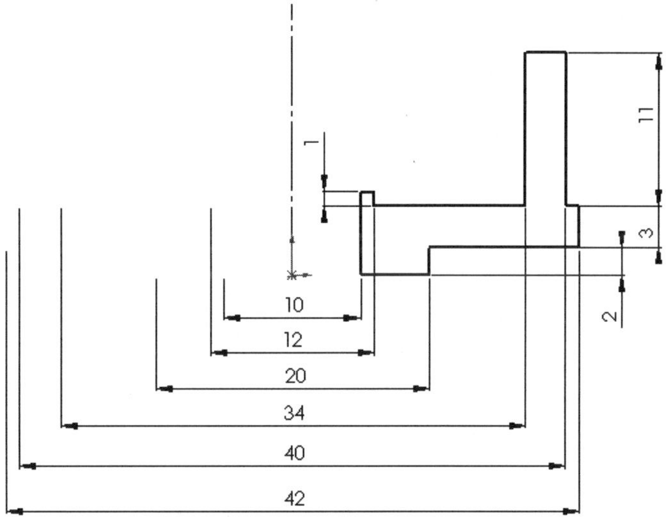

Figure 7-97 Sketch for the base feature

3. Exit the sketching environment and set the value of the **Angle** spinner to **360**.

4. Choose the **OK** button from the **Revolve PropertyManager**. The base feature created after revolving the sketch is shown in Figure 7-98.

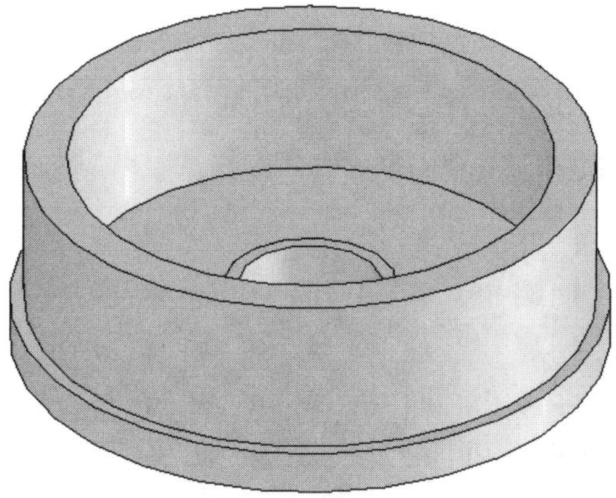

Figure 7-98 Base feature of the model

Creating the Second Feature

The second feature is created by extruding a sketch up to the selected surface.

1. Invoke the **Extruded Boss/Base** tool and select the face shown in Figure 7-99 as the sketching plane.

2. Create the sketch of the second feature and add the required relations and dimensions to the sketch.

3. Exit the sketching environment. Use the **Up To Surface** option to extrude the sketch. The surface to be selected is shown in Figure 7-100.

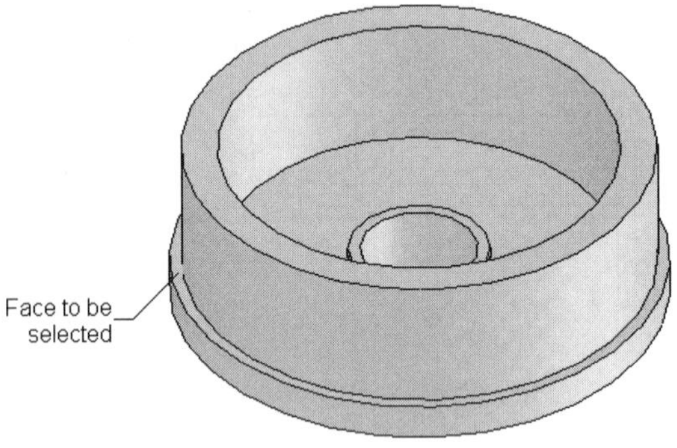

Figure 7-99 Face to be selected

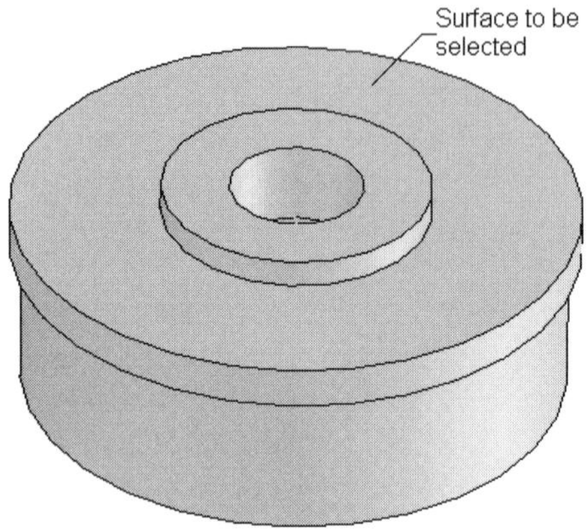

Figure 7-100 Surface to be selected

4. Choose the **OK** button from the **Extrude PropertyManager**.

Creating the Hole Feature

It is evident from Figure 7-96 that a counterbore hole needs to be added to the model by using the **Hole Wizard**. Before invoking this tool, select the placement plane for the hole feature.

1. Select the top face of the second feature as the placement plane for the hole feature.

2. Choose the **Hole Wizard** button from the **Features CommandManager** to invoke the **Hole Definition** dialog box.

   The preview of the hole feature with the default settings in the **Hole Definition** dialog box is displayed in the drawing area.

3. Invoke the **Counterbore** tab, if it is not invoked by default and select the **Ansi Metric** option from the **Standard** drop-down list to specify the standard to be used.

4. Choose the **Socket Button Head Cap Screw - ANSI B18.3.4M** option from the **Screw Type** drop-down list to specify the type of screw.

5. Select the **M3** option from the **Size** drop-down list to specify the size of the fastener to be used in the hole. Leave all the default options as they are and choose the **Next** button from the **Hole Definition** dialog box.

   The **Hole Placement** dialog box is displayed.

6. Invoke the **Add Relations PropertyManager** and apply a concentric relation between the center point of the hole feature and the circular edge of the second feature.

7. Choose the **Finish** button from the **Hole Placement** dialog box to end the feature creation.

Patterning the Features

After creating the second and third feature, you need to pattern them about the temporary axis using the **Circular Pattern** tool.

1. Choose the **Circular Pattern** button from the **Features CommandManager** to invoke the **Circular Pattern PropertyManager**.

You need to define an axis as the direction reference to create a circular pattern. Therefore, display the temporary axes by choosing **View** > **Temporary Axes** from the menu bar.

2. Select the temporary axis that passes through the center of the model and invoke the selection mode in the **Features to Pattern** area. The **Direction 1** callout is displayed.

3. Select the second and third features from the drawing area or from the **FeatureManager Design Tree** that is displayed in the drawing area.

4. Set the value of the **Number of Instances** spinner to **3** and make sure the **Equal spacing** check box is selected. Choose the **OK** button from the **Circular Pattern PropertyManager**.

5. Hide the temporary axes. The model, after patterning the features, is shown in Figure 7-101.

Creating the Rib Feature

The next feature is a rib. The sketch for the rib feature will be created on the **Front Plane**.

1. Choose the **Rib** button from the **Features CommandManager** and select the **Front Plane** from the **FeatureManager Design Tree**.

2. Set the display model to wireframe, create the sketch for the rib feature, and add the required relations, as shown in Figure 7-102.

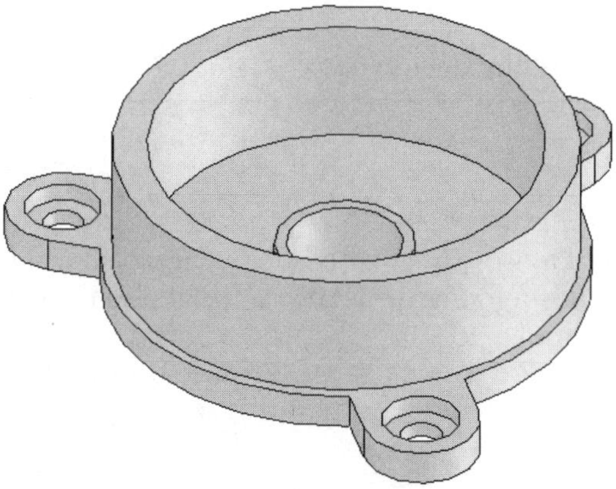

Figure 7-101 *Model after patterning the features*

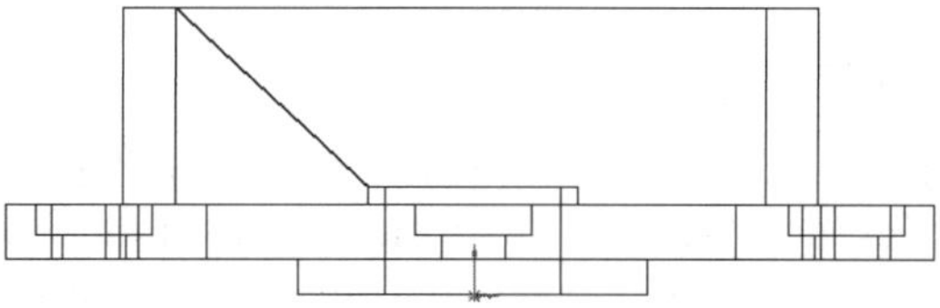

Figure 7-102 *Sketch for the rib feature*

3. Exit the sketching environment and set the value of the **Rib Thickness** spinner to **2**. Leave all the other default options as they are and choose the **OK** button from the **Rib PropertyManager**.

4. Change the model display mode to shaded.

5. Using the **Circular Pattern** tool, create six instances of the rib feature. The final model after, creating all the features, is shown in Figure 7-103.

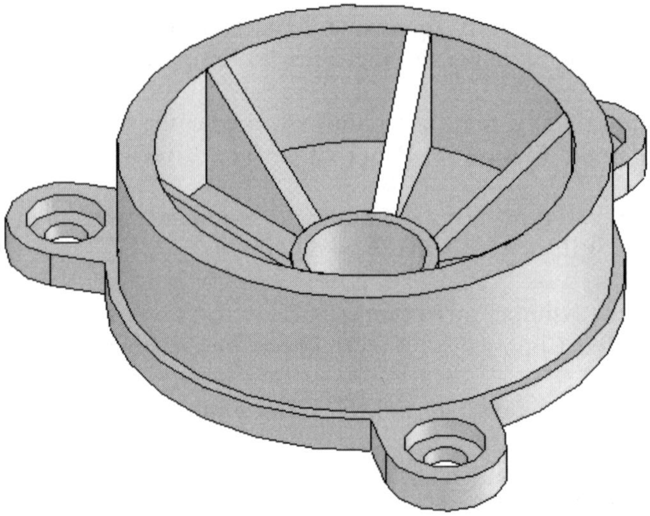

Figure 7-103 *Final solid model*

Saving the Model

1. Choose the **Save** button from the **Standard** toolbar and save the model in the *c07* folder with the name given below.

   \My Documents\SolidWorks\c07\c07tut2.sldprt

2. Choose **File > Close** from the menu bar to close the document.

Tutorial 3

In this tutorial you will create the cylinder head of a two-stroke automobile engine shown in Figure 7-104. The dimensions of the model are shown in Figure 7-105. You will also create a section view of the model using the **Section View** tool.

(Expected time: 1 hr)

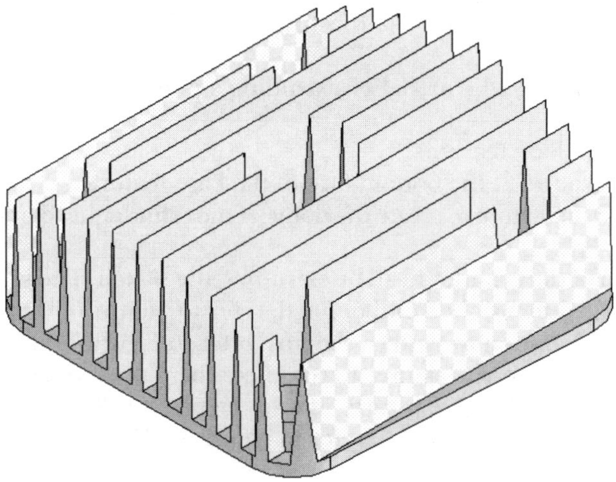

Figure 7-104 Model for Tutorial 3

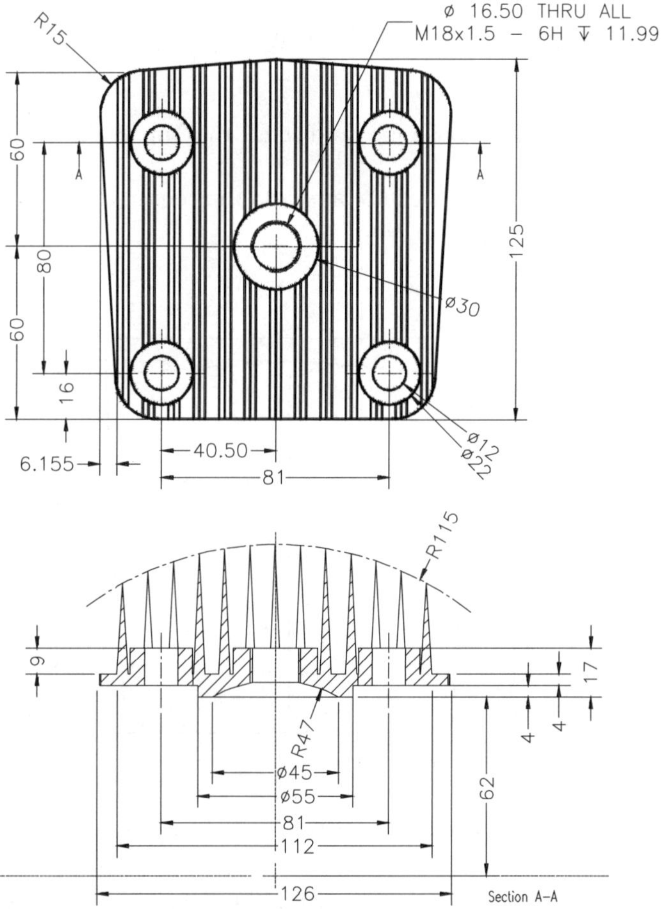

Figure 7-105 Top view and the section front view with dimensions

The steps to be followed to complete this tutorial are discussed next.

a. Create the base feature of the model by extruding a polygon to the given depth, refer to Figure 7-106.
b. Add the fillet to the base feature.
c. Create a circular feature at the bottom face of the base feature.
d. Create the revolve cut feature to create the dome of the cylinder head, refer to Figures 7-107 and 7-108.
e. Create the left fin of the cylinder head by extruding the sketch. The sketch for this feature should be carefully dimensioned and defined, refer to Figure 7-109.
f. Use the **Vary sketch** option to pattern the fins, refer to Figure 7-110.
g. Create other cut and extrude features to complete the model, refer to Figure 7-111.
h. Create a tap hole using the hole wizard, refer to Figure 7-112.
i. Create the section view of the model, refer to Figure 7-113.

Creating the Base Feature

1. Create a new SolidWorks part document.

   The base feature of the model will be created by extruding the sketch created on the **Top** plane.

2. Invoke the **Extruded Boss/Base** tool and select the **Top Plane** as the sketching plane.

3. Create the sketch for the base feature and add the required relations and dimensions to the sketch as shown in Figure 7-106.

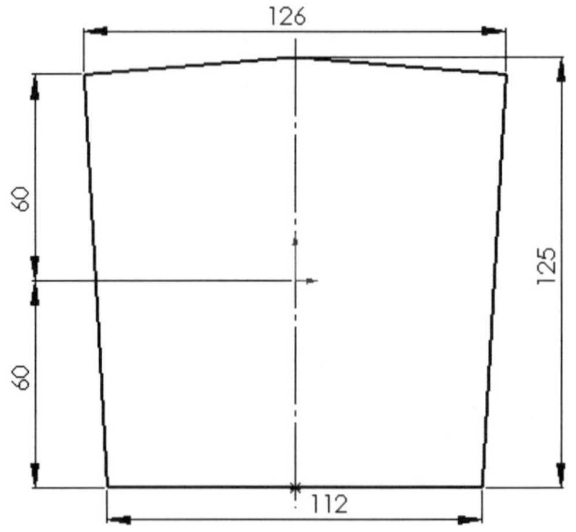

***Figure 7-106** Sketch for the base feature*

4. Exit the sketching environment and extrude the sketch to a depth of 4 mm.

Creating the Second Feature

The second feature of the model is the fillet feature. You need to fillet all the vertical edges of the base feature using the given radius.

1. Invoke the **Fillet** tool and set the value of the **Radius** spinner to **15**. Select all the vertical edges of the base feature to add the fillet feature.

2. Choose the **OK** button from the **Fillet PropertyManager**.

Creating the Third Feature

After creating the base and adding fillet at the corners of the base, you will create the third feature of the model, which is a circular extruded feature. The sketch of the feature will be drawn on the bottom face of the base feature and it will be extruded to the given depth.

1. Invoke the **Extruded Boss/Base** tool and select the bottom face of the base feature as the sketching plane.

2. Create a circle of 55 mm diameter with its center point at origin.

3. Exit the sketching environment and extrude the sketch to a depth of 4 mm.

Creating the Fourth Feature

The fourth feature is a revolved cut feature whose sketch will be drawn on the **Front Plane**. After drawing the sketch, apply the required relations and dimensions to it.

1. Invoke the **Revolved Cut** tool and select the **Front Plane** from the **FeatureManager Design Tree**.

2. Draw the sketch for the revolved cut feature and add the required relations and dimensions as shown in Figure 7-107. You need to apply the vertical relation between the center point of the arc and the origin to fully define it.

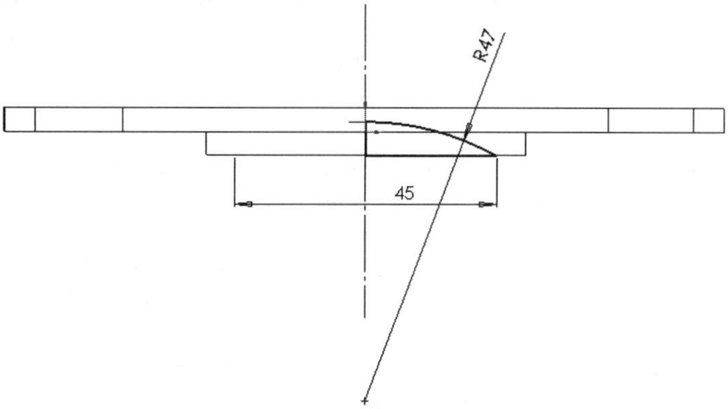

Figure 7-107 *Sketch for the revolve cut feature*

Note

When you draw a sketch for the revolved cut feature for this tutorial, draw a horizontal centerline, such that the start point of the centerline is merged with the upper endpoint of the arc. Also, ensure a tangent relation between the arc and the centerline exists to maintain the tangency of the arc.

3. Select the vertical centerline and exit the sketching environment. Make sure that the **Angle** spinner value is **360**.

4. Choose the **OK** button from the **Cut-Revolve PropertyManager**.

   The rotated model, after creating the fourth feature, is shown in Figure 7-108.

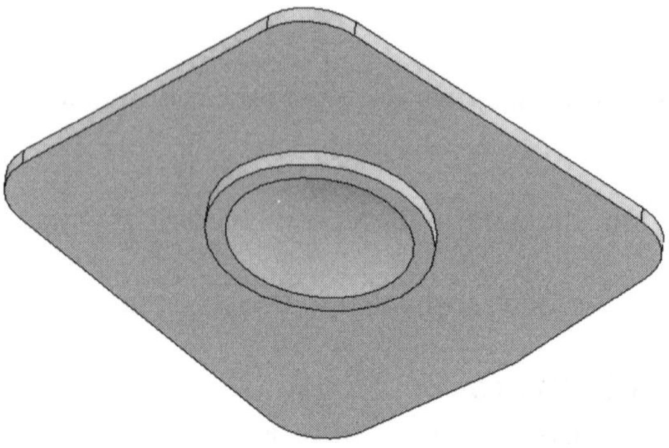

Figure 7-108 *Cut revolve feature added to the model*

Creating the Fifth Feature

Next, you will create the left fin of the cylinder head. It will be created by extruding a sketch to both the directions using the **Through All** option. The sketch of this feature, drawn on the **Front Plane**, will be dimensioned and defined such that the length of the fin is driven by a construction arc and a horizontal dimension. The detailed step by step procedure of drawing, dimensioning, and defining the sketch is discussed next.

1. Invoke the **Extruded Boss/Base** tool and select the **Front Plane** from the **FeatureManager Design Tree**.

2. Using the **Line** tool draw the triangle and then draw a vertical centerline that passes through the upper vertex to the triangle, refer to Figure 7-109.

3. Invoke the **3 Pt Arc** tool and draw the arc as shown in Figure 7-109. Select the arc and select the **For Construction** check box from the **Options** rollout of the **Arc PropertyManager**.

4. Invoke the **Add Relations PropertyManager** and add the coincident relation between the upper vertex of the triangle and the centerline.

5. Add the midpoint relation between the lower endpoint of the centerline and the horizontal line of the triangle. Make sure that the **Coincident** relation between the upper vertex of the triangle and the centerline exists. Also, add the vertical relation to the centerline, if it is missing.

6. Add the coincident relation between the upper vertex of the triangle and the arc.

7. Add the required dimensions and relations to fully define the sketch, as shown in Figure 7-109.

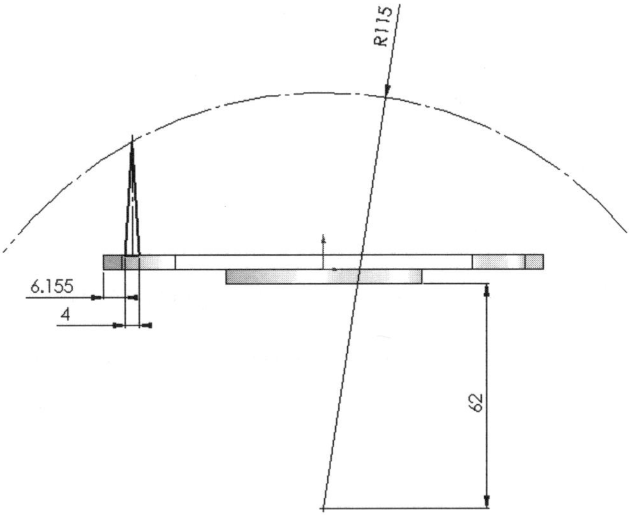

Figure 7-109 *Sketch for the fin of the cylinder head*

Tip. *It is evident from Figure 7-109 that one of the horizontal dimension has a value of 6.155. By default the primary unit precision is set to two decimal places. Therefore, for defining a dimension value with more number of decimal places, you need to select the dimension and set the precision value to the required decimal places from the* ***Primary Unit Precision*** *drop-down list from the* ***Dimension PropertyManager***.

8. Exit the sketching environment and extrude the sketch in both the directions using the **Through All** option. You will notice that the fin extends out of the base feature at both the ends. You will learn how to remove the unwanted material of the fin later in this tutorial.

Patterning the Fifth Feature

You will pattern the fin using the **Vary Sketch** option from the **Linear Pattern** tool. Using the **Vary sketch** option the geometry of each instance of the pattern varies according to the driven dimension and the relation added to the sketch of the feature to be patterned.

1. Choose the **Linear Pattern** button from the **Features CommandManager** to invoke the **Linear Pattern PropertyManager**.

You are prompted to select the directional reference.

2. The fifth feature is selected by default in the **Features to Pattern** selection area. Select the fifth feature from the drawing area if it is not selected.

3. Select the horizontal dimension with the value of **6.155** as the directional reference in the **Pattern Direction** selection area of the **Direction 1** rollout.

4. Set the value of the **Spacing** spinner to **9**. Set the value of the **Number of Instances** spinner to **13**. Choose the **Reverse Direction** button.

5. Invoke the **Options** rollout and select the **Vary Sketch** check box from this rollout.

The preview of the pattern is not displayed in the drawing area.

6. Choose the **OK** button from the **Linear Pattern PropertyManager**.

The model, after adding the pattern feature, is shown in Figure 7-110.

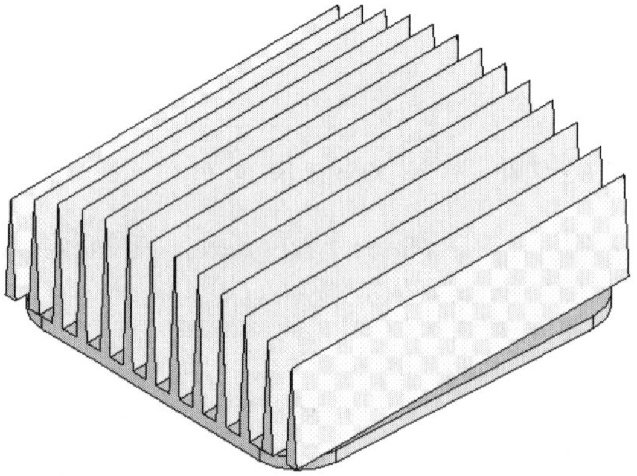

***Figure 7-110** Model after patterning the fin of the cylinder head*

Creating the Cut Feature

The next feature that you will create is a cut feature. Using the **Rotate View** tool rotate the solid model and you will observe that the fins of the cylinder head that you patterned in the last feature extend beyond the boundary of the base feature. Therefore, to trim the extended portion of the fins, you will create a cut feature.

1. Invoke the **Extruded Cut** tool and select the top planar face of the base feature as the sketching plane.

2. Draw the sketch using the standard sketch tools. The sketch for this feature will be the outer profile of the base feature.

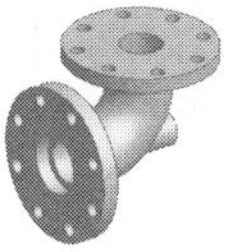

Tip. *You can draw the sketch using the outer profile of the base feature using the* ***Convert Entities*** *tool. Select the lower flat face of the base feature and choose the* ***Convert Entities*** *button from the* ***Sketch*** *toolbar. You will notice that the sketch similar to the outer boundary of the base feature will be placed on the sketching plane.*

3. Exit the sketching environment and choose the **Reverse Direction** button from the **Direction 1** rollout and select the **Through All** option from the **End Condition** drop-down list.

   Since the direction of the side from which the material is to be removed is opposite to the required direction, therefore, you need to flip the direction of material removal.

4. Select the **Flip side to cut** the check box from the **Direction 1** rollout and choose the **OK** button from the **Cut-Extrude PropertyManager**.

5. Using the standard modeling tools shape the model, as shown in Figure 7-111.

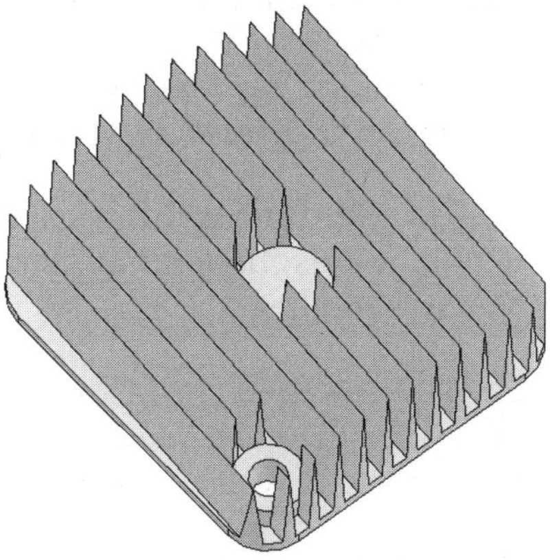

Figure 7-111 *Model after adding other extrude and cut features*

Patterning the Remaining Features

After creating all the features, you need to pattern the cut, extrude, and hole features created at the lower left corner of the model.

1. Invoke the **Linear Pattern PropertyManager** and select the cut, extrude, and hole features, created on the lower left corner of the model.

2. Select the two directional references to pattern the features in both the directions and set the values of the distances between the instances and the number of instances, refer to Figure 7-105.

Tip. *The sketch sharing option is also available in SolidWorks. This allows you to use a sketch used earlier in creating a sketch feature. For sharing the sketch, select it from the* ***FeatureManager Design Tree****, after invoking the feature tool. You need to expand the created sketched feature to select the sketch for sharing from the* ***FeatureManager Design Tree****.*

3. Choose the **OK** button from the **Linear PropertyManager**.

Creating a Tapped Hole

The last feature of the model is a hole feature. You will create a tapped hole using the **Hole Wizard** and then define the placement of the hole.

1. Select the top face of the middle circular extrude feature as the placement plane for the hole feature.

2. Invoke the **Hole Definition** dialog box by choosing the **Hole Wizard** button from the **Features CommandManager** and select the **Tap** tab. Select **ANSI Metric** from the **Standard** drop-down list.

3. Select the **M18x1.5** option from the **Size** drop-down list to define the size of the tap hole.

4. Select the **Through All** option from the **Tap Drill Type & Depth** drop-down list. Also, select the **Through All** option from the **Thread Type & Depth** drop-down list.

5. Select the **Add Cosmetic thread with thread callout** option from the **Add Cosmetic Thread** drop-down list. Choose the **Next** button from the **Hole Definition** dialog box. The **Hole Placement** dialog box is displayed.

   The tapped hole will be placed by default on the placement plane. This default location is not the required position to place the hole. You need to define the placement of the tapped hole concentric with the center circular feature.

6. Invoke the **Add Relations PropertyManager** and add the **Concentric** relation between the center point of the tapped hole and the circular extruded feature of diameter 55 mm.

7. Choose the **Finish** button from the **Hole Placement** dialog box to end the tapped hole feature creation.

   The rotated final model is shown in Figure 7-112.

Displaying the Section View of the Model

Next, you will display the section view of the model. The section view of the model is created using the **Section View PropertyManager**.

1. Orient the model in the isometric view.

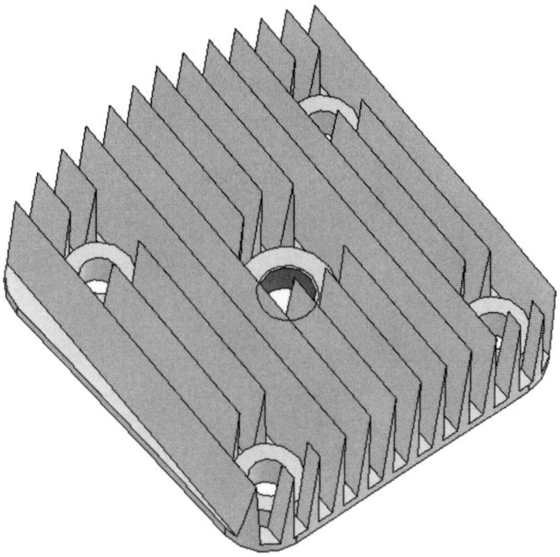

Figure 7-112 *Final solid model*

Tip. *You will observe that a graphic thread is displayed in the tapped hole. You can zoom the area for the better visualization of the graphic thread.*

You will also observe that the cosmetic thread is displayed with the tapped hole feature. On orienting the model in the top view, you will observe the thread convention in the top view. On orienting the model in the front, back, right, or left views, you will view the side convention of the thread.

You can also hide the cosmetic thread by selecting the cosmetic thread from the drawing area and right-clicking to invoke the shortcut menu. Choose the ***Hide*** *option from the shortcut menu to hide the cosmetic thread.*

2. Choose the **Section View** button from the **View CommandManager**.

   The **Front** view is selected as the section plane in the **Section View PropertyManager** by default. The preview of the section view, using the **Front Plane** as the section plane, is displayed in the drawing area.

3. Choose the **OK** button from the **Section View PropertyManager** to view the section of the model.

   The section view of the model is shown in Figure 7-113.

7. Choose the **Section View** button again from the **View CommandManager** to return to the full view mode.

Saving the Model

When you choose the **Save** button from the **Standard** toolbar, the **Save As** dialog box will

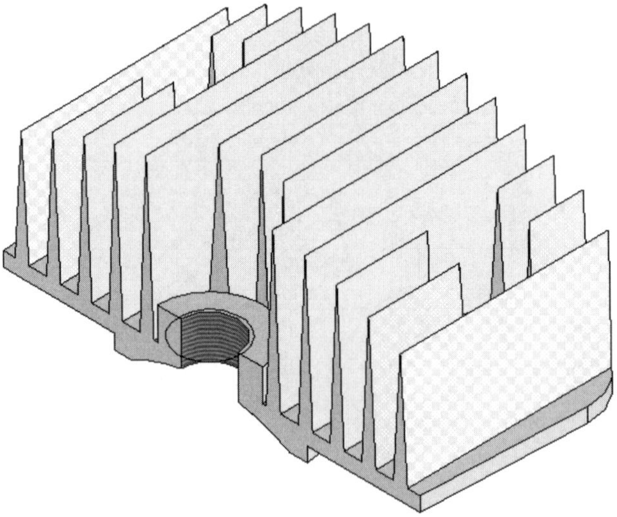

Figure 7-113 Section view of the model

be displayed because the document has not been saved until now. You can enter the name of the document in this dialog box.

1. Choose the **Save** button from the **Standard** toolbar and save the model in the *c07* folder with the name given below.

   \My Documents\SolidWorks\c07\c07tut3.sldprt

2. Choose **File** > **Close** from the menu bar to close the document.

SELF-EVALUATION TEST

Answer the following questions and then compare your answers with those given at the end of this chapter.

1. To invoke the **Mirror PropertyManager**, choose **View** > **Pattern/Mirror** > **Mirror** from the menu bar. (T/F)

2. If you modify the parent feature, then the same change will not reflect on the mirrored feature. (T/F)

3. You cannot preselect the mirror plane and the feature to pattern before invoking the **Mirror** tool. (T/F)

4. You can mirror a single face using the **Mirror** tool. (T/F)

5. You can also pattern a pattern feature. (T/F)

6. The _________ **PropertyManager** is used to view the section view.

7. Using _________ check box you can remove the material using the **Dome** tool.

8. _________ option is used to create a pattern by specifying the coordinates.

9. _________ option is used to create a pattern with respect to the sketched points.

10. Using the _________ rollout you can delete the pattern instances.

REVIEW QUESTIONS

Answer the following questions.

1. The _________ check box is used to accommodate all the instances of a pattern along the selected curve.

2. Enter the coordinates for creating the instances in the _________ area of the **Table Driven Pattern** dialog box.

3. You need to invoke _________ to create a rib feature.

4. _________ check box is selected to create an elliptical dome feature.

5. Using _________ check box from the **Section View PropertyManager**, you can create a section using an invisible plane normal to the eye view as the section plane.

6. When you choose the **Mirror** button from the **Features** tool, which **PropertyManager** is displayed?

   (a) **Mirror Feature PropertyManager** (b) **Mirror All PropertyManager**
   (c) **Mirror PropertyManager** (d) **Copy/Mirror PropertyManager**

7. Which option is used to mirror the exact geometry of the feature independent of the relationships between the geometries?

   (a) **Same Mirror** (b) **Geometry Pattern**
   (c) **Geometry Copy** (d) **Copy Geometry**

8. Which pattern is created along the sketched lines, arcs, or splines?

   (a) Curve-driven pattern (b) Sketch-driven pattern
   (c) Geometry-driven pattern (d) Linear pattern

9. Which dialog box is invoked to create a pattern by specifying the coordinate points?

   (a) **Sketch Driven Pattern** (b) **Table Driven Pattern**
   (c) **Mirror** (d) None of these

10. Which plane is selected by default when you invoke the **Section View PropertyManager** to view a section of the model?

   (a) **Right** (b) **Top**
   (c) **Front** (d) **Plane 1**

EXERCISES

Exercise 1

Create the model shown in Figure 7-114. The dimensions of the model are shown in Figure 7-115. **(Expected time: 1 hr)**

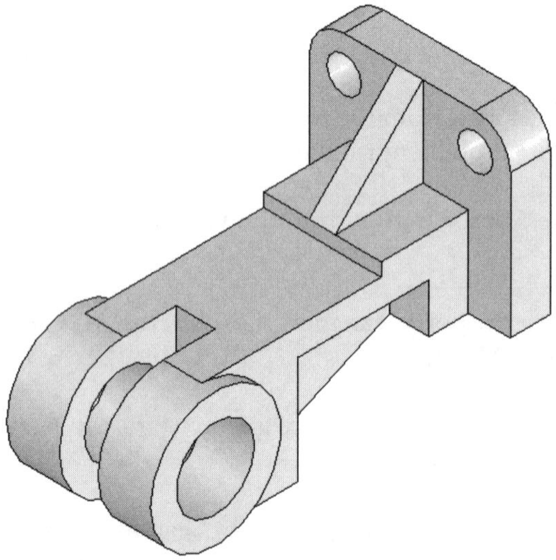

Figure 7-114 *Solid model for Exercise 1*

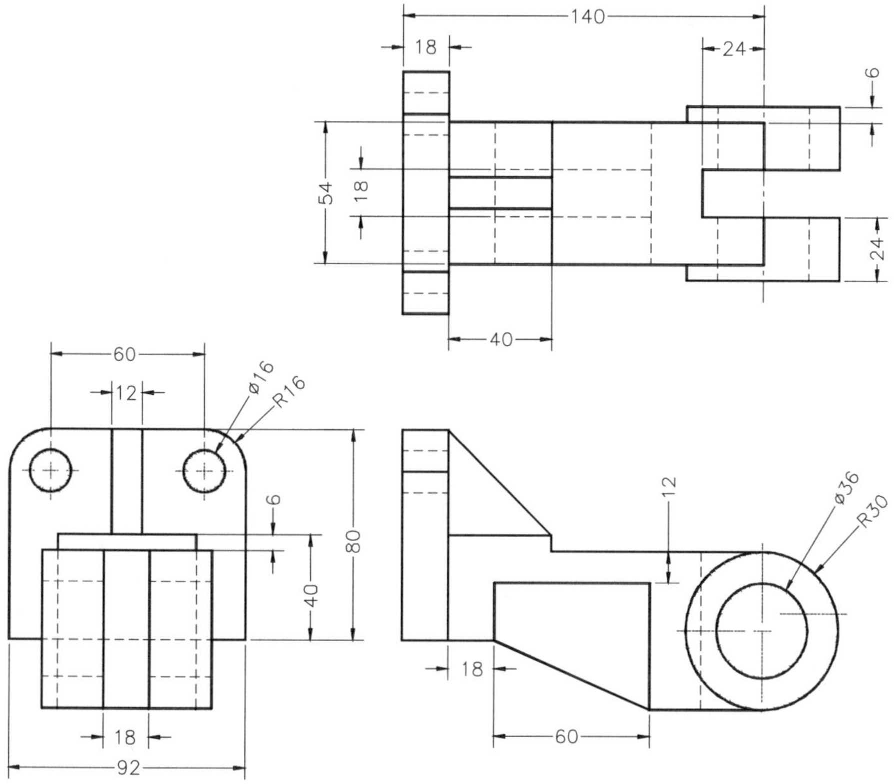

Figure 7-115 Views and dimensions of the model for Exercise 1

Exercise 2

Create the model shown in Figure 7-116. The dimensions of the model are shown in Figure 7-117. **(Expected time: 1 hr)**

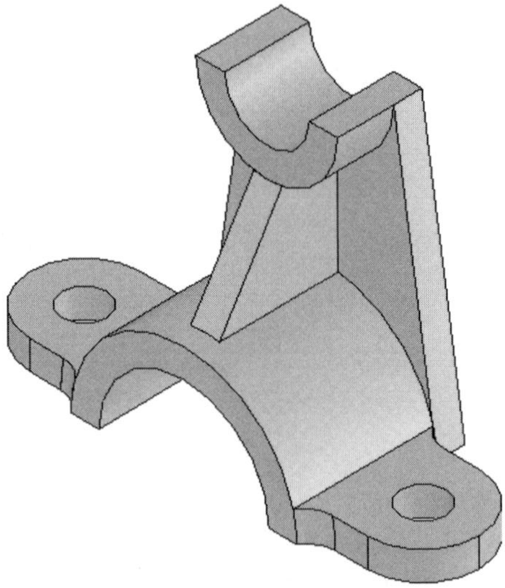

Figure 7-116 *Solid model for Exercise 2*

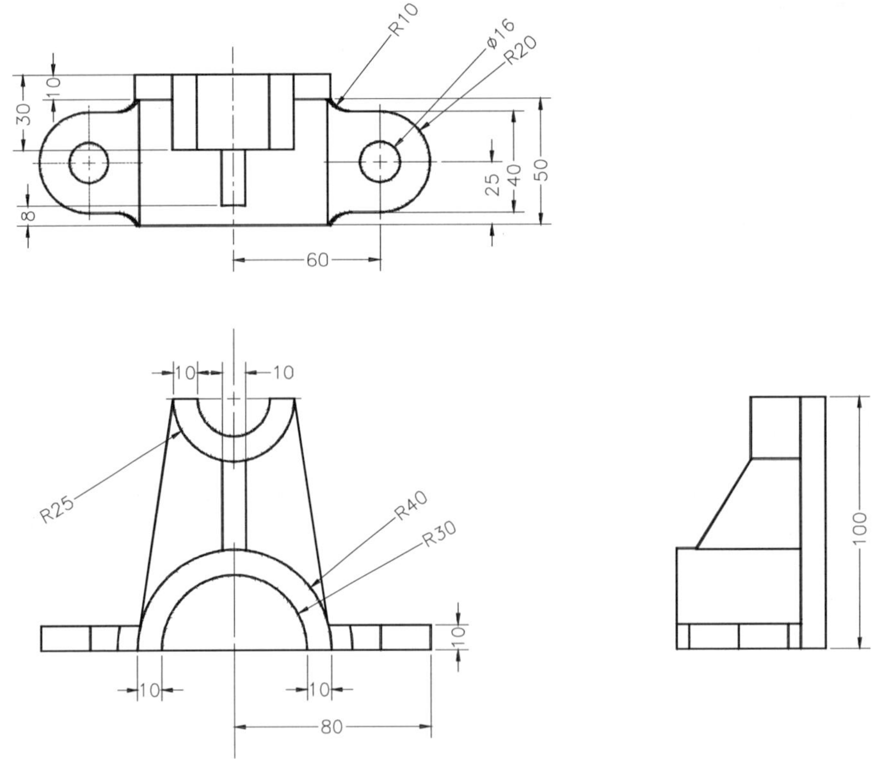

Figure 7-117 *Views and dimensions of the model for Exercise 2*

Exercise 3

Create the model shown in Figure 7-118. Next, create the section view of the model using the **Right Plane**. Figure 7-119 shows the section view of the model whose dimensions are shown in Figure 7-120. **(Expected time: 45 min)**

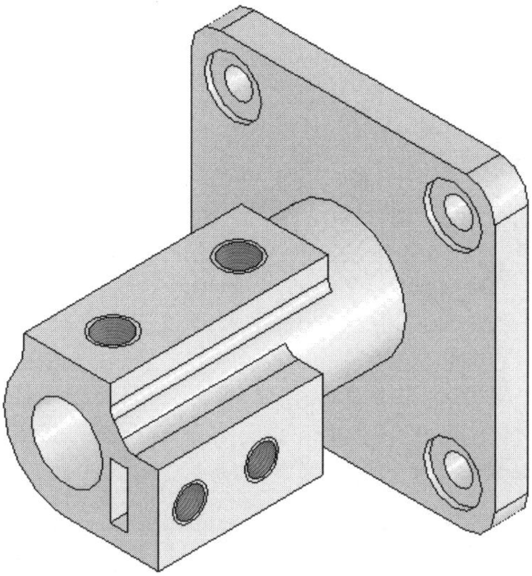

Figure 7-118 *Solid model for Exercise 3*

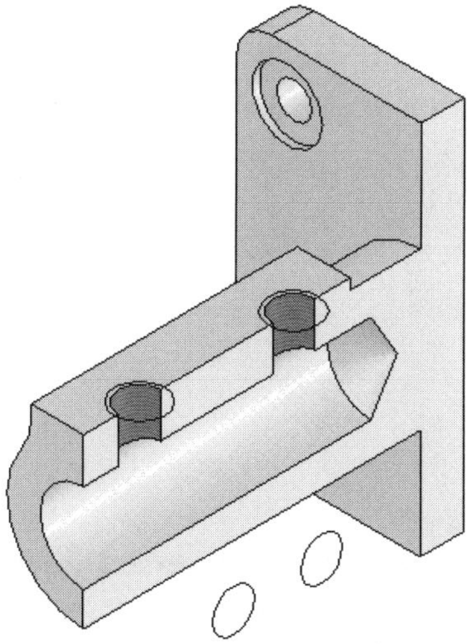

Figure 7-119 *Section view of the model*

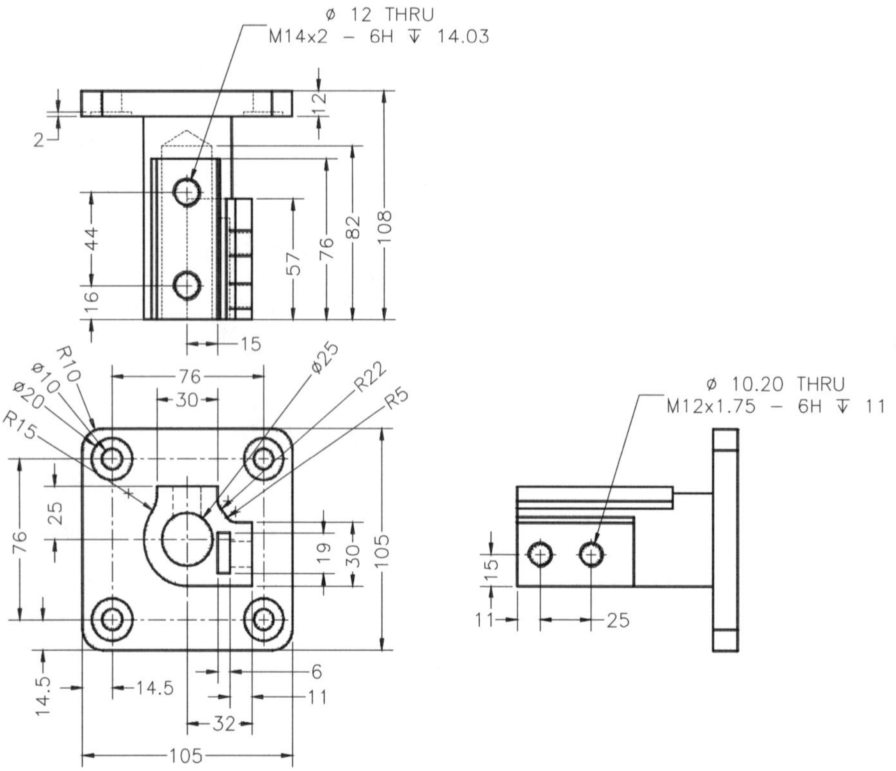

Figure 7-120 *Views and dimensions of the model for Exercise 3*

Answers to Self-Evaluation Test

1. F, **2.** F, **3.** F, **4.** F, **5.** T, **6. Section View**, **7. Reverse Direction**, **8. Table Driven Pattern**, **9. Sketch Driven Pattern**, **10. Instances to Skip**

Chapter 8

Editing Features

Learning Objectives

After completing this chapter you will be able to:

- *Edit Features.*
- *Edit Sketches of the Sketched Features.*
- *Edit the Sketch Plane.*
- *Edit using the Move/Size Features option.*
- *Cut, Copy, and Paste Features and Sketches.*
- *Copy Features using the Drag and Drop Method.*
- *Delete Features.*
- *Delete Bodies.*
- *Suppress and Unsuppress Features.*
- *Move/Copy Bodies.*
- *Reorder Features.*
- *Roll Back the Model.*
- *Rename Features.*
- *Create Folders.*
- *Use What's Wrong? Functionality.*

EDITING THE FEATURES OF THE MODEL

Editing is one of the most important aspects of the product design cycle. Almost all designs require editing either during or after their creation. As discussed earlier, SolidWorks is a feature-based parametric software. Therefore, the design created in SolidWorks is a combination of individual features integrated together to form a solid model. All these features can be edited individually. For example, Figure 8-1 shows a base plate with some holes.

Now, to replace the four inner holes with four counterbore holes, you need to perform one editing operation. Using the editing operation, you will change the drilled holes to the counterbore holes. For editing the holes, you need to select the hole feature and right-click to invoke the shortcut menu. Next, choose the **Edit Feature** option from the shortcut menu to invoke the **Hole Definition** dialog box. Finally, set the new parameters and end the feature creation. The drilled holes will be automatically replaced by counterbore holes. Figure 8-2 shows the base plate with drilled holes modified to the counterbore holes.

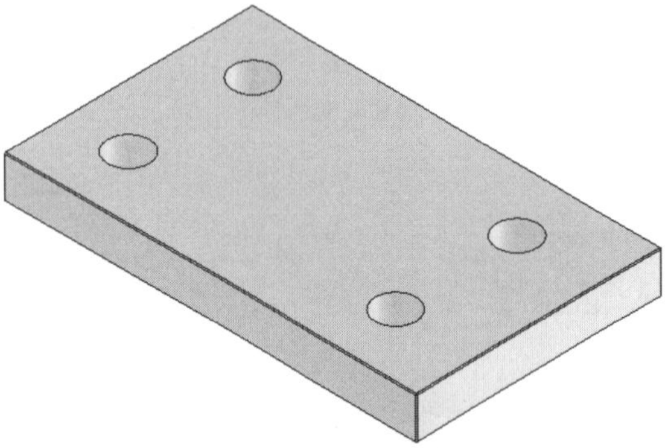

Figure 8-1 *Base plate with drilled holes*

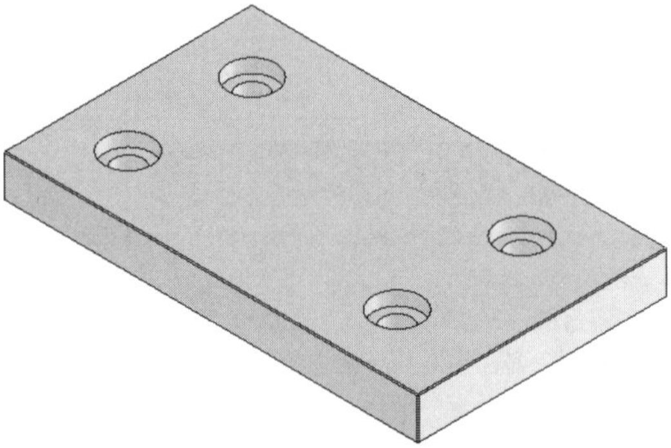

Figure 8-2 *Modified base plate*

Similarly, you can also edit the reference geometry and the sketches of the sketched features. The feature created using the reference geometry also modifies automatically, when you modify the reference geometry. For example, if you create a feature on a plane at some angle and then edit the angle of the plane, the resulting feature is automatically modified. In SolidWorks, you can perform editing tasks using various methods, which are discussed next.

Editing Using Edit Feature Option

Editing, using the **Edit Feature** option, is the most common method in SolidWorks. To edit any feature of the model using this option, select that feature either from the **FeatureManager Design Tree** or from the drawing area. Next, right-click on it to invoke the shortcut menu and choose the **Edit Feature** option, as shown in Figure 8-3. Depending on the feature selected, the **PropertyManager** or dialog box will be invoked, and you can modify the parameters of that feature. The **PropertyManager** will also have the sequence number of the feature. The **Extrude PropertyManager** is displayed in Figure 8-4 with the sequence number 1. After editing the parameters, choose the **OK** button to complete the feature creation. The feature will be modified automatically.

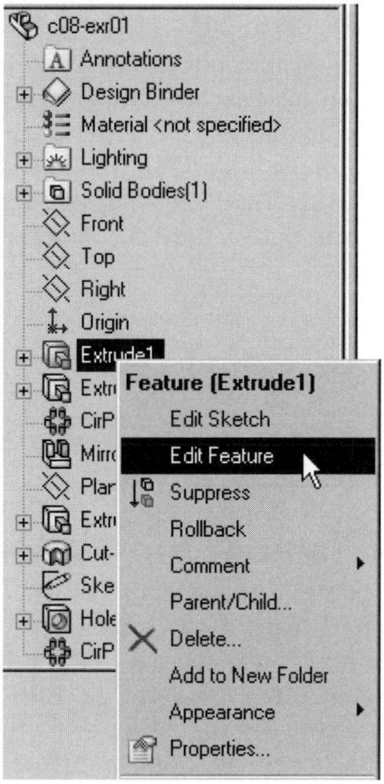

***Figure 8-3** Selecting the **Edit Feature** option from the shortcut menu*

Figure 8-4** The **Extrude PropertyManager

Editing the Sketches of the Sketched Features

In SolidWorks, you can also edit the sketches of the sketched features using the **Edit Sketch** option. Select the feature either from the **FeatureManager Design Tree** or from the drawing area and right-click to invoke the shortcut menu. Choose the **Edit Sketch** option from it and you will enter the sketching environment. Using the standard sketching tools, edit the sketch of the sketched feature and exit the sketching environment. Choose CTRL+B to rebuild the model. You can also select the **Rebuild** button from the **Standard** toolbar to exit the sketching environment and rebuilt the model.

> **Tip**. *You can also use the + sign available on the left of a sketched feature to expand the sketched feature in the **FeatureManager Design Tree**. The sketch icon will be displayed when you expand the sketched feature. Select the sketch icon and invoke the shortcut menu. Select the **Edit Sketch** option from it to enter the sketching environment to edit the sketch.*

Changing the Sketch Plane of the Sketches

You can also change the sketch plane of the sketches of the sketched features. To edit the sketch plane, expand the sketched feature by clicking on the + sign on its left in the **FeatureManager Design Tree**. Select the sketch icon in the **FeatureManager Design Tree**. Right-click to invoke the shortcut menu and choose the **Edit Sketch Plane** option from it. Figure 8-5 shows the **Edit Sketch Plane** option chosen from the shortcut menu. The **Sketch Plane PropertyManager** will be displayed, as shown in Figure 8-6. The name of the current sketch plane is displayed in the **Sketch Plane/Face** selection area. Now, select any other plane or face as the sketching plane. Next, choose the **OK** button from the **Sketch Plane PropertyManager**.

> **Tip**. *If you select a sketch plane on which the relations and dimensions do not find any reference to be placed, the **What's Wrong** dialog box will be displayed. You, will need to undo the last step using the **Undo** button from the **Standard** toolbar. After this, invoke the **Sketch Plane PropertyManager** again, and select the appropriate plane. You will learn more about the **What's Wrong** dialog box later in this chapter.*

Editing by Double-Clicking the Entities and Features

You can also edit a feature, reference geometry, or a sketch by double-clicking the feature either from the **FeatureManager Design Tree** or from the drawing area. For example, if you double-click an extrude feature, all its dimensions, and of the sketch used to create it are displayed in the drawing area. Remember that the dimensions of the sketch will be displayed in black, and the dimensions of the feature will be displayed in blue. Double-click the dimension that you need to modify; the **Modify** dialog box is invoked. Set the new value in the **Modify** dialog box and press the ENTER key on the keyboard, or choose the **Save the current value and exit the dialog** button from the dialog box. You will notice that the value of the dimension is modified but the model is not modified relative to the modified value. Therefore, you need to rebuild the model using the **Rebuild** option. To rebuild the model, choose the **Rebuild** button from the **Standard** toolbar or choose CTRL+B on the keyboard.

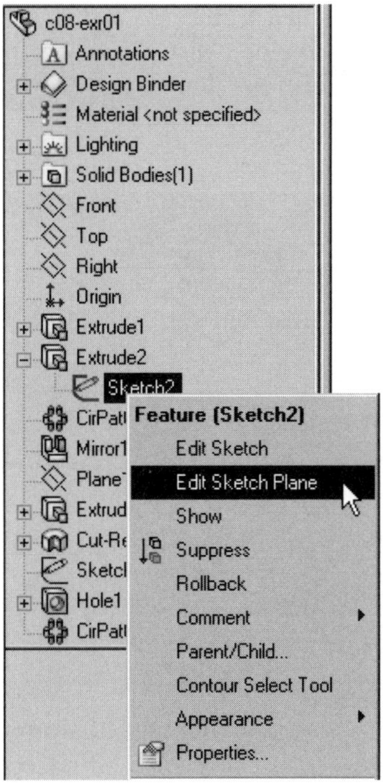

*Figure 8-5 The **Edit Sketch Plane** option being selected*

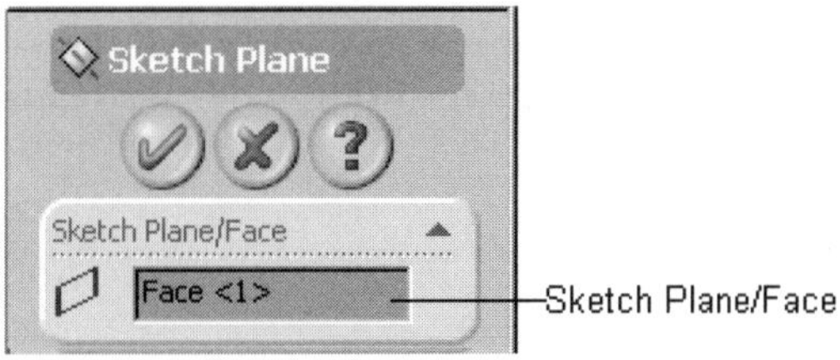

*Figure 8-6 The **Sketch Plane PropertyManager***

Editing Using the Move/Size Features Tool

CommandManager:	Features > Move/Size Features	*(Customize to Add)*
Toolbar:	Features > Move/Size Features	*(Customize to Add)*

You can dynamically modify the feature and the sketch of the sketched feature without invoking the sketching environment by using this option. To edit the feature or sketch, choose the **Move/Size Features** button from the **Features CommandManager**. You will notice that the **Move/Size Features** button is chosen in it. Select any face of the feature to modify. The selected face will be highlighted in green and if the selected feature is a sketched feature then the sketch of that feature will also be highlighted in green. You will be provided with the resize handle, rotate handle, and the move handle. Figure 8-7 shows the move handle, resize handle, and the rotate handle provided for the selected feature.

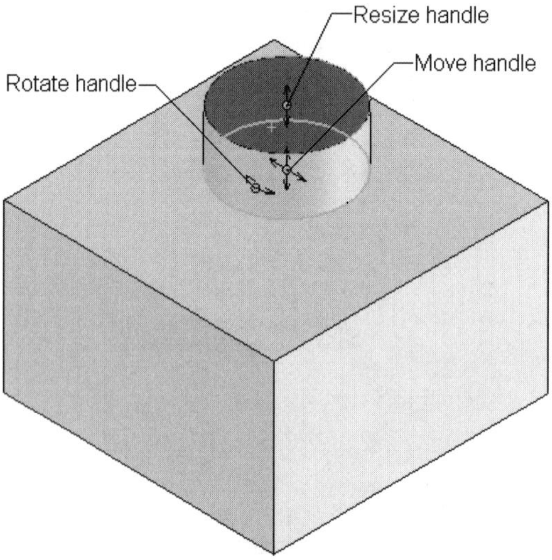

Figure 8-7 *Feature selected with* ***Move/Size Features*** *button selected*

To resize the feature, move the cursor to the resize handle; the select cursor will be replaced by the resize cursor with its name displayed as the tooltip. Press and hold down the left mouse button at this location and drag the cursor to resize the feature. The feature will be resized with an increment of 10. Release the left mouse button when you have resized it. You will notice that the feature will be dynamically resized. Figure 8-8 shows the cursor being dragged to resize the feature. Figure 8-9 shows the resulting modified feature.

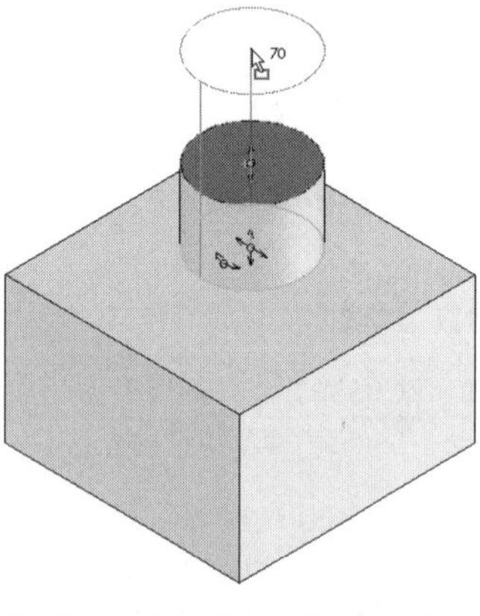

Figure 8-8 *Dragging the resize handle to resize the feature*

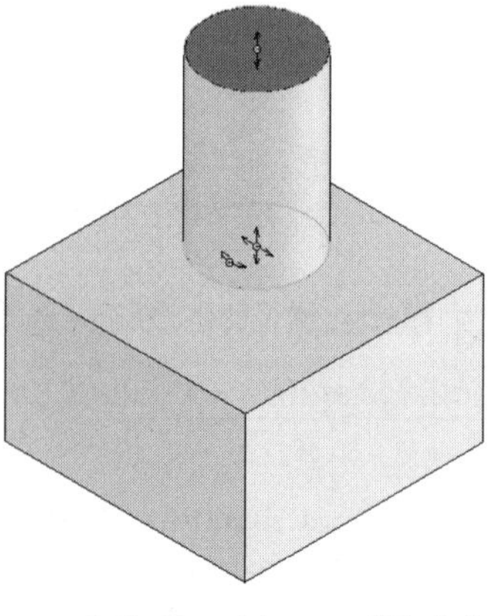

Figure 8-9 *Resulting modified feature*

The rotate handle is used to rotate the selected feature. To rotate the feature using the rotate handle, move the cursor to the rotate handle, press and hold down the left mouse button. Drag the cursor to rotate the feature. You can drag the feature clockwise or counterclockwise. The preview of the feature will be displayed in the drawing area. The feature will be rotated with an increment of 10-degree. Release the left mouse button, after rotating the feature to a specified location.

Note

If you rotate a sketched feature, in which its sketch is fully or partially defined using relations and dimensions, the ***Move Confirmation*** *dialog box will be displayed, as shown in Figure 8-10. This dialog box informs you that the external constraints in the feature are being moved, and asks whether you want to delete those constraints or keep the constraints by recalculating or make them dangling. The relations or dimensions that do not find the external reference after the placement are made dangling.*

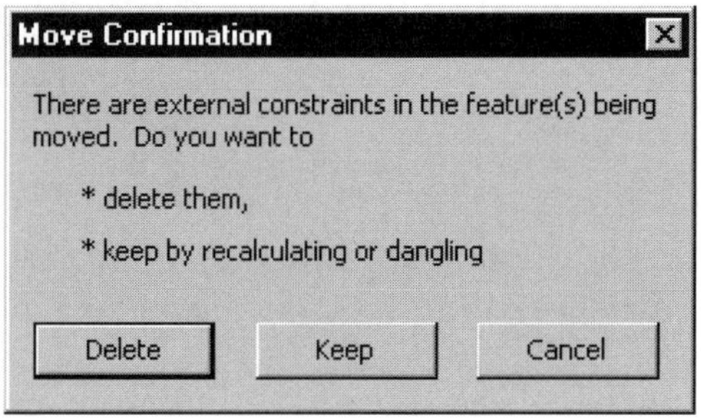

Figure 8-10 *The* ***Move Confirmation*** *dialog box*

After rotating the feature, if you invoke the **Move Confirmation** dialog box, you need to choose either the **Delete** or the **Keep** button, based on the geometric and dimensional conditions. Figure 8-11 shows the preview of rotating feature. Figure 8-12 shows the resulting rotated feature.

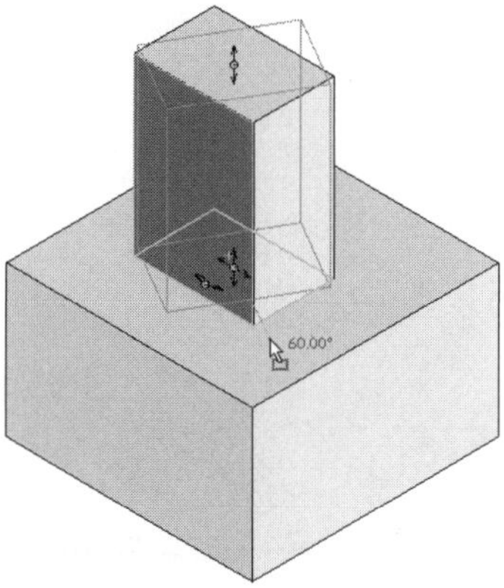

Figure 8-11 *Preview of feature being rotated*

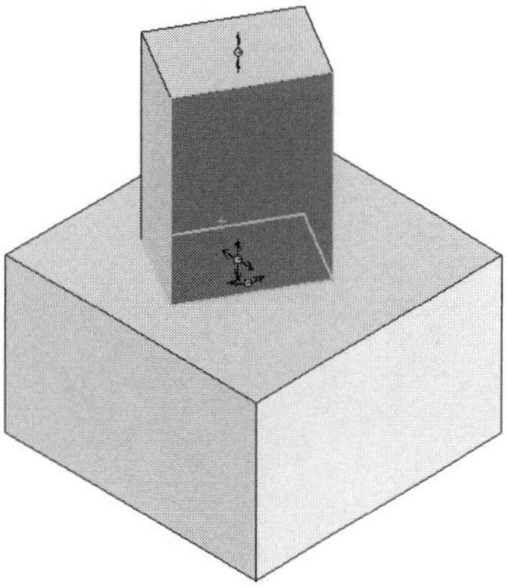

Figure 8-12 *Resulting rotated feature*

To move a feature using the move handle, select a face of the feature and move the cursor to the move handle. Press and hold down the left mouse button, drag the cursor to move the feature to the desired position and then release the left mouse button. If the dimensions or relations of the feature do not remain defined after moving, the **Move Confirmation** dialog box will be displayed. Choose the options available in this dialog box. Using these options, you can also change the placement plane or the sketch plane of the feature. Figure 8-13 shows the feature being moved to another face. Figure 8-14 shows the resulting moved feature.

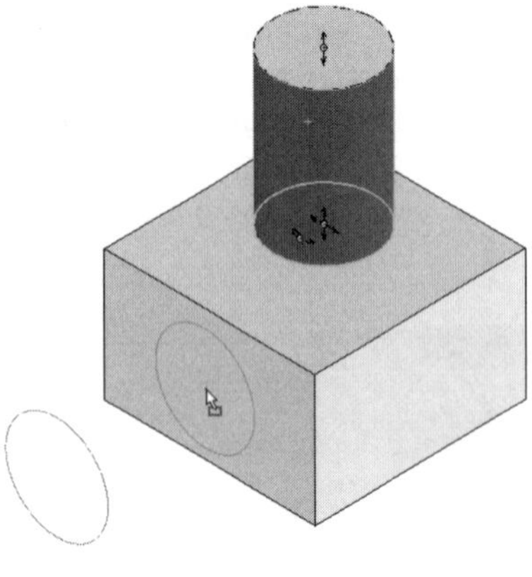

Figure 8-13 *The feature being moved*

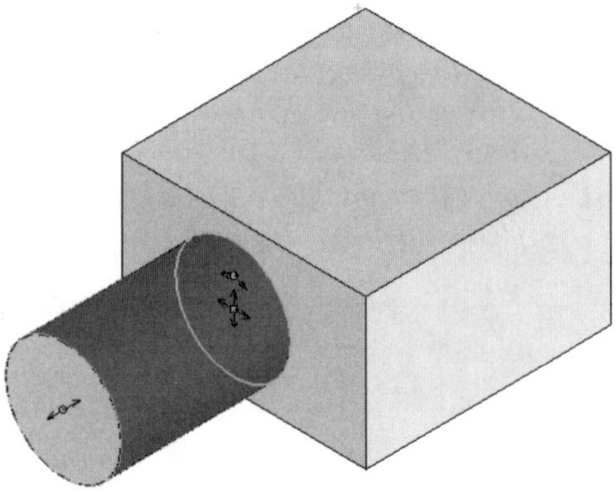

Figure 8-14 *Resulting moved feature*

You can also modify the sketches of the sketched feature using this option. To modify the sketch, move the cursor to the sketch highlighted in light green color. The symbol of the geometry, close to which the cursor is moved, appears on the right of the cursor. Press and hold down the left mouse button, and drag the cursor. After modifying the sketch to the required size, release the left mouse button. Figure 8-15 shows the preview of the sketch being modified. Figure 8-16 shows the resulting modified model.

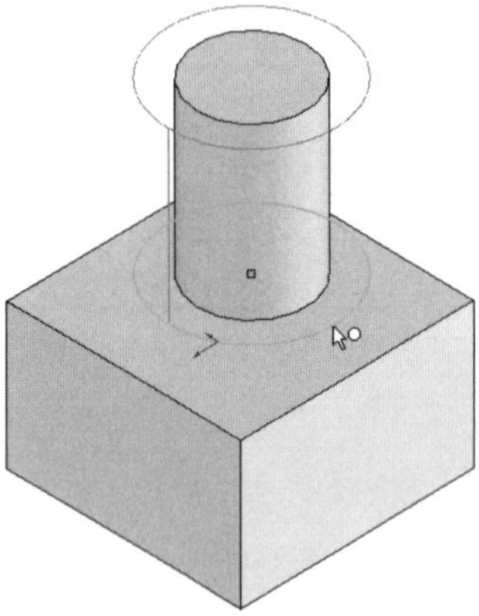

Figure 8-15 *Preview of the sketch being modified*

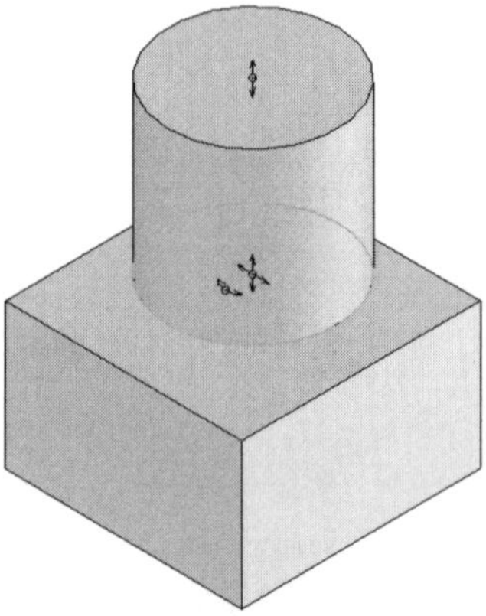

Figure 8-16 *Resulting modified model*

Tip. *When you resize, move, or rotate a feature using the* ***Move/Size Features*** *tool, the feature(s) that has/have a relationship with that feature will also be modified. In other words, if a child-parent relationship is established between the features, the child features will also be modified when the parent feature are modified.*

To clear the selected feature, click once anywhere in the drawing area, or choose the ESC key on the keyboard. You will notice that the ***Move/Size Features*** *button is still chosen in the* ***Features*** *toolbar, indicating that the dynamic move/size features option is still active. To exit this option, choose the* ***Move/Size Features*** *button again.*

Editing the Sketches With the Move/Size Features Tool Active

When the **Move/Size Features** tool is active, and you choose the **Edit Sketch** option to edit the sketch of the sketched feature, you will enter the sketching environment. The feature whose sketch you need to modify will be displayed in transparent yellow temporary graphics, as shown in Figure 8-17.

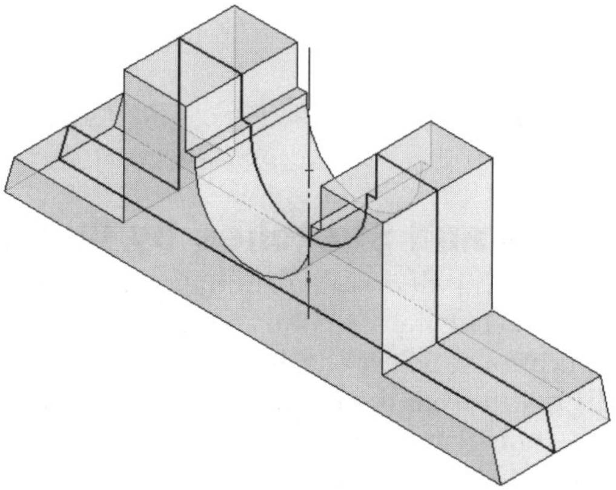

Figure 8-17 *Preview of the sketch being modified*

Tip. *If you want to modify a fully or partially defined sketch, by dragging, the* ***Override Dims on Drag/Move*** *option should be selected. To select this option choose* ***Tools > Sketch Settings > Override Dims on Drag/Move*** *from the menu bar. If this option is not selected, you cannot move or drag a dimensioned sketched entity.*

Now, modify the sketch according to the requirement. Consider a case in which two circles are added to the model and the height of the inclined line is increased. The model shown in temporary graphics will be modified dynamically in the sketching environment. Therefore, you can have a better understanding, that how the model will be displayed after modifying the sketch. Note that the temporary graphic will show the preview of the feature created by the section you add only if you drag one of the entities in the sketch. Otherwise, the new addition will not be displayed as a preview in the temporary graphics. It is recommended that you invoke the **Modify/Size Features** button before modifying the sketch. After modifying the sketch, exit the sketching environment. If the sketch of the model is dimensioned, you can modify it by modifying the dimension. Figure 8-18 shows two circles added to the model and the length of the inclined line being increased. Figure 8-19 shows the resulting modified model in the sketching environment.

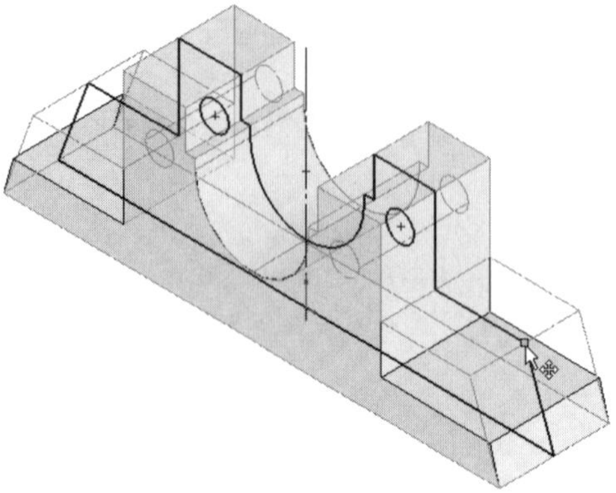

Figure 8-18 *Sketch being modified*

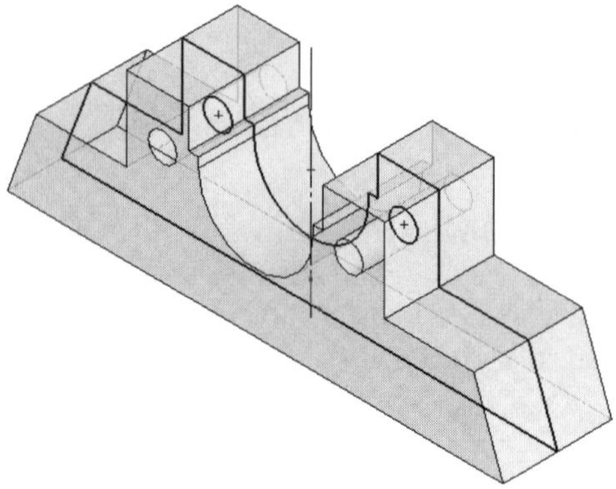

Figure 8-19 *Resulting modified model in the sketching environment*

Editing the Features and Sketches by Cut, Copy, and Paste

SolidWorks allows you to adapt the windows functionality of cut, copy, and paste to copy and paste the features and sketches. The method of using this functionality is the same as used in other windows-based applications. Select the feature or sketch to cut or copy. To cut the selected item, choose **Edit > Cut** from the menu bar or use the shortcut key, CTRL+X. The **Confirm Delete** dialog box is displayed, because when you cut the selected item, it is deleted from the document. Choose the **Yes** button from this dialog box. You will learn more about deleting later in this chapter. After you cut an item, select the placement plane or placement reference where you want to place the feature. Choose **Edit > Paste** from the menu bar or use the shortcut key CTRL+V on the keyboard. Sometimes the **Copy Confirmation** dialog box is displayed, as shown in Figure 8-20, in which, you are prompted either to delete the external constraints or leave them dangling. This dialog box is displayed only when the item to paste has some external references in the form of relations and dimensions. The feature will be pasted on the selected reference.

If you copy and paste an item, the selected item will remain at its position and a copy of that item will be pasted on the selected reference. To copy an item, select the feature or sketch. Choose **Edit > Copy** from the menu bar, or press CTRL+C on the keyboard. Select the reference where you want to paste the selected item and choose **Edit > Paste** from the menu bar, or press CTRL+V on the keyboard to paste. You can paste the selected item any number of times. If you select another item and copy it on the clipboard, the last copied item will be deleted from the memory of the clipboard.

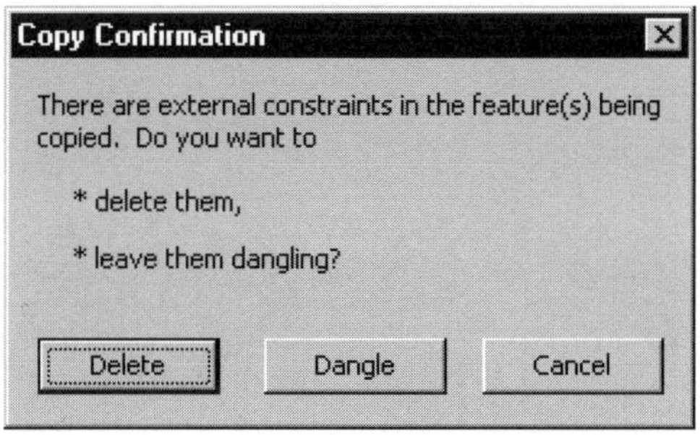

*Figure 8-20 The **Copy Confirmation** dialog box*

Tip. *For pasting a selected sketch, you need to select a plane or a planar face as the reference. For pasting a sketched feature, a simple hole, or a hole created using the hole wizard, you have to select a plane or a planar face as the reference. For pasting chamfers and fillets, you have to select an edge, edges, or face as the references.*

Cut, Copy, and Paste Features and Sketches from One Document to the Other

You can also cut or copy the features and sketches from one document and paste them in another document. For example, you need to copy a sketch created in the current document and paste it in a new document. You need to select the sketch and use CTRL+C on the keyboard to copy the item to the clipboard. Create a new document in the **Part** mode and select the plane on which you want to paste the sketch. Press CTRL+V on the keyboard to paste the sketch on the selected plane. Using the same procedure, you can also copy features from one document to the other.

Copying Features Using Drag and Drop

SolidWorks also provides you with drag drop functionality of windows to copy and paste the item within the document. Press and hold down the CTRL key on the keyboard, and select and drag that item from the drawing area, or from the **FeatureManager Design Tree**. Drag the cursor to a location where you want to paste the item and release the left mouse button on that location. If the item to be pasted is defined using the dimensions or relations, the **Copy Confirmation** dialog box will be displayed to delete or make those constraints dangle. Figure 8-21 shows the feature being dragged. Figure 8-22 shows the resulting pasted feature.

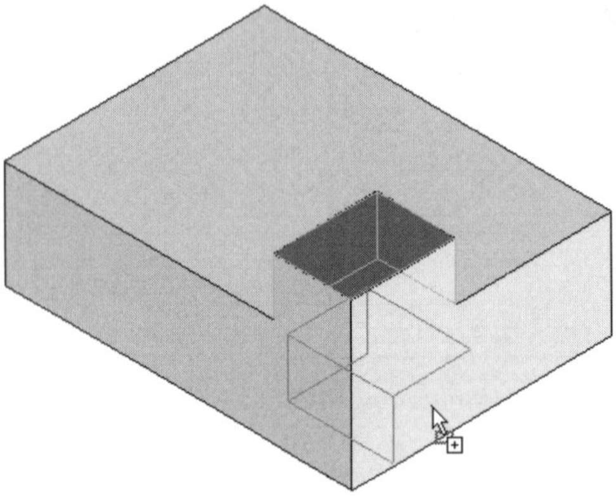

Figure 8-21 Feature being dragged

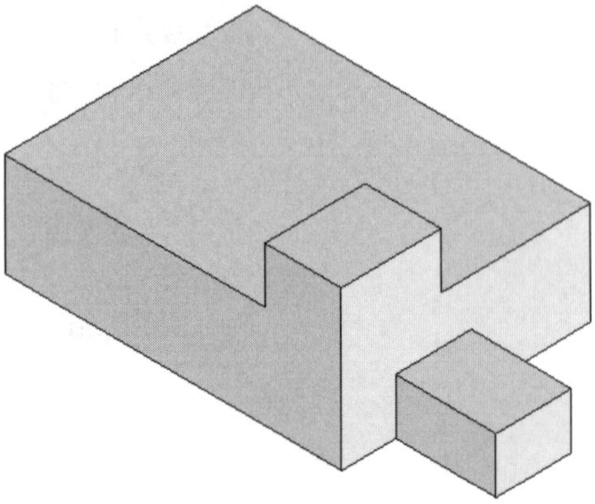

Figure 8-22 Resulting pasted feature

Dragging and Dropping Features from One Document to Other

You can also drag and drop features and sketches from one document to the other. For pasting the items from one document to the other, you should open both the documents in the SolidWorks session. Choose **Windows > Tile Vertical/Tile Horizontal** from the menu bar. Display both the documents at the same time in the SolidWorks window. Press and hold down the CTRL key on the keyboard. Select and drag the feature or sketch to the other document, and place it on the required reference. Figure 8-23 shows the fillet feature being dragged to be applied on the edge of the model in the second document.

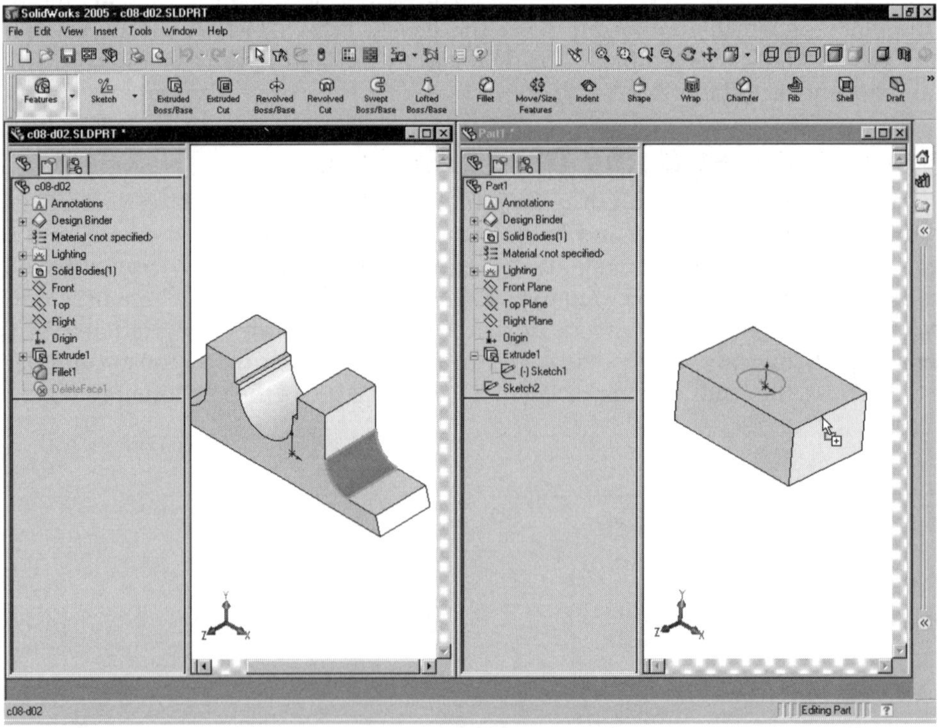

Figure 8-23 Fillet feature being dragged to be pasted in the second document

Deleting the Features

You can delete the unwanted features from the model by selecting the feature either from the **FeatureManager Design Tree** or from the drawing area. After selecting the feature to be deleted, choose the DELETE key on the keyboard, or right-click to invoke the shortcut menu and choose the **Feature** option from the **Feature** area. When you delete a feature, the **Confirm Delete** dialog box is displayed, as shown in Figure 8-24. The features that are dependent on the feature to be deleted are also displayed in the **Confirm Delete** dialog box, which informs you that all the dependent features to the parent feature will also be deleted. You are provided with the **Also delete all child features** check box. If this check box is selected, all the child features related to the parent feature are also deleted. When you delete a sketched feature, the sketches related to it will not be deleted. These sketches are known as absorbed features. To delete the absorbed features along with the parent feature, select the **Also delete absorbed features** check box from the **Confirm Delete** dialog box. Choose the **Yes** button to delete the selected features, or else choose the **No** button to cancel the delete operation. You can also delete a selected feature by choosing **Edit > Delete** from the menu bar.

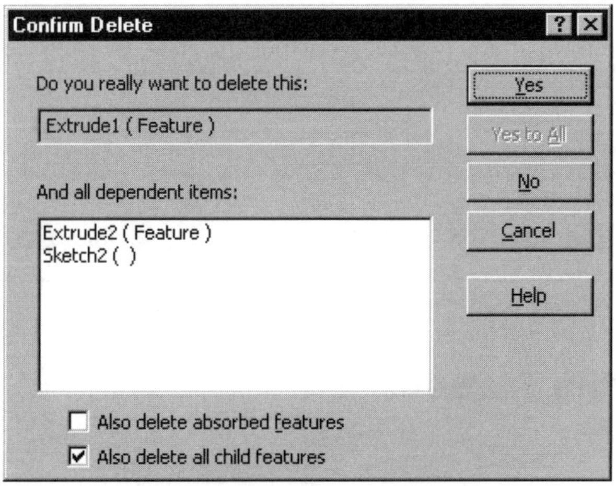

Figure 8-24 *The **Confirm Delete** dialog box*

Deleting the Bodies

CommandManager:	Features > Delete Solid/Surface	*(Customize to Add)*
Menu:	Insert > Features > Delete Body	
Toolbar:	Features > Delete Solid/Surface	*(Customize to Add)*

As discussed earlier, the multibody environment is supported in SolidWorks. Therefore, you can create multiple disjoint bodies in SolidWorks. You can also delete the unwanted bodies. The bodies to be deleted can be either solid bodies or surface bodies. To delete a body, choose **Insert > Features > Delete Body** from the menu bar. You can invoke this tool by selecting the **Delete Solid/Surface** button from the **Features CommandManager** after customizing the **CommandManager**. The **Delete Body PropertyManager** is invoked, as shown in Figure 8-25. You are prompted to select the solid and/or surface bodies to be deleted.

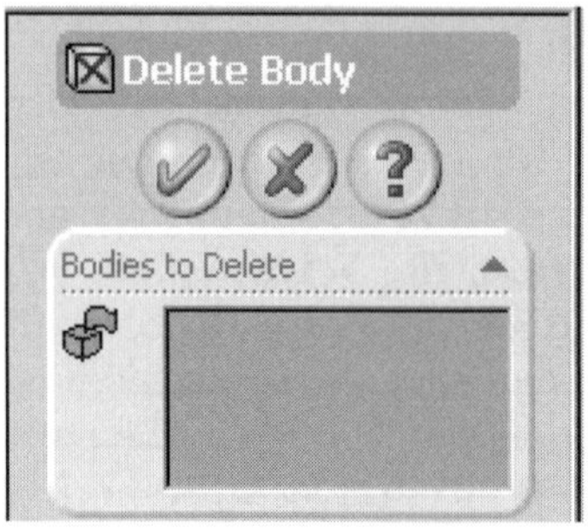

*Figure 8-25 The **Delete Body PropertyManager***

Select the body or bodies to be deleted from the drawing area or from the **Solid Bodies** folder available in the **FeatureManager Design Tree**, which is displayed in the drawing area. The selected body is displayed in green and its name is displayed in the **Bodies to Delete** selection area. Choose the **OK** button from the **Delete Body PropertyManager**. A new item with the name **Body-Delete1** appears in the **FeatureManager Design Tree**. This item stores the deleted bodies. Therefore, at any point of your design cycle, you can delete or suppress this item to resume the deleted body back in your design. You will learn more about suppressing the feature later in this chapter.

Tip. *You can also choose the **Delete Body** option from the shortcut menu. To delete a body using the shortcut menu, select the body and right-click to invoke the shortcut menu. Choose the **Delete** option from **the Body** area of the shortcut menu; the **Delete Body PropertyManager** will be displayed. Choose the **OK** button from the **Delete Body PropertyManager** to delete the body.*

Suppressing the Features

CommandManager:	Features > Suppress	*(Customize to Add)*
Menu:	Edit > Suppress > This Configuration	
Toolbar:	Features > Suppress	*(Customize to Add)*

Sometimes, you do not want a feature to be displayed in the model or in its drawing views. Instead of deleting those features, they can be suppressed. When you suppress a feature, it is neither visible in the model nor in the drawing views. If you create an assembly using that model, the suppressed feature will not be displayed even in the assembly. You can resume the feature anytime by unsuppressing it. When you suppress a feature, the features that are dependent on it are also suppressed. To suppress a feature, select it from the **FeatureManager Design Tree** or from the drawing area. Choose the **Suppress** button from the **Features CommandManager** after customizing it, or right-click and choose the **Suppress** option from the shortcut menu. The suppressed feature will be removed from the display of the model, and the icon of the feature will be displayed in gray in the **FeatureManager Design Tree**. When you suppress a feature, then all the features dependent on that feature are also suppressed.

Unsuppressing the Features

CommandManager:	Features > Unsuppress *(Customize to Add)*
Menu:	Edit > Unsuppress > This Configuration
Toolbar:	Features > Unsuppress *(Customize to Add)*

The suppressed features can be unsuppressed using this option. To resume the suppressed feature, select the suppressed feature from the **FeatureManager Design Tree** and choose the **Unsuppress** button from the **Features CommandManager**. You can also choose this option from the shortcut menu, after selecting the suppressed feature. As discussed earlier, when you suppress a feature, the dependent features are also suppressed. But when you resume a suppressed feature, the dependent features remain suppressed. Therefore, you need to unsuppress all the features independently. The method of resuming the parent feature and the dependent features using a single option is discussed next.

Unsuppressing the Features With Dependents

CommandManager:	Features > Unsuppress with Dependents *(Customize to Add)*
Menu:	Edit > Unsuppress with Dependents > This Configuration
Toolbar:	Features > Unsuppress with Dependents *(Customize to Add)*

Using this option you can resume the suppressed feature along with the dependents of the suppressed parent feature. To resume the suppressed feature using this option, select the suppressed feature from the **FeatureManager Design Tree**. Choose **Unsuppress with Dependents** button from the **Features** toolbar. You will observe that the dependent suppressed features are also unsuppressed.

Hiding the Bodies

While working in the multibody environment, you can also hide the bodies. The hidden body is not displayed in the model, or assembly, or in the drawing views. The display of the dependent bodies is not turned off when you hide a body. To hide a body, expand the **Solid Bodies** folder available in the **FeatureManager Design Tree,** and select the body to hide. Right-click to invoke the shortcut menu and choose the **Hide Solid Body** option from it. The selected body will disappear from the drawing area. The icon of the hidden body is displayed in wireframe in the **Solid Bodies** folder. Select the hidden body from that folder and choose the **Show Solid Body** option from the shortcut menu to turn on the display of the hidden body.

Move/Copy Bodies

CommandManager:	Features > Move/Copy Bodies *(Customize to Add)*
Menu:	Insert > Features > Move/Copy Bodies
Toolbar:	Features > Move/Copy Bodies *(Customize to Add)*

In SolidWorks you can also move or copy the bodies by choosing the **Move/Copy Bodies** button from the **Features CommandManager** after customizing it, or choosing **Insert > Features > Move/Copy Bodies** from the menu bar. The **Move/**

Copy Body PropertyManager will be displayed, as shown in Figure 8-26. The confirmation corner is also displayed in the drawing area.

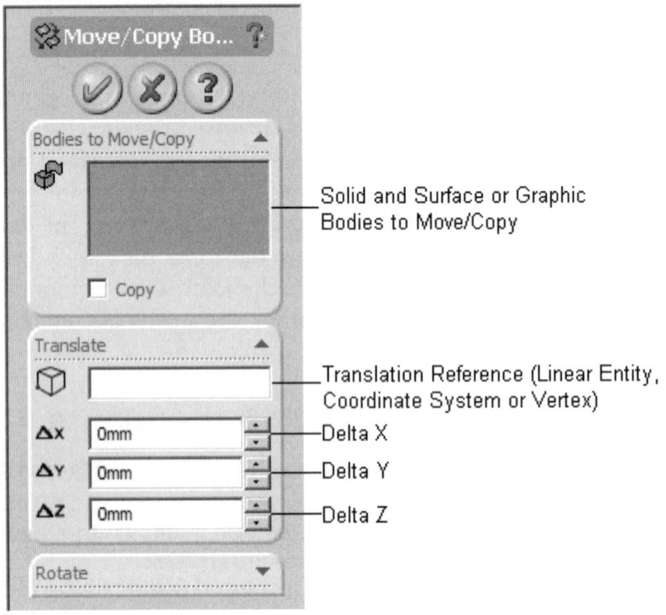

Figure 8-26** The **Move/Copy Body PropertyManager

You will notice a 3D triad is displayed at the centroid of the selected body. It is used to dynamically rotate and move the selected body. The three arrows of this triad denote the three axis, which are X, Y, and Z. Move the cursor close to any one of the arrows of the triad. The select cursor is replaced by the mouse cursor as shown in Figure 8-27. Use the right mouse button to rotate the selected body and the left mouse button to move the selected body along that direction. You can also move the triad and place it at some other location. To move the triad, move the cursor to the spherical ball where the three arrows meet. The cursor is replaced by the move cursor. Drag it to move the triad and then place it at a desired location.

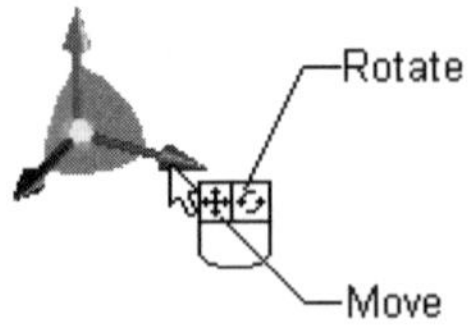

***Figure 8-27** 3D triad along with the cursor*

Various options available in the **Move/Copy Body PropertyManager** are discussed next.

Bodies to Move/Copy

The **Bodies to Move/Copy** rollout is used to define the body to copy or move. On invoking the **Move/Copy Body PropertyManager**, you are prompted to select the bodies to move/copy and set the options. Move the cursor to the body to be selected. The cursor will be replaced by the body selection cursor and the edges of the body will be highlighted in red. The name of the body is also displayed in the tooltip. Use the left mouse button to select the body, which will then be highlighted in green. The name of body is displayed in the **Solid and Surface or Graphics Bodies to Move/Copy** selection area. You can also select the body from the **Solid Bodies** folder after, invoking the **FeatureManager Design Tree** flyout.

Copy

The **Copy** check box is cleared by default. If you select the **Copy** check box, you can create the copies of the selected body. When you select this check box, the **Number of Copies** spinner is displayed below the **Copy** check box. Set the number of copies using this spinner.

Translate

The **Translate** rollout available in the **Move/Copy Body PropertyManager** is used to define the translational parameters to move the selected body. Set the value of the destination in the **Delta X**, **Delta Y**, and **Delta Z** spinners. When you set the value, the preview of the moved body is displayed in temporary graphics in the drawing area. You can also move or copy the selected body between points. To move or copy a body by specifying two points, select the **Translation Reference (Linear Entity, Coordinate System, Vertex)** selection area. The selection mode in this area is active. Select the vertex from which you want the translation to start. When you select the first vertex as translation reference, the **Delta X**, **Delta Y**, and **Delta Z** spinners are replaced by the **To Vertex** selection area. The selection mode in the **To Vertex** selection area is active. Select the second translation reference. You will observe the preview of the body move between the two selected vertices. The placement of the body also depends on the sequence of selection of vertices. Therefore, you need to be very careful, while selecting the two vertices. Figure 8-28 shows the sequence for the selection of references and Figure 8-29 shows the resulting copied body.

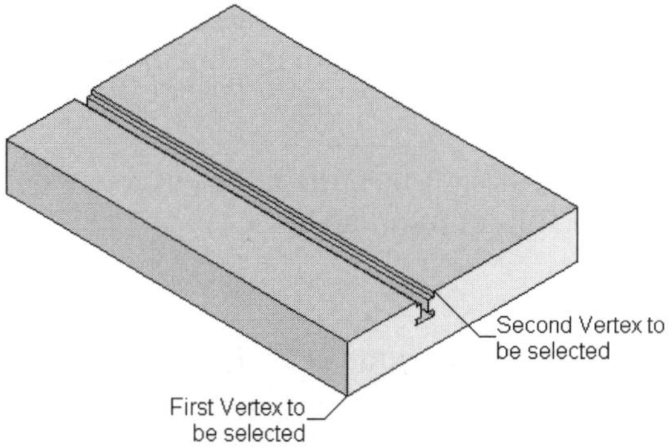

Figure 8-28 *Sequence of selection*

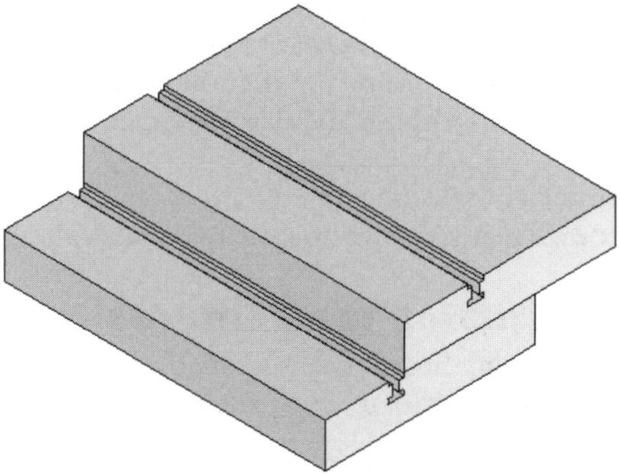

Figure 8-29 *Resulting copied body*

Rotate

The **Rotate** spinner available in the **Move/Copy Body PropertyManager** is used to define the parameters to rotate the body. To open this rollout, click once on the black arrow provided at the right of this rollout. The **Rotate** rollout is shown in Figure 8-30.

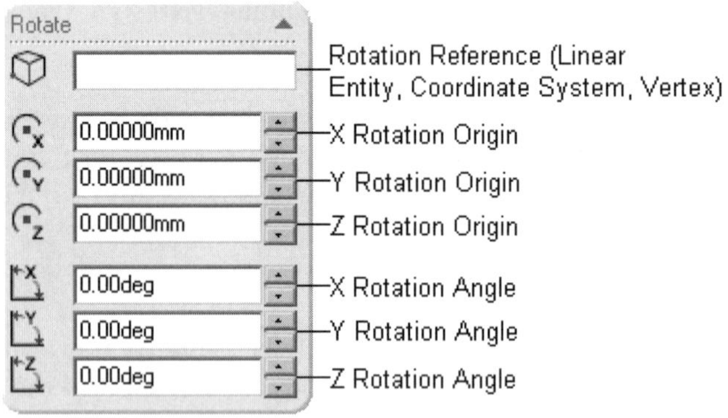

***Figure 8-30** The **Rotate** rollout*

As discussed earlier, a filled square is placed at the origin when you invoke the **Move/Copy PropertyManager**. It is clearly visible, when you hide the origin by choosing **View > Origin** from the menu bar. It indicates the origin along which the selected body will be rotated. You can adjust the position of this temporary moveable origin using the **X Rotation Origin** spinner, **Y Rotation Origin** spinner, and **Z Rotation Origin** spinner. The **X Rotation Angle** spinner is used to set the value of the angular increment to rotate or copy the body along the X axis, the **Y Rotation Angle** spinner is used to rotate or copy the body along the Y axis, and the **Z Rotation Angle** spinner is used to rotate or copy the body along the Z axis.

To rotate or copy the selected body along an edge, click once in the **Rotation Reference (Linear Entity, Coordinate System, Vertex)** selection area to invoke the selection. Select the edge along which you want to rotate the selected body. When you select an edge, all the other spinners disappear from the rollout, and the **Angle** spinner is invoked in the **Rotate** rollout. Set the value of the angular increment in this spinner.

Instead of selecting an edge, you can also select a vertex along which the body will rotate or copy. Then you need to specify the axis along which you want to rotate it.

Reordering the Features

Reordering the features is defined as a process of changing the position of the features created in the model. Sometimes, after creating a model, it may be required to change the order in which its features were created. For reordering the features, the features are dragged and placed before or after another feature in the **FeatureManager Design Tree**.

For reordering the features, select the feature in the **FeatureManager Design Tree,** and drag it to the required position. When you drag the feature to reorder, the bend arrow pointer is displayed, which suggests that feature dragging is possible. If you drag the child feature above the parent feature, the reorder error pointer will be displayed. If you drag a child feature above a parent feature, the **SolidWorks** warning box will be displayed, as shown in Figure 8-31. Choose **OK** from this warning box.

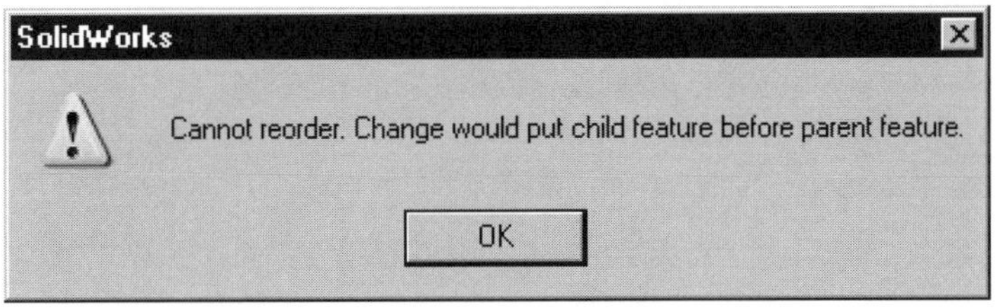

Figure 8-31 *The **SolidWorks** warning box*

Consider a case, in which you have created a rectangular block and a pattern of through holes created on the base feature, as shown in Figure 8-32. Now, if you create a shell feature and remove the top face, the front face, and the right face of the model, it will appear as shown in Figure 8-33.

But this was not the desired result. Hence, you need to reorder the shell feature before the holes. Select the shell feature in the **FeatureManager Design Tree** and drag it above the holes. All the features will be automatically adjusted in the new order, as shown in Figure 8-34.

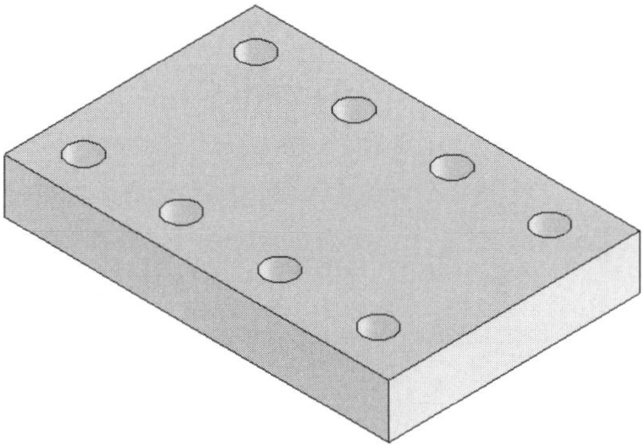

Figure 8-32 *Model created with a pattern of through holes on the base feature*

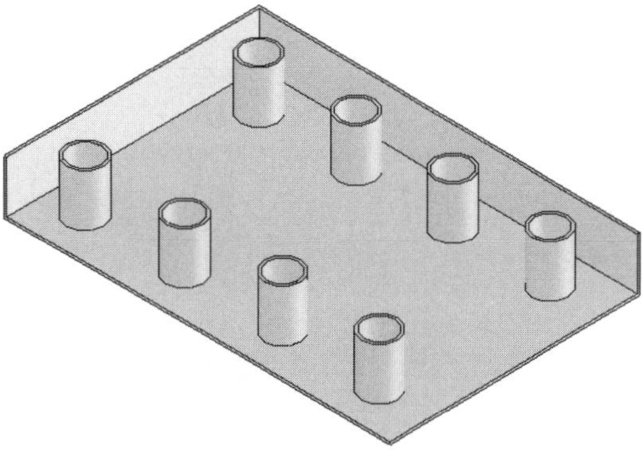

Figure 8-33 *Shell feature added to the model*

Rolling Back the Model

Rolling back the model is defined as a process, in which you rollback the model to an earlier stage. When you rollback a feature or features, those features are suppressed, and you can add new features when the model is in the rollback state. The newly added features are added before the features that are rolled back. While working with a multifeatured model, if you want to edit a feature that was created at the starting of the design cycle of the model, it is recommended that you should rollback the model up to that feature. This is because after

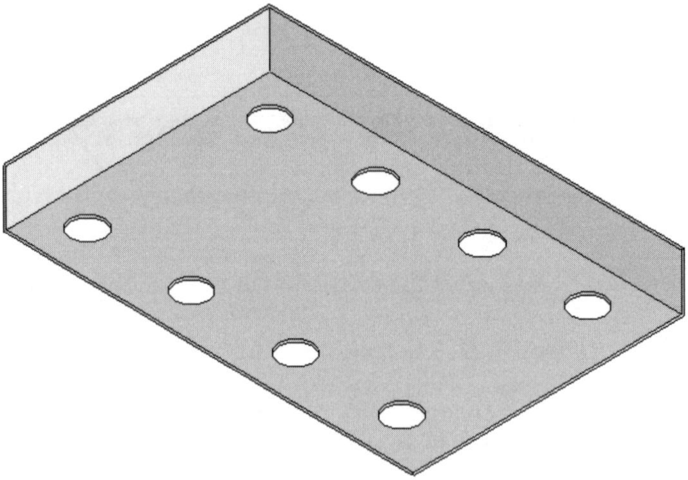

Figure 8-34 *Model after reordering the features.*

each editing operation, the time of regeneration will be minimized. Rolling back is done using the **Rollback Bar** available in the **FeatureManager Design Tree,** as shown in Figure 8-35.

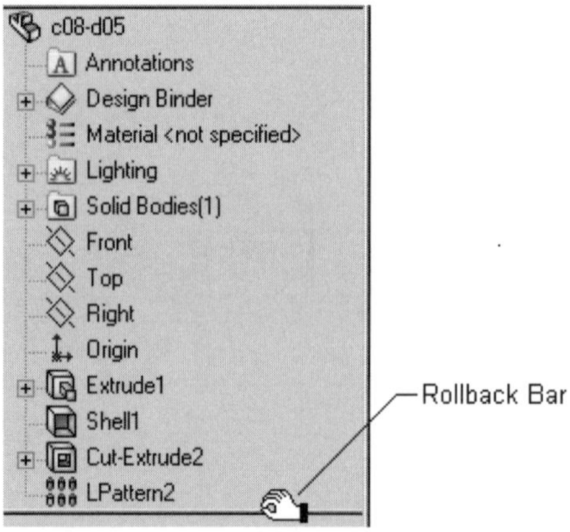

Figure 8-35 *The rollback bar of the* ***FeatureManager Design Tree***

Using the select tool, select the **Rollback Bar**; after selection it will be changed to blue color and the select cursor will be replaced by the hand pointer. Drag the hand pointer to the feature up to which you want to rollback the model, and then release it. To resume or roll the model, drag the **Rollback Bar** to the last feature of the model. You can also rollback using the menu bar. Select the feature up to which you want to rollback the model, and choose **Edit > Rollback** from the menu bar.

Tip. *If you want to rollback the model to the previous step, choose **Edit > Roll to Previous** from the menu bar. To rollback the entire model to its original position, choose **Edit > Roll to End** from the menu bar.*

*You can also choose the **Roll Forward, Roll Previous, Roll End** options from the shortcut menus invoked by selecting the features placed below the **Rollback Bar**. These options are used to control the roll and rollback of the model.*

*You can also rollback the model using the keyboard, by selecting the **Rollback Bar** and using the CTRL+ALT+Up Arrow key to roll forward. Use CTRL+ALT+Down Arrow key to roll backward.*

Renaming the Features

The name of the features is displayed in the **FeatureManager Design Tree**. By default, the naming of the features is done according to the sequence in which they are created. You can also rename the features, according to your convenience, by selecting the feature from the **FeatureManager Design Tree** and then clicking once on the selected feature. An edit box will be displayed in the **FeatureManager Design Tree**. Enter the name of the feature and press the ENTER key or click anywhere on the screen.

Creating Folders in the FeatureManager Design Tree

You can also add the folders in the **FeatureManager Design Tree** and the features displayed in the **FeatureManager Design Tree** are added in the folder. This is done to reduce the length of the **FeatureManager**. Consider a case, in which the base of the model consists of more than one feature. You can add a folder named Base Feature, and add all the features used to create the base, in that folder. To add a folder in the **FeatureManager Design Tree,** select any feature, right-click to invoke the shortcut menu, and choose the **Create New Folder** option. A new folder is created above the selected feature. Specify the name of the folder and click anywhere on the screen. Now, you can drag and drop the features to the newly created folder. You can also rename the folder by selecting it and then clicking it once. Now, enter the name in the edit box and press the ENTER key.

To add the selected feature in a new folder, choose **Add to New Folder** from the shortcut menu. A new folder will be created in the **FeatureManager Design Tree,** and the selected feature will be added to the newly created folder. You can also delete the folder by selecting the folder, and invoking the shortcut menu, and selecting the **Delete** option from it. Using the options in this shortcut menu you can also rollback and suppress the features available in the selected folder.

What's Wrong? Functionality

Sometimes, a model may not rebuild properly after you modify a sketch or a feature because of errors resulting from the modification. Therefore, you are provided with a **What's Wrong** dialog box, as shown in Figure 8-36. The possible errors in the feature are displayed in this dialog box along with their detailed description.

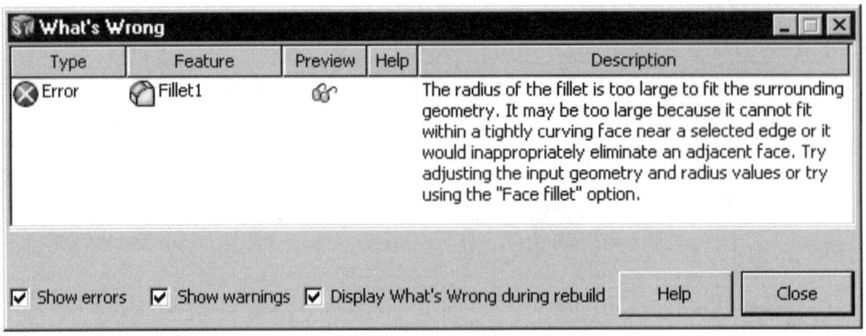

*Figure 8-36 The **What's Wrong** dialog box*

The **Show errors** check box is selected by default to display the errors in the **What's Wrong** dialog box. The **Show warnings** check box is selected by default to display the warnings messages. The **Display What's Wrong during rebuild** check box is selected by default and is used to display the errors at every rebuild of the model unless the error is fixed. After reading the description of the errors from this dialog box, choose the **Close** button to exit it. The errors are also displayed in the **FeatureManager design Tree**. The **FeatureManager Design Tree,** with an error in a feature, is displayed in Figure 8-37.

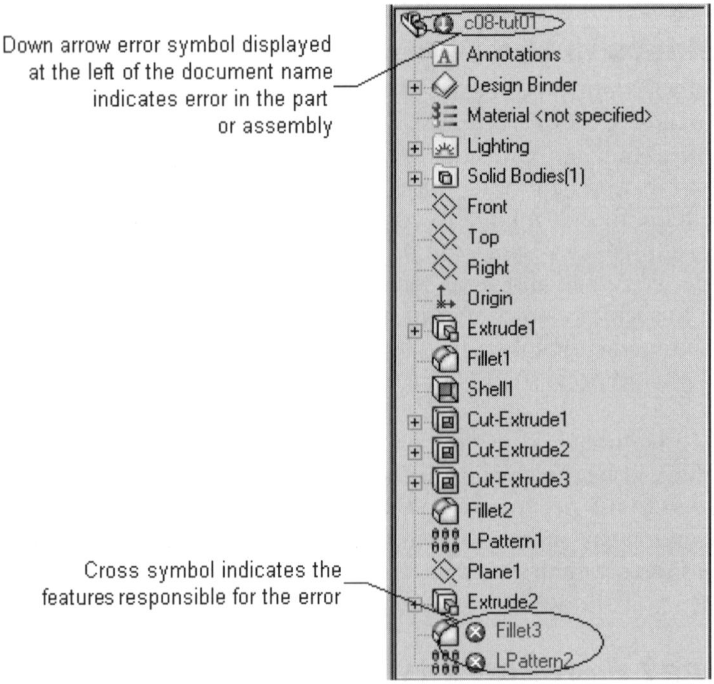

*Figure 8-37 The **FeatureManager Design Tree** with a feature having errors*

Tip. *You can also invoke the* ***What's Wrong*** *dialog box by selecting the feature having errors from the* ***FeatureManager Design Tree*** *and choosing the* ***What's Wrong?*** *option from the shortcut menu. To view all the errors in a model, select the name of the document from the top of the* ***FeatureManager Design Tree*** *and invoke the* ***Rebuild Errors*** *dialog box. All errors in the model are displayed in the* ***Rebuild Errors*** *dialog box.*

If there are some geometrical constraints, while creating the feature, the ***What's Wrong*** *dialog box is displayed.*

If there is an error in a model or in an assembly, the symbol appears on the left of the name of the model or assembly in the **FeatureManager Design Tree**. If a feature has an error, then the symbol appears on the left of the feature in the **FeatureManager Design Tree**. If there is an error in the child feature then the symbol appears on the left of the parent feature and also on the name of the document in the **FeatureManager Design Tree**. If a warning message appears for a feature, then the symbol appears on the left of that feature in the **FeatureManager Design Tree**.

TUTORIALS

Tutorial 1

In this tutorial, you will create the model shown in Figure 8-38. After creating some of its features, you will dynamically modify it, and then undo the modification. The dimensions of the model are shown in Figure 8-39. **(Expected time: 30 min)**

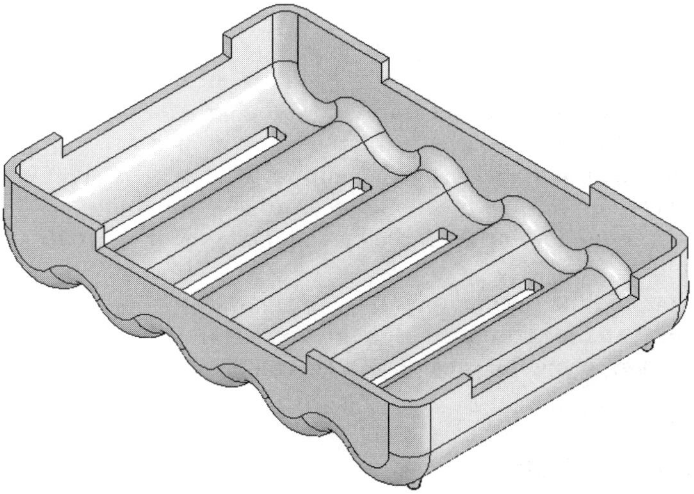

Figure 8-38 *Model for Tutorial 1*

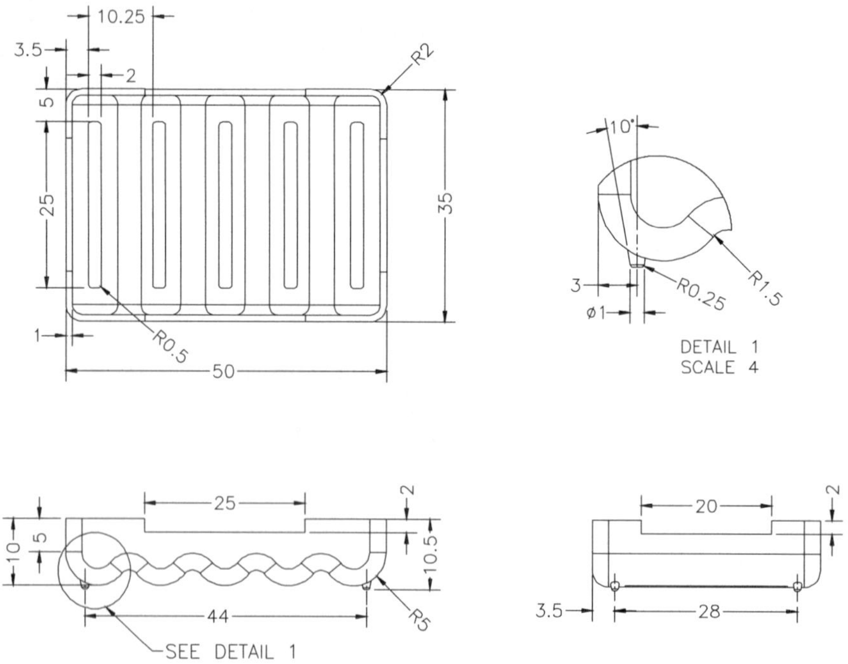

Figure 8-39 Views and dimensions of the model for Tutorial 1

The steps to be followed to create the model are listed next.

a. Create the base feature of the model by extruding the profile to a given distance, refer to Figures 8-40 and 8-41.
b. Add the fillets to the base feature, refer to Figures 8-42 and 8-43.
c. Add the shell feature to the model and remove the top face of the base feature, refer to Figures 8-44 and 8-45.
d. Dynamically modify the model, refer to Figures 8-46 through 8-47.
e. Create the cuts on the sides of the model, refer to Figure 8-49.
f. Create the slots on the lower part of the base and add the fillet to the slots feature, refer to Figure 8-49.
g. Pattern the slots and the fillet feature, refer to Figure 8-49.
h. Create a plane at an offset distance from the **Top Plane**.
i. Create the standoffs using the extrude and **Fillet** tool, and pattern the standoffs, refer to Figure 8-50.

Creating the Base Feature

You will draw the sketch of the base feature on the **Front Plane** and extrude it using the **Mid Plane** option.

1. Start SolidWorks and start a new SolidWorks Part document using the **New SolidWorks Document** dialog box.

2. Invoke the **Extruded Boss/Base** tool and draw the sketch of the base feature on the **Front Plane**. Add the required relations and dimensions to the sketch as shown in Figure 8-40.

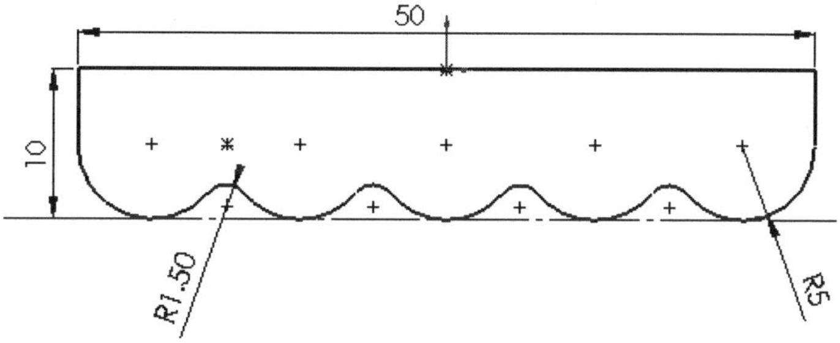

Figure 8-40 *Sketch for the base feature*

3. Extrude the sketch to a distance of 35 mm using the **Mid Plane** option.

   The base feature of the model is shown in Figure 8-41.

Figure 8-41 *Base feature of the model*

Add fillet to the Base Feature

After creating the base feature, you will fillet its lower edges.

1. Invoke the **Fillet PropertyManager**, rotate the model, and select the edges of the base feature, as shown in Figure 8-42.

2. Set the value of the **Radius** spinner to **2.5** and choose the **OK** button from the **Fillet PropertyManager**.

   The model, after adding fillet to its edges, is shown in Figure 8-43.

Adding Shell to the Model

After creating the fillet feature, you need to shell the model using the **Shell** tool.

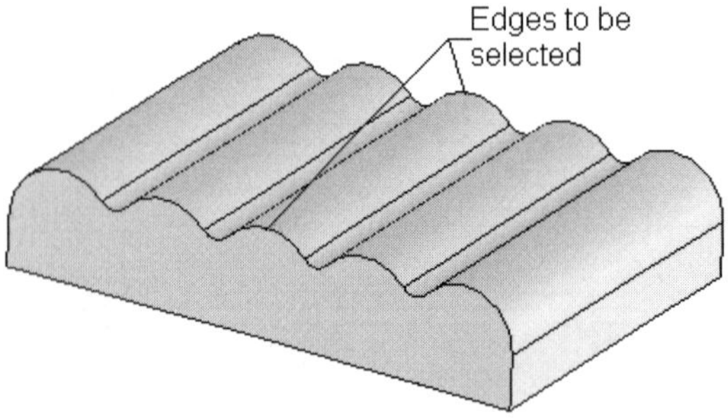

Figure 8-42 Edges to be selected

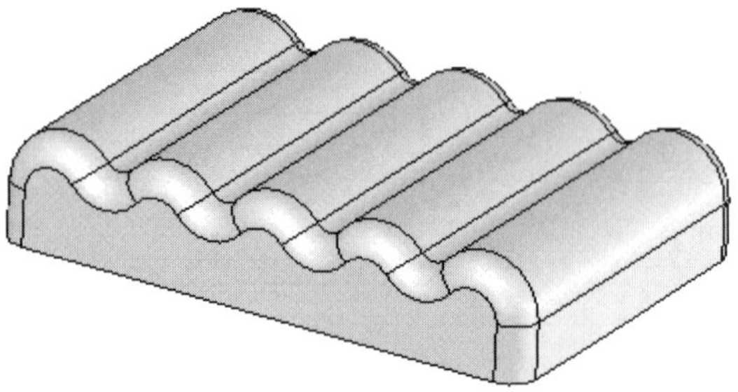

Figure 8-43 Fillet added to the model

1. Orient the model in the isometric view and invoke the **Shell1 PropertyManager**.

2. Select the top planar face of the model, as shown in Figure 8-44.

3. Set the value of the **Thickness** spinner to **1,** and choose the **OK** button from the **Shell PropertyManager**.

   The model, after adding the shell feature, is shown in Figure 8-45.

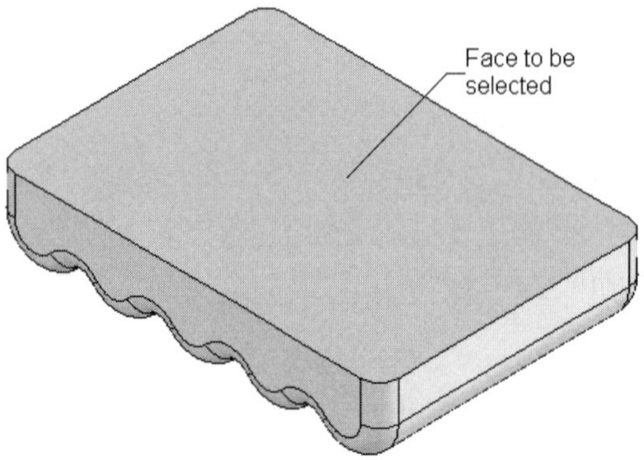

Figure 8-44 Face to be selected

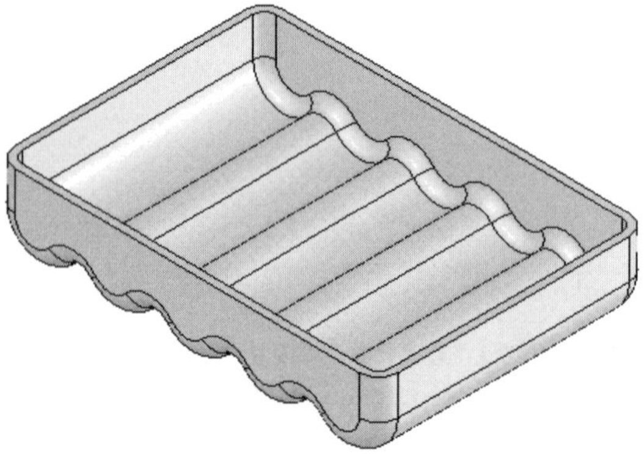

Figure 8-45 Shell feature added to the model

Dynamically Editing the Features

After creating and adding the shell to the base of the model you will learn how to edit the features dynamically using the **Move/Size Features** tool.

1. Choose the **Move/Size Features** button from the **Features CommandManager** after customizing it to invoke the dynamic dragging tool.

2. Select the right planar face of the base feature from the drawing area. The selected face

will be highlighted in green. The sketch of the selected feature is also displayed in the drawing area, along with various editing handles, as shown in Figure 8-46.

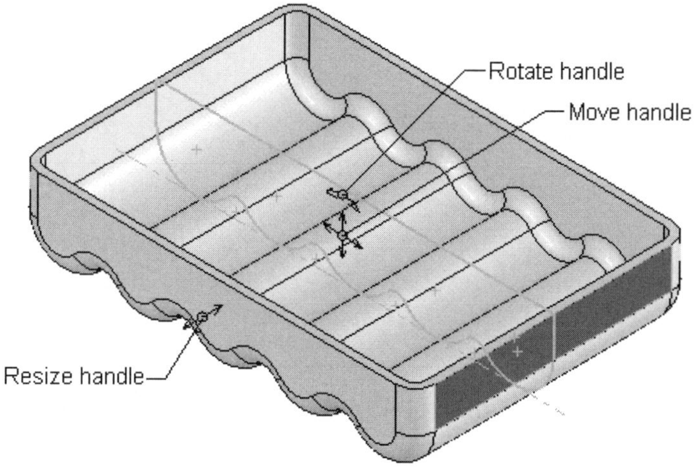

Figure 8-46 *Editing handles for editing the base feature*

Tip. *You can also select the feature to be edited from the **FeatureManager Design Tree**.*

*To edit the sketches using the **Move/Size Features** tool, you need to select the **Override Dimension** option by choosing **Tools > Sketch Settings > Override Dims on Drag/Move** from the menu bar. A check mark is displayed at the left of this option in the menu bar means that the option is already selected.*

3. Select the **Resize** handle from the drawing area and drag the cursor to resize the feature.

   The preview of the resized feature and its dimensions are displayed in the drawing area. As you drag the cursor, the preview and the dimension update automatically.

4. Release the left mouse button after dragging the feature to some distance. Figure 8-47 shows the preview of the dragged feature and Figure 8-48 shows the edited feature.

5. Select the **Move/Size Features** button again to disable the dynamic editing tool and click anywhere in the drawing area to clear all the selections from the selection set.

   You have edited the model by dynamically dragging, but the original depth value of the feature was 35 mm. To bring the base feature back to its original size you need to edit the feature.

6. Double-click the base feature in the **FeatureManager Design Tree,** or from the drawing area. All the dimensions of the feature are displayed in it.

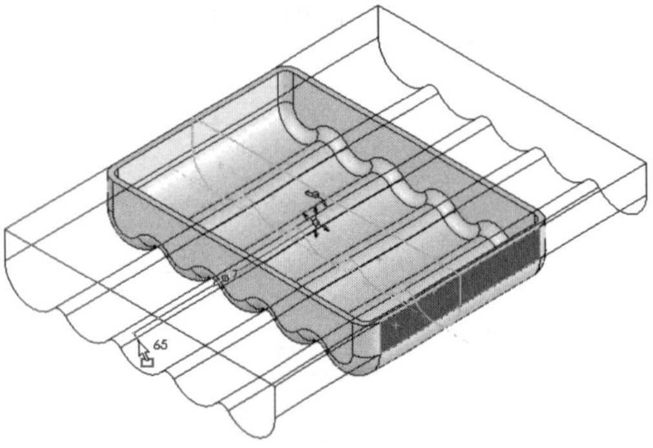

__Figure 8-47__ Dragging the __Resize__ handle

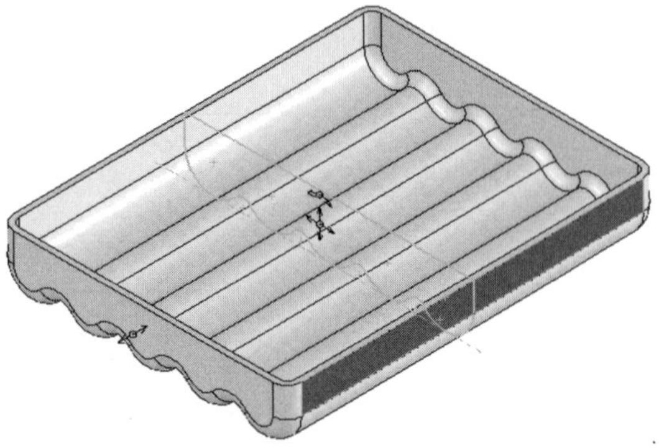

__Figure 8-48__ Resulting edited feature

7. Double-click the dimension that reflects the depth of the base feature that appears in blue.

8. The **Modify** dialog box will be displayed. Set the value of the **Dimension** spinner to **35** and press the ENTER key on the keyboard.

9. Choose the **Rebuild** button from the **Standard** toolbar or CTRL+B on the keyboard to rebuild the model.

   Using the **Extruded Cut**, **Fillet**, and **Linear Pattern** tools create the remaining features of the model. The model, after creating the features using these tools, is displayed in Figure 8-49.

Creating the Standoffs

Now, you need to create the standoffs for the model. It is created by extruding a sketch drawn on a sketch plane at an offset distance from the **Top Plane**. You also need to specify a draft angle, while creating this feature.

1. Create a reference plane at an offset distance of 10.5 mm from the **Top Plane**. You need to select the **Reverse direction** check box from the **Plane PropertyManager**.

2. Select the newly created sketching plane, draw the sketch of the standoff, and apply the required relations and dimensions. The sketch consists of a circle of diameter 1 mm. For more dimensions refer to Figure 8-39.

3. Extrude the sketch using the **Up To Next** option with an outward draft angle of 10-degree. Hide the newly created plane.

   This creates the standoffs of the model.

4. Rotate the model and add a fillet of radius 0.25 to the base of the standoff.

The rotated and zoom view of the complete standoff is displayed in Figure 8-50.

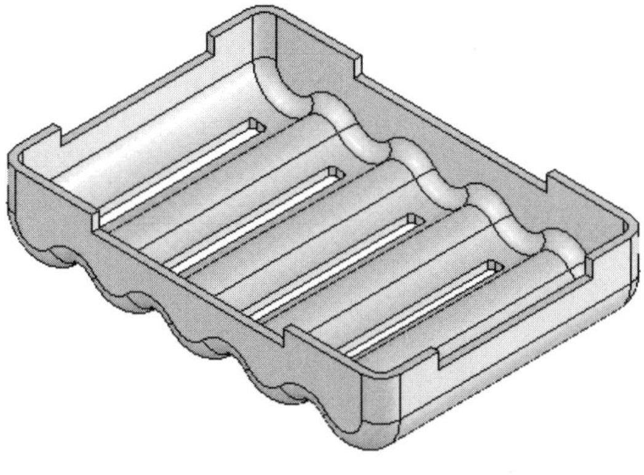

Figure 8-49 *Model after creating other features*

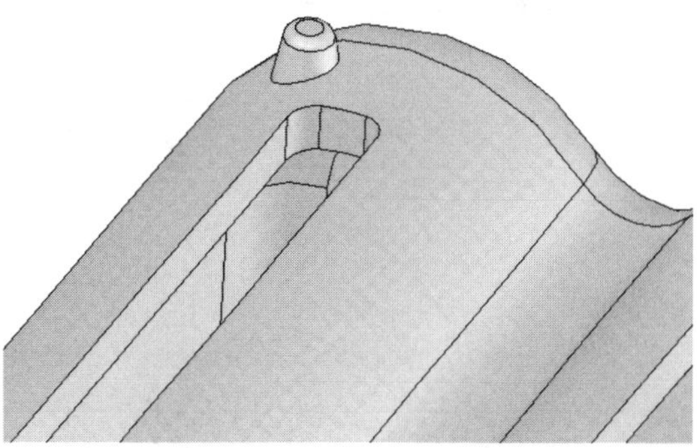

Figure 8-50 *Rotated and zoom view of the model to show the standoff*

5. Pattern the filleted standoff feature using the **Linear Pattern** tool. The isometric view of the final model is shown in Figure 8-51.

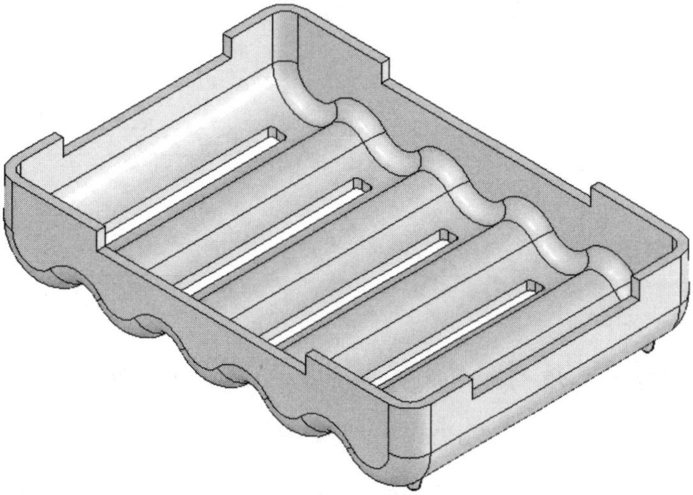

Figure 8-51 *Final model*

Saving the Model

1. Choose the **Save** button from the **Standard** toolbar and save the model with the name given below.

   \My Documents\SolidWorks\c08\c08tut1.sldprt

2. Choose **File > Close** from the menu bar to close the document.

Tutorial 2

In this tutorial you will create the model, shown in Figure 8-52, and then edit it with the **Move/Size Features** option selected. The views and dimensions of the model are shown in the same figure. **(Expected time: 45 min)**

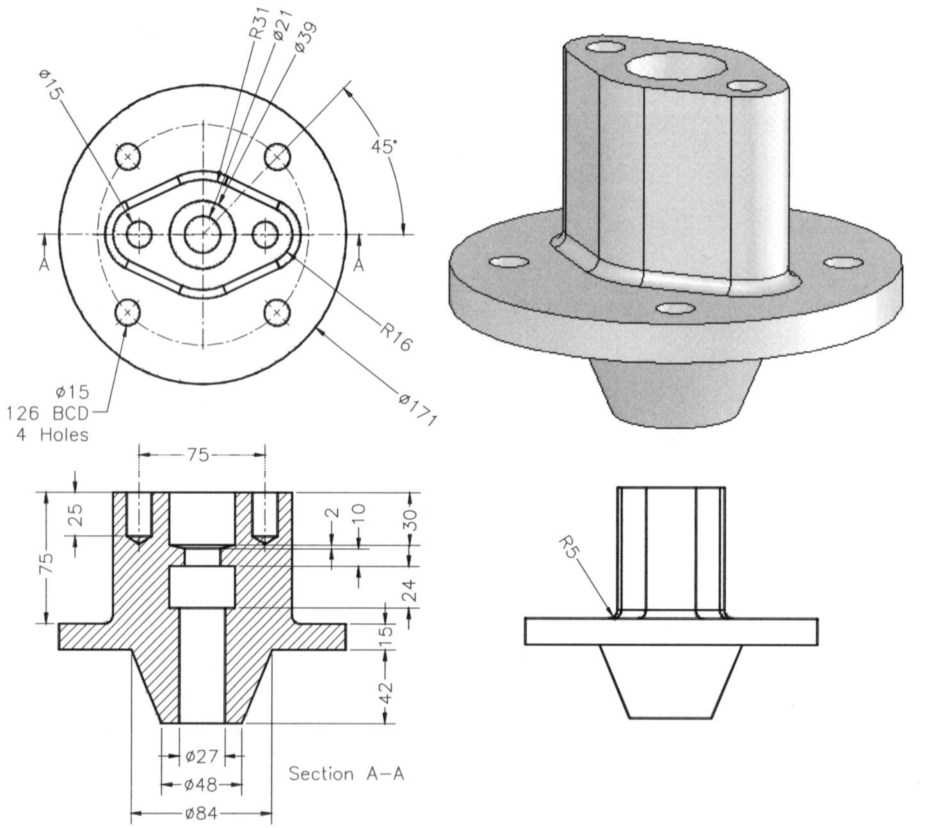

Figure 8-52 *Views and dimensions of the model for Tutorial 2*

The steps to be followed to complete this tutorial are listed next.

a. Create the base feature of the model by revolving the sketch along the central axis of the sketch, refer to Figures 8-53 and 8-54.
b. Draw the sketch of the second feature on the top face of the base feature and extrude it to a given dimension, refer to Figures 8-55 and 8-56.
c. Create the revolve cut feature, refer to Figures 8-57 and 8-58.
d. Create the hole using the **Simple Hole** tool, and then pattern it using the **Circular Pattern** tool, refer to Figure 8-58.
e. Create a drilled hole feature using the **Hole Wizard** tool, refer to Figure 8-58.
f. Mirror the hole feature along the **Right Plane**, refer to Figure 8-58.
g. Apply the fillet, refer to Figure 8-58.
h. Edit the model using the **Move/Size Features** tool, refer to Figures 8-59 through 8-61.

Creating the Base Feature

First, you will create the base feature of the model by revolving the sketch created on the **Front Plane**.

1. Start a new SolidWorks Part document using the **New SolidWorks Document** dialog box.

2. Invoke the **Revolved Boss/Base** tool and draw the sketch of the base feature on the **Front Plane**, Add the required relations and dimensions to the sketch, as shown in Figure 8-53.

3. Exit the sketching environment.

   You do not need to set any parameters in the **Revolve PropertyManager** because the default value in the **Angle** spinner is 360-degree, as required.

4. Choose the **OK** button from the **Revolve PropertyManager**.

   The base feature created, after revolving the sketch, is shown in Figure 8-54.

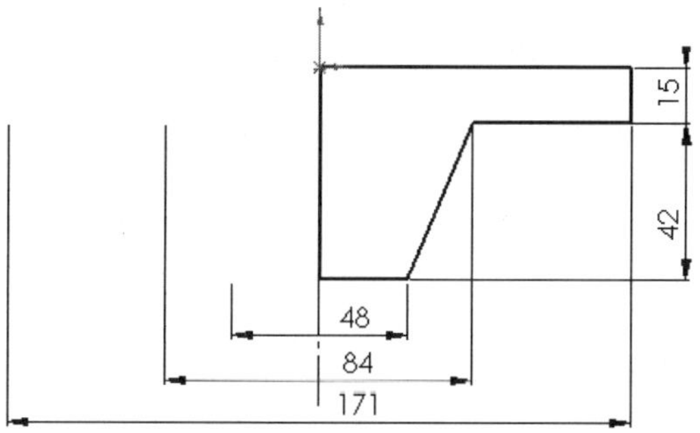

Figure 8-53 *Sketch for the base feature*

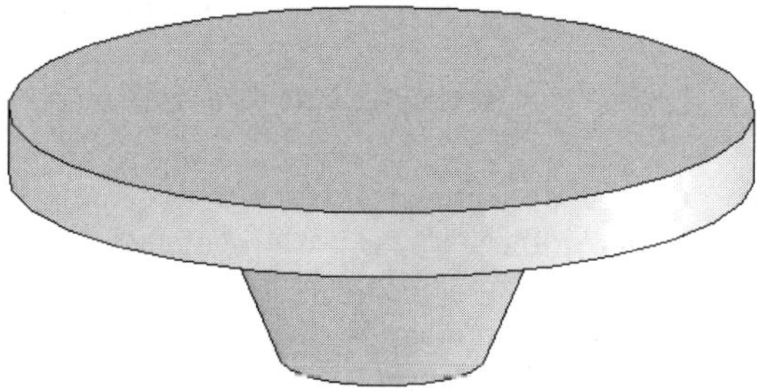

Figure 8-54 *Dimetric view of the base feature*

Creating the Second Feature

The second feature of this model is an extruded feature. It is created by extruding a sketch created on the top planar face of the base feature.

1. Invoke the **Extruded Boss/Base** tool and select the top planar face of the base feature as the sketching plane.

2. Draw the sketch of the second feature and apply the required relations and dimensions to it, as shown in Figure 8-55. Make sure that the sketch is symmetrical about the centerline. If it is not, the sketch will not be properly mirrored.

3. Extrude the sketch to a distance of 75 mm.

The isometric view of the model, after creating the second feature, is displayed in Figure 8-56.

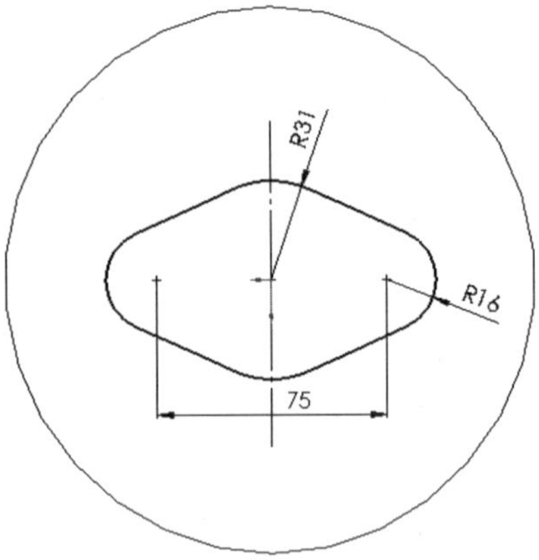

Figure 8-55 Sketch for the second feature

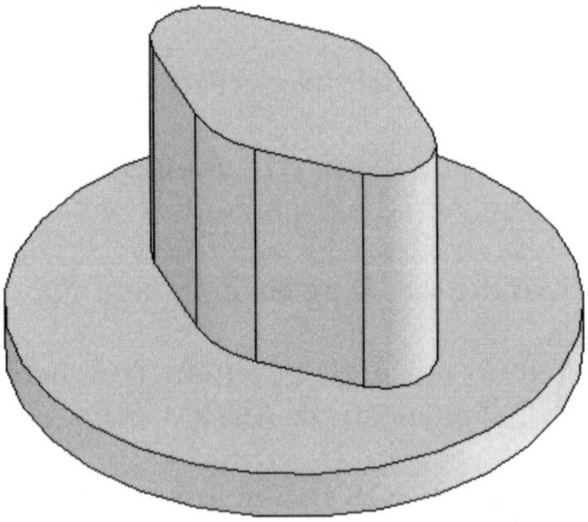

Figure 8-56 Second feature added to the model

Creating the Third Feature

The third feature of the model is created by revolving a sketch using the cut option. The sketch for this feature will be created on the **Front Plane**.

1. Invoke the **Revolved Cut** tool and select the **Front Plane** as the sketching plane.

2. Draw the sketch for the revolved cut feature and apply the required relations and dimensions to it, as shown in Figure 8-57.

3. Exit the sketching environment, and create a revolved cut feature with a default angle value of 360-degree.

Creating the Remaining Features

1. Create the other features of the model by referring to Figure 8-52 and using the **Simple Hole**, **Hole Wizard**, and **Mirror** tools.

   The isometric view of the model, after creating all the other features, is displayed in Figure 8-58.

Editing the Sketch of the Model With the Move/Size Features tool

After creating the model, you will edit the sketch of the second feature with the **Move/Size Features** tool. Therefore, before proceeding further you need to invoke this option.

1. Choose the **Move/Size Features** button from the **Features CommandManager** to invoke the **Move/Size Features** tool.

2. Select the second feature from the drawing area, and right-click to invoke the shortcut menu.

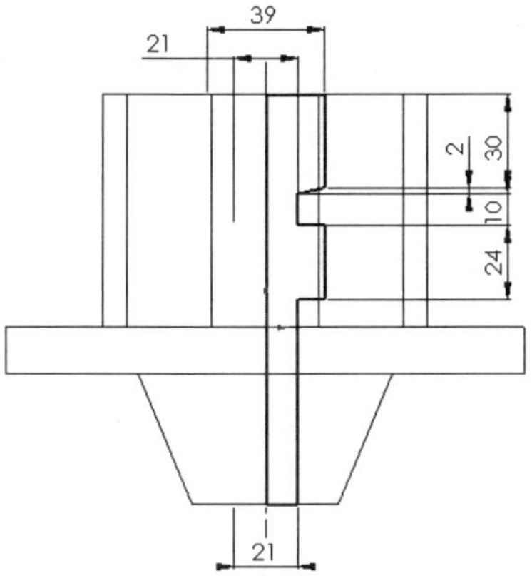

Figure 8-57 Sketch for the revolve cut feature

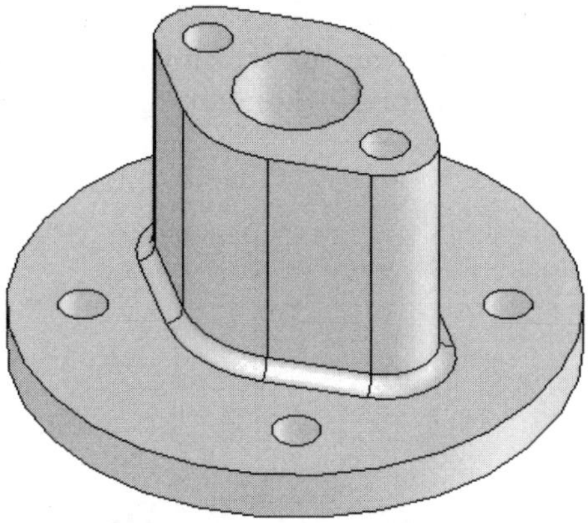

Figure 8-58 Isometric view of the final model

3. Choose the **Edit Sketch** option from the shortcut menu.

   The sketch of the second feature with the preview of the feature in temporary graphics is displayed in the drawing area, as shown in Figure 8-59.

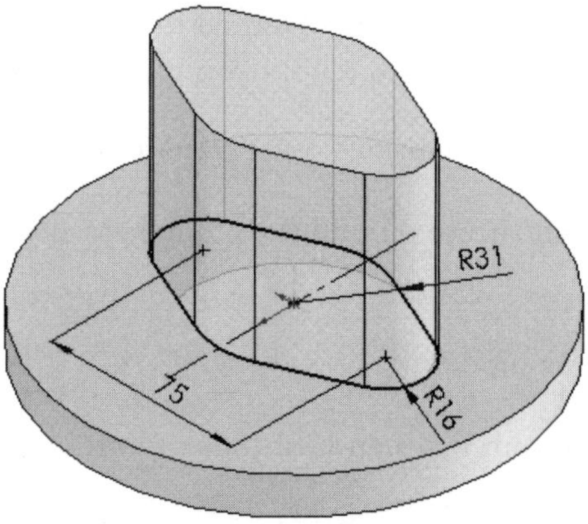

Figure 8-59 Sketch and the preview of the second feature

Now, you will modify the sketch by dragging the sketched entities. You will observe that the preview of the second feature displayed in temporary graphics is also modified.

4. Choose **Tools > Sketch Settings > Override Dims on Drag/Move** option from the menu bar, if it is not selected.

5. Select the center point of the right arc of sketch and drag the cursor toward the right. As you drag the cursor, the preview of the feature is also modified dynamically. Release the left mouse button. The new dimensions appears on the sketch. Continue this process until the value of the distance between the center of the right and left arcs shows a value close to 135.

Figure 8-60 shows the sketch being dragged. Figure 8-61 shows the final dragged sketch. After modifying the sketch by dragging, you need to change the modified dimensions back to the original dimensions.

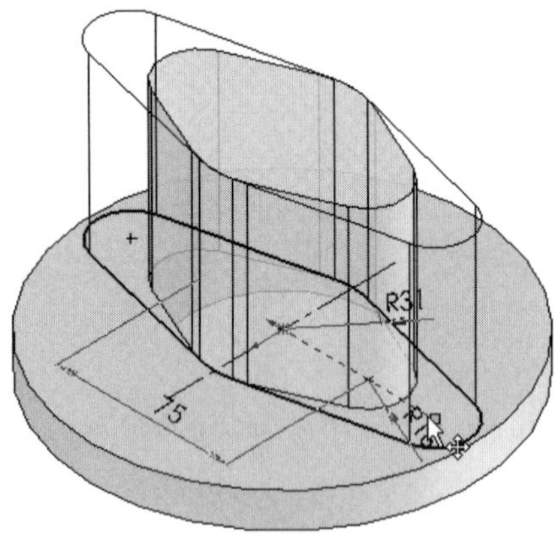

Figure 8-60 *Dragging the center point of right arc*

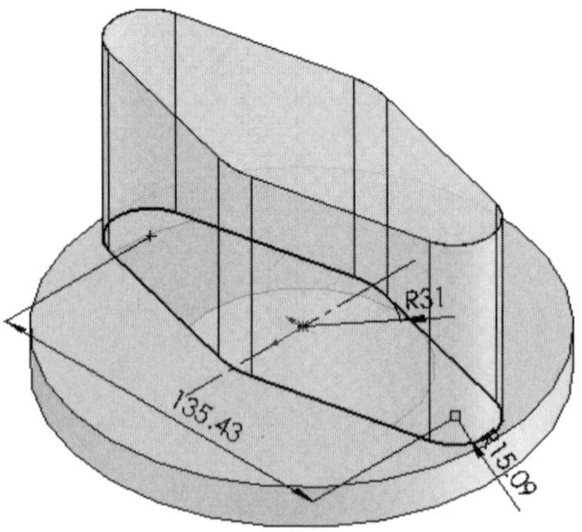

Figure 8-61 *Sketch after dragging*

5. Double-click the linear dimension value between the center points of the right and left arcs, enter **75** in the **Modify** edit box, and press the ENTER key.

   Similarly, edit the other two dimensions using the **Modify** edit box, if necessary.

6. Use CTRL+B on the keyboard to rebuild the model.

Saving the Model

Now, you need to save the model.

1. Choose the **Save** button from the **Standard** toolbar and save the model with the name given below.

   \My Documents\SolidWorks\c08\c08tut2.sldprt.

2. Choose **File** > **Close** from the menu bar to close the document.

Tutorial 3

In this tutorial you will create the model shown in Figure 8-62. While creating it, you will also perform some editing operations on it. The views and dimensions of the model are displayed in Figure 8-63. **(Expected time: 45min)**

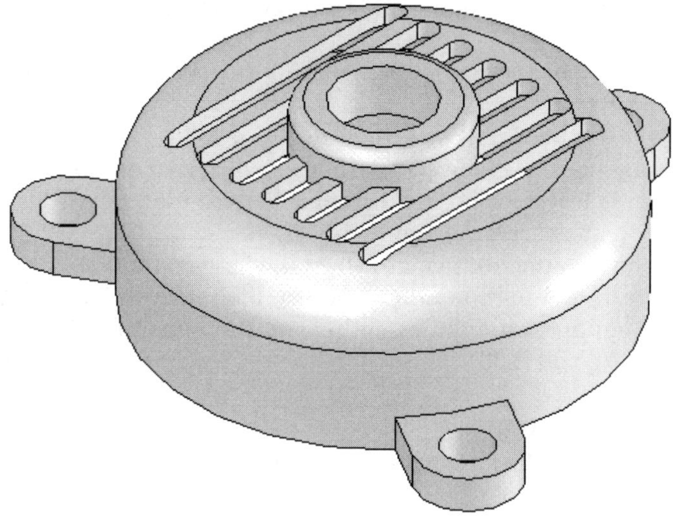

Figure 8-62 *Model for Tutorial 3*

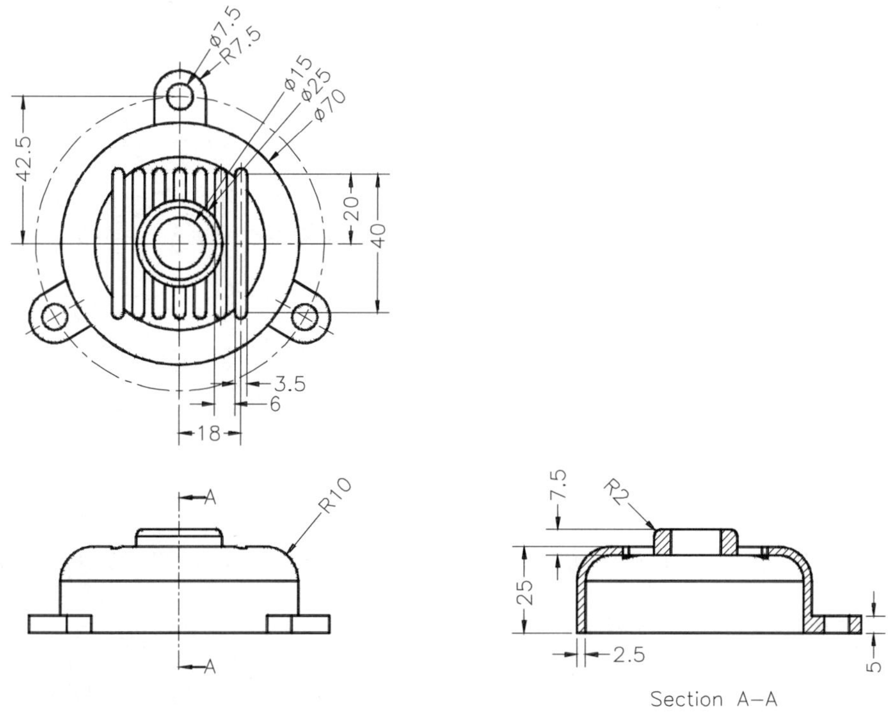

Figure 8-63 *Views and dimensions of the model for Tutorial 3*

The steps to be followed to complete this tutorial are listed next.

a. Create a base feature of the model by revolving the sketch drawn on the **Front Plane**, refer to Figures 8-64 and 8-65.
b. Shell the model using the **Shell** tool, refer to Figure 8-66.
c. Draw the sketch on the **Top Plane** and extrude it to the given distance, refer to Figure 8-67.
d. Pattern the extrude feature using the **Circular Pattern** tool, refer to Figure 8-68.
e. Edit the circular pattern, refer to Figure 8-69.
f. Create a cut feature on the top planar face of the base feature, refer to Figure 8-70.
g. Pattern the cut feature using the **Linear Pattern** tool, refer to Figure 8-71.
h. Unsuppress the suppressed features and create the remaining features of the model, refer to Figures 8-72 and 8-73.

Creating the Base Feature

First, you need to create the base feature of the model by revolving the sketch created on the **Front Plane**.

1. Start a new SolidWorks Part document using the **New SolidWorks Document** dialog box.

2. Invoke the **Revolved Boss/Base** tool and draw the sketch of the base feature on the **Front Plane**. Add the required relations and dimensions to it, as shown in Figure 8-64.

3. Exit the sketching environment and create the base feature of the model, as shown in Figure 8-65.

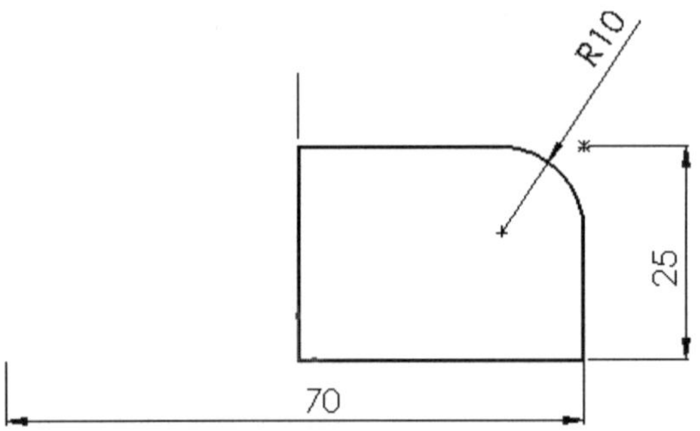

***Figure 8-64** Sketch for the base feature*

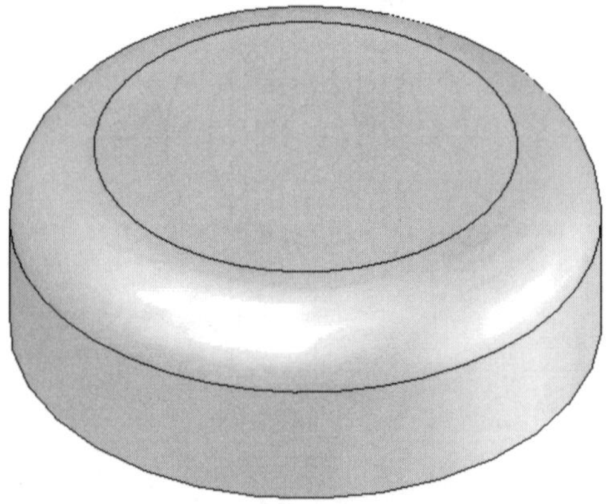

***Figure 8-65** Base feature of the model*

Shelling the Base Feature

After creating the base feature, you need to shell the model using the **Shell** tool. You will also remove the bottom face of the base feature, leaving behind a thin walled model.

1. Invoke the **Shell1 PropertyManager** and set the value of the **Thickness** spinner to **2.5**.

2. Rotate the model and select its bottom face to remove.

3. Choose the **OK** button from the **Shell1 PropertyManager**. The model, after adding the shell feature, is displayed in Figure 8-66.

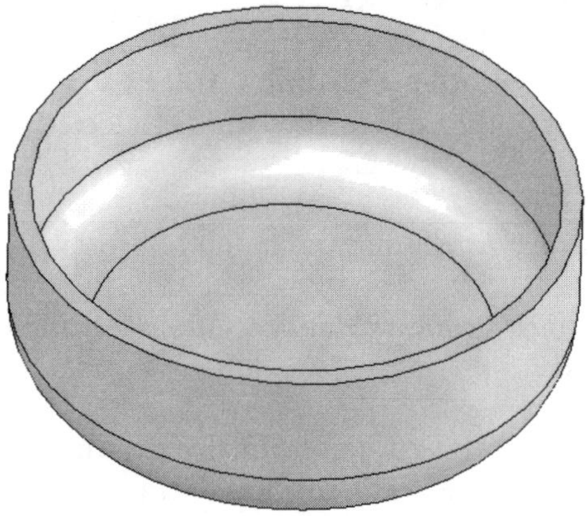

Figure 8-66 *Shell feature added to the model*

Creating the Third Feature

After adding the shell feature to the model, you need to create its third feature, which is an extruded feature. The sketch for this feature will be drawn on the **Top Plane**.

1. Invoke the **Extruded Boss/Base** tool and select the **Top Plane** as the sketching plane.

2. Orient the model in the top view.

3. Draw the sketch of the third feature and add the required relations and dimensions to the sketch as shown in Figure 8-67.

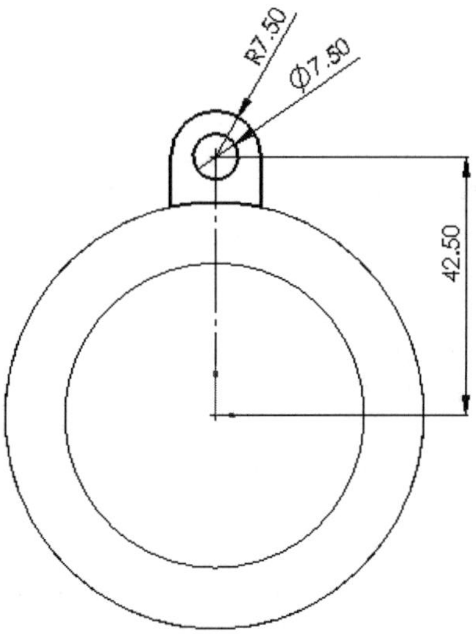

Figure 8-67 *Sketch of the third feature*

4. Exit the sketching environment and extrude the sketch to a depth of 5 mm.

Patterning the Third Feature

You need to pattern the third feature after creating it. This feature will be patterned using the **Circular Pattern** tool. Before proceeding further you need to display the temporary axes. The temporary axis of the base feature will be selected as the central axis to create the circular pattern.

1. Choose **View > Temporary Axes** from the menu bar to display the temporary axes.

2. Invoke the **Circular Pattern PropertyManager** and select the temporary axis of the base feature as the central axis.

3. Select the third feature, created earlier from the drawing area, if not selected in the **Features to Pattern** selection area. The preview of the pattern feature is displayed in the drawing area.

4. Set the value of the **Number of Instances** spinner to **5** and choose **OK** from the **Circular Pattern PropertyManager**.

5. Choose **View > Temporary Axes** from the menu bar to remove the temporary axes from the current display.

   The model, after creating the pattern feature, is displayed in Figure 8-68.

Editing the Pattern Feature

The pattern created is not the same as required, refer to Figure 8-62. As a result, you need to edit it.

1. Select **CirPattern1** from the **FeatureManager Design Tree** or any one of the pattern instance other than the parent instance from the drawing area. Right-click and choose the **Edit Feature** option from the shortcut menu.

   The **CirPattern1 PropertyManager** is displayed. Presently, the number of instances in the pattern feature is 5, but the required number of instances is 3. Therefore, you will edit the number of instances.

2. Set the value of the **Number of Instances** spinner to **3** and choose the **OK** button from the **FeatureManager Design Tree**.

   The model, after editing the features, is shown in Figure 8-79.

Suppressing the Features

As discussed earlier, sometimes you may need to suppress some features to reduce the complications in the model. The suppressed features are actually not deleted; their display is just turned off. When you suppress a feature, the child features associated with that feature are also suppressed.

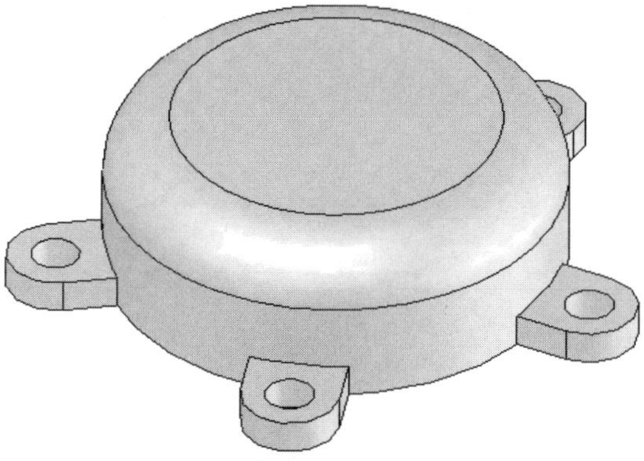

Figure 8-68 Pattern feature added to the model

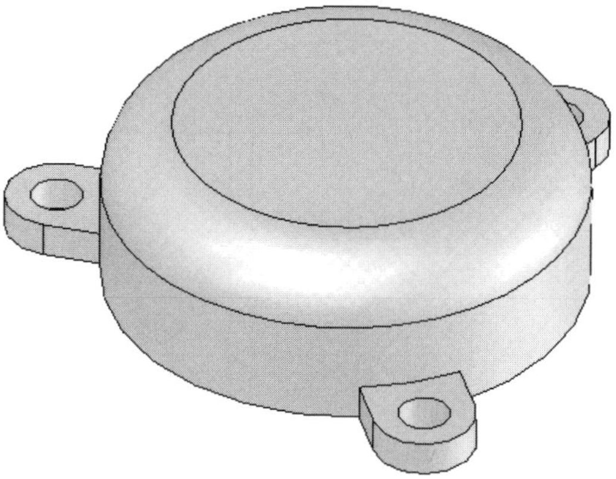

Figure 8-69 The edited pattern feature

1. Select the **Extrude1** feature, which is the third feature of the model, from the **FeatureManager Design Tree**. Right-click and choose the **Suppress** option from the shortcut menu.

   The circular pattern feature is the child feature of the extrude feature. Therefore, it is also suppressed. Both features are not displayed in the drawing area. The **Extrude1** and the **CirPattern1** features are displayed in gray in the **FeatureManager Design Tree**, indicating that both of them are suppressed.

Creating the Cut Feature

The next feature that you are going to create is a cut feature. The sketch for this feature will be drawn on the top planar face of the base feature.

1. Invoke the **Extruded Cut** tool and select the top planar face of the base feature as the sketching plane.

2. Draw the sketch of the cut feature and add the required relations and dimensions to the sketch, as shown in Figure 8-70.

3. Exit the sketching environment and specify the end condition as **Through All** from the **Cut-Extrude PropertyManager**.

4. Choose the **OK** button from the **Cut-Extrude PropertyManager**.

5. Now, using the **Linear Pattern** tool create a linear pattern of the cut feature. You can select the dimension 18 as the directional reference. The model, after creating the linear pattern, is shown in Figure 8-71.

6. Create the other features of the model. For dimensions, refer to Figure 8-63. The model, after creating the other features, is shown in Figure 8-72.

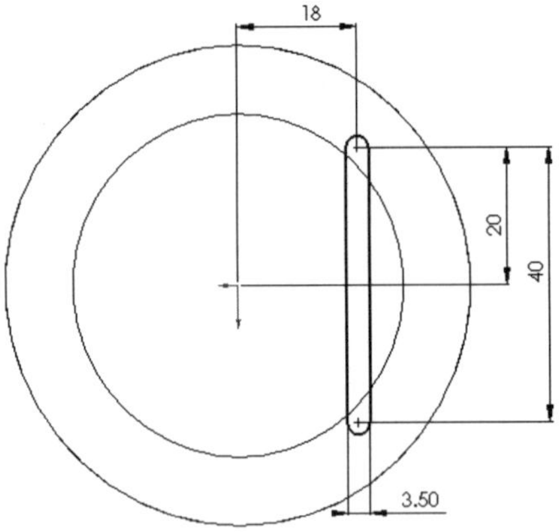

Figure 8-70 Sketch for the cut feature

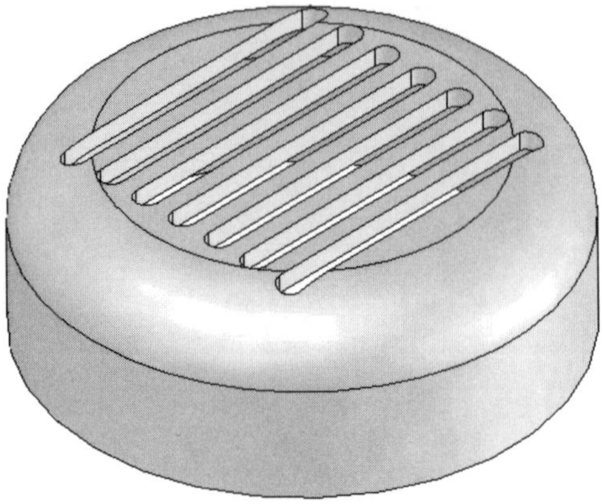

Figure 8-71 Model after patterning the cut feature

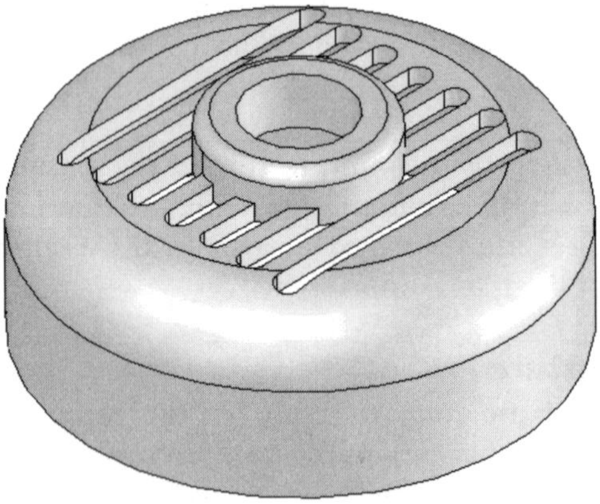

Figure 8-72 Model after creating other features.

Unsuppressing the Features

After completing the model, you need to unsuppress the features that you suppressed earlier.

1. Press and hold down the CTRL key on the keyboard, and select all the suppressed features from the **FeatureManager Design Tree**.

Note

On selecting only the parent suppressed feature and unsuppressing it, the child feature will not be unsuppressed. Therefore, you have to select the parent feature and the suppressed child features.

Instead of selecting all the parent and child suppressed features from the ***FeatureManager Design Tree,*** *select only the parent feature, and choose* ***Edit > Unsuppress with Dependents > All Configurations*** *from the menu bar. You will learn more about the configurations in the later chapters.*

On unsuppressing the child feature, the parent feature will be unsuppressed automatically.

2. Right-click and choose the **Unsuppress** option from the shortcut menu.

   The suppressed features will be restored in the model. The final model, after unsuppressing the features, is shown in Figure 8-73.

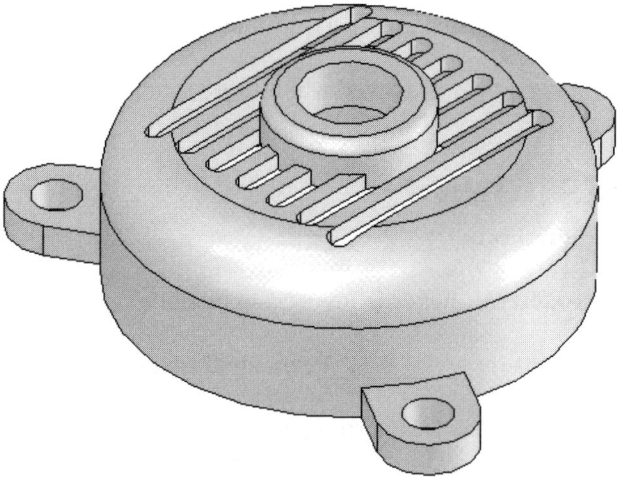

Figure 8-73 *The final model*

Saving the Model

1. Choose the **Save** button from the **Standard** toolbar and save the model with the name given below.

   \My Documents\SolidWorks\c08\c08tut3.sldprt

2. Choose **File** > **Close** from the menu bar to close the document.

SELF-EVALUATION TEST

Answer the following questions and then compare your answers with those given at the end of this chapter.

1. You cannot edit the sketch of a sketched feature. (T/F)

2. The **Edit Feature** option is used to edit any feature. (T/F)

3. You cannot rename the feature in the **FeatureManager Design Tree**. (T/F)

4. You cannot edit the sketch plane of the sketch of a sketched feature. (T/F)

5. You cannot edit the sketches using the **Move/ Size Features** option. (T/F)

6. The __________ dialog box is displayed when you edit a dimension.

7. The process of changing the position of a feature in the **FeatureManager Design Tree** is known as __________.

8. Using __________ **PropertyManager** you can delete the bodies.

9. The __________ **PropertyManager** is used to move or copy the bodies.

10. The __________ dialog box is displayed when there is any error in a feature.

REVIEW QUESTIONS

Answer the following questions.

1. The __________ **PropertyManager** is invoked to delete a body.

2. You can rotate a body using __________ **PropertyManager**.

3. __________ key is used on the keyboard to copy a feature or a sketch.

4. The __________ key is used on the keyboard to cut a feature or a sketch.

5. When __________ tool is active, the preview of the feature is displayed in temporary graphics while editing the sketches.

6. The __________ **PropertyManager** is displayed to edit the sketch plane of a sketch.

7. To add the selected feature in a new folder choose **Add to New Folder** from the shortcut menu. (T/F)

8. For reordering the features, select the feature in the **FeatureManager Design Tree** and drag it to the required position. (T/F)

9. The **Modify** dialog box is invoked, using a single click on the dimension to modify. (T/F)

10. If you want to modify the sketch by dragging the fully or partially defined sketch, the **Override Dims on Drag/Move** option should be selected. (T/F)

EXERCISES

Exercise 1

Create the model that is sectioned and shown in Figure 8-74. The other views and dimensions of the model are also given in the same figure. The complete model is shown in Figure 8-75.

(Expected time: 45 min)

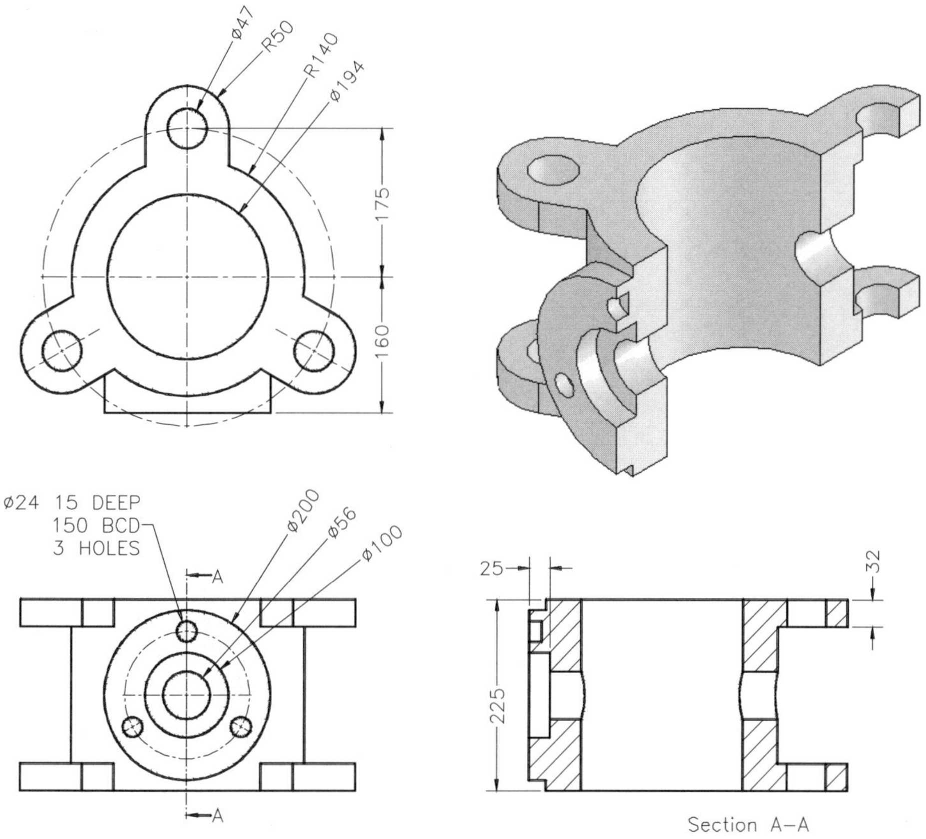

Figure 8-74 *Views and dimensions of the model for Exercise 1*

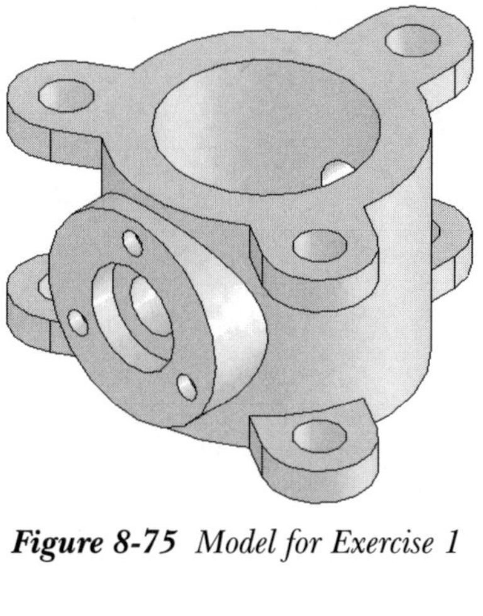

Figure 8-75 *Model for Exercise 1*

Exercise 2

Create the model shown in Figure 8-76. Its dimensions are shown in Figure 8-77.

(Expected time: 30 min)

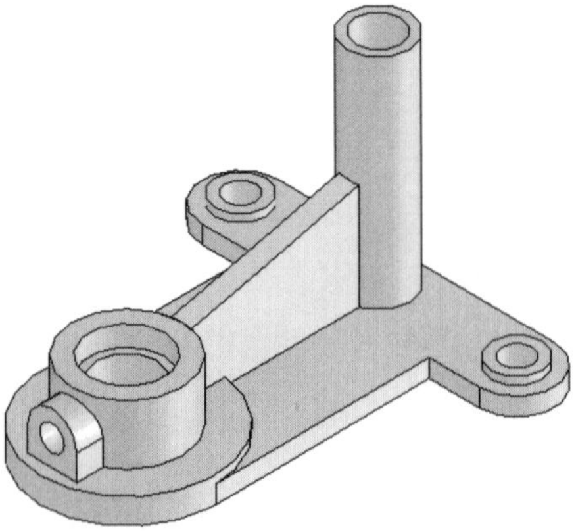

Figure 8-76 *Model for Exercise 2*

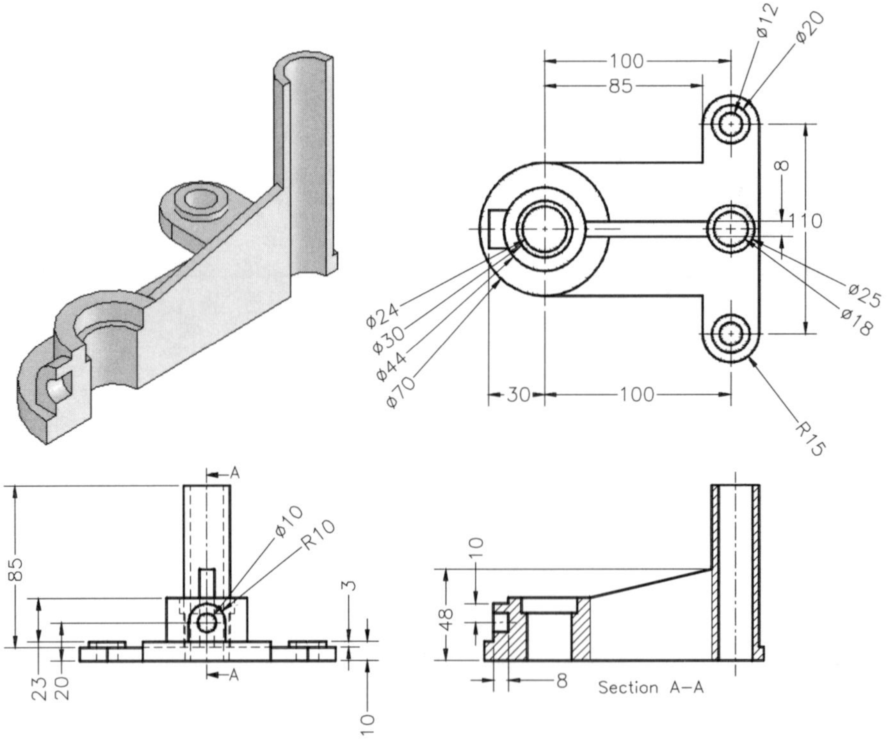

Figure 8-77 *Views and dimensions of the model for Exercise 2*

Answers to Self-Evaluation Test

1. F, **2.** T, **3.** F, **4.** F, **5.** F, **6. Modify**, **7.** Reordering, **8. Delete Body**, **9. Move/Copy Body**, **10. Rebuild Errors**

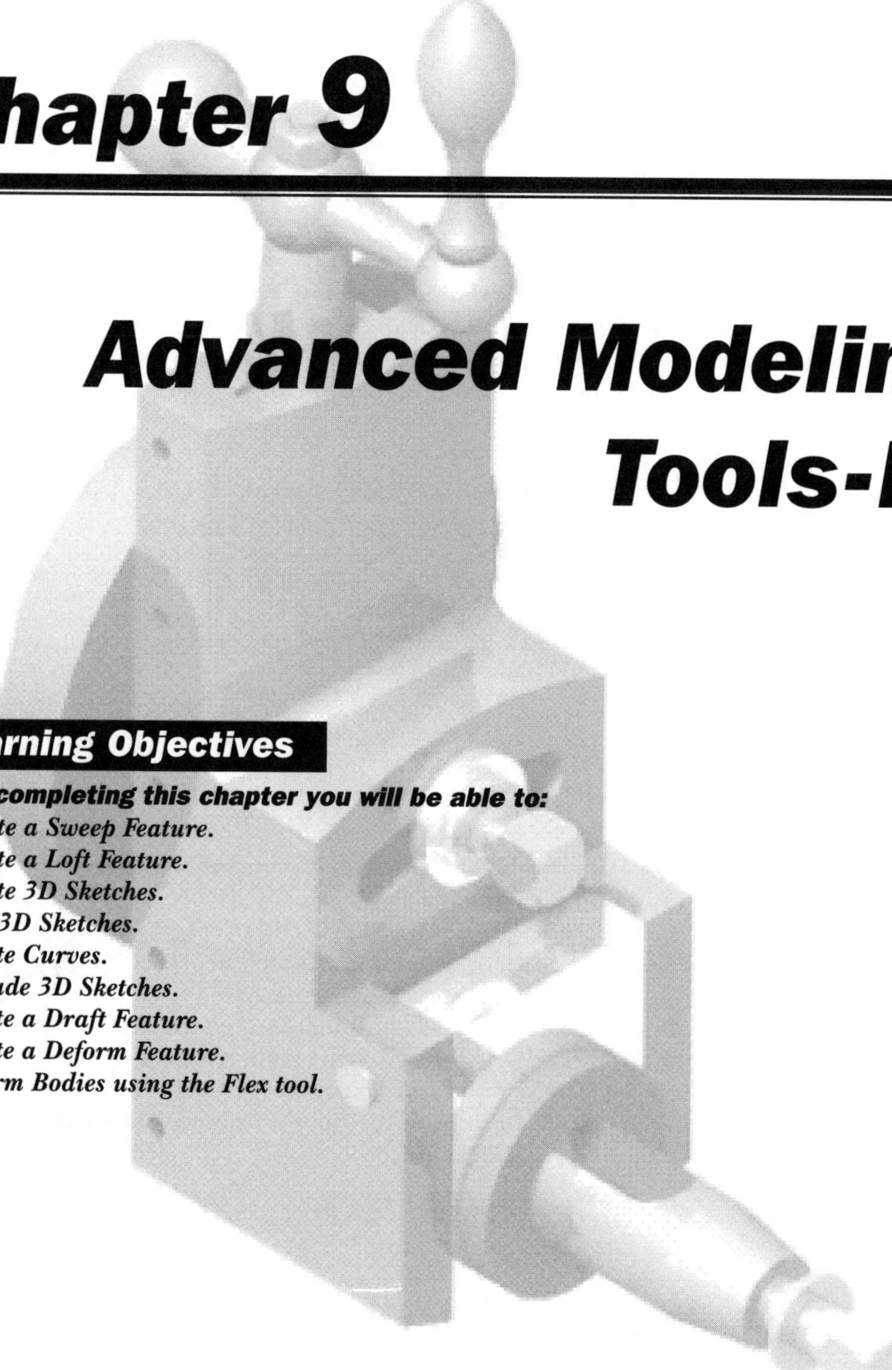

Chapter 9

Advanced Modeling Tools-III

Learning Objectives

After completing this chapter you will be able to:

- *Create a Sweep Feature.*
- *Create a Loft Feature.*
- *Create 3D Sketches.*
- *Edit 3D Sketches.*
- *Create Curves.*
- *Extrude 3D Sketches.*
- *Create a Draft Feature.*
- *Create a Deform Feature.*
- *Deform Bodies using the Flex tool.*

ADVANCED MODELING TOOLS

Some of the advanced modeling tools were discussed in earlier chapters. The remaining advanced modeling tools including sweep, loft, draft, curves, 3D sketches, deform and flex are discussed in this chapter.

Creating the Sweep Feature

CommandManager:	Features > Swept Boss/Base
Menu:	Insert > Boss/Base > Sweep
Toolbar:	Features > Swept Boss/Base

One of the most important advanced modeling tools, is the **Swept Boss/Base** tool. This tool is used to extrude a closed profile along an open or a closed path. Therefore, you need a profile and a path to create a sweep feature. A profile is a section for the sweep feature and a path is the course taken by the profile, while creating the sweep feature. The profile has to be a sketch, but the path can be a sketch, edge, or a curve. You will learn more about how to create the curves later in this chapter. An example of the profile and path for creating the sweep feature, is shown in Figure 9-1.

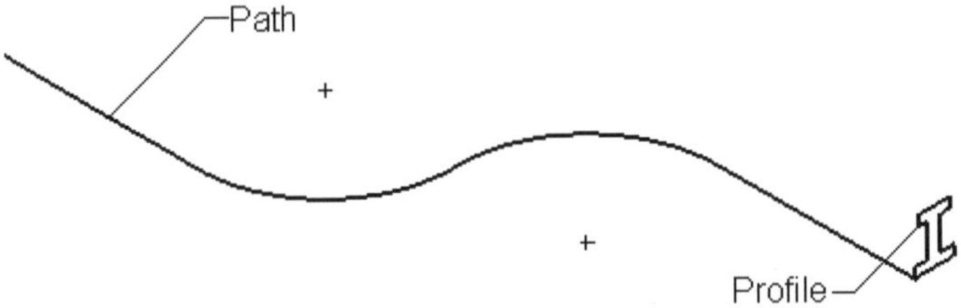

Figure 9-1 *Profile and path to create a sweep feature*

Choose the **Swept Boss/Base** button from the **Features CommandManager** to invoke the **Sweep PropertyManager**, as shown in Figure 9-2.

After invoking the **Sweep PropertyManager**, you are prompted to select the sweep profile. Select the sketch that is drawn as the profile from the drawing area; the sketch is highlighted in green and the **Profile** callout is displayed. Now, you are prompted to select the path for the sweep feature. Select the sketch or an edge to be used as the path, it is highlighted in magenta and the **Path** callout is displayed. The sweep feature is also displayed in temporary graphics in the drawing area. Choose the **OK** button from the **Sweep PropertyManager** to end the feature creation. Figure 9-3 shows a sweep feature.

Tip. *You can also use the **Contour Select Tool** to select a contour as the section for the sweep feature.*

You can also use a shared sketch as the section of the sweep feature.

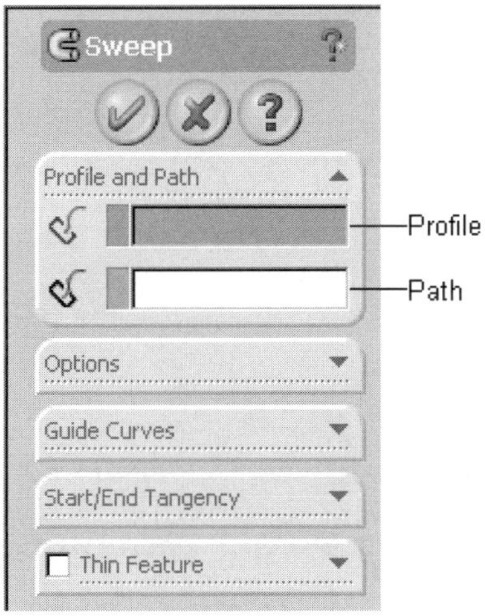

Figure 9-2 *The **Sweep PropertyManager***

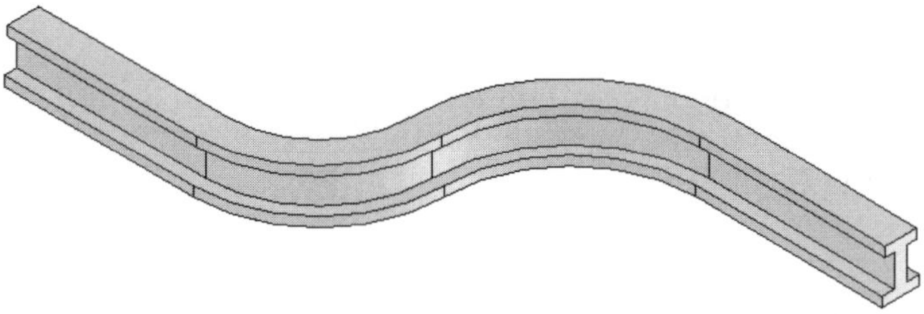

Figure 9-3 *Sweep feature*

It is not necessary that the sketch drawn for the profile of the sweep feature intersects its path. However, the plane on which the profile is drawn should lie at one of the endpoints of the path. Figure 9-4 shows the nonintersecting sketches of profile and path. Figure 9-5 shows the resulting sweep feature. Figure 9-6 shows a sketch of a profile and a closed path. Remember, the plane on which the profile is drawn should intersect the closed path. Figure 9-7 shows the resulting sweep feature.

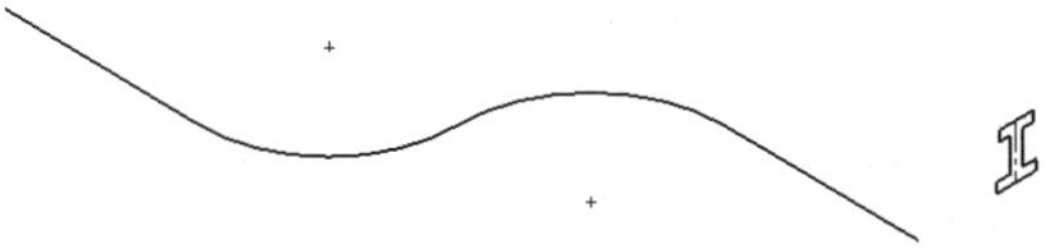

Figure 9-4 *Nonintersecting sketches of profile and path*

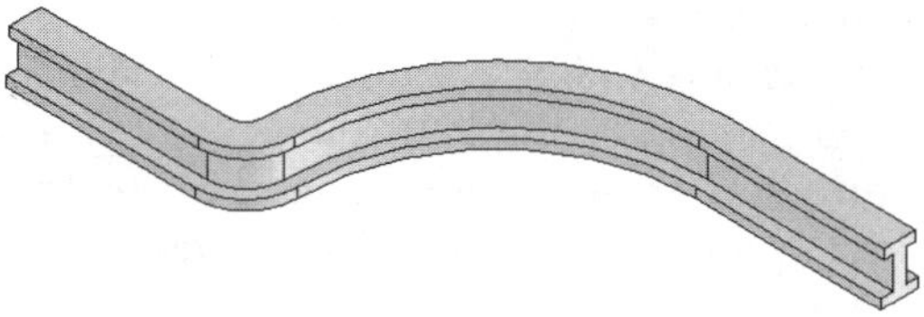

Figure 9-5 *The resulting sweep feature*

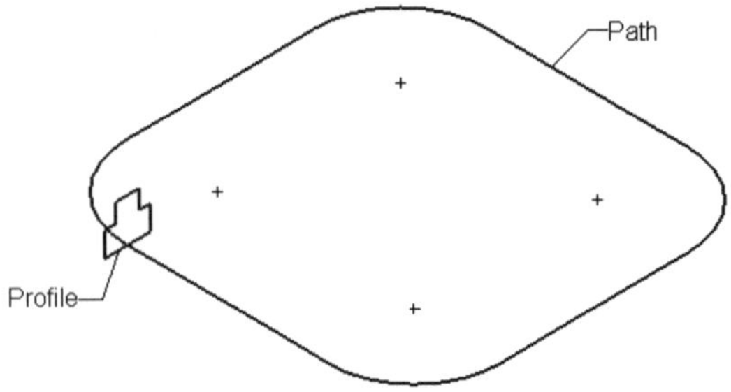

Figure 9-6 Sketch of a profile and a closed path

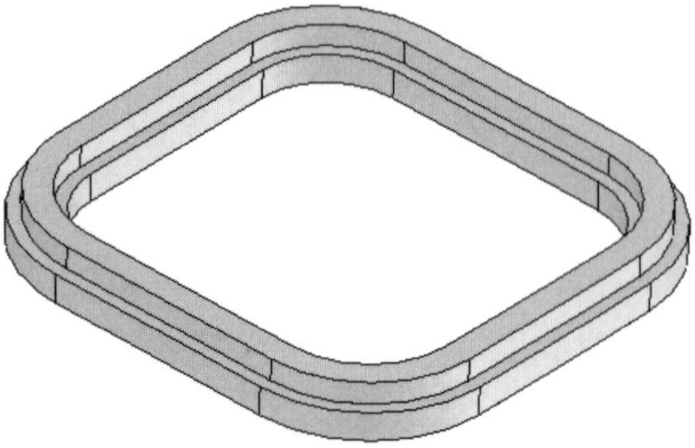

Figure 9-7 The resulting sweep feature

Various other options available in the **Sweep PropertyManager** to create advanced sweep features are discussed next.

Sweep Using the Follow Path and Keep Normal Constant Options

While creating a sweep feature, the **Follow Path** option is selected by default, in the **Orientation/twist type** drop-down list available in the **Options** rollout. The **Options** rollout is displayed in Figure 9-8. The **Path alignment type** drop-down list is also displayed. Using this drop-down list, you can set the options to create an even sweep feature when the curvature along the path fluctuates unevenly.

*Figure 9-8 The **Options** rollout*

While creating a sweep feature using the **Follow Path** option, the section will follow the path to create it. If you select the **Keep normal constant** option from the **Orientation/twist type** drop-down list, the section will be swept along the path with a normal constraint and will not change its orientation along the sweep path. Therefore, the starting and the end face of the sweep feature will be parallel. Figure 9-9 shows the sketches of the path and profile for the sweep feature. Figure 9-10 shows the sweep feature created using the **Follow Path** option. Figure 9-11 shows the sweep feature created using the **Keep normal constant** option. The other options available in the **Orientation/twist type** drop-down list are discussed later in this chapter.

Merge tangent faces

The **Merge tangent faces** check box, available in the **Options** rollout, is used to merge the tangent faces of the profile throughout the swept feature.

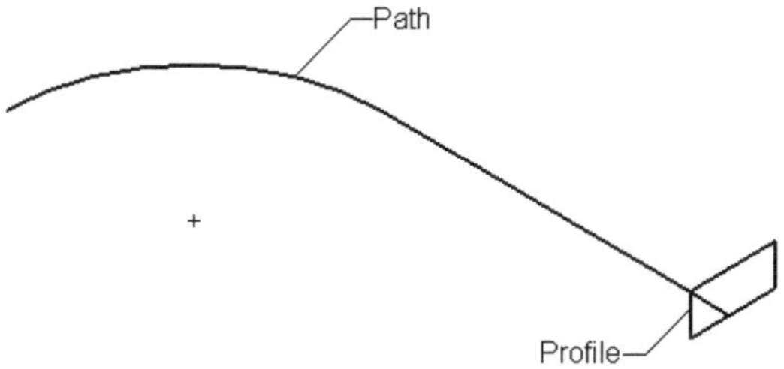

Figure 9-9 *Sketches for the sweep feature*

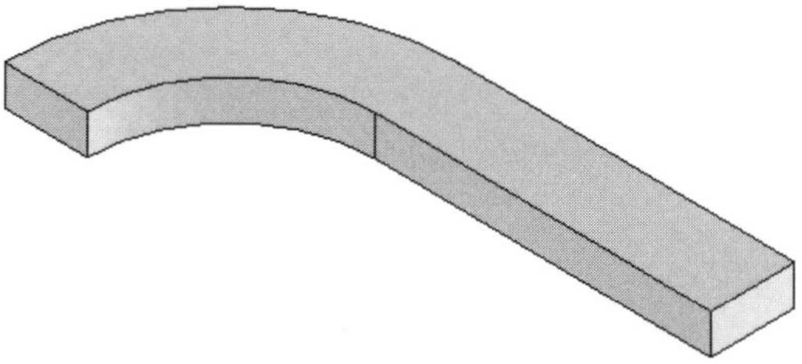

Figure 9-10 *Sweep feature with the* ***Follow Path*** *option selected from the* ***Orientation/twist type*** *drop-down list*

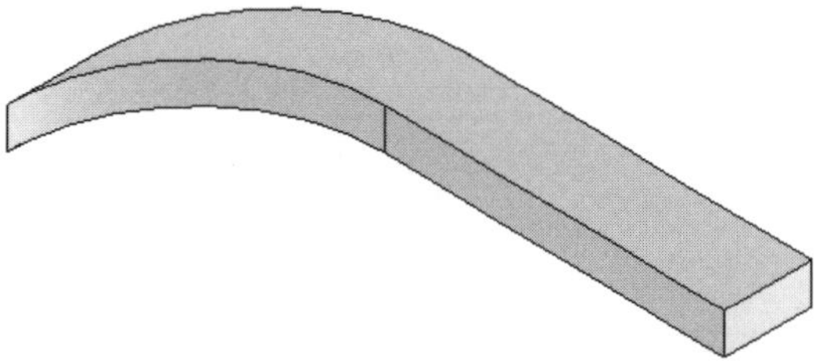

Figure 9-11 *Sweep feature with the* ***Keep normal constant*** *option selected from the* ***Orientation/twist type*** *drop-down list*

Show preview

The **Show preview** check box, available in the **Options** rollout, is used to display the preview of the sweep feature in the drawing area. This check box is selected by default. If you clear it, the preview of the sweep feature will not be displayed in the drawing area.

Merge result

The **Merge result** check box, selected by default, is available only when you have at least one feature in the current document. If you clear the check box, it will result in the creation of the sweep feature as a separate body.

Tip. *On selecting a model edge as the sweep path, the* ***Tangent Propagation*** *check box is displayed in the* ***Options*** *rollout. If this check box is selected, the edges tangent to the selected edge are selected automatically as the path of the sweep feature.*

Align with end faces

The **Align with end faces** check box is available in the **Options** rollout only when at least one feature has already been created in the current document. When this option is selected, the sweep feature is extended or trimmed to align with the end faces. Figure 9-12 shows the profile and path for creating the sweep feature. Figure 9-13 shows the resulting sweep feature created with the **Align with end faces** check box cleared. Figure 9-14 shows the resulting sweep feature created with the **Align with end faces** check box selected.

Note

If the sweep feature does not merge, you need to reduce the size of the profile.

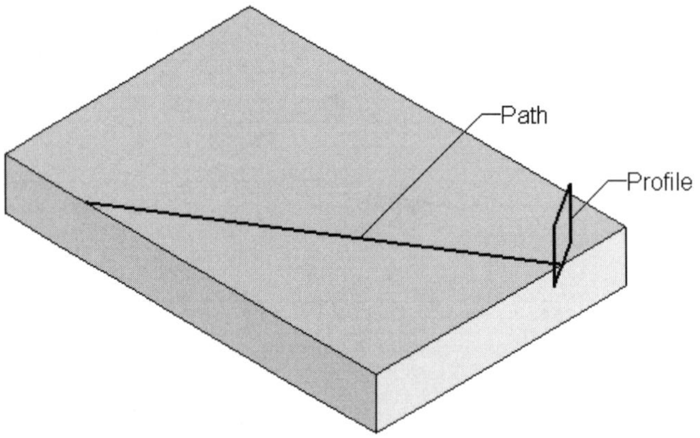

Figure 9-12 Sketches for the sweep feature

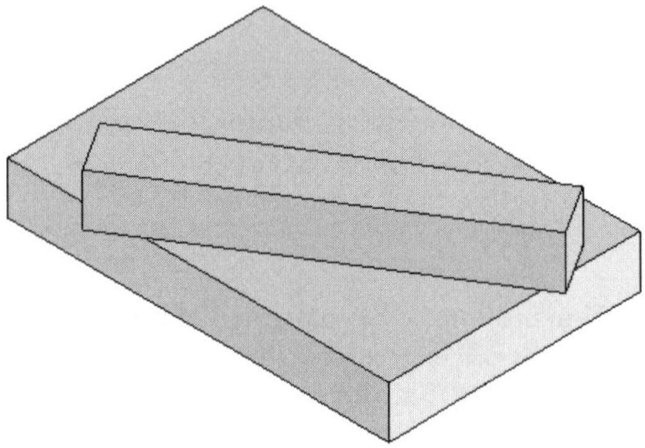

Figure 9-13 Resulting sweep feature with the ***Align with end faces*** check box cleared

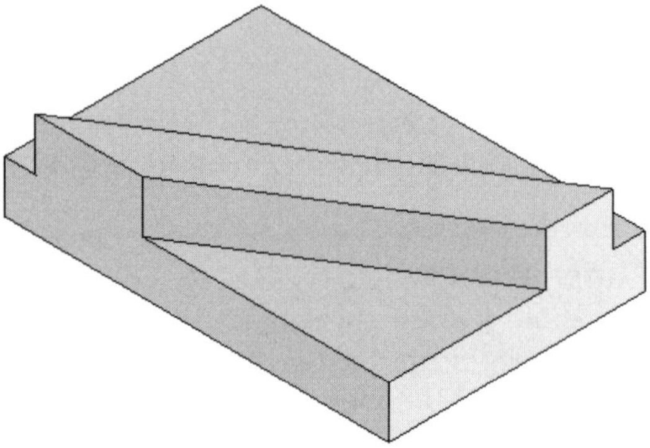

Figure 9-14 Resulting sweep feature with the ***Align with end faces*** check box selected

Applying Twist to the Swept Feature

While creating a sweep feature, you can also apply twist to it by choosing the **Twist Along Path** option from the **Orientation/twist type** drop-down list. The **Defined by** drop-down list and the **Twist Angle** spinner is displayed below the **Orientation/twist type** drop-down list, as shown in Figure 9-15.

Figure 9-15 The ***Options*** rollout

Using the option available in this area, you can apply a twist to the swept feature. By default, the **Degrees** option is selected in the drop-down list provided in this area. With this option selected, you need to specify the twist angle in the **Twist Angle** spinner. You can reverse the direction of twist using the **Reverse Direction** spinner. You can also specify the amount of twist in terms of radians and turns using the options available in the **Defined by** drop-down list. Figure 9-16 shows the swept feature using the **Follow Path** option and Figure 9-17 shows the swept feature after applying the twist.

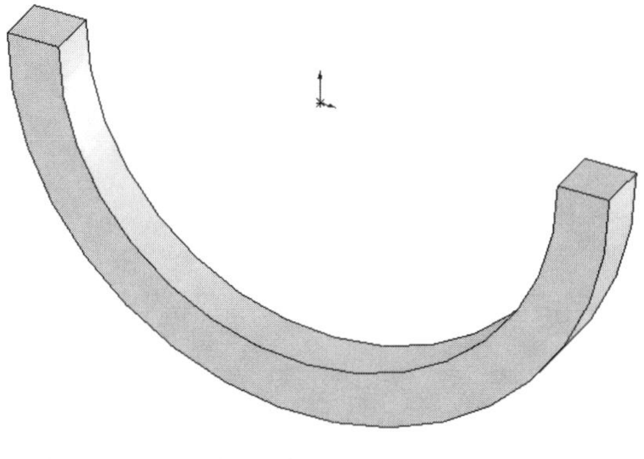

Figure 9-16 *Swept feature created using the **Follow Path** option*

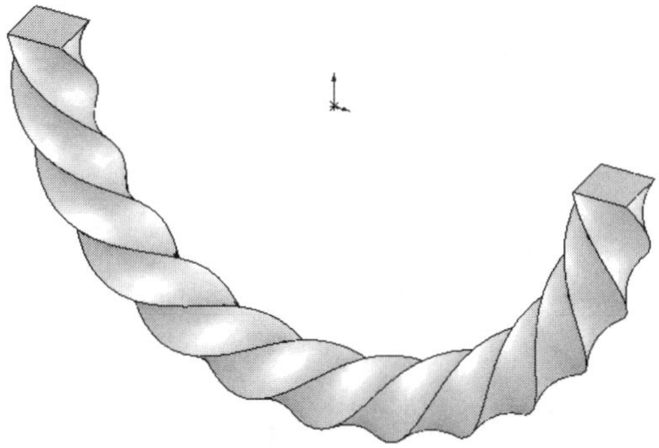

Figure 9-17 *Swept feature after applying twist*

Applying Twist Along the Path Keeping the Normal Constant

You can also apply a twist to the swept feature keeping its end face parallel to the profile and the entire transition normal to the sweep path. To apply this type of twits, choose the **Twist Along Path With Normal Constant** option from the **Orientation/twist type** drop-down list and then set the twist parameters. Figure 9-18 shows the twist applied using the **Twist Along Path** option and Figure 9-19 shows the twist applied using the **Twist Along Path With Normal Constant** option.

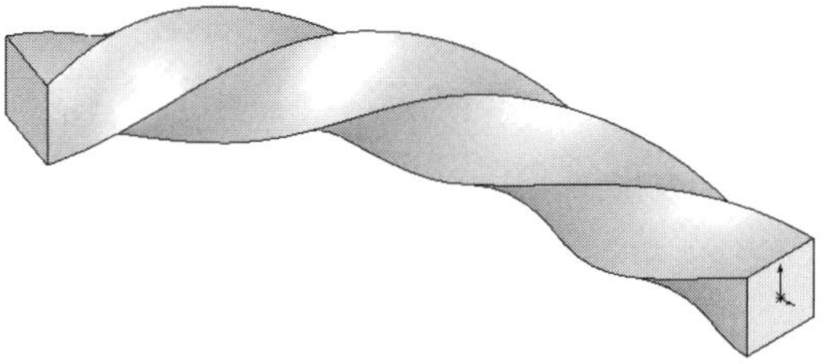

Figure 9-18 *Twist applied using the **Twist Along Path** option*

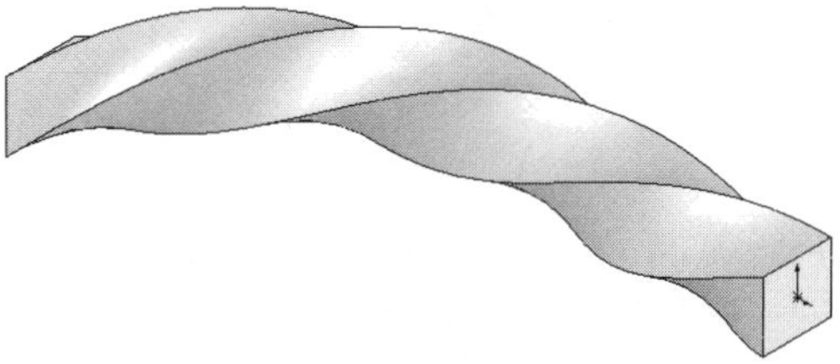

Figure 9-19 *Twist applied using the **Twist Along Path With Normal Constant** option*

Sweep with Guide Curves

The sweep with guide curves is the most important option in the advanced modeling tools. In this sweep feature, the section of the sweep profile varies according to the guide curves along the sweep path. To create this type of feature, you need a profile, path, and the guide curves.

After drawing the sketch of the profile, path, and guide curve, apply the coincident relation between the guide curves and the profile. You need to make sure that the guide curves intersects the profile. Now, invoke the **Sweep PropertyManager**. Select the profile and the path; the preview of the sweep feature is displayed in the drawing area. Click on the blue arrow on the right of the **Guide Curves** rollout to invoke this rollout. The **Guide Curves** rollout is shown in Figure 9-20.

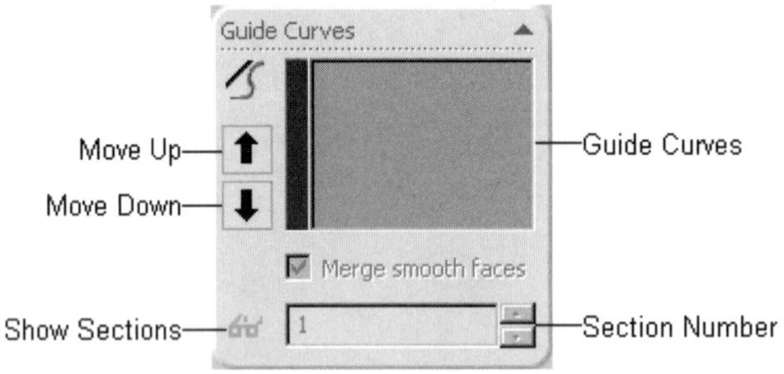

Figure 9-20 *The **Guide Curves** rollout*

Select the sketch of the guide curve; the selected guide curve is displayed in purple with a **Guide Curve** callout it. The preview of the sweep feature is also displayed in the drawing area. Choose the **OK** button from the **Sweep PropertyManager**. Figure 9-21 shows the sketch for the sweep feature with the guide curve. Figure 9-22 shows the resulting sweep feature creation.

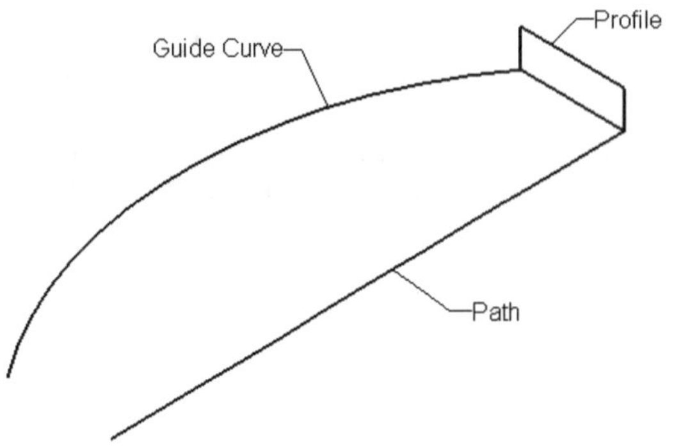

Figure 9-21 *Sketches for the sweep feature with guide curves*

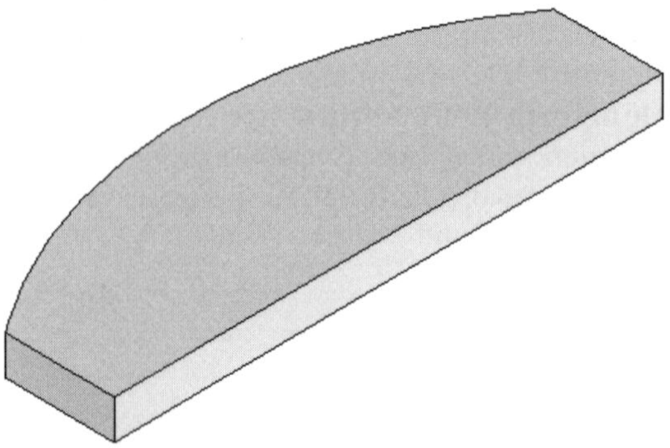

Figure 9-22 *Resulting sweep feature*

In the previous case, the path of the sweep feature was a straight line and the guide curve was an arc. In the next case, an arc is selected as the path of the sweep feature and a straight line is selected as the guide curve. Figure 9-23 shows the sketches for the sweep feature. Figure 9-24 shows the resulting sweep feature.

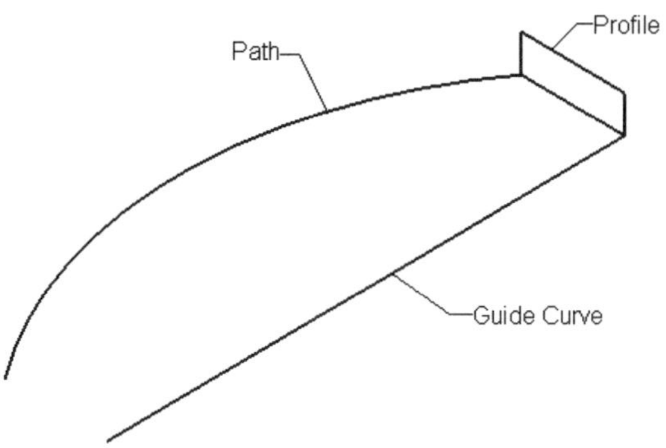

Figure 9-23 Sketches for the sweep feature

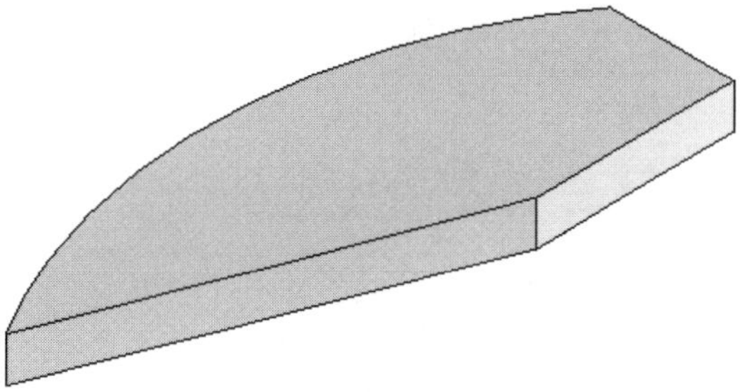

Figure 9-24 Resulting sweep feature

Move Up and Move Down

The **Move Up** button and the **Move Down** button available on the left of the **Guide Curves** selection area are used to change the sequence of the selected guide curve.

Merge smooth faces

The **Merge smooth faces** check box available in the **Guide Curves** rollout is selected by default. This option is used to merge all the smooth faces together, resulting in a smooth sweep feature. After creating the sweep feature, when you edit it and clear the **Merge smooth faces** check box, the **Sweep Preview Warning** dialog box is displayed, as shown in Figure 9-25. In this dialog box you are prompted that the feature you are creating may fail because of change in smooth face option. Choose the **Yes** button from this dialog box to accept the change option. When you create a sweep feature with guide curves and this option cleared, the resulting feature does not merge the smooth faces together. This results in a sweep feature with noncontinuous curvature surface. Figure 9-26 shows a sweep feature created with the **Merge smooth faces** check box selected. Figure 9-27 shows the same sweep feature with the **Merge smooth faces** check box cleared.

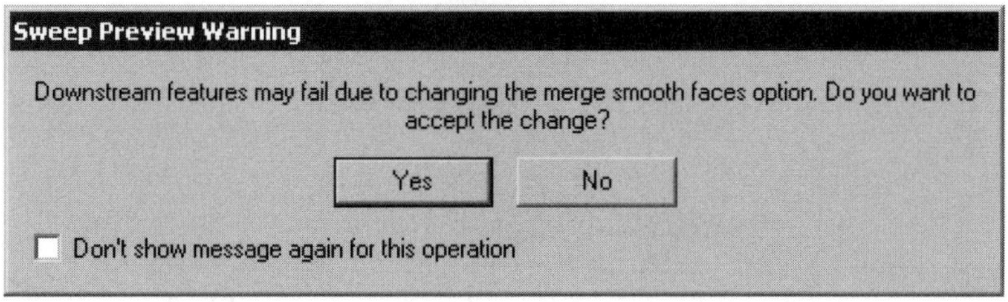

*Figure 9-25 The **Sweep Preview Warning** dialog box*

Note

*Remember that if you create the sweep feature with the **Merge smooth faces** check box cleared, the resulting feature will be generated faster and the adjacent faces and edges will be easily merged. Also, the lines and arcs in the guide curve will match accurately.*

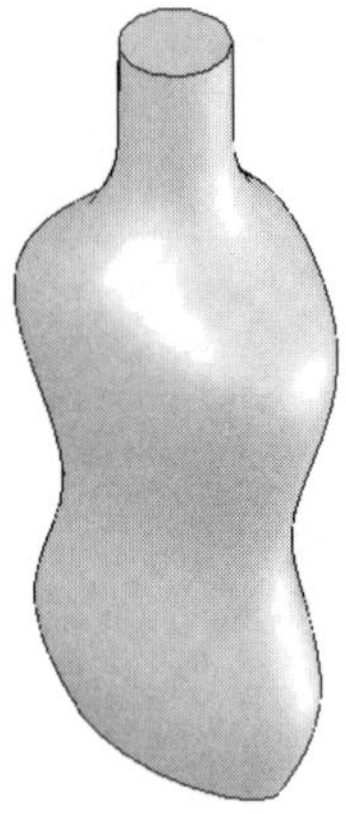

Figure 9-26 *Sweep feature with the* ***Merge smooth faces*** *check box selected*

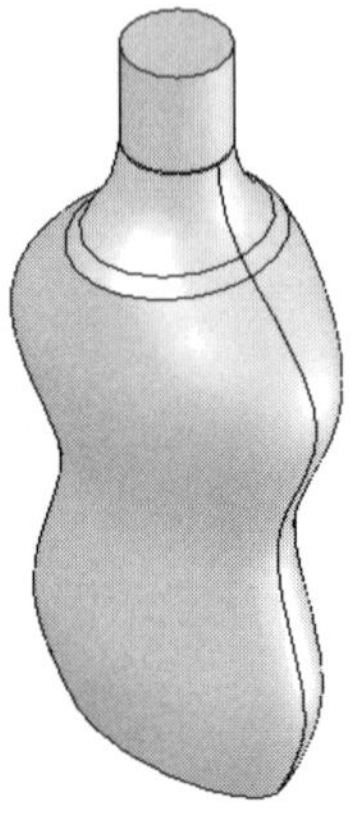

Figure 9-27 *Sweep feature with the* ***Merge smooth faces*** *check box cleared*

Show Sections

The **Show Sections** button available in the **Guide Curves** rollout is used to display the intermediate sections, while creating the sweep feature with guide curves. To display the intermediate profiles or sections along the sweep path, choose the **Show Section** button from the **Guide Curves** rollout. The **Section Number** spinner is invoked. Using this spinner, you can view the sections of the profile along the sweep path. The maximum value of the spinner goes up to the number of sections that fit inside the sweep feature. Figure 9-28 shows a section being displayed using the **Show Section** option and constrained by the guide curves.

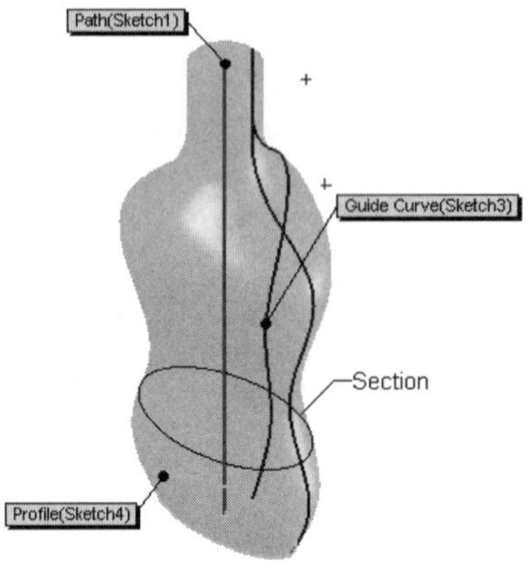

Figure 9-28 *Section being displayed using the* ***Show Section*** *option*

Sweep Feature Using the Follow path and 1st guide curve Option

When you create a sweep feature with a guide curve using the **Follow path and 1st guide curve** option from the **Options** rollout, the profile follows the path and the first guide curve to create the feature. For creating a sweep feature using this option, invoke the **Sweep PropertyManager** and select the profile, path, and guide curve(s). By default, the **Follow**

path option is selected in the **Orientation/twist type** drop-down list of the **Options** rollout. Select the **Follow path and 1st guide curve** option from the **Orientation/twist type** drop-down list. Choose the **OK** button from the **Sweep PropertyManager** to end the creation of the feature.

Sweep Feature Using the Follow 1st and 2nd guide curves Option

Using this option, the profile of the sweep feature follows the first and second guide curves to create the resulting sweep feature. For creating this type of sweep feature, select the **Follow 1st and 2nd guide curves** option from the **Orientation/twist type** drop-down list from the **Options** rollout. Choose the **OK** button from the **Sweep PropertyManager** to end the creation of the feature.

Figure 9-29 shows the sketches of the profile, path, and guide curves for creating a sweep feature. Figure 9-30 shows the sweep feature created using the **Follow path and 1st guide curve** option. Figure 9-31 shows the sweep feature created using the **Follow 1st and 2nd guide curves** option.

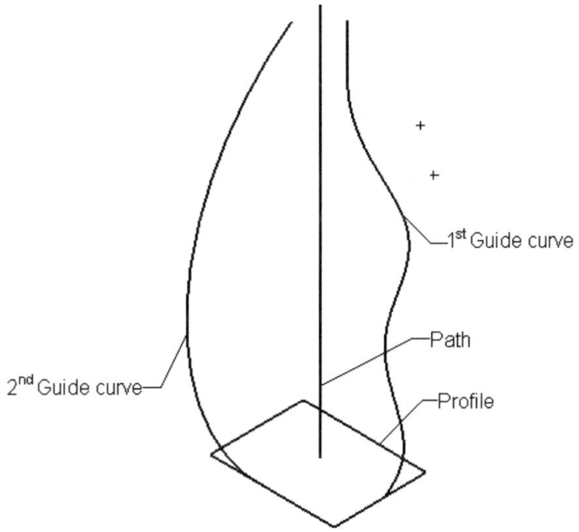

Figure 9-29 *Sketches of profile, path, and guide curves*

Start/End Tangency

The **Start/End Tangency** rollout available in the **Sweep PropertyManager** is used to define the tangency conditions at the start and end of the feature. Invoke the **Start/End Tangency** rollout, as shown in Figure 9-32. The various options available in the **Start/End Tangency** rollout are discussed next.

Start tangency type

The **Start tangency type** drop-down list is used to specify the options to define the tangency at the start of the sweep feature. The various options available in this drop-down list are discussed next.

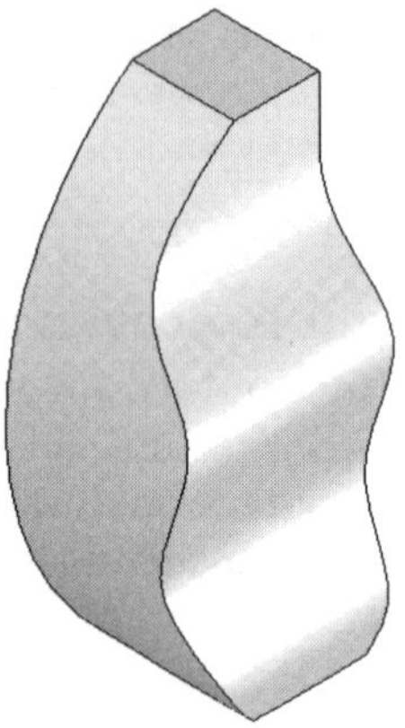

*Figure 9-30 Sweep feature using the **Follow path and 1st guide curve** option*

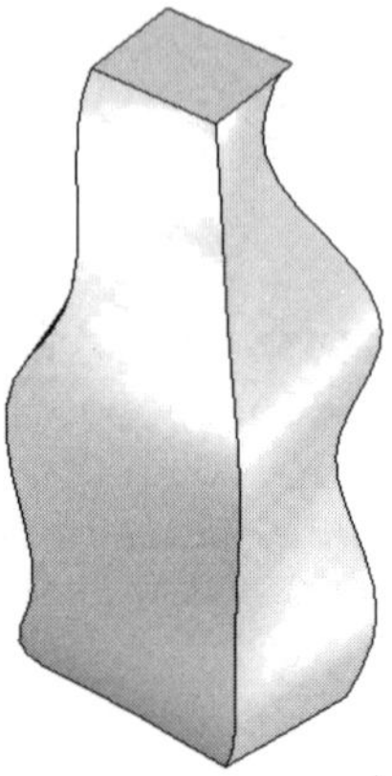

*Figure 9-31 Sweep feature using the **Follow 1st and 2nd guide curves** option*

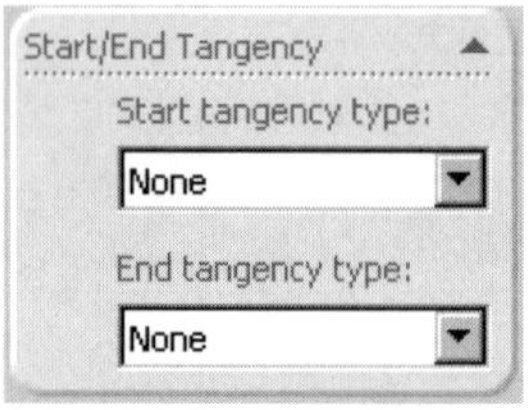

*Figure 9-32 The **Start/End Tangency** rollout*

None
The **None** option is selected by default and is used to create a sweep feature without applying any start tangency.

Path Tangent
The **Path Tangent** option is used to maintain the sweep feature normal to the path at the start.

Direction Vector
When you use the **Direction Vector** option, the starting of the sweep feature will be tangent to a virtual normal created from the selected entity. When you select this option, the **Direction Vector** selection area is also displayed and you need to select a linear edge, axis, planar face, or plane.

All Faces
The **All Faces** option is used to sweep the feature tangent to the adjoining faces of the existing geometry at the start.

The options available in the **End tangency type** drop-down list are the same as those discussed above except the **All Faces** option. The only difference is that the options in this drop-down list are applied to the end of the sweep feature. Figure 9-33 shows the sketches and the references to create the sweep feature with the start and

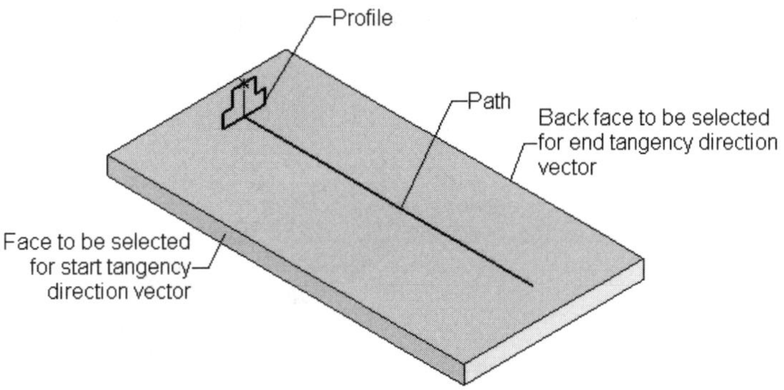

Figure 9-33 *Profile, path, and references for tangency using the* ***Direction Vector*** *option*

end tangency, using the **Direction Vector** option. Figure 9-34 shows the resulting sweep feature.

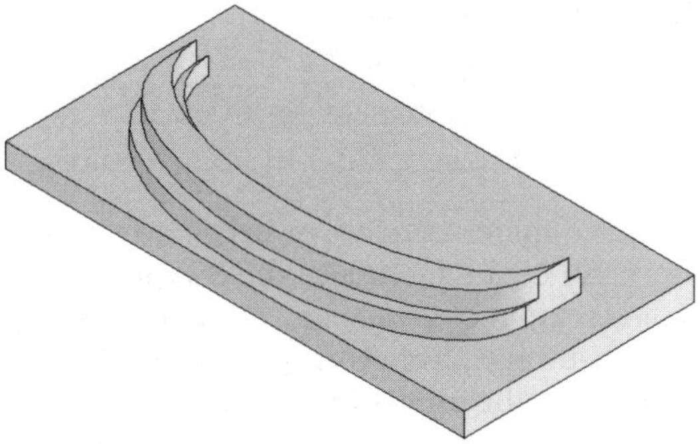

Figure 9-34 *Resulting sweep feature*

Creating a Thin Sweep Feature

You can also create a thin sweep feature by specifying the thickness using the **Thin Feature** rollout, which is invoked by selecting the check box provided at the left of the **Thin Feature** rollout. The **Thin Feature** rollout is shown in Figure 9-35.

Figure 9-35 *The* ***Thin Feature*** *rollout*

The options available in this rollout are the same as those discussed in the earlier chapters, where extruding and revolving of thin features were discussed. Figure 9-36 shows a thin sweep feature.

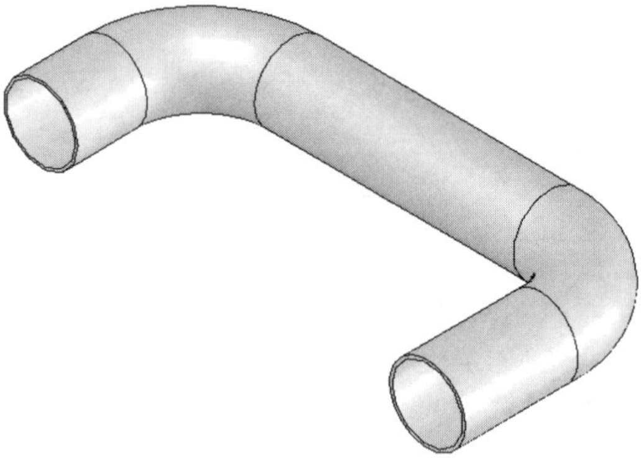

Figure 9-36 *Thin sweep feature*

Creating Cut Sweep Features

CommandManager:	Features > Swept Cut	*(Customize to Add)*
Menu:	Insert > Cut > Sweep	
Toolbar:	Features > Swept Cut	*(Customize to Add)*

You can also remove material from an existing feature or model by creating a cut-sweep feature. To create it, choose the **Swept Cut** button from the **Features CommandManager** after customizing, as shown in Figure 9-37. The options available in the **Cut-Sweep PropertyManager** are the same as those discussed in the **Sweep PropertyManager**, with the only difference being that the options in this case are meant for the cut operation. Figure 9-38 shows the profile and the path. Figure 9-39 shows the resulting cut-sweep feature created using the **Swept Cut** tool.

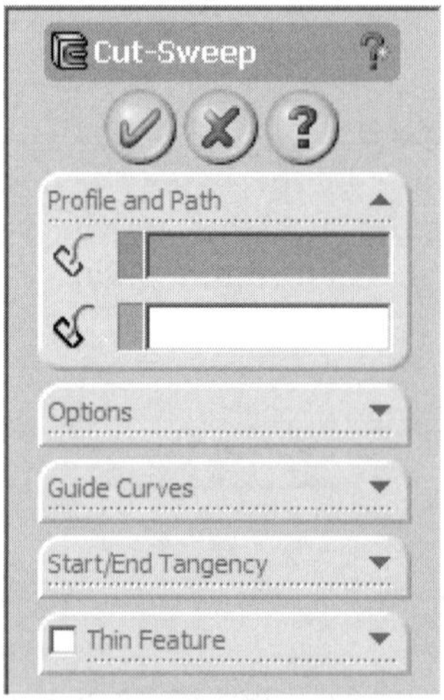

Figure 9-37 *The* ***Cut-Sweep PropertyManager***

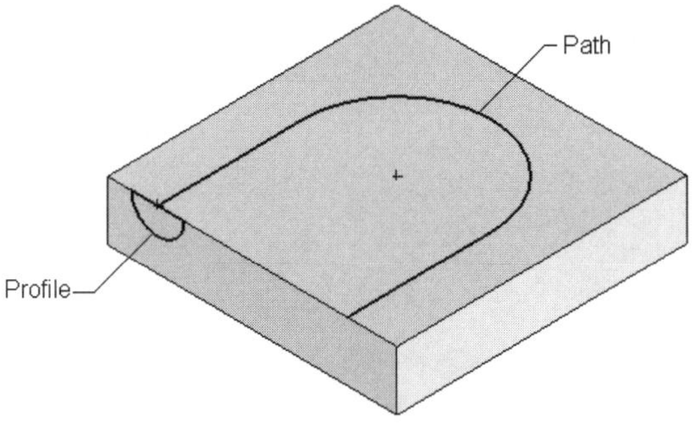

Figure 9-38 Profile and path for the cut sweep feature

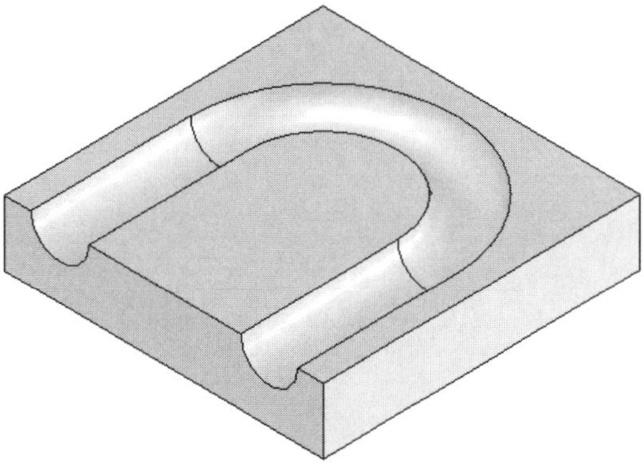

Figure 9-39 Resulting cut-sweep feature

Creating the Loft Feature

CommandManager:	Features > Lofted Boss/Base
Menu:	Insert > Boss/Base > Loft
Toolbar:	Features > Lofted Boss/Base

The lofted features are created by blending more than one similar or dissimilar sections together to get a free form shape. These similar or dissimilar sections may or may not be parallel to each other. Note that the sections for the solid lofts should be closed sketches.

To create a loft feature, invoke the **Loft PropertyManager** by choosing the **Lofted Boss/Base** button from the **Features CommandManager**. A partial view of the **Loft PropertyManager** is shown in Figure 9-40.

On invoking the **Loft PropertyManager**, after drawing the sketches, you are prompted to select at least two profiles. Select the profiles from the drawing area; the preview of the loft feature along with a connector is displayed in it. Choose the **OK** button from the **Loft PropertyManager** to end the creation of the feature. Figure 9-41 shows the preview of the loft feature along with the connector and Figure 9-42 shows the resulting loft feature.

The loft feature can be reshaped using the handles of the connector that appear in cyan in the preview it. Press and hold down the left mouse button on a handle; the handle appears red. Drag the cursor to specify a new location and release the left mouse button to place the connector on it. The process of controlling the loft shape using connectors is known as **Loft**

Tip. *You can also use the **Contour Select Tool** to select a contour that will be used as the profile to create the loft feature. You can also use a shared sketch as a profile that is currently in use by some other sketched feature.*

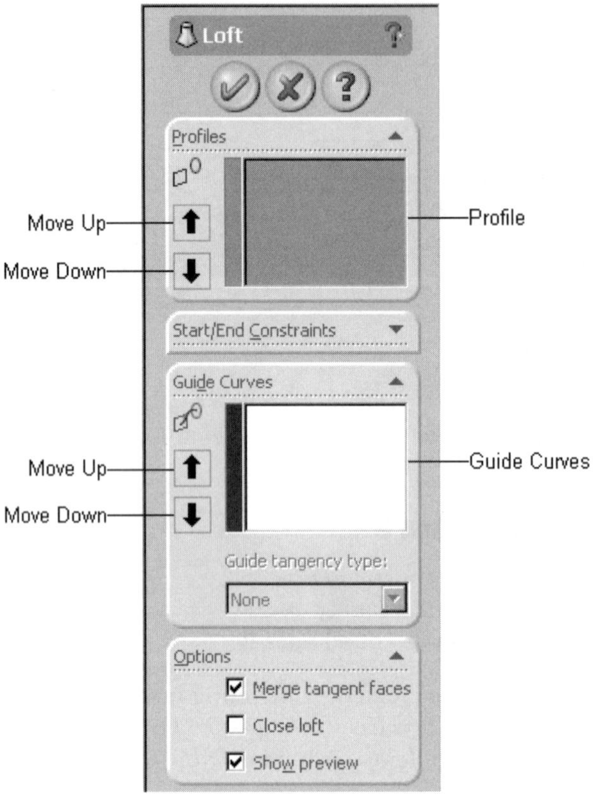

Figure 9-40** Partial view of the **Loft PropertyManager

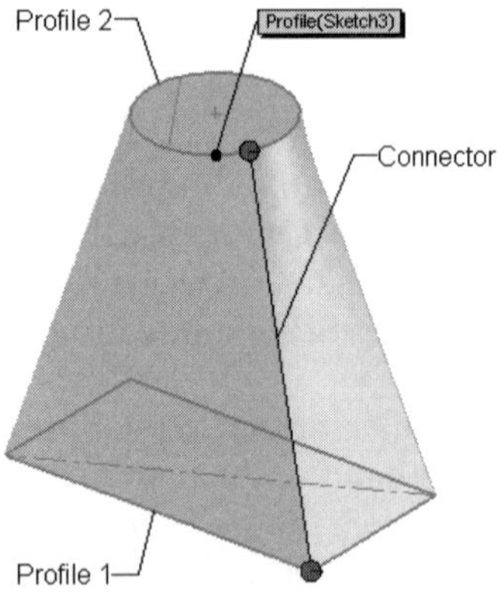

***Figure 9-41** Preview of the loft feature*

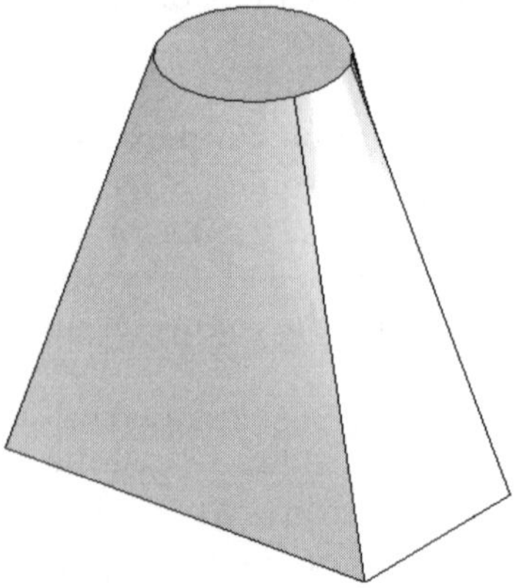

***Figure 9-42** Resulting loft feature*

Synchronization. Figure 9-43 shows the preview of the loft feature, after modifying the location of the handle of the connector. Figure 9-44 shows the resulting loft feature.

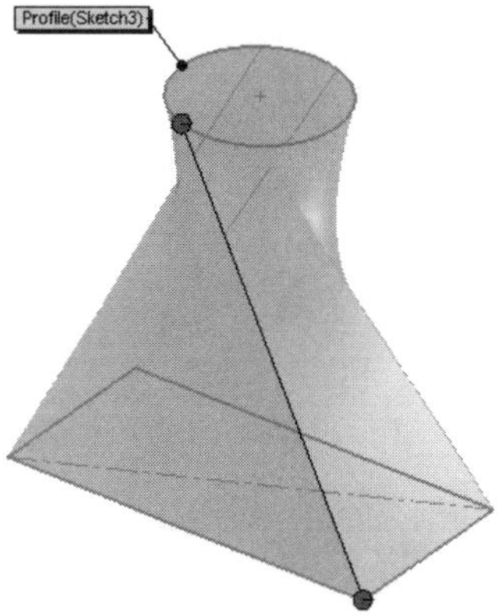

Figure 9-43 *Preview of the loft feature, after modifying the location of the connector*

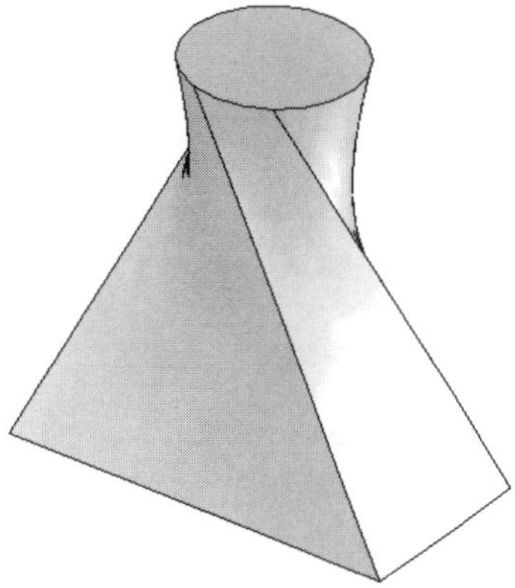

Figure 9-44 *Resulting loft feature*

Note

The reshaping or twisting using the default connector, available while creating the loft feature, is known as global twisting. Global twisting means if you change the location of one connecting point of the profile, other connecting points of the profile automatically change their positions with respect to the modified connecting point.

In global twisting of non-tangent profiles, the handles of the connectors move only from vertex to vertex.

To display all connectors, right-click and choose the **Show All Connectors** option from the shortcut menu. The number of connectors displayed using this option depends on the maximum number of vertices in the start or end loft section. If the start and end loft sections are circular, elliptical, or a closed splined section, then only one controller is displayed.

You can also add additional connectors to manipulate the loft feature. Connectors can be added to straight profiles or curved profiles. To add a connector, right-click on the location in the sketch, where you need to add the connector to invoke the shortcut menu. Choose the **Add Connector** option from the shortcut menu. A connector will be added to the loft feature. You can also add more connectors using this option. You can also modify them by dragging their handles. Figure 9-45 shows the preview of the loft feature after modifying all additional connectors and Figure 9-46 shows the resulting loft feature.

Note

Reshaping or twisting using additional connectors is known as local twisting, because the twisting using one connector does not affect the other connecting points of the profile. Therefore, you can independently modify all connectors simply by dragging the handles along the profile.

Start/End Constraints

The **Start/End Constraints** rollout available in the **Loft PropertyManager** is used to define the constraints at the start and end sections of the loft feature. You can define normal, tangency, or continuity constraints to the loft feature. Invoke this rollout to define the tangency. By

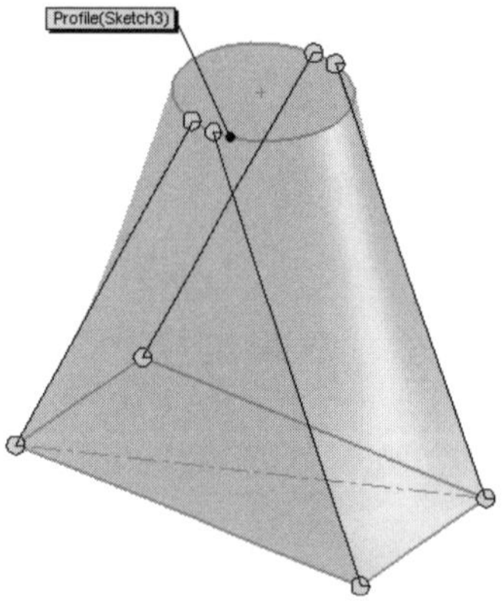

Figure 9-45 *Preview after adding all connectors*

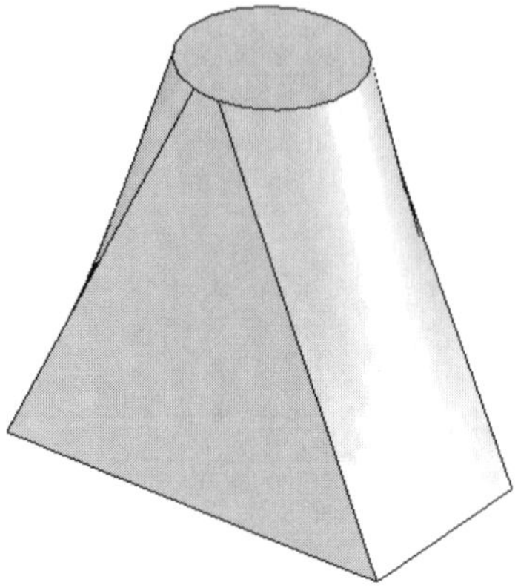

Figure 9-46 *Resulting loft feature*

Tip. *Invoke the shortcut menu and choose* ***Hide All Connectors*** *to hide all connectors. The connectors are not displayed in the preview, but the settings made by them remain in the loft feature. To hide a specific connector, select it and invoke the shortcut menu. Then choose the* ***Hide Connector*** *option.*

If you choose the ***Reset Connectors*** *option from the shortcut menu, all connectors and the settings related to them are deleted from the memory of the feature. Also, the default connector is displayed with the default connection points lying between the profiles of the loft feature.*

default, the **None** option is selected in the **Start constraint** and the **End constraint** drop-down lists. This implies that no constraint is applied to the loft feature. The other options available in these drop-down lists are discussed next.

Normal to Profile

The **Normal to Profile** option is used to define the tangency normal to the profile. When you select this option from the **Start constraints** and **End constraints** drop-down lists, an arrow is displayed at the start and end sections. Also, some additional options are displayed in the **Start/End Constraints** rollout, as shown in Figure 9-47. You can set the length of tangency using the tangency arrows attached to the sections. You can also set the values of the length of tangency using the **Start Tangency Length** and the **End Tangency Length** spinners provided in this rollout. You can also specify a draft angle using the **Draft angle** spinners. You will notice that the **Apply to all** check boxes are selected by default, implying that the tangency is applied evenly to all the vertices of the sections. However, if on clearing these check boxes, you can apply values of tangencies individually to all the vertices. Figure 9-48 shows the preview of the loft feature with the tangency arrows attached to end sections of the loft feature. Figure 9-49 shows the tangency arrows attached to all vertices of the end sections of the loft feature. Figure 9-50 shows the resulting loft feature, after specifying the length of tangency.

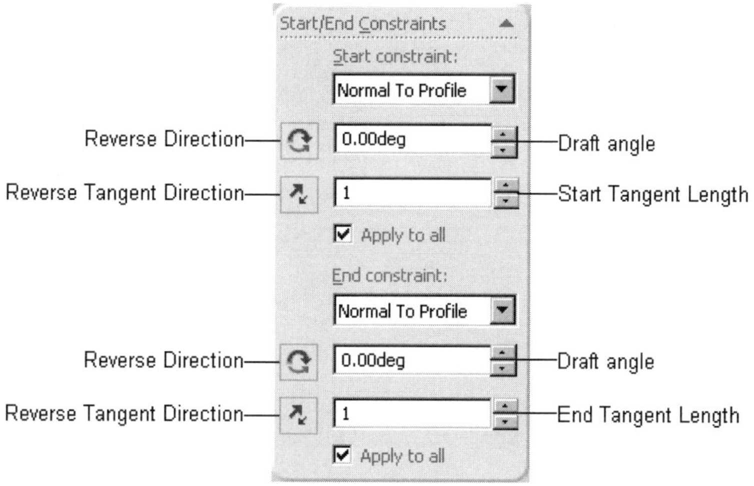

***Figure 9-47** The **Start/End Constraints** rollout with the **Normal to Profile** option selected*

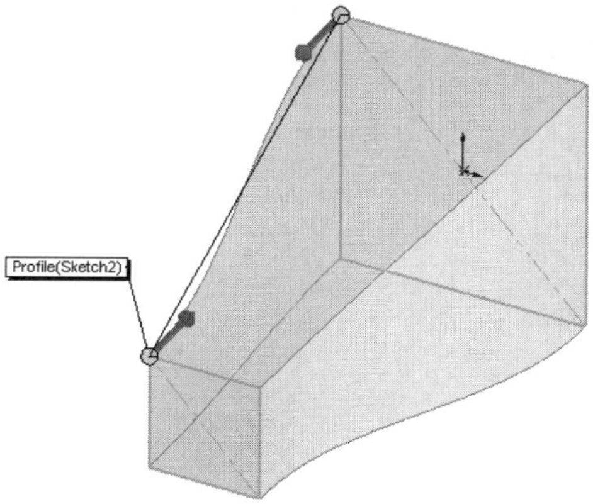

***Figure 9-48** Two arrows attached to the sections*

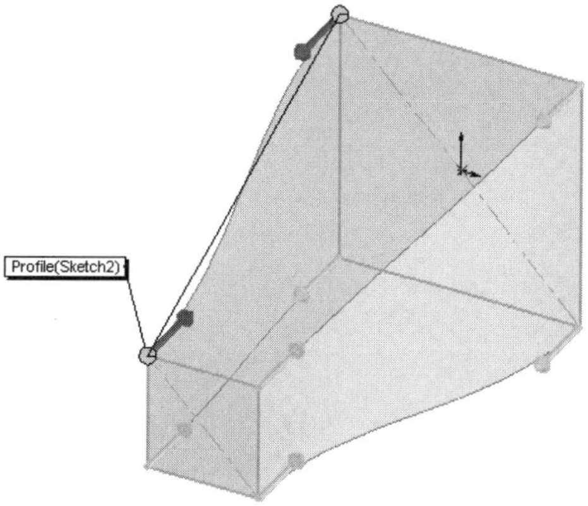

***Figure 9-49** Arrows attached to all vertices*

Direction Vector

The **Direction Vector** option is used to define the tangency at the start and end of the loft feature by defining a direction vector. On invoking this option, you are provided with the **Direction Vector** selection area and the spinners to define the length of tangents and the draft angle. You need to select the direction vectors to specify the tangent at the start and end of the loft feature. You can also specify the length of the tangents, using the spinners provided in the **Start/End Tangency** rollout. The **Start/End Tangency** rollout with the **Direction Vector** option selected is displayed in Figure 9-51.

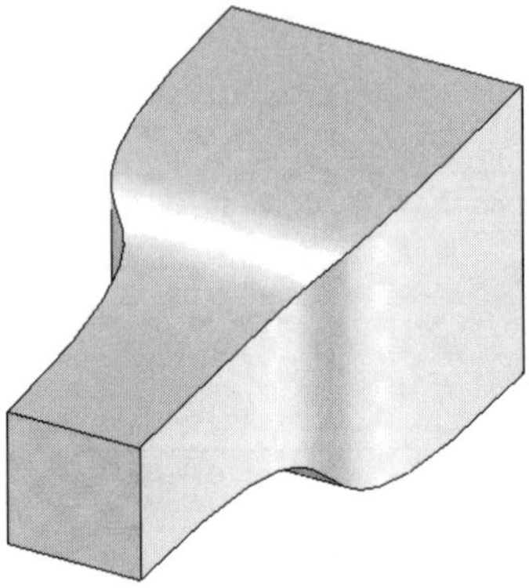

Figure 9-50 *Resulting loft feature*

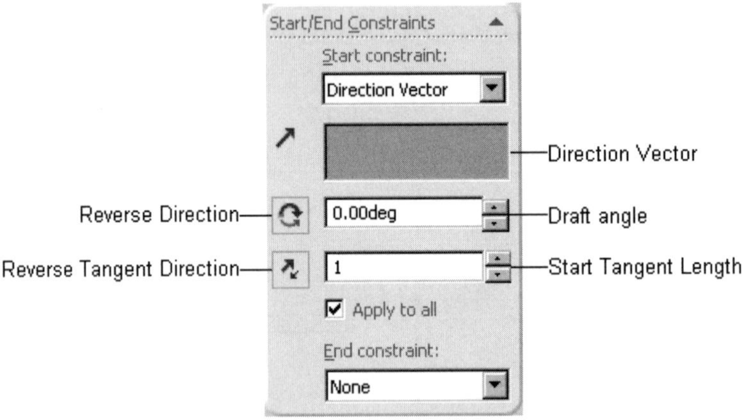

Figure 9-51 *The* ***Start/End Constraints*** *rollout with the* ***Direction Vector*** *option selected*

Figure 9-52 shows the sections for the loft feature. Figure 9-53 shows the initial preview of the loft feature. Figure 9-54 shows the preview of the loft feature with tangent at the start of the loft and Figure 9-55 shows the tangent at the start and at the end of the loft. Figure 9-56 shows the final loft feature.

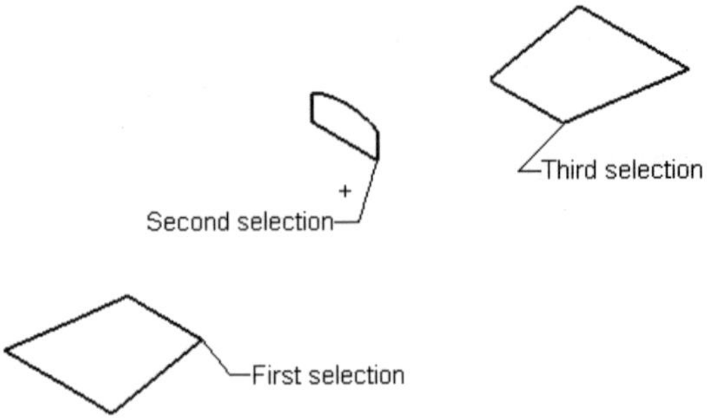

Figure 9-52 *Sections, selection points, and sequence of selection*

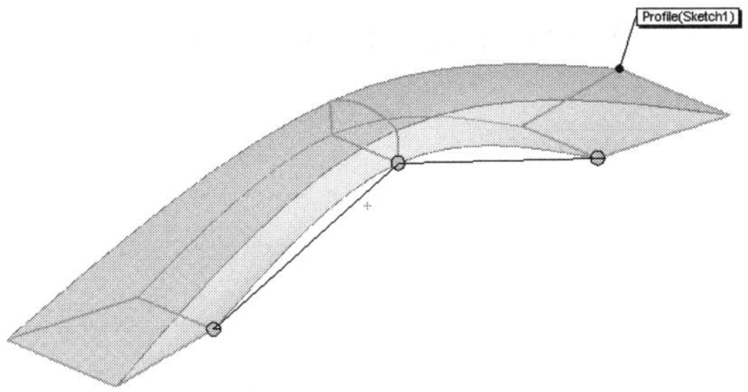

Figure 9-53 Initial preview of the loft feature

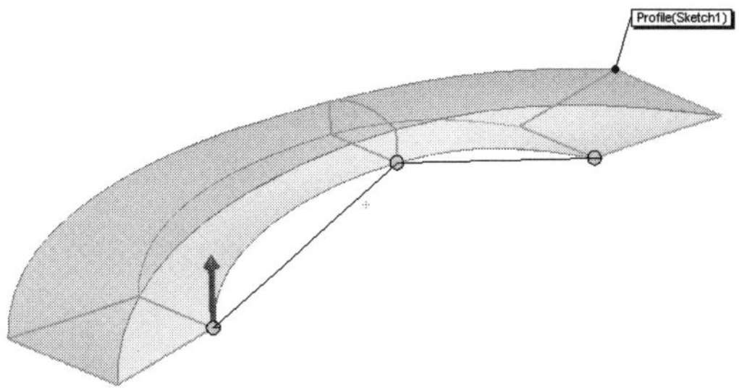

Figure 9-54 Tangent applied at the start

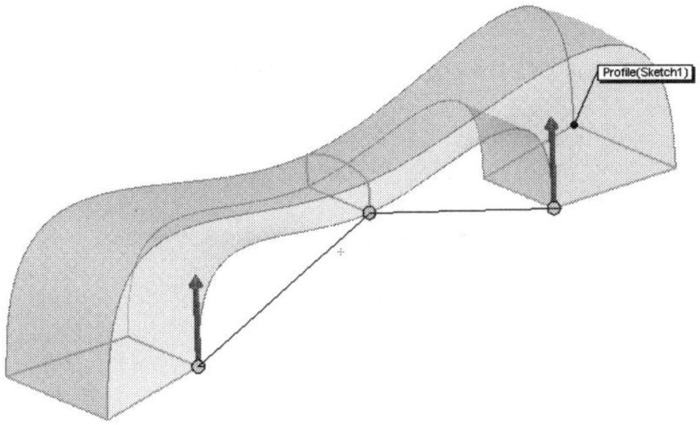

Figure 9-55 Tangent applied at start and end of the loft feature

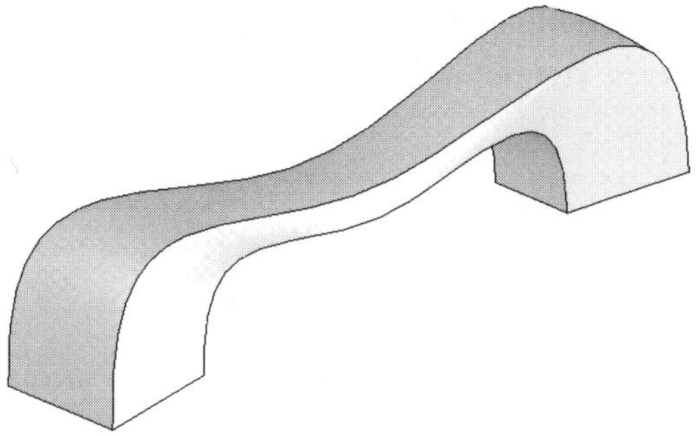

Figure 9-56 The final loft feature

Tangency To Face

If you select the **Tangent To Face** option from the **Start constraint** or **End constraint** drop-down lists, the resulting loft feature will maintain the tangency along the adjacent curved faces. The face, along which the tangency will be maintained, is highlighted in green. You can also switch between the faces along which you need to maintain the tangency using the **Next Face** button. You can also specify the lengths at the start and end tangencies, using the spinners available below it.

Curvature To Face

If you select the **Curvature To Face** option from the **Start constraint** or **End constraint** drop-down lists, the resulting loft feature will maintain the curvature along the adjacent curved faces. The face, along which the curvature will be maintained, is highlighted in green. You can also switch between the faces along which you need to maintain the tangency, using the **Next Face** button.

Guide Curves

SolidWorks enables you to specify the guide curves between the profiles of the loft feature to define the path of transition of the loft feature. The sketches drawn for the guide curve must coincide with the sketches that define the loft sections. The **Guide Curves** rollout is displayed in Figure 9-57.

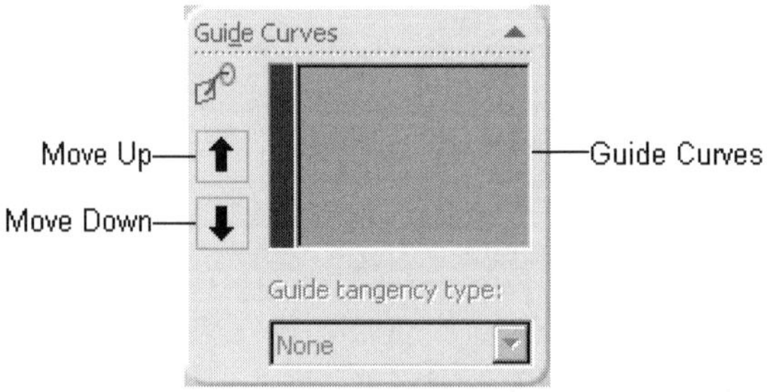

Figure 9-57 *The **Guide Curves** rollout*

Figure 9-58 shows the profiles and the guide curves for creating a loft feature with guide curves. Figure 9-59 shows the resulting loft feature created using the guide curves.

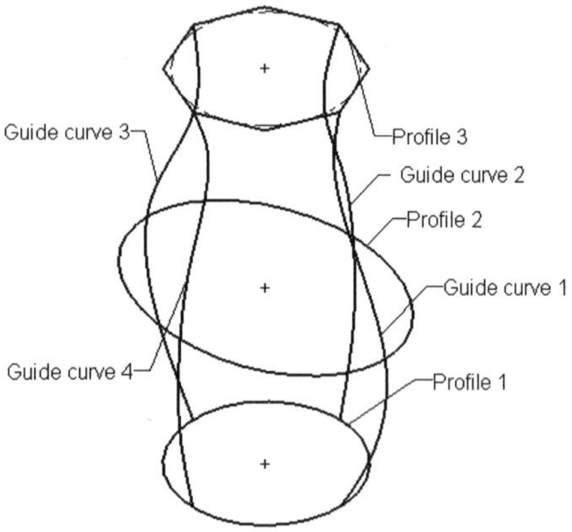

Figure 9-58 *Profiles and guide curves*

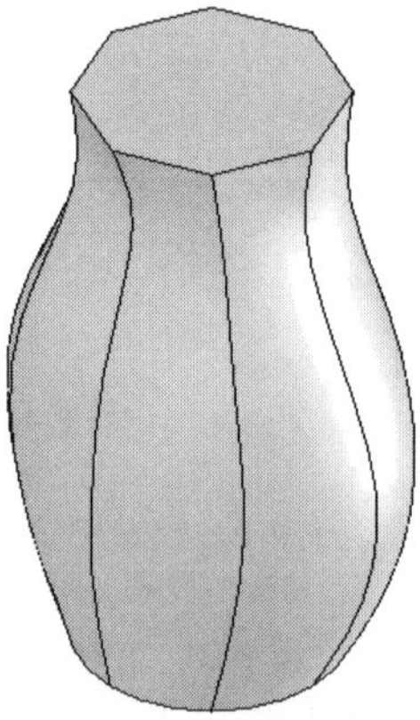

Figure 9-59 *Resulting loft feature*

Options

The **Options** rollout of the **Loft PropertyManager** is provided with options to improve the creation of the loft feature. These options are the same as those discussed in the sweep option. An additional option available with this rollout is discussed next.

Close Loft

A closed loft feature is one in which the end and start of the loft features is joined together. The **Close loft** check box is selected to create a close loft feature. Figure 9-60 shows a loft feature created with the **Close loft** check box cleared. Figure 9-61 shows the loft feature created with the **Close loft** check box selected.

Centerline Parameters

The **Centerline Parameters** rollout is used to create a loft feature by blending two or more than two sections along a specified path. The path that specifies the transition is called

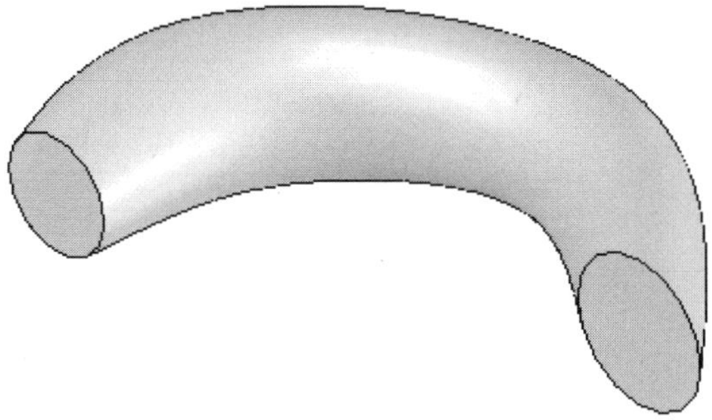

Figure 9-60 Loft feature with the ***Close Loft*** *check box cleared*

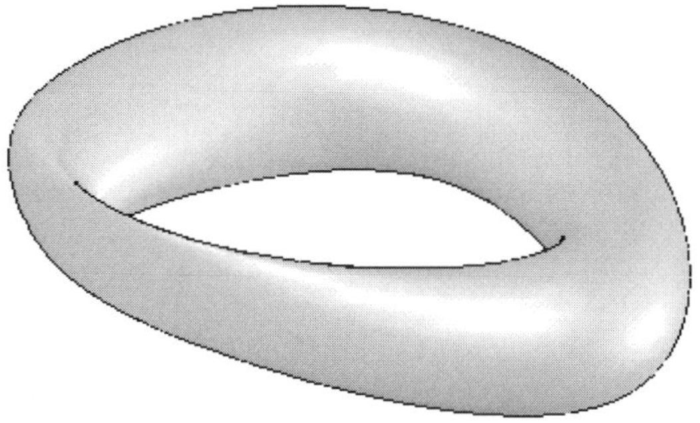

Figure 9-61 Loft feature with the ***Close loft*** *check box selected*

the centerline. From this release of SolidWorks, you can also define guide curves to the loft feature created by using the centerline. The **Centerline Parameters** rollout is displayed in Figure 9-62. The options available in this rollout are discussed next.

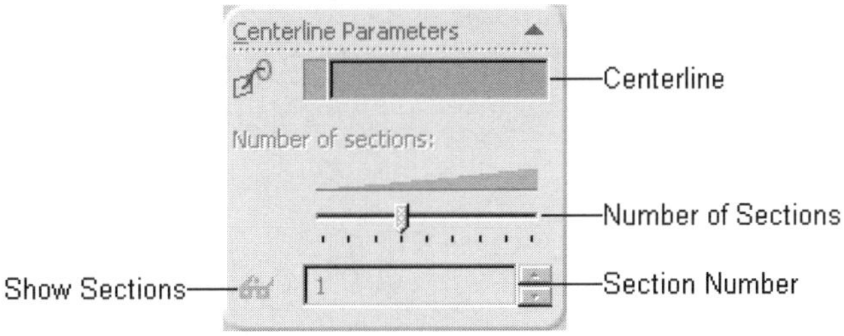

Figure 9-62 The ***Centerline Parameters*** *rollout*

Centerline

You need to specify the centerline after invoking the **Centerline Parameters** rollout. Therefore, select the sketch that defines the centerline for the loft feature; the name of the sketch will be displayed in the **Centerline** selection area.

Number of sections

The **Number of sections** slider bar provided in the **Centerline Parameters** rollout is used to define the number of intermediate sections, which further, define the accuracy and the smoothness of the loft feature.

The **Show Sections** button and the **Section Number** spinner available in the **Centerline Parameters** rollout are used to display the intermediate sections, as discussed earlier. Figure 9-63 shows the sketches of the profiles and the centerline used to create the loft feature. Figure 9-64 shows the resulting loft feature.

You can also create a thin loft feature by defining the thin parameters using the **Thin Feature** rollout. The options available in this rollout are the same as those discussed in the earlier

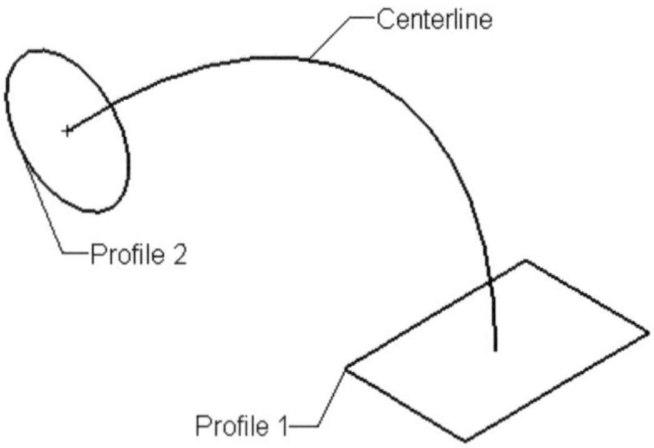

Figure 9-63 Profiles and centerline

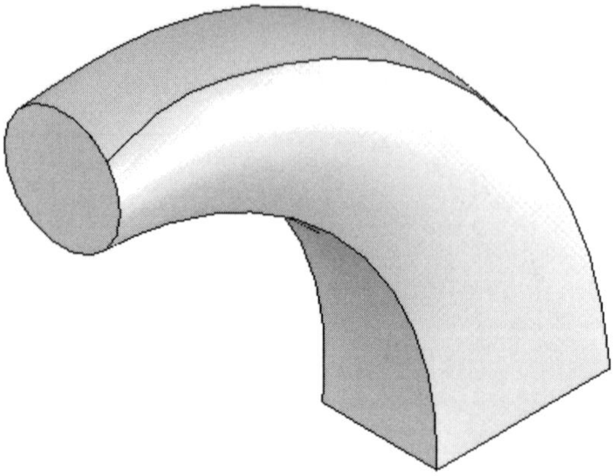

Figure 9-64 The Resulting loft feature

Tip. *If you need to create a smooth loft feature between a circle and a polygon, it is a good practice to split the circle into a number of arcs. The number of arcs that forms the circle should be the same as the number of sides of the polygon. Now, when you create the loft feature the number of connecting points in both the sections are predefined, which results in a smoother loft feature.*

chapters. Figure 9-65 shows a thin loft feature created using the **Thin Features** rollout available in the **Loft PropertyManager**.

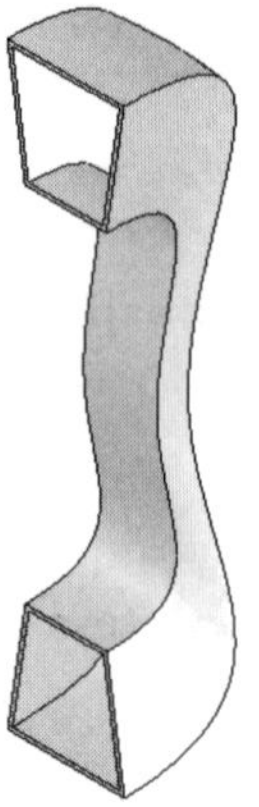

Figure 9-65 A thin loft feature

Tip. *From this release of SolidWorks, you can also display a mesh in the preview of the loft feature by right-clicking to invoke the shortcut menu and choosing* ***Mesh Preview > Mesh All Faces****. The mesh preview will only be displayed on the non-planar faces and not on the planar faces. To display a mesh preview on all faces of the loft preview, you need to change the planar faces to non-planar faces by manipulating the connectors.*

Add a Section in the Loft Feature

After creating a loft feature, you can also add a section to it by selecting one of the side faces of the loft feature and right-clicking to invoke the shortcut menu. Choose the **Add Loft Section** option from it; the **Add Loft Section PropertyManager** is displayed. You are provided with a plane that can be moved or rotated dynamically. To move the plane, place the cursor on the arrows provided on it. The arrows turns red in color. Press and hold down the left mouse button and drag the cursor to move the plane. To rotate the plane, select one of its edges and drag the cursor. Set the position of the plane by dynamically moving and rotating it to define the position to add the sketch. Figures 9-66 and 9-67 show the plane being moved and rotated. After specifying the position of the plane, choose the **OK** button from the **Add Loft Section PropertyManager**. Figure 9-68 shows the section added in the model. This is a closed section created using a spline.

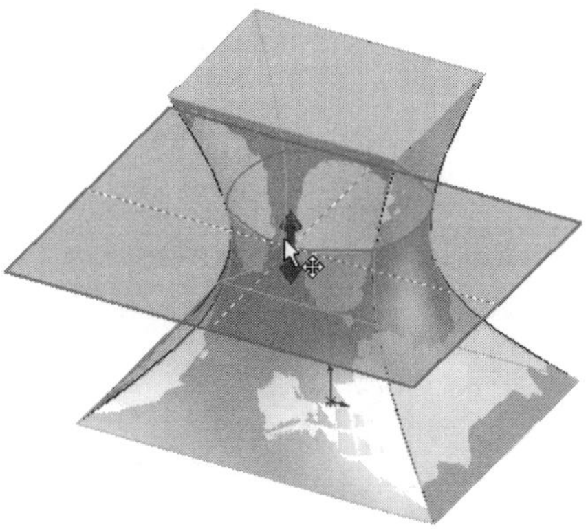

Figure 9-66 Moving the plane

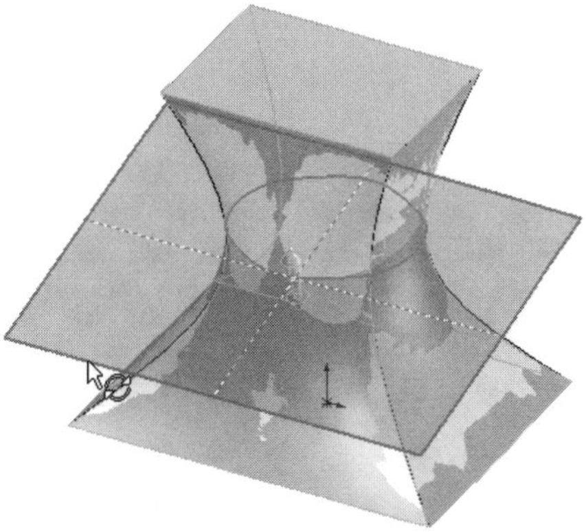

Figure 9-67 Rotating the plane

Expand the **Loft1** feature and select the sketch that is added using the **Add Loft Section** tool. Right-click and choose the **Edit Sketch** option from the shortcut menu. Edit the sketch as shown in Figure 9-69 and rebuild the part. Figure 9-70 shows the resulting modified loft feature.

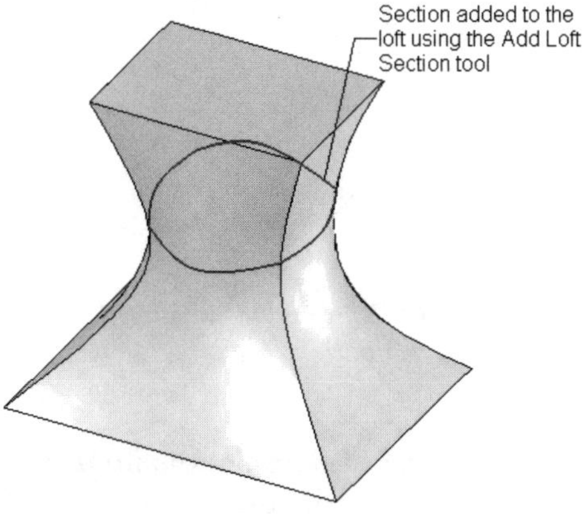

Figure 9-68 Resulting section added to loft feature

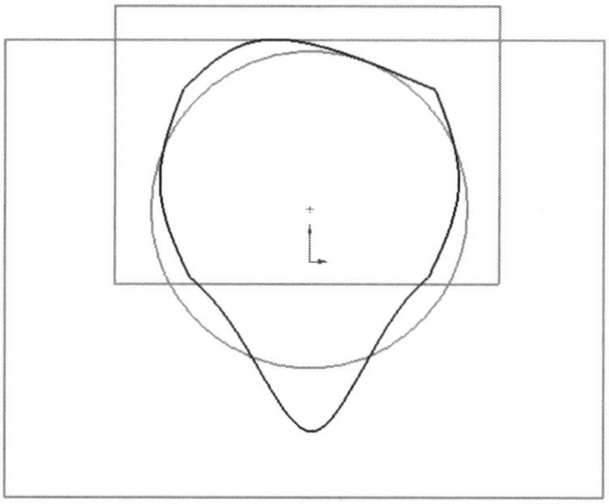

Figure 9-69 Modified loft section

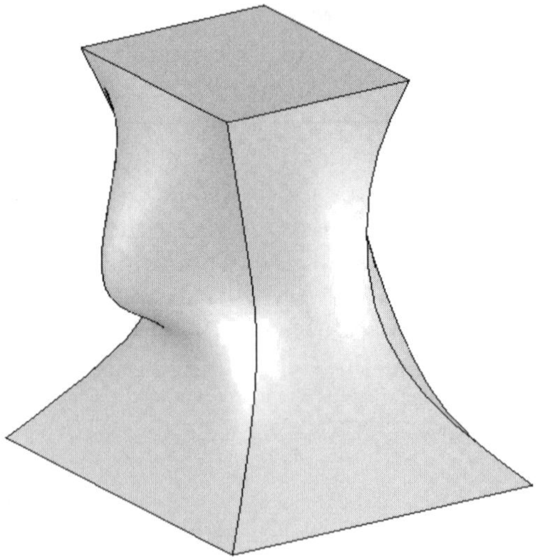

Figure 9-70 Resulting loft feature

The **Cut Loft PropertyManager** is used to create a cut loft feature. This option is invoked by choosing **Insert > Cut > Loft** from the menu bar.

Creating 3D Sketches

CommandManager:	Sketch > 3D Sketch
Menu:	Insert > 3D Sketch
Toolbar:	Sketch > 3D Sketch

In the previous chapters, you learned how to draw 2D sketches in the sketching environment. In this chapter, you will learn how to draw 3D sketches. The 3D sketches are mostly used to create 3D paths for the sweep features, 3D curves, and so on. Figure 9-71 shows a chair frame created by sweeping a profile along a 3D path.

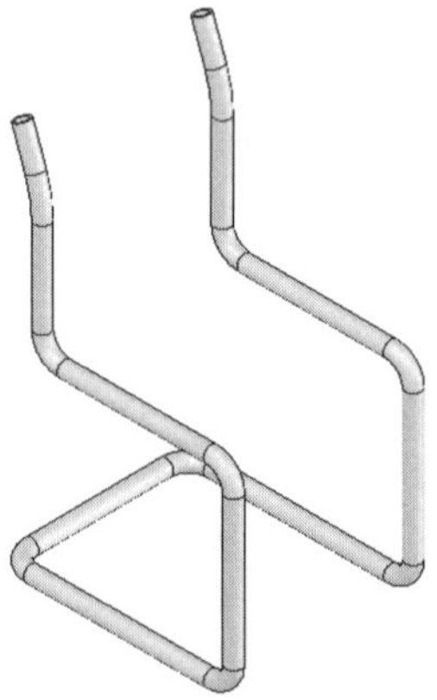

Figure 9-71 A chair frame created by sweeping a profile along a 3D path

To draw a 3D sketch, choose the **3D Sketch** button from the **Sketch CommandManager**. On choosing this option, the 3D sketching environment is invoked and the origin is displayed in red color. You do not need to select a sketching plane for drawing a 3D sketch. When you invoke the 3D sketching environment, some of the sketching tools are activated in the **Sketch CommandManager**. These tools can be used in the 3D sketching environment and are discussed next.

Line

It is better to orient the view to isometric for drawing lines in the 3D sketching environment. Choose the **Line** button from the **Sketch CommandManager** to invoke the **Line** tool; the **Insert Line PropertyManager** is displayed. The **As sketched** radio button selected in the **Orientation** rollout, while the other options are frozen. The select cursor is replaced by the line cursor with **XY** displayed at it bottom. This implies that by default, the sketch will be drawn in the XY plane. The coordinate system is also displayed in the current plane. You can toggle between the default planes using the TAB key on the keyboard. On doing so, the orientation of the coordinate system also modifies with respect to the current plane. Move the cursor to the location from where you want to start sketching. On specifying the start point of the line you are provided with a space handle.

You can also toggle the plane after specifying the start point of the line. The Coordinate system will also change with respect to the current plane. Move the cursor to specify the endpoint of the line. Its length is displayed above the line cursor. Specify the endpoint of the line at this location. You will notice that a rubber-band line segment is attached to the cursor. Toggle the plane using the TAB key, if required. Move the cursor to draw another line and specify its endpoint at the desired location. Right-click to invoke the shortcut menu and select the **Select** option to end drawing the line. Figures 9-72 through 9-74 show sketching in different planes in the 3D sketching environment. Figure 9-75 shows an example of a 3D sketch.

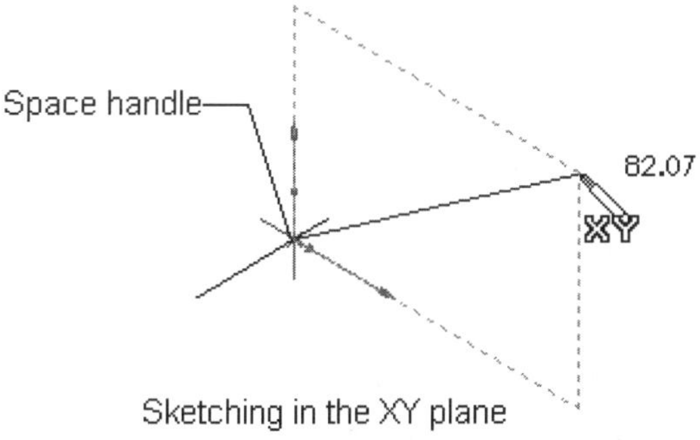

Figure 9-72 *Sketching in the XY plane*

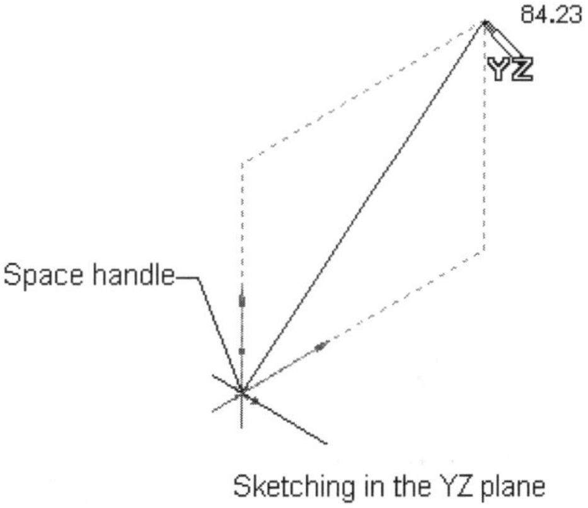

Figure 9-73 *Sketching in the YZ plane*

Spline

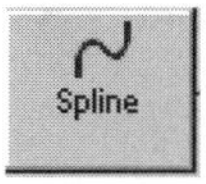

You can also draw a spline in the 3D sketching environment by choosing the **Spline** button from the **Sketch CommandManager**. The select cursor will be replaced by the spline cursor. Move it to the desired location to start the sketch. Specify the start point of the spline; the space handle is displayed. You can toggle between the default planes using the TAB key. Move the cursor to specify the second point of spline and specify the second point of spline. Follow the same procedure to continue drawing the spline. To end the spline creation right-click and select the **Select** or **End Spline** option from the shortcut menu.

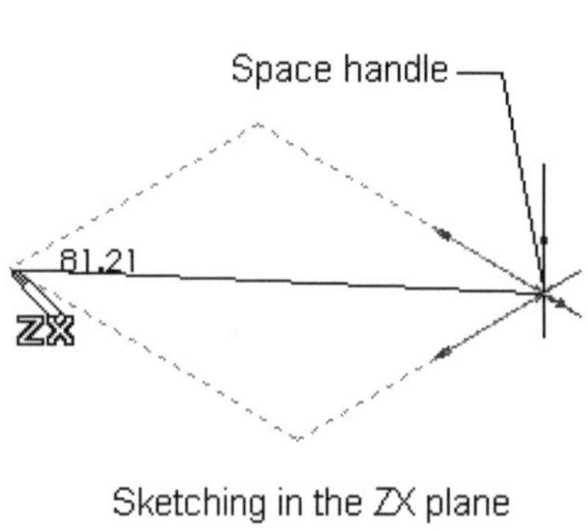

Figure 9-74 Sketching in the ZX plane

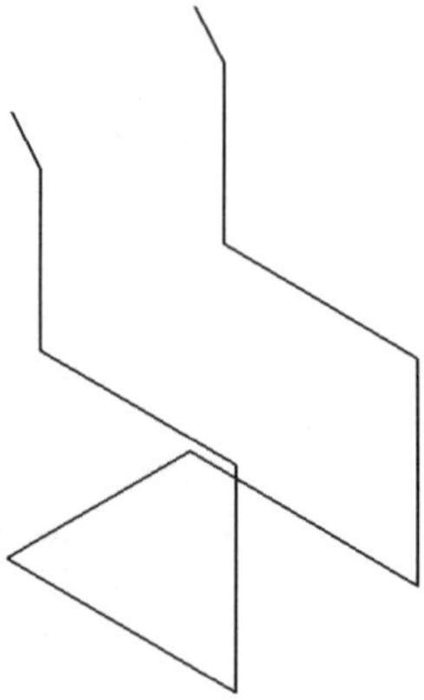

*Figure 9-75 3D sketch created using the **Line** tool*

Point

Choose the **Point** button from the **Sketch CommandManager** to invoke the **Point** tool. The select cursor will be replaced by the point cursor. Use the left mouse button to draw points.

Centerline

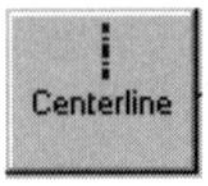

You can also draw centerlines in the 3D sketching environment. Invoke the 3D sketching environment and orient the view to isometric view. Choose the **Centerline** button from the **Sketch CommandManager** to draw a centerline; the select cursor is replaced by the line cursor. The procedure of drawing the centerline is the same as that discussed for drawing lines in the 3D sketching environment.

The dimensioning of 3D sketches is the same as the dimensioning of 2D sketches.

Editing the 3D Sketches

You can perform a number of editing operations on 3D sketches including **Convert Entities**, **Sketch Chamfer**, **Trim Entities**, **Fit Spline**, **Sketch Fillet**, **Extend Entities**, **Construction Geometry**, and **Split Entities**. All these tools have been discussed in the previous chapters.

Creating Curves

You can also create various types of curves in SolidWorks. These curves are mostly used to create complex shapes generally using the **Swept Boss/Base** and **Lofted Boss/Base** tools. The types of curves that can be created in SolidWorks are discussed next.

Creating a Projection Curve

CommandManager:	Curves > Project Curve	*(Customize to Add)*
Menu:	Insert > Curve > Projected	
Toolbar:	Curves > Project Curve	

This option allows you to project a single closed/open sketched entity on one or more than one planar or curved faces. You can also project a sketched entity on another sketched entity to create a 3D curve. To create a projected curve, draw at least two sketches or a single sketch and at least one feature. Next, choose the **Project Curve** button from the **Curves CommandManager**, after customizing it. On choosing the **Project Curve** button, the **Projected Curve PropertyManager** is displayed, as shown in Figure 9-76. The confirmation corner is also displayed in the drawing area. The two different options to create projected curves are discussed next.

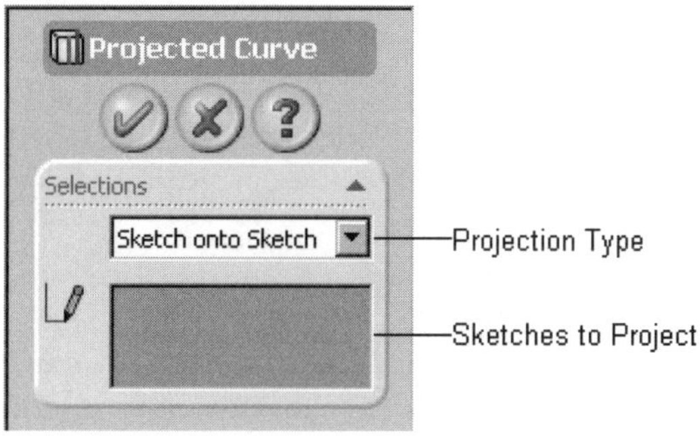

*Figure 9-76 The **Projected Curve PropertyManager***

Sketch onto Sketch

On invoking the **Projected Curve PropertyManager**, you will find the **Sketch onto Sketch** option selected by default in the **Projection Type** drop-down list. You are prompted to select two sketches to project onto one another. Select two sketches that do not lie on the same plane to project them onto one another. When you select the sketches, the names of the sketches are displayed in the **Sketches to Project** selection area. The preview of the projected curve is also displayed in the drawing area. Choose the **OK** button from the **Projected Curve PropertyManager** or choose **OK** from the confirmation corner. Figure 9-77 shows the two sketches selected to create a projected curve. Figure 9-78 shows the preview of the projected curve. Figure 9-79 shows the resulting projected curve.

Sketches onto Face(s)

The **Sketches onto Face(s)** option available in the **Projection Type** drop-down list. This used to project a sketch on a planar or curved face(s). On choosing this option, the **Sketch to Project** and the **Projection Faces** selection area are displayed in the **Selections** rollout. You are also provided with a **Reverse Projection** check box The **Projected Curve PropertyManager** with the **Sketches onto Face(s)** option selected is displayed in Figure 9-80. Next, select the sketch from the drawing area and the face or faces on which you want to project the sketch. The selected sketch is highlighted in green and the selected face is highlighted in magenta. The arrow provided in the drawing area, is used to reverse the direction of the projection. You can also reverse the direction of projection using the

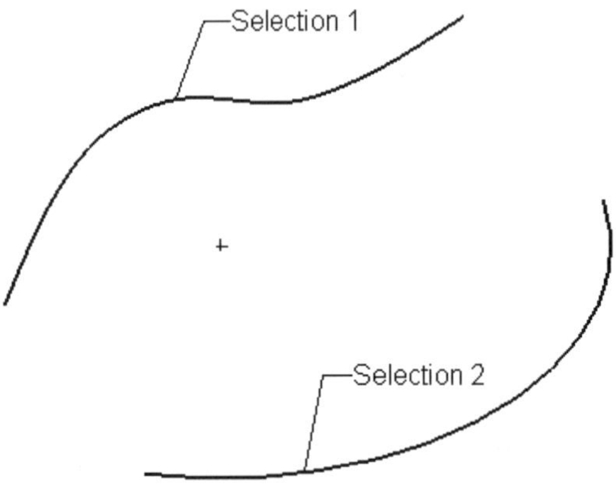

Figure 9-77 *Sketches to be selected*

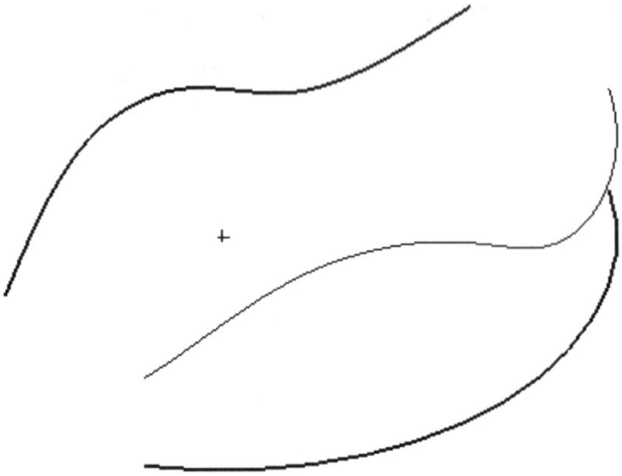

Figure 9-78 *Preview of the projected curve*

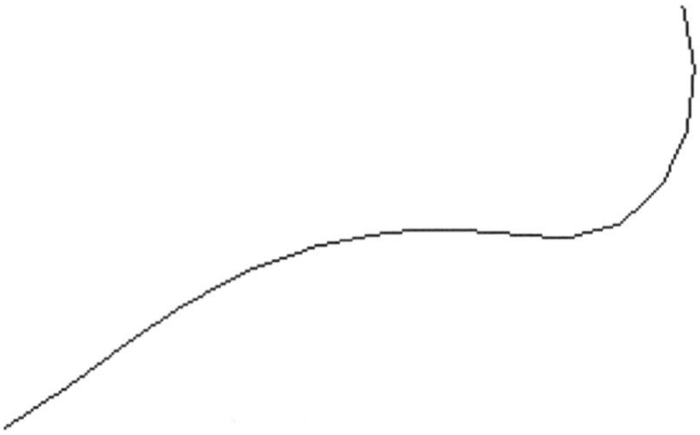

Figure 9-79 *Resulting projected curve*

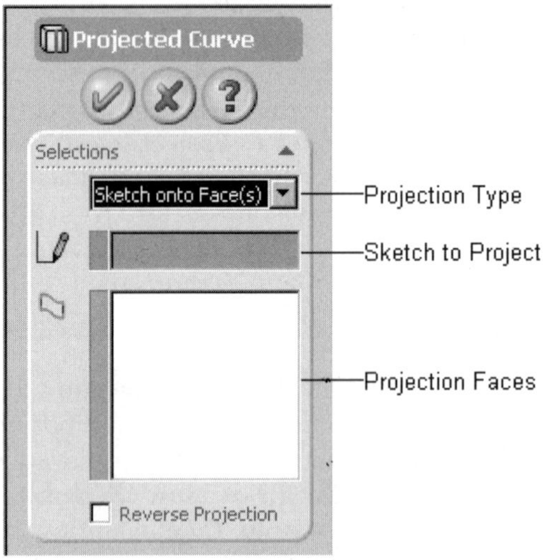

Figure 9-80 *The* ***Projected Curve PropertyManager*** *with the* ***Sketches onto Face(s)*** *option selected*

Reverse Projection check box. Choose the **OK** button from the **Projected Curve PropertyManager** or choose **OK** from the confirmation corner. Figure 9-81 shows the sketch to be selected for projection also the face to be selected on which the sketch will be projected. Figure 9-82 shows the resulting projected sketch.

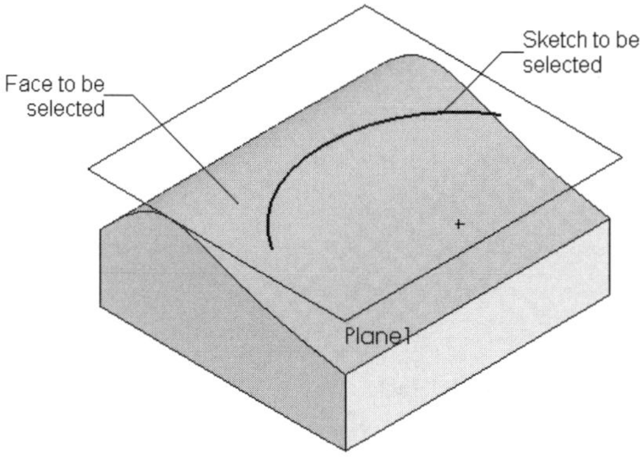

Figure 9-81 *Sketch and the face to be selected*

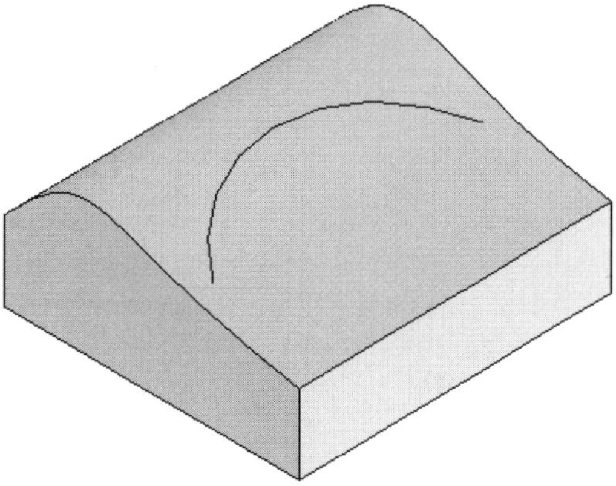

Figure 9-82 *Resulting projected curve*

Creating Split Lines

CommandManager:	Curves > Split Line	*(Customize to Add)*
Menu:	Insert > Curve > Split Line	
Toolbar:	Curves > Split Line	

The **Split Line** tool is generally used to project a sketch on a planar or curved face, which in turn it splits or divides the single face into two or more than two faces. Invoke the **Split Line PropertyManager**. Choose the **Split Line** button from the **Curves CommandManager**, after customizing it to invoke the **Split Line PropertyManager**, as shown in Figure 9-83. The two methods to create a split line are discussed next.

Silhouette

You can split a curved face by creating a silhouette line at the intersection of the projection of direction entity and the curved face. Invoke the **Split Line PropertyManager**; the **Silhouette** option available in the **Type of Split** rollout is selected by default. Change the type of split, or select the direction of pull and the faces to split. You need to select an edge, line, or a plane that defines the direction of pull. Select the direction of pull; the selected entity will be highlighted in magenta color. Next, select the curved face; the selected face will be highlighted in green color. Choose the **OK** button from the **Split Line PropertyManager** or choose the **OK** option from the confirmation corner. The

Tip. *You can also use the **Contour Select Tool** to select the contours for creating a projected curve or a split line. You will learn more about split line later in this chapter.*

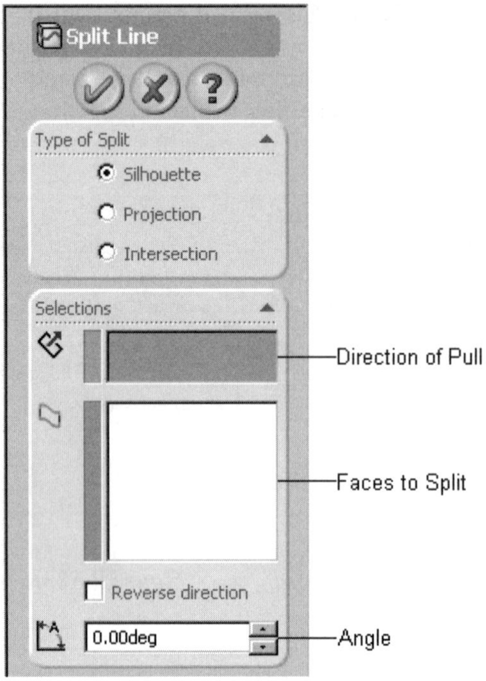

Figure 9-83 *The **Split Line PropertyManager***

curved face will be divided in two or more faces. With this release of SolidWorks you are provided with the **Angle** spinner in the **Selections** rollout to define the draft angle for creating the silhouette line. By default, its value is set to 0-degrees.

Consider a case, in which you need to split a circular face, using this option. Plane will be selected to define the direction of pull. Figure 9-84 shows the plane and the face to be selected. Figure 9-85 shows the resulting split line created to split the selected face.

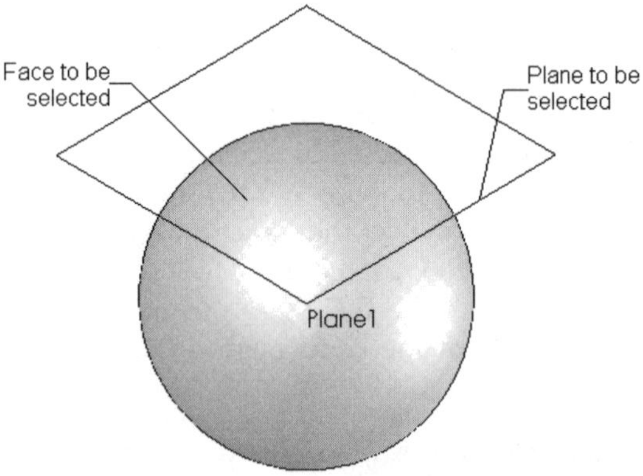

Figure 9-84 *Plane and the face to be selected*

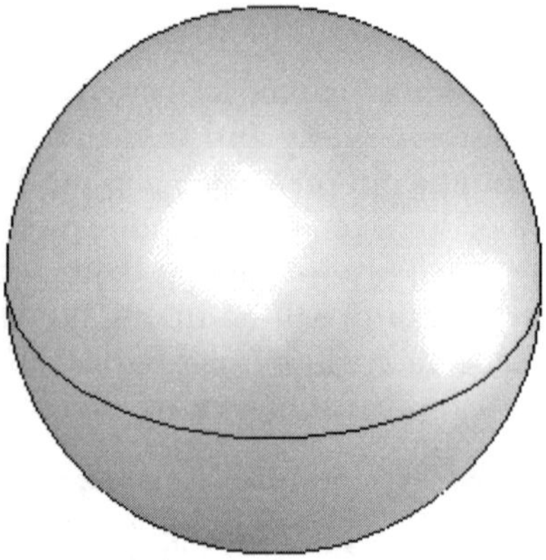

Figure 9-85 *Resulting split line*

Projection

You can project a sketched entity onto a planar or curved face to create a split line on it. The split line splits the selected face on which the sketch is projected. To use this option, invoke the **Split Line PropertyManager**, and choose the **Projection** option from the **Type of Split** rollout. The **Split Line PropertyManager**, with the **Projection** option selected, is shown in Figure 9-86.

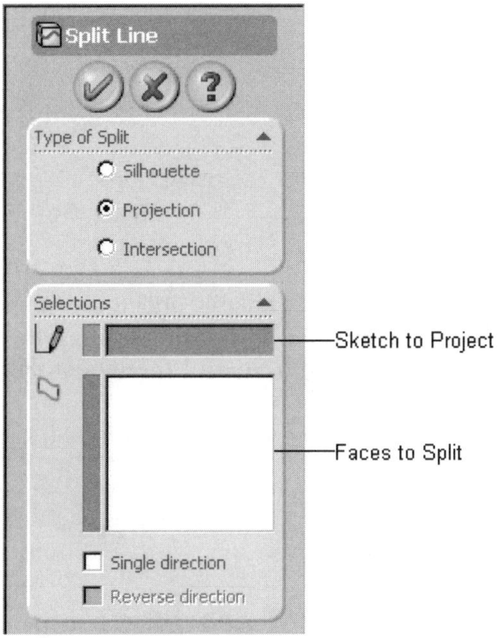

Figure 9-86 *The* ***Split Line PropertyManager*** *with the* ***Projection*** *option selected*

You are prompted to change the type of split, or select the sketch to project, direction, and faces to split. Select the sketch and it will be displayed in magenta. Next, select the face to split; the selected face will be displayed in green. Choose the **OK** button from the **Split Line PropertyManager**. The selected face or faces will split into two or more than two faces, depending upon the sketch that is used to project. Figure 9-87 shows the sketch and the face to be selected. Figure 9-88 shows the resulting split line created to be split the selected face.

The other options available in the **Selections** rollout of the **Split Line PropertyManager** are discussed next.

Single direction

If the sketching plane on which the sketch is created lies within the model, the split line on a cylindrical face will be created on its two sides. The **Single direction** check box available in the **Selections** rollout is used to create the split line only in one direction.

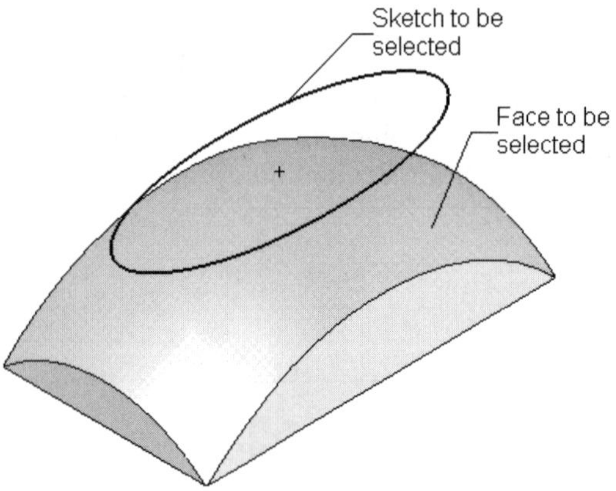

Figure 9-87 *Sketch and the face to be selected*

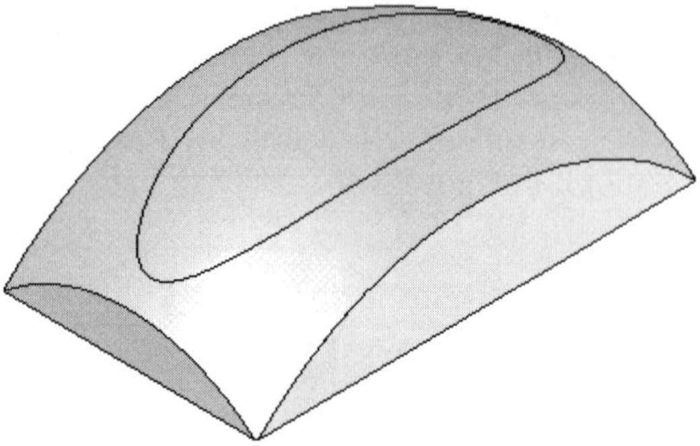

Figure 9-88 *Resulting split line*

Reverse direction

The **Reverse direction** check box, available only if the **Single direction** check box is selected and is used to reverse the direction of the created split line. Figure 9-89 shows the split line created on both sides of the model. Figure 9-90 shows the split line created on a single side of the model.

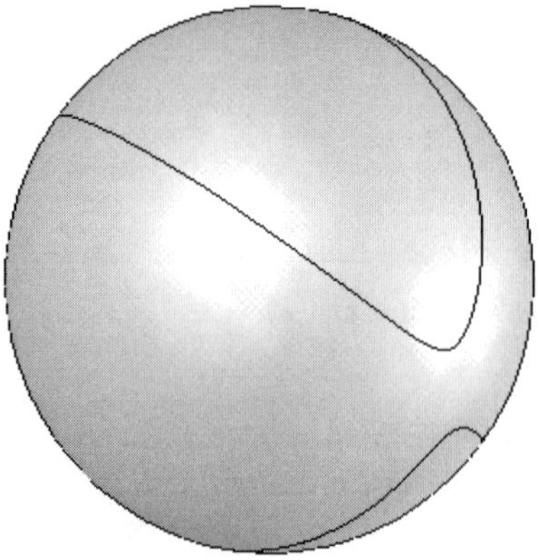

Figure 9-89 *Split line created on both sides*

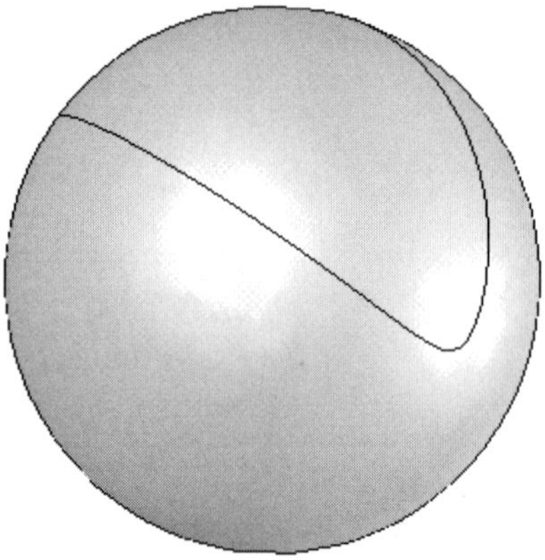

Figure 9-90 *Split line created on single side*

Intersection

The **Intersection** option is used to split the selected bodies or faces using tool bodies, face, or planes. Invoke the **Split Line PropertyManager** and select the **Intersection** radio button from the **Type of Split** rollout; the selection mode in the **Splitting Bodies/Faces/Planes** selection area is active. Select bodies, faces, or planes that will be used as tool bodies. Next, click in the **Faces/Bodies to Split** selection area and select the bodies to be split.

The **Split all** check box provided in the **Surface Split Options** rollout is used to split almost all areas of surfaces using the current selection set. You can also select the **Natural** or **Linear** radio button to define the shape of the split. Figure 9-91 shows the plane to be selected as a split tool and also faces to be split. Figure 9-92 shows the resulting split faces.

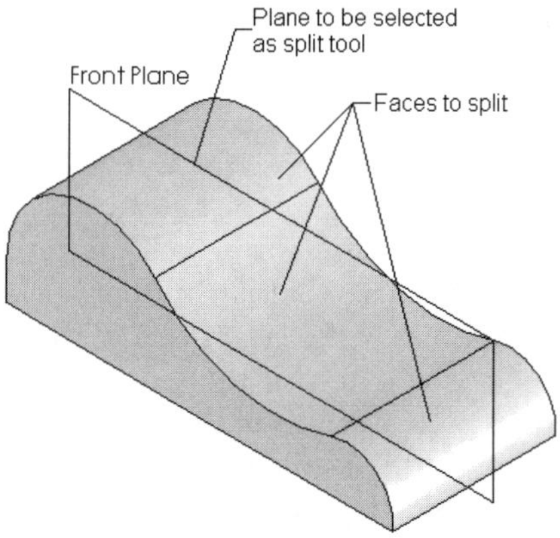

Figure 9-91 Plane and faces to split

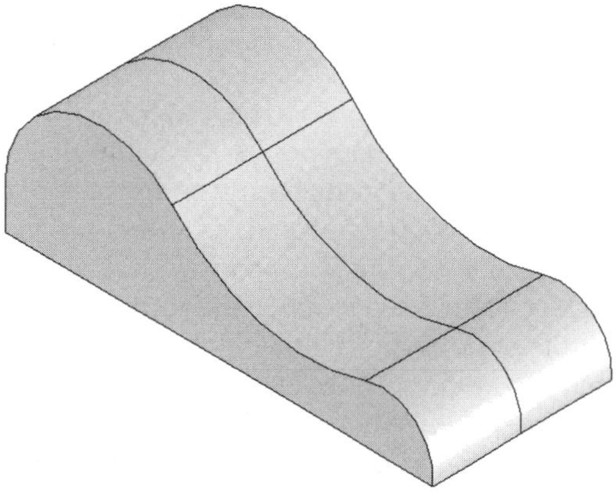

Figure 9-92 Resulting split faces

Creating a Composite Curve

CommandManager:	Curves > Composite Curve
Menu:	Insert > Curve > Composite
Toolbar:	Curves > Composite Curve

The **Composite Curve** option is used to create a curve by combining 2D or 3D curves, sketched entities, and part edges into a single curve. You need to ensure that the selected entities form a continuous chain. The composite curve is mainly used while creating a sweep or a loft feature. To create it, invoke the **Composite Curve PropertyManager**. Choose the **Composite Curve** button from the **Curves CommandManager** to invoke the **Composite Curve PropertyManager**. The **Composite Curve PropertyManager** is shown in Figure 9-93.

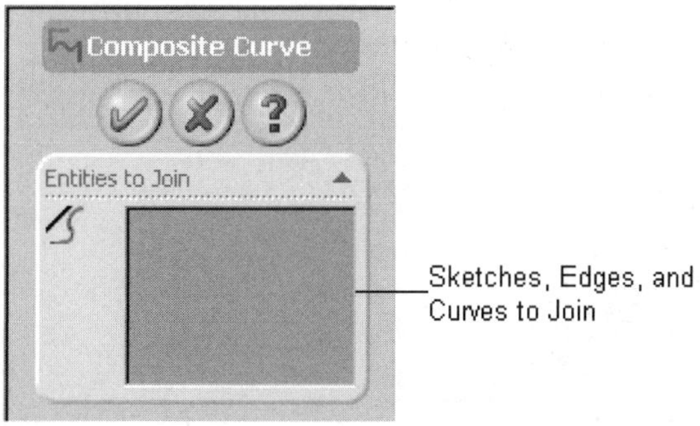

*Figure 9-93 The **Composite Curve PropertyManager***

On invoking the **Composite Curve PropertyManager**, you are prompted to select a continuous set of sketches, edges, and/or curves. Select the continuous edges, curves, or sketched entities to create a composite curve. Choose the **OK** button to end its creation.

Creating a Curve Through XYZ Points

CommandManager:	Curves > Curve Through XYZ Points
Menu:	Insert > Curve > Curve Through XYZ Points
Toolbar:	Curves > Curve Through XYZ Points

The **Curve Through XYZ Points** option is used to create a curve by specifying the coordinate points. To create a curve using this option, choose the **Curve Through XYZ Points** button from the **Curves CommandManager**. The **Curve File** dialog box is displayed, as shown in Figure 9-94. Double-click in the cell under the **X** column to enter the X coordinate as the start point for the curve. Similarly, double-click in the **Y** cell and **Z** cell to enter the respective Y and Z coordinates of the start point of the curve. Double-click in the cell below the first cell to enter the coordinates for the second point for creating the curve. Similarly, specify the coordinates of the other points of the curve, as shown in Figure 9-95. When you enter the coordinates of the points, the preview of the curve is displayed in the drawing area. Choose the **OK** button from the **Curve File** dialog box to complete the creation of the feature, as shown in Figure 9-96.

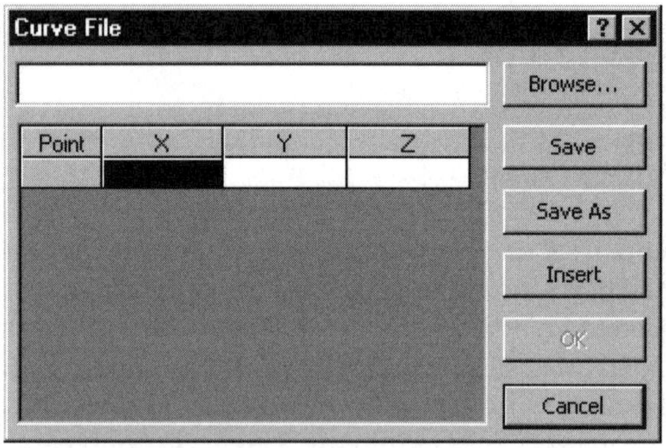

***Figure 9-94** The **Curve File** dialog box*

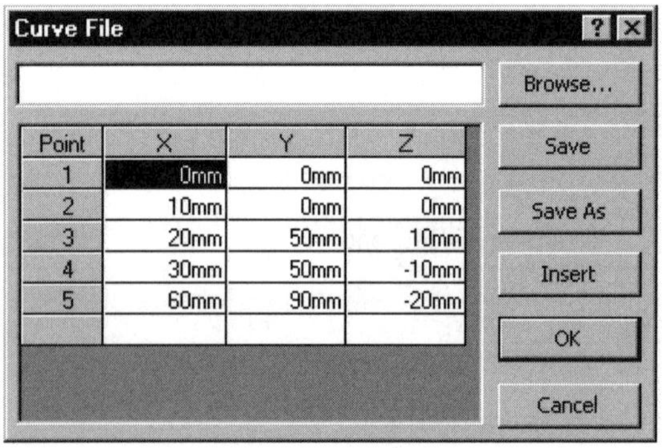

***Figure 9-95** Coordinates entered in the **Curve File** dialog box*

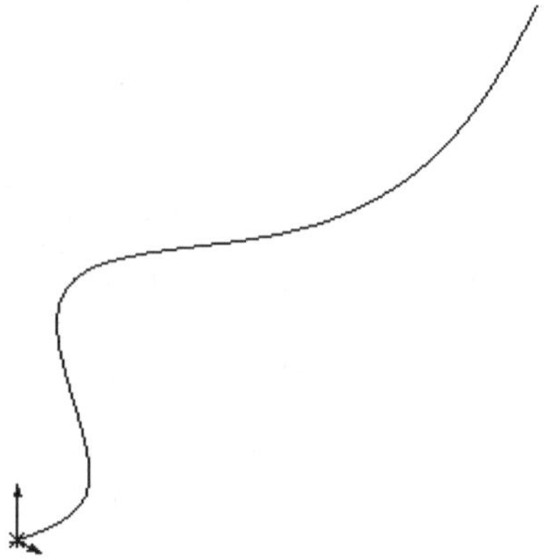

***Figure 9-96** Resulting 3D curve*

> **Tip**. *You can select a row and use the DELETE key on the keyboard to delete it completely. Use the SHIFT key the select the entire row and use the* **Insert** *button, to add a new row between any two rows.*

You can also save the current set of coordinates using the **Save** button available in the **Curve File** dialog box. When choosing this button, the **Save As** dialog box is displayed. Browse the folder to save the coordinates, enter the name of the file in the **File name** edit box and choose the **Save** button. The curve file is saved with extension *.sldcrv*. Using the **Save As** button you can save the current set of coordinates in a file with some other name.

You can use the **Browse** button, to open an existing curve file; the **Open** dialog box is displayed. You can browse the previously saved curve file to specify the coordinate points. You can also write the coordinates in a text (notepad) file and save it. In the **Open** dialog box, choose the **Text Files (*.txt)** option from the **Files of type** drop-down list and browse the text file to specify the coordinates.

Creating a Curve Through Reference Points

CommandManager:	Curves > Curve Through Reference Points
Menu:	Insert > Curve > Curve Through Reference Points
Toolbar:	Curves > Curve Through Reference Points

The **Curve Through Reference Points** option enables you to create a curve by selecting the sketched points, vertices, origin, endpoints, or center points. To create a curve through reference points, you have to invoke the **Curve Through Reference Points PropertyManager**. Choose the **Curve Through Reference Points** button from the **Curves CommandManager** to invoke the **Curve Through Reference Points PropertyManager**, as shown in Figure 9-97.

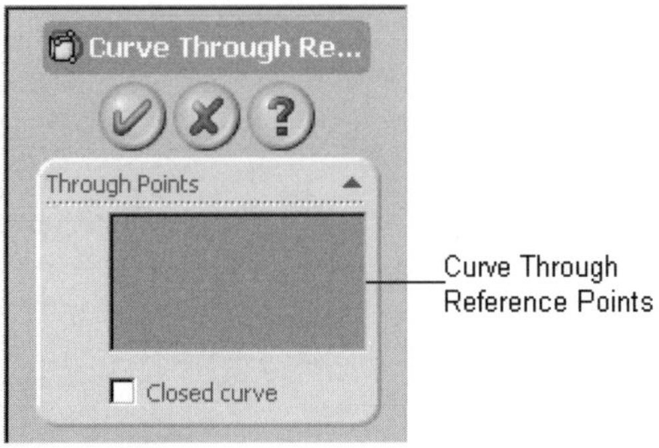

Figure 9-97** The **Curve Through Reference Points PropertyManager

On invoking this tool, you are prompted to select the vertices to define the through points for the curve. Select the points to define the curve. On defining the points, the preview of the resulting curve created using the selected points, is displayed in the drawing area. After specifying all the points, choose the **OK** button. You can select the **Closed curve** check box to create a closed curve. Figure 9-98 shows the vertices to be selected to create the curve through reference points. Figure 9-99 shows the resulting curve.

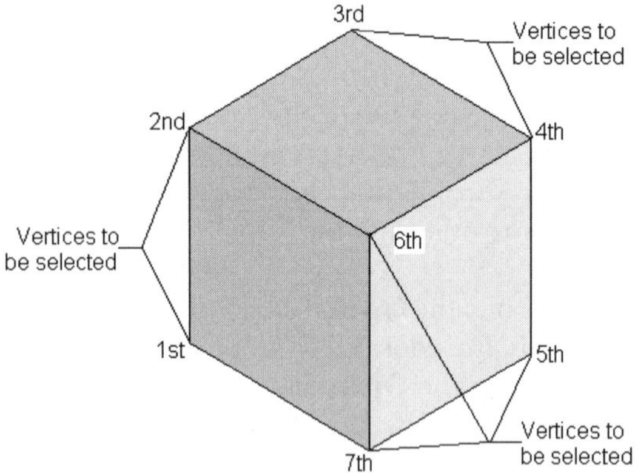

Figure 9-98 *Vertices to be selected*

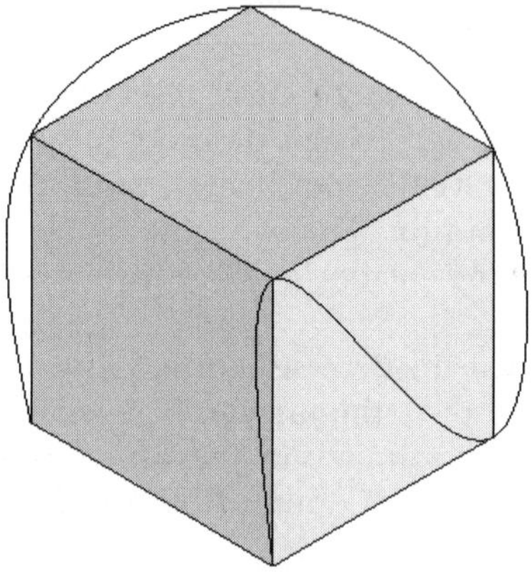

Figure 9-99 *Resulting 3D curve*

Creating a Helical/Spiral Curve

CommandManager:	Curves > Helix and Spiral
Menu:	Insert > Curve > Helix/Spiral
Toolbar:	Curves > Helix and Spiral

The **Helix and Spiral** tool is used to create a helical curve or a spiral curve. This curve is generally used as the sweep path to create springs, threads, spiral coils, and so on. Figure 9-100 shows a spring created by sweeping a circular profile along a helical path. Figure 9-101 shows a spiral coil created by sweeping a rectangular profile along a spiral path.

Figure 9-100 *Spring*

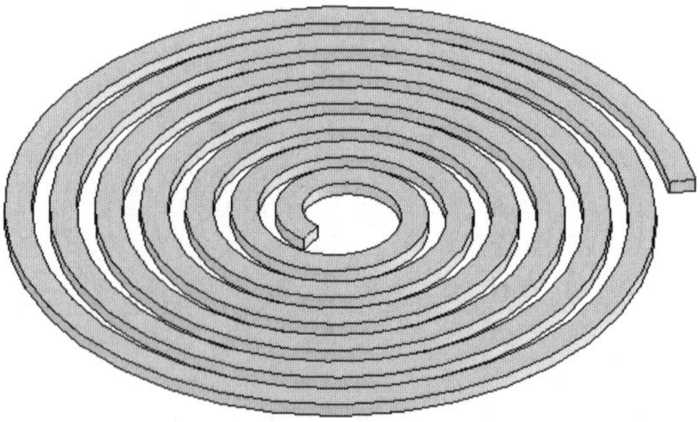

Figure 9-101 *Spiral coil*

To create a helix, choose the **Helix and Spiral** button from the **Curves CommandManager**. The **Helix/Spiral PropertyManager**, is displayed and you are prompted to select a plane, a planar face, or an edge to sketch a circle to define the helix cross-section, or a sketch that contains a single circle. The circle will define the diameter of the spring. If you are creating a spiral, the sketch will define the starting diameter of the spiral curve. The **Helix/Spiral PropertyManager** disappears on selecting a plane and you enter the sketching environment to draw the cross-section sketch for the helix or spiral. Exit the sketching environment after drawing the sketch. The **Helix/Spiral PropertyManager** will appear again, as shown in Figure 9-102.

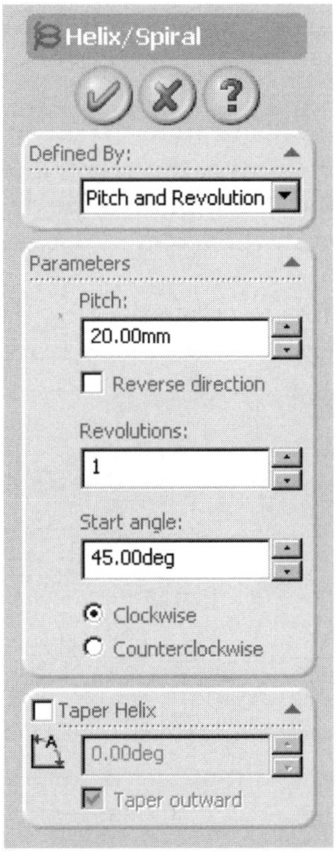

Figure 9-102** The **Helix/Spiral PropertyManager

Two arrows are displayed, one emerging from the center of the circle, and defining the direction of helix. The second arrow is tangent to the circle and defines the rotation of the helix, either clockwise or counterclockwise. The preview of the helix curve, with the default values, is displayed in the drawing area. The various methods to specify the parameters of the helical curve are discussed next.

Pitch and Revolution

The **Pitch and Revolution** option, selected by default in the **Type** drop-down list of the **Defined By** rollout, is be used to specify the pitch of the helical curve and the number of revolutions. When this option is selected, the **Pitch** spinner and the **Revolution** spinner are available in the **Parameters** rollout to define the value of pitch and number of revolutions. You can also select the **Reverse direction** check box, available in the **Parameters** rollout, to reverse the direction of helix creation. You can specify the start angle of the helical curve using the **Start angle** spinner The **Clockwise** and

Counterclockwise radio buttons, available in this rollout, are used to define the direction of rotation of the helix.

Height and Revolution

The **Height and Revolution** option available in the **Type** drop-down list is used to define the parameters of the helix curve in the form of the total helix height and the number of revolutions. On choosing this option, the **Height** and **Revolutions** spinners are displayed in the **Parameters** rollout, along with other options to specify the required parameters.

Height and Pitch

The **Height and Pitch** option available in the **Type** drop-down list is used to define the parameters of the helix curve in terms of the height and the pitch of the helix. When you select this option, the **Height** and **Pitch** spinners are displayed in the **Parameters** rollout, along with other options to specify the required parameters.

When you specify the parameters to create the helix curve, the preview in the drawing area modifies automatically. Figure 9-103 shows a helix curve.

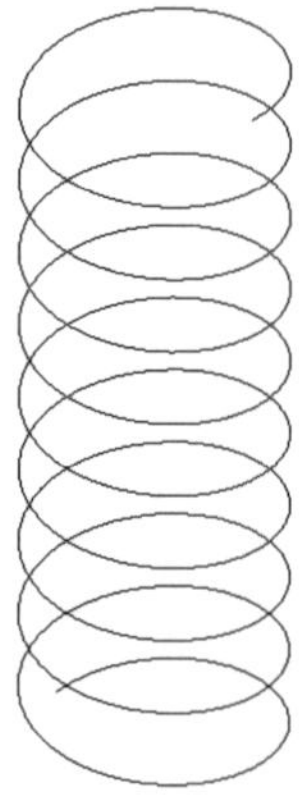

Figure 9-103 *Helix Curve*

You can also create a tapered helix using the **Taper Helix** rollout The procedure for this is discussed next.

Taper Helix

Invoke the **Taper Helix** rollout by selecting the **Taper Helix** check box available in the **Taper Helix** rollout. The **Taper Angle** spinner and the **Taper outward** check boxes are enabled. Using the **Taper Angle** spinner, you can specify the value of the angle between the central axis of the helix and the periphery of the helix. The **Taper outward** check box is selected by default and is used to create an outward taper. When you specify the parameters to create a tapered helical curve, its preview of the helical curve updates automatically in the drawing area. Figure 9-104 shows a tapered helical curve. Figure 9-105 shows a tapered helical curve created with the **Taper outward** check box selected.

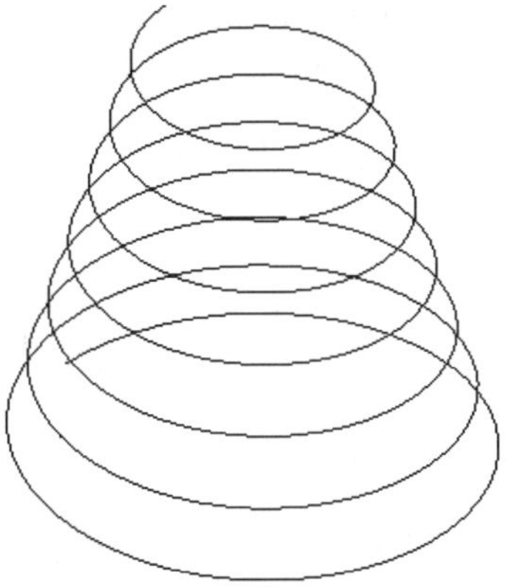

Figure 9-104 *Tapered helical curve*

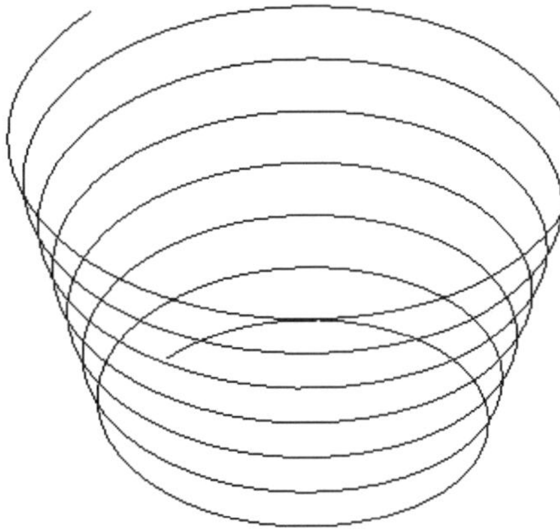

Figure 9-105 *Tapered helical curve with the **Taper outwards** option selected*

By default, the helical curve is created in the clockwise direction. Therefore, the **Clockwise** radio button is selected in the **Parameters** rollout of the **Helix/Spiral PropertyManager**. Select the **Counterclockwise** radio button to create the helical curve in the counterclockwise direction. After setting the parameters, choose the **OK** button from the **Helix/Spiral PropertyManager**.

Creating a Spiral Curve

For creating a spiral curve, select the **Spiral** option from the **Type** drop-down list available in the **Defined By** rollout. The preview of the spiral curve will be displayed in the drawing area. You can define the pitch and the number of revolutions in the **Pitch** and **Revolutions** spinners, respectively. The other options in the **Parameters** rollout, are the same as those discussed earlier. After specifying all the required parameters, choose the **OK** button. Figure 9-106 shows a spiral curve.

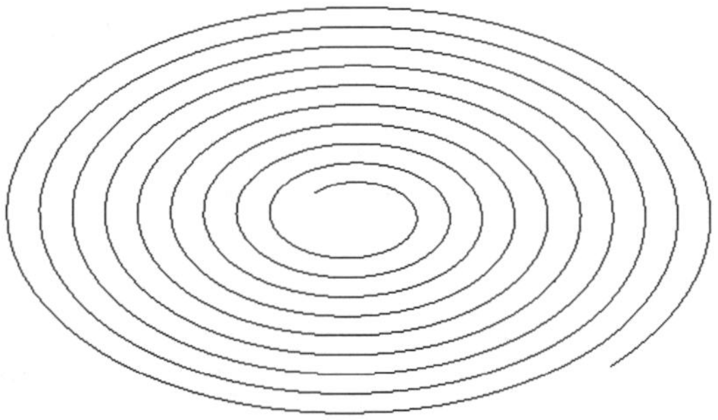

Figure 9-106 *Spiral curve*

Extruding a 3D Sketch

You can also extrude a 3D sketch drawn on the 3D sketching environment. A 3D sketch is always extruded along a direction vector, which can be a line, edge, planar face, or a plane. Invoke the **Extrude PropertyManager** and select the 3D sketch that you need to extrude. You also need to select a direction vector along which the sketch will be extruded. Set the parameters at the start of the extrude feature, using the **From** rollout. Now, set the value of the depth of the extrude feature and choose the **OK** button from the **Extrude PropertyManager**. Figure 9-107 shows a 3D sketch and the direction vector along which it will be extruded. Figure 9-108 shows the resulting extruded feature.

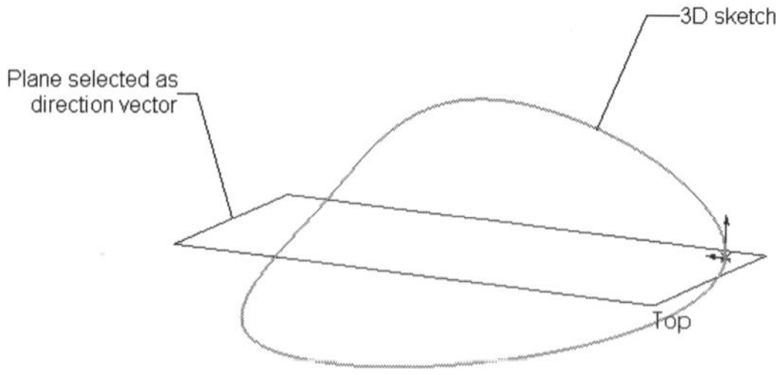

Figure 9-107 3D sketch and the direction vector

Figure 9-108 Resulting extruded feature

Creating Draft Features

CommandManager:	Features > Draft
Menu:	Insert > Features > Draft
Toolbar:	Features > Draft

The **Draft** tool is used to add taper to the selected faces of the model. One of the main applications of creating the draft feature is in the models that are to be moulded or casted so that it is easier to remove them from the mould or die. Invoke the **Draft PropertyManager**. Choose the **Draft** button from the **Features CommandManager** to invoke the **Draft PropertyManager**. The **Draft PropertyManager** is shown in Figure 9-109.

After invoking this tool, you are prompted to select a neutral plane and the faces to draft. You will notice that the **Neutral Plane** option is selected by default in the **Type of Draft** drop-down list in the **Type of Draft** rollout. Therefore, you need to select a neutral plane for creating the draft feature. Select a planar face or a plane that acts as a neutral plane. The selected entity will appear in magenta along with the **Neutral Plane** callout. A **Reverse Direction** arrow is also displayed in the drawing area. The selection mode in the **Faces to Draft** selection area of the **Faces to Draft** rollout is activated. Select the faces to apply the draft. The selected faces will be displayed in green with the **Draft Face** callout. Next, set the value of the draft angle in the **Draft Angle** spinner available in the **Draft Angle** rollout. Choose the **OK** button from the **Draft PropertyManager** or choose **OK** from the confirmation corner.

Figure 9-110 shows the neutral face and the faces to be selected to add the draft. Figure 9-111 shows the resulting draft feature. The other options in the **Draft PropertyManager** are discussed next.

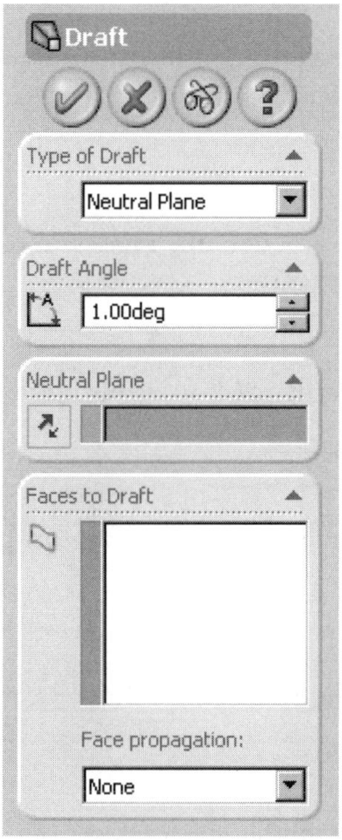

Figure 9-109 *The* ***Draft PropertyManager***

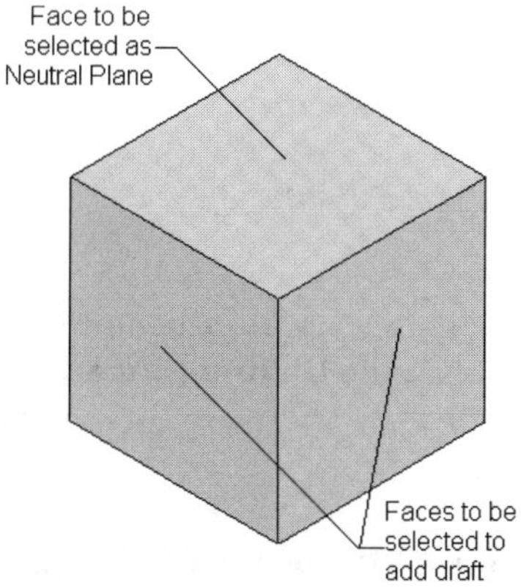

Figure 9-110 *Faces to be selected*

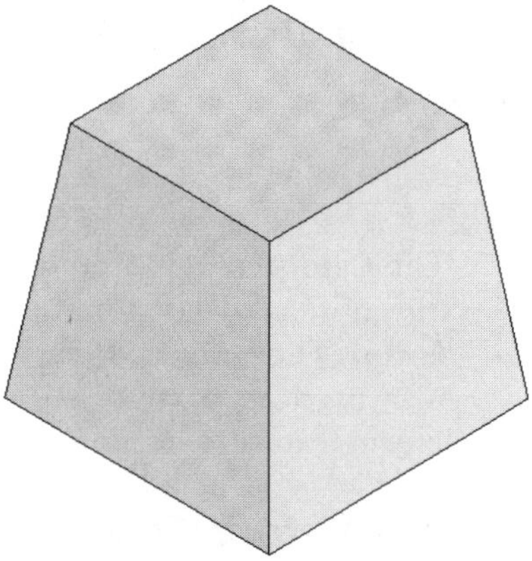

Figure 9-111 *Resulting draft feature*

Reverse Direction

The **Reverse Direction** button, available on the left of the **Neutral Plane** selection area in the **Neutral Plane** rollout, is used to reverse the direction of creation of the draft. You can also reverse the direction of draft creation by selecting the **Reverse Direction** arrow from the drawing area.

Face propagation

The options available in the **Face propagation** drop-down list in the **Faces to Draft** rollout are used to extend the draft feature to the other faces. The options available in this drop-down list are discussed next.

None

The **None** option is selected by default. This option is used when you do not need to apply any type of face propagation.

Along Tangent

The **Along Tangent** option is used to apply the draft to the faces tangent to the selected face.

All Faces

The **All Faces** option is used to apply the draft to all the faces attached to the neutral plane or face.

Inner Faces

This option is used to draft all the faces inside the model that are attached to the neutral plane or face.

Outer Faces

This option is used to draft all the outside faces of the model that are attached to the neutral plane or face.

Creating the Deform Feature

Enhanced

CommandManager:	Features > Deform	*(Customize to Add)*
Menu:	Insert > Features > Deform	
Toolbar:	Features > Deform	*(Customize to Add)*

The **Deform** tool is used to create free-style designs by manipulating the shape of the entire model or a particular portion of it. Choose the **Deform** button from the **Features CommandManager**, after customizing it; the **Deform PropertyManager** is invoked. Its partial view is shown in Figure 9-112. There are three methods of creating a deform feature, which are discussed next.

Creating the Deform Feature Using the Curve to curve Method

To create a deform feature using the **Curve to curve** method, invoke the **Deform PropertyManager**. Select the **Curve to curve** radio button, if it is not selected by default. The options available to create the deform feature using this method are discussed next.

Deform Curves

The **Deform Curves** rollout is used to define the curves that are used to deform the shape of the model. On invoking this tool, you are prompted to select the geometry to define the deformation and fixed geometry for the deform operation. You first need to select the initial curve. The initial curve defines the reference from where the deformation will start. Select an edge, or a chain of edges, or curves to define the initial curve. Next, you

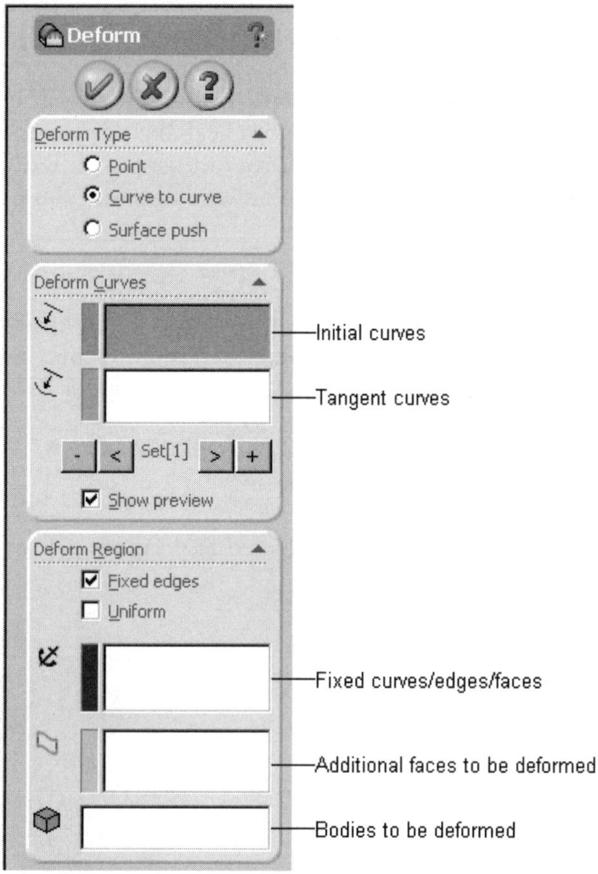

*Figure 9-112 The **Deform PropertyManager***

need to select the target curves. Click once in the **Target curves** selection area and select the target curve, which can be a sketch, an edge, or a curve. The profile of the target curve will define the shape of the deform feature, refer to Figure 9-113 and 9-114.

When you select the initial curve and the target curve, a connector connecting both curves is displayed. You can position the connector by using the handles provided at its sides. To define additional connectors, invoke the shortcut menu and choose the **Add Connector** option from it.

Deform Region

The **Deform Region** rollout is used to define the options to deform a specific area of the model. By default, the entire model is deformed, when you specify the initial curves and the target curves. The **Fixed edges** check box is selected by default. Therefore, the **Fixed curves/edges/faces** selection area is invoked. Using this selection area, you can select the curves, edges, or faces that you need to keep fixed, while creating the deform feature. The **Additional faces to be deformed** selection area is used to specify the additional faces that you need to deform. If you clear the **Fixed edges** check box, the **Fixed curves/edges/faces** and **Additional faces to be deformed** selection areas are not displayed.

The **Uniform** check box available in the **Deform Region** rollout is used to uniformly deform the model. If you clear the **Fixed edges** check box and select the **Uniform** check box from the **Deform Region** rollout, the **Deform radius** spinner is displayed. The **Bodies to be deformed** selection area is used to define the bodies that you need to deform. This option is used in a multibody model.

Shape Options

The options available in this rollout are used to define the intensity of stiffness, shape, accuracy, and intensity of weight along the fixed reference or moving reference.

Note

*The **Weight** slider bar is available in the **Shape Options** rollout if the **Uniform** check box available in the **Deform Region** rollout is cleared.*

Figure 9-113 shows the references to be selected and Figure 9-114 shows the resulting preview of the deform feature.

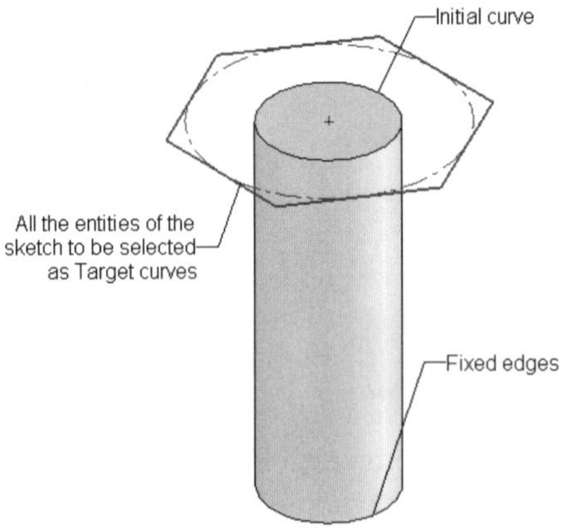

Figure 9-113 *References to be selected*

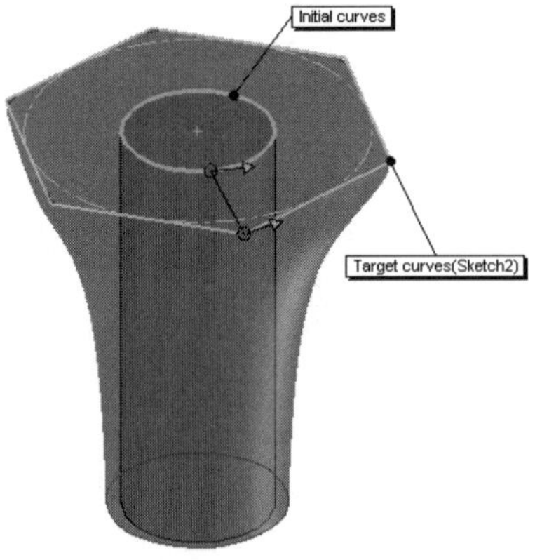

Figure 9-114 *Preview of the deform feature*

After setting all the parameters, choose the **OK** button from the **Deform PropertyManager**. Figure 9-115 shows the resulting deform feature.

Creating the Deform Feature Using the Point Method

To create the deform feature using the **Point** method, select the **Point** radio button from the **Deform Type** rollout, as shown in Figure 9-116. You are prompted to select the geometry to define the region, deformation point, and fixed geometry for the deform operation. Also, select a point on the model face, vertex, or edge from where you need to deform the model. On selecting any of the above-mentioned entities from the model, the model will be deformed at the selected point, using the default value of deform distance and deform radius. You can set the value of the deform distance and the deform radius using the **Deform distance** and

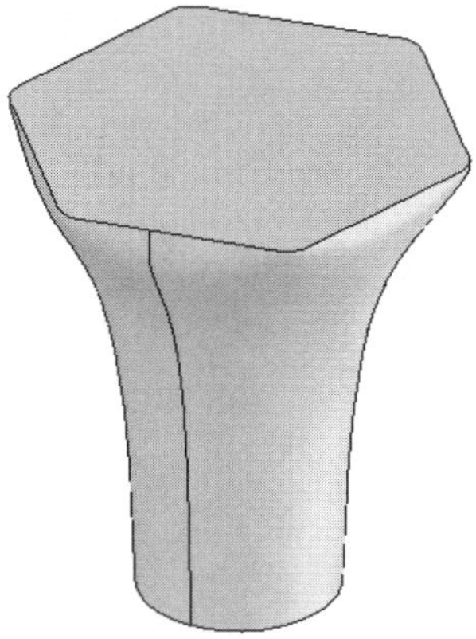

Figure 9-115 Resulting deform feature

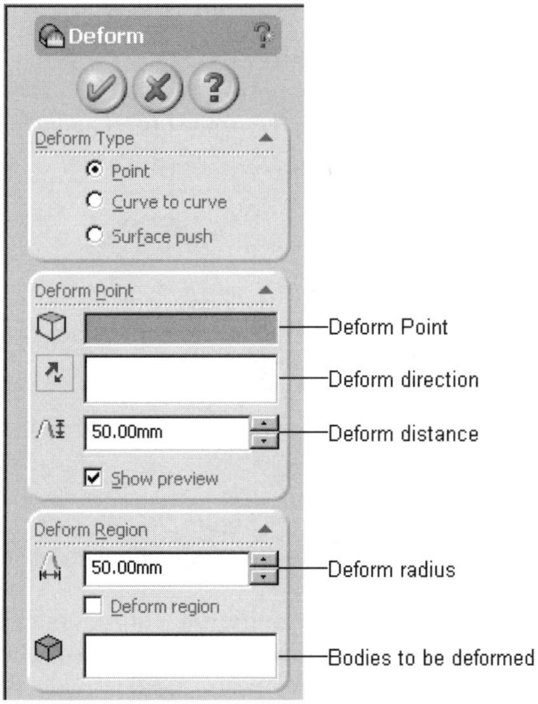

Figure 9-116 Deform PropertyManager

Deform radius spinners from the **Deform Point** and **Deform Region** rollouts, respectively. You can reverse the direction of the deform feature by using the **Reverse deform direction** button in the **Deform point** rollout.

You can also specify a deform direction to create a deform feature by clicking once in the **Deform direction** selection area and selecting the direction to create the deform feature. By default, the entire model is deformed using this tool. You can also specify a particular portion by selecting the **Deform region** check box. The options available to specify a region for deformation are the same as discussed earlier.

Figure 9-117 shows the point where you need to select the face of the model. Figure 9-118 shows the resulting deform feature. Figure 9-119 shows the deform feature created using the **Point** option and the **Deform region** check box selected. Figure 9-120 shows the deform feature with the **Deform region** check box selected and the front face of the base feature selected as the additional faces to deform.

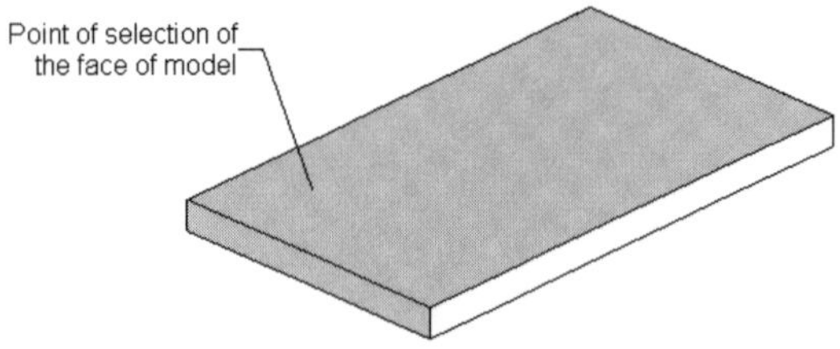

Figure 9-117 *References to be selected*

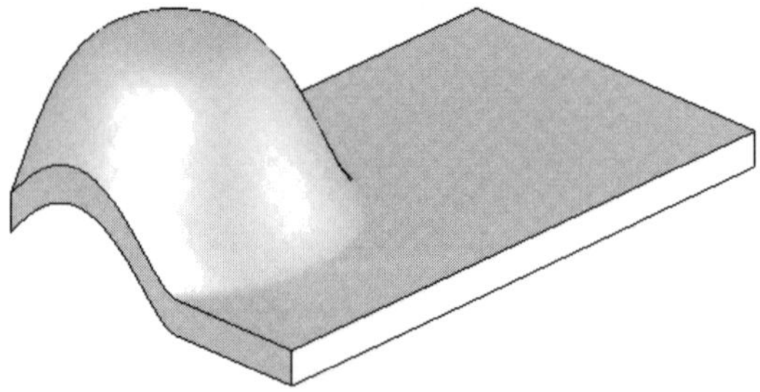

Figure 9-118 *Resulting preview of deform feature*

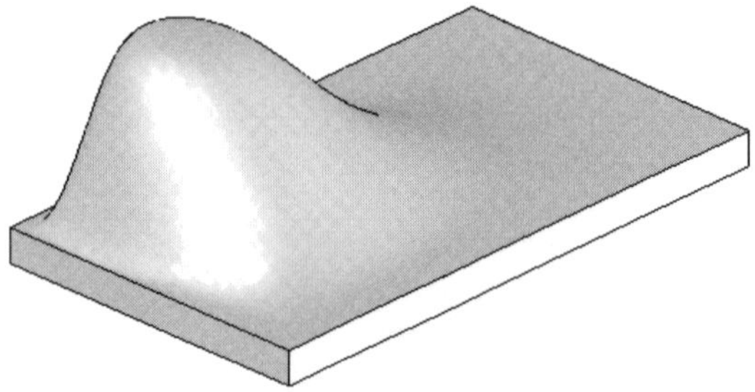

Figure 9-119 *Deform feature with the **Point** option and the **Deform region** check box selected*

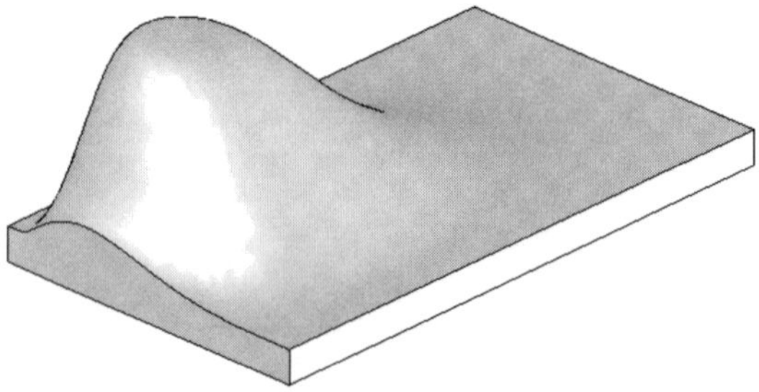

Figure 9-120 *Deform feature with front face selected as the additional face to deform*

You can also define the deform axis for deforming the entire model by specifying it in the **Deform axis** selection area available in the **Shape Options** rollout. Other options available are the same as discussed earlier.

Creating the Deform Feature Using the Surface push Method

One of the major enhancements in the free style modeling using SolidWorks 2005 is the **Surface push** option in the **Deform** tool. Using this tool, you can deform the selected body or bodies by pushing the tool bodies inside them. To deform a body using this tool, invoke the **Deform PropertyManager** and select the **Surface push** radio button from the **Deform Type** rollout. The partial view of the resulting **PropertyManager** is shown in Figure 9-121.

A 3D triad is displayed placed coincident to the part origin. A transparent spherical ball is

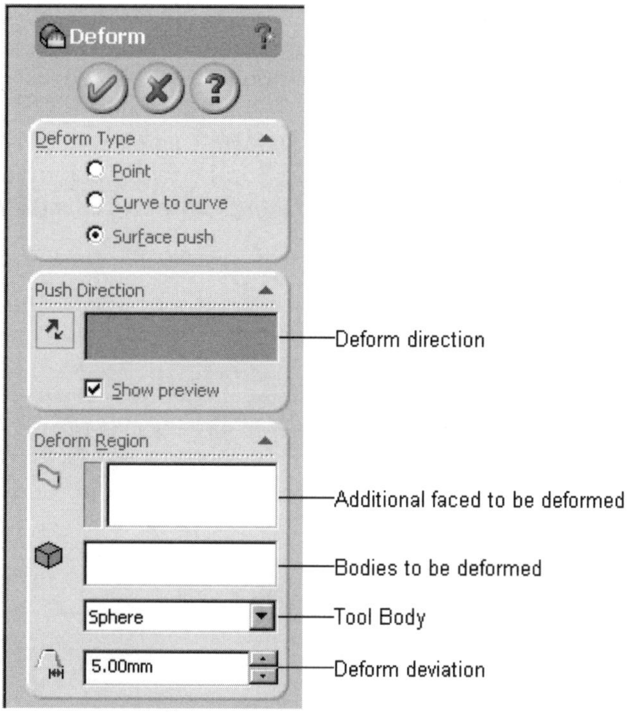

***Figure 9-120** The **Deform PropertyManager** with the **Surface push** radio button selected*

also displayed placed coincident to the 3D triad. This spherical ball is the default tool using which the selected bodies will be deformed.

By default, the spherical geometry is used as the tool body. However, you can also used other type of geometries as the tool body using the options available in the **Tool Body** drop-down list. Other type of geometries that can be used as tool body are ellipse, ellipsoid, polygon, and rectangle. You can also select a separately created body as the tool body by choosing the **Select Body** option from the **Tool Body** drop-down list and selecting the body required as the tool body. You will notice that with each type of a default tool body, a callout is displayed, using which you can modify the size of the geometry of the tool body.

Next, you need to select the direction in which you need to push the target body. Select an edge, sketched line, or a face to define the push direction reference. Choose the **Reverse push direction** button to flip the default pushing direction, if required.

Click once in the **Bodies to be deformed** selection area and select the body or bodies to be deformed. The preview of the resulting deformed body is displayed in the drawing area.

You can modify the position of the tool body using the triad. To modify the location of tool body, move the cursor close to any of the arrows of the cursor along which you need to move or rotate the tool body. The cursor will be replaced by cursor. Press and hold down the left mouse button and drag the cursor to move the tool body in the direction pointing the arrow.

Press and hold down the right mouse button and drag the cursor to rotate the tool body along the direction of the arrow of triad. You can also use the options available in the **Tool Body Position** rollout to modify the position of the tool body.

The **Deform deviation** spinner provided in the **Deform Region** rollout is used to define the fillet radius where the tool body intersects the face of the target body. Figure 9-122 shows a deformed body with the lower value of deviation and Figure 9-123 shows a deformed body with a higher value of deviation.

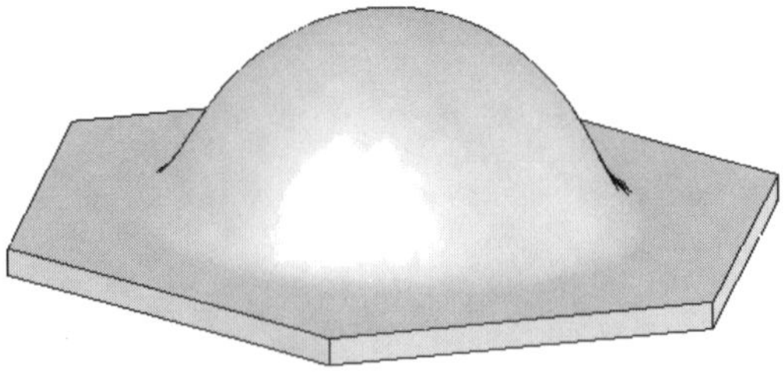

***Figure 9-122** Deform feature with lower value of deviation*

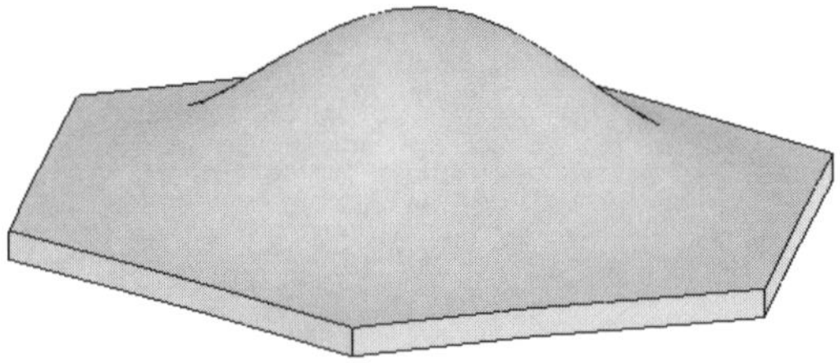

***Figure 9-123** Deform feature with higher value of deviation*

Creating Flex Features

CommandManager:	Features > Flex	*(Customize to Add)*
Menu:	Insert > Features > Flex	
Toolbar:	Features > Flex	*(Customize to Add)*

The most awaited enhancement in the **Part** mode of SolidWorks is made in this release of the software. This enhancement is the **Flex** tool, which enables you to perform free form bending, twisting, tapering, and stretching of a selected body. The portion of the bodies to be deformed is defined by two planes that are known as trimming planes. You will learn more about these planes later in this chapter. To deform a body using this tool, choose the **Flex** button from the **Features CommandManager**; the **Flex PropertyManager** is displayed. The partial view of the **Flex PropertyManager** is shown in Figure 9-124.

The **Bending** radio button is selected by default in the **Flex Input** rollout; you are prompted to select the bodies for flex and set the options. Select the body from the drawing area. You will notice that the selected body is displayed in transparent and two trimming planes are displayed at its extremes. A bending axis is also displayed passing through the selected body along which the body will be bent. A 3D triad is displayed placed coincident to the part origin. Figure 9-125 shows the selected body along with the trimming planes, 3D triad, and the bending axis.

You can modify the position and location of the trimming planes and the bending axis using the 3D triad. Move the cursor on the triad arrow, along which you need to move the bending axis, or about which you need to rotate the trimming planes. The select cursor is replaced by cursor. Press and hold the left mouse button and drag the cursor to move the location of

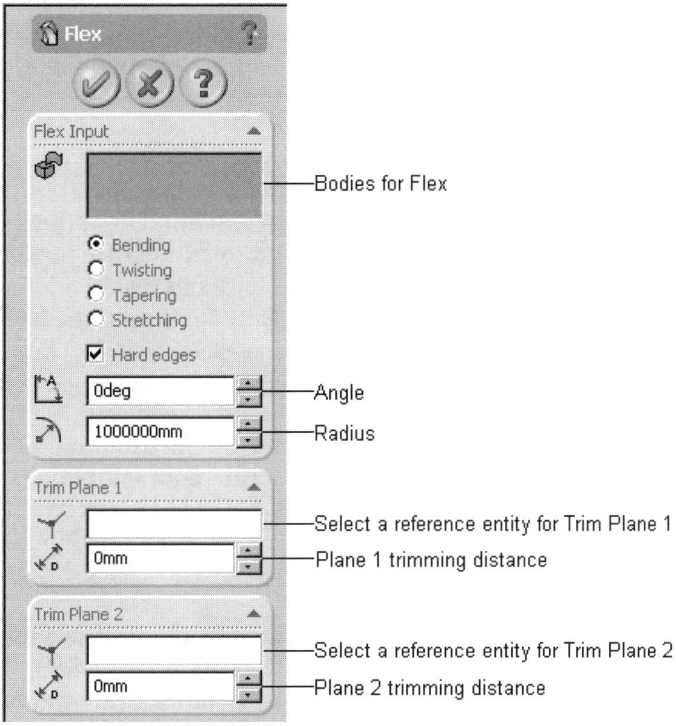

Figure 9-124** The **Flex PropertyManager

the bending axis, along the arrow on which the cursor was moved. Press and hold the right mouse button and drag the cursor to rotate the trimming planes and the bending axis. Figure 9-126 shows the preview of the flex feature, after modifying the value of the bending axis and the trimming planes.

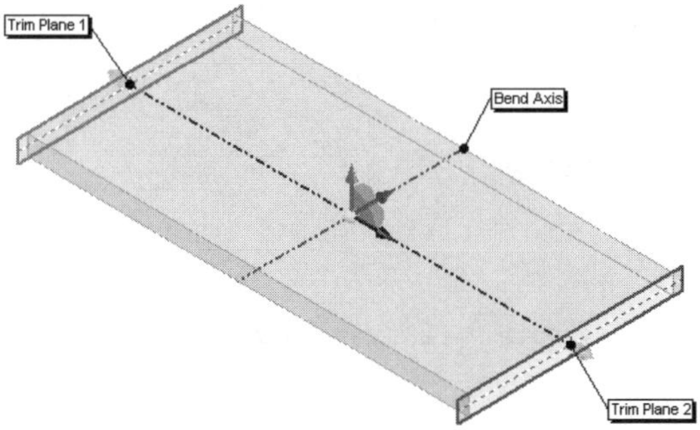

***Figure 9-125** Selected body along with the trimming planes, 3D triad, and the bending axis*

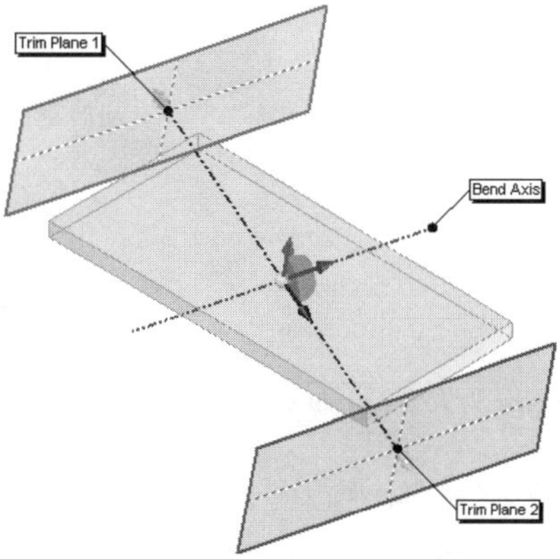

***Figure 9-126** Body after modifying the bending axis and the trimming planes*

You can also modify the location of the 3D triad. To do this, move the cursor to the origin of the triad, press and hold down the left mouse button and drag the cursor to change the location of the triad. You can also use the **Triad** rollout to modify the position of triad and rotation angle of the bending axis and trimming planes.

As discussed earlier, by default the 3D triad is placed coincident to the part origin. Therefore, the body is deformed about the part origin. If you need to fix a face of the body while deforming it, move the cursor to the origin of the triad and press and hold down the left mouse button. Drag the cursor and place it on the face that you need to fix. Release the left mouse button, when the boundary of the face on which the cursor is moved is displayed in red. You can also set the position of the triad using the **Triad** rollout. Note that you can coincide the origin of the triad with an existing geometry such as a vertex, an edge, or a sketched entity.

The **Trim Plane 1** and the **Trim Plane 2** rollouts are used to specify an offset value of the current trimming planes. You can modify the gap between the trimming planes by dragging them using the arrows attached to them. The **Select a reference entity for Trim Plane 1** and the **Select a reference entity for Trim Plane 2** areas are used to select a vertex along which you need to align the trimming planes. The slider bar provided in the **Flex Options** rollout is used to increase the accuracy of the flex tool.

Next, move the cursor on the boundary of any of the trimming planes; the select cursor is replaced by cursor. Drag the cursor to dynamically define the bending angle and radius of the bend. You can also define them using the **Angle** and **Radius** spinners, respectively. Figure 9-127 shows the preview of a body being bent and Figure 9-128 shows the body after bending using the **Flex** tool.

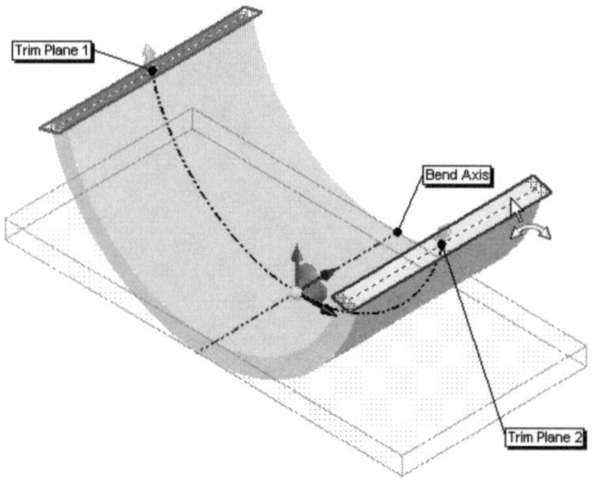

Figure 9-127 *Body being bent dynamically*

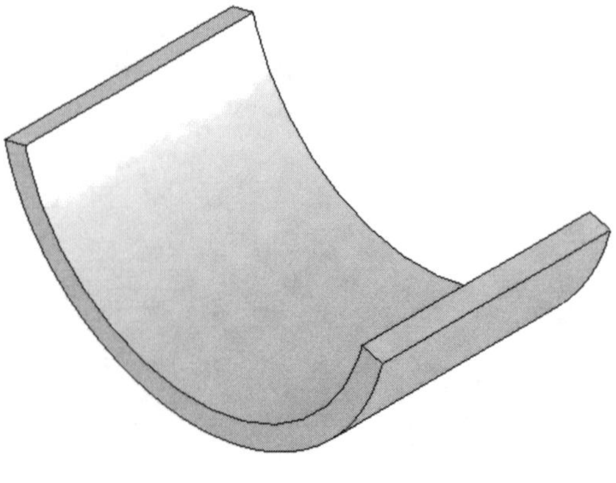

Figure 9-128 *Body after bending*

Twisting a Body

To twist a body using the **Flex** tool, invoke the **Flex PropertyManager** and select the **Twist** radio button from the **Flex Input** rollout. Select the body or bodies that you need twist; the 3D triad and the trimming planes are displayed. Set the location of the trimming planes and the triad, if required. Move the cursor on the boundary of any of the trimming planes. The select cursor is replaced by cursor. Drag the cursor to dynamically twist the selected body, or set the value of the twisting angle in the **Angle** spinner provided. Figure 9-129 shows the body being twisted with the triad placed coincident to the left face of the body and Figure 9-130 shows the resulting twisted body.

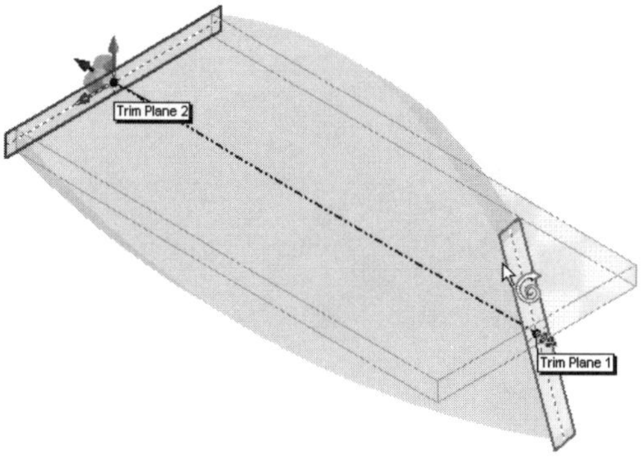

Figure 9-129 *Body being twisted*

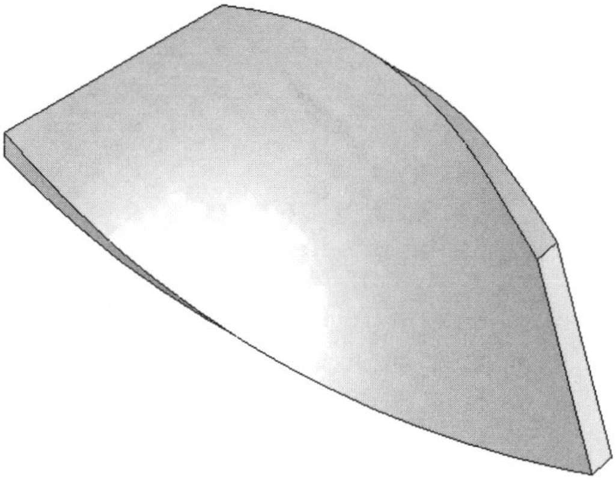

Figure 9-130 *The resulting twisted body*

Tapering a Body

To taper a body using the **Flex** tool, invoke the **Flex PropertyManager**, and select the **Taper** radio button from the **Flex Input** rollout. Select the body and set the position of the 3D triad and the trimming planes, if required. Move the cursor close to the boundary of any of the trimming planes. The select cursor is replaced by the cursor. Drag the cursor to specify the taper factor. You can also set the taper factor using the **Taper factor** spinner provided in the **Flex Input** rollout. Figure 9-131 shows the body being tapered and Figure 9-132 shows the resulting tapered body.

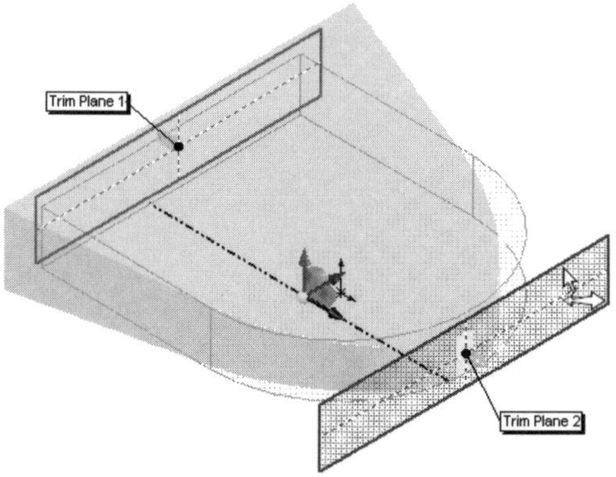

Figure 9-131 *Body being tapered*

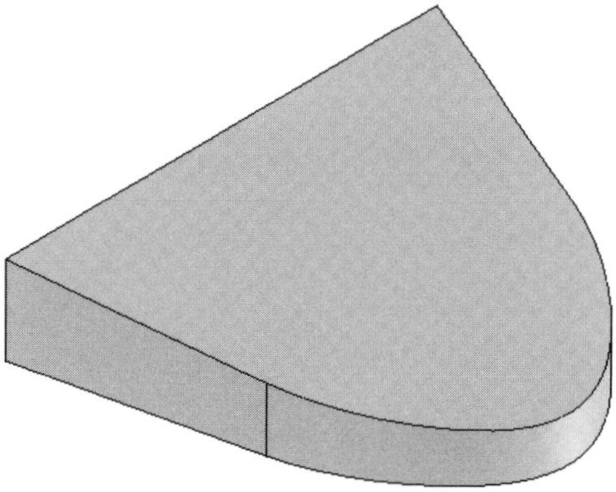

Figure 9-132 *Resulting tapered body*

Stretching a Body

You can also stretch a selected body using the **Flex** tool. To do this, invoke the **Flex PropertyManager** and select the **Stretching** radio button from it. Select the body and move the cursor on the boundary of any one of the trimming planes. The select cursor is replaced by the ⇔ cursor. Drag the cursor to specify the stretching distance. You can also specify it using the **Stretch distance** spinner. Figure 9-133 shows the body being stretched and Figure 9-134 shows the resulting stretched body.

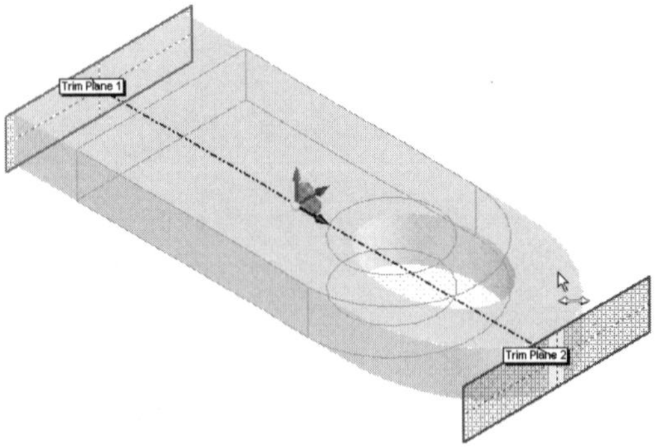

Figure 9-133 *Body being stretched*

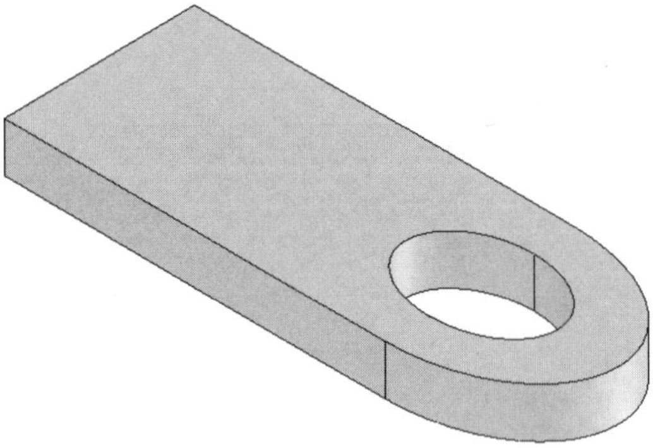

Figure 9-134 *Resulting stretched body*

TUTORIALS

Tutorial 1

In this tutorial, you will create the model shown in Figure 9-135. The dimensions of the model are shown in the same figure. **(Expected time: 45 min)**

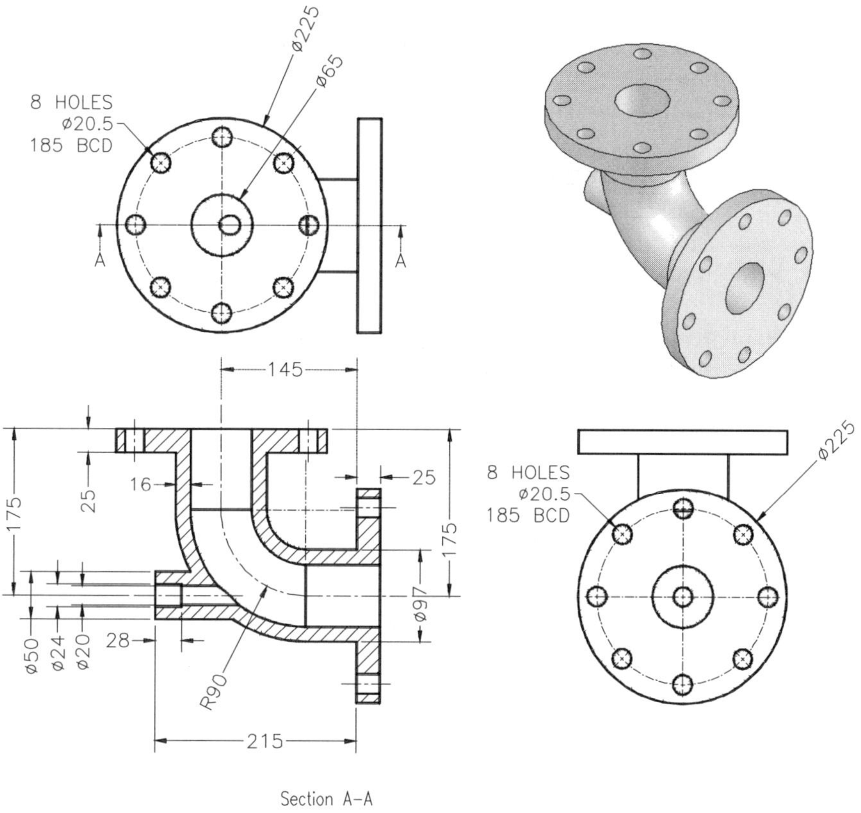

Figure 9-135 *Views and dimensions of the model for Tutorial 1*

The steps to be followed to complete this tutorial are listed next.

a. The base feature of the model is a sweep feature. First, you need to create the path for the sweep feature on the **Front Plane**. Next, you need to create a plane normal to this path. Select the newly created plane as the sketching plane and create the profile for the sweep feature. You will create a thin sweep feature, because the base feature of the model is a hollow feature, refer to Figures 9-136 through 9-138.
b. Create the extrude features on both ends of the sweep feature, refer to Figure 9-139.
c. Create a plane at an offset distance from the right face of the model. Create the circular feature by extruding the feature using the **Up To Next** option.
d. Create the hole using the **Simple Hole** tool, refer to Figure 9-139.
e. Create the pattern of the hole feature, refer to Figure 9-139.
f. Create the counterbore hole using the cut revolve option, refer to Figure 9-139.

Creating the Path for the Sweep Feature

As discussed earlier, the base feature of the model is a sweep feature. To create the sweep feature, you first need to create its path. This path will be created on the **Front Plane**.

1. Start a new SolidWorks Part document using the **New SolidWorks Document** dialog box.

2. Draw the sketch of the path for the sweep feature on the **Front Plane** and add the required relations and dimensions to the sketch as shown in Figure 9-136. Exit the sketching environment and change the view to isometric.

Creating the Profile for the Sweep Feature

After creating the path for the sweep feature, you will create its profile. For creating the profile, first you need to create a reference plane normal to the path. The newly created plane will be selected as the sketching plane for creating the profile for the sweep feature.

1. Invoke the **Plane PropertyManager** and using the **Normal to Curve** option, create a plane normal to the path, as shown in Figure 9-137.

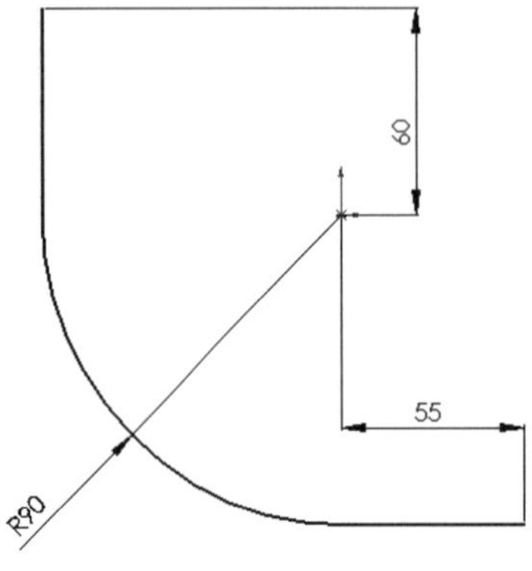

Figure 9-136 *Sketch of the path*

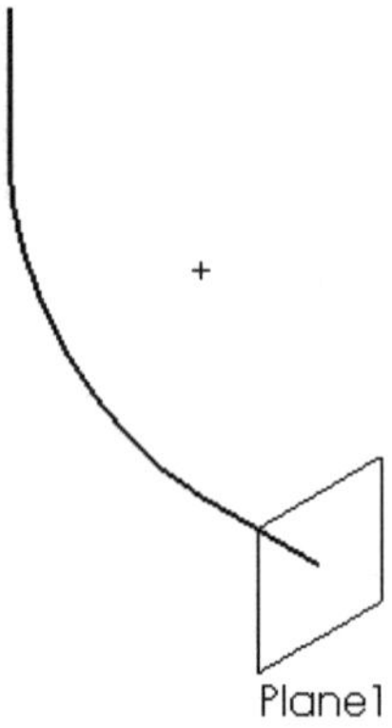

Figure 9-137 *Plane created normal to path*

2. Invoke the sketching environment by selecting the newly created plane as the sketching plane.

3. Draw the sketch of the profile of the sweep feature using the **Circle** tool. The center of the circle is at the origin and the diameter of the circle is 97 mm.

4. Exit the sketching environment after drawing the profile for the sweep feature.

Creating the Sweep Feature

The sweep feature that you are going to create is a thin sweep feature. You will use the **Thin Feature** rollout to specify its parameters.

1. Choose the **Swept Boss/Base** button from the **Features CommandManager**. The **Sweep PropertyManager** is invoked and you are prompted to select the sweep profile.

2. Select the profile for the sweep feature. The selected profile will be highlighted in green and the **Profile** callout will also be displayed.

   You are prompted to select the sweep path after selecting the profile.

3. Select the path for the sweep feature. The selected path will be highlighted in magenta and the **Path** callout is also displayed. The preview of the sweep feature is also displayed in the drawing area.

4. Select the **Thin Feature** check box available on the left of the **Thin Feature** rollout; the **Thin Feature** rollout is invoked.

5. Set the value of thickness in the **Thickness** spinner to **16**.

   Since the wall thickness added to the model is reverse to the required direction, therefore, you need to reverse the direction of the creation of thin feature.

6. Choose the **Reverse Direction** button from the **Thin Features** rollout. Choose the **OK** button from the **Sweep PropertyManager**, or choose **OK** from the confirmation corner.

   The base feature created by sweeping a profile along a path is shown in Figure 9-138.

Tip. *Instead of creating a thin sweep feature, you can also create a solid sweep feature and then add a shell feature to hollow the base feature.*

Creating the Remaining Features

1. Create the remaining join features of the model using the extrude option.

2. Using the **Simple Hole PropertyManager**, create the holes and pattern them using the **Circular Pattern** tool. Create the counterbore hole using the **Hole Wizard** or the revolve cut option.

   The final solid model of Tutorial 1 is shown in Figure 9-139.

Saving the Model

1. Choose the **Save** button from the **Standard** toolbar and save the model with the name given below

   \My Documents\SolidWorks\c09\c09tut1.sldprt

2. Choose **File** > **Close** from the menu bar to close the file.

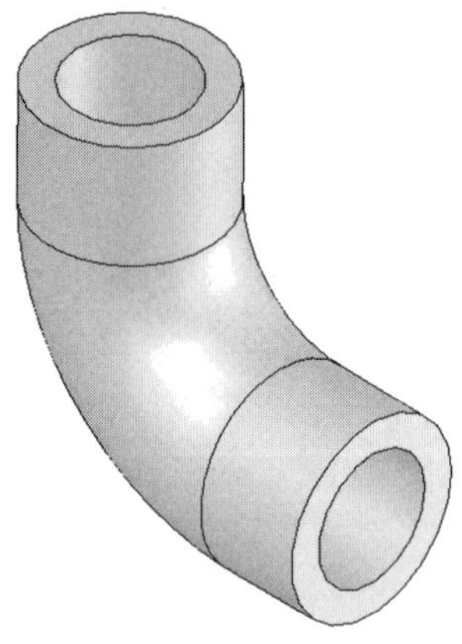

Figure 9-138 Base feature of the model

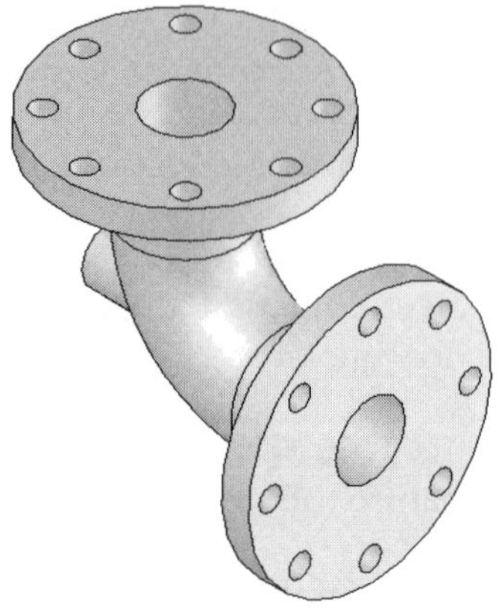

Figure 9-139 Final model of Tutorial 1

Tutorial 2

In this tutorial, you will create the chair frame shown in Figure 9-140. The dimensions of the chair frame are also shown in the same figure. **(Expected time: 30 min)**

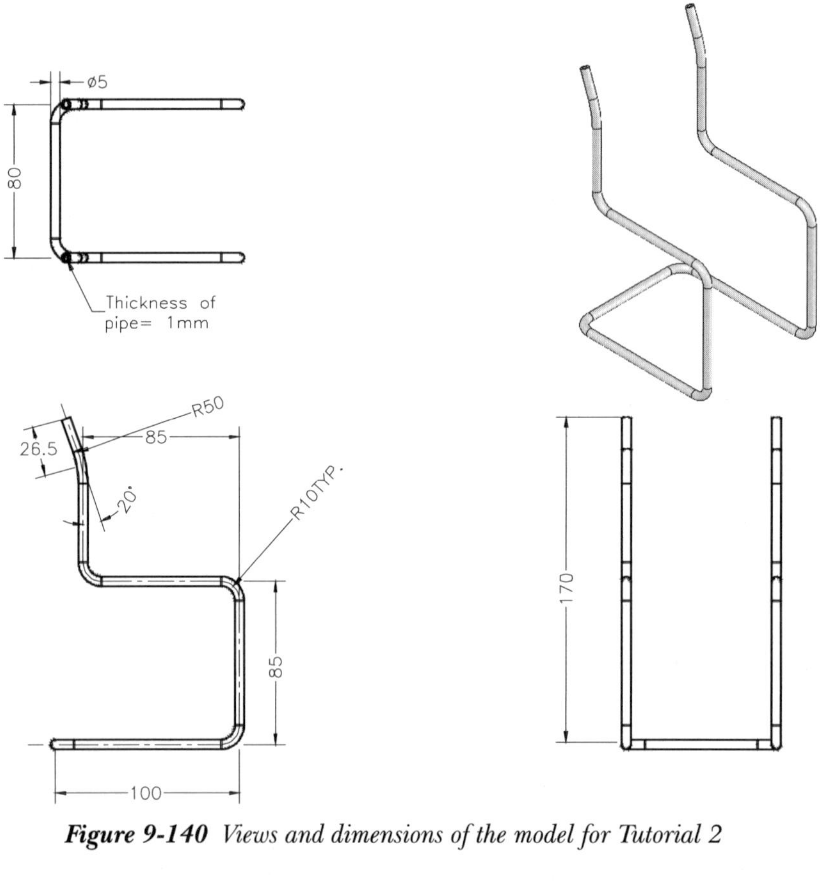

Figure 9-140 Views and dimensions of the model for Tutorial 2

The steps to be followed to complete this tutorial are listed next.

a. The chair frame is created by sweeping a profile along a 3D path. The 3D path will be created in the 3D sketching environment. Therefore, you need to invoke the 3D sketching environment and then draw the sketch of the 3D path. You will create only the left half of the 3D path in the 3D sketching environment, refer to Figure 9-141.
b. Create a plane normal to the 3D path. Select the newly created plane as the sketching plane and draw the sketch of the profile.
c. Sweep the profile along the 3D path using the **Thin Feature** option, refer to Figure 9-142.
d. Mirror the sweep feature using the **Front Plane**, refer to Figure 9-143.

Creating the Path of Sweep Feature Using 3D Sketching Environment

It is evident from Figure 9-140 that the model is created by sweeping a profile along a 3D path. Therefore, you need to create a path of the sweep feature in the 3D sketching environment.

1. Start a new SolidWorks part document.

2. Create a plane at an offset distance of 40 mm from the **Front Plane**.

3. Change the current view to isometric.

4. Choose **Insert > 3D Sketch** from the menu bar to invoke the 3D sketching environment. You can also use the **3D Sketch** button from the **Sketch CommandManager** to invoke the 3D sketching environment.

The 3D sketching environment is invoked and the sketch origin is displayed in red. You are also provided with the confirmation corner on the top right of the drawing area. The sketching tools that can be used in the 3D sketching environment are also invoked.

5. Invoke the **Line** tool; the select cursor is replaced by the line cursor. The XY that is displayed below the line cursor suggests that by default, the line will be sketched in the XY plane.

The first line you need to draw is in the ZX plane. Therefore, you will toggle the plane before you start creating the sketch.

6. Press the TAB key on the keyboard twice to switch to ZX plane.

7. Move the line cursor to the origin. When a red dot is displayed, specify the start point of the line.

8. Move the cursor in the positive Z direction of the triad. A small triad with Z appears below the cursor indicating that you are drawing the line in the **Z** direction. Specify the end point of the line, when a value close to 40 is displayed above the cursor.

A rubber-band line is displayed attached to the cursor.

9. Move the cursor toward the right along the infinite line indicating the X axis. Specify the end point of the line when a value close to 100 is displayed above the cursor.

10. Press the TAB key to switch over to the XY plane. Move the cursor vertically upwards.

11. Specify the end point of the line when the value above the cursor shows a value close to 85.

14. Similarly, draw the remaining sketch and add the required relations and dimensions to it. The final sketch is displayed in Figure 9-141.

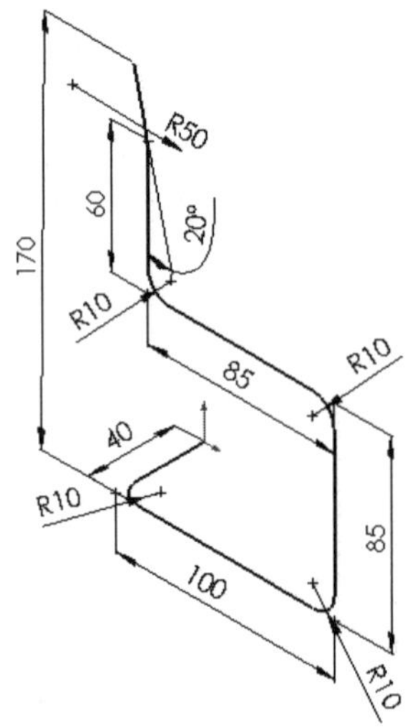

Figure 9-141 *Sketch of the 3D path*

15. Add the **Coincident** relation between the upper endpoint of the 3D sketch and **Plane1**.

16. Exit the 3D sketching environment using the confirmation corner.

Creating the Profile of the Sweep Feature

As discussed earlier, the sweep feature will be created by sweeping the profile along a 3D path. After creating the path, you need to create the profile for the sweep feature.

1. Select the **Front Plane** as the sketching plane and invoke the sketching environment.

   You do not need to create a reference plane, because the **Front Plane** is normal to the 3D path.

2. Draw a circle of diameter 5 mm, that will be used as the sketch of the profile of the sweep feature, refer to Figure 9-140.

3. Exit the sketching environment.

Sweeping the Profile Along the 3D Path

After creating the 3D path and the profile for the sweep feature, you need to sweep the profile along the 3D path using the **Swept Boss/Base** tool.

1. Choose the **Swept Boss/Base** button from the **Features CommandManager**. The **Sweep PropertyManager** is invoked.

2. Make sure that the sketch of the profile is selected. The selected profile will be displayed in green. The profile callout will also be displayed.

   You are prompted to select the path for the sweep feature.

3. Select the path for the sweep feature. The path will be displayed in magenta color. The path callout will also be displayed. The preview of the sweep feature will also be displayed in the drawing area.

   As evident from Figure 9-140, the frame of the chair is made from a hollow pipe. Therefore, you need to create a thin sweep feature to create a hollow chair frame.

4. Invoke the **Thin Feature** rollout and set the value of the **Thickness** spinner to **1**.

5. Choose the **Reverse Direction** button from the **Thin Features** rollout to reverse the direction of the thin feature creation.

6. Choose the **OK** button from the **Sweep PropertyManager** to end the creation of the feature.

   The model, after creating the sweep feature, is displayed in Figure 9-142.

7. Mirror the sweep feature about the **Front Plane** using the **Mirror** tool. The model, after mirroring is shown in Figure 9-143.

Saving the Model

1. Choose the **Save** button from the **Standard** toolbar and save the model with the name given below.

   \My Documents\SolidWorks\c09\c09tut2.sldprt

2. Close the document.

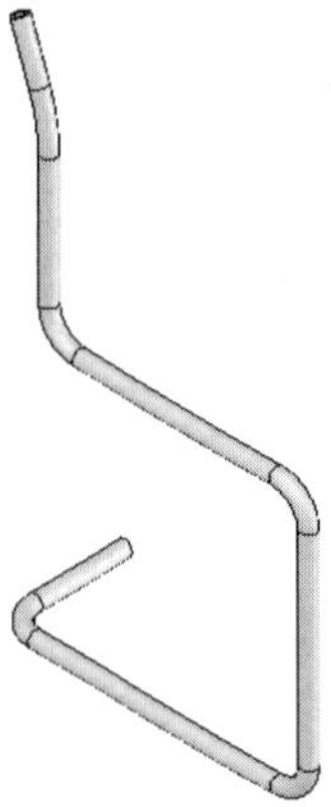

Figure 9-142 *Sweep feature created by sweeping a profile along a 3D path*

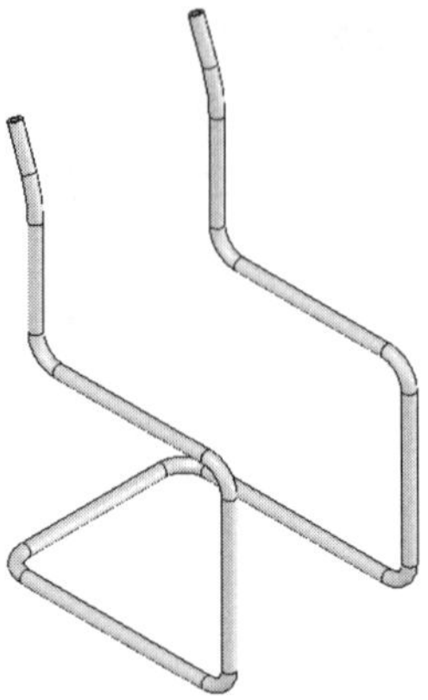

Figure 9-143 *The final model*

Tutorial 3

In this tutorial, you will create the spring shown in Figure 9-144. The dimensions of the spring are shown in Figure 9-145. **(Expected time: 45 min)**

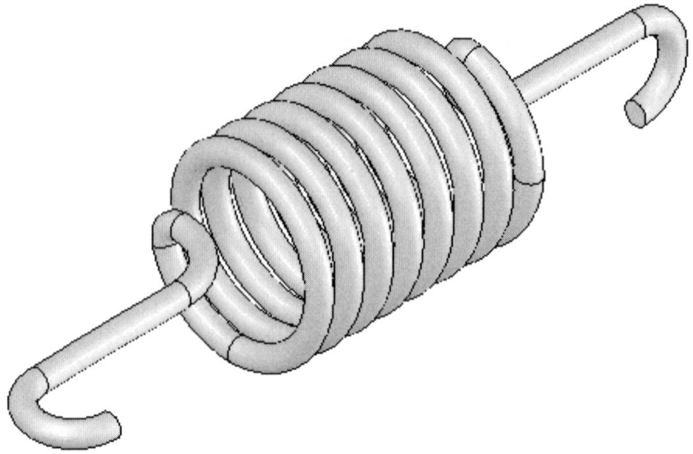

Figure 9-144 *Model for Tutorial 3*

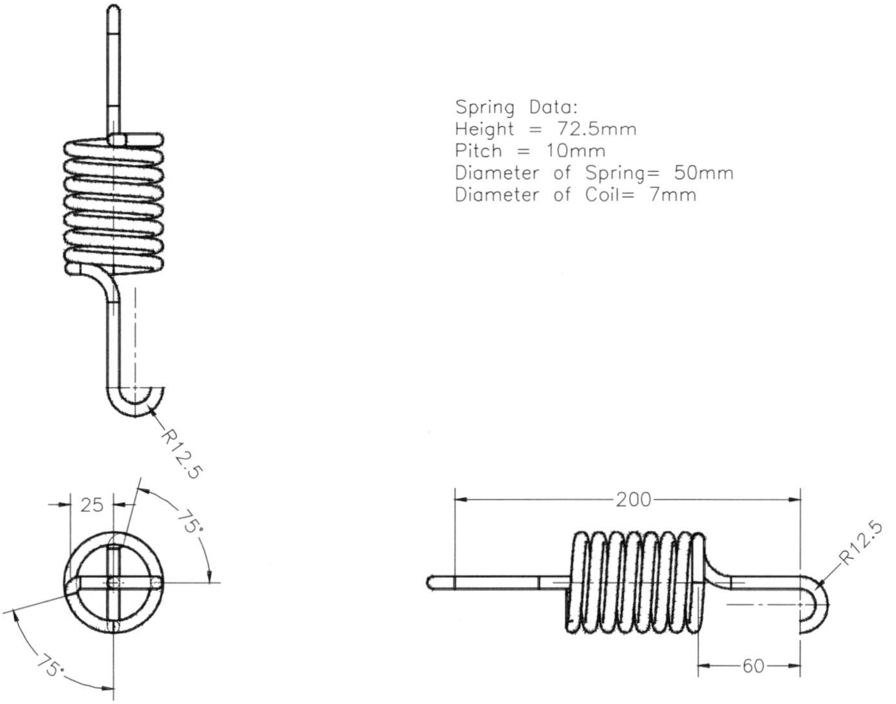

Figure 9-145 *Views and dimensions of the model for Tutorial 3*

The steps to be followed to complete this tutorial are listed next.

a. Create the helical path of the spring, refer to Figure 9-146.
b. Create the end clips of the spring, refer to Figure 9-149.
c. Combine the end clips and the helical curve to create a single curve using the **Composite Curve** option, refer to Figure 9-150.
d. Create the profile on the plane normal to the curve and sweep it along the curve, refer to Figures 9-151 and 9-152.

Creating the Helical Curve

For creating this model first you need to create the helical curve of the spring. For creating the helical curve first you need to create a circular sketch. This sketch will define the diameter of the spring.

1. Start a new SolidWorks Part document.

2. Choose the **Helix and Spiral** button from the **Curves CommandManager** and select the **Front Plane** as the sketching plane.

3. Draw a circle of diameter 50. Change the view to isometric view and exit the sketching environment.

The **Helix/Spiral Curve PropertyManager** is displayed and the preview of the helical curve with the default values is displayed in the drawing area.

4. Select the **Height and Pitch** option from the **Type** drop-down list available in the **Defined By** rollout.

5. Set the value of the **Height** spinner to **72.5** and the value of the **Pitch** spinner to **10**.

   You will observe that the preview of the helical curve is updated automatically when you modify the values in the spinners.

6. Set the value of the **Starting angle** spinner to **0** and choose the **OK** button from the **Helix/Spiral PropertyManager**.

   The helical curve, created using the above steps, is shown in Figure 9-146.

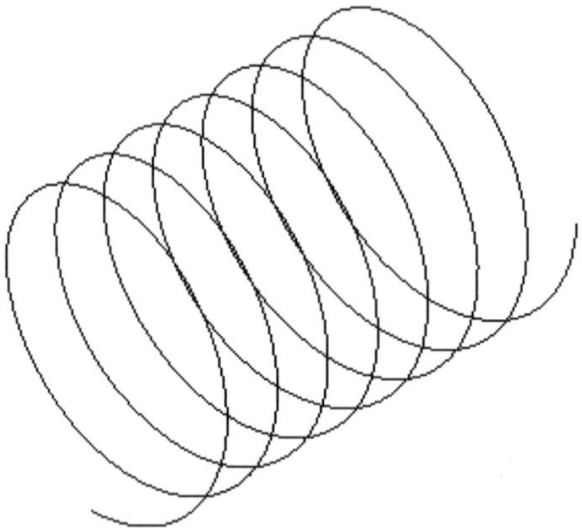

Figure 9-146 *Helical curve*

Drawing the Sketch of the End Clips of the Spring

After creating the helical curve, you need to draw the sketch that defines the path of the end clips. There are two end clips in this spring, and each end clip is created using two sketches. First, you will create the right end-clip and then the left end-clip.

1. Select the **Front Plane** as the sketching plane and invoke the sketching environment.

   The first sketch of the right end-clip consists of two arcs. The first arc will be drawn using the **Centerpoint Arc** tool and the second arc will be drawn using the **3 Point Arc** tool.

2. Choose the **Centerpoint Arc** button from the **Sketch CommandManager** and specify the centerpoint of the arc at the origin.

3. Move the cursor toward the right and specify the start point of the arc when the radius of the reference circle shows a value close to **25**.

4. Move the cursor in the counterclockwise direction. Specify the endpoint of the arc when the value of the angle above the cursor shows a value close to **75**.

5. Choose the **3 Point Arc** button from the **Sketch CommandManager**. Specify the start point of the arc on the upper endpoint of the previous arc. Create the arc as shown in Figure 9-147.

6. Add the **Pierce** relation between the lower endpoint of the first arc and the helical curve. Add the other relations and the dimensions to fully define the sketch. The fully defined sketch is shown in Figure 9-147.

7. Exit the sketching environment.

   After drawing the first sketch of the right end-clip, you need to draw the second sketch of the right end-clip. The second sketch of the right end-clip will be drawn on the **Right Plane**.

8. Select the **Right Plane** and invoke the sketching environment.

9. Draw the sketch as shown in Figure 9-148. You need to apply the **Pierce** relation between the upper arc and the previous sketch and the left end point of the left arc of the current sketch.

10. Add a **Tangent** relation between the left arc of the current sketch and the upper arc of the sketch previously drawn. Add the required relations and dimensions to the current sketch. The sketch, after applying all the relations and dimensions, is shown in Figure 9-148.

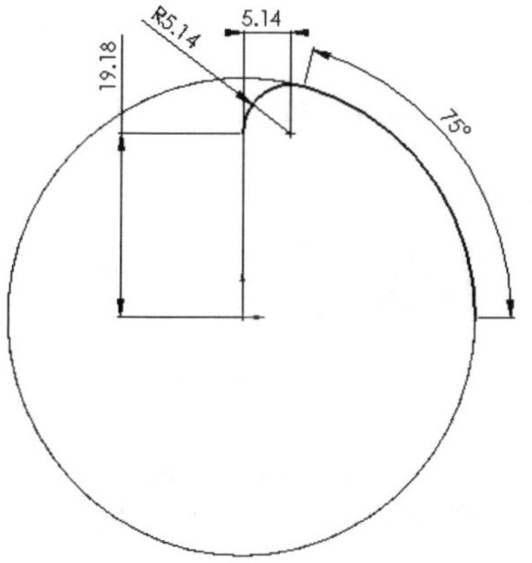

Figure 9-147 *First sketch of right end-clip*

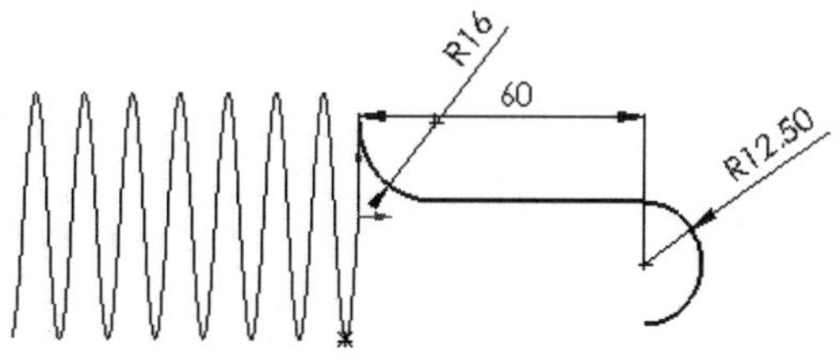

Figure 9-148 *Second sketch of right end-clip*

Similarly, draw the sketch of the left end-clip. Figure 9-149 shows the path of the spring after creating the helical curve, right end-clip, and the left end-clip.

Creating the Composite Curve

After creating all the required sketches and the helical curve, you need to combine them so that they form a single curve. This is done because the path of the sweep feature has to be a single curve or sketch.

1. Choose the **Composite Curve** button from the **Curves CommandManager** or choose **Insert > Curve > Composite** from the menu bar. The **Composite Curve PropertyManager** is displayed and the confirmation corner is also available.

2. Select both the end-clips and the helical curve from the drawing area or from the **FeatureManager Design Tree**.

3. Choose the **OK** button from the **Composite Curve PropertyManager** or choose **OK** from the confirmation corner.

   The composite curve is created and is displayed in Figure 9-150.

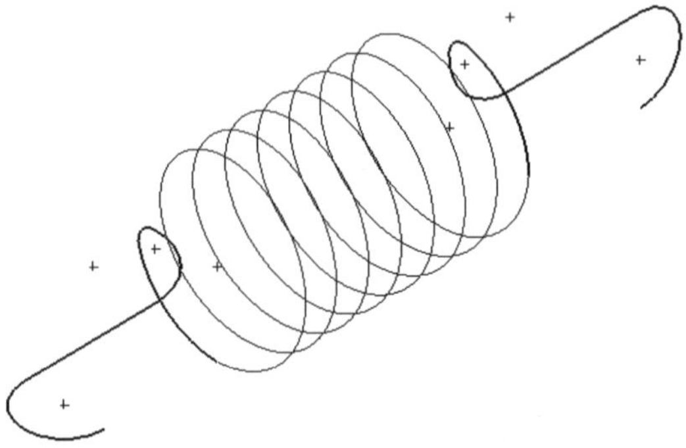

Figure 9-149 *Path for the spring*

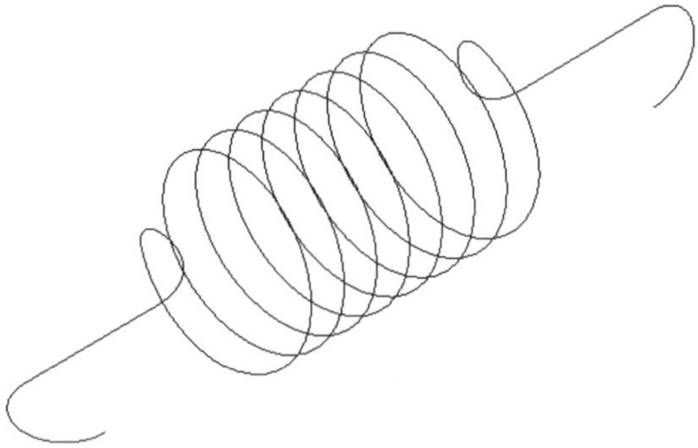

Figure 9-150 *Composite curve*

Creating the Profile for the Sweep Feature

Next, you need to create the profile for the sweep feature. The sketch of the profile will be drawn on a plane normal to the curve on the endpoint of the right end-clip.

1. Create a plane normal to the path and at the endpoint of the right end-clip.

2. Draw the profile of the sweep feature and add the **Pierce** relation between the center of the circle and the composite curve. Add the required dimensions to the sketch.

3. Exit the sketching environment. Figure 9-151 shows the profile and the path for the sweep feature.

Creating the Sweep Feature

You will create a sweep feature to complete the creation of spring.

1. Invoke the **Sweep PropertyManager** and you are prompted to select the sweep profile.

2. Select the sweep profile from the drawing area.

   You are prompted to select the sweep path.

3. Select the path and choose the **OK** button from the **Sweep PropertyManager**.

   The spring created, after sweeping the profile along the path, is shown in Figure 9-152.

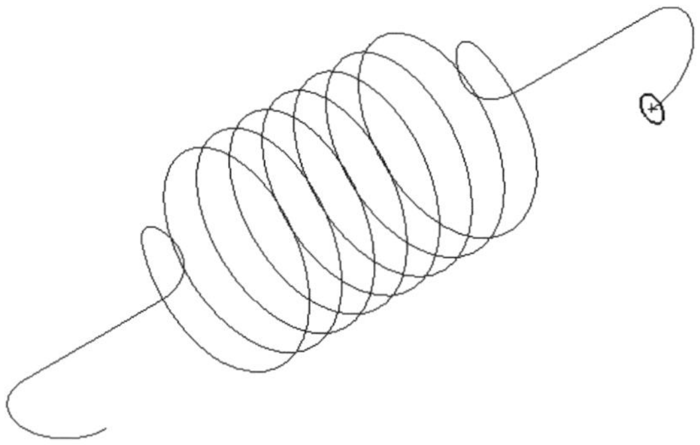

Figure 9-151 *Profile and path of the sweep feature*

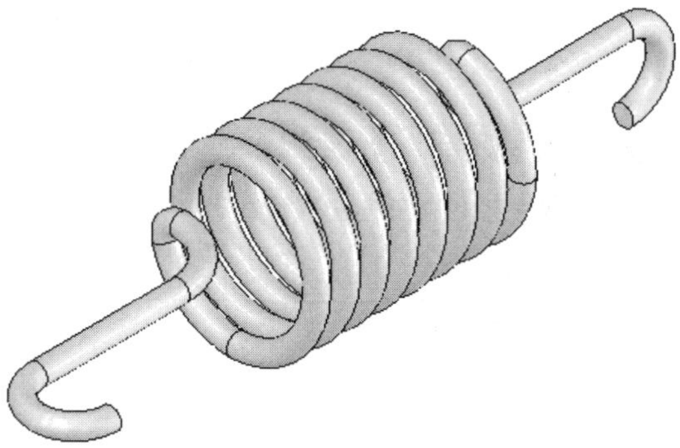

Figure 9-152 *Final model*

Saving the Model

1. Choose the **Save** button from the **Standard** toolbar and save the model with the name given below.

   \My Documents\SolidWorks\c09\c09tut3.sldprt

2. Close the document.

SELF-EVALUATION TEST

Answer the following questions and then compare your answers with those given at the end of this chapter.

1. You need a profile and a path to create a sweep feature. (T/F)

2. At least two sections are required to create a loft feature. (T/F)

3. You cannot sweep a closed profile along a closed path. (T/F)

4. You cannot create a thin sweep feature. (T/F)

5. You can also create a loft feature using open sections. (T/F)

6. The __________ **PropertyManager** is used to create a loft feature.

7. The __________ option is used to create a single curve by joining the continuous chain of existing sketches, edges, or curves.

8. You need to apply __________ relation between the sketch and the guide curve, while sweeping a profile along a path using guide curves.

9. __________ option is used to create a curve by defining the coordinates.

10. __________ **PropertyManager** is invoked to create a cut-sweep feature.

REVIEW QUESTIONS

Answer the following questions.

1. Using the __________ rollout, you can define the tangency at the start and end in the sweep feature.

2. The __________ rollout is used to create a thin loft feature.

3. You need to invoke __________ dialog box to create a spiral curve.

4. The __________ **PropertyManager** is invoked to create a curve projected on a surface.

5. The __________ dialog box is used to specify the coordinates to create the curve.

6. Which button is selected to invoke the 3D sketching environment?

   (a) **2D Sketch** (b) **3D Sketching Environment**
   (c) **3D Sketch** (d) **Sketch**

7. Which rollout, available in the **Sweep PropertyManager**, is used to define the tangency?

   (a) **Start/End Tangency** (b) **Tangency**
   (c) **Options** (d) None of these

8. Which button, available in the **Features PropertyManager**, is used to invoke the **Draft PropertyManager**?

   (a) **Draft** (b) **Taper Angle**
   (c) **Draft Feature** (c) **Draft Angle**

9. In which rollout would you define the pull direction, while creating a draft feature using the parting line option?

   (a) **Pull Direction** (b) **Direction of Pull**
   (c) **Reference Direction** (d) **Options**

10. Which button available in the **Guide Curves** rollout is used to display the sections, while creating the sweep feature with guide curves?

    (a) **Preview Sections** (b) **Show Sections**
    (c) **Sections** (d) **Preview**

EXERCISES

Exercise 1

Create the model of the Upper Housing shown in Figure 9-153. The dimensions of the model are shown in Figure 9-154. **(Expected time: 1 hr)**

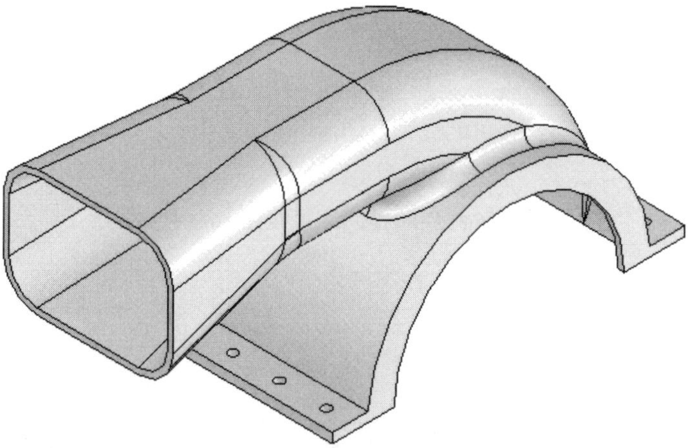

Figure 9-153 *Model of the Upper Housing*

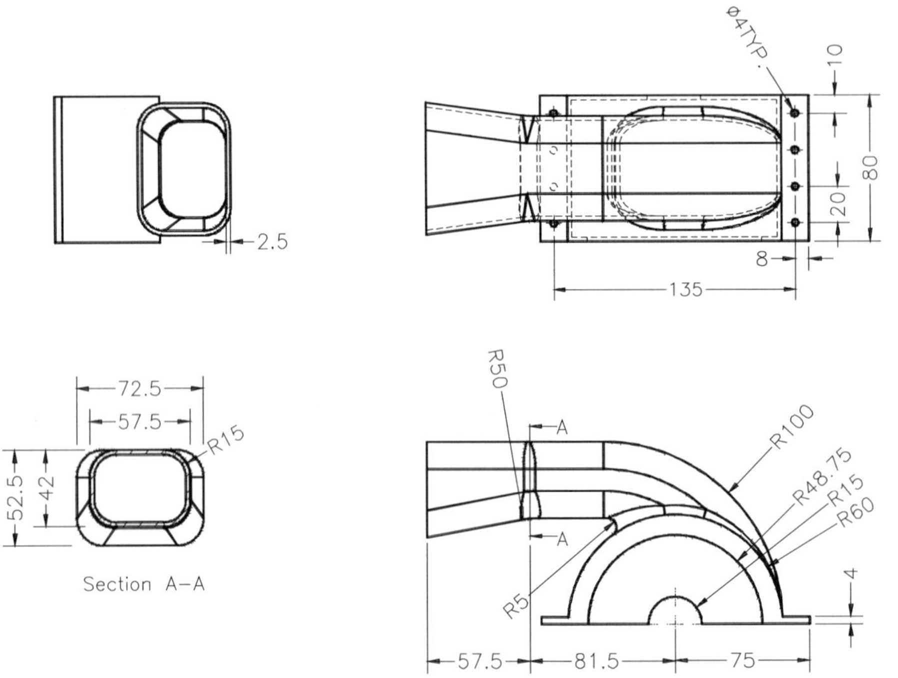

***Figure 9-154** Dimensions of the Upper Housing*

Tip. *This model is divided in three major parts. The first part is the base and is created by extruding the sketch to a distance of 80 mm using the **Mid Plane** option.*

*The second part of this model is the right portion of the discharge venturi and is created using the sweep feature. The path of the sweep feature will be created on the **Right Plane**. You need to create a plane at the left endpoint of the path and normal to the path.*

The third part of this model is the left portion of the discharge venturi created using the loft feature. You need to create the first section of the loft feature on the planar face of the sweep feature created earlier. The second section will be created on a plane at an offset distance from the planar face of the sweep feature created earlier. Create a loft feature using the two sections created earlier. The other features needed to complete the model are fillets, hole, circular pattern, and so on.

Exercise 2

Create the model shown in Figure 9-155. The dimensions of the model are shown in Figure 9-156. **(Expected time: 1 hr)**

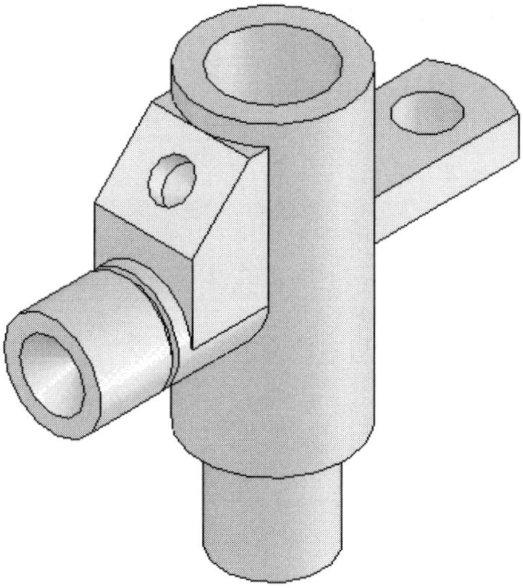

Figure 9-155 *Model for Exercise 2*

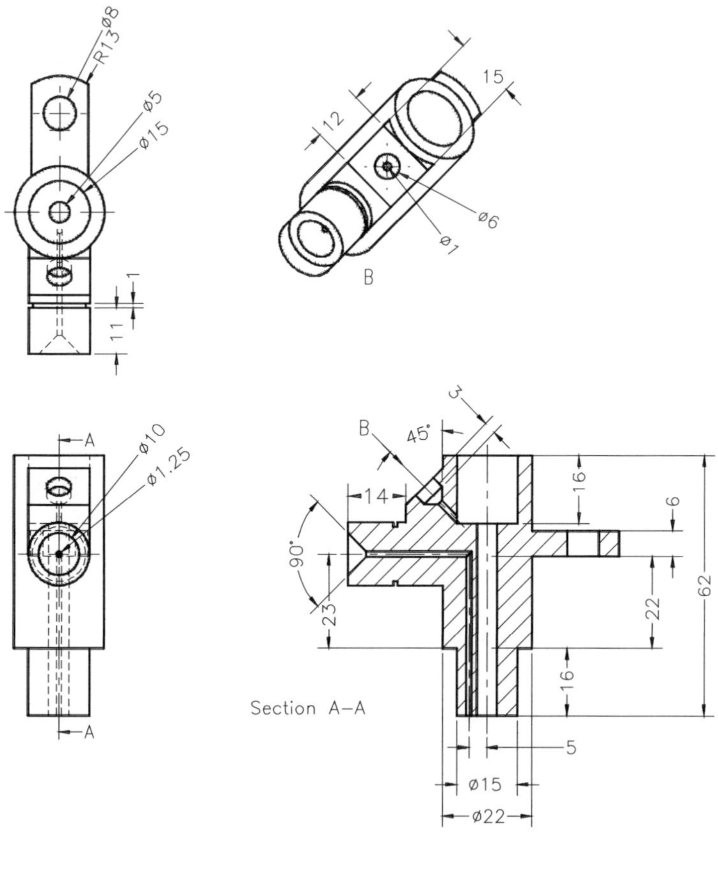

Figure 9-156 *Views and dimensions of the model for Exercise 2*

Answers to Self-Evaluation Test

1. T, **2.** T, **3.** F, **4.** F, **5.** T, **6. Loft**, **7. Composite Curve**, **8. Coincident**, **9. Curve Through Free Points**, **10. Cut-Sweep**

Chapter 10

Assembly Modeling-I

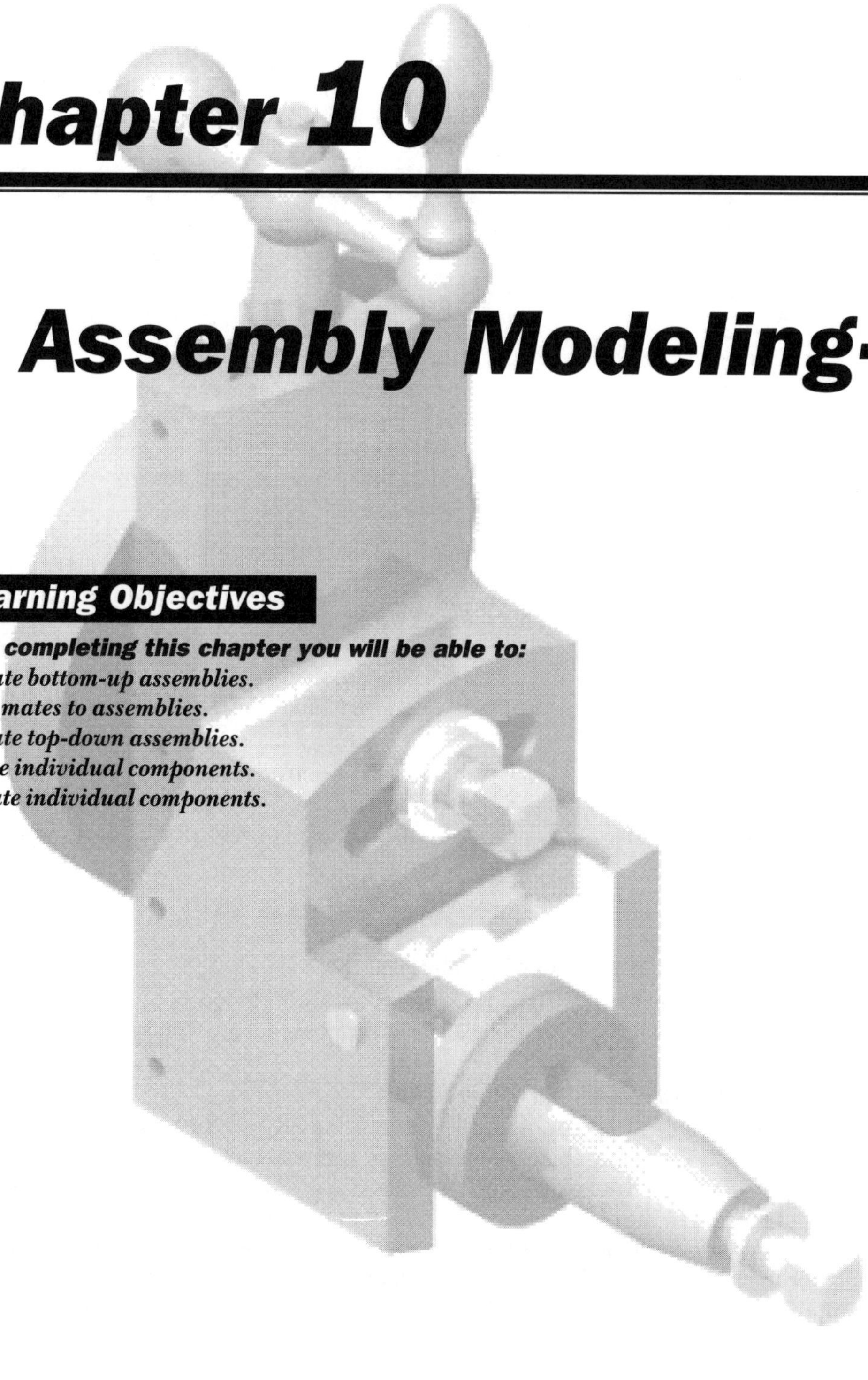

Learning Objectives

After completing this chapter you will be able to:

- *Create bottom-up assemblies.*
- *Add mates to assemblies.*
- *Create top-down assemblies.*
- *Move individual components.*
- *Rotate individual components.*

ASSEMBLY MODELING

An assembly design is a design consisting of two or more components assembled together at their respective work positions using the parametric relations. In SolidWorks, these relations are called mates. These mates allow you to constrain the degrees of freedom of the components at their respective work positions. To proceed to the **Assembly** mode of SolidWorks, invoke the **New SolidWorks Document** dialog box and choose the **Assembly** button as shown in Figure 10-1.

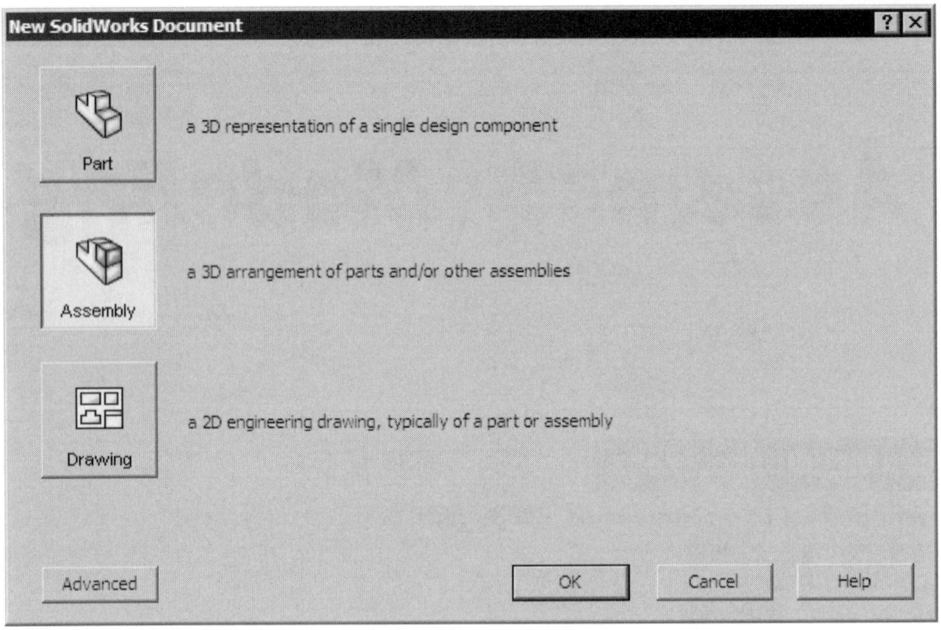

***Figure 10-1** The **New SolidWorks Document** dialog box*

Choose the **OK** button to create a new assembly document; new SolidWorks document will be started in the **Assembly** mode and the **Insert Component PropertyManager** will be invoked, as shown in Figure 10-2.

Types of Assembly Design Approach

In SolidWorks, the assemblies are created using two types of design approaches: bottom-up approach and the top-down approach. These design approaches are discussed next.

Bottom-up Assembly Design Approach

The bottom-up assembly design approach is the traditional and the most widely preferred approach of assembly design. In this assembly design approach, all components are created as separate part documents and are placed and referenced in the assembly as external components. In this type of approach, the components are created in the **Part** mode and saved as the *.sldprt* documents. After creating and saving all components of the assembly, you will start a new assembly document (*.sldasm*) and insert the components using the tools provided in the **Assembly** mode. After inserting you will assemble the components using the assembly mates.

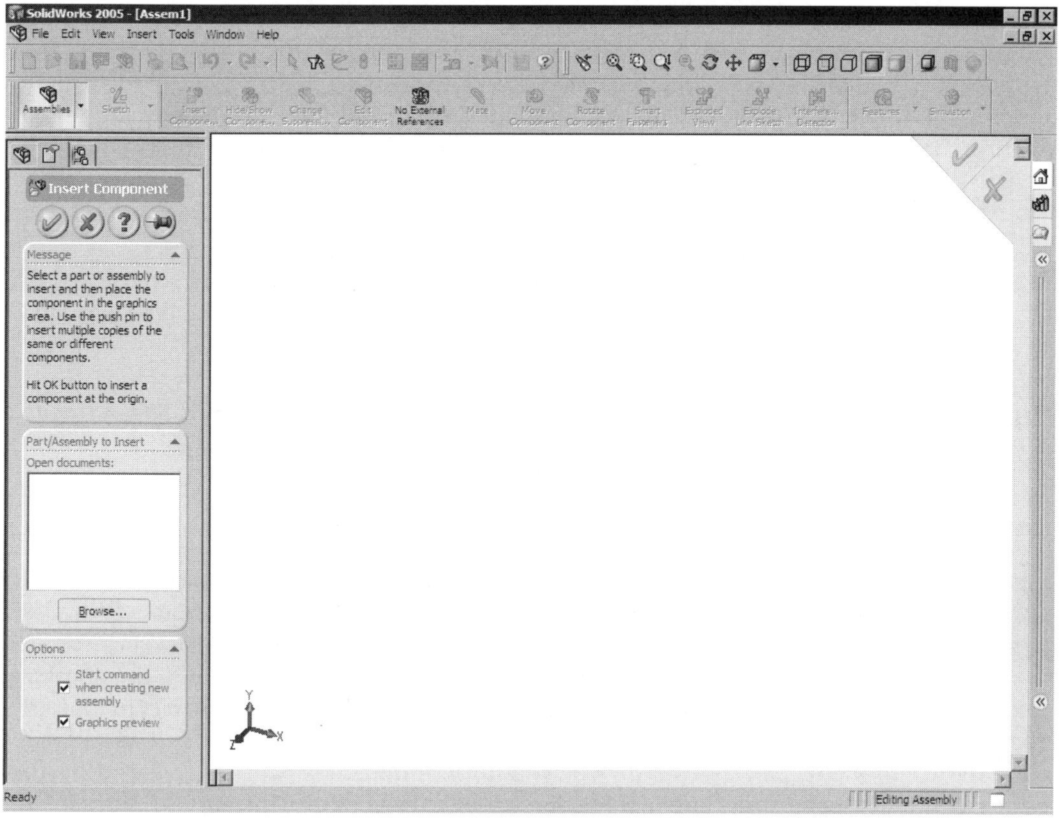

Figure 10-2 *Screen display of SolidWorks in the* ***Assembly*** *mode*

The main advantage of this assembly design approach is that because there is only a single part in the current file, the view of the part is not restricted. Hence, this approach allows you to pay more attention and focus more on the complex individual features. This approach is preferred while handling large assemblies or the assemblies with complex parts.

Top-down Assembly Design Approach

In the top-down assembly design approach, the components are created in the same assembly document, but saved as separate part files. Therefore, the top-down assembly design approach is entirely different from the bottom-up design approach. In this approach you will start your work in the assembly document and the geometry of one part will help you define the geometry of the other.

Note

You can also create an assembly with a combination of the bottom-up and the top-down assembly approaches.

CREATING BOTTOM-UP ASSEMBLIES

As mentioned earlier, the bottom-up assemblies are the assemblies in which the components are created as separate part documents in the **Part** mode. After creating the components, they are inserted in the assembly and then assembled using the assembly mates. For starting an assembly design with this approach, you first need to insert the components in the assembly. It is recommended that the first component should be placed at the origin of the assembly document. By doing this the default planes of the assembly and the part will coincide and the component will be in the same orientation as it was in the **Part** mode. When you place the first component in the assembly, that component will be fixed at its placement position. The techniques used to place the components in the assembly file are discussed next.

Placing Components in the Assembly Document

In SolidWorks, there are various options to place the components in the assembly. These options are discussed next.

Placing Components Using the Insert Component PropertyManager

CommandManager:	Assemblies > Insert Components
Menu:	Insert > Component > Existing Part/Assembly
Toolbar:	Assembly > Insert Components

When you start a new SolidWorks document in the **Assembly** mode, the **Insert Component PropertyManager** will be displayed, as shown in Figure 10-3. The **Message** rollout available in the **Insert Component PropertyManager** prompts you to select a part or an assembly and then place the component in the graphics area. When you choose the **Browse** button available in the **Part/Assembly to Insert** rollout, the **Open** dialog box will be displayed. Browse the location where the component is saved and then select the component and choose the **Open** button. The cursor will be replaced by the component cursor and graphic preview of the component will be also displayed. The name of the selected component will be displayed in the **Open documents** selection area of the **Part/Assembly to Insert** rollout. It is recommended that the origin of the first component should be aligned with the assembly origin. To place the component origin on the assembly origin choose the **OK** button from the **Insert Component PropertyManager**.

To place the second component, invoke the **Insert Component PropertyManager** again and choose the **Browse** button from the **Part/Assembly to Insert** rollout. Select the component from the **Open** dialog box; the cursor will be replaced by the component cursor and the preview of the component will also be displayed in the drawing area. Select a point anywhere in the drawing area to place the second component.

When you select a component to insert in the assembly, the **Thumbnail Preview** rollout will be displayed in the **Insert Component PropertyManager**. You can invoke this rollout to view the thumbnail preview of the selected component.

The **Start command when creating new assembly** check box available in the **Options** rollout is selected by default and is used to invoke the **Insert Component PropertyManager** automatically when you start a new SolidWorks assembly document. The **Graphics preview** check box available in the **Options** rollout is selected by default and is used to display the graphic preview of the component selected to insert.

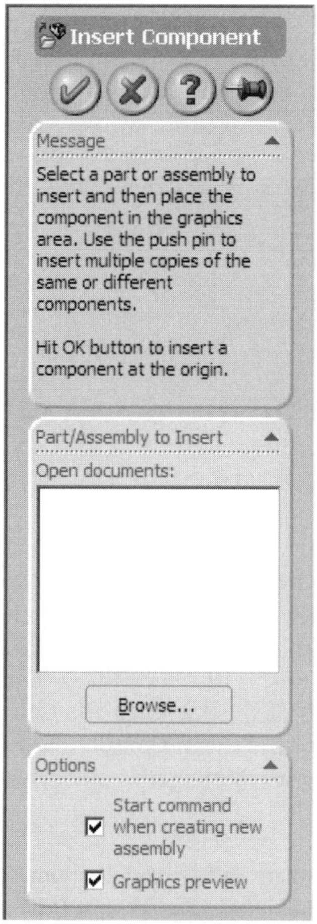

Figure 10-3 The ***Insert Component PropertyManager***

Tip. *If you need to place multiple components or multiple instances of the same component, choose the* ***Keep Visible*** *button from the* ***Insert Component PropertyManager*** *and select the placement points in the drawing area to place the multiple components. The* ***Keep Visible*** *button is available on the right of the* ***Help*** *button on top of the* ***Message*** *rollout.*

If some part documents are opened in the SolidWorks window and you start an assembly document, those parts will be *listed in the* ***Open documents*** *selection area. You can directly select the names of the parts from the* ***Open documents*** *selection area and place them in the drawing area.*

Starting an Assembly From Within the Part Document

Menu:	File > Make Assembly from Part
Toolbar:	Standard > Make Assembly from Part/Assembly

Tip. *When you insert a component in the assembly, it* will be *displayed in the* ***FeatureManager Design Tree****. The convention of naming the first component is (f)* ***Name of Component <1>****. In this convention (f) implies that the component is fixed. You cannot move a fixed component. You will learn more about fixed and floating components later in this chapter. After this the name of the component is displayed. After the name of the component,* ***<1>*** *symbol is displayed. This implies the serial number of the same component in the entire assembly.*

The (-) symbol implies that the component is floating and is under defined. You need to apply the required mates to the component to fully define it. You will learn more about assembly mates later in this chapter.

The (+) symbol implies that the component is over defined.

If no symbol appears before the name of the component, the component will be *fully defined.*

You can also start an assembly document from within the part document. If the part document of the base component of the assembly is opened, choose the **Make Assembly from Part/Assembly** button from the **Standard** toolbar or choose **File > Make Assembly from Part** from the menu bar. You can also use CTRL+A on the keyboard as the shortcut key. If the **New SolidWorks Document** dialog box is invoked, choose the **OK** button from this dialog box. An assembly document will be started and the **Insert Component PropertyManager** will be invoked. Choose the **OK** button from this **PropertyManager** to place the component at the origin.

Note

If the last time you invoked the ***New SolidWorks Document*** *dialog box in the novice mode, you just need to select the* ***OK*** *button from the* ***New SolidWorks Document*** *dialog box if it is displayed while creating an assembly from a part document. If you used the* ***New SolidWorks Document*** *in the advanced mode last time, you first need to select the assembly template and then choose the* ***OK*** *button. When you choose the* ***Make Assembly from Part/Assembly*** *button and if a* ***SolidWorks*** *warning box is displayed, choose* ***No*** *from the warning box. The* ***New SolidWorks Document*** *dialog box will be displayed in the advanced mode.*

Placing Components Using the Opened Document Window

Another most widely used method of placing components in the assembly is the use of currently opened part or assembly documents. For example, the assembly that you are going to create consists of three components. Open the part document of the part that you want to insert and then start a new assembly document. Close the **Insert Component PropertyManager**. Now, choose **Window > Tile Horizontally** or **Tile Vertically** from the menu bar. All SolidWorks document windows will be tiled vertically or horizontally, depending on the option selected. You first need to place the first component in the assembly and this needs to be placed at the origin of the assembly. If the origin is not displayed by default, choose **View > Origins** from the menu bar. Now, move the cursor on the name of the first component in the **FeatureManager Design Tree** of the window of that component. Press and hold the left mouse button down on the name and then drag the cursor to the assembly origin in the assembly window. When the

Tip. *Only the information about the mates is stored in the assembly file. The feature information of parts is stored in the individual part files. Therefore, the size of the assembly file is small.*

It is recommended that all parts of an assembly should be saved in the same folder in which the assembly is saved. This is because if the location of the parts is changed, the component will not be displayed in the assembly and assembly file will show errors.

coincident symbol appears below the component cursor, release the left mouse button to place the component in the assembly. Similarly, place the other components in the assembly. Figure 10-4 displays placing the first component from an opened document window to the assembly document window. If another existing assembly document is opened, you can also drag and drop the part from that assembly document.

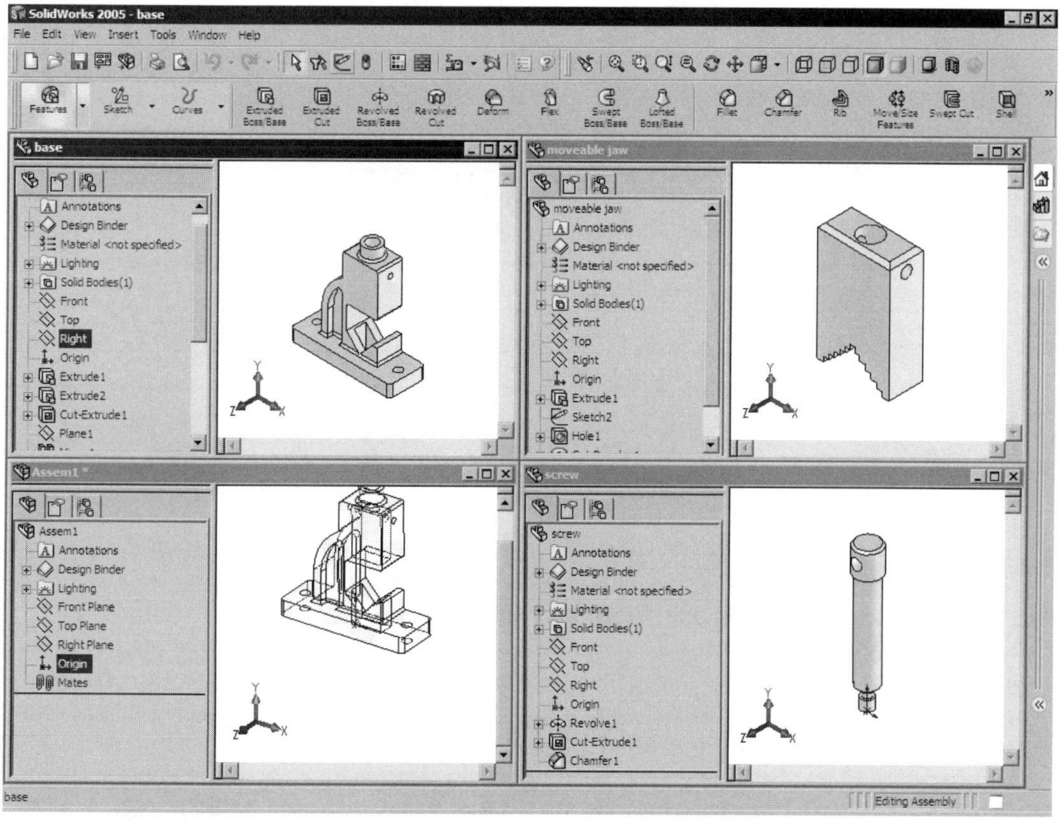

Figure 10-4 Placing a component in the assembly file from an existing window

Placing Components by Dragging From Windows Explorer

You can also place components in the assembly document by dragging them from Windows Explorer. Open Windows Explorer and browse the location where the part documents are saved. Tile the Windows Explorer window and the SolidWorks window such that you can view both windows. Move the cursor on the icon of the part document in Windows Explorer. Press and hold the left mouse button down and drag the cursor to the assembly document window. Release the left mouse button at the origin of the assembly document to coincide the origin of the part with that of the assembly document. Figure 10-5 shows the part being dropped in the assembly window.

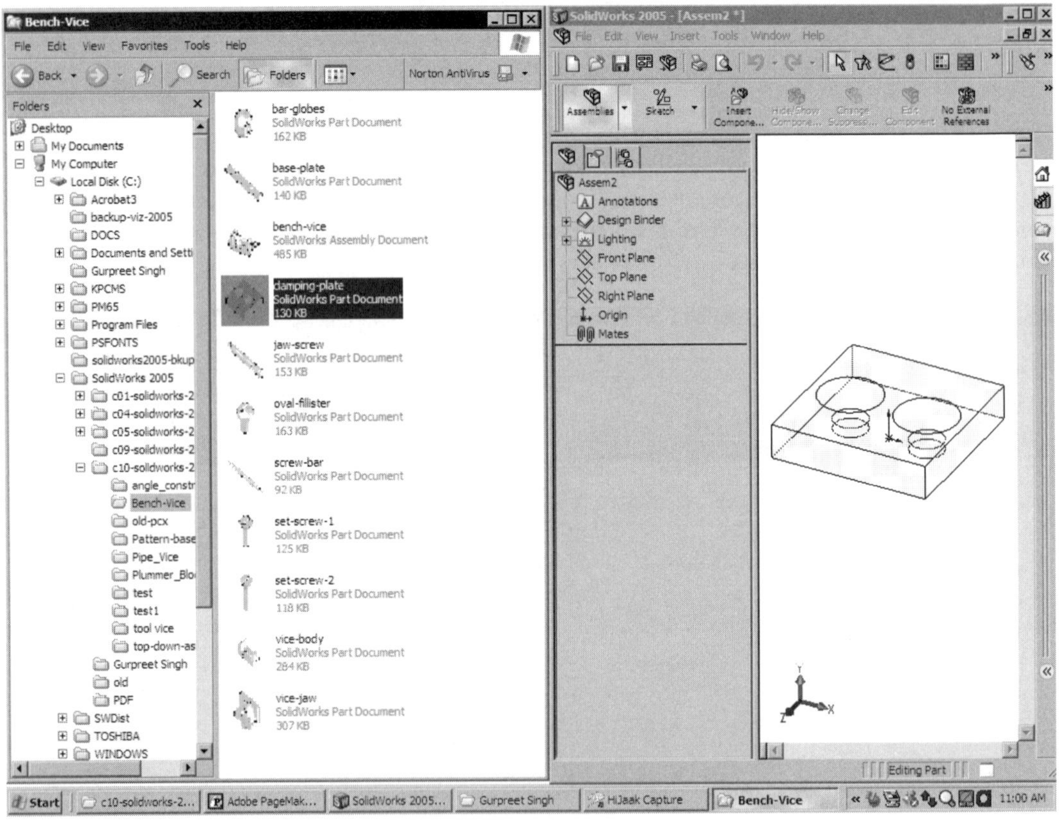

Figure 10-5 Dropping the part from Windows Explorer to the assembly window

Placing Components From Internet Explorer

You can also place the components from Internet Explorer. You need Internet Explorer 4.0 or later for that. Browse the location of the SolidWorks part file link on the Web. Drag the hyperlink and drop it in the drawing area of the assembly document. The **Save As** dialog box will be displayed; save the part document at the desired location.

Placing Additional Instances of Existing Component in the Assembly

Sometimes you need more than one instance of the component to be placed in the assembly document. To do so, press and hold the CTRL key down on the keyboard. Now, select the component in the assembly file and drag the cursor to the location where you want to place the instance of the selected component. Release the left mouse button to drop the new instance of the component. Similarly, you can place as many copies of the existing component as you want by following the above-mentioned procedure.

Placing the Component Using the Feature Palette Window

You can also place a component in the assembly file from the **Feature Palette** window. If you have saved the component in the **Feature Palette** window, you can place that component by dragging and dropping from this window. The **Feature Palette** window is invoked by choosing **Tools > Feature Palette** from the menu bar.

Assembling the Components

After placing the components in the assembly document, you need to assemble them. By assembling the components, you will constrain the degrees of freedom of the components. As mentioned earlier, the components are assembled using mates. Mates help you precisely place and position the component with respect to the other components and the surroundings in the assembly. You can also define the linear and rotatory movement of the component with respect to the other components. In addition, you can create a dynamic mechanism and check the stability of the mechanism by precisely defining the combination of mates. There are two methods of adding mates to the assembly. The first method is using the **Mate PropertyManager** and the second and the most widely used method of adding mates to the assembly is using the **Smart Mates**. Both methods are discussed next.

Assembling Components Using the Mate PropertyManager

CommandManager:	Assemblies > Mate
Menu:	Insert > Mate
Toolbar:	Assembly > Mate

In SolidWorks, mates can be applied using the **Mate PropertyManager**. Choose the **Mate** button available in the **Assemblies CommandManager** or choose **Insert > Mate** from the menu bar. The **Mate PropertyManager** will be invoked, as shown in Figure 10-6.

Select a planar face, curved face, axis, or point on the first component and then select the entity from the second component. The selected entities will be highlighted in green. The names of the selected entities will be displayed in the **Entities to Mate** selection area of the **Mate Selections** rollout. The **Mate** pop-up toolbar will be invoked, as shown in Figure 10-7. The most suitable mates to be applied to the current selection set are displayed in the **Mate** pop-up menu toolbar and in the **Standard Mates** rollout of the **Mate PropertyManager**. The most appropriate mate is selected by default. The preview of the assembly using most appropriate mate is displayed in the drawing area. You can also select the mates from the given list of the mates that can be applied to the current selection set. As you select some other mate from the **Mate** pop-up toolbar the preview of the assembly will be displayed using the newly selected mate. Now, choose the **Add/Finish Mate** button from the **Mate** pop-up toolbar; the **Mate PropertyManager** will be still displayed, and you can add other mates to the assembly. After adding all mates choose the **OK** button from the **Mate PropertyManager**. Various types of mates that can be applied are discussed next.

Coincident

The **Coincident** mate is applied to make two planar faces coplanar. But you can also apply the **Coincident** mate to other entities. The details of the geometries on which

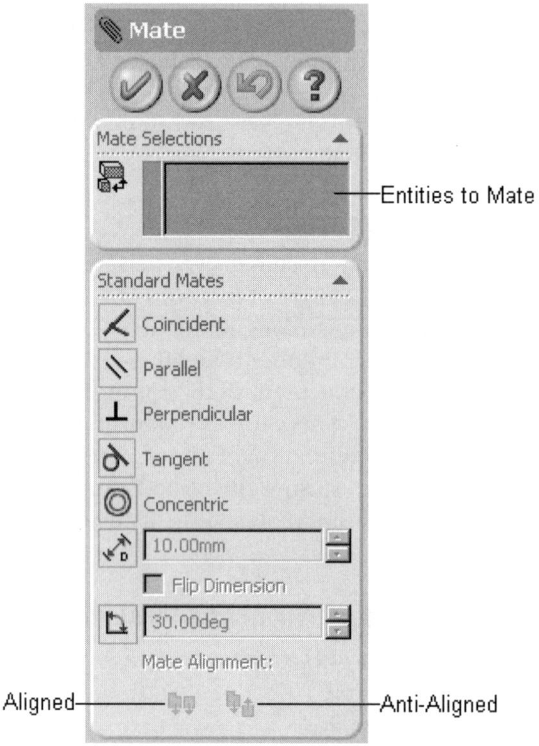

Figure 10-6 *The* ***Mate PropertyManager***

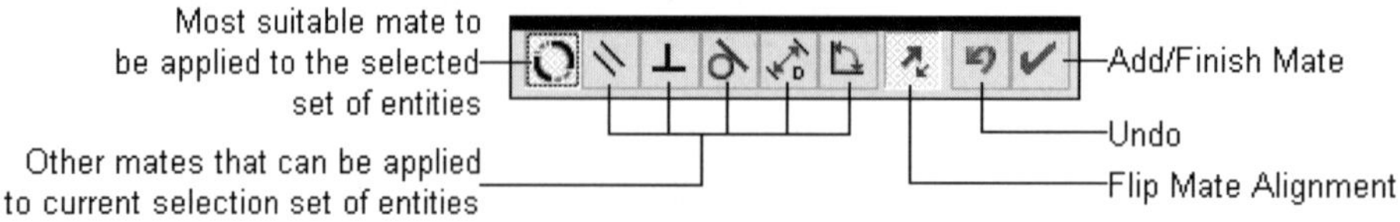

Figure 10-7 *The* ***Mate*** *pop-up toolbar*

the **Coincident** mate can be applied are shown in Figure 10-8.

When you choose the **Coincident** button from the **Mate** pop-up toolbar, the preview of the model will be displayed according to the current selection of the mate and the model will be assembled in aligned or anti-aligned direction, depending on the current orientation of the model. Depending on the preview of the assembled component the **Aligned** or **Anti-Aligned** button is selected in the **Standard Mates** rollout. You can choose the alignment of the assembled component by choosing the **Flip Mate Alignment** button from the **Mate** pop-up toolbar or switch between **Aligned** and **Anti-Aligned** buttons in the **Mate PropertyManager.** Figure 10-9 displays the faces to be selected to apply the

ASSEMBLY MATE COMBINATIONS (USING COINCIDENT MATE)		Second Component									
		Cone	Cylinder	Line	Point	Sphere	Circular/ Arc Edge	Extrusion	Surface	Plane	Cam
First Component	Cylinder	✗	✗	✓	✓	✗	✓	✗	✗	✗	✗
	Sphere	✗	✗	✗	✓	✗	✗	✗	✗	✗	✗
	Cone	✓	✗	✗	✓	✗	✓	✗	✗	✗	✗
	Circular/ Arc Edge	✗	✓	✗	✗	✗	✓	✗	✗	✓	✗
	Line	✗	✓	✓	✓	✗	✗	✗	✗	✓	✗
	Point	✗	✓	✓	✓	✓	✗	✓	✓	✓	✓
	Extrusion	✗	✗	✗	✓	✗	✗	✗	✗	✗	✗
	Surface	✗	✗	✗	✓	✗	✗	✗	✗	✗	✗
	Plane	✗	✗	✓	✓	✗	✓	✗	✗	✓	✗

Figure 10-8 *Table displaying the combinations for applying the* ***Coincident*** *mate*

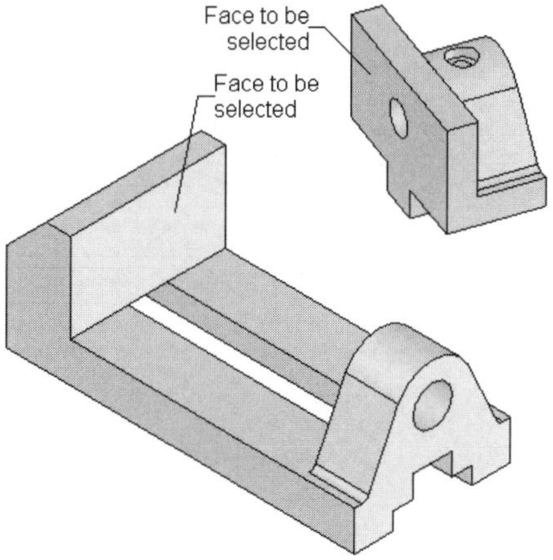

Figure 10-9 *Faces to be selected*

coincident mate. Figure 10-10 shows the resulting mate applied with the default option, that is, **Anti-Aligned** selected. Figure 10-11 shows the coincident mate applied with the **Aligned** button selected.

Concentric

The **Concentric** mate is generally used to align the central axis of one component with that of the other. You generally need to select the circular faces to apply the concentric mate. You can also apply the **Concentric** mate between a point and a circular face or edge. The other combinations of applying the **Concentric** mate are displayed in the table given in Figure 10-12. For applying a **Concentric** mate, invoke the **Mate PropertyManager**. Select two entities from two different components; the names of the selected entities will be displayed in the **Entities to Mate** selection area. The **Concentric** button will be chosen in the **Mate** pop-up toolbar. If this button is not chosen by default, you

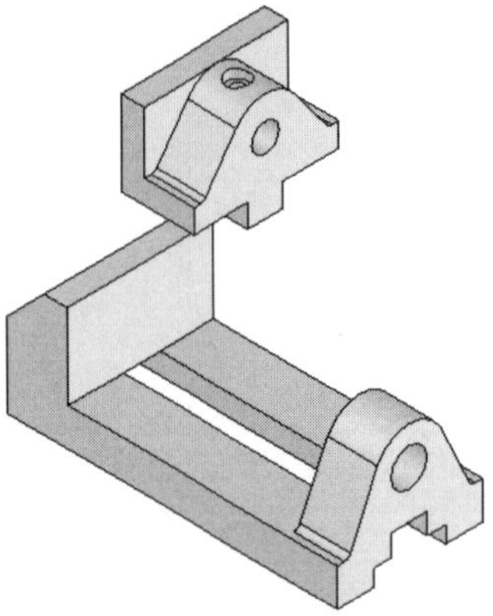

Figure 10-10 ***Coincident*** *mate applied with the* ***Anti-Aligned*** *button selected*

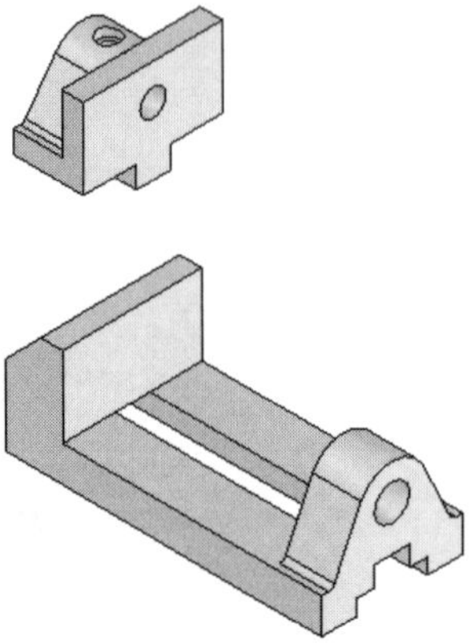

Figure 10-11 ***Coincident*** *mate applied with the* ***Aligned*** *button selected*

ASSEMBLY MATE COMBINATIONS (USING CONCENTRIC MATE)		Second Component									
		Cone	Cylinder	Line	Point	Sphere	Circular/ Arc Edge	Extrusion	Surface	Plane	Cam
First Component	Cylinder	✓	✓	✓	✓	✓	✓	X	X	X	X
	Sphere	X	✓	✓	✓	✓	X	X	X	X	X
	Cone	✓	✓	✓	✓	X	✓	X	X	X	X
	Circular/ Arc Edge	X	✓	✓	X	X	✓	X	X	X	X
	Line	✓	✓	X	X	✓	✓	X	X	X	X
	Point	✓	✓	X	X	✓	X	X	X	X	X
	Extrusion	X	X	X	X	X	X	X	X	X	X
	Surface	X	X	X	X	X	X	X	X	X	X
	Plane	X	X	X	X	X	X	X	X	X	X

Figure 10-12 *Table displaying the combinations for applying the* ***Concentric*** *mate*

need to manually choose it. The preview of the models being assembled after applying this mate will be displayed in the graphics area. You can choose the **Aligned** and **Anti-Aligned** buttons on the basis of the design requirement. Choose the **Add/Finish Mate** button from the **Mate** pop-up toolbar.

Figure 10-13 shows the faces to be selected to apply the concentric mate. Figure 10-14 shows the **Concentric** mate applied with the **Aligned** button chosen and Figure 10-15 shows the **Concentric** mate applied with the **Anti-Aligned** button chosen.

Distance

The **Distance** button is chosen to apply the **Distance** mate between two components. To apply this mate, invoke the **Mate PropertyManager** and select the entities from both components. Choose the **Distance** button from the **Mate** pop-up toolbar; the

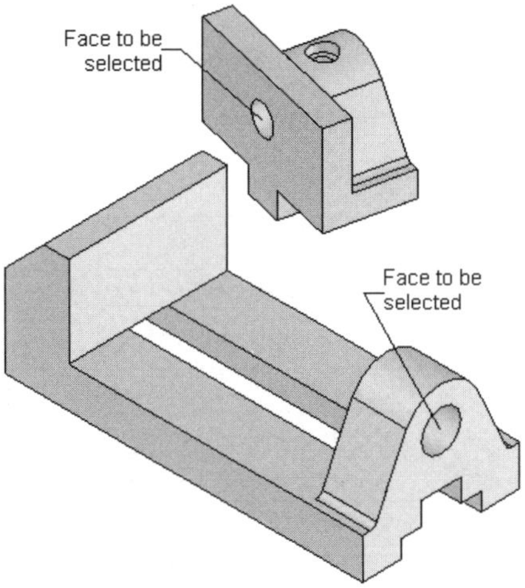

Figure 10-13 *Faces to be selected*

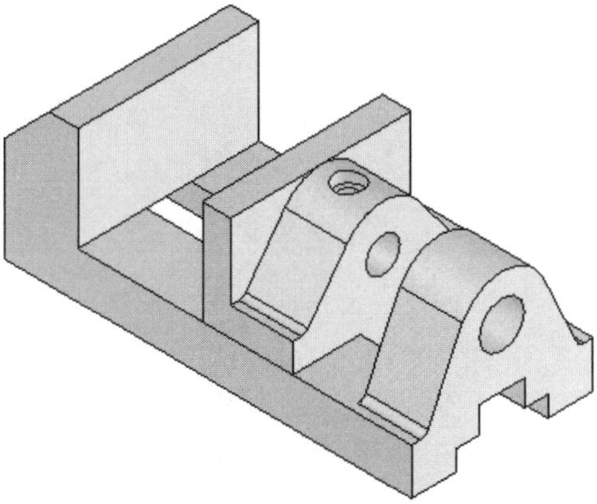

Figure 10-14 ***Concentric*** *mate applied with the* ***Aligned*** *button selected*

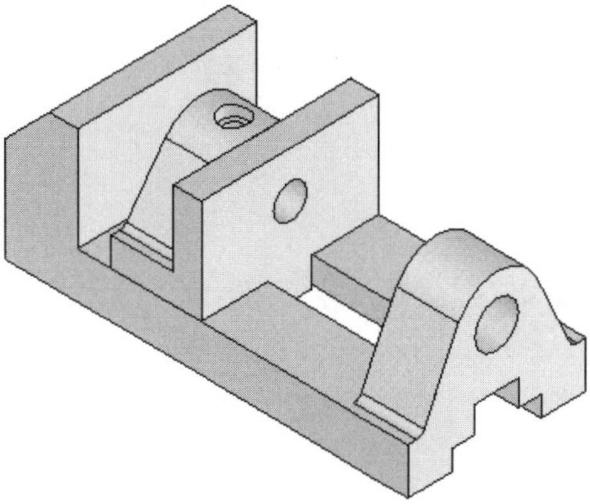

Figure 10-15 ***Concentric*** *mate applied with the* ***Anti-Aligned*** *button selected*

Distance spinner will be displayed in the **Mate** pop-up toolbar. Set the value of the distance in the **Distance** spinner. The preview of the assembly will be updated automatically after you set the value of the distance. Using the **Flip Dimension** button available on the left of the **Distance** spinner, you can specify a negative distance value. If needed, you can choose the **Aligned** button or the **Anti-Aligned** button. Figure 10-16 shows the combinations of components to apply the **Distance** mate. Figure 10-17 shows the faces to be selected and Figure 10-18 shows the **Distance** mate applied between two components.

Angle

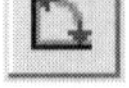

The **Angle** button is used to apply the **Angle** mate between two components. This mate is used to specify the angular position between the selected plane, planar face, or edges of the two components. To apply this mate, invoke the **Mate PropertyManager** and select the entities from the two components. Choose the **Angle** button from the **Mate** pop-up toolbar. The preview of the models will be modified with the

ASSEMBLY MATE COMBINATIONS (USING DISTANCE MATE)		Second Component									
		Cone	Cylinder	Line	Point	Sphere	Plane	Extrusion	Surface	Circular/ Arc Edge	Cam
First Component	Cylinder	✗	✓	✓	✓	✗	✓	✗	✗	✗	✗
	Sphere	✗	✗	✓	✓	✓	✗	✗	✗	✗	✗
	Cone	✓	✗	✗	✗	✗	✗	✗	✗	✗	✗
	Plane	✗	✓	✓	✓	✓	✓	✗	✗	✗	✗
	Line	✗	✓	✓	✓	✓	✓	✗	✗	✗	✗
	Point	✗	✓	✓	✓	✓	✓	✗	✗	✗	✗
	Extrusion	✗	✗	✗	✗	✗	✗	✗	✗	✗	✗
	Surface	✗	✗	✗	✗	✗	✗	✗	✗	✗	✗
	Circular/ Arc Edge	✗	✗	✗	✗	✗	✗	✗	✗	✗	✗

Figure 10-16 *Table displaying the combinations for applying the* ***Distance*** *mate*

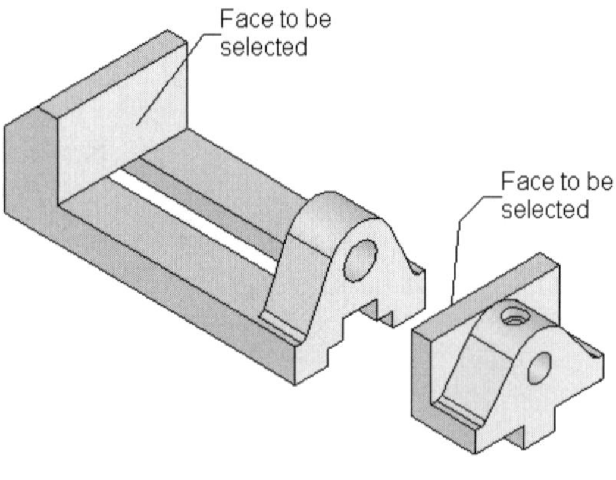

Figure 10-17 *Selecting the faces to apply the* ***Distance*** *mate*

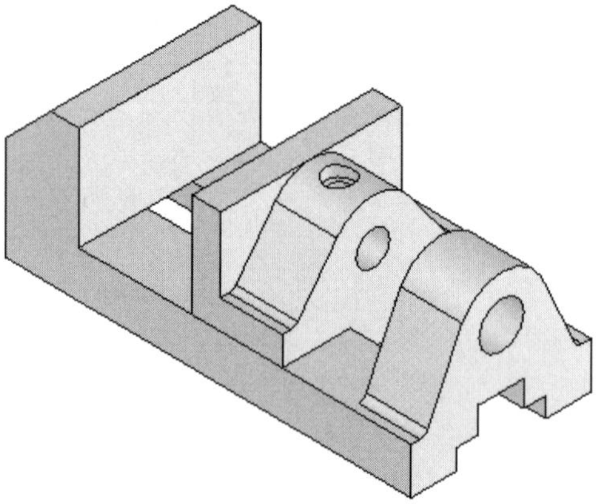

Figure 10-18 ***Distance*** *mate applied to the selected faces*

default value of the angle. Also, the **Angle** spinner will be invoked and you can set the value of the angle in this spinner. You can also change the angle direction using the **Flip Dimension** button provided on the right of the angle spinner. You can choose the **Aligned** button and the **Anti-Aligned** button on the basis of the design requirement. After adding the mate, choose the **Add/Finish Mate** button from the **Mate** pop-up toolbar. Figure 10-19 shows the table of combinations for applying the **Angle** mate.

Figure 10-20 shows the faces to be selected to apply the **Angular** mate and Figure 10-21 shows the assembly after applying the **Angular** mate with the angle value of 90-degrees.

Parallel

The **Parallel** button available in the **Mate** pop-up toolbar is used to apply the **Parallel** mate between two components. To apply the **Parallel** mate, invoke the **Mate**

ASSEMBLY MATE COMBINATIONS (USING ANGLE MATE)		Second Component									
		Cylinder	Extrusion	Line	Plane	Sphere	Circular/ Arc Edge	Cone	Surface	Point	Cam
First Component	Cylinder	✓	✓	✓	X	X	X	X	X	X	X
	Extrusion	✓	✓	✓	X	X	X	X	X	X	X
	Line	✓	✓	✓	X	X	X	X	X	X	X
	Plane	X	X	X	✓	X	X	X	X	X	X
	Sphere	X	X	X	X	X	X	X	X	X	X
	Circular/ Arc Edge	X	X	X	X	X	X	X	X	X	X
	Cone	✓	✓	✓	X	X	X	✓	X	X	X
	Point	X	X	X	X	X	X	X	X	X	X
	Surface	X	X	X	X	X	X	X	X	X	X

Figure 10-19 *Table displaying the combinations for applying the* ***Angle*** *mate*

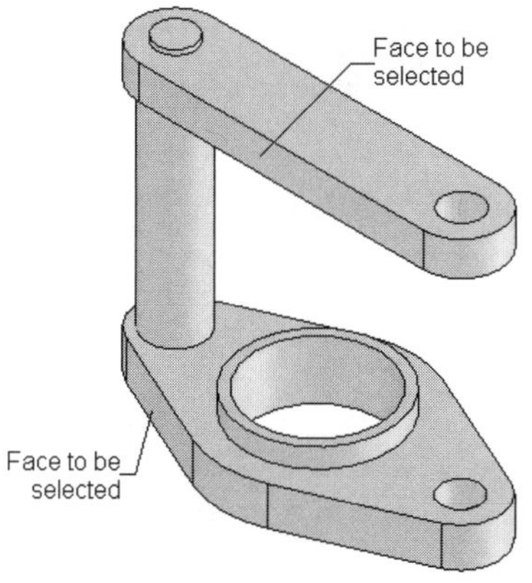

Figure 10-20 *Faces selected to apply the* ***Angle*** *mate*

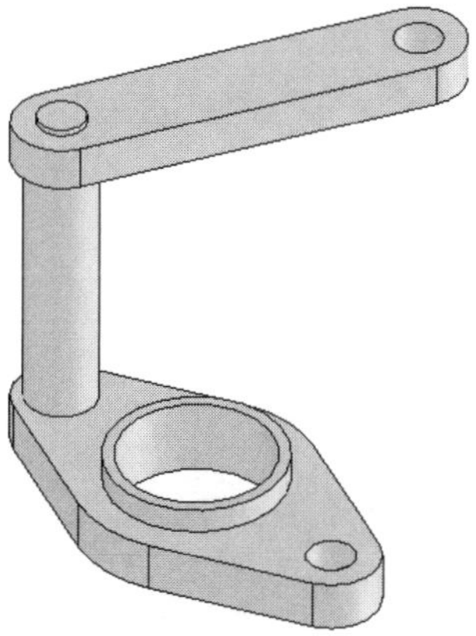

Figure 10-21 *Assembly after applying the* ***Angle*** *mate*

PropertyManager and select the two entities from the two components. Choose the **Parallel** button from the **Mate** pop-up menu to apply the mate. You can also choose the **Align** button and the **Anti-Aligned** button. After applying the mate, choose the **Add/Finish Mate** button from the **Mate** pop-up toolbar. Figure 10-22 shows the combinations to components on which you can apply the **Parallel** mate. Figure 10-23 shows the entities to be selected to apply the **Parallel** mate and Figure 10-24 shows the assembly after applying the **Parallel** mate.

Perpendicular

The **Perpendicular** button available in the **Mate Settings** rollout is used to apply the **Perpendicular** mate between two components. Invoke the **Mate PropertyManager** and select two entities from the two components. Choose the **Perpendicular** button from the **Mate** pop-up toolbar. You can also toggle between the **Aligned**

ASSEMBLY MATE COMBINATIONS (USING PARALLEL MATE)		Second Component									
		Cylinder	Extrusion	Line	Plane	Sphere	Circular/ Arc Edge	Cone	Surface	Point	Cam
First Component	Cylinder	✓	✓	✓	✗	✗	✗	✗	✗	✗	✗
	Extrusion	✓	✓	✓	✗	✗	✗	✗	✗	✗	✗
	Line	✓	✓	✓	✓	✗	✗	✗	✗	✗	✗
	Plane	✗	✗	✓	✓	✗	✗	✗	✗	✗	✗
	Circular/ Arc Edge	✗	✗	✗	✗	✗	✗	✗	✗	✗	✗
	Sphere	✗	✗	✗	✗	✗	✗	✗	✗	✗	✗
	Cone	✓	✓	✓	✗	✗	✗	✓	✗	✗	✗
	Surface	✗	✗	✗	✗	✗	✗	✗	✗	✗	✗

Figure 10-22 *Table displaying the combinations for applying the* ***Parallel*** *mate*

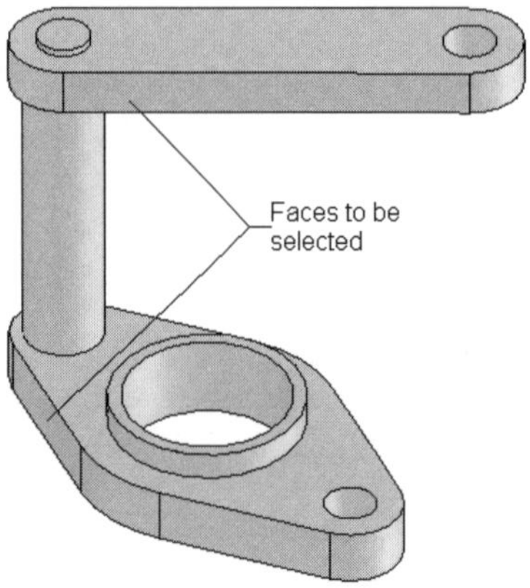

Figure 10-23 *Faces selected to apply the* ***Parallel*** *mate*

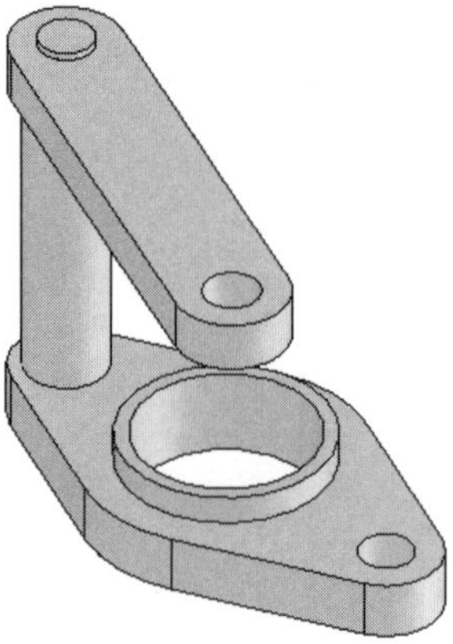

Figure 10-24 *Assembly after applying the* ***Parallel*** *mate*

button and the **Anti-Aligned** button. Choose the **Add/Finish Mate** button from the **Mate** pop-up toolbar. Figure 10-25 shows the table displaying the combinations for applying the **Perpendicular** mate. Figure 10-26 shows the entities to be selected and Figure 10-27 shows the **Perpendicular** mate applied to the assembly.

ASSEMBLY MATE COMBINATIONS (USING PERPENDICULAR MATE)		Second Component									
		Cylinder	Extrusion	Line	Plane	Sphere	Circular/ Arc Edge	Cone	Surface	Point	Cam
First Component	Cylinder	✓	✓	✓	X	X	X	X	X	X	X
	Extrusion	✓	✓	✓	X	X	X	X	X	X	X
	Line	✓	✓	✓	✓	X	X	X	X	X	X
	Plane	X	X	✓	✓	X	X	X	X	X	X
	Circular/ Arc Edge	X	X	X	X	X	X	X	X	X	X
	Sphere	X	X	X	X	X	X	X	X	X	X
	Cone	✓	✓	✓	X	X	X	✓	X	X	X
	Surface	X	X	X	X	X	X	X	X	X	X

Figure 10-25 *Table displaying the combinations for applying the* ***Perpendicular*** *mate*

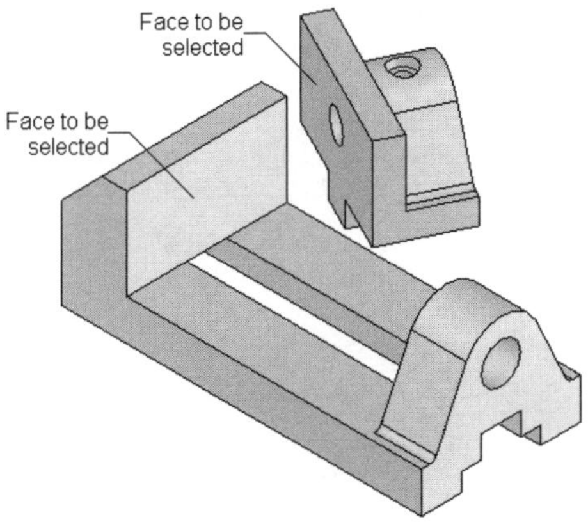

Figure 10-26 *Faces selected to apply the* ***Perpendicular*** *mate*

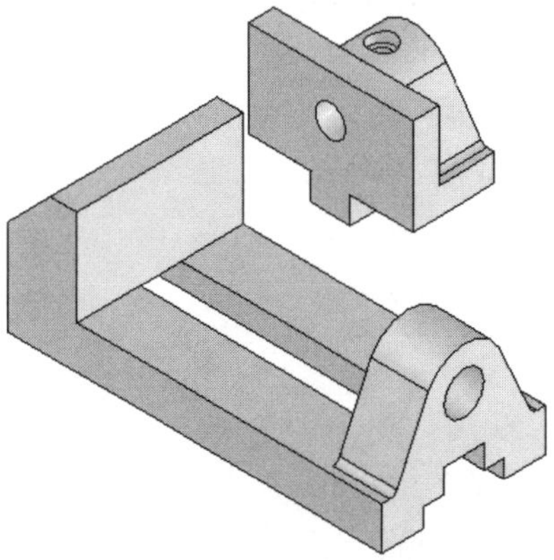

Figure 10-27 *Assembly after applying the* ***Perpendicular*** *mate*

Tangent

The **Tangent** button available in the **Mate** pop-up toolbar is used to apply the **Tangent** mate between two components. Figure 10-28 shows the table displaying the combinations for applying the tangent mate. Figure 10-29 shows the entities to be selected to apply the **Tangent** mate and Figure 10-30 shows the **Tangent** mate applied to the assembly.

Note

The ***Advanced Mates*** *are discussed in Chapter 11.*

ASSEMBLY MATE COMBINATIONS (USING TANGENT MATE)		Second Component								
		Cone	Cylinder	Line	Point	Sphere	Plane	Surface	Cam	Extrusion
First Component	Cylinder	✓	✓	✓	✗	✓	✓	✓	✓	✓
	Sphere	✓	✓	✓	✗	✓	✓	✗	✗	✗
	Cone	✗	✗	✗	✗	✓	✓	✗	✗	✗
	Plane	✗	✓	✗	✗	✓	✗	✓	✓	✓
	Line	✗	✓	✓	✓	✓	✓	✓	✓	✓
	Extrusion	✗	✓	✗	✗	✗	✓	✗	✗	✗
	Surface	✗	✓	✗	✗	✗	✓	✗	✗	✗
	Cam	✗	✗	✗	✗	✗	✗	✗	✗	✗

Figure 10-28 *Table displaying the combinations for applying the* ***Tangent*** *mate*

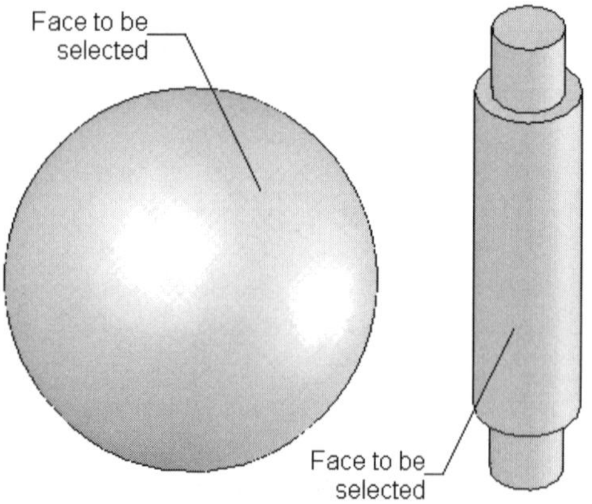

Figure 10-29 *Entities to be selected to apply the* ***Tangent*** *mate*

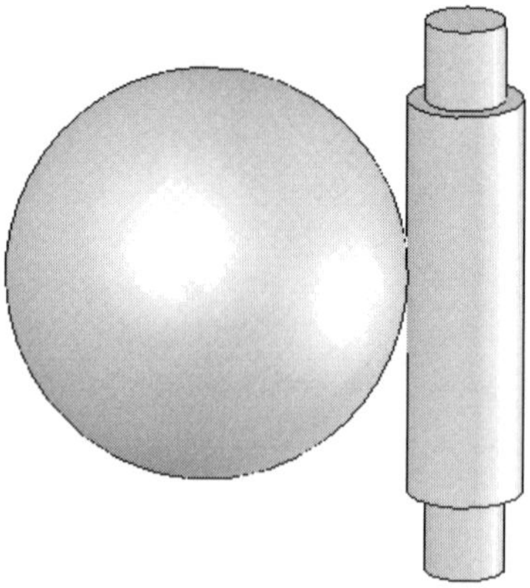

Figure 10-30 *Assembly after applying the* ***Tangent*** *mate*

The options available in the **Mate PropertyManager** are discussed next.

Mate

The **Mate** rollout is used to display the mates that are applied between the selected entities.

Options

The **Add to new folder** check box available in this rollout is used to add the currently applied mate in a folder. This folder is placed in the **Mates** mategroup in the **FeatureManager Design Tree**. You can also drag and drop the other mates from the **Mates** mategroup in the newly created folder. The **Show popup dialog** check box is selected by default and is used to display the **Mate** pop-up toolbar. The **Show preview** check box is selected by default and is used to display the dynamic preview of the assembly as you apply the mates to the components. The **Use for positioning only** check box is used only to define the position of the component by applying the mate. This mate is displayed in the **Mate** rollout, but when you exit the **Mate PropertyManager**, this mate will not be displayed in the **Mates** mategroup in the **FeatureManager Design Tree**.

Assembling Components Using the Smart Mates

Smart Mates is the most attractive feature of the assembly design environment of SolidWorks. The **Smart Mates** technology speeds up the design process in the assembly environment of SolidWorks. To add smart mates to the components, choose the **Move Component** button from the **Assembly CommandManager**; the **Move Component PropertyManager** will be displayed. Now, choose the **SmartMates** button available in the **Move** rollout; the **Move Component PropertyManager** will be replaced by the **SmartMates PropertyManager,** as shown in Figure 10-31. Also, the select cursor will be replaced by the move cursor.

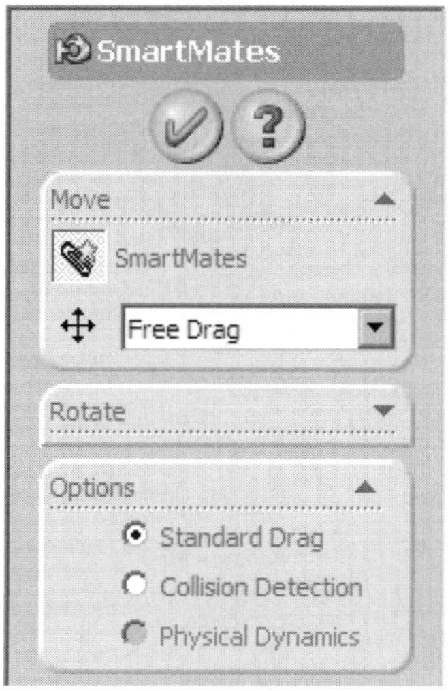

Figure 10-31** The **SmartMates PropertyManager

Next, double-click on the entity of the first component to add a mate; the component will appear transparent and the cursor will be replaced by the smart mates cursor. Press and hold the left mouse button down on the selected entity and drag the cursor to the entity with which you want to mate the previously selected entity. The symbol of the constraint that can be applied between the two entities will be displayed below the smart mates cursor. You can use the TAB key to toggle between the aligned and anti-aligned options while applying **Smart Mates**. When the symbol of the mate is displayed below the cursor, release the left mouse button; the **Mate** pop-up toolbar will be displayed. Choose the **Add/Finish Mate** button from this toolbar. Figure 10-32 shows the face to be selected to apply a smart mate and Figure 10-33 shows the component being dragged.

Figure 10-34 shows that the concentric symbol appears below the smart mates cursor when the cursor is placed near a circular face of the other component.

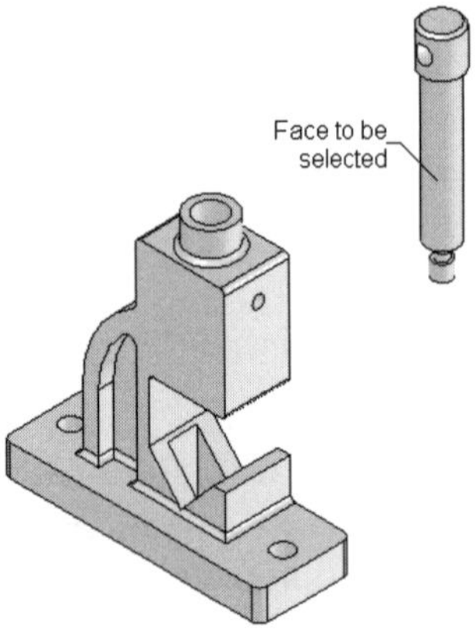

Figure 10-32 Face to be selected to apply a smart mate

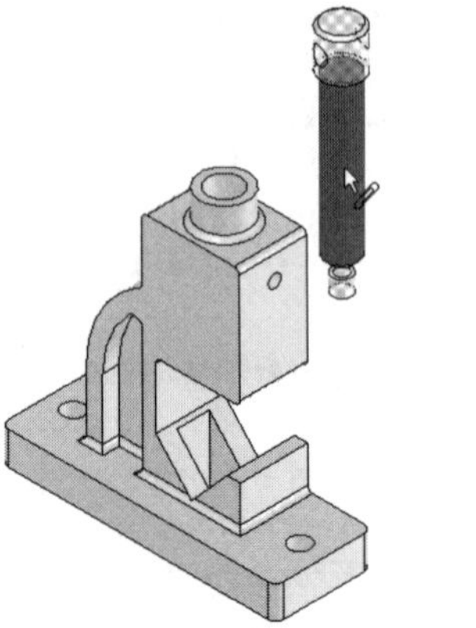

Figure 10-33 Component being dragged to apply a smart mate

Figure 10-34 Concentric symbol appears below the smart mates cursor

Figure 10-35 shows a planar face selected to apply a smart mate. You can use the TAB key to toggle between the aligned and anti-aligned mates. Figure 10-36 shows that the coincident symbol appears below the cursor and Figure 10-37 shows the assembly after applying the **Coincident** mate using the smart mate.

Note

To apply smart mates without invoking the ***SmartMates*** *tool, press and hold the ALT key down and select the entity of the first component. Drag the component to the entity of the other component to which you need to apply the mate. As you release the left mouse button, the* ***Mate*** *pop-up toolbar will be displayed. Choose the* ***Add/Finish Mate*** *button from this toolbar.*

When you drag a component for applying a smart mate, the selected entity of the first component will snap all corresponding entities of the second component. You can press the ALT key on the keyboard to exit the snap. To enter the snap mode again press the ALT key on the key board.

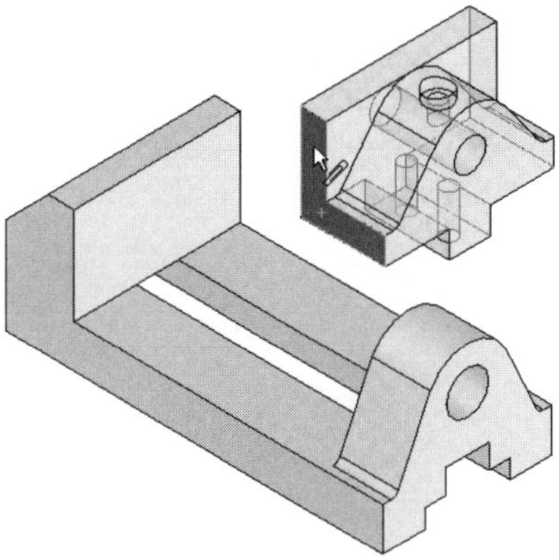

Figure 10-35 Planar face to be selected to apply a mate using smart mates

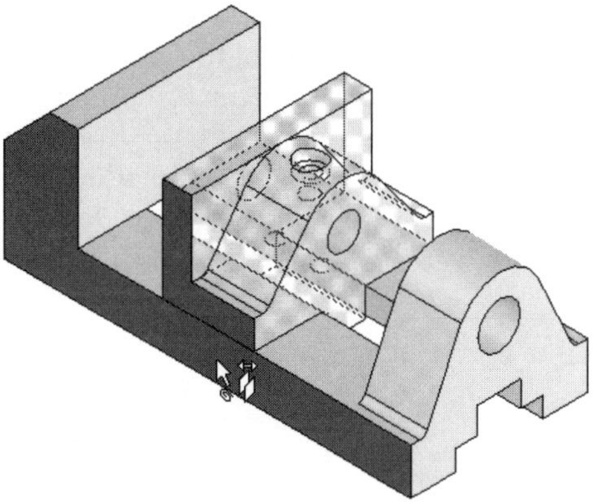

Figure 10-36 Coincident symbol appears after dragging the component near another planar face

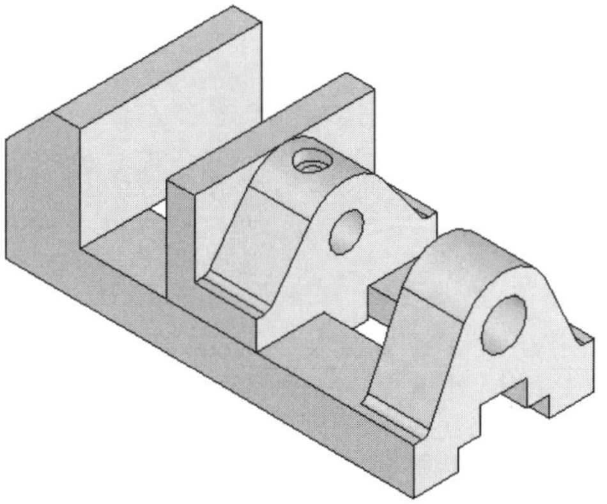

***Figure 10-37 Coincident** mate applied to the assembly using smart mates*

Tip. *You can also add smart mates without dragging the component. To add a smart mate without dragging, invoke the* ***Smart Mates*** *tool. Double-click on the entity from the first component. The component will be displayed in transparent. Now, select the entity from the second component. The preview with the most appropriate mate* will be *displayed and the* ***Mate*** *pop-up toolbar is invoked.*

Geometry-based Mates

In the assembly design environment of SolidWorks, you can also add geometry-based mates. Geometry-based mates are also a type of smart mate and are applied while you are placing a component in the assembly environment. Consider a case in which the first component is already placed in the assembly environment. Now, open the part document of the second component. Choose **Window** > **Tile Horizontal** or **Tile Vertical** from the menu bar.

Suppose you need to insert the revolved feature of the second component in the circular slot of the first component and at the same time you need to align the larger bottom face of the second component with the upper face of the first component. To do so, press and hold the left mouse button down on the edge of the second component, as shown in Figure 10-38. Drag the cursor to the assembly window near the upper edge of the circular slot of the first component; the second component mated with the first component will be displayed in temporary graphics, as shown in Figure 10-39. The coincident symbol will also be displayed below the smart mates cursor.

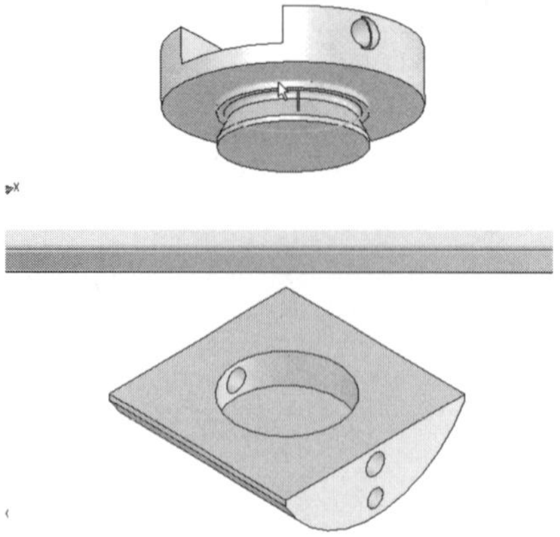

Figure 10-38 *Edge of the second component to be selected.*

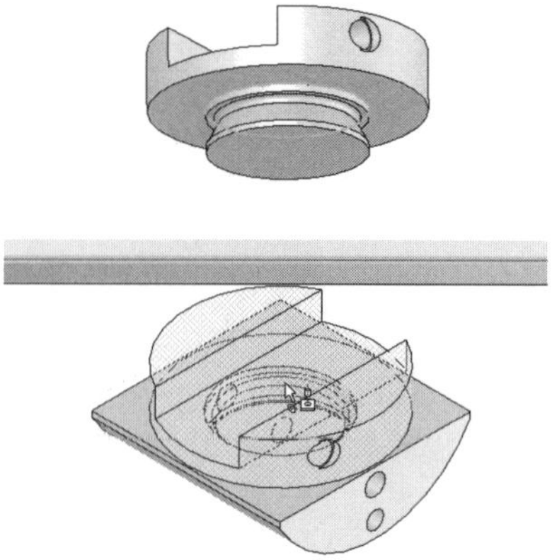

Figure 10-39 *Component being dragged into the assembly window for applying geometry-based mates*

You can also toggle the direction of placement of the component. Figure 10-40 shows that the direction of placement is flipped using the TAB key. Again press the TAB key to return to the default direction. Release the left mouse button to place the component. If you expand the **Mates** option from the **FeatureManager Design Tree**, you will notice that two mates are applied to the assembly; one is the coincident mate and the other is the concentric mate. Figure 10-41 shows the assembly after adding the geometry-based smart mates.

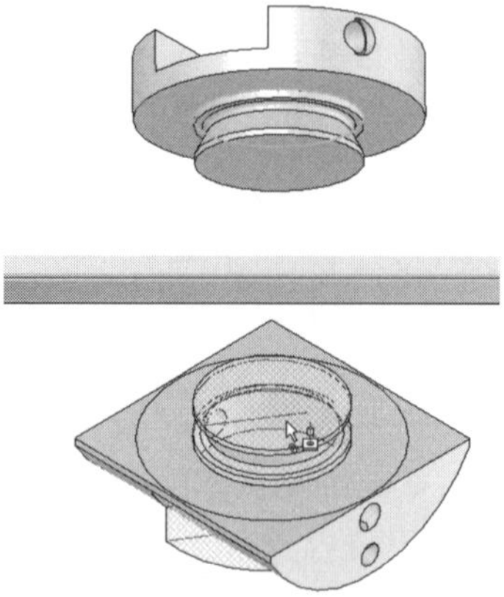

Figure 10-40 *The placement direction of component is flipped using the TAB key*

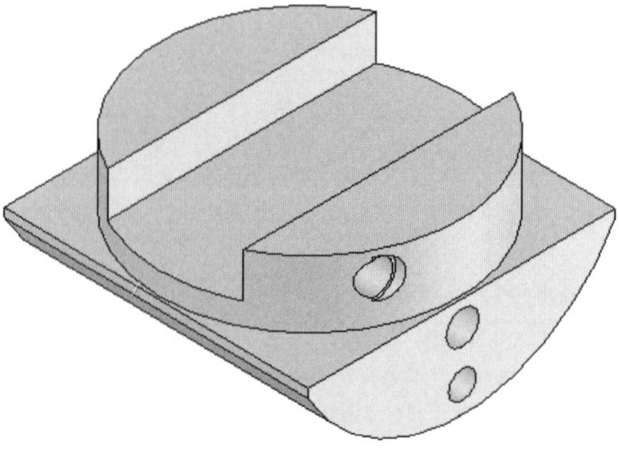

Figure 10-41 *Assembly after adding a geometry-based mate*

Tip. *You can add the geometry-based mates between two linear edges, two planar faces, two vertices, two conical faces, two axes, between an axis and a conical face, and between two circular edges.*

Feature-based Mates

In the assembly design environment of SolidWorks, you can also add feature-based mates. For adding feature-based mates, one of the features of the first component must be a circular base or boss feature and the second component must have a hole or a circular cut feature. The feature can be an extruded or a revolved feature. Also, in the assembly document, one of the parts must be placed earlier. Open the part document of the component to be placed using feature-based mates and tile both document windows. In the **FeatureManager Design Tree** of the part document, select the extruded or revolved feature and drag it to the assembly window. Place the cursor at a location where you need to place the component. You can also change the alignment or direction of placement using the TAB key. Release the left mouse button to drop the component. Figure 10-42 shows the component being dragged by selecting the revolved feature from the **FeatureManager Design Tree** of the part document. Figure 10-43 shows the resulting component assembled using feature-based mates.

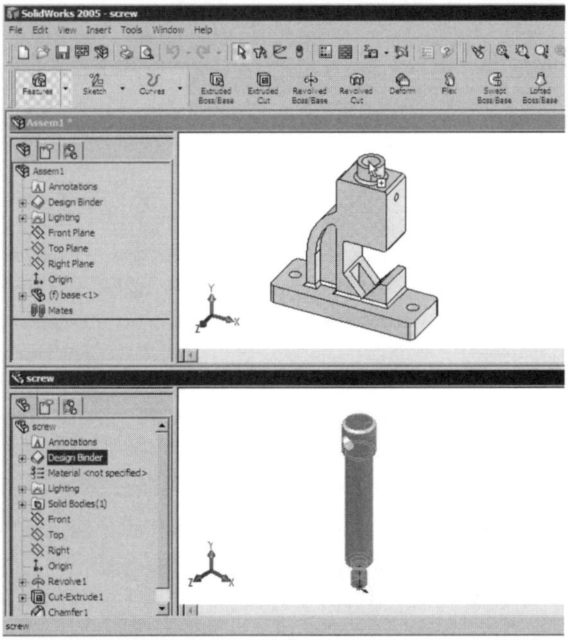

***Figure 10-42** Component being dragged by selecting the revolved feature*

Pattern-based Mates

Pattern-based mates are used to assemble the components that have a circular pattern created on the circular feature. The best example for these type of components is the flange or a shaft coupling. Remember that to create pattern-based mates, all components that will be assembled must have a circular pattern on the mating faces. To create pattern-based mates, select the outer edge of the

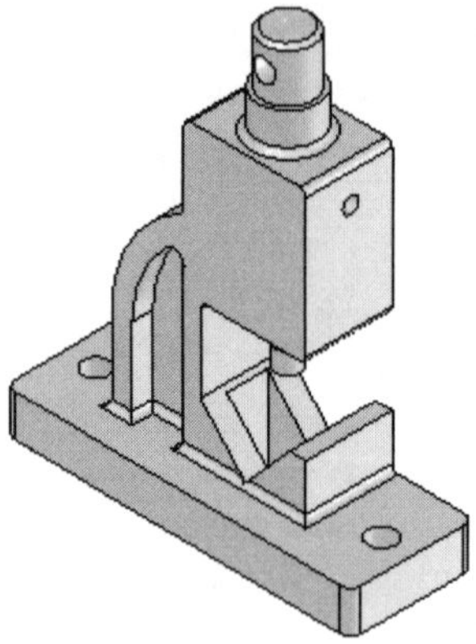

Figure 10-43 Component assembled using feature-based mates

Tip. *The feature-based mates are applied only to the components having cylindrical or conical features. You cannot add feature-based mates using features other than cylindrical or conical geometry. If you are adding feature-based mates to a component having a conical face, the second component also must have a conical face. You cannot add a feature-based mate if the geometry of the feature of one component is cylindrical and that of the other is conical.*

If you are adding feature-based mates using the features having conical geometry, there must be a planar face adjacent to the conical face of both features.

second component and drag it to the circular edge of the first component that is already placed in the assembly document. The preview of the component assembled with the first component will be displayed. Using the TAB key, you can switch the part with respect to the pattern instances. Release the left mouse button to drop the part. Figure 10-44 shows the component being dragged to the assembly document window. Figure 10-45 shows the preview of the component being assembled and Figure 10-46 shows the component assembled using pattern-based mates.

Assembling Components Using Mate Reference

CommandManager:	Reference Geometry > Mate Reference	*(Customize to Add)*
Menu:	Insert > Geometry > Mate Reference	
Toolbar:	Reference Geometry > Mate Reference	*(Customize to Add)*

In SolidWorks, you can define the mate reference to the part in the **Part** mode or in the **Assembly** mode. The mate references allow you to define the mating references such as planar surfaces, axis, edges, and so on before assembling the component.

It is recommended that the mate reference should be added to the component in the **Part** mode. The mate references are created using the **Mate Reference PropertyManager**, which is invoked by choosing the **Mate Reference** button from the **Reference Geometry Command** after customizing it. The **Mate Reference PropertyManager** is shown in Figure 10-47.

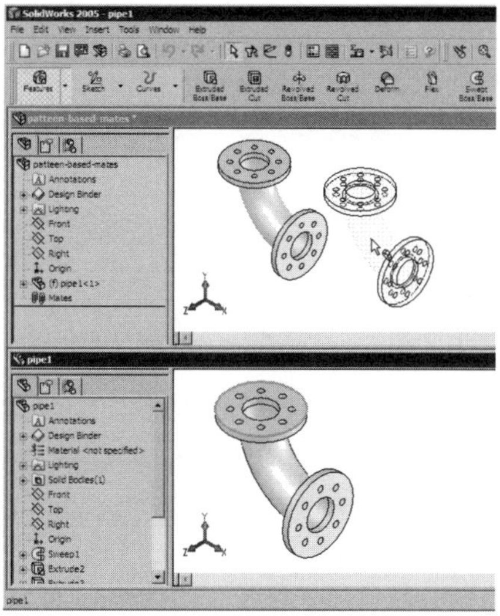

Figure 10-44 Component being dragged into the assembly document

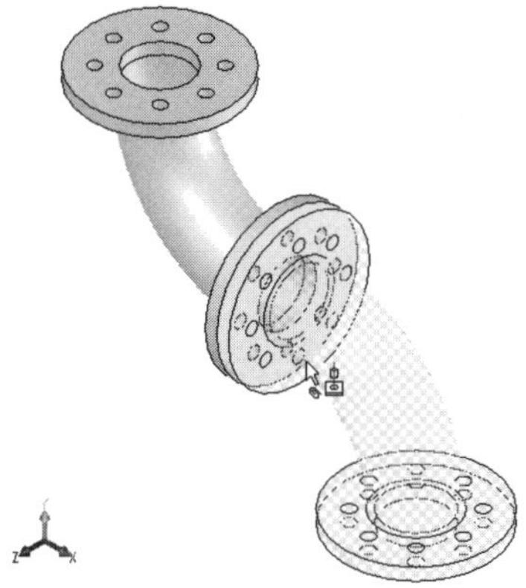

Figure 10-45 Preview of the component being assembled using pattern-based mates

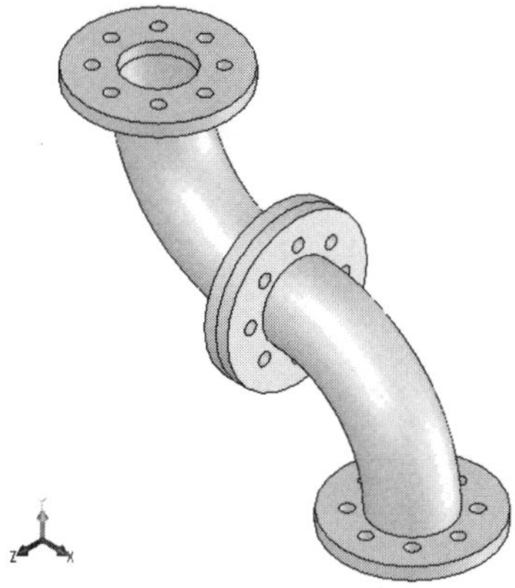

Figure 10-46 Component assembled using pattern-based mates

The **Mate Reference Name** edit box is used to define the name of the mate reference. The **Primary Reference Entity** rollout is used to define the primary mate reference. The **Mate Reference Type** drop-down list is used to define the type of mate. The **Mate Reference Alignment** drop-down list is used to define the type of alignment.

The **Secondary Reference Entity** rollout is used to define the secondary mate reference. The **Tertiary Reference Entity** rollout is used to define the tertiary mate reference.

To assemble a component using the mate reference, you need to define the mate references for two components. Also, the names of the mate references should be the same for both components. After defining the mate references for both components, place the first component coincident with the origin in the assembly document. Now, drag the second component in the assembly

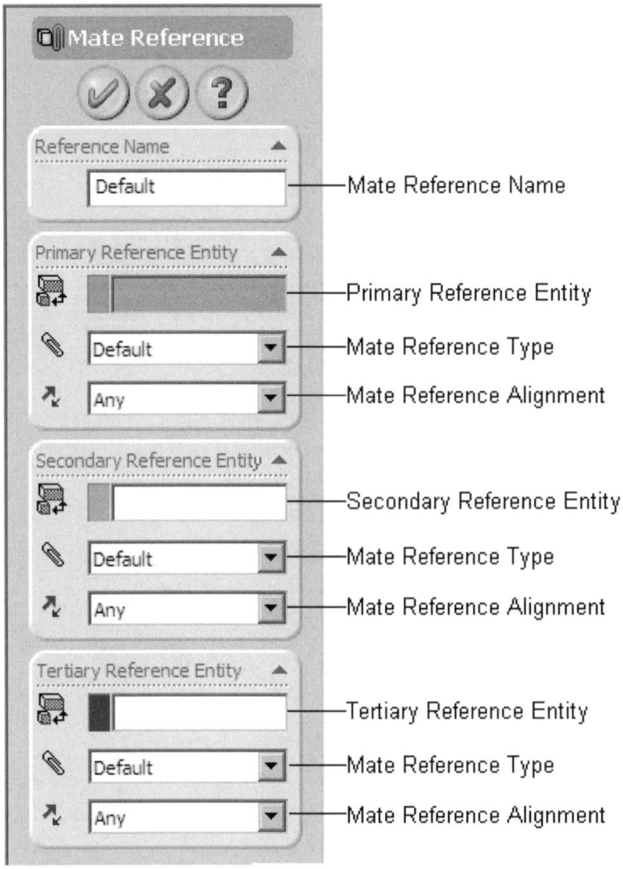

Figure 10-47** The **Mate Reference PropertyManager

document and you will notice that the second component is aligned to the references that were defined as mate references. Therefore, you do not need to apply mates in the assembly environment.

CREATING TOP-DOWN ASSEMBLIES

As mentioned earlier, top-down assemblies are the assemblies in which all components are created in the same assembly file. However, to create the components, you require an environment in which you can draw the sketches or the sketched features and the environment where you can convert the sketches into features. In other words, you need a sketching environment and a part modeling environment in the assembly file. In SolidWorks, you can invoke the sketching environment and the part modeling environment in the assembly document itself. The basic procedure for creating the components in the assembly or in other words the procedure for creating the top-down assembly is discussed next.

Creating the Component in the Top-Down Assembly

CommandManager:	Assemblies > New Part	*(Customize to Add)*
Menu:	Insert > Component > New Part	
Toolbar:	Assembly > New Part	*(Customize to Add)*

Before creating the first component in the top-down assembly, you first need to save the assembly document. After creating a new assembly document, choose the **Save** button from the **Standard** toolbar to save the assembly file. It is recommended that you create a new folder and save the assembly file and the other referenced file in the same folder.

Now, choose the **New Part** button from the **Assemblies CommandManager** after customizing it. The **Save As** dialog box will be displayed where you need to specify the name and location to save the new component. Enter the name of the new component in the **File name** edit box and choose the **Save** button. The select cursor will be replaced by the new component cursor. Now, you need to place the new component in the assembly document. Using the left mouse button place the component on the **Front** assembly plane, which is displayed in the **FeatureManager Design Tree**. The plane on which you place the component is selected as the sketching plane and the sketching environment is invoked automatically. You will notice that the **Edit Part** button available in the **Assemblies CommandManager** is selected. This means that the part modeling environment is invoked in the assembly document. You can draw the sketch of the base feature in the current sketching environment or you can also exit the sketching environment and select any other sketching plane to create the sketch. After creating the sketch of the base feature, exit the sketching environment. Now, as the **Features CommandManager** is not available by default, you need to invoke it and use the tools available in the **Features CommandManager** to convert the sketch into a model. Similarly, create the remaining features in the model.

After creating all features, choose the **Edit Part** button from the **Assembly** toolbar to exit the part modeling environment. The newly created component will have an **Inplace** mate with the default assembly plane on which it was placed earlier. Therefore, the newly created component is fixed. Using the procedure mentioned above, create other components.

Whenever you create a component in a top-down assembly, the component is fixed using the **Inplace** mate. You can delete this mate by selecting the **Inplace** mate by expanding the **Mates** option available in the **FeatureManager Design Tree**. After selecting the mate, press the DELETE key. Now, this component is floating and you can move this component. You can also assemble this component according to your requirement. You will learn more about fixed and floating components later in this chapter.

Note

As discussed earlier, when you place the first component in the assembly in bottom-up assembly design approach, that component is fixed by default. Therefore, you cannot apply any mates to a fixed component. If you want to add some mates to the fixed component, you first need to float the component. To do so, select the component from the drawing area or from the ***FeatureManager Design Tree****. Right-click to invoke the shortcut menu and choose the* ***Float*** *option.*

The components placed after the first component are floating components by default. If you need to fix a floating component, select the component and invoke the shortcut menu. Choose the ***Fix*** *option from the shortcut menu.*

MOVING THE INDIVIDUAL COMPONENTS

In SolidWorks you can move the individual unconstrained components in the assembly document without affecting the position and location of the other components. There are three methods of moving an individual component. Two methods of moving the individual component are discussed next. The third method is discussed later in this chapter.

Moving the Individual Components by Dragging

One of the major enhancements in the **Assembly** mode of SolidWorks 2005 is that you can move the component placed in the assembly without invoking any tool. To move an individual component, simply select the component and drag the cursor to move the component. Release the left mouse button to place the component at the desired location.

Moving Individual Components Using the Move Component Tool

CommandManager:	Assemblies > Move Component
Toolbar:	Assembly > Move Component

You can also move an individual component using the **Move Component** tool. Choose the **Move Component** button from the **Assemblies CommandManager t**o invoke the **Move Component PropertyManager** is invoked. You will notice that the **Free Drag** option is selected in the **Move** drop-down list in the **Move** rollout. You will be prompted to select a component and drag to move it. The select cursor will be replaced by the move cursor; select the component and drag the cursor to move the component. Release the left mouse button to move the component to the desired location. The other options available in the **Move** drop-down list to move the component are discussed next.

Along Assembly XYZ

Using the **Along Assembly XYZ** option from the **Move** drop-down list, you can move the component dynamically along the X, Y, and Z axes of the assembly document. Select the **Along Assembly XYZ** option from the **Move** drop-down list in the **Move** rollout. An assembly coordinate system in magenta will be displayed in the drawing area and you will be prompted to select a component and drag parallel to an assembly axis to move along that axis. Select the component and drag the cursor to move the component along any one of the assembly axes.

Along Entity

The **Along Entity** option is used to move the component along the direction of the selected entity. When you invoke this option, the **Selected item** selection area will be displayed. You will be prompted to select an entity to drag along, then select a component and drag to move it. Select an entity to define the direction in which you need to move the component. The name of the selected entity is displayed in the **Selected item** selection area. Now, select the component and drag the cursor to move the component along the direction of the selected entity.

By Delta XYZ

The **By Delta XYZ** option available in the **Move** drop-down list is used to move the selected component to a given distance in a specified direction. When you select this option, the **Delta X**, **Delta Y**, and **Delta Z** spinners will be invoked and you will be prompted to select a component and enter the distance to move in the PropertyManager. An assembly coordinate system will also be displayed in magenta. Select the component to move and specify the distance in the respective direction spinners in the X, Y, or Z direction. Choose the **Apply** button to move the component.

To XYZ Position

The **To XYZ Position** option is used to specify the coordinates of the origin of the part where the component will be placed after moving. When you select this option, the **X Coordinate**, **Y Coordinate**, and **Z Coordinate** spinners will be invoked and you will be prompted to select a component and enter the XYZ coordinates for the part's origin. An assembly coordinate system will also be displayed in magenta. Select the component and enter the respective coordinates of the part origin in the spinners and choose the **Apply** button to move the component.

ROTATING INDIVIDUAL COMPONENTS

In SolidWorks you can rotate an individual unconstrained component in the assembly document without affecting the position and location of the other components. The **Rotate Component** tool is used to rotate the component. There are three methods of rotating an individual component. Two methods of rotating an individual component are discussed next. The third method is discussed later in this chapter.

Rotating the Individual Components by Dragging

In SolidWorks 2005, you can rotate the component placed in the assembly without invoking any tool. To rotate an individual component, simply select the component using the right mouse button and drag the cursor to rotate the component. Release the right mouse button after attaining the desired orientation of the individual component.

Rotating Individual Components Using the Rotate Component Tool

CommandManager:	Assemblies > Rotate Component
Toolbar:	Assembly > Rotate Component

You can also rotate an individual component using the **Rotate Component** tool. To rotate an individual component using this tool choose the **Rotate Component** button from the **Assemblies CommandManager**; the **Rotate Component PropertyManager** will be invoked. You will notice that the **Free Drag** option is selected in the **Rotate** drop-down list in the **Rotate** rollout. Therefore, you will be prompted to select a component and drag it to rotate. The select cursor will be replaced by the rotate cursor; select the component and drag the cursor to rotate the component. The other options available in the **Rotate** drop-down list to rotate the component are discussed next.

About Entity

The **About Entity** option available in the **Rotate** drop-down list is used to rotate the component with respect to a selected entity. The selected entity is defined as the rotational axis. When you invoke this option, the **Selected item** selection area will be displayed and you will be prompted to select an axis entity to rotate about. Select an entity to define the rotational axis. The name of the selected entity will be displayed in the **Selected item** selection area. Now, select the component and drag the cursor to rotate the component about the selected axis.

By Delta XYZ

The **By Delta XYZ** option available in the **Rotate** drop-down list is used to rotate the selected component by a given incremental angle along the specified axis. When you select this option, the **Delta X**, **Delta Y**, and **Delta Z** spinners will be invoked and you will be prompted to select a component and enter the desired rotation in the PropertyManager. Select the component to rotate and specify the rotation angle in the respective spinners in the direction in which you need to rotate the component. Choose the **Apply** button to rotate the component.

Tip: *You can toggle between the **Move PropertyManager**, **Rotate PropertyManager**, and **SmartMates PropertyManager**. If you need to invoke the **Move PropertyManager**, invoke the **Rotate** rollout to switch to the **Rotate PropertyManager**. If you select the **SmartMates** button, the system will switch you to the **SmartMates PropertyManager**. From the **Rotate** or **SmartMates PropertyManager**, if you invoke the **Move** rollout, the system will switch you to the **Move PropertyManager***

MOVING AND ROTATING AN INDIVIDUAL COMPONENT USING THE TRIAD

To move an individual component using the triad, you first need to select the component and then right-click to invoke the shortcut menu. Choose the **Move with Triad** option from the shortcut menu; the component will appear transparent and the triad will be displayed on it. Using this triad you can move and rotate a component. The different components of the triad are displayed in Figure 10-48.

Press and hold the left mouse button down on the X arm; the select cursor will be replaced by the move/rotate cursor. Use the left mouse button to drag the selected component in the X direction. If you use the right mouse button and drag the cursor, the selected component will rotate about the X axis. Similarly, you can select the Y or the Z arm and drag the cursor to move/rotate the component in the Y or Z direction.

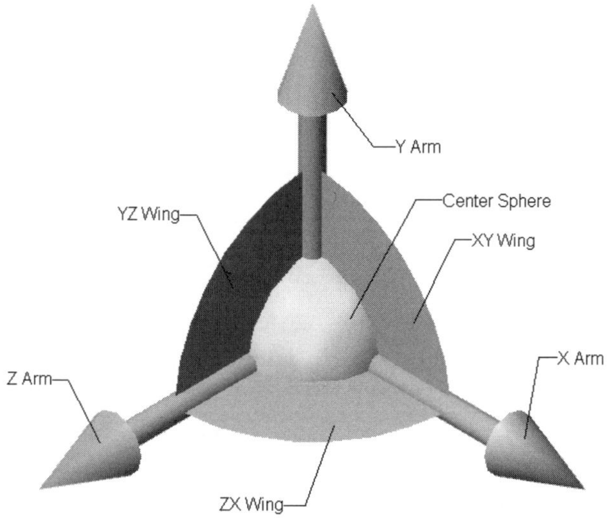

Figure 10-48 *Triad*

You can also move a component in the XY plane. To do so, select the XY wing of the triad and drag the cursor to move the component. To move the component in the YZ plane, select the YZ wing of the triad and drag the cursor to move the component. Similarly, by selecting the ZX wing and dragging the cursor, you can move the component in the ZX plane.

If you select the center sphere and invoke the shortcut menu, various options will be available, which are discussed next.

Show Translate XYZ Box

The **Show Translate XYZ Box** option available in the shortcut menu is used to display the **Translate XYZ** box. Using this box you can specify the value of the X, Y, and Z coordinates of the destination where you need to place the selected component. Specify the X, Y, and Z coordinates in the respective edit boxes and then choose **OK** from the box. When you move the component by dragging, the values of the X, Y, and Z spinners will change automatically.

Show Translate Delta XYZ Box

The **Show Translate Delta XYZ Box** option available in the shortcut menu is used to display the **Translate Delta XYZ** box. Using this box you can specify the incremental value by which you need to move the selected component in the X, Y, or Z direction. Set the incremental value in the X, Y, or Z edit box and then choose **OK** from the box.

Show Rotate Delta XYZ Box

The **Show Rotate Delta XYZ Box** option available in the shortcut menu is used to display the **Rotate Delta XYZ** box. Using this box, you can specify the incremental value by which the selected component will rotate in the X, Y, or Z direction. Set the incremental value in the X, Y, or Z edit box and then choose **OK** from the box.

Align to Component

The **Align to Component** option available in the shortcut menu is used to align the triad with respect to the current component origin.

TUTORIALS

Tutorial 1

In this tutorial, you will create all components of the Bench Vice and then assemble them. The Bench Vice assembly is shown in Figure 10-49. The dimensions of various components of the assembly are given in Figures 10-50 through 10-53. **(Expected time: 2 hrs 45 min)**

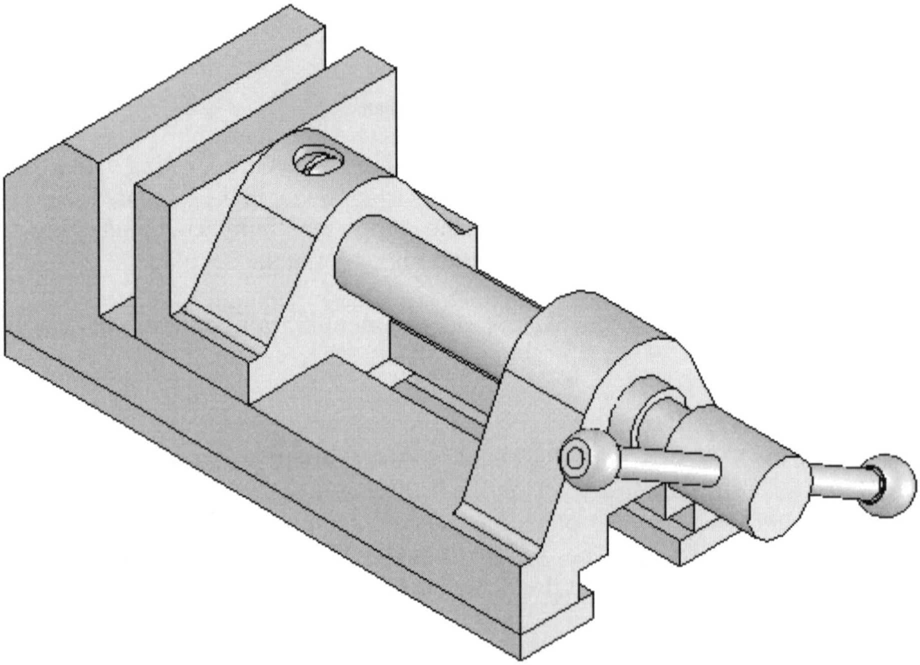

Figure 10-49 *Bench Vice assembly*

The steps to be followed to complete the assembly of this tutorial are listed below.

a. Create all components in individual part documents and save them. The part documents will be saved in the *\My Documents\SolidWorks\c10\Bench Vice* folder.
b. Open Vice Body and Vice Jaw part documents and define the mate references in both part documents, refer to Figures 10-54 and 10-55
c. Create a new assembly document and open all part documents. Place the first component, which is Vice Body, by dragging and dropping from the part document window. Now, drag and drop the Vice Jaw in the assembly document. It will automatically assemble with the

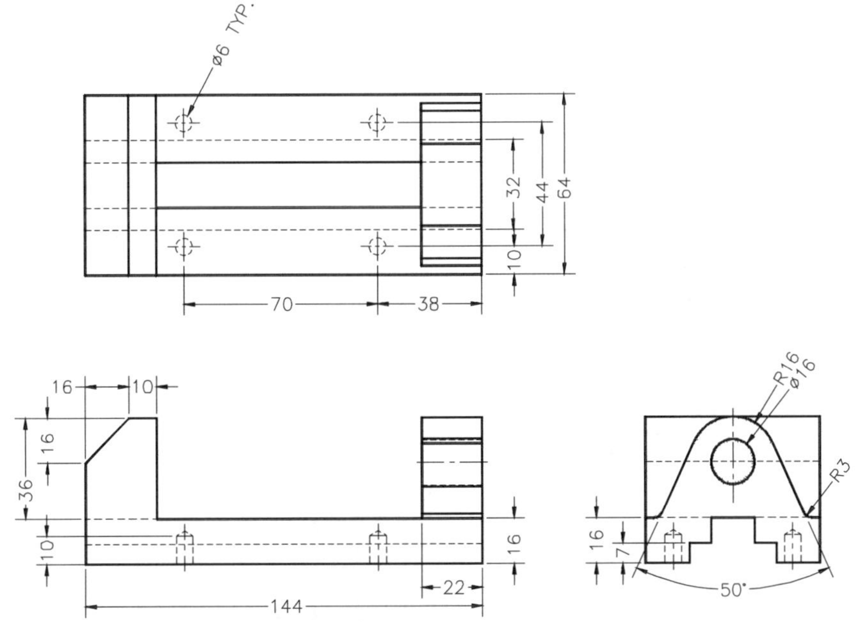

Figure 10-50 *Views and dimensions of the Vice Body*

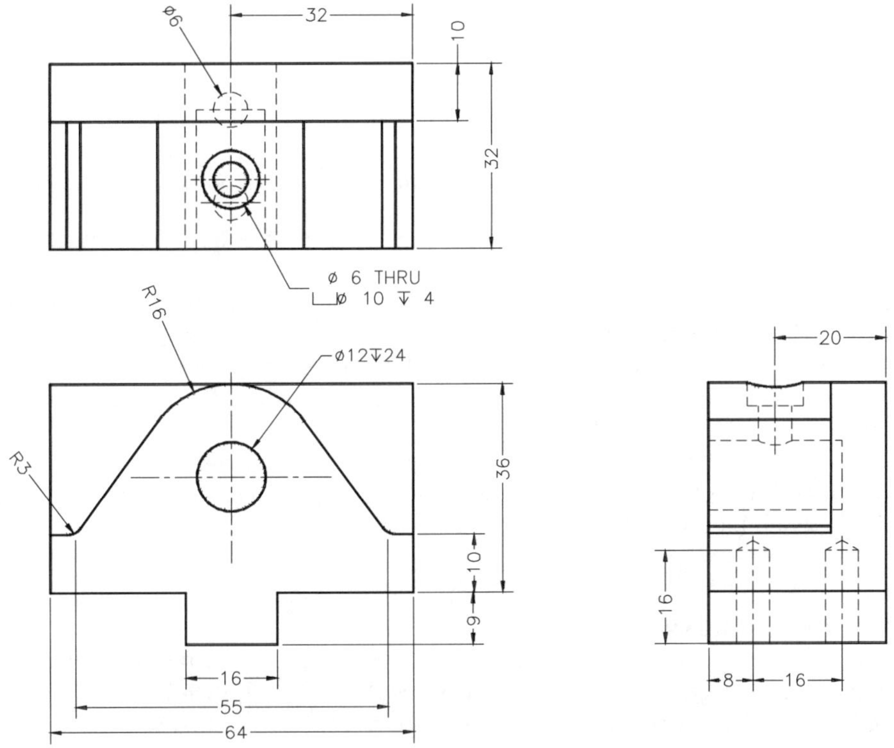

Figure 10-51 *Views and dimensions of Vice Jaw*

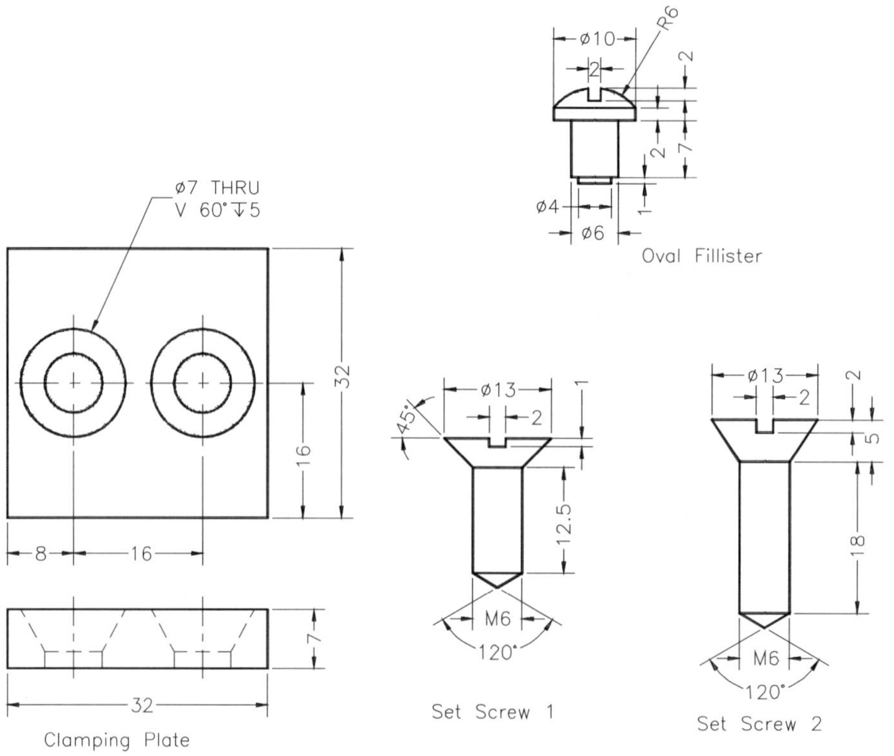

Figure 10-52 *Views and dimensions of Clamping Plate, Oval Fillister, Set Screw 1, and Set Screw 2*

Vice Jaw because the mate references are already defined in both part documents, refer to Figures 10-56 and 10-57.

d. Drag and drop the jaw screw in the assembly document and apply the required mates, refer to Figures 10-59 through 10-64.
e. Next, analyze the assembly for degrees of freedom of the components.
f. After analyzing the assembly, apply the required mates to constrain all degrees of freedom, refer to Figures 10-65 and 10-66.
g. Next, assemble the Clamping Plate, refer to Figures 10-67 through 10-70.
h. Next, assemble the Oval Fillister using the feature-based mates, refer to Figure 10-71.
i. Similarly, assemble the other components.

Creating the Components

1. Create all the components of the Bench Vice assembly as separate part documents. Specify the names of the documents as shown in Figures 10-50 through 10-53. The files should be saved in the folder *\My Documents\SolidWorks\c10\Bench Vice*. Make sure that the *Bench Vice* is your current folder.

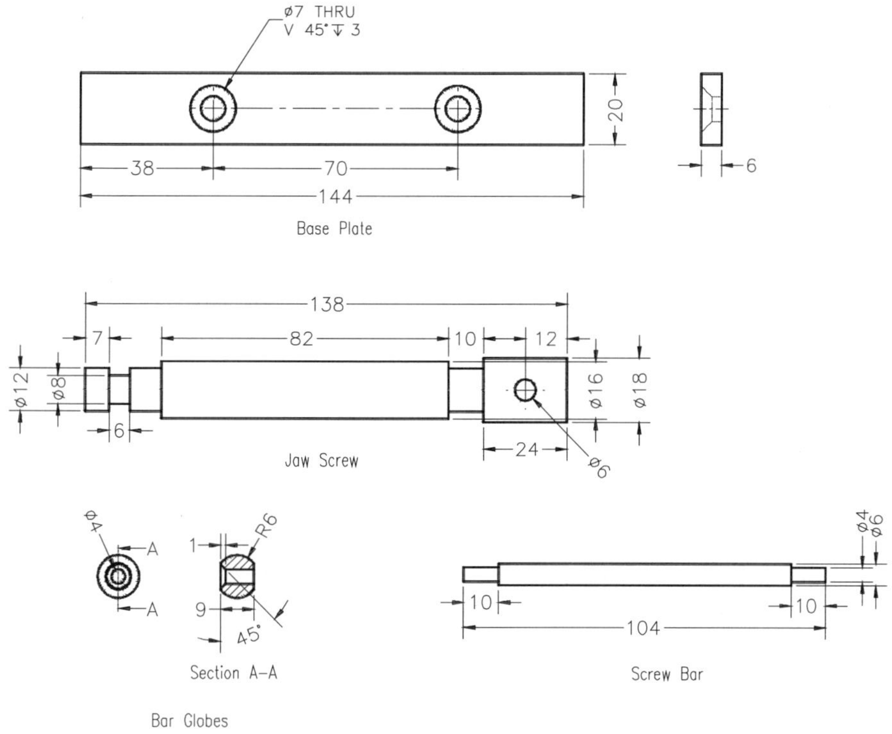

***Figure 10-53** Views and dimensions of Base Plate, Jaw Screw, Screw Bar, and Bar Globes*

Creating the Mate References

In this tutorial, you need to assemble the first two components of the assembly using the mate references. For assembling the components using the mate references, first you need to create the mate reference. Therefore, you need to open the part documents in which you will add the mate references.

1. Invoke the **Open** dialog box from the **Standard** toolbar.

2. Double-click on the Vice Body; the vice-body part document is opened in the SolidWorks window.

3. Choose **Insert > Reference Geometry > Mate Reference** from the menu bar; the **Mate Reference PropertyManager** is invoked. The selection mode in the **Primary Reference Entity** selection area is active.

4. Select the planar face of the model shown in Figure 10-54 as the primary reference. The selected planar face is highlighted in green.

5. Select the **Coincident** option from the **Mate Reference Type** drop-down list in the **Primary Reference Entity** rollout. The selection mode in the **Secondary Reference Entity** selection area is active.

6. Select the planar face of the model shown in Figure 10-54 of the model as the secondary reference. The selected face is highlighted in magenta.

7. Select the **Coincident** option from the **Mate Reference Type** drop-down list in the **Secondary Reference Entity** rollout.

   The selection mode in the **Tertiary Reference Entity** selection area is active.

8. Select the planar face of the model shown in Figure 10-54 as the tertiary reference. The selected face is highlighted in blue.

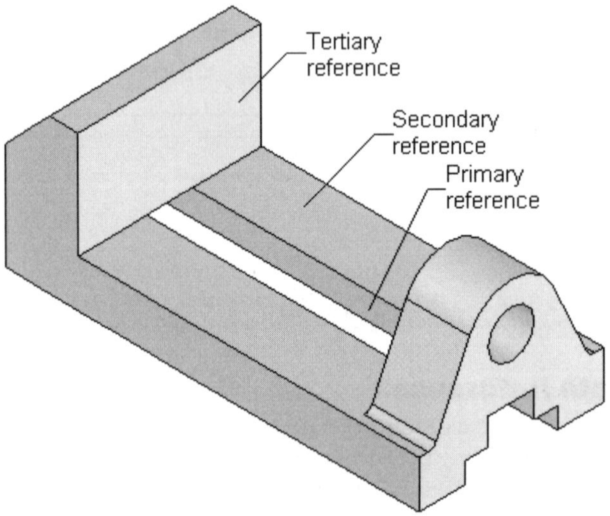

Figure 10-54 *Faces to be selected as mate references*

9. Select the **Parallel** option from the **Mate Reference Type** drop-down list in the **Tertiary Reference Entity** rollout.

10. Enter **Vice Mate Reference** as the name of mate reference in the **Mate Reference Name** edit box available in the **Reference Name** rollout.

11. Choose the **OK** button from the **Mate Reference PropertyManager**.

12. Similarly, create the mate reference in the Vice Jaw part document. The faces to be selected as reference are displayed in Figure 10-55. The name of the mate reference in the **Reference Name** rollout should be the same in both part documents.

13. Close all part documents, except *Vice Body.sldpart* and *Vice Jaw.sldprt*, if they are opened.

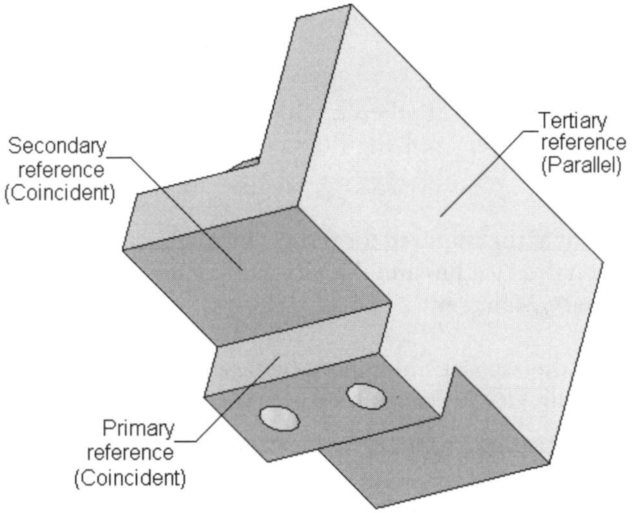

Figure 10-55 *Faces to be selected as mate references*

Assembling the First Two Components of the Assembly

After creating the mate references in the part documents, you need to assemble the components. To do so, you need to start a new SolidWorks assembly document.

1. Start a new SolidWorks assembly document; the **Insert Component PropertyManager** is invoked automatically and the names of components that were opened are displayed in the **Open documents** selection area.

2. Select Vice Body from the **Open documents** selection area; the preview of the Vice Body is displayed along with the component cursor.

3. Choose the **Keep Visible** button from the **Insert Component PropertyManager** because after placing this component, you also need to place the second component that is displayed in the **Open documents** selection area.

   It is recommended that the first component of the assembly should be placed at the assembly origin.

4. Choose the **OK** button from the **Insert Component PropertyManager** to place the first component coincident to the origin.

   As the **Keep Visible** button is selected, the preview of the Vice Body is again displayed with the component cursor in the drawing area.

5. Change the current view to isometric.

   Next, you need to place the second component in the assembly. As discussed earlier, the second component of the assembly, which is the Vice Jaw, is assembled with the Vice Body using the mate references.

6. Select the Vice Jaw from the **Open documents** selection area; the preview of the Vice Jaw is available in the drawing area.

   When you move the cursor close to the Vice Body in the assembly document, the preview of the Vice Jaw assembled after applying the mates with the Vice Body is displayed in the assembly document.

7. Place the component at the required location. The mates specified in the mate references are applied between the Vice Jaw and the Vice Body. Choose the **Keep Visible** button and exit the **Mate PropertyManager**.

   Figure 10-56 shows the second component being placed in the assembly document and Figure 10-57 shows the Vice Jaw assembled with the Vice Body.

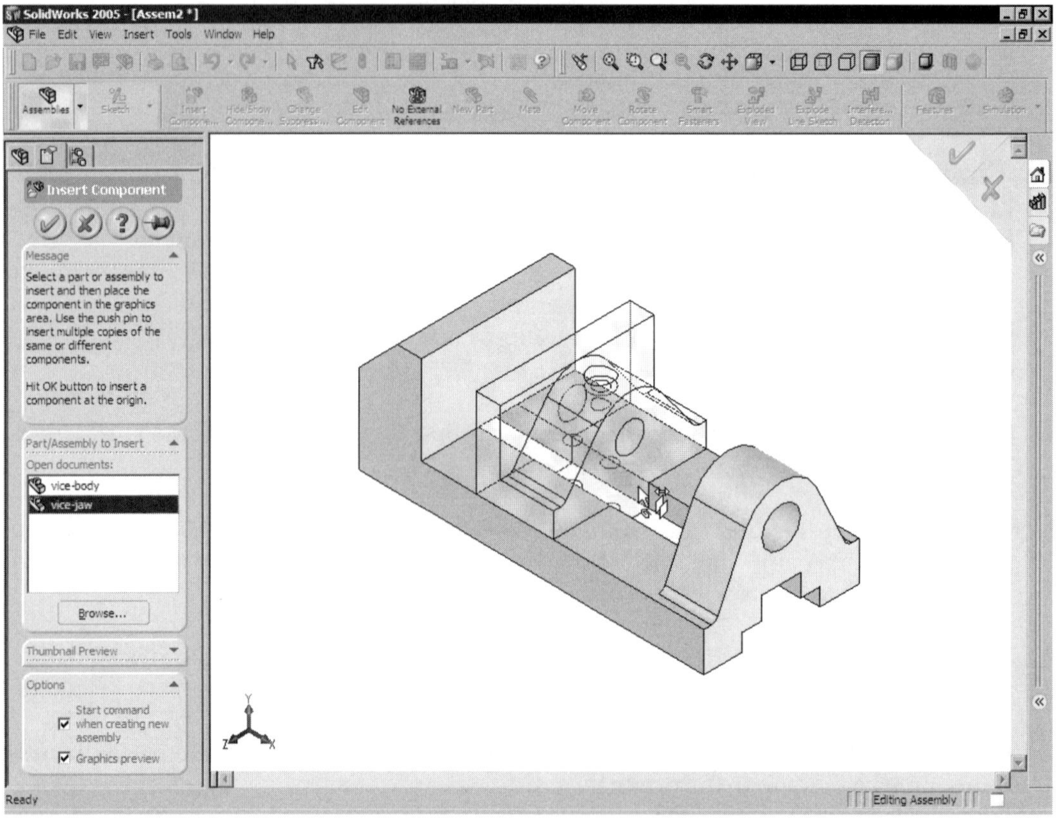

Figure 10-56 *Second component being dragged*

Before proceeding further, it is recommended that you close all part documents.

8. Close the part document windows of all parts that are placed in the assembly document.

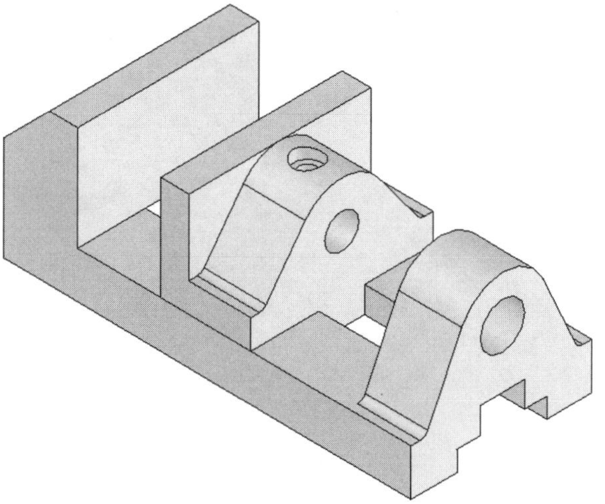

Figure 10-57 *Vice Jaw assembled with the Vice Body*

Tip. *The mates that are defined in the mate reference are applied to the components when you place the components in the assembly. You can view the mates applied to both components by expanding the* ***Mates*** *option from the* ***FeatureManager Design Tree*** *of the assembly document.*

Assembling the Jaw screw

Now, you need to place the Jaw Screw in the assembly document.

1. Choose the **Insert Component** button from the **Assemblies CommandManager**.

2. Choose the **Browse** button from the **Part/Assembly to Insert** rollout to display the **Open**.

3. Double-click on the Jaw Screw to open the part document of Jaw Screw. The preview of the Jaw Screw is displayed in the drawing area.

4. Click anywhere in the drawing area to place the Jaw Screw.

   Figure 10-58 shows the Jaw Screw placed arbitrarily in the assembly document. Next, you need to add the assembly mates to assemble the components placed in the assembly.

5. Press and hold the ALT key down on the keyboard and then select the face from the location shown in Figure 10-59.

6. Drag the Jaw Screw to the hole in the Vice Body, as shown in Figure 10-60; the Jaw Screw appears transparent and the select cursor is replaced by the smart mates cursor. Also, the symbol of concentric mate is displayed below the cursor.

7. Release the left mouse button at this location; the **Mates** pop-up toolbar is displayed and the **Concentric** button is chosen by default, which suggests that the **Concentric** mate is the most

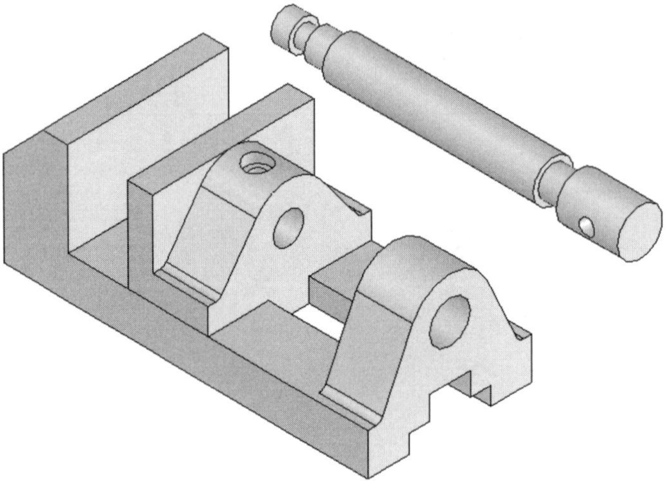

Figure 10-58 Jaw Screw placed in the assembly document

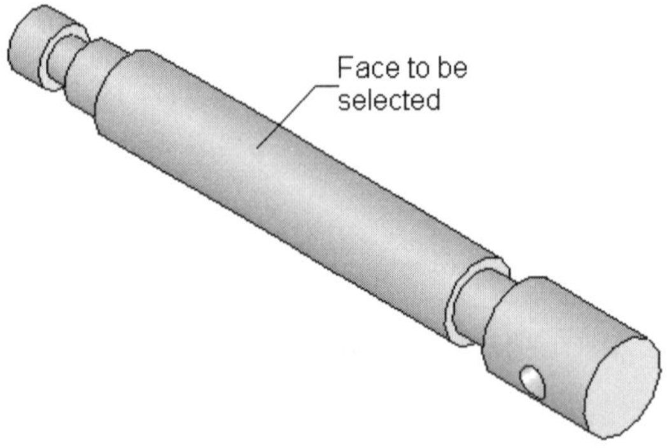

Figure 10-59 Face to be selected

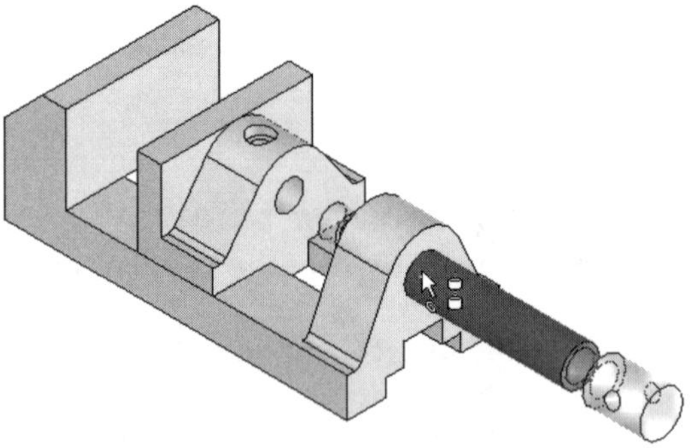

Figure 10-60 Jaw Screw being dragged

appropriate mate to be applied. Choose the **Add/Finish Mate** button from the **Mates** pop-up toolbar. Figure 10-61 shows the Jaw Screw assembled to the Vice Body.

Next, you need to apply the coincident mate between the planar faces of the Jaw Screw and the Vice Jaw.

8. Choose the **Mate** button from the **Assemblies CommandManager** to invoke the **Mate PropertyManager**.

9. Rotate the assembly view and select the face of the Vice Jaw, as shown in Figure 10-62. Next, select the face of the Jaw Screw, as shown in Figure 10-63.

As soon as you select the faces, the **Mates** pop-up toolbar is invoked and the preview of the assembly with coincident mate is displayed in the drawing area.

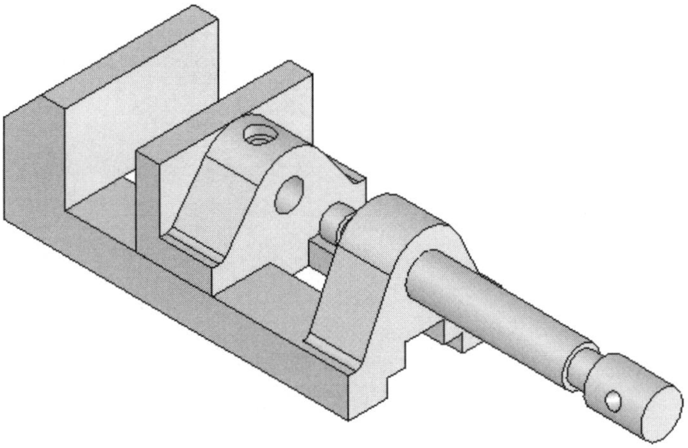

Figure 10-61 ***Concentric*** *mate applied between the Jaw Screw and the Vice Body*

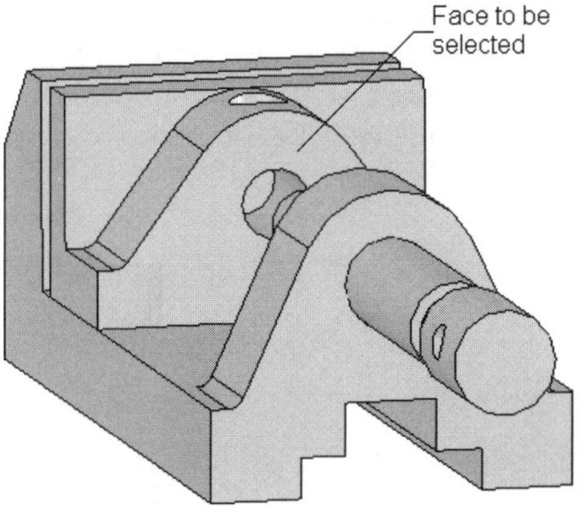

Figure 10-62 *Face to be selected*

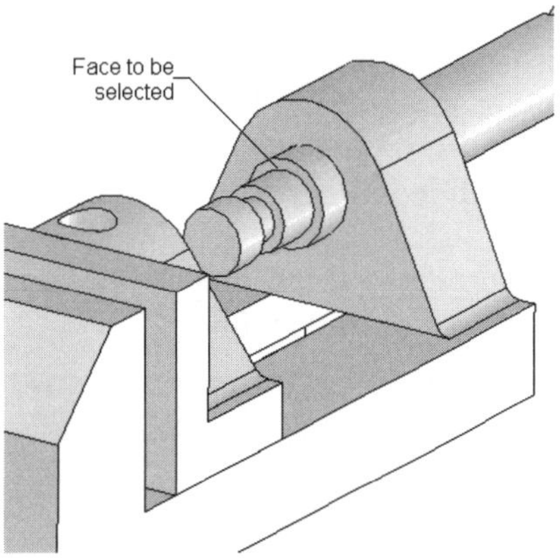

Figure 10-63 *Face to be selected*

10. Choose the **Add/Finish Mate** button from the **Mate** pop-up toolbar. The assembly after adding the **Coincident** mate is displayed in Figure 10-64.

    In real world, there are two types of assemblies. The first is the fully defined assembly in which all degrees of freedom of all components are restricted. The other type of assembly is that in which some degrees of freedom of the components are left free so that they can be moved or rotated. These type of assemblies are used for mechanism that you will learn in the next chapter.

    After adding the **Coincident** mate, you need to move the assembly to analyze the degree of freedom of the components of the assembly. After analyzing the components of an assembly you need to add the mates to constrain that degree of freedom.

11. Select the circular face of the Jaw Screw using the left mouse button and drag the cursor.

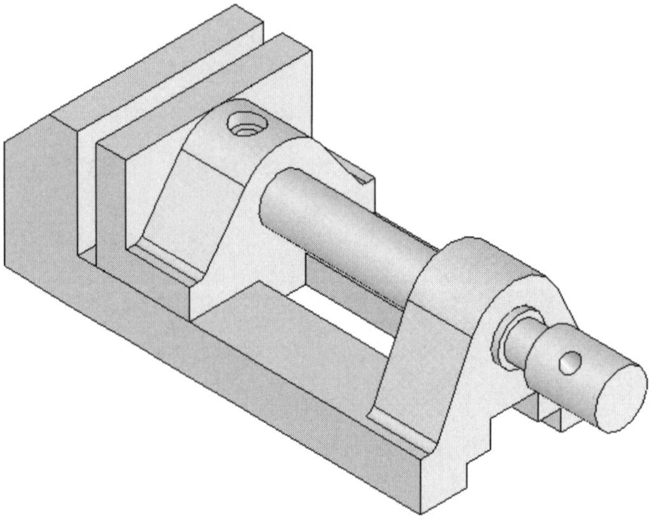

Figure 10-64 *Assembly after applying the* ***Coincident*** *mate to the Jaw Screw*

You will notice that the Jaw Screw is rotating about its axis and is moving along the X axis. Also, it is forcing the Vice Jaw to move along the X axis. Originally, this degree of freedom of Vice Jaw and Jaw Screw needs to be left free so that the assembly can function in the mechanism. But in this chapter, you need to restrict this degree of freedom so as to create a fully defined assembly.

12. Select the two faces, one of the Vice Body and the other of the Jaw Screw, as shown in Figures 10-65 and 10-66.

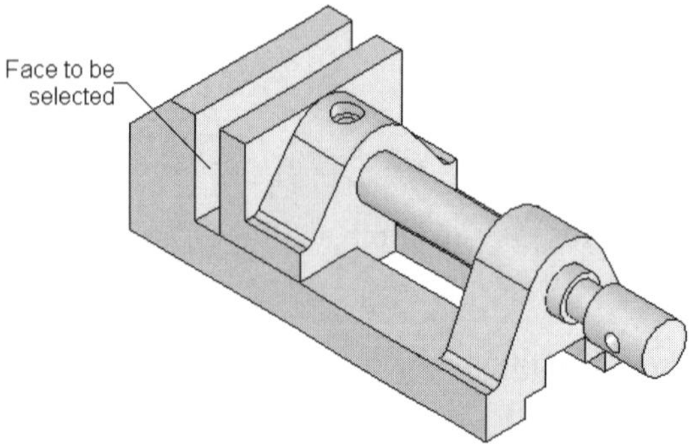

Figure 10-65 *Face to be selected*

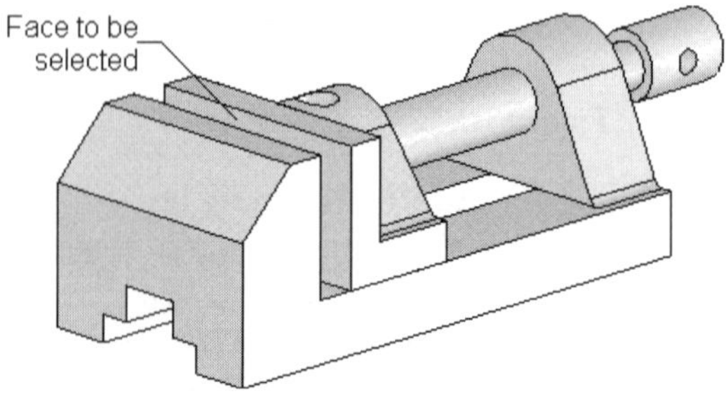

Figure 10-66 *Face to be selected*

13. Choose the **Distance** button from the **Standard Mates** rollout and set the **Distance** spinner value to **10**. As you choose the **Distance** button, the **PropertyManager** changes to the **Distance 1 PropertyManager**.

14. Choose the **OK** button from the **Distance 1 PropertyManager** to again display the **Mate PropertyManager**.

15. Expand the **FeatureManager Design Tree** displayed in the drawing area and select the **Top Plane**. Now, expand the **Jaw Screw** from the **FeatureManager Design Tree** and select the **Top Plane**. The **Mates** pop-up toolbar is displayed.

16. Choose the **Angle** button from the **Mate** pop-up toolbar.

17. Set the value of the **Angle** spinner to **45** and choose the **Add/Finish Mate** button from the **Mates** pop-up toolbar.

18. Choose the **OK** button from the **Mate PropertyManager** to exit it.

Assembling the Clamping Plate

Next, you need to assemble the Clamping Plate with the assembly.

1. Invoke the **Insert Component PropertyManager** and choose the **Browse** button to invoke the **Open** dialog box. Open and place the Clamping Plate in the drawing area.

2. Rotate the assembly such that the bottom face of the assembly is displayed as shown in Figure 10-67.

3. Select the Clamping Plate using the right mouse button and drag the cursor to rotate the Clamping Plate as shown in Figure 10-68.

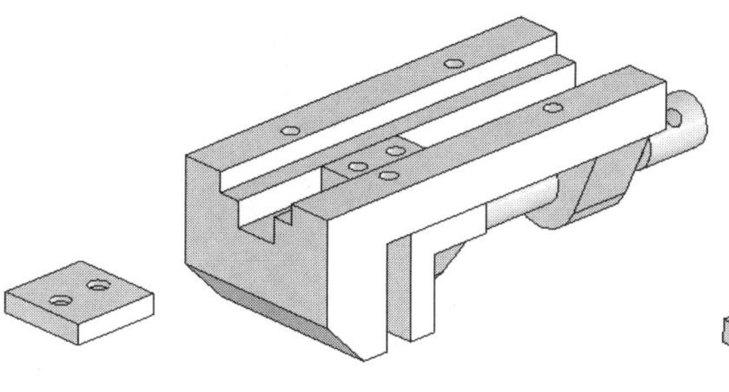

Figure 10-67 *Rotated assembly*

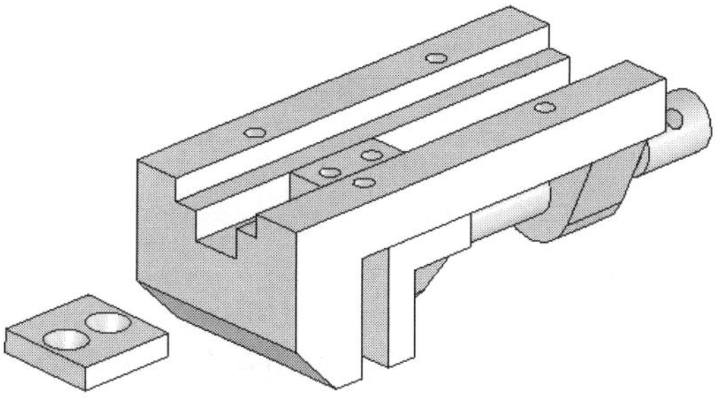

Figure 10-68 *Clamping Plate after rotating*

4. Apply the **Concentric** mate between the two cylindrical faces of the clamping plate and the two holes of the Vice Jaw, refer to Figure 10-69. You may have to move the Clamping Plate after applying the first mate.

5. Move the Clamping Plate by dragging. Now, apply the **Coincident** mate between the faces of the Clamping Plate and the Vice Jaw, as shown in Figure 10-70.

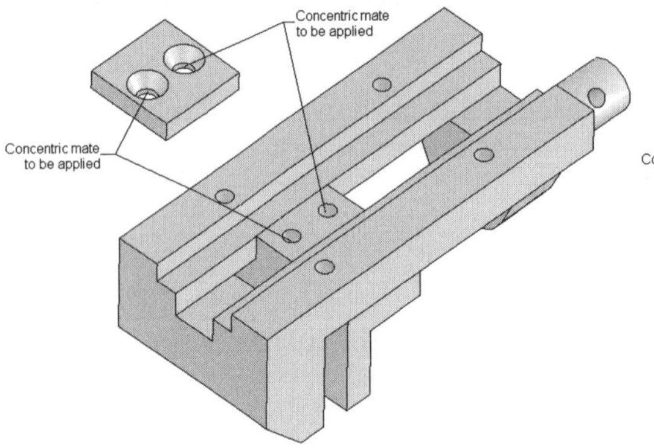

Figure 10-69 Faces to be selected to apply mate

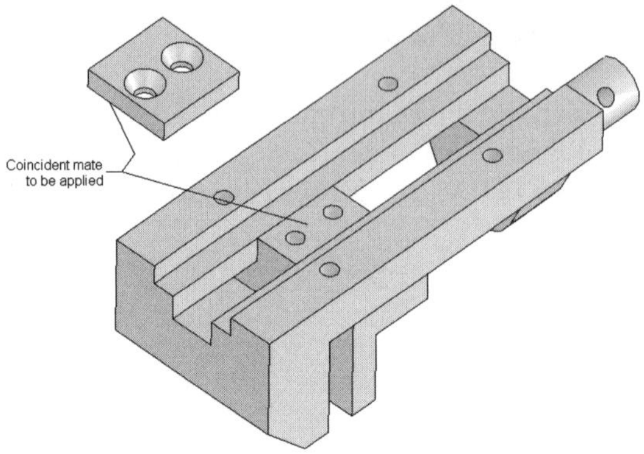

Figure 10-70 Faces to be selected to apply mate

6. Similarly, assemble the Screw Bar, Support Plates, and Bar Globes. The assembly after assembling all these components is shown in Figure 10-71.

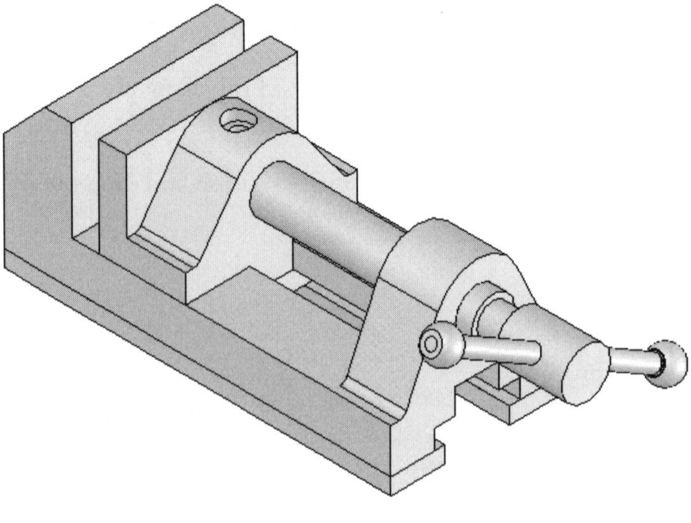

Figure 10-71 Assembly after assembling the Vice Body, Vice Jaw, Jaw Screw, Screw Bar, Clamping Plate, Base Plate, and Bar Globes

Assembling the Remaining Components

Next, you need to assemble the Oval Fillister, Set Screw 1, and Set Screw 2. These fasteners are assembled using feature-based mates.

1. Open the part documents of Oval Fillisters, Set Screw 1, and Set Screw 2.

2. Choose **Window > Tile Horizontally** from the menu bar to rearrange the windows.

3. Select the **Revolve1** feature from the **FeatureManager Design Tree** of the Oval Fillister part document. Drag the cursor to place the component in the assembly, as shown in Figure 10-72.

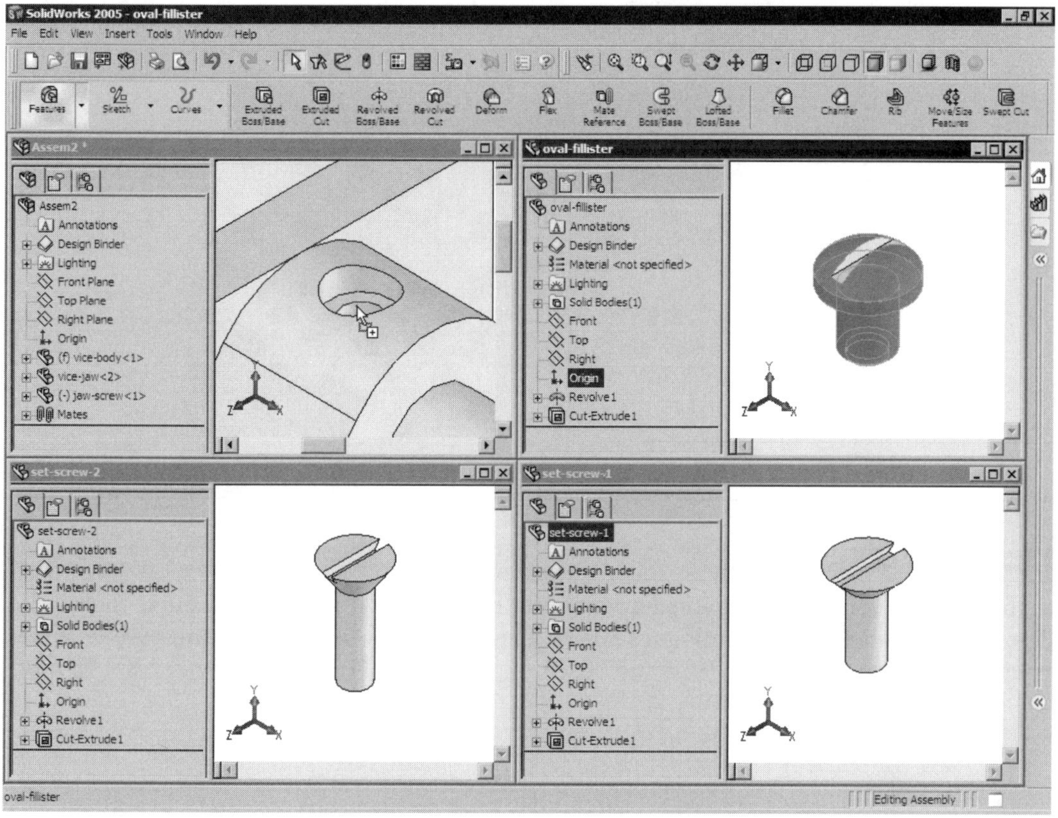

Figure 10-72 *Dragging the Oval Fillisters in the assembly using feature-based mates*

4. Drop the component at this location.

   Similarly, assemble Set Screw 1 and Set screw 2 using feature-based mates. Use the TAB key on the keyboard to reverse the direction. The rotated view of the assembly after assembling Set Screw 1 and Set Screw 2 is shown in Figure 10-73.

5. Choose the **Save** button to save the assembly with the name *Bench Vice* in the folder *\My Documents\SolidWorks\Bench Vice*.

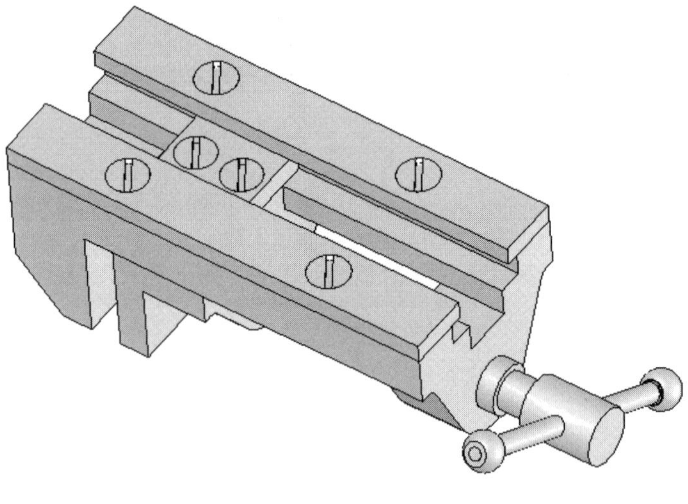

Figure 10-73 *Viewing the final assembly from the bottom*

Tutorial 2

In this tutorial, you will create all components of the Pipe Vice and then assemble them. The Pipe Vice assembly is shown in Figure 10-74. The dimensions of various components of this assembly are given in Figures 10-75 and 10-76. **(Expected time: 2 hrs 45 min)**

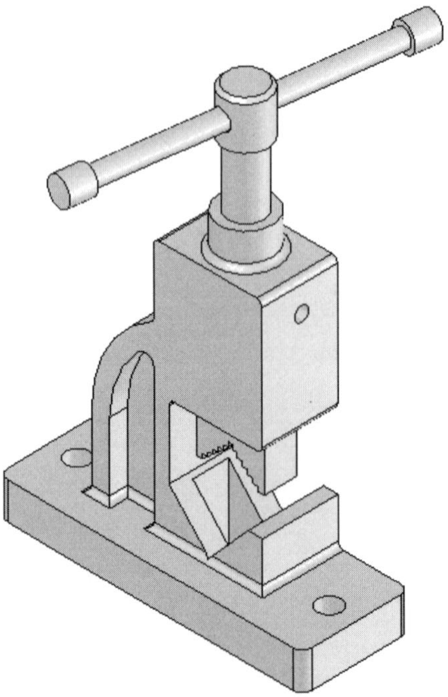

Figure 10-74 *Pipe Vice assembly*

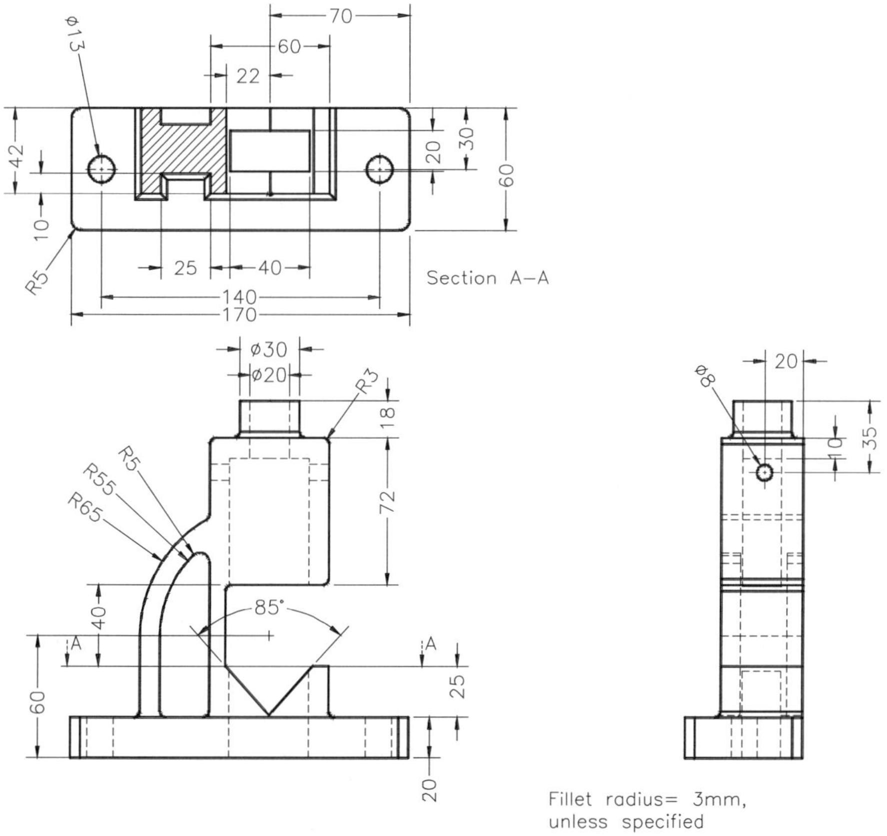

***Figure 10-75** Views and dimensions of the base*

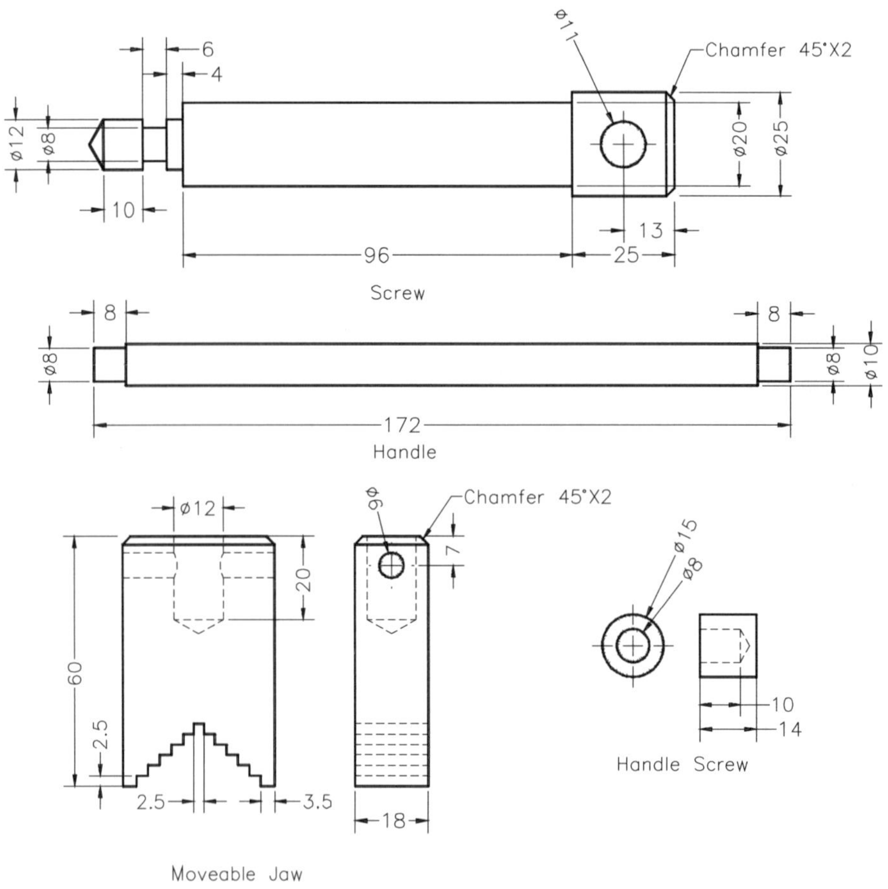

***Figure 10-76** Views and dimensions of the Screw, Handle, Moveable Jaw, and Handle Screw*

You need to create all components of the Pipe Vice assembly as separate part documents. After creating the parts, you will assemble them in the assembly document. Therefore, in this tutorial, you need to use the bottom-up approach for creating the assembly.

The steps to be followed to complete the assembly of this tutorial are listed below.

a. Create all components in the individual part documents and save them. The part documents will be saved in *\My Documents\SolidWorks\c10\Pipe Vice*.
b. Place the base at the origin of the assembly.
c. Place the Moveable Jaw and the Screw in the assembly. Apply the mates between Moveable Jaw and the Screw, refer to Figures 10-77 through 10-79.
d. Assemble the Screw with the Base.
e. Place the other components in the assembly and apply the required mates to the assembly, refer to Figure 10-80.

Creating the Components

1. Create all components of the Pipe Vice assembly as separate part documents. Specify the names of the files, as shown in Figures 10-75 and 10-76. The documents should be saved in the folder */My Documents/SolidWorks/c10/Pipe Vice*.

Inserting the First Component in the Assembly

After creating all components of the Pipe vice Assembly, you need to start a new SolidWorks assembly document.

1. Start a new SolidWorks assembly document; the **Insert Component PropertyManager** is invoked by default.

2. Choose the **Browse** button from the **Part/Assembly to Insert** rollout to display the **Open** dialog box. Double-click on Base.

3. Choose the **OK** button from the **Insert Component PropertyManager** to place the Base origin coincident to the origin of the assembly document.

4. Change the view orientation to isometric.

Inserting and Assembling the Moveable Jaw and the Screw

After placing the first component in the assembly document, you need to place the Moveable Jaw and the Screw in the assembly document. After placing these components you need to apply the required mates.

1. Choose the **Insert Components** button from the **Assemblies CommandManager**. Then, choose the **Keep Visible** button from the **PropertyManager** and invoke the **Open** dialog box by choosing the **Browse** button from the **Part/Assembly to Insert** rollout.

2. Double-click on the Moveable Jaw. Place the component anywhere in the assembly document such that it does not interfere with existing component.

3. Similarly, place the Screw in the assembly document and choose the **OK** button from the **Insert Component PropertyManager**. Figure 10-77 shows the Moveable Jaw, Screw, and Base placed in the assembly document.

First, you need to assemble the Screw with the Moveable Jaw. Therefore, for assembling the Screw with the Moveable Jaw, you need to fix the Moveable Jaw.

4. Select the Moveable Jaw from the drawing area or from the **FeatureManager Design Tree**. Right-click to invoke the shortcut menu.

5. Choose the **Fix** option from the shortcut menu. The Moveable Jaw is fixed and you cannot move or rotate the Moveable Jaw.

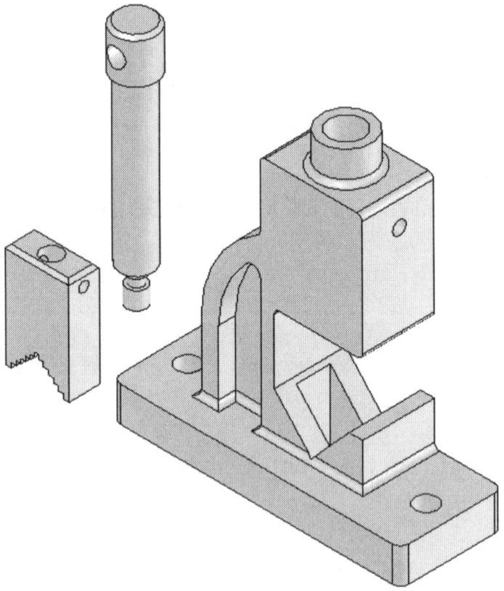

Figure 10-77 The Moveable Jaw, Screw, and Base placed in the assembly document

6. Invoke the **Move PropertyManager** and choose the **SmartMates** button from the **Move** rollout. Double-click on the lowermost cylindrical face of the Screw; the Screw appears transparent.

7. Drag the cursor to the hole located on the top of the Moveable Jaw. Release the left mouse button as soon as the concentric symbol is displayed below the cursor. Choose the **Add/Finish Mate** button from the **Mates** pop-up toolbar.

8. Select the Screw and move it up so that it is not inside the Moveable Jaw.

9. Right-click in the drawing area and choose **Clear Selections** to clear the current selection.

10. Rotate the assembly and double-click on the lower flat face of the Screw; the Screw appears transparent.

11. Again, rotate the model and select the top planar face of the Moveable Jaw. The **Coincident** relation is applied between the two selected faces. Choose the **Add/Finish Mate** button from the **Mates** pop-up toolbar.

12. Choose the **OK** button from the **SmartMates PropertyManager**. Figure 10-78 shows the Screw after applying the mates.

    Next, you need to assemble the Screw and the Moveable Jaw with the Base.

13. Select the Moveable Jaw and invoke the shortcut menu. Choose the **Float** option from the shortcut menu.

    Now, the Moveable Jaw and the Screw assembled to it can be moved.

14. Using smart mates, add a **Concentric** mate between the cylindrical face of the Screw and the hole created on the top face of the Base.

15. Invoke the **Mate PropertyManager** and select the front planar face of the Moveable Jaw and the front planar face of the Base.

16. Choose the **Parallel** button from the **Mates** pop-up toolbar and choose the **Add/Finish Mate** button to add a parallel mate between the selected faces.

17. Next, select the faces as shown in Figure 10-79.

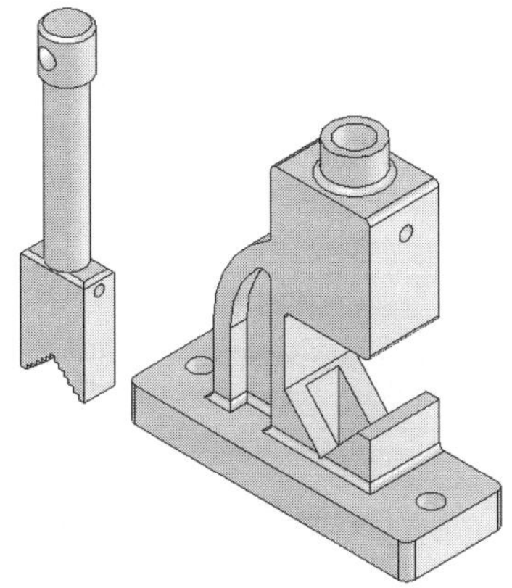

Figure 10-78 The Screw assembled with the Moveable Jaw

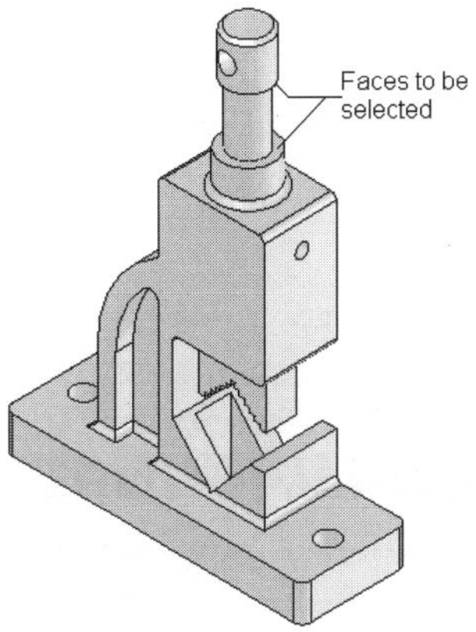

Figure 10-79 Faces to be selected

18. Choose the **Distance** button from the **Mates** pop-up toolbar and set the value of the **Distance** spinner to **35**.

19. Choose the **Add/Finish Mate** button from the **Mates** pop-up toolbar and then choose the **OK** button from the **Mate PropertyManager**.

20. Similarly, assemble the other components of Pipe Vice. Figure 10-80 shows the final Pipe Vice assembly.

21. Choose the **Save** button to save the assembly document in the *\My Documents\SolidWorks\Pipe Vice* folder.

Figure 10-80 *Final Pipe Vice assembly*

SELF-EVALUATION TEST

Answer the following questions and then compare your answers with those given at the end of this chapter.

1. The bottom-up assembly design approach is the traditional and the most widely preferred approach of assembly design. (T/F)

2. In the top-down assembly design approach, all components are created in the same assembly document. (T/F)

3. The **Coincident** mate is generally applied to make two planar faces coplanar. (T/F)

4. The most suitable mates that can be applied to the current selection set are displayed in the **Mate Selections** rollout of the **Mate PropertyManager**. (T/F)

5. Feature-based mates are applied only to the components having cylindrical features. (T/F)

6. Pattern-based mates are used to assemble the components that have a circular pattern created on the circular feature. (T/F)

7. Choose the __________ button from the **Assemblies CommandManager** to invoke the **Rotate Component PropertyManager**.

8. The __________ option available in the **Rotate** drop-down list is used to rotate the selected component by a given incremental angle about the specified axis.

9. The __________ mate is generally used to align the central axis of one component with that of the other.

10. The __________ option available in the **Move** drop-down list is used to move the component along the direction of the selected entity.

REVIEW QUESTIONS

Answer the following questions.

1. The names of the selected entities are displayed in the __________ selection area of the **Mate PropertyManager**.

2. Choose __________ from the menu bar to place a component in the assembly document.

3. Using the __________ option from the **Move** drop-down list, you can move the component dynamically along the X, Y, and Z axes of assembly document.

4. The __________ button available in the **Standard Mates** rollout is used to make the two selected entities normal to each other.

5. The __________ option is used to specify the coordinates of the origin of the part where the component will be placed after moving.

6. If you are adding feature-based mates using the features having conical geometry, then there must be a __________ face adjacent to the conical face of both features.

7. The most widely used method of adding the mates to the components in the assembly in SolidWorks is

   (a) **Smart Mates** (b) **Mate PropertyManager**
   (c) By dragging from part document (d) None of these

8. Which button is used to make the **Mate PropertyManager** available after applying a mate to the selected entities?

   (a) **Help** (b) **OK**
   (c) **Keep Visible** (d) **Cancel**

9. Which option available in the **Rotate** drop-down list is used to rotate the component with respect to the selected entity?

   (a) **Along Entity** (b) **Selected Edge**
   (c) **Reference Entity** (d) None of these

10. Which option is used to specify the coordinates of the origin of the part where the component will be placed after moving?

(a) **To XYZ Position** (b) **Reference Position**
(b) **Along Entity** (c) None of these

EXERCISE

Exercise 1

Create the Plummer Block assembly as shown in Figure 10-81. The dimensions of the components of this assembly are shown in Figures 10-82 through 10-84. **(Expected time: 1 hr)**

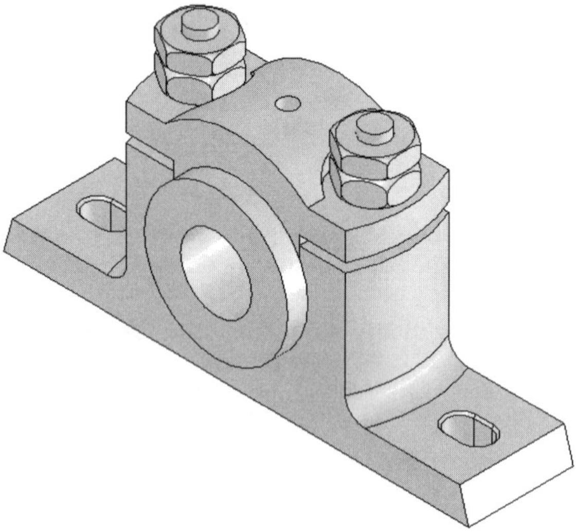

Figure 10-81 *Plummer Block assembly*

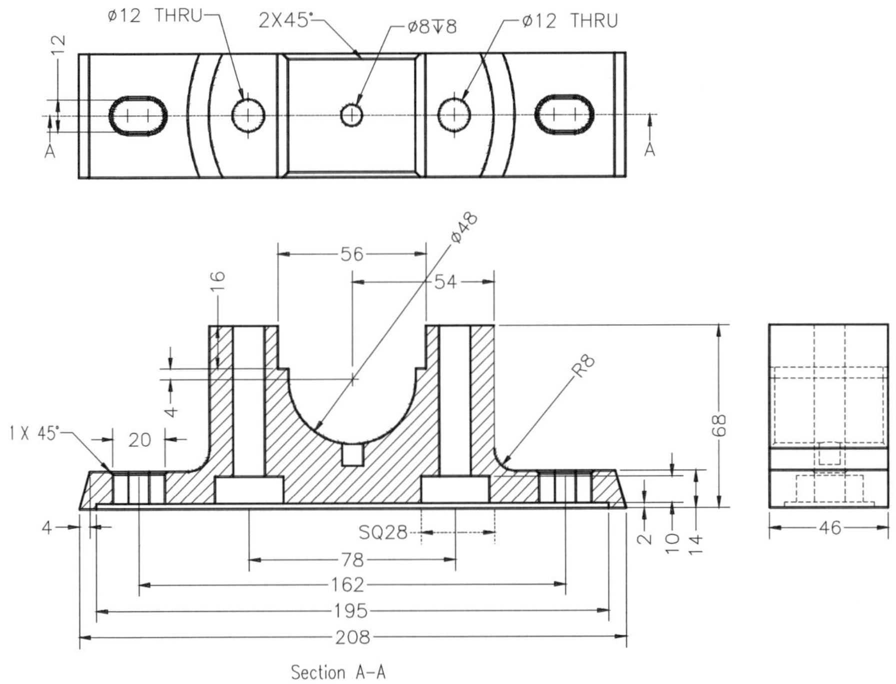

Figure 10-82 Views and dimensions of Casting

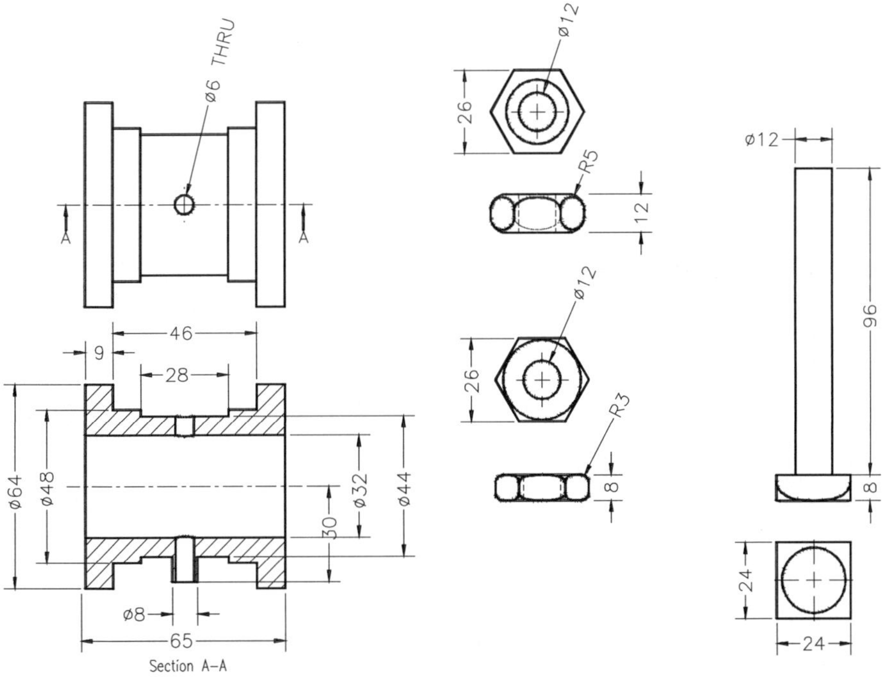

Figure 10-83 Views and dimensions of Brasses, Nut, Lock Nut, and Bolt

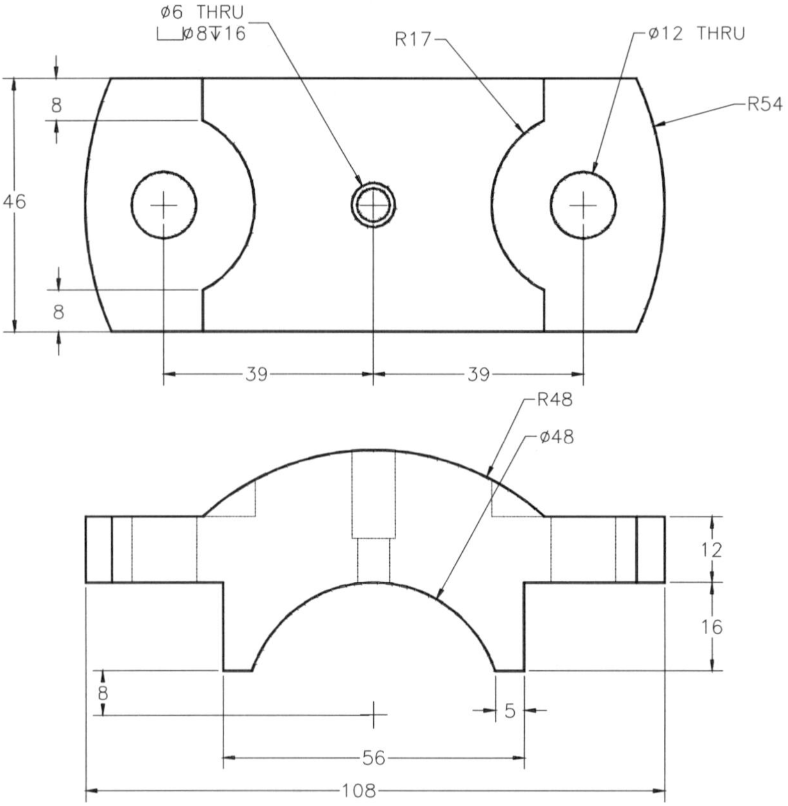

Figure 10-84 *Views and dimensions of Cap*

Answers to Self-Evaluation Test

1. T, **2.** T, **3.** T, **4.** T, **5.** T, **6.** T, **7. Rotate Component**, **8. By Delta XYZ**, **9. Coincident**, **10. Along Entity**

Chapter 11

Assembly Modeling-II

Learning Objectives

After completing this chapter you will be able to:

- *Apply advanced mates.*
- *Create subassemblies.*
- *Delete components and subassemblies.*
- *Edit assembly mates.*
- *Replace mate entities.*
- *Edit components and subassemblies.*
- *Dissolve subassemblies.*
- *Replace components in assemblies.*
- *Create patterns of components in the assembly.*
- *Create mirrored components.*
- *Hide and suppress components in assemblies.*
- *Change the transparency condition of assembly.*
- *Check the interference in an assembly.*
- *Create assemblies for mechanism.*
- *Detect collision while the assembly is in motion.*
- *Create the exploded state of an assembly.*
- *Create the explode line sketch.*

ADVANCED ASSEMBLY MATES

In the previous chapter, you learned how to place the components in an assembly document and also how to apply assembly mates to them. In this chapter you will learn how to apply advanced mates to the components by invoking the **Mate PropertyManager** and then the **Advanced Mates** rollout available in it. The **Advanced Mates** rollout is displayed in Figure 11-1.

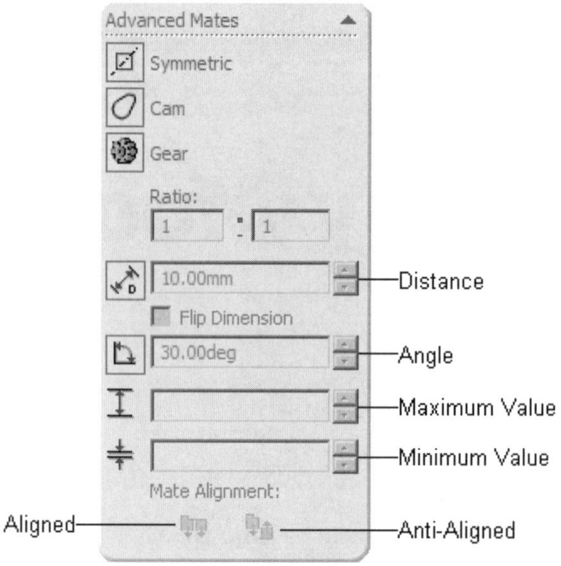

*Figure 11-1 The **Advanced Mates** rollout*

The advanced mates that you will learn in this chapter are the **Symmetric** mate, **Cam** mate, **Gear** mate, and **Limit** mate. All these mates are discussed next.

Applying the Symmetric Mate

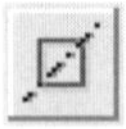

The **Symmetric** mate is applied to create a symmetric relation between two components. To apply this mate, choose the **Symmetric** button available in the **Advanced Mates** rollout; the **Symmetry Plane** selection area is displayed in the **Mate Selections** rollout. Select two similar entities from two components to which you need to apply the symmetric mate. The similar entities can be edges, vertices, planar faces, and so on. Next, you need to select a plane or a planar face that will act as the symmetry plane. Click once in the **Symmetry Plane** selection area to invoke the selection mode and select the symmetry plane; the **Symmetry Plane** callout is attached to the selected plane. Choose the **OK** button from the **Symmetric1 PropertyManager**. Figure 11-2 shows the entities and the plane to be selected to apply the **Symmetric** mate and Figure 11-3 shows the **Symmetric** mate applied to the assembly.

Applying the Cam Mate

You can add a **Cam** mate between a face that is formed by closed and continuous extruded faces from a single profile and a face of the follower. The face of the follower

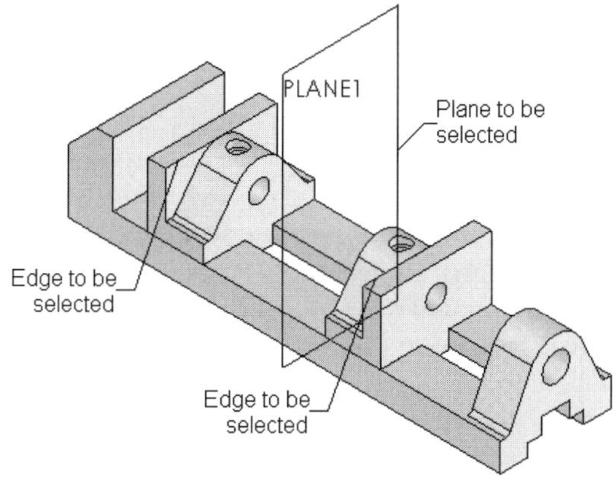

***Figure 11-2** Entities and the plane to be selected to apply the symmetric mate*

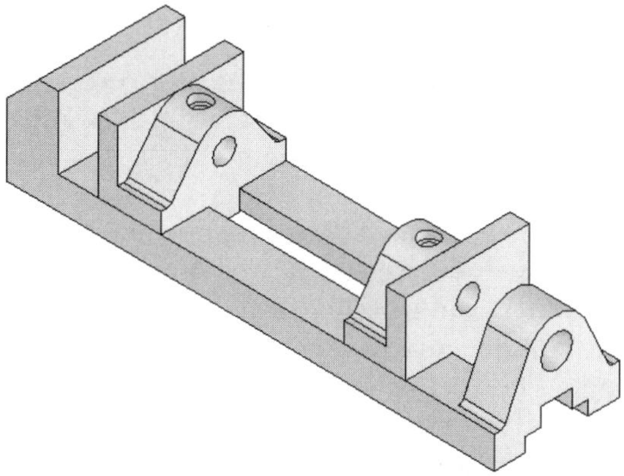

***Figure 11-3** Assembly after applying the symmetric mate*

could be cylindrical or planar, or it could be a vertex. To apply a **Cam** mate, you need to select a chain of continuous tangent faces that will act as surface of the cam. Next, select a curved face, a planar face, or a vertex from the follower, that will remain in contact with the selected chain of tangent faces. To apply a **Cam** mate, invoke the **Advanced Mates** rollout of the **Mate PropertyManager**. Choose the **Cam Follower** button from this rollout; the **Cam Follower** selection area is displayed in the **Mate Selections** rollout. Select the chain of continuous faces that will act as the cam surface. The selected faces are highlighted in green. Next, click once in the **Cam Follower** selection area and select the face from the second component that will act as the surface of the follower. The follower will touch the surface of the Cam and the **Cam Follower** callout will be attached to the selected face. Choose the **OK** button from the **CamMate Tangent1 PropertyManager**. On rotating the cam, the follower will attain the respective linear motion, that is, following the cam surface. Figure 11-4 shows the faces to be selected and Figure 11-5 shows the preview of the assembly, after applying the **Cam** mate.

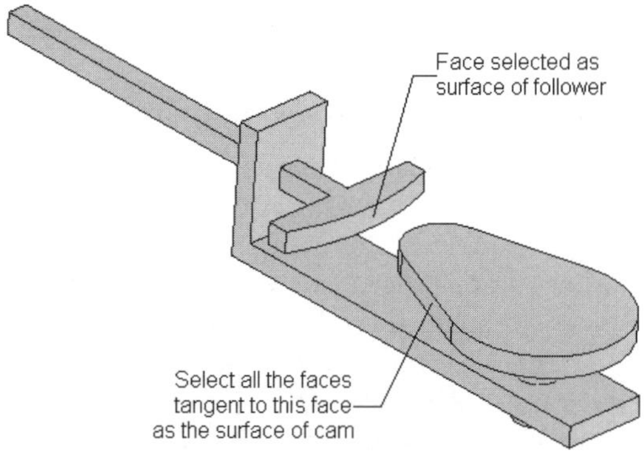

***Figure 11-4** Faces selected to apply cam mate*

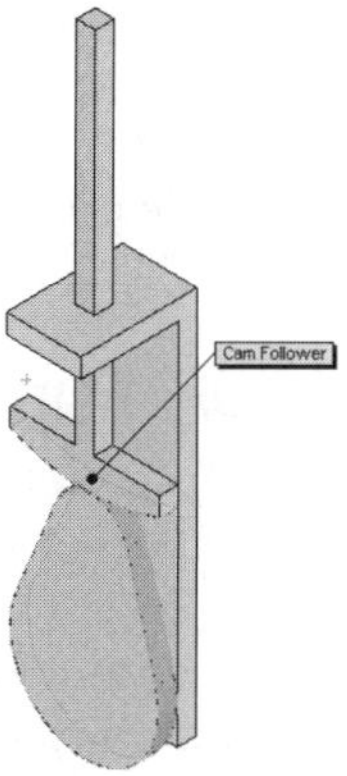

***Figure 11-5** Preview of the assembly with cam mate*

Note

For a better understanding of this mate, make sure that the front faces of the cam and follower are coplanar. Also, ensure that a proper degree of freedom of the components of the cam and follower assembly should be kept free.

Applying Gear Mate

The **Gear** mate allows you to rotate two components, such as gears, with respect to each other about a selected axis. You can also define a specific gear ratio between the selected components. To apply this mate between two components, invoke the **Advanced Mates** rollout from the **Mate PropertyManager**. Choose the **Gearmate** button from the **Advanced Mates** rollout; the **Ratio** edit boxes and the **Reverse** check box are also invoked. Next, select the references along which the two components will rotate. These references can be cylindrical faces, conical faces, axes, circular edges, or linear edges. Select the required references; the names of references are displayed in the **Entities to Mate** selection area. The **Teeth/Diameter** callouts are also displayed attached to the selected references. You can set the ratio of rotation using the **Ratio** edit boxes or using the **Teeth/Diameter** callouts. The **Reverse** check box is selected to change the direction of rotation of one component with respect to the other. Choose the **OK** button from the **GearMate1 PropertyManager**. After adding this mate, rotate any one of the component and you will observe the other component rotating with the specified gear ratio. Generally, this mate is used to display the rotation of the gears and direction. Figure 11-6 shows the cylindrical faces that you need to select to apply the gear mate. Figure 11-7 shows the **Teeth/Diameter** callouts displayed, after selecting the respective faces.

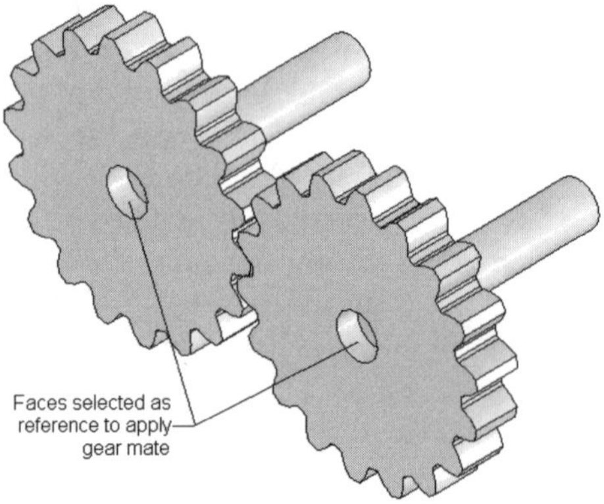

Figure 11-6 *Faces to be selected to apply gear mate*

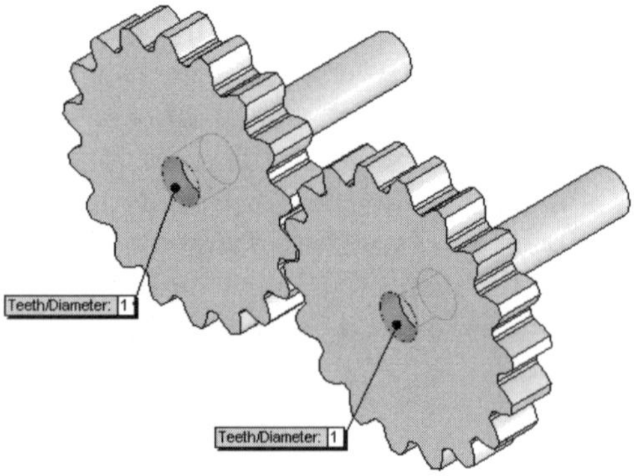

Figure 11-7 ***Teeth/Diameter*** *callouts attached to the selected faces*

Applying Limit Mate

The introduction of the limit mate is another major enhancement in the recent releases of SolidWorks. It is used to apply a distance mate or angle mate to the selected components, but the difference between a simple distance or angle mate and limit mate is that in limit mate you can move the component within the specific range of values. To apply a limit mate, invoke the **Advanced Mates** rollout from the **Mate PropertyManager** and then apply the distance or the angle mate, as required. Next, specify the maximum and minimum values of the limit mate in the **Maximum Value** and the **Minimum Value** spinners respectively. Choose the **OK** button from the **PropertyManager**.

CREATING SUBASSEMBLIES

In the previous chapter, you learned to place the components in the assembly document and apply the assembly mates to them. In this chapter you will also learn how to create the subassemblies and place them in the main assembly.

For placing the subassemblies in the main assembly, there are two approaches that are followed in the assembly design environment. These two approaches are discussed next.

Bottom-up Subassembly Design

In the bottom-up subassembly design approach, the subassembly is created in the assembly environment and then saved as an assembly file. To place the subassembly in the main assembly, open the main assembly document and invoke the **Insert Component PropertyManager**. On choosing the **Browse** button, the **Open** dialog box is displayed; select **Assembly** (**.asm, *.sldasm*) from the **Files of Type** drop-down list. All assemblies saved in the current location will be displayed in the selection area. Select the subassembly from the selection area and choose the **Open** button from the **Open** dialog box. Select a point in the drawing area to place the subassembly. Now, using the assembly mates, assemble the subassembly with the main assembly. Figure 11-8 shows a subassembly of a piston and an articulated rod. Figure 11-9 shows the main assembly of the piston and master rod. Figure 11-10 shows the main assembly, after assembling the subassembly with it.

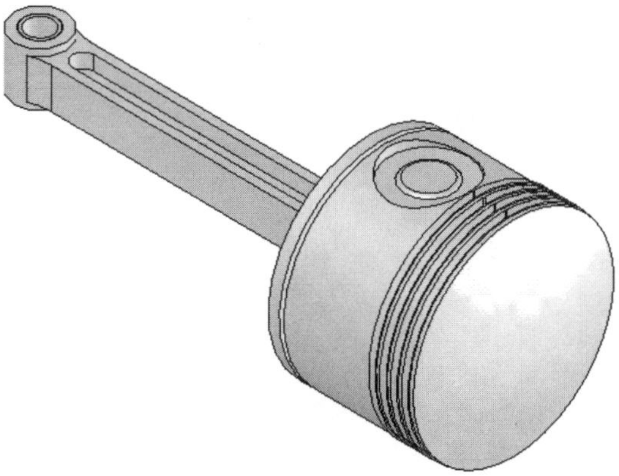

Figure 11-8 *The subassembly*

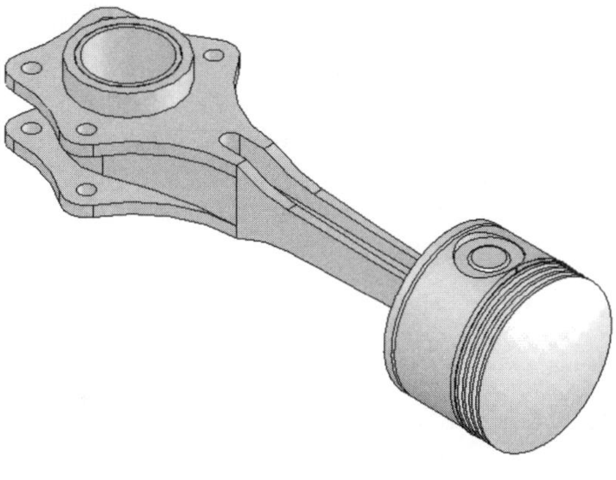

Figure 11-9 *The main assembly*

Top-down Subassembly Design

The top-down subassembly design approach is the most flexible subassembly design approach. In this, you create a new subassembly in the main assembly document. The approach is generally used in conceptual design or while managing a large assembly. To create a new subassembly in the assembly document, choose **Insert > Component > New Assembly** from the menu bar. The **Save As** dialog box is displayed; save the subassembly in the current location. You can drag and place the components in the new subassembly from the **FeatureManager Design Tree**. You will learn more about this approach, while discussing the editing of assemblies and subassemblies.

Tip. *You can also place a subassembly in the main assembly using the drag and drop method that was discussed in the previous chapter.*

When you place a subassembly in the main assembly, an assembly icon will be displayed, with the name of the subassembly in the ***FeatureManager Design Tree****. If you expand the subassembly in the* ***FeatureManager Design Tree****, all the parts assembled in the subassembly will be displayed.*

You can also create a subassembly of the components that are already placed in an assembly file. Press and hold down the CTRL key and select the components either from the drawing area or from the ***FeatureManager Design Tree****. Invoke the shortcut menu and choose either* ***Form New Sub-assembly Here*** *or* ***Assembly from [Selected] Components****. The* ***Save As*** *dialog box is displayed; you need to enter the name of the subassembly and save it in the current location. You will observe that the selected components will be combined and an assembly icon will be displayed in the* ***FeatureManager Design Tree****.*

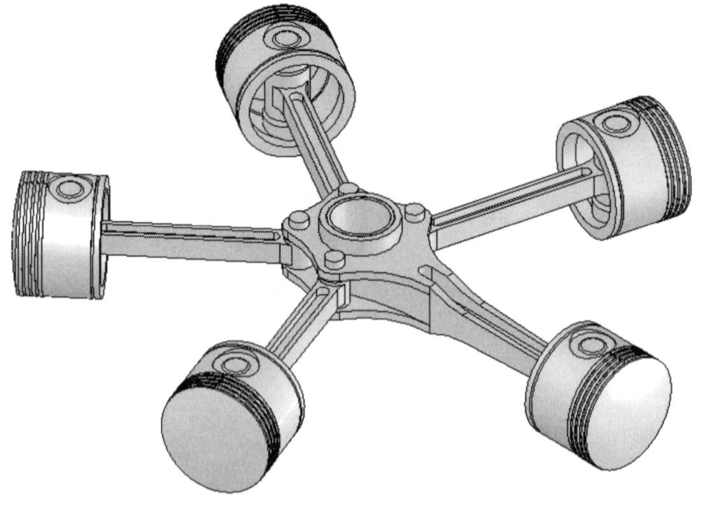

Figure 11-10 *Main assembly after assembling the subassemblies*

DELETING COMPONENTS AND SUBASSEMBLIES

After creating the assembly, at a certain stage of your design cycle you may need to delete any component or the subassembly. To delete a component of the assembly, select the component either from the drawing area or from the **FeatureManager Design Tree**. Invoke the shortcut menu and choose the **Delete** option from it. You can also delete the selected component by pressing the DELETE key on the keyboard. When you delete a component, the **Confirm Delete** dialog box is displayed. The name of the component and the items dependent on it are displayed in this dialog box. Choose the **Yes** button from the **Confirm Delete** dialog box.

To delete the subassembly, select it from the **FeatureManager Design Tree** and press the DELETE key. The **Confirm Delete** dialog box is displayed; choose the **YES** button from it. Note that on deleting the subassembly, all the components of the subassembly are also deleted.

EDITING ASSEMBLY MATES

You may need to edit the assembly mates, after creating the assembly or during the process of assembling the components. The editing operations that can be performed include modifying the type of the assembly mate, or angle and offset values, changing the component to which the mate was applied, and so on. To edit the mates in SolidWorks, you first need to expand the **Mates** option available at the bottom of the **FeatureManager Design Tree**. Next, select the mate that you need to modify, invoke the shortcut menu and choose the **Edit Feature** option. The **Mate PropertyManager** and the **Mate** pop-up toolbar are displayed. The name of the **Mate PropertyManager** will depend on the name and sequence of the mate applied. Figure 11-11 shows a partial view of the **Mate PropertyManager** to edit the **Coincident** mate. You can edit the entities to mate, type of mate, value of offset, value of angle, and so on using this **PropertyManager**.

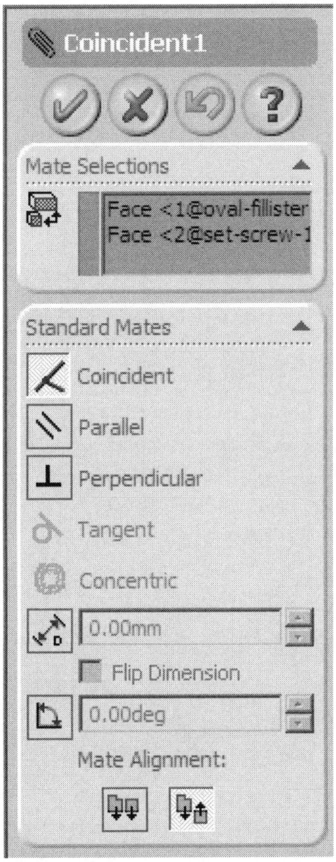

Figure 11-11 *Partial view of the* ***Mate PropertyManager*** *to edit the* ***Coincident*** *mate*

Tip. *The entities used in the mate are highlighted in red in the drawing area, when you move the cursor on a mate in the* ***FeatureManager Design Tree****. On selecting the mate from the* ***FeatureManager Design Tree****, the entities used in the selected mate will be highlighted on the drawing area in green.*

Replacing the Mated Entities

CommandManager:	Assemblies > Replace Mate Entities	*(Customize to Add)*
Toolbar:	Assembly > Replace Mate Entities	*(Customize to Add)*

As discussed, you can edit the mate entities using the **Mate PropertyManager**. They can also be modified using the **Mated Entities PropertyManager**. Select the **Mates** option from the **FeatureManager Design Tree,** invoke the shortcut menu and choose the **Replace Mate Entities** option. The **Mated Entities PropertyManager** is displayed, as shown in Figure 11-12. You can customize the **Assemblies CommandManager** and then select the specific mate whose entities you need to replace.

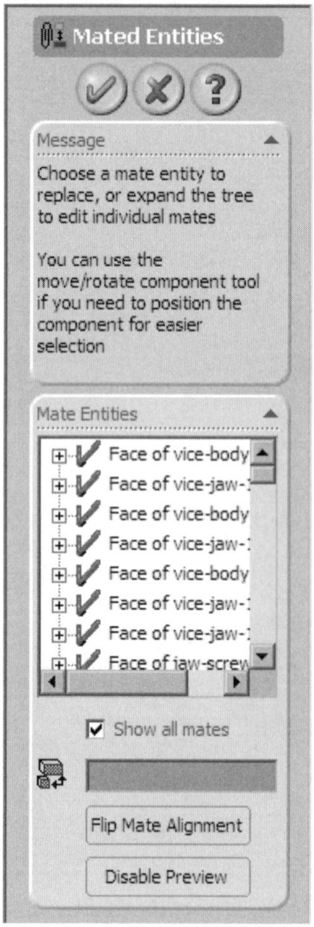

Figure 11-12 The ***Mated Entities PropertyManager***

Next, choose the **Replace Mate Entities** button from the **Assemblies CommandManager**. On invoking the **Mated Entities PropertyManager**, you are prompted to select the entity to be replaced. Select the entity from the **Mate Entities** selection area. You can also expand the entity tree to edit an individual mate. After selecting, the selected face will be highlighted on the drawing area in green its name will be displayed in the **Replacement Mate Entity** selection area. You are prompted to select the entity to be mated. Select the entity that will replace the previously selected entity. The SolidWorks warning box, informing the possible cause of error, will be displayed if the selected entity overdefines the mate, or the mating is not possible between the entities.

The **Flip Mate Alignment** button is used to flip the direction of mate. The **Disable Preview** button is used to disable the preview of the assembly after replacing the mate entity. The **Show All** check box is used to display all the mated entities.

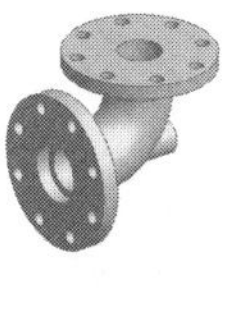

Tip. *Select the **Mates** option from the **FeatureManager Design Tree**, and invoke the shortcut menu. Choose the **Parent/Child Relationship** option from the shortcut menu. The **Parent Child Relationship** dialog box will be displayed. You can display the child and parent relationship of any component placed in the assembly using the **Parent Child Relationship** dialog box.*

EDITING COMPONENTS

CommandManager:	Assemblies > Edit Component
Toolbar:	Assembly > Edit Component

At some stage of your design cycle, you may need to edit the components, after placing and mating them in the assembly document. This may include editing the features, sketches, and sketch planes. To edit components, you first need to select the component and invoke the part modeling environment in the assembly document. To do so, select the component from the drawing area or from the **FeatureManager Design Tree** and choose the **Edit Component** button from the **Assembly CommandManager**. The part modeling environment will be invoked and the entire assembly, except the component to be edited, is displayed in transparent. If not, choose the **Assembly Transparency** button available in the **Assemblies CommandManager**; a flyout is invoked and the options available in it can be used to set the transparency of the assembly. On choosing the **Opaque** option, all components of the assembly are set to opaque. The **Maintain Transparency** option retains the default transparency settings of the individual components. If the **Force Transparency** option is selected, all components of the assembly, except the component being edited, are set as transparent. The name of the component to be edited will be displayed in blue in the **FeatureManager Design Tree**. Select the desired feature to edit, and invoke the respective **PropertyManager**. The **PropertyManager** corresponding to that feature will be displayed and you can easily edit the feature parameters. If required, you can also add new features to the component. This type of editing is technically termed as **Editing in the Context of Assembly**. After editing the component, again choose the **Edit Component** button from the **Assembly CommandManager** to return to the assembly environment.

Tip. *You can also modify the dimensions of a component assembled or those placed in the assembly, by double-clicking on the desired feature of that component. All dimensions of that feature will be displayed in the drawing area. Invoke the* ***Modify*** *dialog box by double-clicking on the dimension to modify. Enter the new dimension in the* ***Modify*** *dialog box and press the ENTER key on the keyboard; the dimension will be modified, but the geometry of the feature will not changed. You need to rebuild the assembly to reflect the modifications. To do so, choose the* ***Rebuild*** *button from the* ***Standard*** *toolbar or use the CTRL+B key on the keyboard.*

While in the part editing mode in the assembly document, you can use the ***Move/Size Features*** *tool to edit the features dynamically using editing handles.*

Note

To edit components separately in their part documents, select the component, invoke the shortcut menu and select the ***Open Part*** *option from it. The part document of the selected component will be opened. You can edit the component individually in the part document. After editing, save the part, close the part document and return to the assembly document. The* ***SolidWorks 2005*** *dialog box will be displayed, as shown in Figure 11-13, which prompts you that models contained within the assembly have changed. Would you like to rebuild the assembly now? Choose the* ***Yes*** *button from this dialog box.*

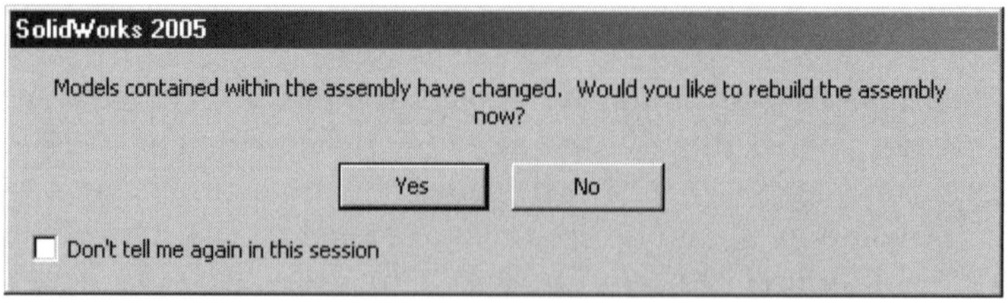

Figure 11-13 *The* ***SolidWorks 2005*** *dialog box*

EDITING SUBASSEMBLIES

CommandManager:	Assemblies > Edit Component
Toolbar:	Assembly > Edit Component

To edit subassemblies, select it from the **FeatureManager Design Tree** and choose the **Edit Component** button from the **Assembly CommandManager**. You can set the transparency setting using the **Assembly Transparency** button available in the **Assemblies CommandManager**. You can add components in the subassembly, modify the mates, and replace the components while in the editing mode. After editing the subassembly, choose the **Edit Component** button to exit the editing mode.

Note

*For editing a component of the subassembly, select the component from the drawing area, invoke the shortcut menu and choose the **Edit Part** option from it. The part editing mode will be invoked in the assembly document.*

*You need to expand the desired subassembly from the **FeatureManager Design Tree** to select its components. All the components assembled in that subassembly will be displayed once the subassembly is expanded.*

DISSOLVING SUBASSEMBLIES

Dissolving the subassembly means the components of the subassembly become the components of the next higher level subassembly. When you dissolve a subassembly, the subassembly is removed from the assembly and the components of the subassembly become the components of the assembly or the subassembly in which it was inserted. To dissolve a subassembly, select the subassembly from the **FeatureManager Design Tree,** invoke the shortcut menu and select the **Dissolve Sub-assembly** option from the shortcut menu. The subassembly will be removed from the **FeatureManager Design Tree** and its components will be displayed as the components of the assembly or the subassembly in the **FeatureManager Design Tree**.

REPLACING COMPONENTS

CommandManager:	Assemblies > Replace Components	*(Customize to Add)*
Toolbar:	Assembly > Replace Components	*(Customize to Add)*

Sometimes, in the assembly design cycle, you may need to replace a component of the assembly with some other component. To replace a component, select the component to be replaced, invoke the shortcut menu, expand it and select the **Replace Components** option to display the **Replace PropertyManager**, as shown in Figure 11-14. You can also invoke the **Replace PropertyManager** by choosing the **Replace Components** button from the **Assemblies CommandManager** after customizing it.

On invoking the **Replace PropertyManager**, you are prompted to select components to be replaced. If the component is already selected before invoking this tool, then it will remain selected. The name of the selected component is displayed in the **Components to be Replaced** selection area. Next, you need to specify the replacement component. Choose the **Browse** button from the **Selection** rollout; the **Open** dialog box will be displayed. Select the replacement component and choose the **Open** button from the **Open** dialog box. The name and location of the replacement component is displayed in the **Replacement Component** area.

The **All instances** check box available in the **Selection** rollout is used to replace all the instances of the selected component. The options available in the **Configuration** area are used to define the selection procedure of the configurations. You will learn more about configurations later.

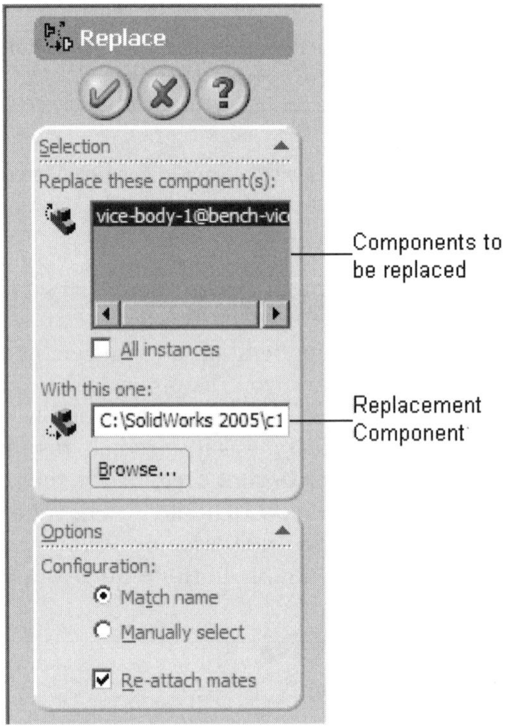

Figure 11-14** The **Replace PropertyManager

The **Re-attach mates** check box is selected by default. On exiting the **Replace PropertyManager** with this check box selected, the **What's Wrong** dialog box is displayed. Choose **Close** from this dialog box. The **Mated Entities PropertyManager** is displayed. You can replace the mate entities by using the **Mated Entities PropertyManager**.

If the **Re-attach mates** check box is cleared, the **What's Wrong** dialog box is displayed after you exit the **Replace PropertyManager**. This dialog box informs you the name of the mates that contain errors. Therefore, you need to redefine the mates. To do so, expand the **Mates** option from the **FeatureManager Design Tree**. Select the mate that contains symbol on the left. Invoke the shortcut menu and choose the **Edit Feature** option. The **Mate PropertyManager** will be displayed and you can edit the mate entities.

Figure 11-15 shows the assembly in which the bolt is to be replaced by a pin. Figure 11-16 shows the faces of the pin to be used as mate entities, after replacing the component. Figure 11-17 shows the isometric view of the assembly after bolts are replaced by pins.

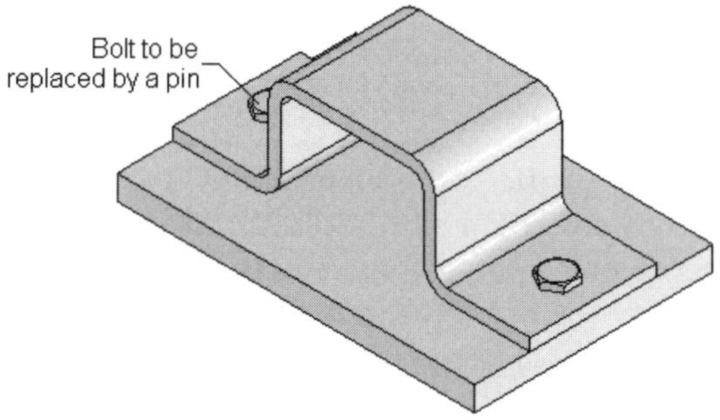

Figure 11-15 Bolt to be replaced by a pin

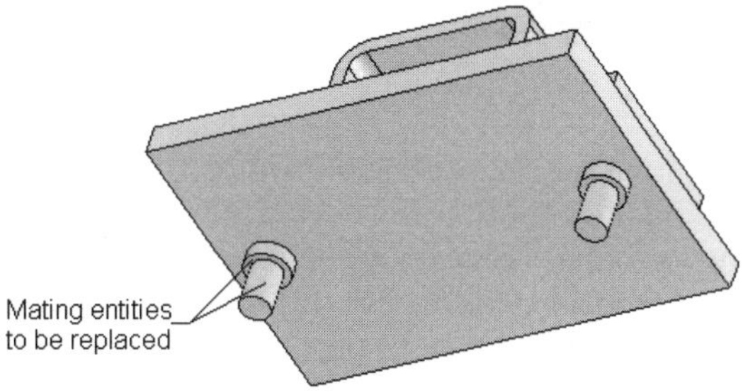

Figure 11-16 Mating entities to be replaced

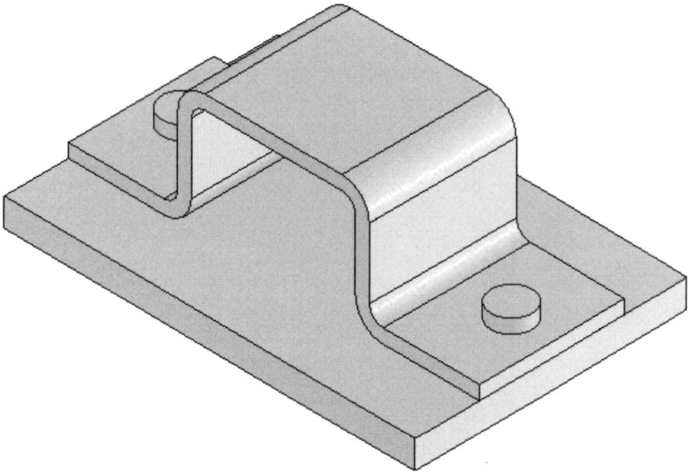

Figure 11-17 Bolt replaced by pin in the assembly

Tip. *The ⚠ symbol displayed on the* ***Mates*** *option in the* ***FeatureManager Design Tree****, means that the mates group has errors in it. Expand the mates group; the mate that displays ⚠ sign will have some errors. Select that mate, invoke the shortcut menu and choose the* ***Whats Wrong?*** *option to display the* ***What's Wrong*** *dialog box. The possible cause of the error will be displayed in it.*

If you choose the ***Mate Diagnostics*** *option from the shortcut menu, the* ***Diagnostics PropertyManager*** *will be displayed. The* ***Diagnose*** *button in this* ***PropertyManager*** *is used to display the entities that are the cause of errors in the mate. Right-click on the mate and choose the* ***Edit Mates*** *option from the shortcut menu to edit the mate with errors.*

CREATING PATTERNS OF COMPONENTS IN AN ASSEMBLY

Menu: Insert > Component Pattern > Linear Pattern/ Circular Pattern/ Feature Driven

You may need to assemble more than one instance of the component about a specified arrangement, while working in the assembly design environment of SolidWorks. Consider the case of a flange coupling where, you have to assemble eight instances of the nut and bolt to fasten the coupling. In this case, you need to make the instances of the nut and bolt, and then assemble the eight nuts and the eight bolts manually. However, this is a very tedious and time-consuming precess. Therefore, to reduce time, SolidWorks has provided a tool to create the patterns of the components. The three types of component patterns provided in SolidWorks are discussed next.

Feature-driven Pattern

The feature driven pattern is used to pattern the instances of the components using an existing pattern feature. To create a derived pattern, choose **Insert > Component Pattern > Feature Driven** from the menu bar; the **Feature Driven PropertyManager** is displayed, as shown in Figure 11-18.

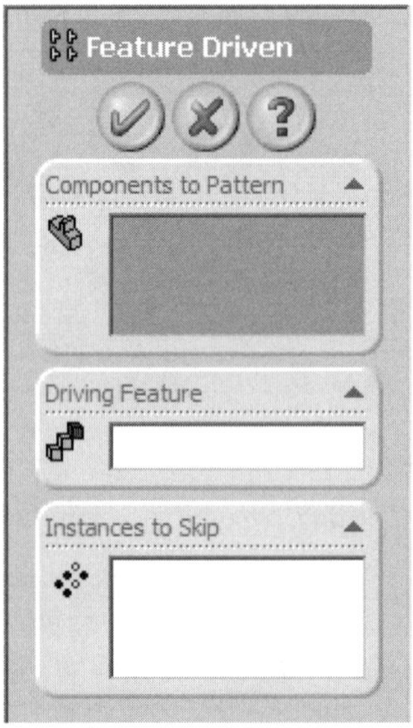

Figure 11-18** The **Feature Driven PropertyManager

You are prompted to select the driving feature pattern and the components to pattern. Select the components to pattern from the drawing area. The selected components will be displayed in green. Click once in the **Driving Feature** selection area and select a pattern feature that will drive the component pattern. The preview of the resulting pattern is displayed in the drawing area. Choose the **OK** button from the **Feature Driven PropertyManager**.

Figure 11-19 shows the components to be selected to pattern and an instance of an existing pattern feature to be selected. Figure 11-20 shows the assembly after creating the derived pattern feature.

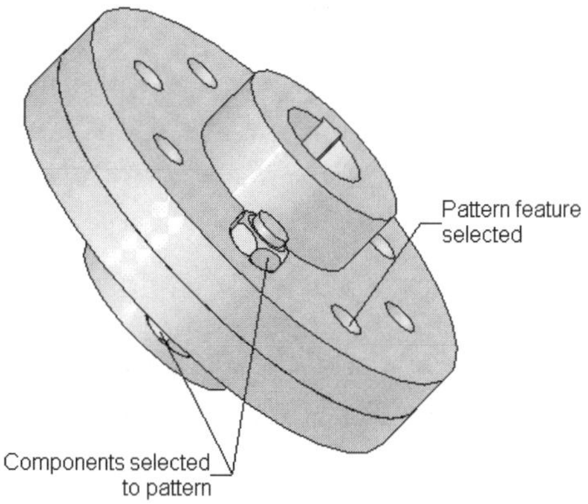

***Figure 11-19** Components and the instance of the pattern feature to be selected*

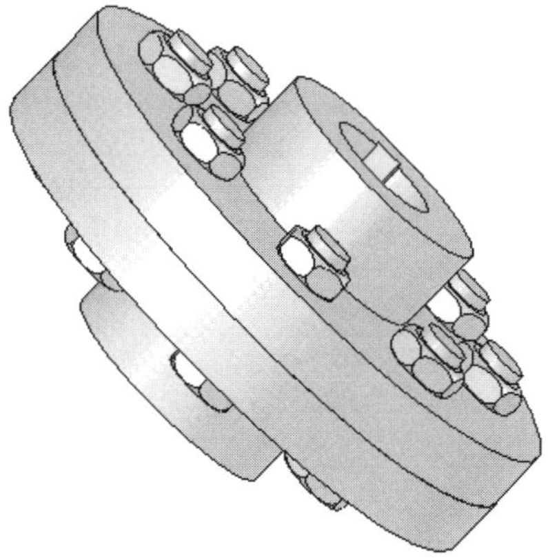

***Figure 11-20** Assembly after creating the derived component pattern*

In order to skip some of the instances of the pattern after it has been created, select the derived pattern feature from the **FeatureManager Design Tree** and invoke the shortcut menu. Choose the **Edit Feature** option from the shortcut menu, the resulting **PropertyManager** is displayed. Click once in the **Instances to Skip** selection area. Filled dots in magenta are displayed with the preview of the instances of the pattern. Select the dot on the instance that you need to skip; the selected instances disappear from the display and the color of the dot changes to red. To retain the instances again, select the dot. After selecting the instances to skip, choose the **OK** button from the **PropertyManager**.

Note

The number of instances of the component pattern are automatically modified when you change the number of instances of the pattern feature, using which the derived component pattern is created. This indicates the associative nature of the derived component pattern.

*You may have to choose the **Rebuild** button, after editing the number of entities in the pattern feature.*

Remember that you need to modify the mates if the feature that was used to assemble the seed component is deleted while reducing the number of features in the pattern. This is because the original feature on which the mates were applied does not exist any more and so the mates give an error.

Local Pattern

You can also create the patterns of the components individually, even if there is no pattern feature. This type of a component pattern is known as a local pattern. You can create two types of local pattern: first is the linear pattern and the other is the circular pattern. Both the types of local patterns are discussed next.

Tip. *You will observe that after creating a derived circular pattern, the* ***DerivedCirPattern1*** *feature will be displayed in the* ***FeatureManager Design Tree****. The name of the feature will be* ***DerivedLPattern1,*** *if the derived pattern creates a rectangular pattern of the components. The number at the end of the feature displays the sequence number of the derived pattern feature.*

On expanding the component pattern feature, all instances of the patterned component are displayed. Select the instance to be skipped from the ***FeatureManager Design Tree*** *and press the DELETE key. The* ***Confirmation*** *dialog box will be displayed; choose the* ***Yes*** *button from it.*

To restore the skipped pattern instance, select the derived pattern from the ***FeatureManager Design Tree*** *and invoke the shortcut menu. Choose the* ***Edit Feature*** *option. Select the skipped instance from the* ***Instances to Skip*** *rollout and invoke the shortcut menu. Choose the* ***Delete*** *option; the instance is restored.*

Linear Pattern

Choose **Insert > Component Pattern > Linear Pattern** from the menu bar, for creating the local linear pattern. The **Linear Pattern PropertyManager** is displayed, and you are prompted to select an edge or axis for direction reference, and also the components to pattern. Select the direction reference and then the component to pattern from the drawing area. Other options available in this **PropertyManager** are the same as those discussed, while creating a linear pattern of the features, faces, and bodies in Chapter 7.

Circular Pattern

For creating the local circular pattern, choose **Insert > Component Pattern > Circular Pattern** from the menu bar. The **Circular Pattern PropertyManager** is displayed, and you are prompted to select an edge or an axis for direction reference, and also the components to pattern. Select the direction reference and then the component to pattern from the drawing area. Other options available in this **PropertyManager** are the same as those discussed, while creating circular pattern of the features, faces, and bodies in Chapter 7.

Tip. *If one instance of the component pattern is modified or edited, the other instances of the component will also be modified.*

You can also create the component pattern of a component pattern feature.

MIRRORING COMPONENTS

Menu: Insert > Mirror Components

In the **Assembly** mode of SolidWorks, you can also mirror a component to place the new instance of the component in the assembly document. To mirror the component, choose **Insert > Mirror Components** from the menu bar. The **Mirror Components PropertyManager** is displayed as shown in Figure 11-21.

Figure 11-21** Partial view of the **Mirror Components PropertyManager

Select the planar face or plane that will act as a mirror plane, as shown in Figure 11-22. Next, select the component to mirror; you will observe that the name of the component with a check box is displayed in the **Components to Mirror** area. Leave this check box cleared and choose the **Next** button from the **Mirror Components PropertyManager**. The preview of the mirrored component is displayed, as shown in Figure 11-23. The **Orientation** rollout is displayed and using the options available in this rollout, you can change the orientation of the mirrored component. Choose the **OK** button from the **Mirror Components PropertyManager**. The mirrored instance of the selected component is displayed in the drawing area, as shown in Figure 11-24.

If you select the check box available with the name of the selected component in the **Components to Mirror** area, a new component, mirrored along the selected mirror plane and saved in a specified location, is created. Select the check box available along the name of the selected component and choose the **Next** button. The **Filenames** rollout is displayed;

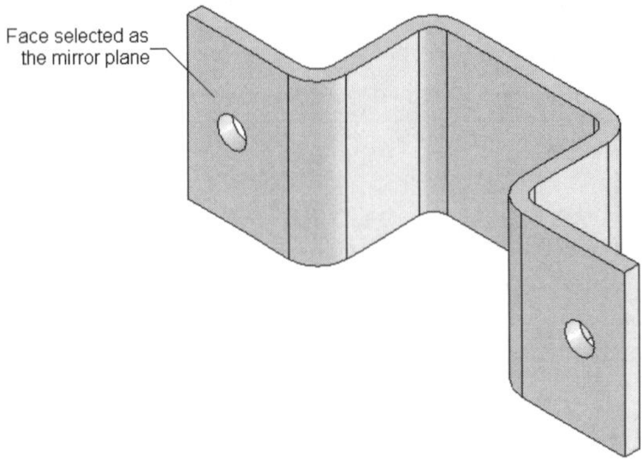

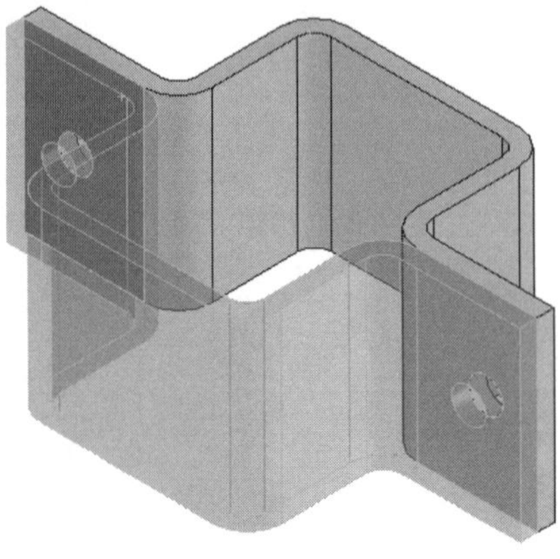

Figure 11-22 *Face to be selected as mirror plane* ***Figure 11-23*** *Preview of the mirrored component*

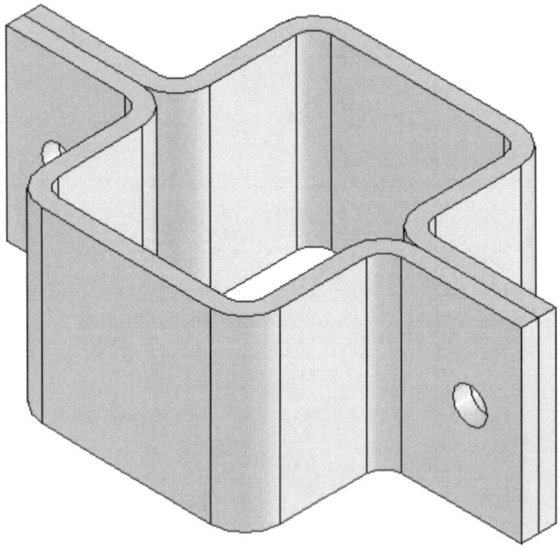

Figure 11-24 *Instance created by mirroring the component*

choose the **Browse** button available in the **Filenames** rollout. This button is used to specify the name and location where the file will be saved. You can save all the mirrored instances in one folder by selecting the **Place files in one folder** check box. You can also add a suffix or a prefix to the file name by using the **Add** drop-down list.

SIMPLIFYING ASSEMBLIES USING THE VISIBILITY OPTIONS

When you are assembling the components in an assembly, whether it be a large assembly or a small assembly, you may need to simplify the assembly using the visibility options. By simplifying, you can hide the components and or set its transparency at any stage of the design cycle. You can also suppress and unsuppress the components at any stage of the design cycle. The various methods of simplifying the assembly are discussed next.

Hiding Components

CommandManager: Assemblies > Hide/Show Components
Menu: Edit > Hide > This Configuration
Toolbar: Assembly > Hide/Show Components

In order to hide the component placed in the assembly, select the component from the drawing area or from the **FeatureManager Design Tree**. You can select more than one component to hide by using the CTRL key. Right-click to invoke the shortcut menu and select the **Hide** option from it; the component will disappear from the drawing area. The icon of the hidden component is displayed in transparent in the **FeatureManager Design Tree**. To turn on the display of the hidden component, select the icon of the component from the **FeatureManager Design Tree** and invoke the shortcut menu. Select the **Show** option from the shortcut menu. The hidden component will be redisplayed in the drawing area. You can also use the **Hide/Show Components** button from the **Assembly CommandManager** to hide or show the components.

Tip. *You will observe that the* ***Lightweight*** *check box is provided in the* ***Open*** *dialog box. On selecting this check box before opening the assembly file, the assembly will be opened only with the lightweight components.*

A lightweight component is one in which the feature information is available in the part document, and only the graphical representation of the component is displayed in the assembly document. Therefore, the assembly environment becomes light. An icon of a lightweight component is displayed as a feather attached to the component icon in the ***FeatureManager Design Tree****.*

To get the feature information of the lightweight component, you need to resolve the component to the normal state. Therefore, select the component from the drawing area or from the ***FeatureManager Design Tree*** *and invoke the shortcut menu. Select the* ***Set to Resolved*** *option from this shortcut menu. To set a resolved component to a lightweight component, select the component and invoke the shortcut menu. Select* ***Set to Lightweight*** *from the shortcut menu.*

Suppressing Components

CommandManager: Assemblies > Change Suppression State
Menu: Edit > Suppress > This Configuration
Toolbar: Assembly > Change Suppression State

You can also suppress the components placed in the assembly to simplify the assembly representation. To do so, select the component from the drawing area or from the **FeatureManager Design Tree,** invoke the shortcut menu, and choose **Suppress**. The component will not be displayed in the assembly document and the icon of the suppressed component will be displayed in grey in the **FeatureManager Design Tree**. To unsuppress the suppressed component, select the component to be resolved from the **FeatureManager Design Tree,** invoke the shortcut menu and choose the **Set to Resolve** option from it.

You can also suppress the component using the **Change Suppression State** button from the **Assembly CommandManager**. You can not only suppress the selected component but also set the selected component to lightweight by using the **Change Suppression State** button. When you select the component and choose the **Change Suppression State** button from the **Assembly CommandManager**, a flyout is displayed. This flyout has three options: **Suppress** option, **Lightweight** option, and **Resolve** option. The **Resolve** option is used to set the suppressed or lightweight components to resolve state.

Changing the Transparency Conditions

In SolidWorks, you can change the transparency of the components or selected faces to simplify the assembly. First, you will learn how to change the transparency of the components. Select the component to change its transparency and invoke the shortcut menu. Choose the **Properties** option from the **Component** area of the shortcut menu; the **Component Properties** dialog box will be displayed. Choose the **Color** button available in this dialog box; the **Assembly Instance Color** dialog box is displayed, as shown in Figure 11-25.

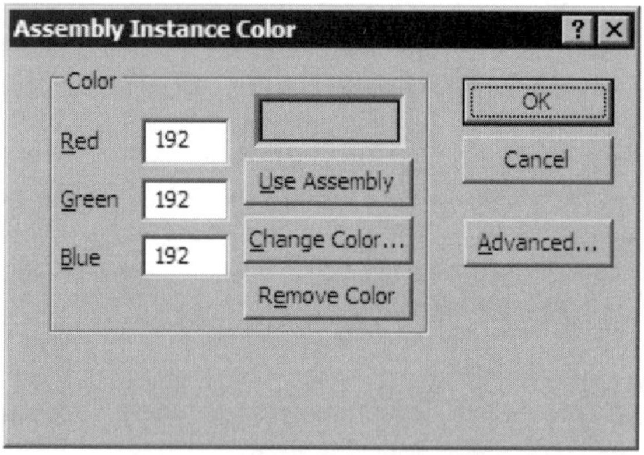

*Figure 11-25 The **Assembly Instance Color** dialog box*

Choose the **Advanced** button from this dialog box; the **Advanced Properties** dialog box is displayed, as shown in Figure 11-26. You can set the transparency of the component from the **Transparency** slider. You can also set the various other advanced color settings such as **Ambient**, **Diffuse**, **Specularity**, **Shininess**, and **Emission** from this dialog box.

You can also change the transparency of a face. To do so, select the face, invoke the shortcut menu and choose the **Properties** option from the **Face** area of the shortcut menu; the **Entity Property** dialog box is displayed. Choose the **Advanced** button from this dialog box to invoke the **Advanced Properties** dialog box that can be used to set the transparency of the selected face.

Tip. *The face that has been made transparent cannot be selected by clicking. You need to select some other faces, and then choose the **Select Other** option from the shortcut menu. Right-click until the desired face is highlighted and then left-click to select it. Now, right-click anywhere in the drawing area to display the shortcut menu.*

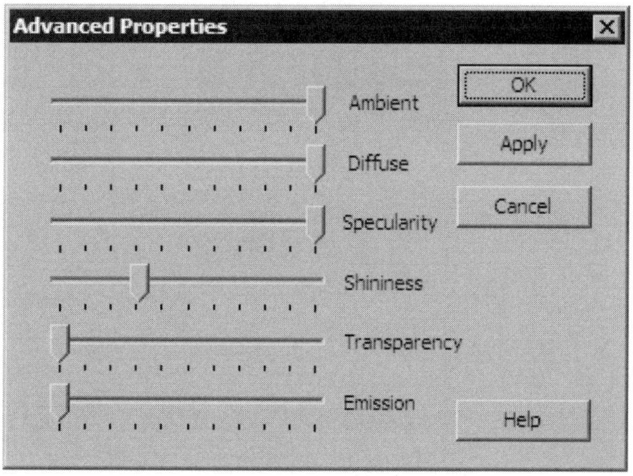

***Figure 11-26** The **Advanced Properties** dialog box*

CHECKING INTERFERENCES IN AN ASSEMBLY

CommandManager:	Assemblies > Interference Detection
Menu:	Tools > Interference Detection
Toolbar:	Assembly > Interference Detection

After creating the assembly design, the first and the most essential step is to check the interference between the components of the assembly. If there is interference, the components may not assemble properly after they roll out from the machine shop or tool room. Therefore, before sending the part and assembly files for the detailing and drafting purpose, it is essential to check the interference. To do so, choose the **Interference Detection** button from the **Assemblies CommandManager**; the **Interference Detection PropertyManager** is displayed. The name of the current assembly is displayed in the **Components to Check** selection area in the **Selected components** rollout. You can also check the interference between two or more than two components. To select components, first clear the current selection set and select the components from the assembly. Next, choose the **Calculate** button from this **Selected components** rollout to check the interference. If there is any interference between the components in the assembly, it will be displayed in the assembly and also the interfering components will be displayed in the **Results** rollout, as shown in Figure 11-27. Expand the names of the interfering components to view the cause of interference. You can select a component/interference from the **Results** rollout and choose the **Ignore** button to ignore that particular component/interference while calculating the interference.

The **Treat coincidence as interference** check box available in the **Options** rollout considers the **Coincident** mates also as interference. The **Show ignored interferences** check box is selected to show the interferences that you ignored while using the **Ignore** button in the **Results** rollout. If you select the **Treat subassemblies as components** check box, all subassemblies in the current assembly will be treated as single components and any interference in the subassembly will be ignored. The **Include multibody part interferences** check box is selected to analyze the multibody parts for interference. If the **Make interfering**

Figure 11-27** Interfering parts displayed in the **Interference Detection PropertyManager

parts transparent check box is selected, the interfering parts are made transparent in the drawing window. The **Create Fasteners Folder** check box is selected to place the interfering fasteners in a separate folder in the **Results** rollout.

The **Non-interfering Components** rollout is used to select the display option for the non-interfering parts. After analyzing the assembly you can edit or modify the part.

CREATING ASSEMBLIES FOR MECHANISM

As mentioned earlier, there are two types of assemblies. The first one is a fully defined assembly in which the relative movement of all the components is contained. The second type is the one in which the components are not fully defined and some degree of freedom is kept unconstrained. As a result, they can move in a certain direction with respect to the surroundings of the assembly. This flexibility in turn helps you to create mechanisms and then you can move the assembly to check the mechanisms that you have designed. Consider the case of a Bench Vice, in which you are assembling the Vice Jaw with the Vice Body. For this assembly to work, the linear movement of the Vice Jaw, when placed on the Vice Body, should be free. Therefore, while creating this assembly for mechanism, you should not apply the mates for constraining the linear motion of the Vice Jaw with respect to the Vice Body. Figure 11-28 shows the degree of freedom that is required to be free to create an assembly for motion.

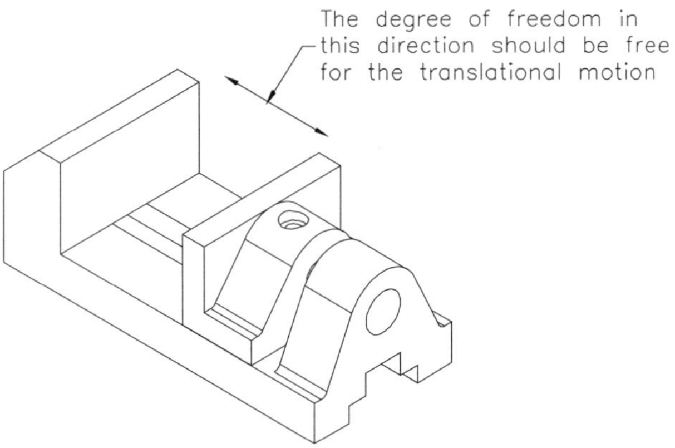

***Figure 11-28** Direction in which the degree of freedom should be free*

After creating the assembly for mechanism by defining minimum mates, invoke the **Move Component** tool. Select one of the faces of the component that you need to move, and drag the cursor to move the assembly. The options available for analyzing the assembly, while moving the assembly for mechanism design, are discussed next.

Analyzing Collisions Using the Collision Detection Tool

In SolidWorks, you can also analyze any collision between the components of the assembly while the assembly is in motion. Invoke the **Move Component PropertyManager** and then the **Options** rollout. Select the **Collision Detection** radio button, as shown in Figure 11-29; the **Check between** area is displayed. The options available in this area are used to specify the components between which the collision will be detected. These options available in the **Check between** area are discussed next.

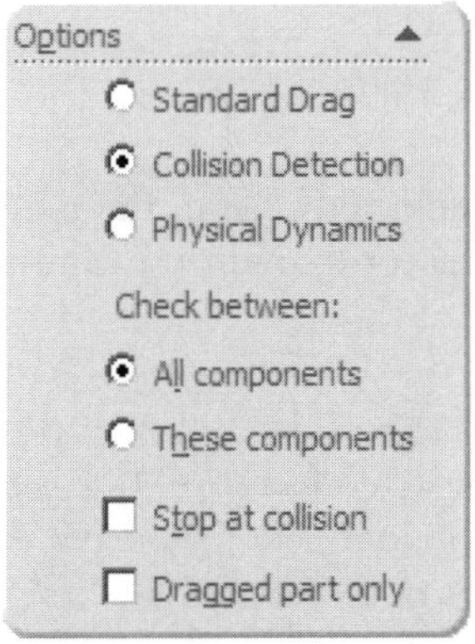

Figure 11-29** The **Options** rollout of the **Move Component PropertyManager

All components

This radio button is selected to check the collision between all components of the assembly.

These components

The **These components** radio button is selected to check the collision only between the selected components when the assembly is in motion. On selecting this radio button, the **Components for Collision Check** selection area and the **Resume Drag** button is displayed in the **Options** rollout, as shown in Figure 11-30. The name of the components, between which the collision is to be detected, is displayed in the **Components for Collision Check** selection area. After selecting the components, choose the **Resume Drag** button from the **Options** rollout, and drag the cursor to move the assembly.

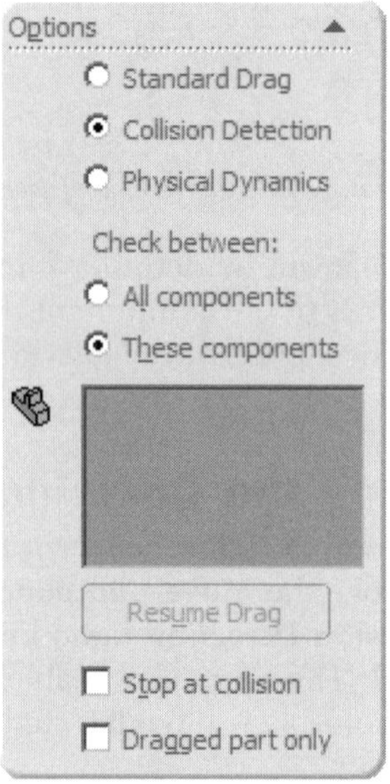

*Figure 11-30 The **Options** rollout with the **These components** radio button selected*

Stop at collision

The **Stop at collision** check box is selected to stop the motion of the assembly when one of the component collides with another component during the assembly motion.

Dragged part only

The **Dragged part only** check box is to be selected when you need to detect the collision only between the components that you selected to move. If this check box is cleared, the collision is determined between the components that you have selected to move and any other component that moves because of mates with the selected components.

After setting the options in the **Options** rollout, drag the assembly either using the **Move Component** tool or using the **Rotate Component** tool. If a component of the assembly collides with another component while the assembly is in motion, the faces of the components that collide with each other, will be displayed in green. If the **Stop at collision** check box is selected, the motion of the assembly will be stopped, when one of the components collides with another component of the assembly.

Consider the assembly shown in Figure 11-31. In this assembly, you need to move the slider in the given direction. Invoke the **Move PropertyManager** and select the **Collision Detection** radio button from the **Options** rollout. Drag the slider to move in the given direction. Figure 11-32 shows that the slider collides with the extrusion feature created in the vertical column of the base component. The faces of the components that collide are displayed in green. If the **Stop at collision** check box is selected, you cannot move the component further, after it collides with one of the components of the assembly.

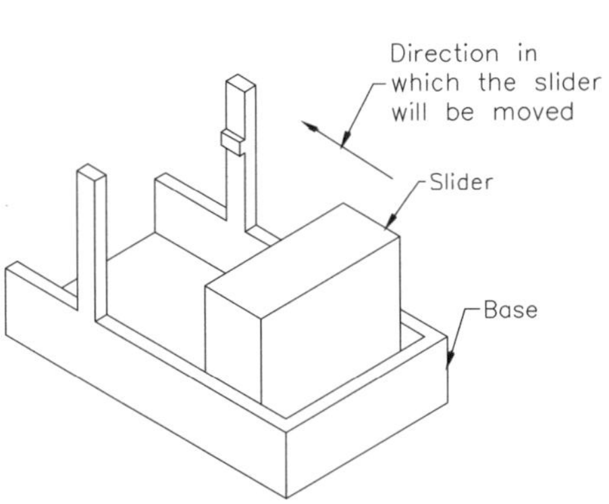

Figure 11-31 *Direction in which the slider will be moved inside the base*

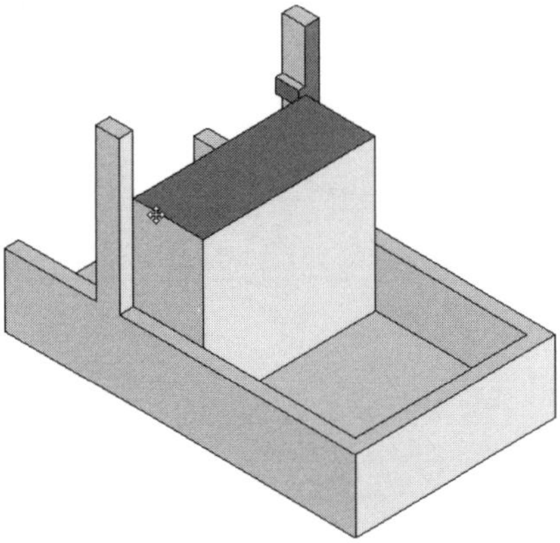

Figure 11-32 *The faces of the components highlighted in green after collision*

Once the collision is detected in the assembly, you can edit and modify the components that collide during the assembly motion.

CREATING THE EXPLODED STATE OF AN ASSEMBLY

Enhanced

CommandManager:	Assemblies > Exploded View
Menu:	Insert > Exploded View
Toolbar:	Assembly > Exploded View

In SolidWorks, you can create an exploded state of the assembly using the **Explode PropertyManager**. To invoke this **PropertyManager**, choose the **Exploded View** button from the **Assemblies CommandManager**. You can also invoke this **PropertyManager** by first invoking the **ConfigurationManager** and then selecting the **Default** option from the **ConfigurationManager** for invoking the shortcut menu. Choose the **New Explode View** option from it. The **Explode PropertyManager** will be displayed, as shown in Figure 11-33.

The **Explode PropertyManager** allows you to explode an assembly using the manipulator. After invoking the **Explode PropertyManager**, select the component to be exploded; the manipulator is displayed on the component and the name of the component is displayed in the **Component(s) of the explode step** selection area of the **Settings** rollout. Now, to explode the component along any of the axes of the manipulator, move the cursor over the arrowhead

Figure 11-33 The ***Explode PropertyManager***

of that axis. When the cursor changes into the move cursor, press and hold the left mouse button down and drag the cursor to explode the component. You can also set the exact numeric value of the explode distance. To do this, select the direction from the manipulator and set the value of the explosion distance in the **Explode distance** spinner provided in the **Settings** rollout. Now, choose the **Apply** button provided below this spinner. You can select an edge or a face to define the direction of explosion.

If the **Auto-spec components after drag** check box is selected in the **Options** rollout, the explosion is displayed as **Chain 1** in the **Existing explode steps** area of the **Explode Steps** rollout. However, if the **Auto-spec components after drag** check box is cleared, the explosion is displayed as **Explode Step 1**. You will learn more about the **Auto-spec components after drag** check box later in this chapter. To edit the **Chain 1** or **Explode Step 1**, select it and invoke the shortcut menu. Choose the **Edit Step** option from the shortcut. Edit the position of the selected component.

You can delete an explode step by selecting it and choosing the **Delete** option from the shortcut menu. After exploding a component, select another component to explode. Similarly, you can create as many explode steps as you want.

You can also perform the auto-explosion by selecting all the components of the assembly together. You can drag a window across the assembly to select all its components. Now, select the direction from the manipulator, along which you need to explode all the components. Set the value of the explode distance in the **Explode distance** spinner. Choose the **Apply** button.

The options available in the **Options** rollout, shown in Figure 11-34, are used to specify the explode options. The **Auto-space components after drag** check box is used to specify an automatic spacing between the components, after exploding them using the auto-explode method. The **Adjust the spacing between the chain components** slider bar is used to adjust the spacing between the chain of components in order to define the automatic spacing between the components after dragging. The **Select sub-assembly's parts** check box is used to also select the components of the subassembly for exploding.

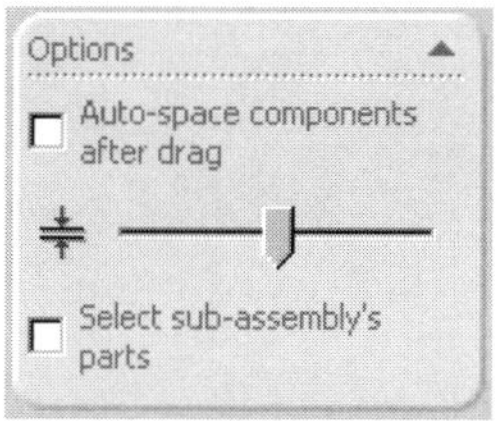

Figure 11-34** **Options** rollout of the **Explode PropertyManager

The **Re-use Sub-assembly Explode** button is chosen to reuse the explode steps of a subassembly also in the current assembly.

Figures 11-35 shows the assembly exploded using the auto-explode method. Figure 11-36 show assemblies exploded using the **Explode PropertyManager**.

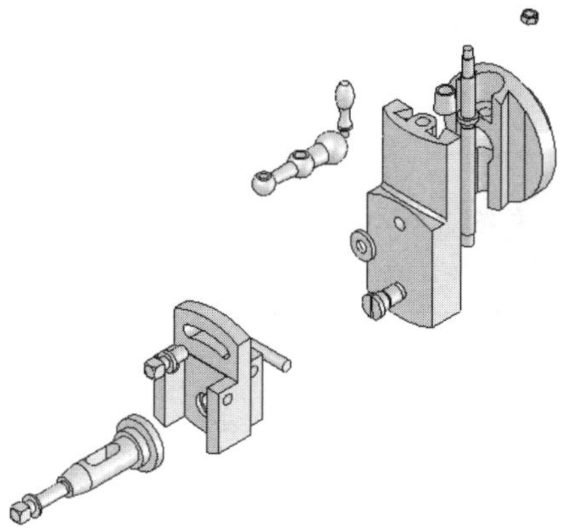

***Figure 11-35** Auto-explosion of an assembly*

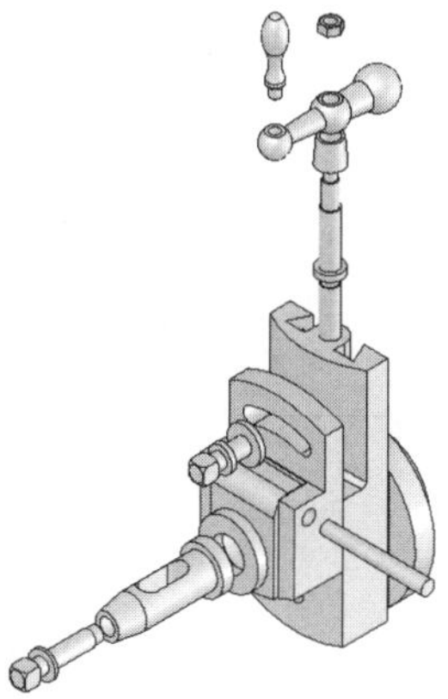

***Figure 11-36** A systematic exploded state of an assembly*

Creating the Explode Line Sketch

CommandManager:	Assemblies > Explode Line Sketch
Menu:	Insert > Explode Line Sketch
Toolbar:	Assemblies > Explode Line Sketch

Explode lines are the parametric axes that display the direction of explosion of the components in the exploded state. Figure 11-37 shows an exploded assembly with explode lines.

To create an explode line sketch, choose the **Explode Line Sketch** button from the **Assembly CommandManager**. The sketching environment in invoked and the **Explode Sketch** toolbar is displayed. Also, the **Route Line PropertyManager** is displayed, as shown in Figure 11-38. You are prompted to select a cylindrical face, planar face, vertex, point, arc, or line entities. One by one, select the cylindrical faces of the two components to create an explode line between them. For example, in order to create an explode line between the Oval Fillister and the Vice Jaw, select the cylindrical face of the Oval Fillister that goes inside the Vice Jaw. Now, select the cylindrical hole of the Vice Jaw. The preview of the explode line appears. Choose

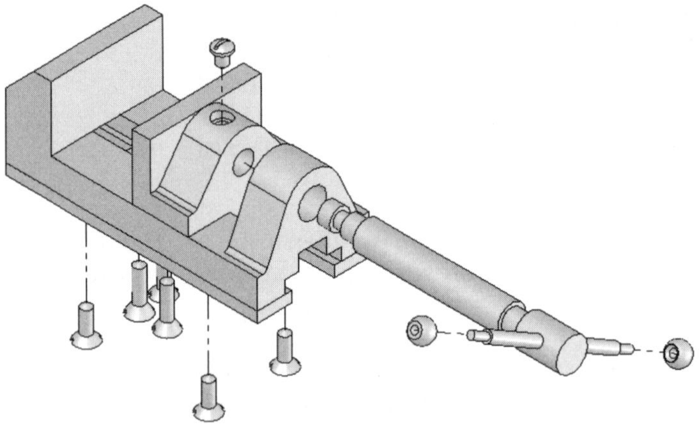

Figure 11-37 Explode line sketch created on an exploded assembly

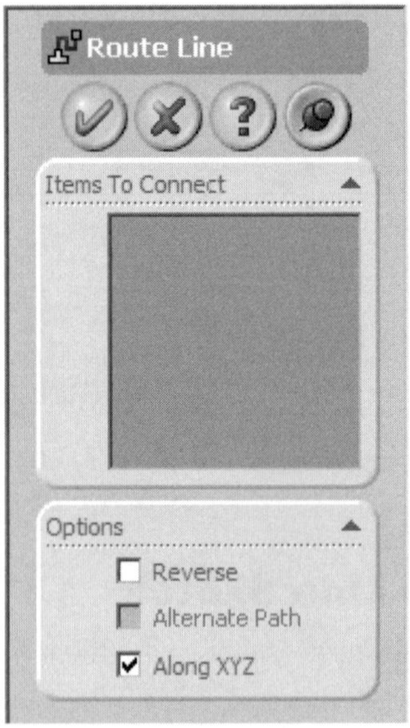

*Figure 11-38 The **RouteLine PropertyManager***

OK to create the explode line. The names of the selected faces are displayed in the **Items To Connect** selection area. When you select an entity to create an explode line, an arrow is also displayed with the line. You can use that arrow or select the **Reverse** check box from the **Options** rollout to reverse the direction of the explode line creation. Next, choose the **OK** button from the **Route Line PropertyManager** and exit the sketching environment.

TUTORIALS

Tutorial 1

In this tutorial, you will create the radial engine assembly shown in Figure 11-39. This assembly will be created in two parts: one will be the subassembly and the second will be the main assembly. You will also create the exploded state of the assembly and then create the explode line sketch. The exploded state of the assembly is displayed in Figure 11-40. The views and dimensions of all the components of this assembly are displayed in Figures 11-41 through 11-44. **(Expected time: 3 hrs)**

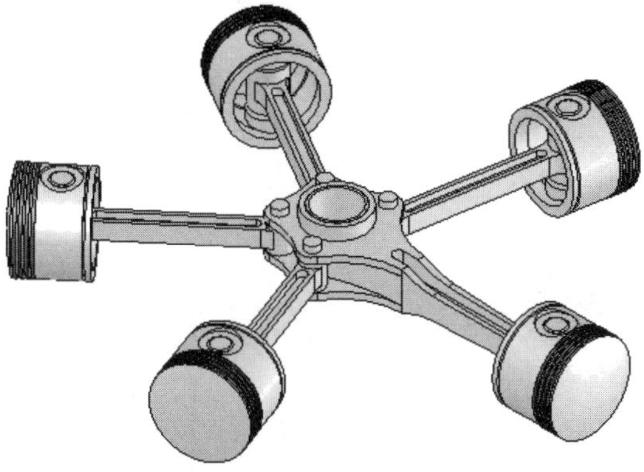

Figure 11-39 *The radial engine assembly*

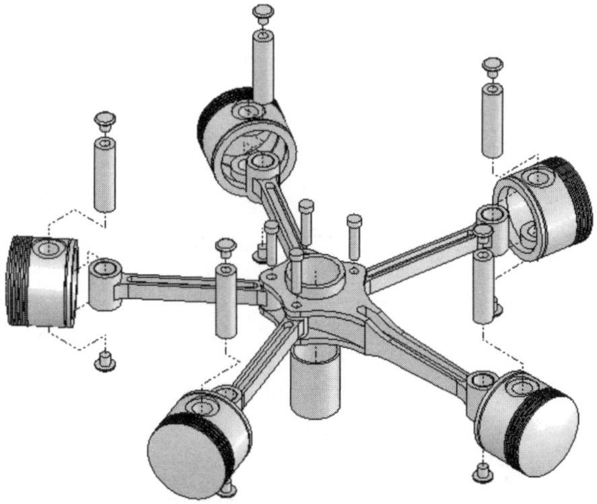

Figure 11-40 *Exploded view of the assembly*

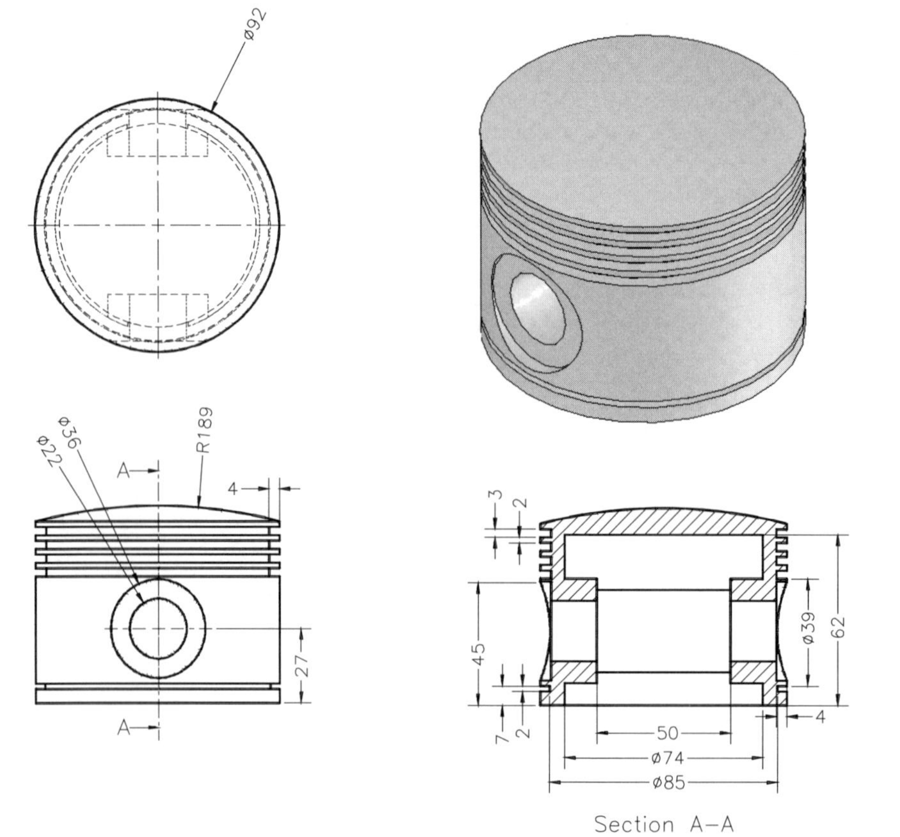

Figure 11-41 *Views and dimensions of the Piston*

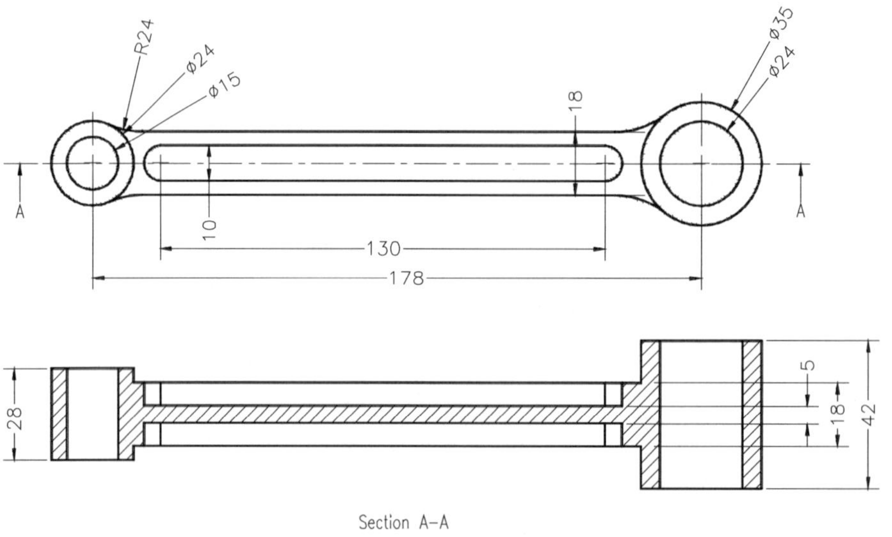

Figure 11-42 *Views and dimensions of the Articulated rod*

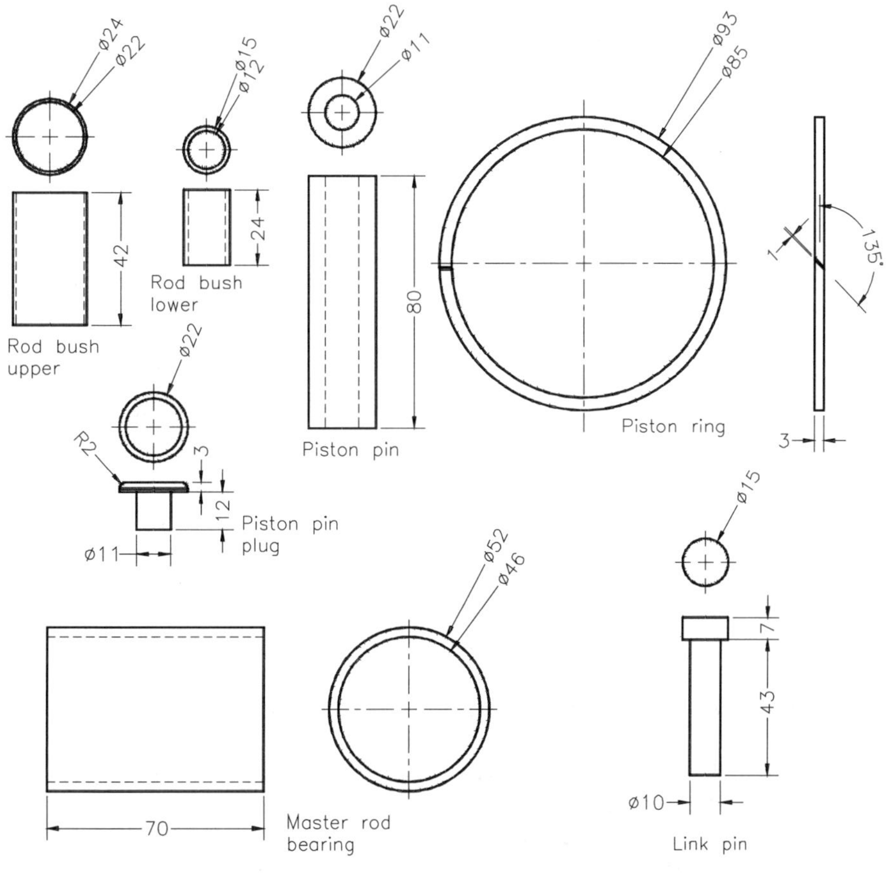

Figure 11-43 Views and dimensions of other components

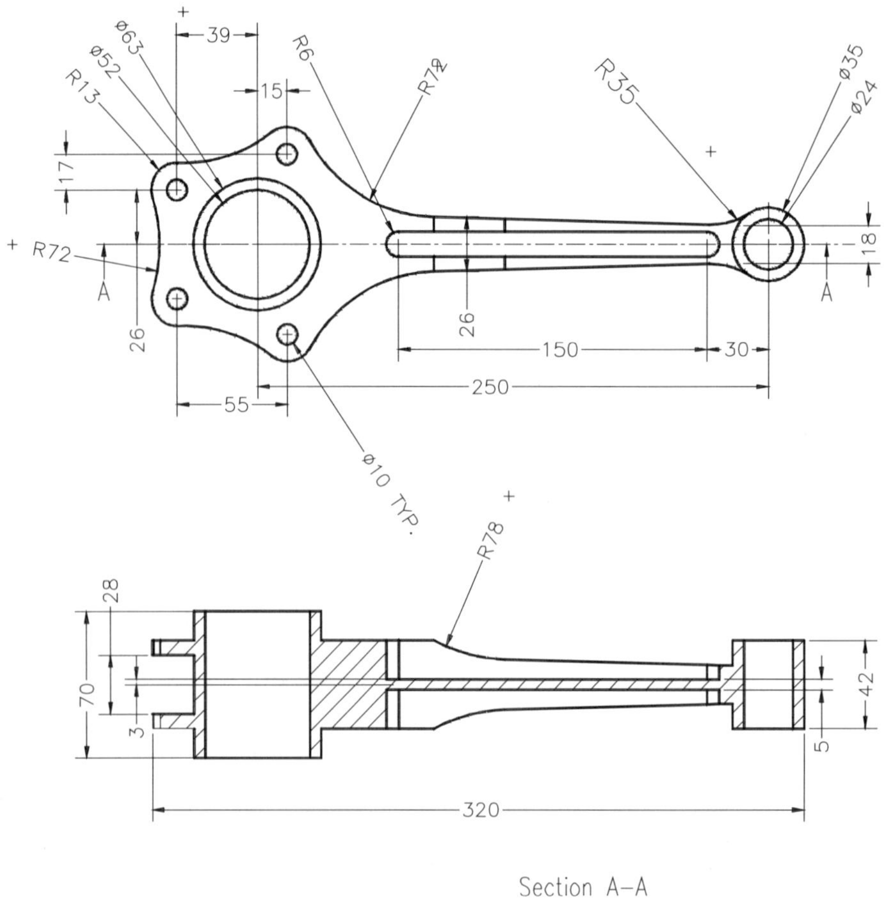

Figure 11-44 *Views and dimensions of Master rod*

The steps to be followed to complete this tutorial are listed next.

You need to break this assembly in two steps because it is a large assembly. One will be the subassembly and the other will be the main assembly. First, you need to create the subassembly, consisting of Articulation Rod, Piston, Piston Rings, Piston Pin, Rod Bush Upper, Rod Bush Lower, and Piston Pin Plug. Next, create the main assembly by assembling the Master Rod with the Piston, Piston Rings, Piston Pin, Rod Bush Upper, and Piston Pin Plug. Finally, you will assemble the subassembly with the main assembly.

a. Create all components of the assembly in the **Part** mode and save the components in a folder named *Radial Engine Assembly*.
b. Start a new assembly document and assemble components to complete the subassembly, refer to Figures 11-45 through 11-47.
c. Start a new assembly document and assemble components of the main assembly, refer to Figure 11-48.

d. Assemble the subassembly in the main assembly, refer to Figures 11-49 through 11-53.
e. Create the exploded view of the assembly and then create the explode line sketch, refer to Figures 11-54 through 11-56.

Creating the Components

1. Create a folder with the name *Radial Engine Assembly* in the *\My Documents\SolidWorks\c11* folder. Create all components in the individual part documents and save them in this folder.

Note

While creating the Master Rod, make sure that the holes on the left of the Master Rod should be created using the sketch-driven pattern. This is done because while assembling the Link Pin you will create the derived pattern of the Link Pin using the sketch-driven pattern feature.

Creating the Subassembly

As discussed earlier, you will first create the subassembly and then assemble it with the main assembly.

1. Start a new SolidWorks assembly document and exit the **Insert Component PropertyManager**. Next, save it with the name **Piston Articulation Rod Subassembly** in the same folder in which the parts are created.

2. First place the Articulated Rod at the origin of the assembly and then place the other components such as Piston, Piston Pin, Piston Pin Plug, Rod Bush Upper, and Rod Bush Lower in the assembly document.

3. Apply the required mates to assemble these components. Figure 11-45 shows the sequence in which you need to assemble the components. The exploded view and the explode line sketch is given only for your reference. The assembly, after assembling the Articulated Rod, Piston, Piston Pin, Piston Pin Plug, Rod Bush Upper, and Rod Bush Lower, is shown in Figure 11-46.

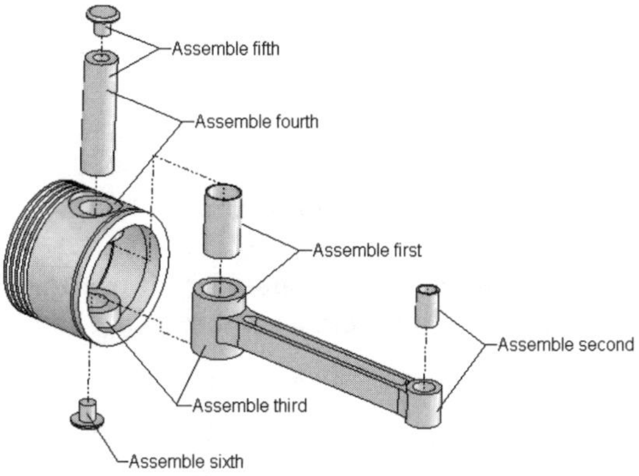

***Figure 11-45** Assembly of the Articulated Rod, Piston, Piston Pin, Piston Pin Plug, Rod Bush Upper, and Rod Bush Lower*

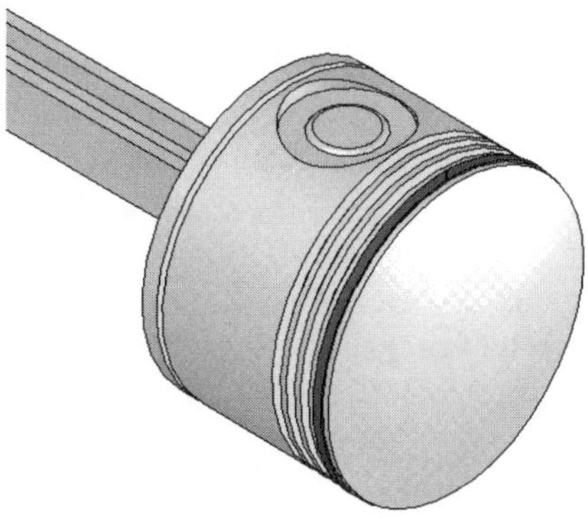

***Figure 11-46** First instance of the Piston Ring assembled with the Piston*

It is clear from the assembly that you need to assemble four instances of the Piston Ring. You will assemble only one instance of the Piston Ring at the uppermost groove of the ring and then create a local linear pattern.

4. Insert the Piston Ring in the assembly document and assemble the Piston Ring at the uppermost groove of the Piston using the assembly mates, see Figure 11-46. In this figure, the color of the Piston Ring is changed by selecting it and choosing the **Component Properties** option from the shortcut menu. Choose the **Color** button from the **Component Properties** dialog box, to change the color.

   Next, you need to create the local linear pattern of the Piston Ring.

5. Choose **Insert > Component Pattern > Linear Pattern** from the menu bar; the **Linear Pattern PropertyManager** is displayed.

6. Select any one of the horizontal edges of the Articulated Rod to define the direction of pattern creation.

7. Click once in the **Components to Pattern** selection area and select the Piston Ring from the drawing area.

   The preview of the linear pattern with the default settings is displayed in the drawing area.

8. Select the **Reverse Direction** button to reverse the direction of pattern creation, if required.

9. Set the value of the **Spacing** spinner to **5** and the value of the **Number of Instances** spinner to **4**.

10. Choose the **OK** button from the **Linear Pattern PropertyManager.**

    The subassembly, after patterning the Piston Ring, is shown in Figure 11-47

    The subassembly is completed. Save and close the assembly document.

Creating the Main Assembly

Next, you will create the main assembly and then assemble the subassembly with it.

1. Start a new SolidWorks assembly document and exit the **Insert Component PropertyManager**. Now, save it with the name **Radial Engine Subassembly** in the same folder in which the parts are created.

2. First, place the Master Rod at the origin of the assembly and then place the Piston, Piston Pin, Piston Pin Plug, Piston Ring, Rod Bush Upper, and Master Rod Bearing in the current assembly document.

3. Using the assembly mates and the local linear pattern option, assemble all components of the main assembly.

   The components, after assembling in the main assembly, are displayed in Figure 11-48.

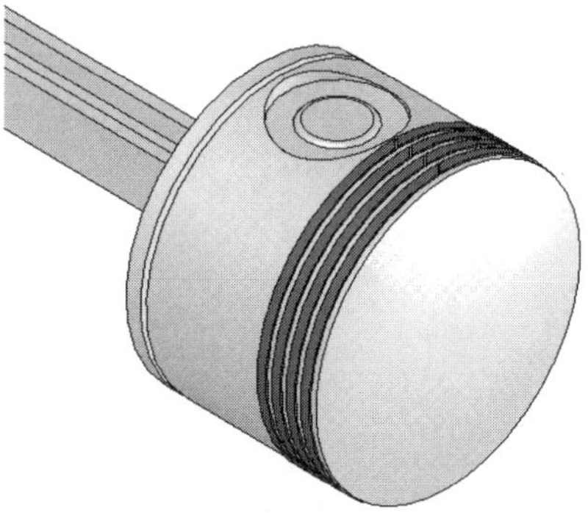

Figure 11-47 *Subassembly after patterning the Piston ring*

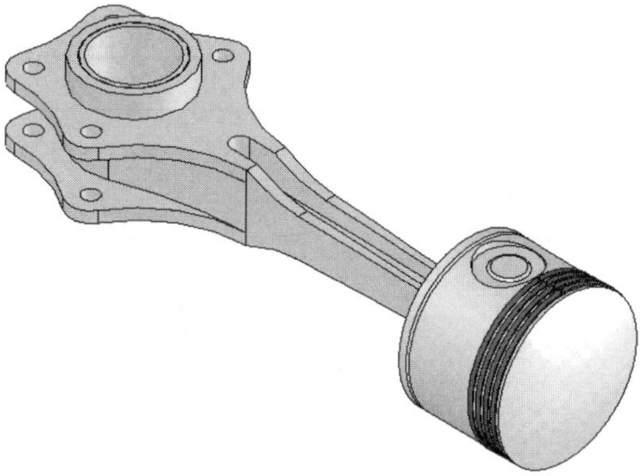

Figure 11-48 *Components assembled in the Main assembly.*

Assembling the Subassembly With the Main Assembly

Next, you will place the subassembly in the main assembly and then assemble them together.

1. Choose the **Insert Components** button from the **Assemblies CommandManager**; the **Insert Component PropertyManager** is displayed.

2. Choose the **Browse** button available in the **Part/Assembly to Insert** rollout; the **Open** dialog box is displayed.

3. Choose **Assembly** (**.asm, *.sldasm*) from the **Files of type** drop-down list.

4. Double-click on *Piston Articulated Rod Subassembly* and place the subassembly in the main assembly. Figure 11-49 shows the subassembly and the main assembly placed together.

5. Using the assembly mates, assemble the subassembly with the main assembly. Refer to Figure 11-50, which shows the assembly structure to help you in assembling the instances of the subassembly.

Tip. *You can create more than one instance of the subassembly by holding down the CTRL key; select and drag the subassembly from the* ***FeatureManager Design Tree****. Release the left mouse button to place the assembly in the current assembly document.*

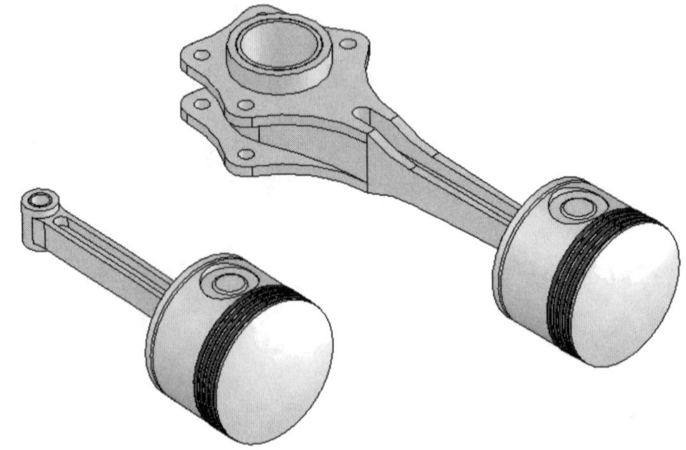

Figure 11-49 Subassembly and the main assembly placed together

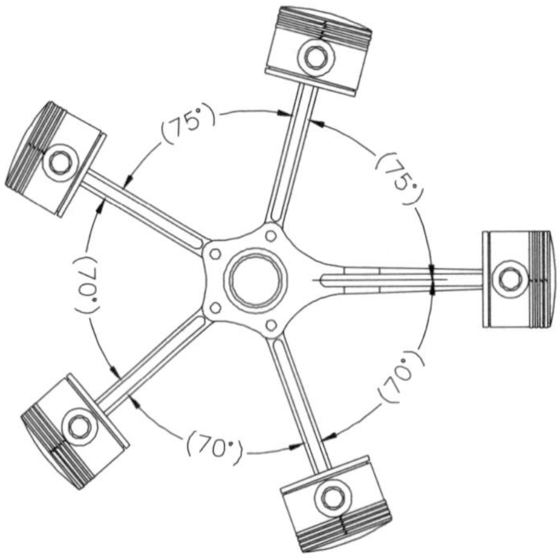

Figure 11-50 Assembly structure

Figure 11-51 shows all instances of the subassembly assembled with the main assembly.

Assembling the Link Pin

After assembling the subassembly with the main assembly you need to assemble the Link Pin with the main assembly.

1. Place the Link Pin in the current assembly document and using the assembly mates, assemble the Link Pin with the main assembly. Figure 11-52 shows the first instance of Link Pin assembled with the main assembly.

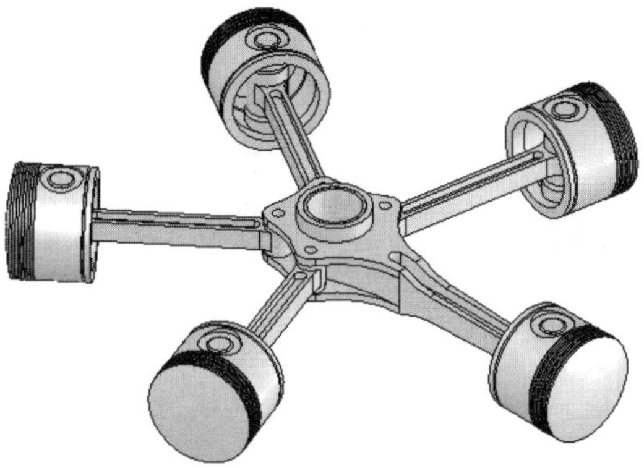

Figure 11-51 *Subassembly assembled to the main assembly*

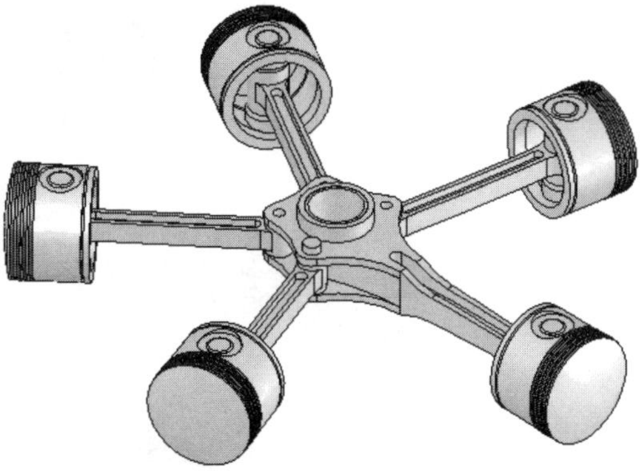

Figure 11-52 *First instance of the Link pin assembled with the main assembly*

As discussed earlier, the other instances of the Link Pin will be assembled using the sketch-driven pattern feature of the holes created on the left of the master rod.

2. Choose **Insert** > **Component Pattern** > **Feature Driven** from the menu bar. The **Feature Driven PropertyManager** is displayed.

3. Select the Link Pin from the main assembly; its name is displayed in the **Components to Pattern** selection area.

4. Click once in the **Driving Feature** selection area to activate the selection mode.

5. Select any one of the hole instances from the master rod. The name of the sketch pattern feature will be displayed in the **Driving Feature** selection area and the preview of the resulting pattern is also displayed.

6. Choose the **OK** button from the **Feature Driven PropertyManager**.

   Figure 11-53 displays the final assembly.

Exploding the Assembly

After creating the assembly you need to explode it using the **Assembly Exploder** tool.

1. Choose the **Exploded View** button from the **Assemblies CommandManager**; the **Explode PropertyManager** is displayed.

2. Click once in the **Explode direction** selection area to highlight it. Select any vertical edge from the assembly to define the direction in which the components will be exploded. The **Component(s) of the explode step** selection area is automatically highlighted.

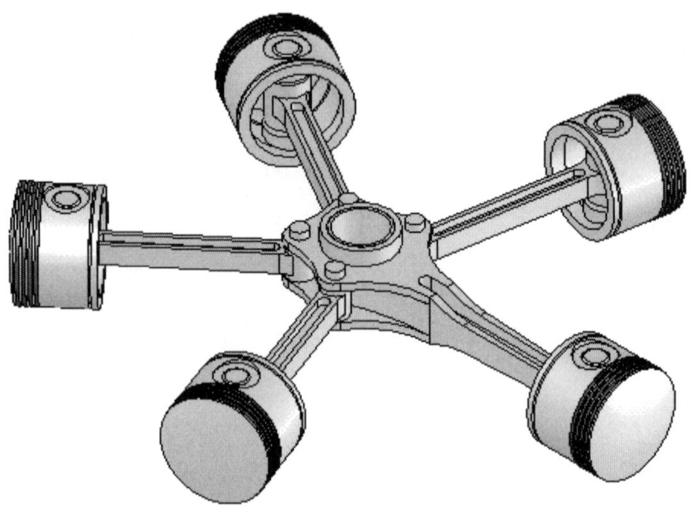

Figure 11-53 *The final assembly*

The **Select sub-assembly's parts** check box has to be selected, if it is cleared in the **Options** rollout. If this check box is cleared then selecting a component selects the complete sub-assembly. On selecting this check box, you can select the individual components of the sub-assembly.

3. Select the **Select sub-assembly's parts** check box. From the assembly, select all the Piston Pin Plugs that are assembled on the upper hole of the piston.

   Ensure that the **Auto-space components after drag** check box is cleared in the **Options** rollout.

4. Set the value of the **Explode distance** spinner to **170**, choose the **Apply** button from the **Settings** rollout and then choose **Done**. The selected instances of the Piston Pin Plug are exploded and the components removed from the selection set. Also, this sequence of explosion is displayed as **Chain 1** in the **Existing explode steps** area of the **Explode Steps** rollout.

5. Click once in the **Explode direction** selection area and again select the same vertical edge to define the direction of explosion. Choose the **Reverse direction** button available on the left of the **Explode direction** selection area to reverse the direction of explosion.

6. Rotate the view of the assembly and select all instances of the Piston Pin Plug that are assembled with the lower holes of the Pistons.

   The explosion value in the **Explode distance** spinner shows a value of 60. You can accept this value to explode the lower Piston Pin Plugs.

7. Choose **Apply** and then choose **Done** to explode the selected components.

8. Click in the **Explode direction** selection area again and select the same vertical edge to define the explosion direction. Now, select all instances of the Piston Pin from the assembly.

9. Set the value of the **Distance** spinner to **140**, choose **Apply** and then **Done**.

10. Similarly, explode the other components of the assembly and then choose **OK** to exit the **Explode PropertyManager**. The assembly, after exploding the components, is shown in Figure 11-54.

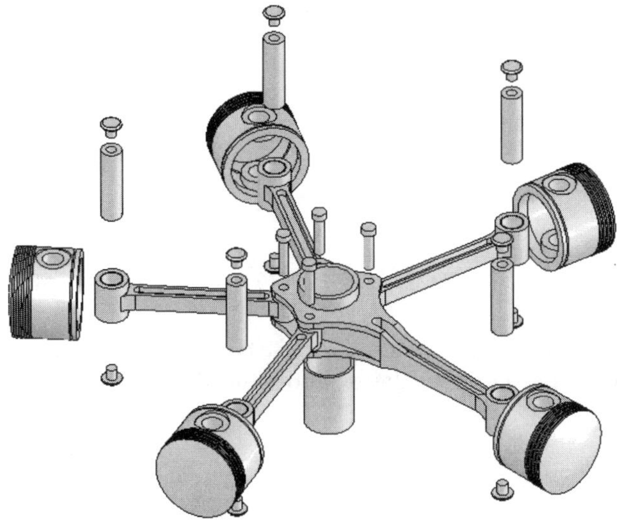

Figure 11-54 *Final exploded assembly*

Creating the Explode Line Sketch

After exploding the assembly, you need to create the explode line sketch of the exploded state of the assembly.

1. Choose the **Explode Line Sketch** button from the **Assemblies CommandManager**; the **Route Line PropertyManager** is displayed and you are prompted to select a cylindrical face, planar face, vertex, point, arc, or line entities.

2. Choose the **Keep Visible** button, if not chosen automatically, from the **Route Line PropertyManager**, to keep it visible on the screen.

3. Select the cylindrical face, as shown in Figure 11-55, as the first selection; the name of the selected face will be displayed in the **Items To Connect** selection area. Also, the preview of the explode line sketch is displayed at the center of the selected face.

4. Refer to Figure 11-55 and select the other cylindrical faces to create the explode line sketch.

5. Next, choose the **OK** button, an exploded line will be created.

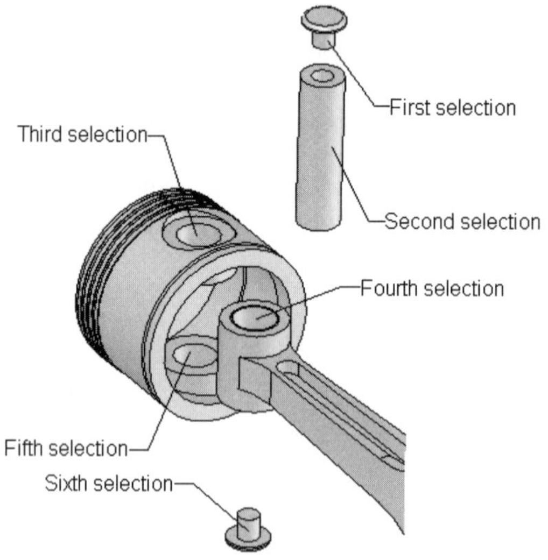

Figure 11-55 *Faces to be selected to create explode line sketch*

6. Similarly, create explode lines between other parts of the exploded assembly. Figure 11-56 shows the assembly after creating the explode line sketch.

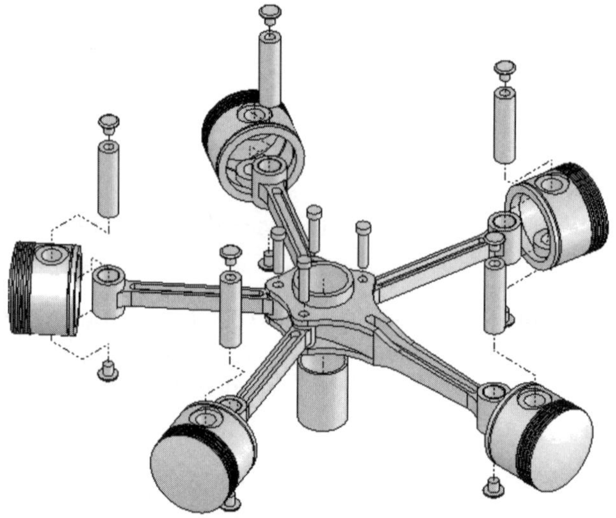

Figure 11-56 *Explode line sketch created to the exploded state of the assembly*

6. Invoke the **ConfigurationManager** and expand the **Default** option. Select **ExplView1** and invoke the shortcut menu. Choose the **Collapse** option from the shortcut menu to switch back to the collapse state of the assembly.

7. Save and close the assembly document.

Tutorial 2

In this tutorial, you will modify the assembly created in Tutorial 1 (Bench Vice) of Chapter 10. You will modify the design of the components of the assembly and then suppress some mates that enable it to move along a particular degree of freedom. Next, check the assembly for collision detection when the assembly is in motion and then modify the assembly and check the interference. **(Expected time: 1 hr)**

The steps that will be followed to complete this tutorial are listed next.

a. Save the Bench Vice assembly folder in *c11* directory and then open the Bench Vice assembly.
b. Modify the design of the components within the context of the assembly, refer to Figures 11-57 through 11-61.
c. Suppress the mate to enable the Vice Jaw to move along the slide ways of the Vice Body.
d. Check the new assembly design for collision detection when the assembly is in motion. Modify the design, if there is any collision between the components, refer to Figures 11-62 and 11-63.
e. Check the interference in the modified assembly.

Opening the Bench Vice Assembly

The assembly created in Tutorial 1 of Chapter 10 is the Bench Vice assembly. You need to copy and save it in the current folder of Chapter 11.

1. Copy the folder in which the Bench vice assembly is saved and paste it in the *c11* folder.

2. Invoke the **Open** dialog box and browse the Bench Vice assembly document. Double-click on it to open the assembly document.

Modifying the Design of the Components of Bench Vice Assembly

You need to modify the components in the context of the assembly because of some alteration in the design of some of the components.

Before you start modifying the components, it is recommended that you hide some of the components. This will simplify the assembly and facilitate in selection of components while editing and modifying them.

1. Press and hold down the CTRL key on the keyboard and select the Clamping Plate, Base Plate, all four Set Screw 1, all four Set Screw 2, Oval Fillister, Screw Bar, and Jaw Screw from the **FeatureManager Design Tree**.

2. Invoke the shortcut menu and choose the **Hide** option from the **Component** area of the shortcut menu.

The display of the selected components is turned off and is not displayed in the drawing area of the assembly document.

The design alteration includes creating a through slot on the right face of the Vice Jaw. To modify its design, you first need to enable the part editing environment.

3. Select the Vice Jaw from the assembly and choose the **Edit Component** button from the **Assemblies CommandManager**. The part modeling environment is invoked in the assembly document.

4. Make sure that the Vice Body is transparent. If not, choose the **Assembly Transparency** button from the **Assemblies CommandManager**; a flyout is displayed. Choose the **Force Transparency** option from the flyout. The Vice Body is displayed in transparent, as shown in Figure 11-57.

5. Select the right face of the Vice Jaw and invoke the sketching environment.
6. Create the sketch of the slot, as shown in Figure 11-58.

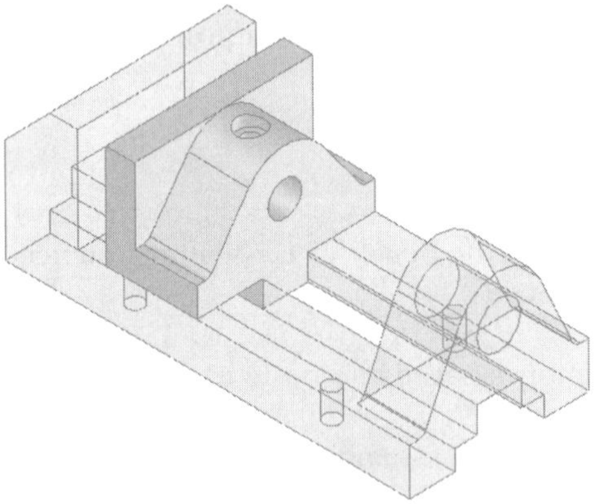

Figure 11-57 *Vice Jaw in the part edit mode in the assembly document*

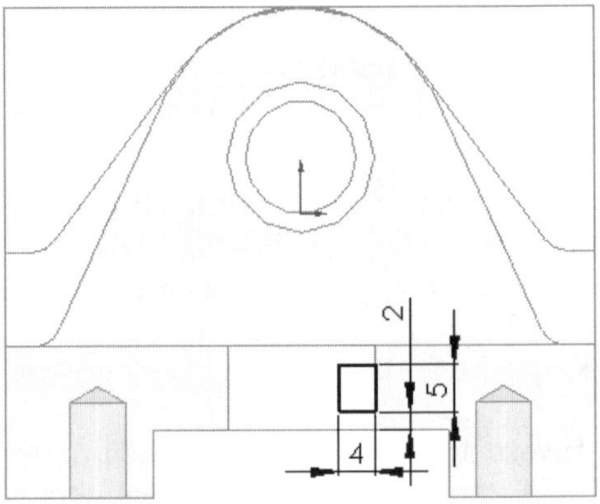

Figure 11-58 *Sketch of the slot*

7. Invoke the **Cut-Extrude PropertyManager**.
8. Create the cut feature using the **Through All** option.
9. Choose the **Edit Component** button to exit the part editing environment.

   Figure 11-59 shows the assembly after modifying the design of the Vice Jaw.

10. Similarly modify the design of the Vice Body. You need to create a blind extruded boss feature up to 60 mm depth on the right face of the component. You also need to reverse the direction of feature creation.

    The sketch of the feature is shown in Figure 11-60. Figure 11-61 shows the assembly, after exiting the part editing environment.

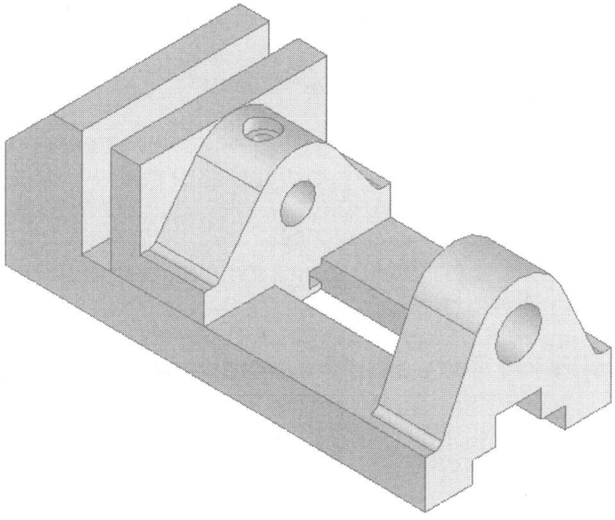

Figure 11-59 *Modified Vice Jaw*

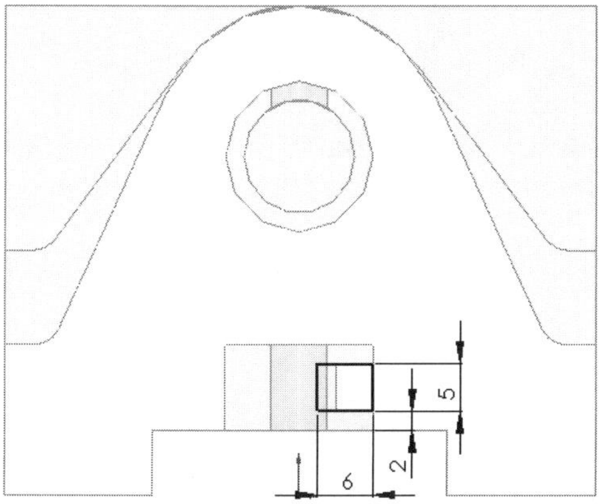

Figure 11-60 *Sketch of the extruded boss feature*

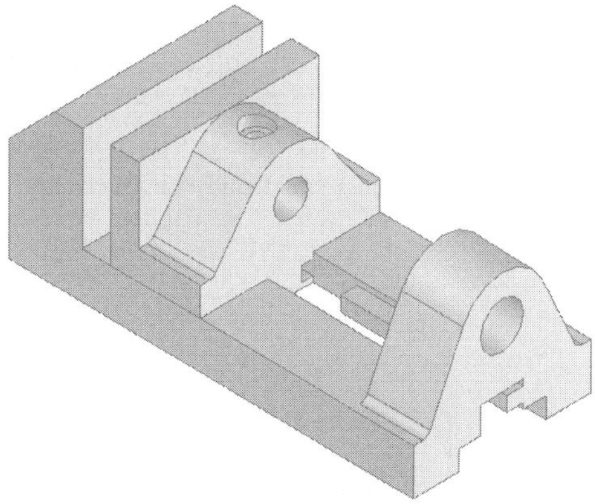

Figure 11-61 *Modified assembly*

11. Choose the **Save** button from the **Standard** toolbar. The **SolidWorks** information box is displayed and you are informed that some models referenced in the document are modified and they must be saved. Choose **Yes** to save the referenced models also.

Suppressing the Mate to Make the Movement of Vice Jaw Free in a Specified Direction

To analyze the movement of the Bench vice assembly, you need to make the movement of the Vice Jaw free in the X direction. By doing so the Vice Jaw will slide on the sideways of the Vice Body.

1. Expand the **Mates** mategroup from the **FeatureManager Design Tree** and select the **Distance1** mate. The planar faces of the Vice Jaw and the Vice Body, to which this mate is applied, are highlighted in green.

2. Invoke the shortcut menu. Choose the **Suppress** option from the shortcut menu.

   Now, the degree of freedom in the X direction is free.

3. Choose the **Move Component** button from the **Assemblies CommandManager** and select a horizontal edge of the Vice Jaw. On dragging the cursor, you will observe that you can move the Vice Jaw in the X direction.

4. Drag the Vice Jaw back to its original position and choose the **OK** button from the **Move Component PropertyManager**.

Analyzing the Collision Between the Components When the Assembly is in Motion

Next, you will analyze the collision between the components of the assembly when the assembly is in motion.

1. Again, choose the **Move Component** button from the **Assemblies CommandManager**. Select the **Collision Detection** radio button from the **Options** rollout.

2. Select Vice Jaw and drag the cursor to move it in the direction as shown in Figure 11-62.

3. On moving the Vice Jaw in the specified direction, you will observe that the right face of the Vice Jaw and the newly created extrusion feature of the Vice Body are highlighted in green as shown in Figure 11-63.

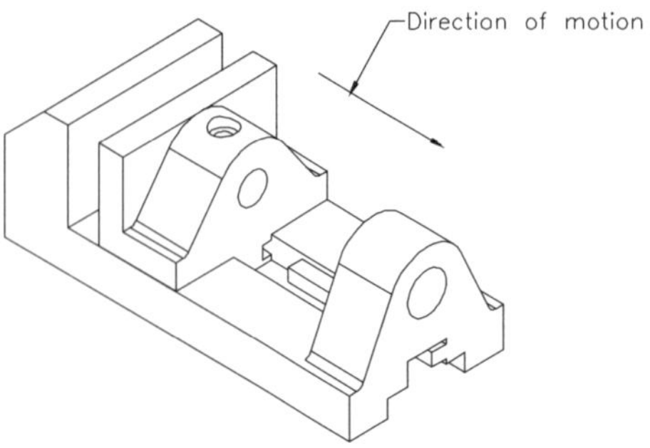

Figure 11-62 *Direction in which the Vice Jaw will be moved*

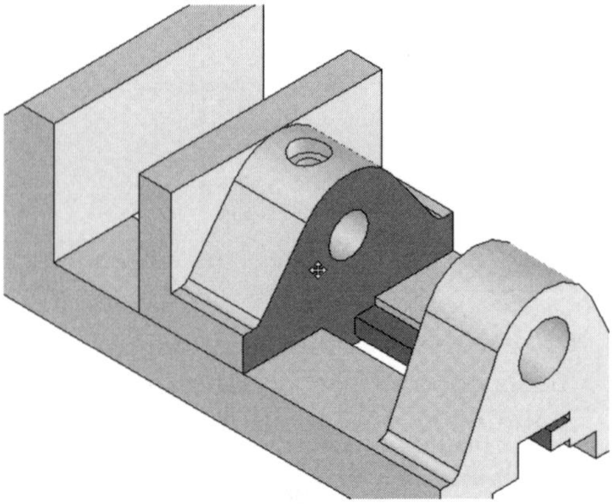

Figure 11-63 *Faces of the Vice Jaw and the Vice Body highlighted in green*

This indicates that the Vice Jaw collides with the Vice Body. Leave the assembly at this location.

4. Choose the **OK** button from the **Move Component PropertyManager**.

The collision is detected in the assembly, and so you need to modify the design of one of the components. In this case, you will modify the dimensions of the extruded boss feature.

5. Double-click the newly created extrusion feature of the Vice Body; the dimensions of the newly created feature will be displayed.

6. Double-click the dimension that has a of value **6**; the **Modify** dialog box is displayed; set the value of the dimension to **4** and press the ENTER key.

7. Press CTRL+B on the keyboard to rebuild the entire assembly.

8. Choose the **Interference Detection** button from the **Assemblies CommandManager**. The **Interference Detection PropertyManager** is displayed.

9. You will observe that **No Interferences** is displayed in the **Interference Results** selection area in the **Results** rollout of this **PropertyManager**.

10. Choose the **Cancel** button from the **Interference Detection PropertyManager**.

Next, you need to show all the components of this assembly.

11. Press and hold the CTRL key, select the hidden components from the **FeatureManager Design Tree,** and invoke the shortcut menu. Choose the **Show** option from the shortcut menu.

12. Save the assembly document and all the referenced part documents.

SELF-EVALUATION TEST

Answer the following questions and then compare your answers with those given at the end of this chapter.

1. You can create subassemblies in the assembly environment of SolidWorks. (T/F)

2. You cannot create a subassembly of the components that are already placed in an assembly document. (T/F)

3. When you move the cursor on a mate in the **FeatureManager Design Tree**, the entities used in the mate are highlighted in red in the drawing area. (T/F)

4. You cannot edit the assembly mates. (T/F)

5. While in the part editing mode in the assembly document, you can use the **Move/Size Features** tool to edit the features dynamically using editing handles. (T/F)

6. The component patterns created individually without the use of any existing pattern feature are known as __________ patterns.

7. The component patterns created using an existing pattern feature are known as __________ patterns.

8. In __________ component, the feature information is available in the part document and only the graphical representation of the component is displayed in the assembly document.

9. After selecting the component, choose the _________ option from the shortcut menu to change the transparency condition of the selected component.

10. To create the explode line sketch, choose the __________ button from the **Assemblies CommandManager**.

REVIEW QUESTIONS

Answer the following questions.

1. Which option is used to open a component separately in the part document?

   (a) **Modify** (b) **Edit**
   (c) **Open Part** (d) None of these

2. Which option is used to define whether a component collides with another component of the assembly or not?

   (a) **Collision Detection** (b) **Interference Detection**
   (c) **Mass Properties** (d) None of these

3. Which check box is selected in the **Open** dialog box to open an assembly with lightweight parts?

   (a) **Lightweight** (b) **Open Lightweight**
   (c) **Lightweight parts** (d) **Lightweight assembly**

4. Which button available in the **Assembly CommandManager** is used to suppress a component?

   (a) **Change Suppression State** (b) **Suppress**
   (c) **Hide/Show Component** (d) **Move Component**

5. Which button is selected from the **Assemblies CommandManager** to create an exploded view?

   (a) **Exploded View** (b) **Assembly Exploder**
   (c) **Mate** (d) None of these

6. The _________ radio button is used to create a local linear pattern.

7. To show the hidden component, select the icon of the component from the **FeatureManager Design Tree,** invoke the shortcut menu and choose the _________ option from it.

8. The _________ option is used to pattern the instances of the components using an existing pattern feature.

9. The explode state of the assembly is created using the _________ dialog box.

10. The _________ check box is selected to stop the motion of the assembly, when one of the component collides with another component, when the assembly is in motion.

EXERCISE

Exercise 1

Create the assembly as shown in Figure 11-64. Ensure that the back plate is fixed and the entire assembly can move in the Y direction with respect to the back plate. Keep the rotational degree of freedom of the screw rod free so that it can also rotate on its axis. After creating the assembly, explode it and create the explode line sketch. The exploded view of the assembly with explode line sketch is shown in Figure 11-65. The dimensions of the model are given in Figures 11-66 thought 11-70. **(Expected time: 4 hrs)**

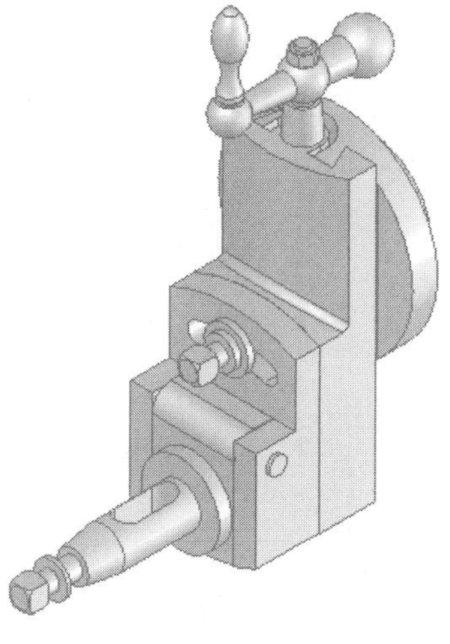

Figure 11-64 *Shaper tool holder assembly*

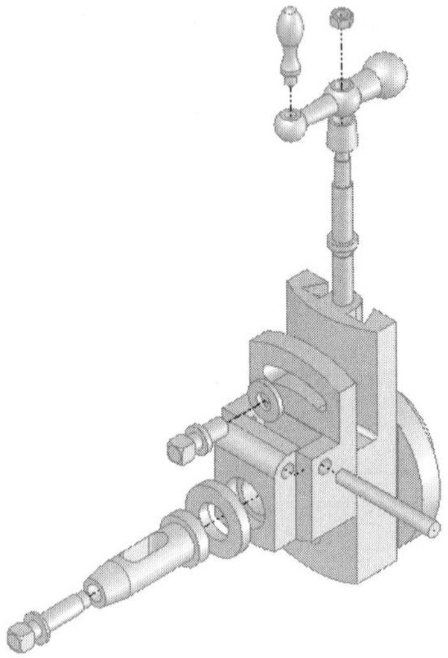

Figure 11-65 Exploded view of the Shaper tool holder assembly with explode lines

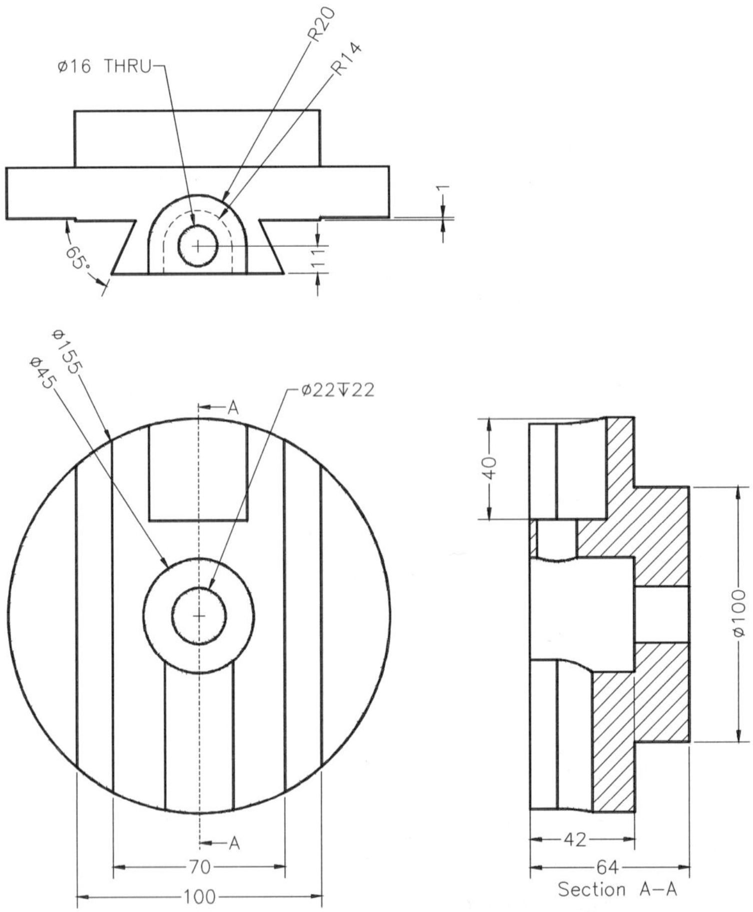

Figure 11-66 Views and dimensions of Back Plate

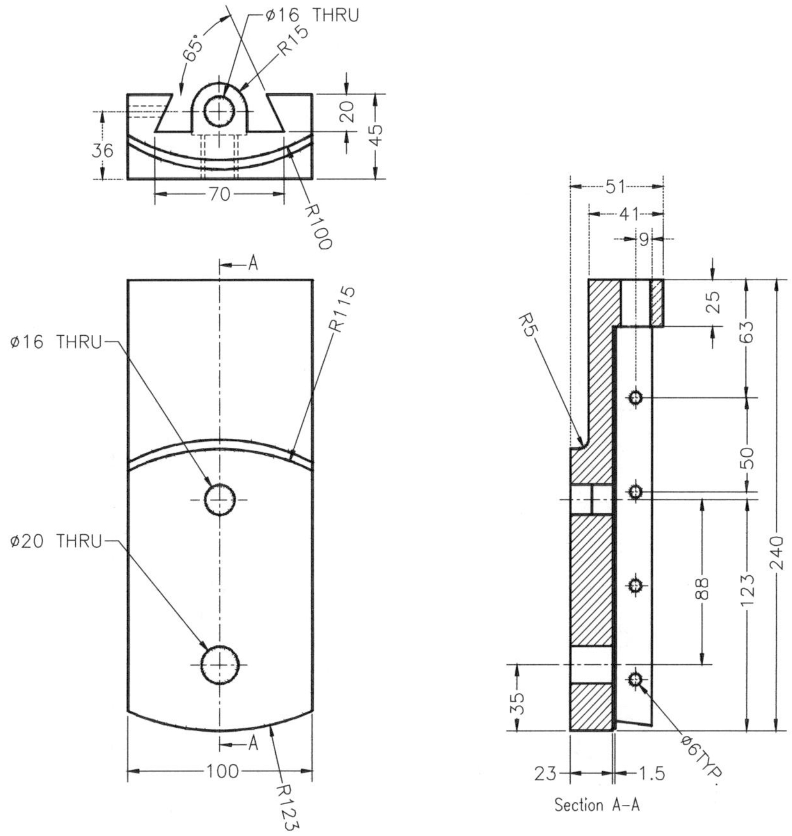

Figure 11-67 Views and dimensions of Vertical Slide

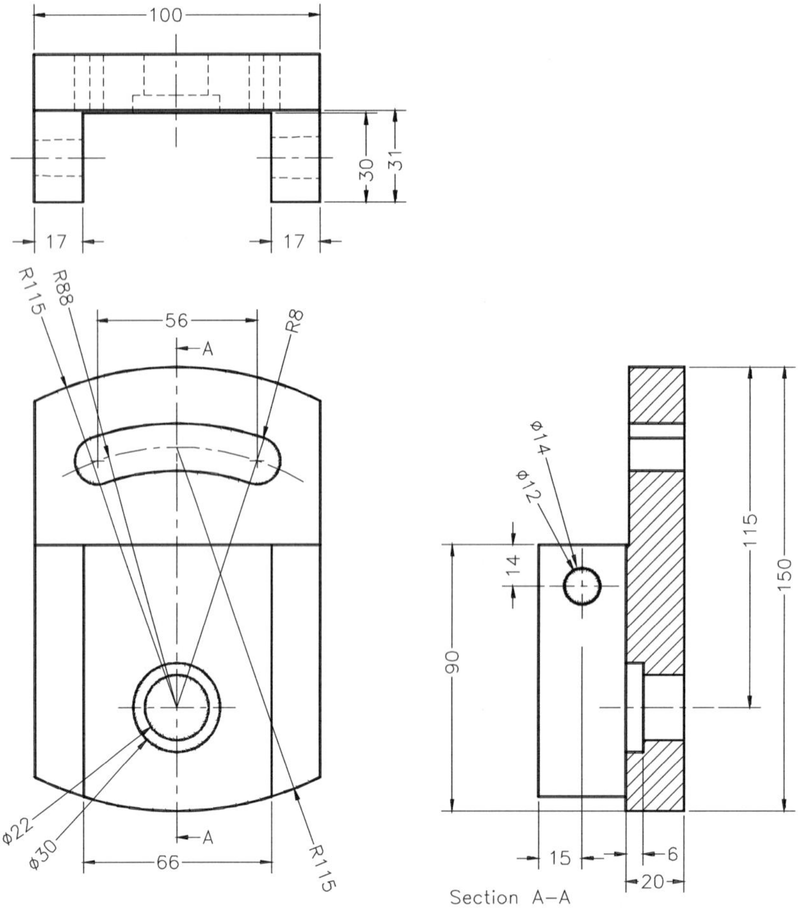

Figure 11-68 *Views and dimensions of Swivel Plate*

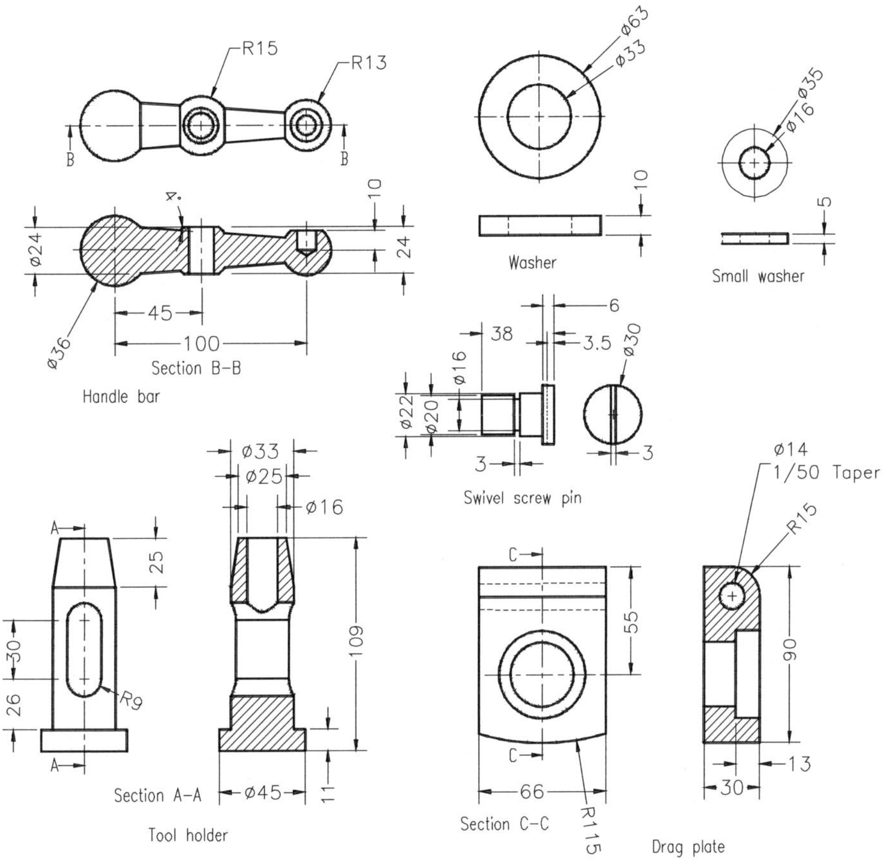

Figure 11-69 *Views and dimensions of the components*

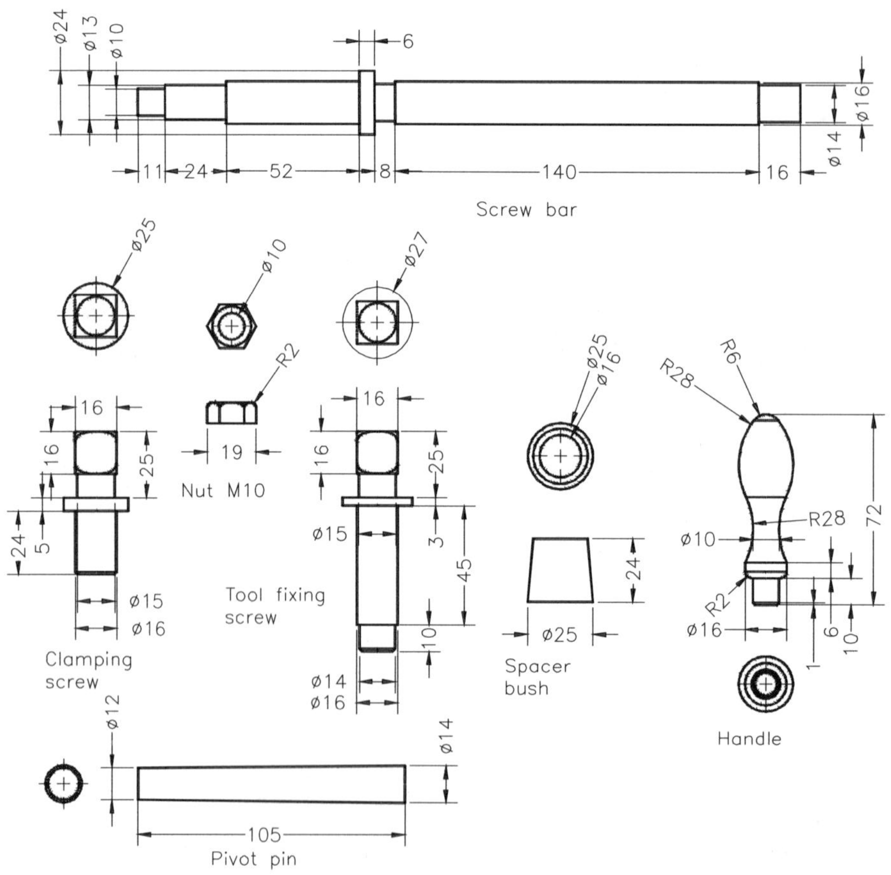

Figure 11-70 Views and dimensions of the components

Answers to Self-Evaluation Test

1. T, **2.** F, **3.** T, **4.** F, **5.** F, **6. Local**, **7. Derived**, **8. Lightweight**, **9. Component Properties**, **10. Explode Line Sketch**

Chapter 12

Working With Drawing Views-I

Learning Objectives

After completing this chapter you will be able to:

- ***Generate standard three views.***
- ***Generate Model views.***
- ***Generate Relative views.***
- ***Generate Predefined views.***
- ***Generate Projected views.***
- ***Generate Section views.***
- ***Generate Aligned Section views.***
- ***Generate Broken-out Section views.***
- ***Generate Auxiliary views.***
- ***Generate Detail views.***
- ***Generate Crop views.***
- ***Generate Broken views.***
- ***Generate Alternate Position views.***
- ***Generate the view of an assembly in the exploded state.***
- ***Work with interactive drafting.***
- ***Edit the drawing views.***
- ***Change the scale of the drawing views.***
- ***Delete drawing views***
- ***Modify the hatch pattern of the Section Views.***

THE DRAWING MODE

After creating the solid models of the parts, or assemblies, you need to generate the two-dimensional (2D) drawing views. These views are the lifeline of all the manufacturing systems because at the shop floor or machine floor, the machinist mostly needs the 2D drawing for manufacturing. SolidWorks has provides a specialized environment, known as the **Drawing** mode, that has all the tools required to generate and modify the drawing views, and add dimensions and annotations to them. In other words, you can get the final shop floor drawing using this mode of SolidWorks. You can also sketch the 2D drawings in the **Drawing** mode of SolidWorks using the sketching tools provided in this mode.

In other words, there are two types of drafting methods available in SolidWorks: Generative drafting and Interactive drafting. Generative drafting is a technique of generating the drawing views using a solid model or an assembly. Interactive drafting is a technique of using the sketching tools to sketch a drawing view in the **Drawing** mode. In this chapter, you will learn about generating the drawing views of parts or assemblies.

One of the major advantages of working in SolidWorks is that this software is bidirectionally associative in nature. This property ensures that the modifications made in a model in the **Part** mode are reflected in the **Assembly** mode and the **Drawing** mode, and vice versa.

STARTING A DRAWING DOCUMENT

To generate the drawing views, you need to start a new drawing document. There are two methods of starting a drawing document in SolidWorks 2005. You can use the **New SolidWorks Document** dialog box or the option available in the part of assembly document to start a drawing document. Both these methods are discussed next.

Starting a New Drawing Document Using the New SolidWorks Document Dialog box

To start a new drawing document for generating the drawing views, invoke the **New SolidWorks Document** dialog box. Choose the **Drawing** button, as shown in Figure 12-1, and choose the **OK** button. A new drawing document is started and the **Sheet Format/Size** dialog box is also displayed. Figure 12-2 shows the initial screen of the drawing document with the **Sheet Format/Size** dialog box. Select the available drawing template file from this dialog box. A new drawing document will be started.

The Model View PropertyManager is invoked automatically when you start a new drawing document. Its appearance will depend on whether any part or assembly document was opened or not when you started the new drawing document.

Tip. *If you start a new drawing document by choosing the* ***drawing*** *template from the* ***Tutorial*** *tab of the* ***New SolidWorks Document*** *dialog box, when used in the* ***Advanced*** *mode, the* ***Sheet Format/Size*** *dialog box is not displayed.*

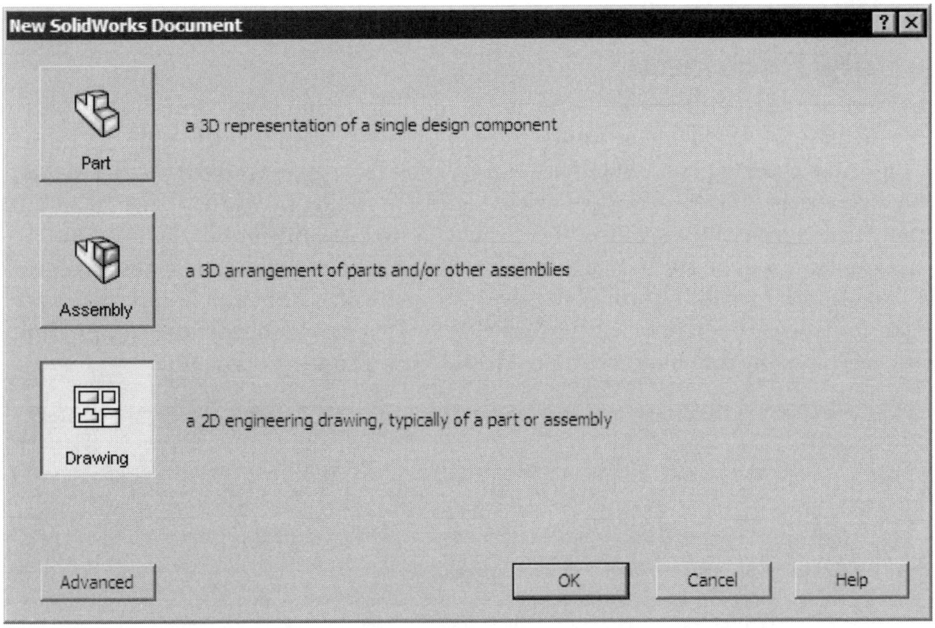

***Figure 12-1** The **New SolidWorks Document** dialog box*

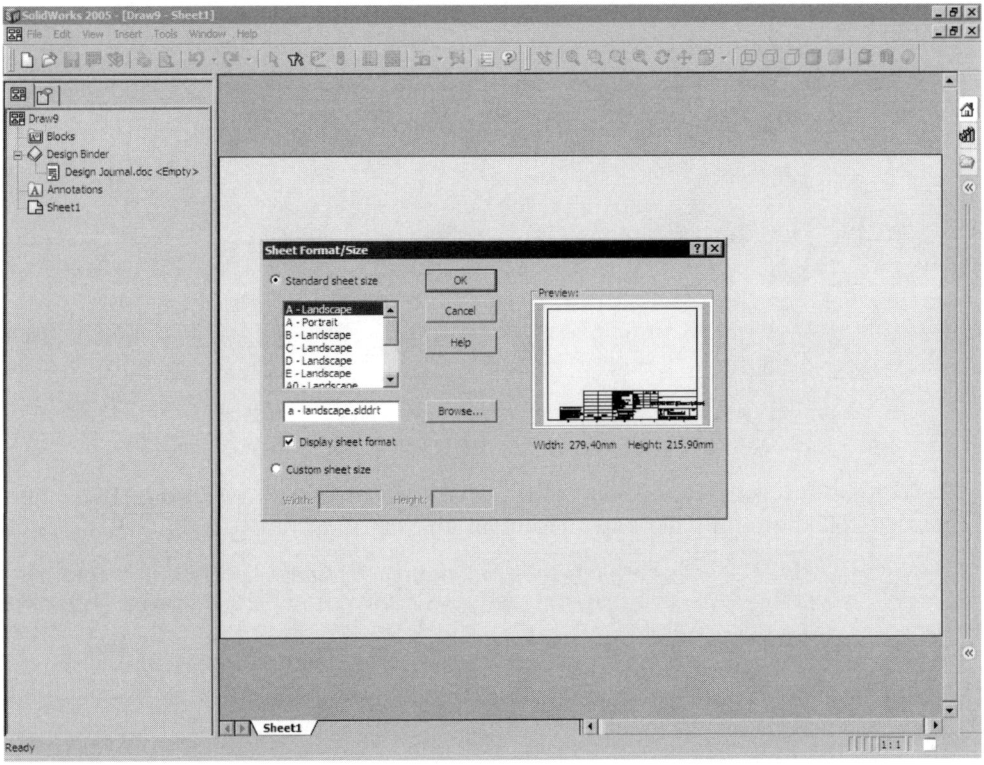

***Figure 12-2** The initial screen of the drawing document with the **Sheet Format/Size** dialog box*

Starting a New Drawing Document From Within the Part/ Assembly Document

This method of starting a new drawing document, is recommended when the part or the assembly document of which you want to generate the drawing views, is open. In this case, choose the **Make Drawing from Part/Assembly** button from the **Standard** toolbar of the part or assembly document. The **New SolidWorks Document** dialog box will be displayed if you are using it in the advanced mode. Select the drawing template and choose the **OK** button. A new drawing document will be started and the **Sheet Format/Size** dialog box is displayed. You can select the required format and size of the sheet from the **Sheet Format/Size** dialog box. A new drawing document will be started and the **Model View PropertyManager** is displayed. Figure 12-3 shows the new drawing document with the **Model View PropertyManager**.

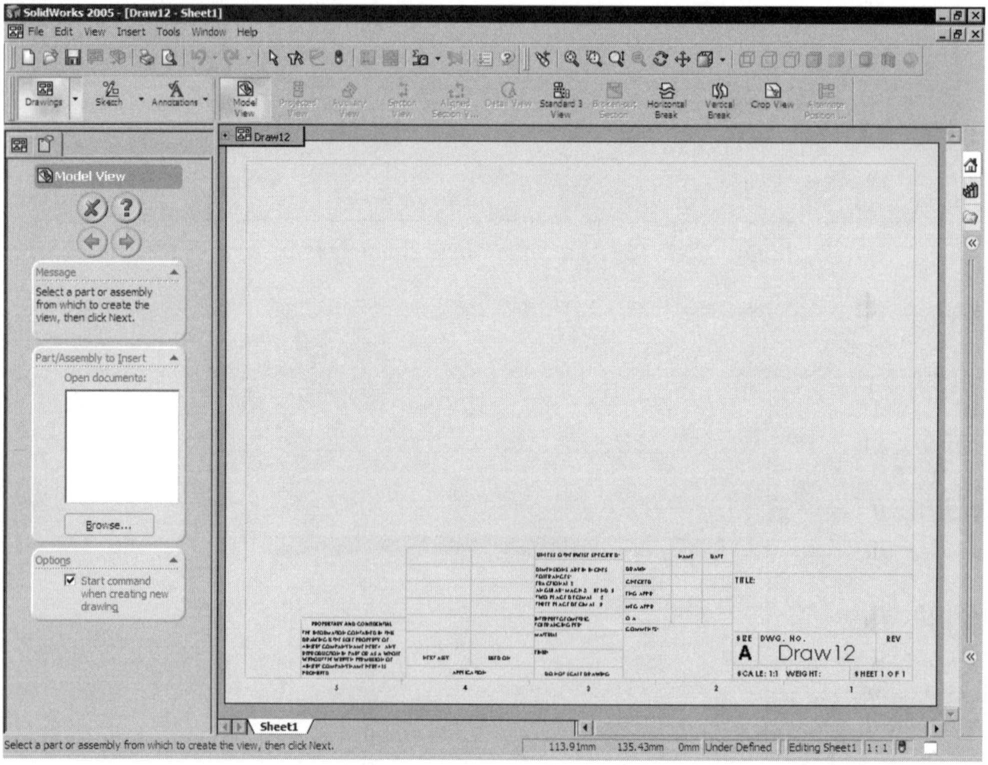

Figure 12-3 *A new drawing document with the* ***Model View PropertyManager***

Tip. *If you choose the* ***Cancel*** *button from the* ***Sheet Format/Size*** *dialog box, a blank custom sheet of size 431.80 mm x 279.40 mm will be inserted in the drawing document.*

TYPE OF VIEWS

You can generate nine type of views in SolidWorks. Generally, you first need to generate a standard view, such as the top view or the front view, and then use it to derive the remaining generate or derive the following views from the standard view.

Model View

The model view is used to create the base view in the drawing sheet. You can generate orthogonal views such as the front, top, left, and so on as the model view. You can also generate isometric, trimetric, or dimetric views as the model view.

Projected View

The projected view is generated by taking an existing view as the parent view. It is generated by projecting the lines normal from the parent view or at an angle. The resulting view will be an orthographic view or a 3D view.

Section View

A section view is generated by chopping a part of an existing view using a plane and then viewing the parent view from a direction normal to the section plane. In SolidWorks, the section plane is defined using one or more sketched line segments.

Aligned Section View

An aligned section view is used to section the features that are created at a certain angle to the main section planes. Align sections straighten these features by revolving them about an axis that is normal to the view plane. Remember, that the axis about which the feature is straightened should lie on the cutting planes.

Auxiliary View

An auxiliary view is generated by projecting the lines normal to a specified edge of an existing view.

Detail View

A detail view is used to display the details of a portion of an existing view. You can select the portion whose detailing has to be shown in the parent view. The portion that you have selected will be magnified and placed as a separate view. You can control the magnification of the detail view.

Broken View

A broken view is the one in which a portion of the drawing view is removed from in between, keeping the ends of the drawing view intact. This type of view is used to display the components whose length to width ratio is very high. This means that either the length is very large as compared to the width or the width is very large as compared to the length. The broken view will break the view along the horizontal or vertical direction such that the drawing view fits the required area.

Broken-out Section View

A broken-out section view is used to remove a part of the existing view and display the area of the model or the assembly that lies behind the removed portion. This type of view is generated using a closed sketch associated with the parent view.

Crop View

A crop view is used to crop an existing view enclosed in a closed sketch associated to that view. The portion of the view that lies inside the associated sketch is retained and the remaining portion is removed.

Alternate Position View

The alternate position view is used to create a view in which you can show both the maximum and minimum range of motion of the assembly. The main position is displayed in the drawing view in continuous lines and the alternate position of the assembly is shown in the same view in dashed lines (phantom lines).

GENERATING STANDARD DRAWING VIEWS

A standard view is generally the first view that you generate in the current drawing sheet. There are a number of methods of generating the standard drawing views. All these methods are discussed next.

Generating Model Views

Command Manager:	Drawing > Model View
Menu:	Insert > Drawing View > Model
Toolbar:	Drawing > Model View

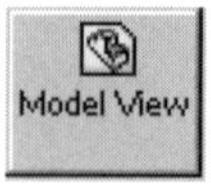

As mentioned earlier, the model views can be used to generate the base view in the drawing sheet. Whenever you start a new drawing document, the **Model View PropertyManager** is automatically invoked.

If you start the new drawing document from within the part or the assembly document, the part or the assembly is automatically selected and you can place the view. However, if you start the new drawing document using the **New SolidWorks Document** dialog box, a message will be displayed in the **Model View PropertyManager** and you will be prompted to select a part or assembly to generate the drawing view. If any part or assembly document is opened, it will be displayed in the selection area of the **Part/Assembly to Insert** rollout. You can preview the part or the assembly document by opening the **Thumbnail Preview** rollout, as shown in Figure 12-4.

You can also choose the **Browse** button and use the **Open** dialog box to select the document. After selecting the required part or assembly document, double-click on it in the **Part/Assembly to Insert** a rollout. You can also choose the **Next** button available above the **Message** rollout in the **Model View PropertyManager**.

The **Model View PropertyManager** is automatically modified and now provides the options related to generating the standard views, as shown in Figure 12-5. The various options available in this **PropertyManager** are discussed next.

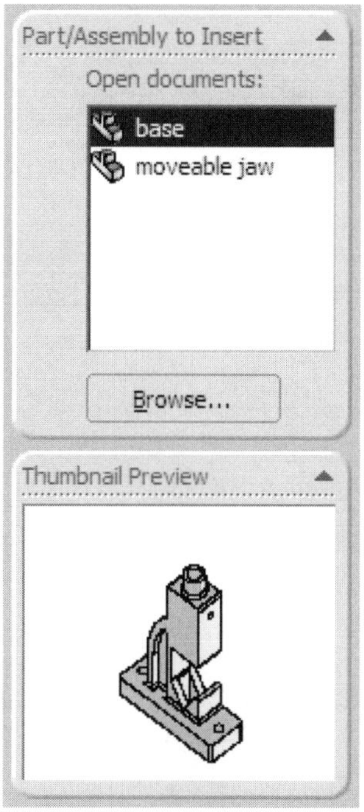

***Figure 12-4** Selecting the document to generate the model views*

Orientation Rollout

The list box available in the **Orientation** rollout is used to specify the orientation of the view. The selection of the **Preview** check box allows you to preview the drawing view before it is placed.

Note

*You can change its orientation of the model view, even after placing it. To do this, place the view and make sure that it is still selected. This can be confirmed by the box that appears around the model view. The view is selected, if the box is displayed in green with grip point. Now, double-click on the required view orientation from the **Orientation** rollout in the **Model View PropertyManager**. The orientation of the view is automatically modified.*

Options Rollout

The **Auto-start projected view** check box available in this rollout is used to automatically invoke the **Projected View** tool to generate the projected view immediately after placing the model view. The model view that you generate will be automatically taken as the parent view to generate the projected view.

Display Style Rollout

The options available in this rollout are used to specify the display styles for the model view. These display styles are similar to those available in the **View** toolbar to display parts or assemblies in the part or assembly documents.

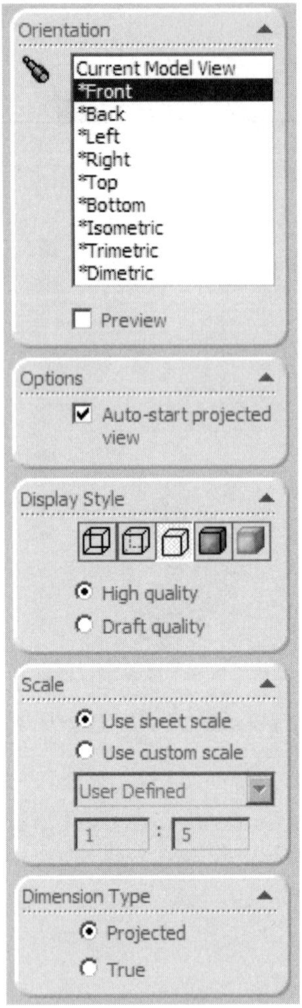

Figure 12-5 *Partial view of the* ***Model View PropertyManager*** *after selecting the document to generate the drawing views*

Scale Rollout

A default scale, used to generate the drawing views is automatically defined, when you select a template to generate the drawing views. This is the reason the **Use sheet scale** radio button is selected in the **Scale** rollout. If you want to define a custom scale for the model view, select the **Use custom scale** radio button and specify the scale factor in the edit boxes available below this radio button.

Dimension Type Rollout

The radio buttons available in this rollout are used to specify whether the model view will have true dimensions or projected dimensions. The true dimensions are the exact model dimensions that were specified while creating the model. The projected dimensions are reduced dimensions that are used in case of isometric, dimetric, or trimetric view. Generally, the value of the projected dimension is about 81.6% of the value of true dimension.

Tip. *The drawing views will be generated depending on the default projection type of the current sheet. If the sheet is configured for first angle projection, the drawing views will be generated according to that. If the drawing sheet is configured for third angle projection, the views will be generated accordingly.*

To change the projection type of the current sheet, right-click on ***Sheet Format1*** *in the* ***FeatureManager Design Tree*** *and choose* ***Properties*** *from the shortcut menu to display the* ***Sheet Properties*** *dialog box. Set the required projection type using the* ***Type of projection*** *area.*

Generating the Three Standard Views

CommandManager:	Drawing > Standard 3 View
Menu:	Insert > Drawing View > Standard 3 View
Toolbar:	Drawing > Standard 3 View

You can generate three default orthographic views of the specified part or assembly by using the **Standard 3 View** option. To create the three standard views, choose the **Standard 3 View** button from the **Drawing CommandManager**; the **Standard 3 View PropertyManager** will be displayed. If any part or assembly document is opened in the current session of SolidWorks, it will be displayed in the list box available in the **Part/Assembly to Insert** rollout, as shown in Figure 12-6.

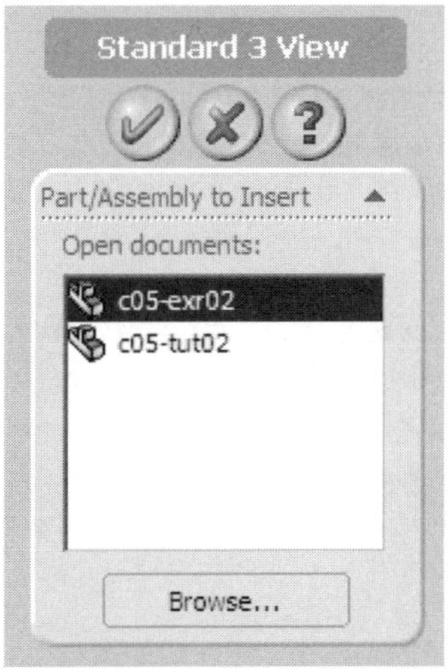

Figure 12-6 Standard 3 View PropertyManager

You can select the document from this list box. You can use the **Browse** button to select the part or assembly document, if no documents are opened. As soon as you select a document, three standard views are generated based on the default scale of the current sheet. Figure 12-7 shows the three standard views of a model generated in the third angle projection using the **Standard 3 View** tool.

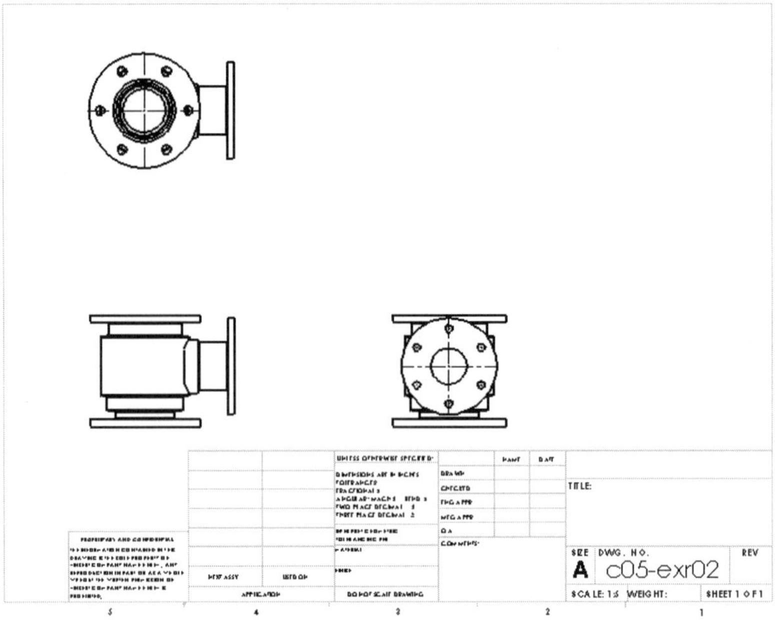

Figure 12-7 *Three standard views generated using the* ***Standard 3 View*** *tool*

Note

You will observe that the name of the part document, whose drawing views are generated, is displayed in the ***DWG NO.*** *text box of the title block. The size of the sheet is also displayed at the lower right corner of the title block. Try changing the sheet format, if these parameters are not displayed.*

You will observe the that center marks are automatically created on generating the drawing views. Otherwise, you can set the option to automatically create the center marks. To do this, invoke the ***System Options*** *dialog box and choose the* ***Document Properties*** *tab. The* ***Detailing*** *tab is chosen by default. Select the* ***Center marks*** *check box from the* ***Auto insert on view creation*** *area. You can also set auto-insertion of centerlines, balloons, and drawing using the options provided in this dialog box.*

Tip. *You need to move the generated view if it overlaps the title block. Place the cursor over the view to move it; the bounding box of the view is displayed in dashed red lines. At this point, click to select the view. Next, move the cursor to the boundary of the selected view; the cursor is replaced by the move cursor. Press and hold down the left mouse button and drag the cursor to move the view. Remember that on moving the parent view, all the views generated using it are also moved.*

Generating Standard Views Using the Relative View Tool

CommandManager:	Drawing > Relative View	*(Customize to Add)*
Menu:	Insert > Drawing View > Relative To Model	
Toolbar:	Drawing > Relative View	*(Customize to Add)*

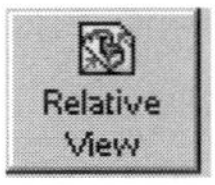

The **Relative View** tool is used to generate an orthographic view. The orientation of the view is defined by selecting two reference planes or the planar faces of the model. This option is very useful if you need the orientation of the parent view other than the default orientations.

Open the part or the assembly document and tile it vertically or horizontally with the drawing document in order to create a relative view. Invoke the **Relative View** tool to display the **Relative View PropertyManager**. You are prompted to select a planar face of the model. Click in the window of the part or assembly document; the **Relative View PropertyManager** will be displayed, as shown in Figure 12-8.

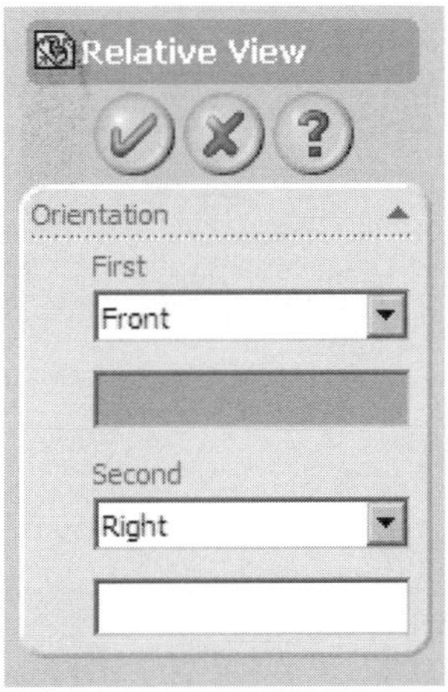

Figure 12-8 Relative View PropertyManager

Select the orientation for the first plane or planar face from the drop-down list available in the **First** area. Then select the plane or planar face of the model to be oriented in that direction. For example, you can select the **Top** option from this drop-down list and then select the top planar face of the model.

The selection area in the **Second** area is highlighted on specifying the first reference. Select the orientation for the second reference from the drop-down list and then select a plane or planar face. Next, choose the **OK** button from the **Relative View PropertyManager**. You will return to the drawing document. Place the view at the required location. Figure 12-9 shows the faces of the model selected to generate a standard view and Figure 12-10 shows the resulting view.

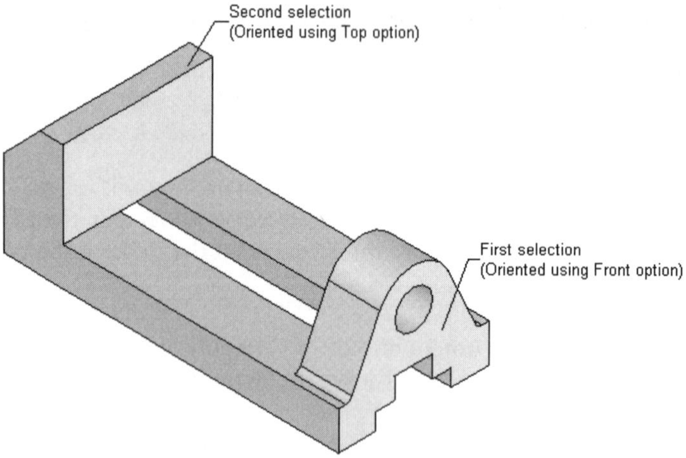

Figure 12-9 *Faces to be selected*

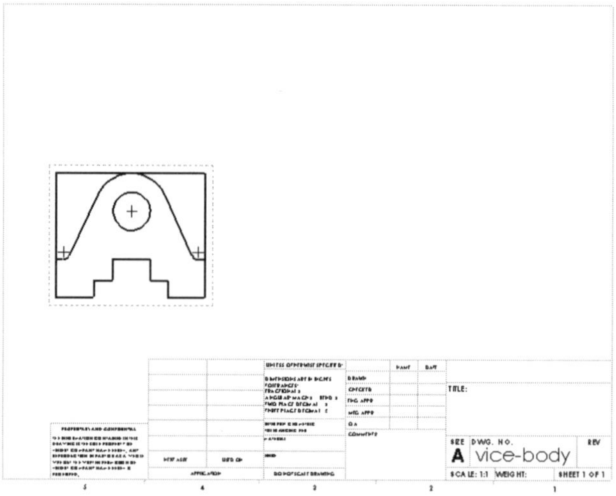

Figure 12-10 *Resulting view*

Generating Standard Views Using the Predefined View Tool

Toolbar:	Drawing > Predefined View	*(Customize to Add)*
Menu:	Insert > Drawing View > Predefined	
Toolbar:	Drawing > Predefined View	*(Customize to Add)*

The **Predefined View** tool is used to create empty views with the predefined orientation. After their creation, you can populate them by dragging a part from the other window by holding the name of the part document in the **FeatureManager Design Tree**. All the predefined views will be populated. This option adds the empty views in the drawing document and then saves it as a drawing template. You just need to drag and drop a part from another window, in the drawing document opened using the template, to populate it with all the predefined views. To create predefined views, choose the **Predefined View** button from the **Drawing CommandManager**. An empty view will be attached to the cursor. Specify a point in the drawing document to place the predefined view. The view will be placed in the drawing document and the **Drawing View PropertyManager** will be displayed, as shown in Figure 12-11.

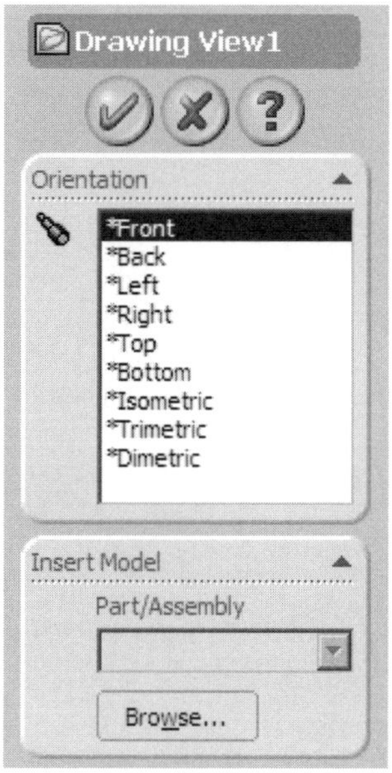

Figure 12-11 *The* ***Drawing View PropertyManager***

Select the view orientation from the **View Orientation** area of the **View Orientation** rollout and choose the **OK** button from the **Drawing View PropertyManager**.

You need to define the alignment option while placing the next predefined view. To create additional predefined views, invoke the **Predefined View** tool and place the view in the drawing sheet. The bounding box of the view is displayed. Right-click inside the bounding box and choose **Alignment > Align Horizontal by Center/Align Vertical by Center**. Next, select the previous predefined view to align the corresponding view. Similarly, add the other predefined views using this tool.

After creating all the predefined views, open the part or assembly document and tile the document windows horizontally or vertically. Next, press and hold the left mouse button down on the name of the part or assembly in the **PropertyManager** and drag the component or the assembly in the drawing document; all the predefined views will be populated. You can also populate individual predefined views by selecting them and choosing the **Browse** button from the **Insert Model** rollout of the **Drawing View PropertyManager**.

Figure 12-12 shows the selected predefined views with the orientation in which the views are created, and Figure 12-13 shows the drawing document after populating the drawing views.

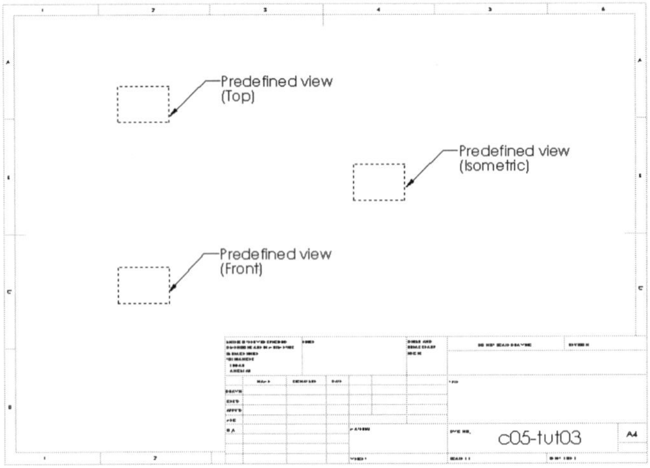

Figure 12-12 *Various predefined views*

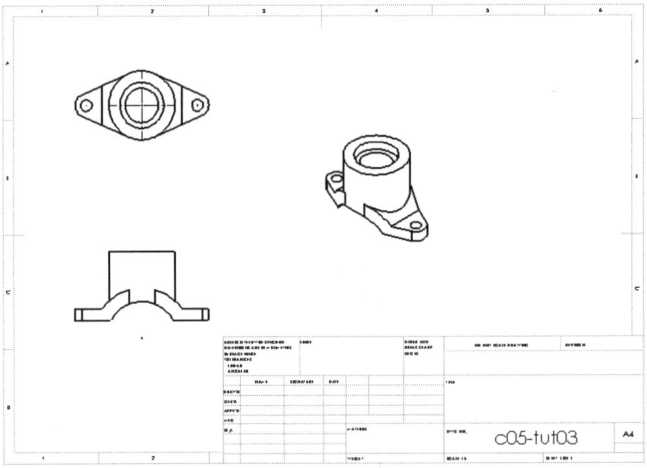

Figure 12-13 *Views created after populating the predefined views*

Note

*The views generated in the **Drawing** mode of SolidWorks, are automatically scaled, depending on the size of the sheet.*

The views will also be scaled automatically, if the drawing contains more than one predefined view.

*One predefined view, placed in the drawing document, will be scaled with respect to the **Custom Scale** value, if specified. Otherwise, it will be scaled with the default scale factor of the drawing sheet. You can change the view scale using the **Sheet Properties** dialog box. You will learn more about scaling the views in the next chapter.*

GENERATING THE DERIVED VIEWS

All views, generated from a view already placed in the drawing document, are known as derived views. These include:

1. Projected view
2. Section view
3. Aligned Section view
4. Broken-out Section view
5. Auxiliary view
6. Detail view
7. Crop view
8. Broken view
9. Alternate Position view

The methods of generating the various derived drawing views are discussed next.

Generating Projected Views

Enhanced

CommandManager:	Drawing > Projected View
Menu:	Insert > Drawing View > Projected
Toolbar:	Drawing > Projected View

As mentioned earlier, the projected views are generated by projecting the normal lines from an existing view, or at an angle from an existing view. To generate a projected view, choose the **Projected View** button from the **Drawing CommandManager**. The **Projected View PropertyManager** is displayed and you are prompted to select a drawing view to project the normal lines. The select cursor is replaced by the view cursor. Select the parent view and move the cursor vertically to generate the top view or the bottom view, or move the cursor horizontally to create right or left view. If you move the cursor at an angle, a 3D view will be created. Specify a point on the drawing sheet to place the view. For generating more than one projected views, choose the **Keep Visible** button to pin the **Projected View PropertyManager**. Figure 12-14 shows the front view generated from the top view.

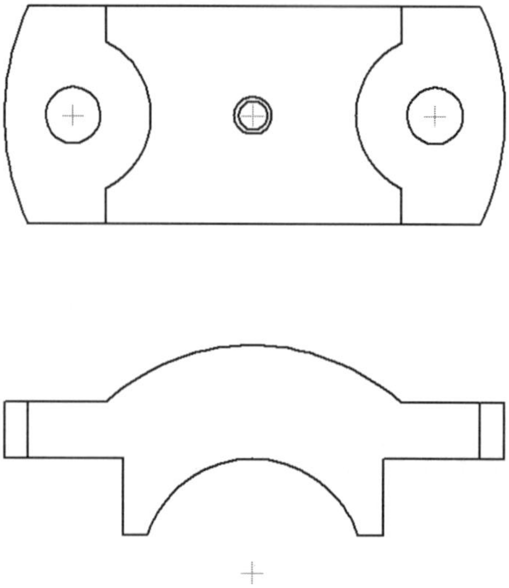

***Figure 12-14** Front view generated from the top view*

Tip. *The drawing view is aligned to the parent view, when you generate a projected drawing view. To place the projected view that is not in alignment with the parent view, press and hold down the CTRL key before placing it. Next, move the cursor to the desired location and place the view.*

All the standard and derived views such as projected views, section view, detailed view, and so on are linked to their parent view by a Parent-Child relationship. If you select the child view, the bounding box of the parent view will also be displayed.

*Select the child view, invoke the shortcut menu, and choose the **Jump to Parent View** option from it. The parent view will be selected automatically.*

Generating Section Views

CommandManager:	Drawing > Section View
Menu:	Insert > Drawing View > Section
Toolbar:	Drawing > Section View

As mentioned earlier, section views are generated by chopping a portion of an existing view, using a cutting plane (defined by sketched lines), and then viewing the parent view from a direction normal to the cutting plane.

In SolidWorks, you can use the **Section View** tool to create a full section view, or a half section view, as shown in Figure 12-15 and Figure 12-16 respectively. A full section view is defined using a single line segment, but a half section view is defined using three line segments. Note that the section plane for a full section view can be defined after invoking the **Section View** tool. But to generate a half section view, you need to draw the line segments to define the section plane before invoking the **Section View** tool. To do this, select the drawing view that you want to use as

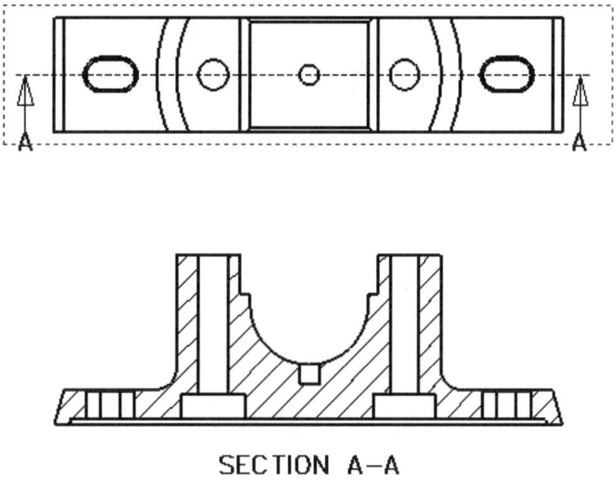

Figure 12-15 *Full section view*

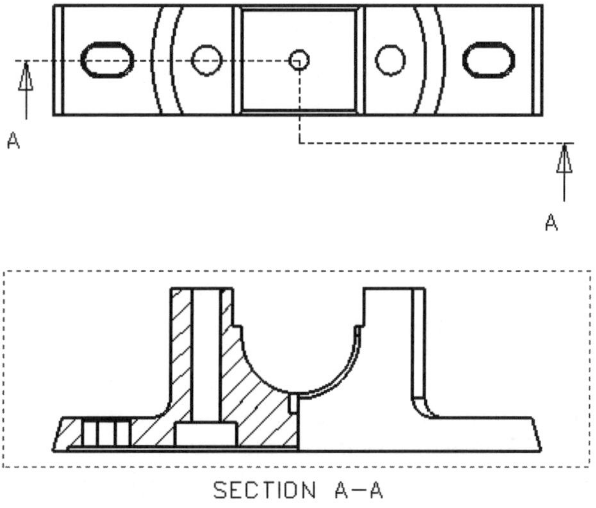

Figure 12-16 *Half section view*

the parent view and choose the **Sketch** button from the **CommandManager**; the sketching environment will be invoked. You can use the inferencing lines to draw the lines for the section plane.

To create a full section view, activate the view in which you need to draw the section line. The view symbol will be displayed below the cursor and the bounding box of the view will also be displayed. Now, choose the **Section View** button from the **Drawing CommandManager**. The **Section View PropertyManager** is displayed and you will be prompted to sketch a line to continue view creation.

After activating the view, draw a line that will define the section plane. On specifying the endpoint of the section line, the view is defined and is attached to the cursor. Also, the other options in the **Section View PropertyManager** are displayed.

To generate a half section view, select all the line segments that define the section plane and then invoke the **Section View** tool. The section view is defined and is attached to the cursor.

Move the cursor and specify a point on the drawing sheet to place the section view. The name and the scale factor of the drawing view is displayed below the section view and the **Section View PropertyManager** is invoked, as shown in Figure 12-17.

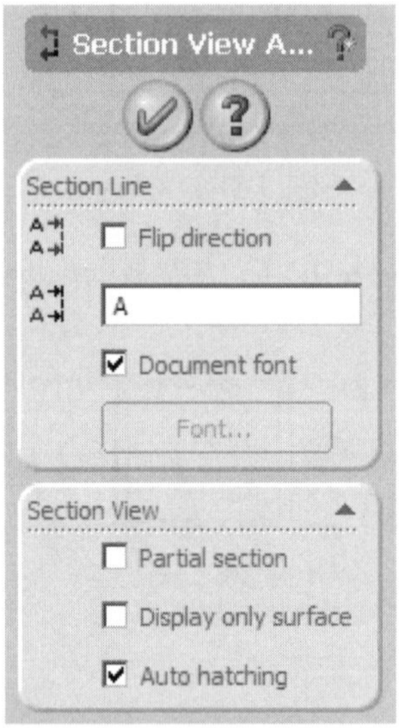

Figure 12-17 *Partial view of the* ***Section View PropertyManager***

You can use the **Flip direction** check box to flip the direction of the section view. The view will be automatically modified in the drawing sheet. You can also flip the direction of the viewing section view using the TAB key on the keyboard before placing the section view. The **Scale with model** check box is used to scale the drawing view, if the model is scaled in the part document. You will learn more about scaling the model in the next chapter.

Tip. *On creating a section view and moving the cursor to place the section view, you will observe that the view is aligned to the direction of arrows on the section line. In order to remove this alignment to place the section view, press and hold down the CTRL key and move the view to the desired location. Select a point in the drawing sheet to place the view.*

Note

The default hatch pattern in the section view depends on the material assigned to the model. Also, you may need to increase the spacing of the hatch pattern, if it is not correct. You will learn more about editing the hatch pattern later in this chapter.

Some other options available in the **Section View PropertyManager** to a create partial section view and the surface section view, are discussed next

Creating the Partial Section View

If the section line does not cut through the model, the **SolidWorks** information box will be displayed. This dialog box prompts you that the section line does not completely cut through the bounding box of the model in this view. Do you want this to be a partial section cut? For creating the partial section view, choose the **Yes** button from this dialog box. If you choose the **No** button from this dialog box, then the complete section view will be created. Figure 12-18 shows a partial section view generated from the top view.

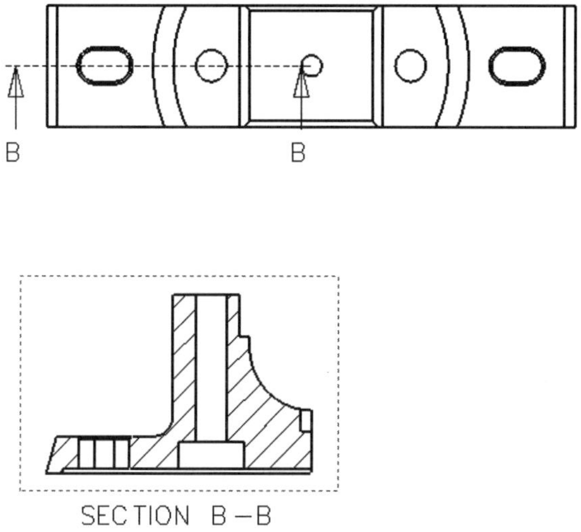

Figure 12-18 *A partial section view*

Creating the Surface Section View

A surface section view is the one in which only the sectioned surface is displayed in the section view. To create a surface section view, you first need to create the section view and then select the **Display only surface** check box from the **Section View PropertyManager**. Figure 12-19 shows a surface section view.

Tip. *Sometimes, the sectioned view is generated upside down, even if you have set the projection type to third angle. In such cases, you need to flip the direction of the section line from the* ***Section View PropertyManager****.*

Generating the Section View of an Assembly

According to the drawing standards, when you create the section view of an assembly, some components, such as fasteners, shafts, keys, and so on should not be sectioned. Therefore, when you create the section view of an assembly, the **Section View** dialog box is displayed, as shown in Figure 12-20.

This dialog box allows you to select the components that will be excluded from the section cut. You can also select the components from the parent view. But if the components are not visible

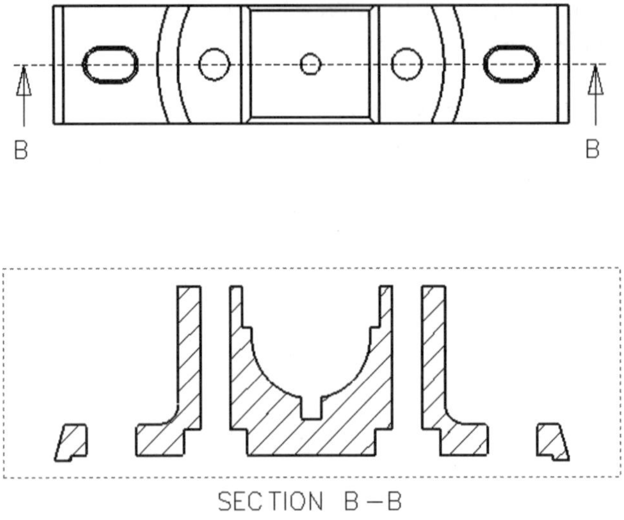

Figure 12-19 A surface section view

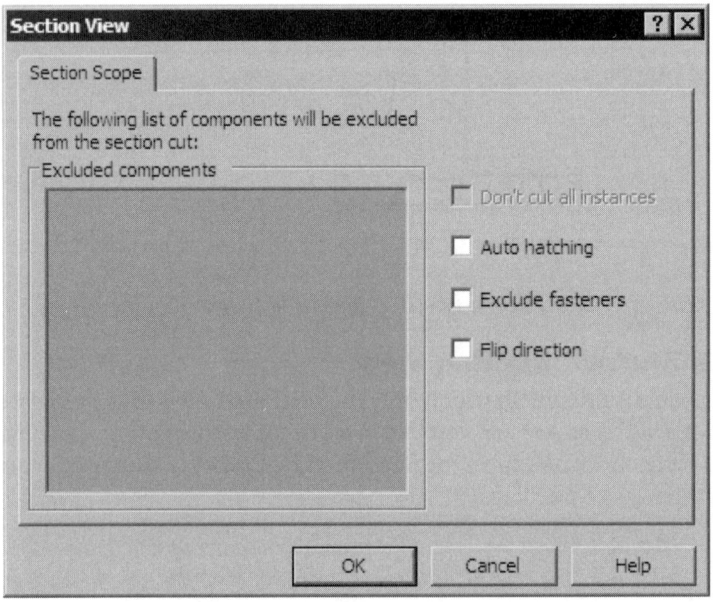

*Figure 12-20 The **Section View** dialog box*

in the parent view, you can invoke the **FeatureManager Design Tree** and expand the parent drawing view. Next, expand the assembly tree view to display all the components of the assembly. Select the components that are not required to be sectioned. The name of the selected component is displayed in the **Excluded components** selection area.

The **Auto hatching** check box is used to automatically define the hatch patterns. You can even change them if required. The method of changing hatch patterns is discussed later. This release of SolidWorks provides you with an option to exclude the fasteners that are inserted in the assembly, using the Toolbox application. Toolbox, one of the add-ins of SolidWorks, is used to insert standard fasteners to the assembly. To exclude the fasteners that are inserted using this option, select the **Exclude fasteners** check box from the **Drawing View Properties** dialog box.

The **Flip direction** check box is used to flip the direction of viewing of the section view.

In case you have more than one instance of the component in the assembly, and you need to exclude all the instances of the component from the section view, select the component from the drawing sheet and also the name of the component from the **Exclude components** selection area. Select the **Don't cut all instances** check box from the **Section Scope** dialog box. All instances of the selected component will be excluded from the section view. Figure 12-21 shows an assembly section view with, fasteners excluded from the cut.

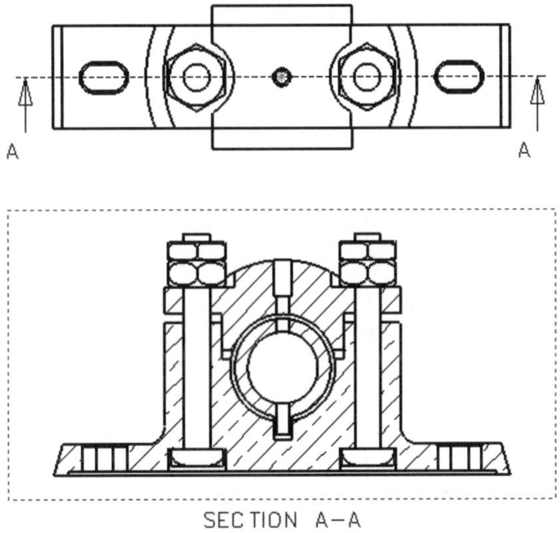

Figure 12-21 Section view of an assembly with some of the components excluded from the cut

Tip. *You can add or remove the components that are sectioned by, right-clicking the drawing view and choosing **Properties** from the shortcut menu. Next, choose the **Section Scope** tab and add or remove the components.*

Generating Aligned Section Views

CommandManager: Drawing > Aligned Section View
Menu: Insert > Drawing View > Aligned Section
Toolbar: Drawing > Aligned Section View

This tool is used to generate a section view of the component in which at least one of the feature is at an angle. In the aligned section view, the sectioned portion revolves about an axis normal to the view such that it is straightened. For an example, refer to Figure 12-22.

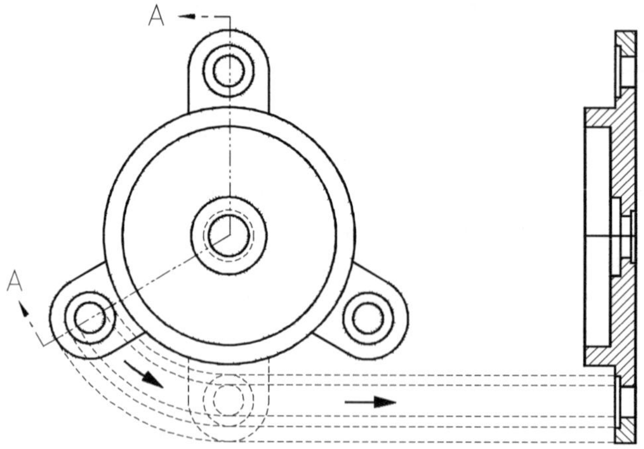

Figure 12-22 *Aligned section view*

This figure explains the concept of an aligned section view of a model. Notice that the inclined feature, sectioned in this view is straightened. As a result, the section view is longer than the parent view. Activate the view to create the aligned section view. Choose the **Aligned Section View** button from the **Drawing CommandManager**. Draw the sketch that defines the section plane. The aligned section view will be attached to the cursor; place the view at an appropriate location in the drawing sheet. Note that the resulting view will be projected normal to the line drawn at the end in the section sketch. Therefore, to get the aligned section view similar to that shown in Figure 12-22, the inclined line in the section sketch should be drawn first, followed by the vertical line. Figure 12-23 shows the aligned section view in which the vertical line in the section sketch is drawn first. This is the reason the section view is projected normal to the inclined line that is drawn last. On the other hand, Figure 12-24 shows the view in which the inclined line is drawn first.

With this release of SolidWorks, you can also create a section view and aligned section views from a crop view, detail view, and an orthogonal exploded views.

Note

You can also create a sketch associated to a view. This sketch can be selected as the section plane for generating the section view. To create an associated sketch, activate the view and draw the sketch that defines the section plane, using the ***Line*** *tool.*

If you create a sketch to define the section plane for the aligned section view before invoking the ***Aligned Section View*** *tool, the view will be projected normal to the line that you select last. However, if you select the sketch, by dragging a window around it, the view will be projected normal to the line that was drawn last.*

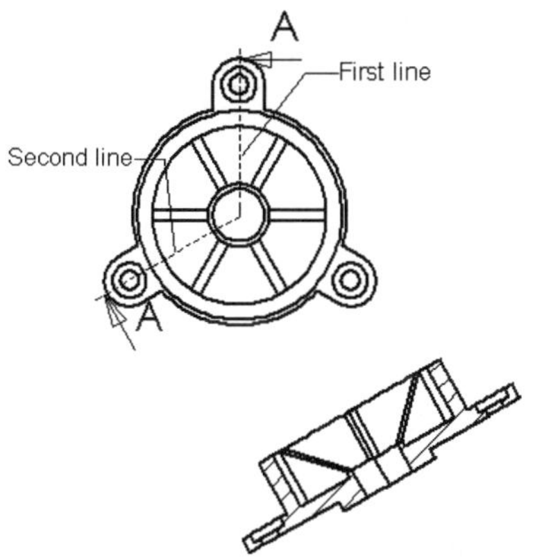

Figure 12-23 Aligned section view

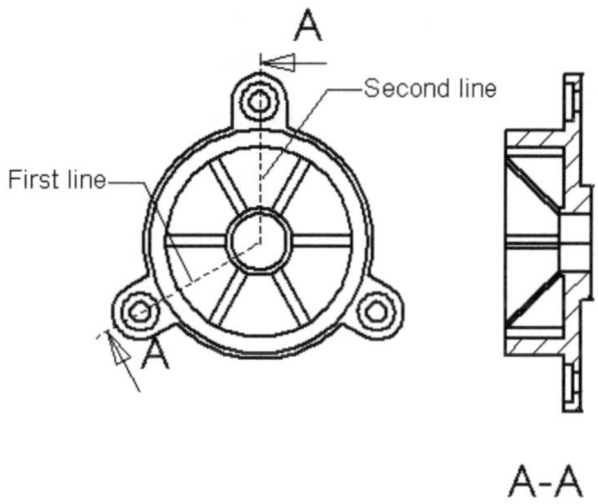

Figure 12-24 Aligned section view

Tip. *With this release of SolidWorks you can also use more than two lines to create an aligned section view. To do this, you need to draw the lines prior to invoke the* ***Aligned Section View*** *tool.*

Generating Broken-out Section Views

CommandManager:	Drawing > Broken-out Section
Menu:	Insert > Drawing View > Broken-out Section
Toolbar:	Drawing > Broken-out Section

This tool is used to create a broken-out section view ,which is used to remove a part of the existing view and display the area of the model or the assembly behind the removed portion. This view is generated using a closed sketch that is associated with the parent view. To create a broken-out section view, activate the view on which you need to create the broken-out section view. Choose the **Broken-out Section** button from the **Drawing CommandManager**. The **Broken-out Section PropertyManager** is displayed and it prompts you to create a closed spline to continue the section creation. The cursor will be replaced by the spline cursor. Draw a closed sketch using the spline cursor. If you do not want a spline profile, select a closed profile before choosing the **Broken-out Section** button. Figure 12-25 shows an associated sketch created for creating a broken-out section view.

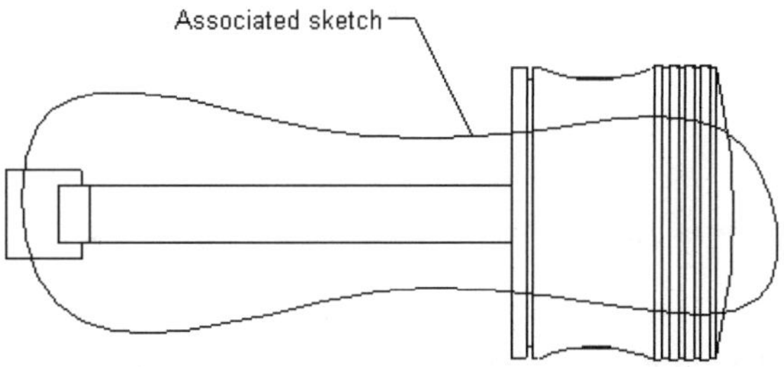

Figure 12-25 Sketch for creating a broken-out section view

When you draw a closed sketch, the options are displayed in the **Broken-out Section PropertyManager**, as shown in Figure 12-26, and you are prompted to specify the depth of the broken-out section.

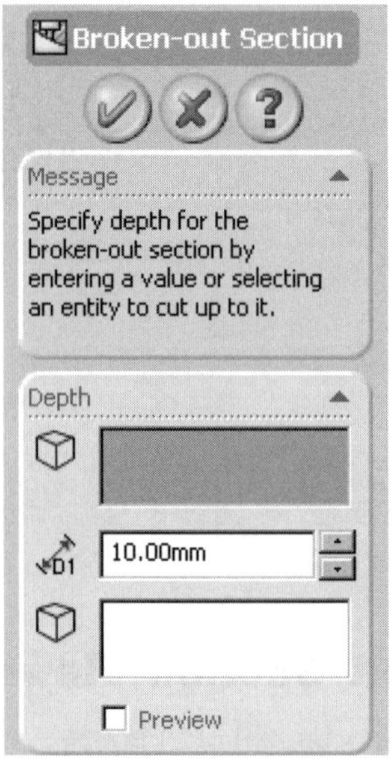

Figure 12-26 *The* ***Broken-out Section PropertyManager***

Select the **Preview** check box to preview the broken-out section view. The **Auto hatching** check box is used to automatically define the hatch pattern to the section drawing view of the assembly. Note that the **Auto hatching** check box is not available in the **Broken-out Section PropertyManager,** if you are creating the broken-out section view of a part. Figure 12-27 shows the preview of the broken-out section view.

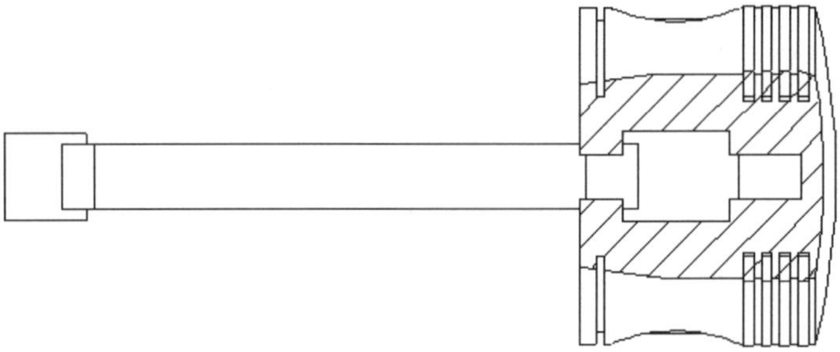

Figure 12-27 *Preview of the broken-out section view*

Set the value of the depth of the broken-out section in the **Depth** spinner. The preview of the section will be modified dynamically in the drawing view. After setting the value of the depth of the broken-out section, choose the **OK** button from the **Broken-out Section PropertyManager**. Figure 12-28 shows a broken-out section view with a different depth value.

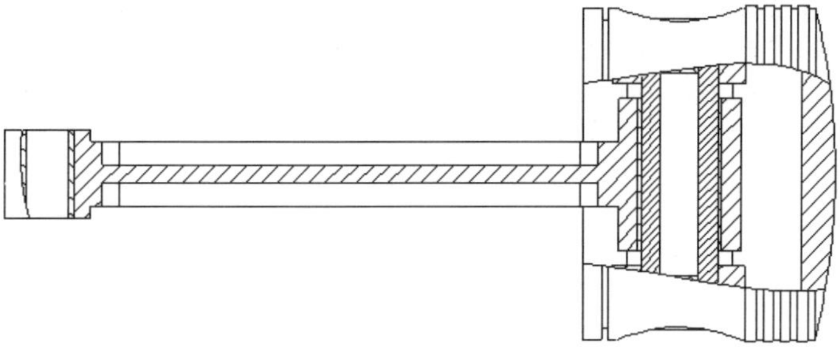

Figure 12-28 Broken-out section view

Generating Auxiliary Views

CommandManager:	Drawing > Auxiliary View
Menu:	Insert > Drawing View > Auxiliary
Toolbar:	Drawing > Auxiliary View

An auxiliary view is a drawing view that is generated by projecting the lines normal to a specified edge of an existing view. SolidWorks also allows you to create a line segment associated to the view that can be used to generate the auxiliary view. For this, the associated line segment needs to be created before invoking this tool.

To create an auxiliary view, choose the **Auxiliary View** button from the **Drawing CommandManager**. The **Auxiliary View PropertyManager** is displayed and you are prompted to select a reference edge to continue. Select the edge or the associated sketch; a view will be attached to the cursor and some options will be displayed in the **Auxiliary View PropertyManager**, as shown in Figure 12-29. Also, you will be prompted to specify the location to place the view.

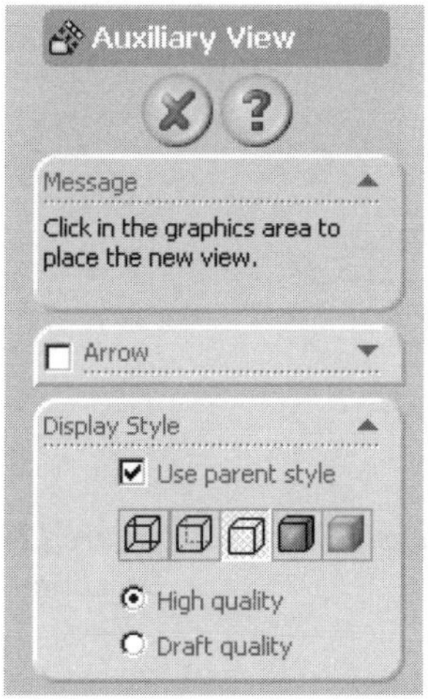

Figure 12-29 Partial view of the ***Auxiliary View PropertyManager***

The check box available in the **Arrow** rollout is used to display the arrow of the viewing direction in the drawing views. The name of the auxiliary view is specified in the **Label** edit box. Using the **Flip Direction** check box, you can flip the viewing direction for creating the auxiliary view. Figure 12-30 shows the reference edge to be selected to create the auxiliary view.

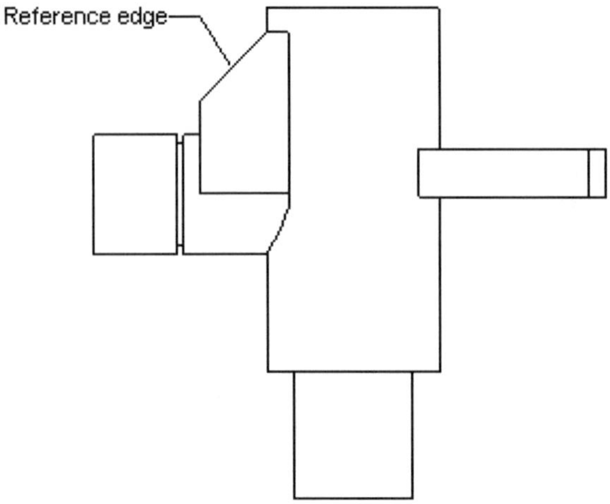

***Figure 12-30** Reference edge to be selected to create the auxiliary view*

Figure 12-31 shows the auxiliary view created with the default viewing direction. Figure 12-32 shows the auxiliary view created with the **Flip Direction** check box selected.

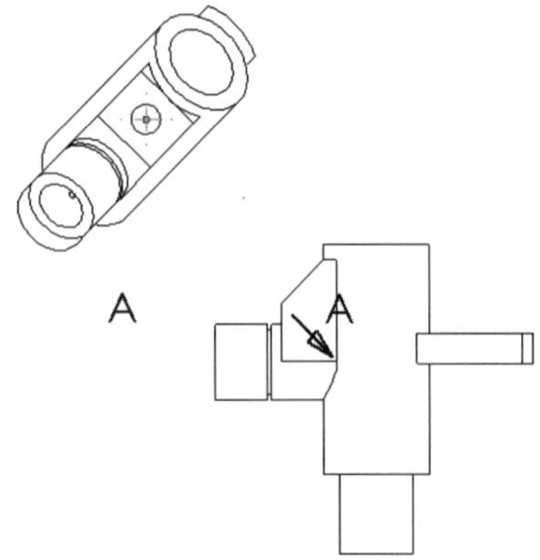

***Figure 12-31** Auxiliary view created with the **Flip Direction** check box cleared*

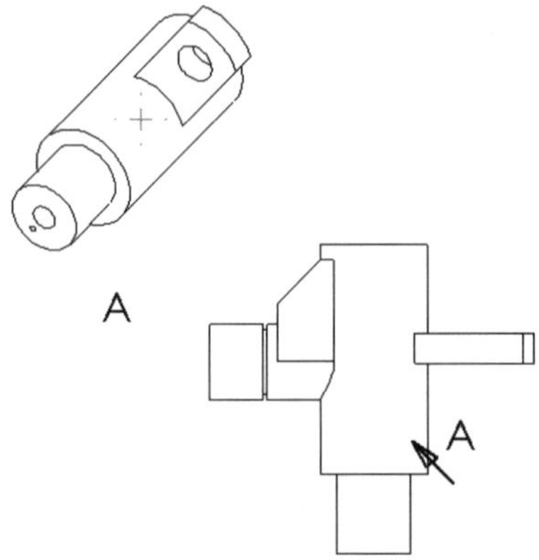

***Figure 12-32** Auxiliary view created with the **Flip Direction** check box selected*

Generating Detail Views

CommandManager: Drawing > Detail View
Menu: Insert > Drawing View > Detail
Toolbar: Drawing > Detail View

A detail view is used to display the details of a portion of an existing view. You can select the portion whose detailing has to be shown in the parent view. The portion that you select will be magnified and placed as a separate view. You can control the magnification of the detail view. To create a detail view, activate the view from which you will generate the detail view. Next, choose the **Detail View** button from the **Drawing CommandManager**; the **Detail View PropertyManager** is displayed and you are prompted to sketch a circle to continue view creation. The cursor is replaced by a circle cursor. Create the circle on the portion of the view that is to be displayed in the detail view. If you want to use a profile other than the circle, you need to create it associated to the sketch before invoking the **Detail View** tool.

As soon as you draw the circle, the detail view is attached to the cursor and the options are displayed in the **Detail View PropertyManager**, as shown in Figure 12-33. You are also prompted to select a location for the new view. Specify a point on the drawing sheet to place the view. The options available in the **Detail View PropertyManager** are discussed next.

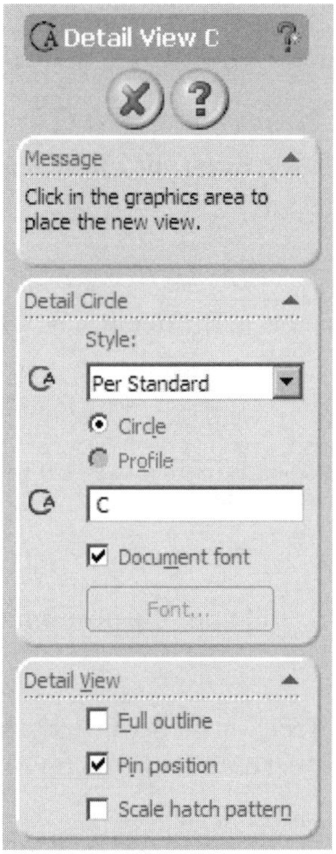

*Figure 12-33 The **Detail View PropertyManager***

Detail Circle

This rollout is used to define the options to display the circle of the detail view. You can also apply the leader to the detail view, using the options available in the rollout, which are discussed next.

Style

The **Style** area has the **Style** drop-down list to specify the style of a closed profile. By default, the **Circle** radio button is selected below the **Style** drop-down list. Therefore, the portion of the parent view that is shown in the detail view is highlighted in the circle. Select the **Profile** radio button, if you have already created a closed profile for defining the portion to be shown in the detail view. The options available in the **Style** drop-down list are discussed next.

Per Standard. The **Per Standard** option is used to create the detail view as per the default standards.

Broken Circle. The **Broken Circle** option is used to display the area of the parent view to be displayed in the detailed view in a broken circle.

With Leader. The **With Leader** option is used to add the leader to the callout of the detail view.

No Leader. The **No Leader** option is used to remove the leader from the callout of the detail view.

Connected. This option is used to create a line that connects the detail view with the closed profile in the parent view.

Detail View

This rollout is used to set the parameters of the detail view. The various options available in this rollout are discussed next.

Full outline

The **Full outline** check box is used to display the complete outline of the closed profile in the detail view.

Pin position

The **Pin position** check box is used to pin the position of the detail view.

Scale hatch pattern

While creating a detail view of a section view, the **Scale hatch pattern** check box is used to scale the hatch pattern with respect to the scale factor of the detail view.

If you create a detail view with another detail view or a crop view as the parent view, the default scale factor of the resulting detail view is twice of the immediate parent view. Figure 12-34 shows the detail view created using the **Detail View** tool.

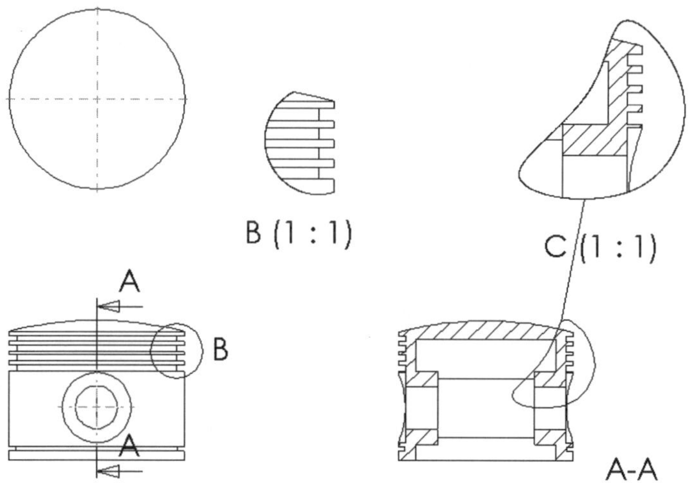

Figure 12-34 Detail views generated using the existing views

Tip. *When you create a detail view, by default it is scaled as 1:1. You can define the default scale factor in the* ***System Options*** *dialog box so that the detail view, it will be created with the scaling factor provided by you. To specify the scale factor for the detail view, invoke the* ***System Options*** *dialog box and select the* ***Drawings*** *option from its left. Set the value of the scale factor of the detail view in the* ***Detail view scaling*** *edit box and choose the* ***OK*** *button. Hence forth, the detail view will be created of the scale factor defined in the* ***System Options*** *dialog box.*

Cropping Drawing Views

CommandManager:	Drawing > Crop View
Menu:	Insert > Drawing View > Crop
Toolbar:	Drawing > Crop View

This tool is used to crop an existing view using a closed sketch associated to it. The portion of the view that lies inside the associated sketch is retained and the remaining portion is removed. To crop the view, you first need to create a closed profile that defines the area of the view that will be displayed. The area of the view outside this closed profile will not be displayed when you crop the view. Select the closed profile and choose the **Crop View** button from the **Drawing CommandManager**. Figure 12-35 shows the closed profile used to crop the view and Figure 12-36 shows the cropped view.

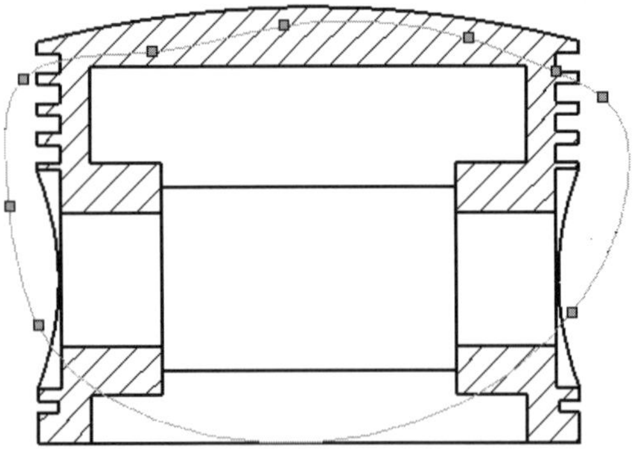

Figure 12-35 *Closed profile to crop the view*

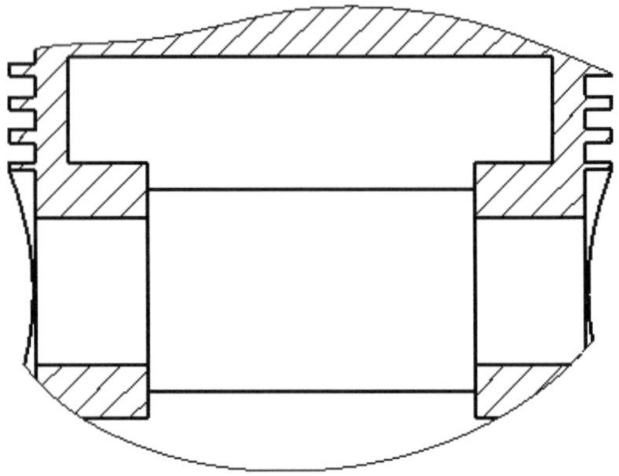

Figure 12-36 *Resulting crop view*

Tip. *You can remove the cropping of view by selecting the crop view and invoking the shortcut menu. Choose **Crop View > Remove Crop** from the shortcut menu. The initial view will be displayed in the drawing sheet.*

*To edit the closed profile of the crop view, select the crop view and invoke the shortcut menu and choose **Crop View > Edit Crop** from it. The sketch of the closed profile and the complete view is displayed in the drawing sheet. Edit the closed profile and choose the **Rebuild** button from the **Standard** toolbar or use CTRL+B on the keyboard.*

Generating Broken View

A broken view is the one in which a portion of the drawing view is removed from in between, keeping the ends of the drawing view intact. This view is used for displaying the components whose length to width ratio is very high. This means that either the length is very large as compared to the width or the width is very large as compared to the length. The broken view will break the view along the horizontal or vertical direction such that the drawing view fits the area you require. To create a broken view, you first need to define the break line. Select the view you need to break and choose the **Horizontal Break/Vertical Break** buttons from the **Drawings CommandManager**, depending on the direction in which you need to break the component. Two break lines will be displayed on the selected view, as shown in Figure 12-37.

Tip. *If you suppress the features of a model whose drawing views have been generated, the suppressed features will not be displayed in the drawing views. The feature, on suppressing, will be displayed in the drawing views.*

When you hide or suppress the components of an assembly, the hidden or suppressed components are not displayed in the drawing views.

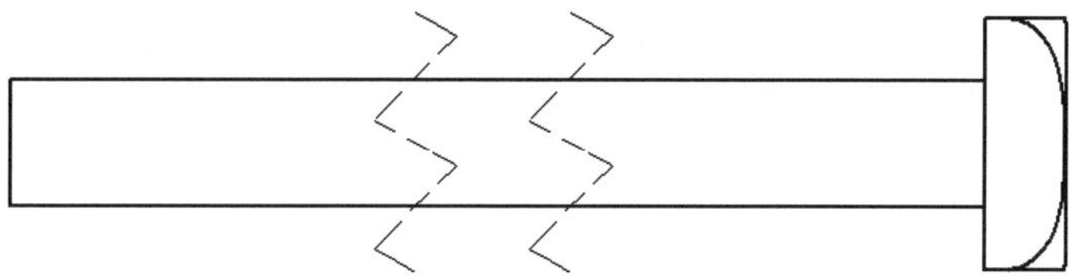

Figure 12-37 *Break lines added to the view*

After adding the break lines, you need to move the break lines to define the gap in the broken view. Select the break lines and move them away from each other, as shown in Figure 12-38. Now, select the view and invoke the shortcut menu. Choose the **Break View** option from the shortcut menu. The broken view will be created, as shown in Figure 12-39.

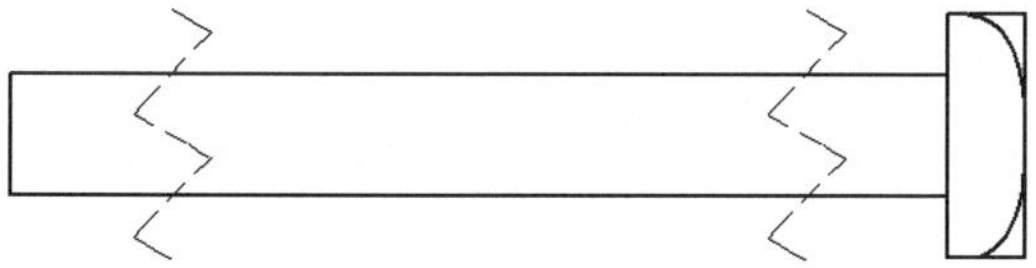

Figure 12-38 *Extended gap between break lines*

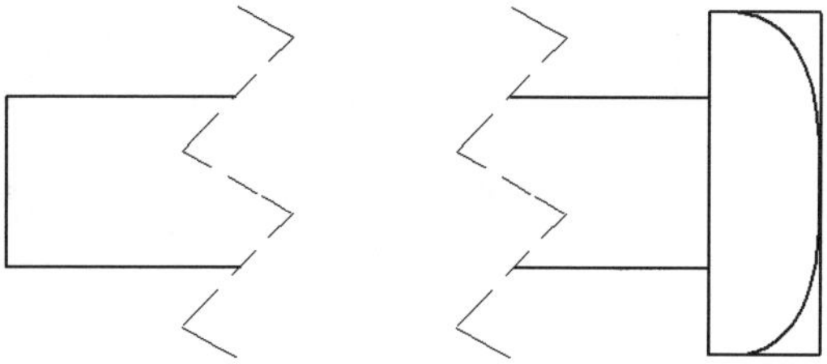

Figure 12-39 *Resulting broken view*

You can also break an isometric view; the procedure of breaking an isometric view or any 3D view is the same as discussed earlier. Figure 12-40 shows a broken isometric view.

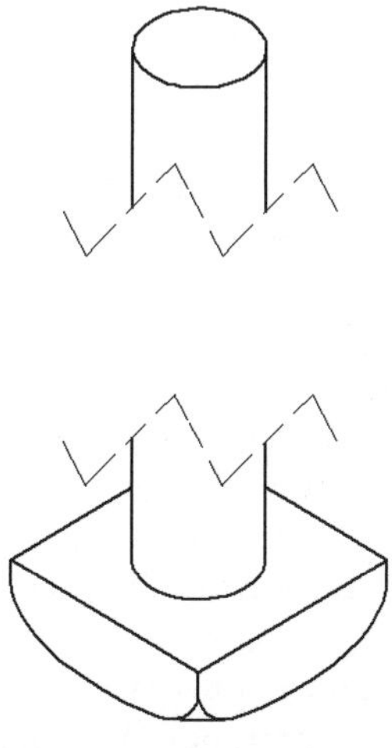

Figure 12-40 *A broken isometric view*

Note

Select the break line to increase or decrease the gap between the broken view. On moving the break line, the broken view is modified dynamically.

If you generate a projected view from a broken view, the resulting projected view is also a broken view.

If you break a 3D view placed horizontally, the two parts of the view, as a result of the ***Broken View*** *tool, will lose their alignment.*

Tip. *You can change the style of the break line by selecting it and invoking the shortcut menu. The various break line styles available are straight cut, curve cut, zig zag cut, and small zig zag cut.*

To unbreak the broken view, select the view, invoke the shortcut menu and choose the ***Un-Break View*** *option from it.*

Select the view, invoke the shortcut menu and choose the ***Break View*** *option to again break the view.*

If you select the break lines and press the DELETE key on the keyboard, the broken view will be replaced by the parent view.

Alternate Position View

CommandManager:	Drawing > Alternate Position View
Menu:	Insert > Drawing View > Alternate Position
Toolbar:	Drawing > Alternate Position View

The alternate position view is used to create a view in which you can show the maximum and minimum range of the motion of an assembly. The main position is displayed with continuous lines in the drawing view, while the alternate position of the assembly is shown in the same view with dashed (phantom) lines. To create an alternate position view, activate and select the view of the assembly drawing on which you need to create the alternate position view. Choose the **Alternate Position View** button from the **Drawing CommandManager**. The **Alternate Position PropertyManager** is displayed, as shown in Figure 12-41.

The **Alternate Position PropertyManager** prompts you to select a new configuration, click OK or enter and define the new configuration parameters. If you have not created any configurations, the **New Configuration** radio button will be automatically selected to create one. Enter the name of the configuration in the edit box given below and choose the **OK** button from the **Alternate Position PropertyManager**. Choose **OK** from the **Tangent Edge Display** dialog box, if displayed.

The assembly document is opened and the **Move Component PropertyManager** is displayed in the assembly document. The **Move Component PropertyManager** prompts you to move the desired components to the position to be shown in the alternate view. Note that the component

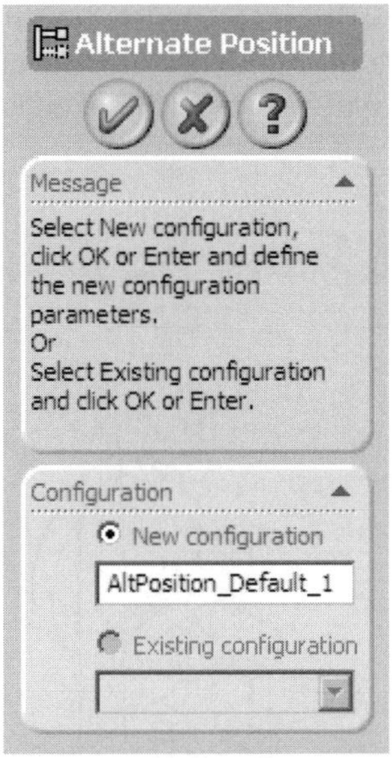

Figure 12-41** The **Alternate Position PropertyManager

or components that you need to move should have that particular degree of freedom free. These components should not be fully defined in the assembly. Select and drag the cursor to move the components to the desired location. After defining the alternate position of the components, choose the **OK** button from the **Move Component PropertyManager**. You will return to the drawing document automatically. The alternate position of the components that are moved will be displayed in the phantom lines in the drawing view, as shown in Figure 12-42.

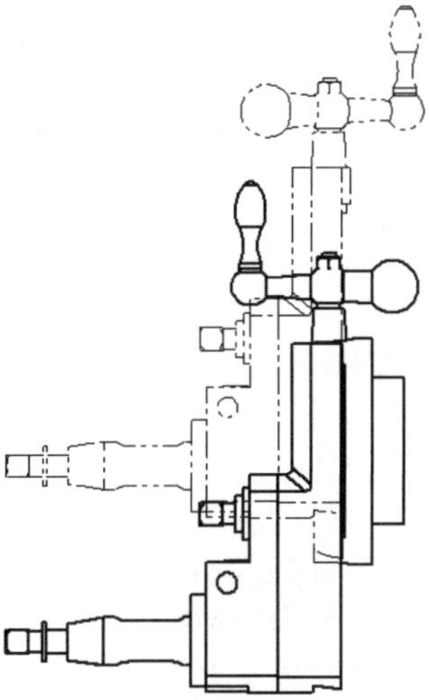

***Figure 12-42** Alternate position view*

You can also create the alternate position view of an isometric view or of any 3D view. The procedure of creating the alternate position view of a 3D view is the same as that discussed earlier. Figure 12-43 shows the alternate position view of an isometric view.

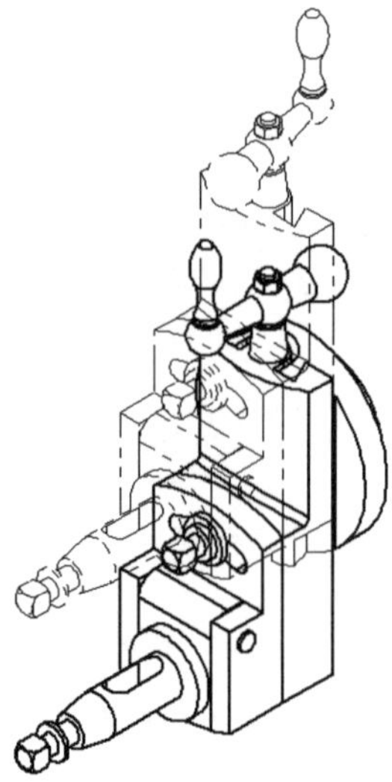

Figure 12-43 *Alternate view of an isometric view.*

Note

On creating an alternate view of an assembly, a new configuration is created inside the assembly document with the same name that is specified to the configuration, while creating the alternate position view. Open the assembly document and invoke the ***ConfigurationManager****; you will observe that a new configuration has been created along with the default configuration. By default, the newly created configuration is selected. Therefore, the assembly is displayed with the components moved to their extreme positions. To switch back to the default configuration, select* ***Default*** *from the* ***ConfigurationManager****, invoke the shortcut menu and choose the* ***Show Configuration*** *option from it. You will observe that the assembly will be displayed with the moved components back at their original positions. If the* ***Show Configuration*** *option is not available in the shortcut menu, the assembly is at the default configuration.*

Creating the Drawing View of the Exploded State of an Assembly

You can create the drawing view of the exploded state of the assembly. For this, you need to have an exploded state defined in the assembly document. Generate the isometric view of the assembly on the drawing sheet. Select the view, invoke the shortcut menu, and choose the **Properties** option from it. The **View Properties** tab of the **Drawing View Properties** dialog box is displayed. Select the **Show in exploded state** check box from the **Configuration information** area and choose the **OK** button. Figure 12-44 shows the drawing view of the exploded state of assembly with explode lines.

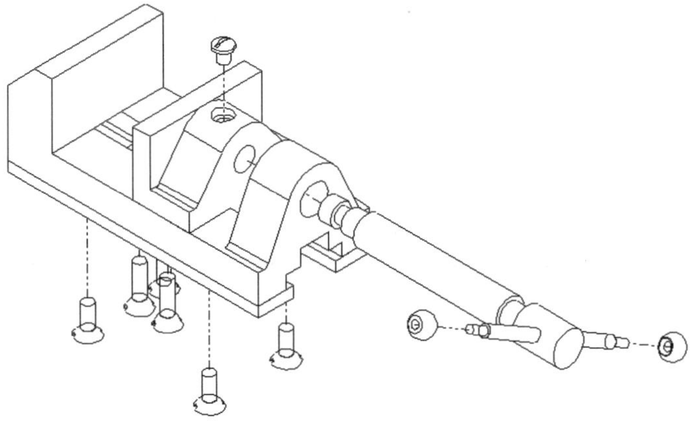

Figure 12-44 Drawing view of the exploded state with explode lines

Tip. *All the views of the assembly will be generated in the exploded state, if the assembly in the assembly document is in the exploded state and you drag and drop the assembly to generate the drawing views.*

To the collapse state in the drawing view, select the view and clear the ***Show in exploded state*** *check box from the* ***View Properties*** *tab of the* ***Drawing View Properties*** *dialog box.*

WORKING WITH INTERACTIVE DRAFTING IN SolidWorks

As mentioned earlier, you can also sketch the 2D drawings in the drawing document of SolidWorks. In technical terms, sketching 2D drawings is known as interactive drafting. Before starting the drawing, it is recommended that you insert an empty view. To create an empty view, choose **Insert > Drawing View > Empty** from the menu bar. An empty view is attached to the cursor. Select a point at the desired location to place an empty view. Now, select the empty view to activate and use the tools in the **Sketch CommandManager** to sketch the view.

The sketched entities will also move if you move the empty view by selecting and dragging it. This is because the sketch that you draw is associated to the empty view.

EDITING AND MODIFYING THE DRAWING VIEWS

In SolidWorks, you can perform various kinds of editing operations and modifications on the drawing views. For example, you can change the orientation of the view, or the view scale, or you can delete the view. All these operations are discussed next.

Changing the View Orientation

You can change the orientation of the views generated using the **Model View** or the **Predefined View** option. To change the orientation, select the view; the **Drawing View PropertyManager** is displayed in both the cases. Double-click on the view orientation that you want as the current one, from the **View Orientation** rollout. The orientation of the selected view will be modified. Choose the **OK** button from the **Drawing View PropertyManager**.

All the derived views will also change their orientation, when you change the orientation of the parent view.

Changing the Scale of Drawing Views

In SolidWorks, you can also change the scale of the drawing views. For doing so, select the drawing view and then select the **Use custom scale** radio button from the **Scale** rollout. Set the new scale of the drawing view in the edit boxes available below this radio button. You can also change the scale of the derived views. However, the scale of the parent view will not be changed if you change the scale of a derived view.

Deleting Drawing Views

The unwanted views are deleted from the drawing sheet using the **FeatureManager Design Tree** or directly from the drawing sheet. Select the view to be deleted from the **FeatureManager Design Tree**, invoke the shortcut menu and choose the **Delete** option from it. The **Confirm Delete** dialog box will be displayed. Choose the **Yes** button from this dialog box. You can also delete a view by selecting it directly from the drawing sheet and pressing the DELETE key on the keyboard. The **Confirm Delete** dialog box is displayed; choose the **Yes** button from this dialog box. On deleting a parent view, the projected views are not deleted. However, if you delete the parent view from which a section, auxiliary, or a detail view is generated, the name of the dependent view will also be displayed in the **Confirm Delete** dialog box. If you choose **Yes**, the dependent views will also be deleted.

Rotating Drawing Views

SolidWorks allows you to rotate a drawing view in the 2D plane. Select the view and choose the **Rotate View** button from the View toolbar. The **Rotate Drawing View** dialog box is displayed, as shown in Figure 12-45. You can enter the value or rotation angle in this dialog box or you can also dynamically rotate the drawing view by dragging the mouse. If you select the **Dependent views update to change in orientation** check box, the views dependent on the rotated view will also change their orientation.

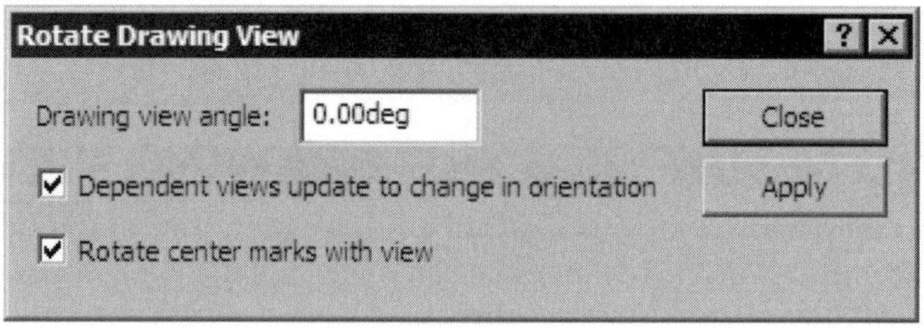

***Figure 12-45 Rotate Drawing View** dialog box*

Tip. *You can also copy and paste the drawing view in the drawing sheet. To do so, select the view to copy and press CTRL+C on the keyboard. Now, click anywhere on the drawing sheet to select the sheet and press CTRL+V on the keyboard to paste the drawing view.*

MODIFY HATCH PATTERN IN SECTION VIEWS

As discussed earlier, when you generate a section view of an assembly or a component, a hatch pattern is applied to the component or components. This hatch pattern is based on the material assigned to the components in the part document. If you need to modify the default hatch pattern, select the hatch pattern from the section view; the **Area Hatch/Fill PropertyManager** is displayed, as shown in Figure 12-46. The options available in this dialog box are discussed next.

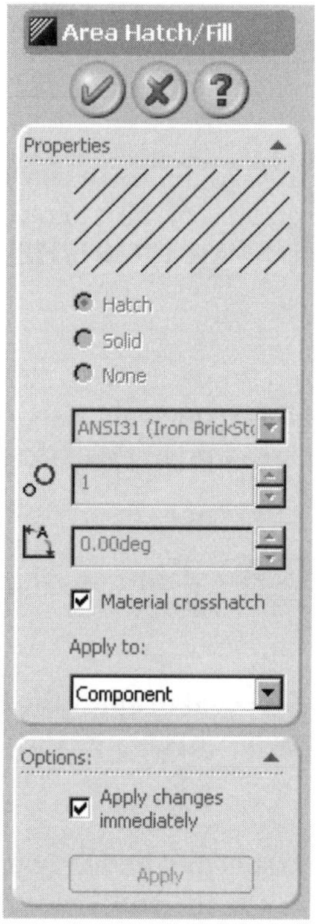

Figure 12-46** The **Area Hatch/Fill PropertyManager

Properties Rollout

The **Properties** rollout is used to define the type of hatch pattern and its properties. Some of the options in this area are not available by default. This is because by default, the material-dependent hatch pattern is applied to the component. If you want to make the other options also available, clear the **Material crosshatch** check box. The options available in this area are discussed next.

Preview
The **Preview** area displays the preview of the hatch pattern with the current setting.

Hatch
The **Hatch** radio button is selected to apply the standard hatch patterns to the section view. On selecting this button, some options available in the dialog box are invoked to define the properties of the hatch pattern. The options available to define the properties of the hatch pattern are discussed next.

Solid
The **Solid** radio button is used to apply the solid filled hatch pattern to the section view. By default, black color is applied to the solid filled hatch pattern.

None
Select the **None** radio button, if you do not need to apply any hatch pattern in the section view.

Hatch Pattern
The **Hatch Pattern** drop-down list is used to define the style of the standard hatch pattern you need to apply to the section view. The preview of the hatch pattern selected from this drop-down list is displayed in the **Preview** area of the **Area Hatch/Fill** dialog box.

Hatch Pattern Scale
The **Hatch Pattern Scale** spinner is used to specify the scale factor of the standard hatch pattern selected from the **Pattern** drop-down list. When you change the scale factor using this spinner, the preview displayed in the **Preview** area updates dynamically.

Hatch Pattern Angle
The **Hatch Pattern Angle** spinner is used to define an angle to the selected hatch pattern.

Material crosshatch
The **Material crosshatch** check box is selected to apply the hatch pattern based on the material assigned to the model. Clear this check box to change the type of the hatch pattern.

Apply to
The **Apply to** drop-down list is used to specify whether you need to apply this hatch pattern to the selected component, region, body, or to the entire view. Note that some of these options are available only while modifying the hatch pattern of an assembly section view.

Options Rollout
The **Options** rollout provides the **Apply changes immediately** check box, which when selected applies the changes immediately on the view and the preview is modified dynamically. If you clear this check box, you need to choose the **Apply** button, after making the changes to reflect the changes in the preview.

TUTORIALS

Tutorial 1

In this tutorial, you will generate the front view, top view, right view, aligned section view, detail view, and isometric view of the model created in Tutorial 2 of Chapter 7. Use the Standard A4 Landscape sheet format for generating the views. **(Expected time: 30 min)**

The steps to be followed to complete this tutorial are listed next.

a. Copy the part document of Tutorial 2 of Chapter 7 in the folder of the current chapter.
b. Open the copied part document and start a new drawing document from within the part document.
c. Select the standard A4 landscape sheet format and generate the parent view using the **Model View** tool, refer to Figure 12-47.
d. Generate the projected views using the **Projected View** tool, refer to Figure 12-47.
e. Generate the aligned section view using the **Aligned Section View** tool, refer to Figures 12-48 and 12-49.
f. Generate the detail view, refer to Figure 12-50.
g. Save and close the drawing document.

Copying and Opening the Part Document

1. Create a folder with the name *c12* in the *SolidWorks* directory and copy *c07tut2.sldprt* from the *\My Document\SolidWorks\c07* folder to this folder.

2. Start SolidWorks and open the part document of Tutorial 2 of Chapter 7 that you copied in the folder of the current chapter.

Starting the New Drawing Document

As mentioned earlier, one of the latest enhancements of SolidWorks 2005 is its ability to let you start a new drawing document from within the part document. This way, the model in the part document is automatically selected and you can generate its drawing views.

1. Choose the **Make Drawing from Part/Assembly** button from the **Standard** toolbar.

If you are in the practice of using the advanced form of the **New SolidWorks Document** dialog box, this dialog box will be displayed every time you choose the **Make Drawing from Part/Assembly** button.

2. From the **New SolidWorks Document** dialog box, select the drawing template from the **Template** tab and choose **OK**.

Remember that if you are not using the advanced form of the **New SolidWorks Document** dialog box, this dialog box is not displayed. Instead, a new drawing document is started directly and the **Sheet Format/Size** dialog box is displayed.

3. Select the **A4 - Landscape** sheet from the list box available in this dialog box and choose the

OK button. The new drawing document is started with the standard A4 sheet and the **Model View** tool is invoked automatically. The model of Tutorial 2 of Chapter 7 is selected by default for generating the drawing views. A view is also attached to the cursor.

Generating the Parent View and the Projected Views

Before you proceed with generating the drawing views, you need to confirm whether the projection type for the current sheet is set to the third angle.

1. Press the ESC key to exit the **Model View** tool. Select **Sheet1** from the **FeatureManager Design Tree** and then right-click on it. Choose the **Properties** option from the shortcut menu to display the **Sheet Properties** dialog box.

2. Select the **Third angle** radio button from the **Type of projection** area, if it is not already selected. Close this dialog box.

Note

*This option can also be selected **in the Sheet Format/Size** dialog box, which is displayed when you start a new drawing document.*

3. Now, choose the **Model View** button from the **CommandManager** to invoke the **Model View PropertyManager**.

4. Select the **Front** option from the list box available in the **Orientation** rollout of the **Model View PropertyManager**, if not already selected. Also, select the **Preview** check box; the preview of the front view of the model is displayed.

5. Select the **Auto-start projected view** check box in the **Options** rollout, if it is not selected. Move the cursor to the middle left of the drawing sheet just above the title block and specify a point at this location to place the front view, refer to Figure 12-47.

   Specify the location of the view; the front view will be generated and placed at this location. This view is generated at a 1:1 scale. Note that because you selected the option to start projected views immediately after generating the front view, the **Projected View PropertyManager** is invoked and the preview of the projected view is attached to the cursor. This view is being generated by referencing the front view as the parent view.

6. Move the cursor above the front view and specify a point to place the top view, refer to Figure 12-47. The top view of the model is generated and the preview of another projected view with the front view as the base view, is attached to the cursor.

7. Move the cursor to the right of the front view and place the right view, refer to Figure 12-47.

8. Similarly, move the cursor horizontally toward the right and then move it upward, the preview of the isometric view is displayed. Specify a point to place the isometric view. Right-click to exit the **Projected View PropertyManager**.

   The current location of the isometric is such that it will interfere with the aligned section view that you need to place next. Therefore, you need to move the isometric view close to the top right corner of the drawing sheet.

9. Move the cursor over the isometric view; the bounding box of the view is displayed in red.

10. Click to select the view; the border of the view is displayed in green.

11. Move the cursor on one of the border lines of the view; the cursor changes to the move cursor.

12. Press and hold the left mouse button and drag the view close to the upper right corner of the drawing sheet. The drawing sheet, after generating and moving the drawing view, is shown in Figure 12-47.

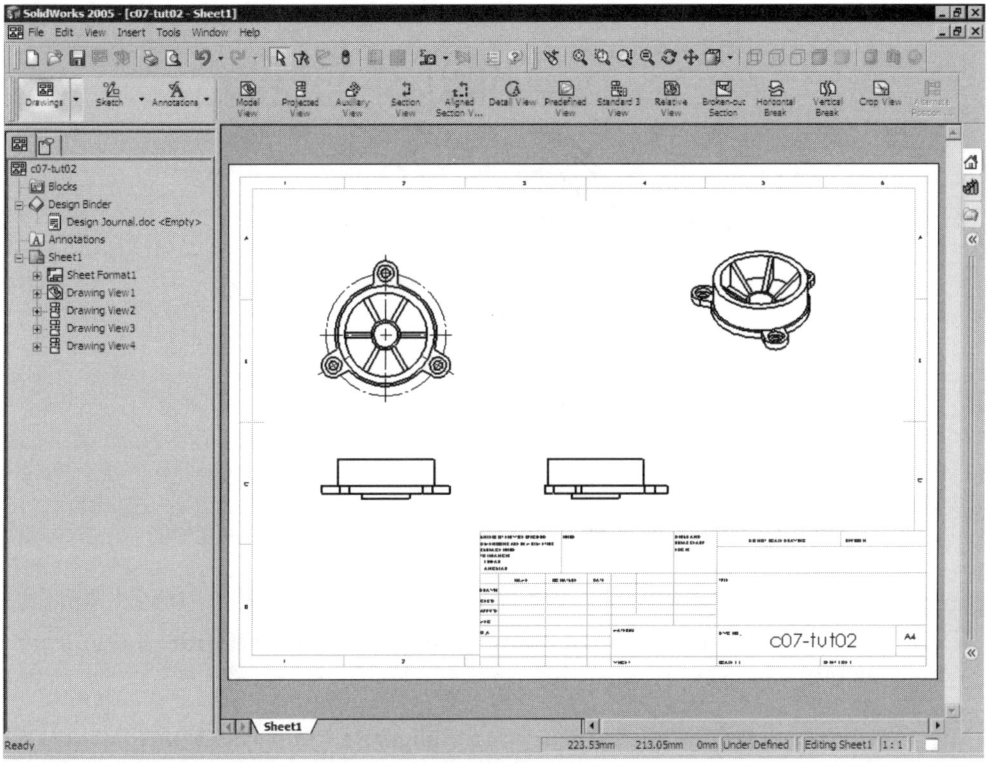

***Figure 12-47** Drawing sheet after generating the front, top, right, and isometric views*

Tip. *You can turn off the origins displayed in the drawing views by using the* ***View*** *menu.*

Center marks are automatically created in the drawing views of the circular features in the model.

Generating the Aligned Section View

Next, you need to generate the aligned section view. The line segments, used to generate this view, will be drawn before invoking the **Aligned Section View** tool. Remember that the view is projected normal to the line selected last, irrespective of the last line drawn. This means that you can draw any line first. In this tutorial, you will first select the inclined line and then the vertical line.

1. Click on the top view to activate the view.

2. Choose the **Sketch** button from the **CommandManager** to invoke the sketching tools. Using the **Line** tool, draw the sketch and apply the relations and dimensions to the sketch, as shown in Figure 12-48.

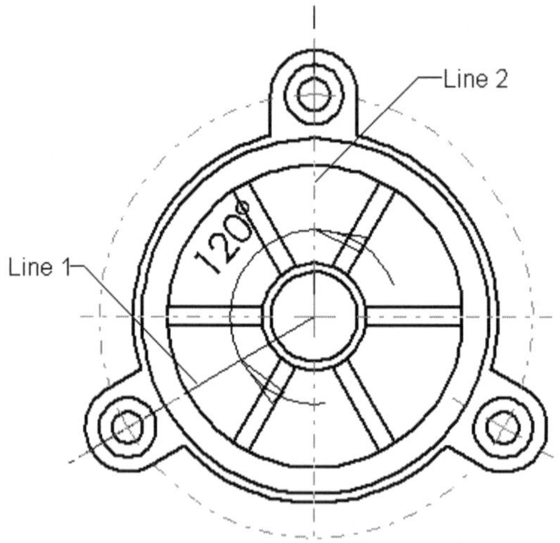

Figure 12-48 *Sketch to be used as section sketch for the aligned section view*

3. Select the dimension and invoke the shortcut menu. Choose the **Hide** option from the shortcut menu.

   Next, you need to select the lines to generate the aligned section view. Note that the vertical line should be selected last to generate the view normal to this line. It will be difficult to select the line because the vertical line coincides with the center marks. You will use the **Select Other** option to select this line.

4. Select the inclined line. Make sure you do not select any segment of the center mark.

5. Next, press and hold the CTRL key down and move the cursor to the vertical line. Right-click on the vertical line and choose the **Select Other** option from the shortcut menu. The **Select Other** list box is displayed.

6. Select the **Line** option from the **Select Other** list box. Also, because the CTRL key was pressed, the inclined line will still be in the current selection set.

7. Now, choose the **Aligned Section View** button from the **Drawing CommandManager**. The aligned section view will be attached to the cursor.

The view generated is normal to the vertical line. If the direction of viewing the aligned section view is reversed, you need to flip it after placing the view.

8. Move the cursor to the right of the top view and place the aligned section view. The **Section View PropertyManager** is displayed. Select the **Flip direction** check, if the direction of viewing is not as required. Click anywhere on the sheet to exit the **PropertyManager**. The sheet, after generating the aligned section view, is shown in Figure 12-49.

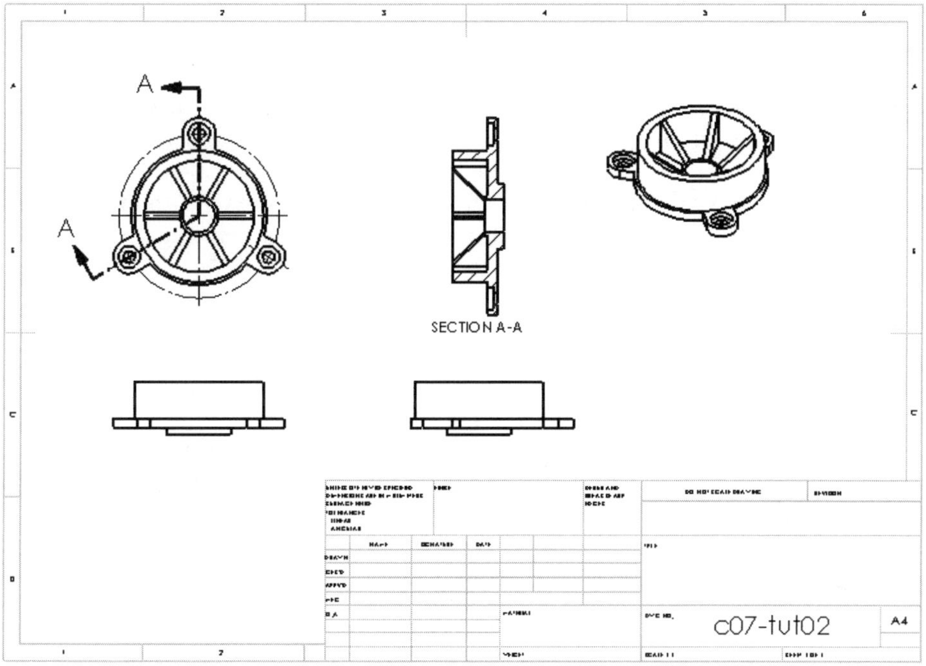

***Figure 12-49** Sheet after generating the aligned section view*

Modifying the Hatch Pattern of the Aligned Section View

The gap between the hatching lines in the aligned section view is large. Therefore, you need to modify the spacing.

1. Select the hatch pattern from one of the sections in the aligned section view; the **Area Hatch/Fill PropertyManager** is displayed.

2. Clear the **Material crosshatch** check box; the **Hatch Pattern**, **Hatch Pattern Scale**, and

Hatch Pattern Angle options are available. Set the value of the **Hatch Pattern Scale** spinner to **2** and select **View** from the **Apply to** drop-down list. Choose **OK** to close the dialog box.

Generating the Detail View

Next, you need to generate the detail view of the right circular feature of the model. Before doing so, you need to activate the view from which you will drive the detail view.

1. Activate the top view and choose the **Detail View** button from the **Drawing CommandManager**; the **Detail View PropertyManager** is displayed and you are prompted to sketch a circle to continue to view the creation. Also, the cursor is replaced by the circle cursor.

2. Draw a small circle on the right circular feature of the model in the top view, refer to Figure 12-50. As you draw the circle, the detail view is attached to the cursor.

4. Place the view on the right side of the drawing sheet above the title block, refer to Figure 12-50.

5. Set the value of the scale factor of the detail view to **3:1** and choose the **OK** button from the **Detail View PropertyManager**.

You may need to move the drawing view and its label so that the view is not overlapping the title block. Figure 12-50 shows the final drawing sheet, after generating the detail view from the top view.

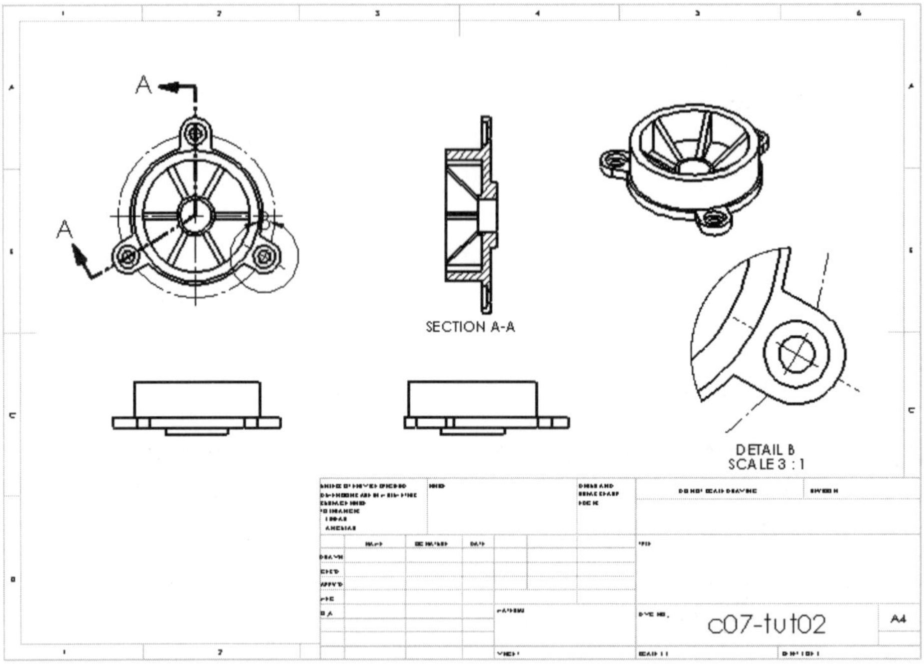

Figure 12-50 The detail view derived from the top view

Saving the Drawing

1. Choose the **Save** button from the **Standard** toolbar and save the drawing document with the name given below.

   \My Documents\SolidWorks\c12\c12tut1.slddrw

2. Choose **File > Close** to close this document. Also, close the part document of Tutorial 2.

Tutorial 2

In this tutorial, you will generate the drawing view of the Bench Vice assembly created in Chapter 10. You will generate the top view, sectioned front view, right view, and isometric view of the assembly in the exploded state. **(Expected time: 45 min)**

The steps to be followed to complete this tutorial are listed next.

a. Copy the folder of the Bench Vice assembly from Chapter 10 to the *c12* folder.
b. Create the exploded state of the Bench Vice assembly, refer to Figure 12-51.
c. Start a new drawing document from within the assembly document using A4 landscape sheet format. Generate the top view using the **Model View** tool, refer to Figure 12-52.
d. Generate the section view using the **Section View** tool, refer to Figure 12-53.
e. Generate the right view using the **Projected View** tool, refer to Figure 12-54.
f. Generate the isometric view and change the state of the isometric view to the exploded state, refer to Figure 12-55.
g. Save and close the drawing and assembly documents.

Copying the Folder of the Bench Vice Assembly

First, you need to copy the folder of the Bench Vice assembly to *c12* folder.

1. Copy the folder of the Bench Vice assembly from the *\My Document\SolidWorks\c10* folder to the *c12* folder.

Creating the Exploded View of the Assembly

Before proceeding further to generate the drawing views of the assembly, you need to create the exploded state of the assembly in the **Assembly** mode.

1. Open the Bench Vice assembly and create the exploded state and the explode lines, as shown in Figure 12-51.

   It is recommended that whenever you create an exploded state of an assembly, you must revert to the collapsed state. On saving the assembly in the exploded state, whenever you generate the drawing views of the assembly, it will generate views with the exploded state.

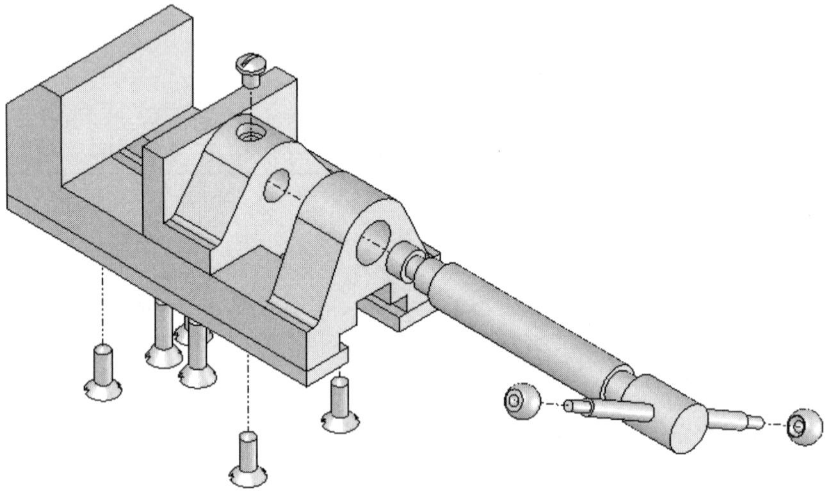

Figure 12-51 Exploded view of the assembly with explode lines

2. Right-click **Bench Vice Configutarion(s)** > **Default** in the **ConfigurationManager** and choose **Collapse** to unexplode the assembly.

3. Save the assembly.

Starting a New Drawing Document From Within the Assembly Document

As mentioned earlier, you can also start a drawing document from within the assembly document.

1. Choose the **Make Drawing from Part/Assembly** button from the **Standard** toolbar. Start a new drawing document from the **Templates** tab of the **New SolidWorks Document** dialog box, if it appears.

A new SolidWorks drawing document is started and the **Sheet Format/Size** dialog box is displayed.

2. Select the **A4 - Landscape** sheet and make sure that the projection type for the current sheet is set to the third angle. Choose the **OK** button to close this dialog box.

The **Model View PropertyManager** is automatically invoked and you are prompted to select a named view and place the drawing view.

Generating the Top View

1. Select the **Top** option from the list box available in the **Orientation** rollout of the **Model View PropertyManager**.

2. Select the **Use custom scale** radio button from the **Scale** rollout.

3. Set the value of the scale factor to **1:2** and then select the **Preview** check box from the **Orientation** rollout. The preview of the top view of assembly is displayed.

4. Make sure the **Auto-start projected view** check box in the **Options** rollout is cleared.

5. Place the view close to the top left corner of the drawing sheet, refer to Figure 12-52. Click anywhere on the sheet to exit the **PropertyManager**.

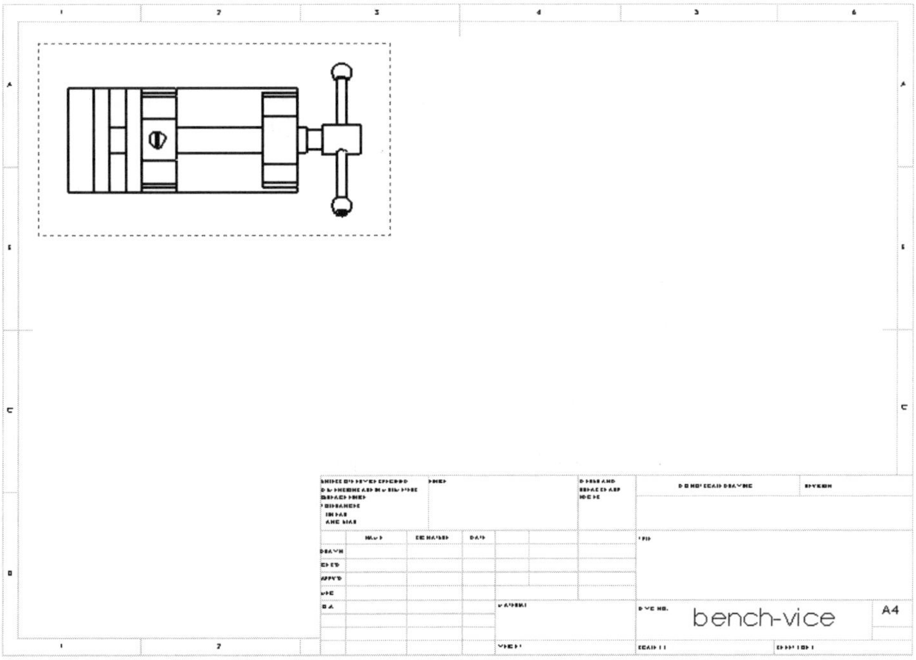

***Figure 12-52** Top view generated using the **Model View** tool*

Creating the Sectioned Front View

Next, you need to generate the sectioned front view that is derived from the top view.

1. Activate the top view and choose the **Section View** button from the **Drawings CommandManager**.

The **Section View PropertyManager** is displayed and you are prompted to sketch a line in order to continue to view the creation. The cursor will be replaced by the line cursor.

2. Draw a horizontal line such that it passes through the center of the Bench Vice assembly.

On specifying the endpoint of the line, the **Section View** dialog box is displayed. This dialog box is used to exclude the components from the section cut.

3. Click on the + sign located on the left of the **FeatureManager Design Tree** and continue expanding it until Bench Vice assembly name is no longer displayed. Next, expand the assembly.

4. Select Screw Bar, Bar Globes, Jaw Screw, Oval Fillister, Set Screw1, and Set Screw2. Next, activate the **Section View** dialog box and select the **Don't cut all instances** check box individually for all the selected items.

5. Select the **Auto hatching** check box and choose the **OK** button from the **Section View** dialog box.

   The preview of the section view is displayed in the drawing sheet as you move the cursor up and down.

   If the direction of viewing of the section view is not what is required, flip the direction by selecting the **Flip direction** check box from the **Section Line** rollout.

6. Place the section view below the top view. Click anywhere on the sheet to exit the **PropertyManager**.

7. Modify the hatch scale for the components. Figure 12-53 shows the section view generated using the **Section View** tool, after modifying the hatch scale.

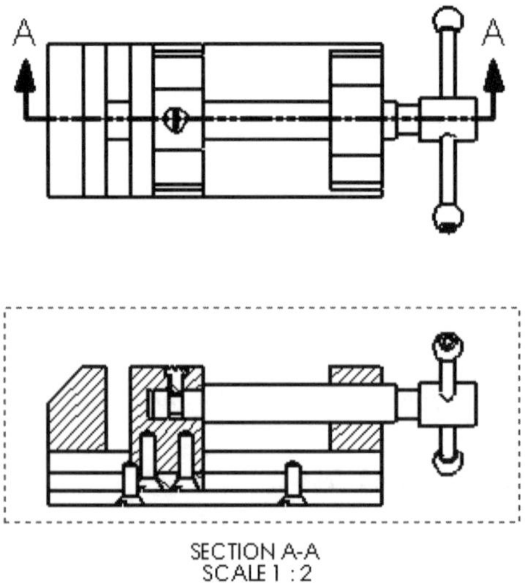

Figure 12-53 *Section view generated using the* ***Section View*** *tool*

Generating the Right-Side View

The next view that you need to generate is the right-side view derived from the sectioned front view and it will be generated using the **Projected View** tool.

1. Select the sectioned front view and invoke the **Projected View** tool. The **Projected View PropertyManager** is displayed and a projected view is attached to the cursor.

2. Move the cursor to the right of the sectioned front view and place the view on the right of the sectioned front view.

3. Choose **OK** from the **PropertyManager**. The sheet after generating the projected view is shown in Figure 12-54.

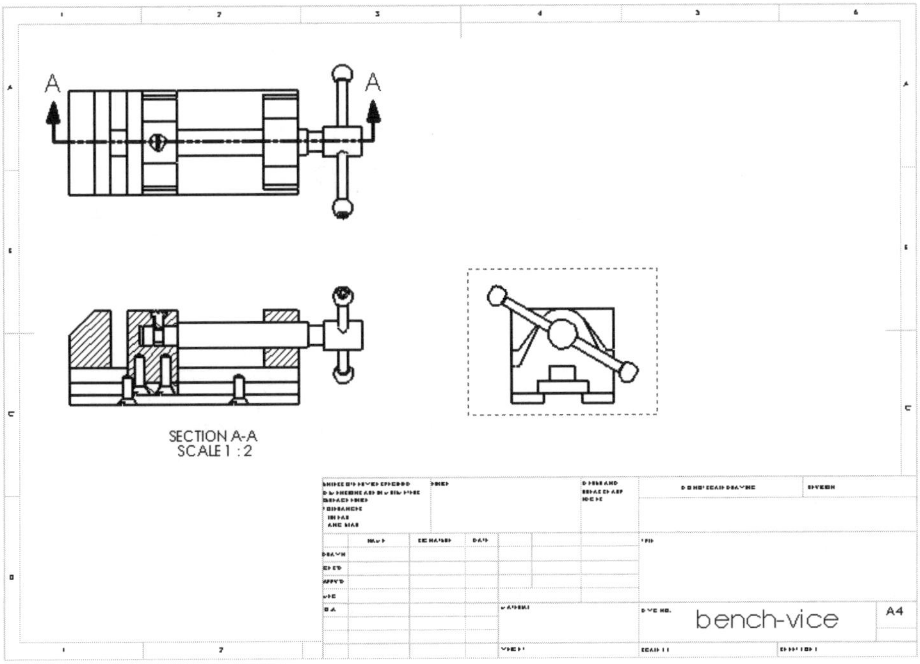

***Figure 12-54** Right-side view generated using the **Projected View** tool*

Creating the Isometric View of the Assembly in the Exploded State

The last view to be generated is the isometric view of the assembly in the exploded state.

1. Using the **Model View** tool, generate the isometric view and place the view close to the upper right corner of the drawing sheet. Choose **Yes** from the **SolidWorks** information box.

2. Set the scale factor of the drawing view to **1:2**.

3. Right-click on the view to invoke the shortcut menu. Choose the **Properties** option from the shortcut menu. The **Drawing View Properties** dialog box is displayed.

4. Select the **Show in exploded state** check box from the **View Properties** tab of the **Drawing View Properties** dialog box and choose the **OK** button.

5. Move the views to place all of them in the drawing sheet. Figure 12-55 shows the final drawing sheet after generating all the drawing views.

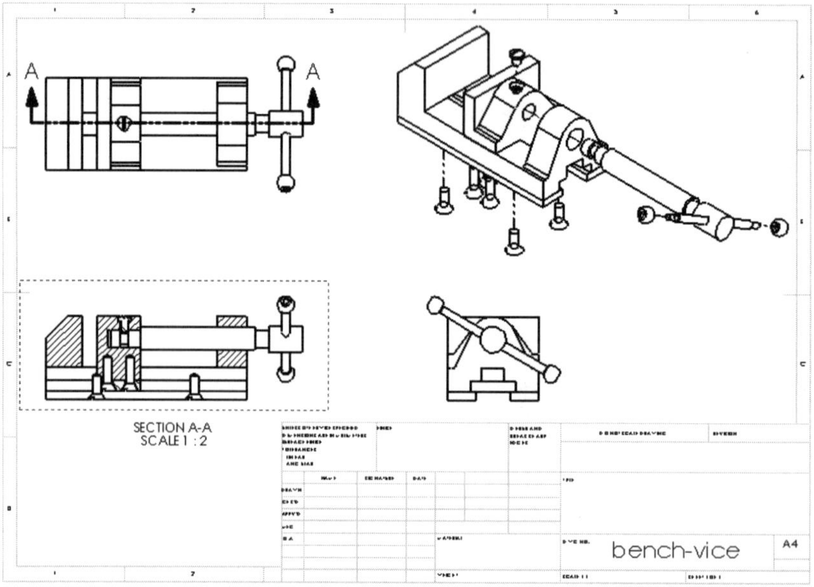

Figure 12-55 Drawing sheet after generating all the views

Saving the Drawing

Next, you need to save the drawing document.

1. Choose the **Save** button from the **Standard** toolbar and save the drawing document with the name given below.

   \My Documents\SolidWorks\c12*c12tut2.slddrw*

2. Close the drawing and assembly documents.

SELF-EVALUATION TEST

Answer the following questions and then compare your answers with those given at the end of this chapter.

1. The **Standard sheet size** radio button is selected by default in the **Sheet Format/Size** dialog box. (T/F)

2. The **Display sheet format** check box is selected if you want to use the empty sheet without any margin lines or a title block. (T/F)

3. The **Relative View** tool is used to generate an orthographic view by defining the orientation of the view, using reference planes or planar faces of the model. (T/F)

4. An auxiliary view is a drawing view that is generated by projecting the lines normal to a specified edge of an existing view. (T/F)

5. You cannot change the style of the break line in a broken view. (T/F)

6. In technical terms, creating a 2D drawing in the drawing document is known as __________.

7. To start a new drawing document from within the part document, choose the __________ button from the **Standard** toolbar.

8. The __________ check box available in the **Detail View** rollout of the **Detail View PropertyManager** is used to display the complete outline of the closed profile in the detail view.

9. To change the scale of the drawing views, select the drawing view and select the __________ radio button from the **Scale** rollout.

10. To rotate a drawing view, select the view and choose the __________ button from the **View** toolbar.

REVIEW QUESTIONS

Answer the following questions.

1. Choose the __________ button from the **Drawings CommandManager** to create an alternate position view.

2. The __________ dialog box is used to modify the hatch pattern of a section view.

3. The __________ check box needs to be cleared to modify the scale of the hatch pattern.

4. A __________ view is a section view in which only the sectioned surface is displayed in the section view.

5. The __________ dialog box is displayed to confirm the deletion of the views.

6. The views that are generated from a view already placed in the drawing sheet are known as

   (a) Child views (b) Derived views
   (c) Predefined views (d) Empty views

7. By default, the detail view boundary is in which shape?

   (a) Circle (b) Ellipse
   (c) Rectangle (d) None

8. In which edit box is the name of the auxiliary view specified?

(a) **Label** (b) **Arrow**
(c) **Name** (d) **Detail view label**

9. From which rollout can you select the view orientation in the **Named View PropertyManager**?

(a) **View Orientation** (b) **Define View**
(c) **Specify View** (d) **Scale View**

EXERCISE

Exercise 1

In this exercise you will generate the front view, section right view, isometric view, and the alternate position view on the isometric view of Exercise 1 of Chapter 11. You need to scale the parent view to the scale factor of **1:3**. The views that you need to generate are shown in Figure 12-56.

(Expected time: 30 min)

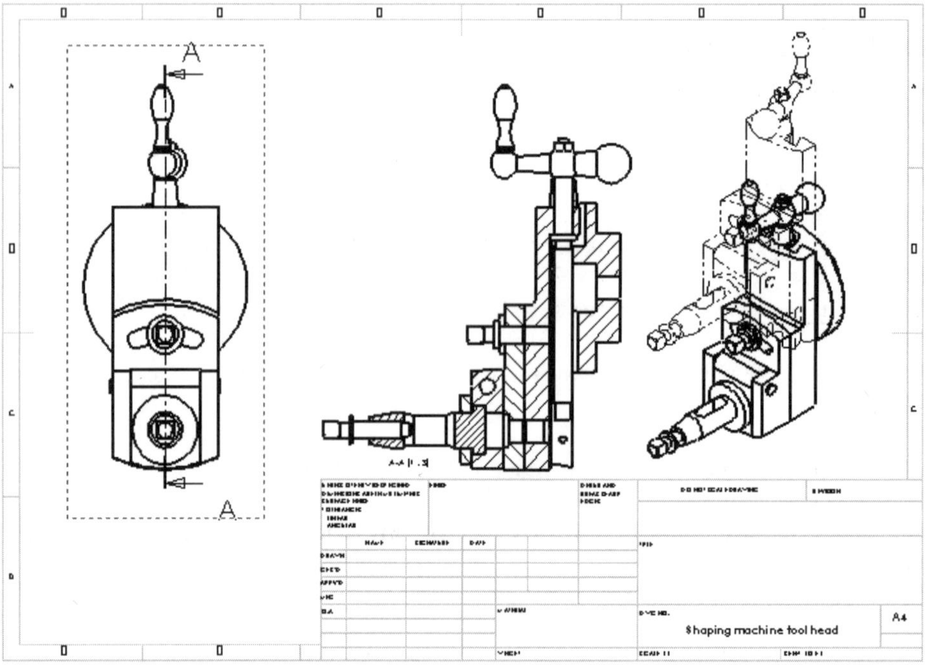

Figure 12-56 Views of Exercise 1

Answers to Self-Evaluation Test

1. T, **2.** T, **3.** T, **4.** T, **5.** F, **6.** Interactive drafting, **7. Make Drawing from Model/Assembly**, **8. Full outline**, **9. Use custom scale**, **10. Rotate View**

Chapter 13

Working With Drawing Views-II

Learning Objectives

After completing this chapter you will be able to:

- *Add annotations in the drawing views.*
- *Add reference dimensions.*
- *Add notes to the drawing views.*
- *Add Surface Finish symbols to the drawing views.*
- *Add Datum Feature symbols to the drawing views.*
- *Add geometric tolerance to the drawing views.*
- *Add Datum Target symbols to the drawing views.*
- *Add Center Marks and Centerlines to the drawing views.*
- *Add Hole Callouts to the drawing views.*
- *Add Cosmetic Threads to the drawing views.*
- *Add Weld symbols to the drawing views.*
- *Add Multi-jog Leader to the drawing views.*
- *Add Dowel Pin Symbol to the drawing views.*
- *Edit the annotations.*
- *Add BOM*
- *Add Balloons to the assembly.*
- *Add new sheets in the drawing document.*
- *Edit the sheet format.*
- *Create a user-defined sheet format.*

ADDING ANNOTATIONS TO THE DRAWING VIEWS

After generating the drawing views, you need to generate the dimensions in the drawing views and add other annotations, such as notes, surface finish symbols, geometric tolerance, and so on. There are two types of annotations that can be displayed in the drawing views. The first type are the generative annotations which are added while creating the part in the **Part** mode, for example, the dimensions that you add to the sketch and features of the part. The second type of annotations are added manually to the geometry of drawing views such as reference dimensions, notes, surface finish symbols, and so on. Both these types of annotations are discussed next.

Generating Annotations Using the Model Items Tool

Enhanced

CommandManager:	Annotations > Model Items
Menu:	Insert > Model Items
Toolbar:	Annotation > Model Items

The **Model Items** tool is used to generate the annotations that were added while creating the model in the **Part** mode. To invoke this tool, choose the **Model Items** button from the **Drawing Annotation CommandManager**. The **Model Items PropertyManager** is displayed, as shown in Figure 13-1. The options available in this **PropertyManager** are discussed next.

Import from Rollout

The drop-down list available in the **Import from** rollout defines the options from where the annotation are imported. These options are discussed next.

Entire model

When a feature of a model is selected from the **FeatureManager Design Tree** in the **Drawing** mode, the **Entire model** option is selected to import the annotations from the entire model. In case of assemblies, the annotations from the entire assembly are imported, even if a single component of the assembly is selected.

Selected component

The **Selected component** option is available only when you generate the drawing views of an assembly. This option is selected to import the annotations from only the selected component.

Selected feature

The **Selected feature** option is selected to import the annotations from only the selected feature or features.

Only Assembly

The **Only Assembly** option is available when, you generate drawing views of an assembly. This option is selected to import the annotations that are applied to the assembly in the **Assembly** mode, such as offset distance and so on.

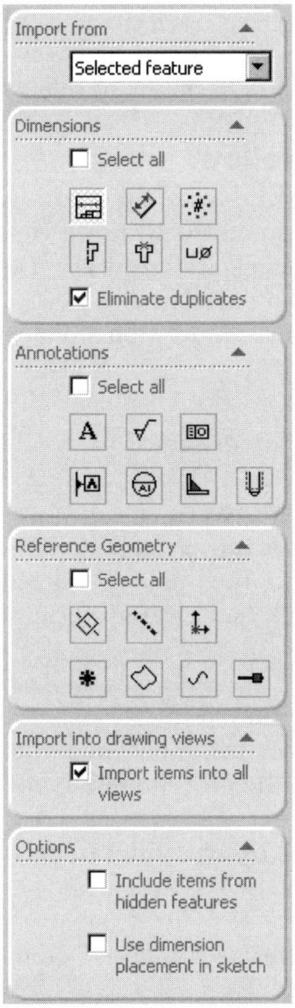

Figure 13-1** The **Insert Model Items PropertyManager

Dimensions Rollout

This rollout is used to select the type of dimensions, pattern annotations, and the hole annotations that you need to generate in the drawing views. You can choose the button of the required annotation type. You can also select the **Select all** check box to select all options. The **Eliminate duplicates** check box is selected to remove duplicate instances of annotations.

Annotations Rollout

The **Annotations** rollout is used to select the annotations that are to be generated in the drawing views. Using the buttons available in this rollout, you can generate the cosmetic thread, datums, datum targets, dimensions, geometric tolerances, notes, surface finish, and weld symbols. You can also select the **Select all** check box to select all options.

Reference Geometry Rollout

The **Reference Geometry** rollout is used to generate the reference geometries that were used in creating the part. You can generate axes, curves, planes, surfaces, and so on. Select the check box of the type of reference geometry that you need to generate in the drawing views. You can also select the **Select all** check box to select all options.

Import into drawing views Rollout

The **Import items into all views** check box available in this rollout is selected to automatically display and distributed annotations in all the drawing views. Select the view to display the annotations in it. This check box is cleared and the **Import to views** selection area is displayed in this rollout. You can also preselect the views in which you need to display the annotations before invoking the **Insert Model Items** dialog box.

Options Rollout

The options available in this rollout are discussed next.

Include items from hidden features

The **Include items from hidden features** check box is selected to display the annotation that belongs to a hidden feature of the model. By default, this check box is cleared. It is recommended to keep this check box cleared because it will eliminate the display of unwanted annotations.

Use dimension placement in sketch

This check box is selected to place the dimension at the exact location with respect to the sketch, in which it was placed in the part modeling environment.

After setting all the parameters in the **Insert Model Items** dialog box, choose the **OK** button to display the annotations.

Tip. *You can toggle the display of annotations when the **Model Items PropertyManager** is displayed by right-clicking on the annotation.*

The annotations mostly overlap each other when they are generated. Therefore, you may need to move the annotations after generating them. To move an annotation, move the cursor on the annotation; the annotation is highlighted in red. Press and hold down the left mouse button and drag the cursor to place the annotation at the desired location. Release the left mouse button when the cursor is placed at the desired location.

The dimensions generated, while generating the annotations, are the same as the ones used to create the model. These dimensions are linked to the model because of the bidirectional associativity in SolidWorks.

*Double-click on the dimension to modify it. The **Modify** dialog box is displayed, using which, you can modify the value of the dimension. Next, rebuild the drawing views using the **Rebuild** button. The change in the dimension will also be made in the original model. If the model is used in an assembly, the changes will also be reflected in the assembly.*

Adding Reference Annotations

You can add reference annotations in the drawing views in SolidWorks. These include reference dimensions, notes, surface finish symbols, datum feature symbol, geometric tolerance, and so on. The method of adding reference annotations is discussed next.

Adding Reference Dimensions

CommandManager:	Annotations > Smart Dimension
Menu:	Tools > Dimensions > Smart
Toolbar:	Annotations > Smart Dimension

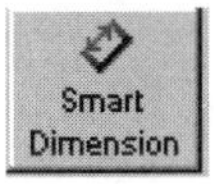

You can use the **Smart Dimension** tool to add reference dimensions to the drawing views in the **Drawing** mode of SolidWorks. This tool is similar to the **Smart Dimension** tool discussed in the sketching environment of SolidWorks.

You can also use the **Dimension/Relations CommandManager** to add dimensions. For example, you can use this **CommandManager** in the **Drawing** mode to add chamfer dimensions. The technique of adding chamfer dimensions to the drawing views in the **Drawing** mode of SolidWorks is discussed next.

Chamfer Dimension

The **Chamfer Dimension** tool is used to add the reference dimension to the chamfers in the drawing view. To add a chamfer dimension, choose **Chamfer Dimension** from the **Dimension/Relations CommandManager**. The cursor is replaced by the chamfer dimension cursor. Now, select the inclined chamfered edge and then select a horizontal or vertical edge. The chamfer dimension will be attached to the cursor. Select a point on the sheet to place the dimension. Figure 13-2 shows a chamfer dimension created using the **Chamfer Dimension** tool.

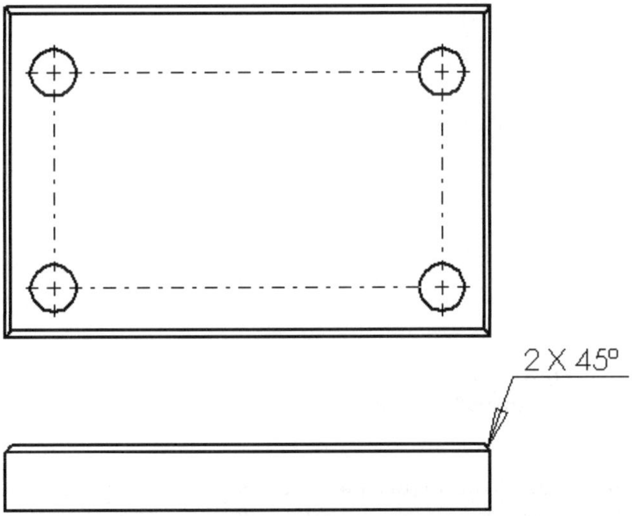

Figure 13-2 *Chamfer dimension added using the* ***Chamfer Dimension*** *tool*

Adding Notes to the Drawing Views

CommandManager:	Annotations > Note
Menu:	Insert > Annotations > Note
Toolbar:	Annotation > Note

In SolidWorks, you can add notes to the drawing views in the **Drawing** mode. To do so, choose the **Note** button from the **Annotations CommandManager**. The partial view of the **Note PropertyManager** is displayed, as shown in Figure 13-3.

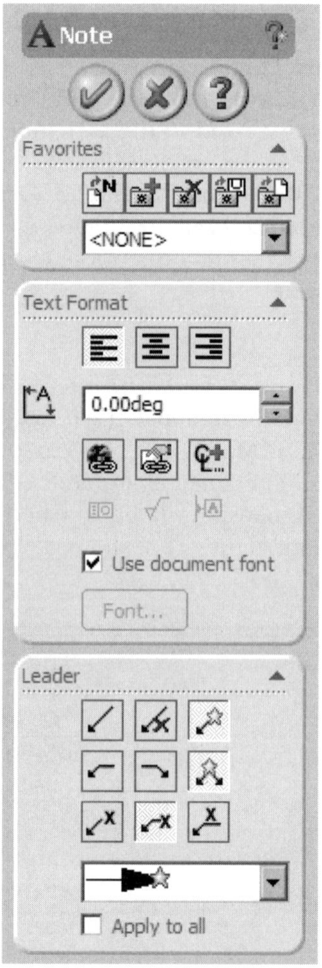

Figure 13-3** Partial view of the **Note PropertyManager

On invoking the **Note PropertyManager**, a text box is attached to the cursor. If you move the cursor close to an edge in a drawing view, a leader will be displayed with the text box. This is because the **Auto Leader** button is chosen in the **Leader** rollout. Place the text box at the desired location; the text edit box will be invoked in the drawing area and you can enter the text. Next, choose **OK** from the **Note PropertyManager**. The other options available in the **Note PropertyManager** are discussed next.

Favorites

You can save a note as a favorite using the options available in the **Favorites** rollout. The options available in this rollout are the same as those discussed in Chapter 3.

Text Format

The **Text Format** rollout is used to set the format of the text such as font, size, justification, and rotation of the text. You can also add symbols and hyperlinks to the text using the options available in this rollout.

Leader

The options available in the **Leader** rollout are used to define the style of arrows and leaders that are displayed in the notes.

Border

The options available in the **Border** rollout are used to define the border in which the note text will be displayed. You can use various types of borders from the **Style** drop-down list. The **Size** drop-down list available in this rollout is used to define the size of the border in which the text will be placed.

Adding Surface Finish Symbols to the Drawing Views

CommandManager:	Annotations > Surface Finish
Menu:	Insert > Annotations > Surface Finish Symbol
Toolbar:	Annotation > Surface Finish

You can add the surface finish symbols to the edges or faces in the drawing views using the **Surface Finish** tool. To do so, choose the **Surface Finish** button from the **Annotations CommandManager**. The **Surface Finish PropertyManager** is displayed, as shown in Figure 13-4. When you invoke this tool, a surface finish symbol is attached to the cursor. The options available in the **Surface Finish PropertyManager** dialog box are discussed next.

Favorites

With this release of SolidWorks, you can use the options available in the **Favorites** rollout to save a surface finish as a favorite. The options available in this rollout are the same as those discussed in Chapter 3.

Symbol

The **Symbol** rollout is used to define the type of surface finish symbol that you need to add to the drawing views. The surface finish symbols include the **Basic** surface finish symbol, **Machining Required**, **Machining Prohibited**, **JIS Basic**, **JIS Machining Prohibited**, **JIS Machining Prohibited**, and so on.

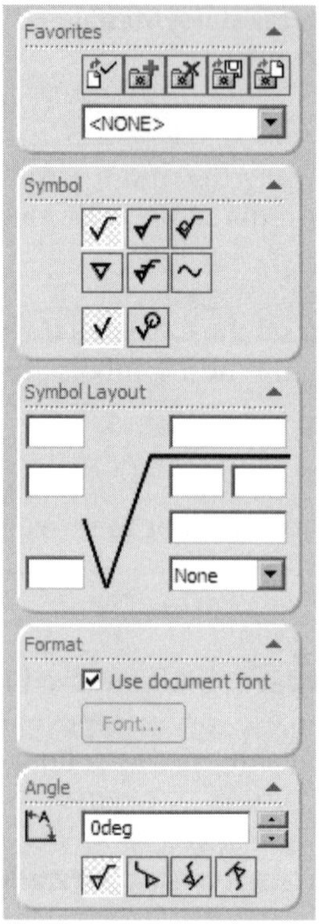

*Figure 13-4 Partial view of the **Surface Finish PropertyManager***

Symbol Layout

The options available in the **Symbol Layout** rollout are used to define the parameters of the surface finish symbol.

Format

The options available in the **Format** rollout are used to define the font size and the orientation of the surface finish symbol.

Angle

The options available in the **Angle** rollout are used to define the angle of the surface finish symbol. You can enter the angle value in the spinner provided in this rollout or choose the buttons available below the spinner.

The other options available in the **Surface Finish PropertyManager** are the same as those discussed while adding notes.

Adding Datum Feature Symbol to the Drawing Views

CommandManager:	Annotations > Datum Feature
Menu:	Insert > Annotations > Datum Feature Symbol
Toolbar:	Annotation > Datum Feature

The **Datum Feature** tool is used to add a datum feature symbol to an entity in the drawing view. The datum feature symbols are used as datum references, while adding the geometric tolerances in the drawing view. To add the datum feature symbol, choose the **Datum Feature** button from the **Annotations CommandManager**. The **Datum Feature PropertyManager** is displayed and a datum feature symbol with default parameters is attached to the cursor. The **Datum Feature PropertyManager** is displayed in Figure 13-5. The options available in the **Datum Feature PropertyManager** are discussed next.

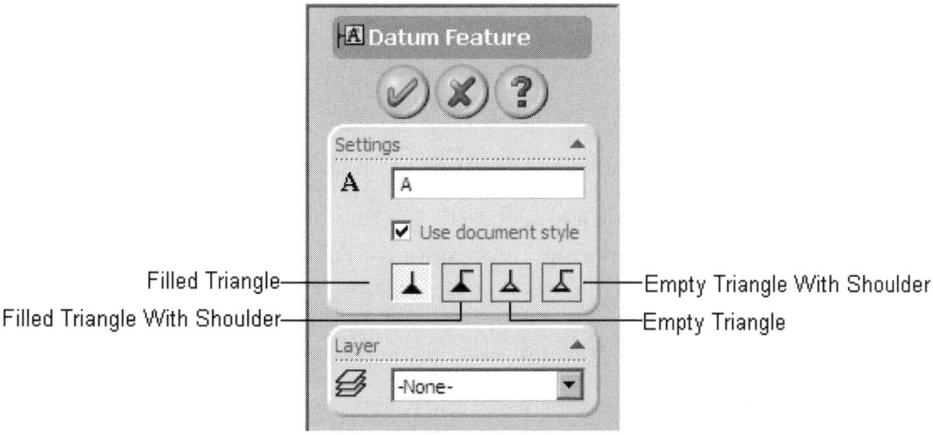

*Figure 13-5 The **Datum Feature PropertyManager***

Label

The **Label** edit box in the **Settings** rollout is used to define the label to be used in the datum feature symbol. You can use alphabets or numeric characters as labels.

Use document style

This check box is selected to use the datum feature style defined in the document to display the datum feature symbol.

Square

The **Square** button is displayed below the **Use document style** check box, when you clear it. This button is used to place the text of the datum feature inside a square. By default, the text of the datum feature is placed using this option. Therefore, by using the buttons available below the **Square** button you can set the type of datum feature such as filled triangle, filled triangle with shoulder, empty triangle, empty triangle with shoulder.

Round (GB)

The **Round (GB)** button is displayed below the **Use document style** check box, when you clear it. This button is used to place the text of the datum feature inside a circle. To do so, you first need to clear the **Use document style** check box and then choose the **Round (GB)** button. On choosing this button, additional buttons are displayed below the **Round (GB)** button. These buttons are used to set the style of the datum feature.

After defining all parameters of the datum feature symbol, specify a point on an existing entity in the drawing sheet. Next, move the cursor to define the length and the placement of the datum feature symbol. As soon as you place one datum feature symbol, another datum feature symbol is attached to the cursor. Therefore, you can place as many datum feature symbols as you want using the **Datum Feature PropertyManager**. The sequence of the names of datum feature symbols automatically follows the order based on the labels, as you place multiple datum feature symbols.

Adding Geometric Tolerance to the Drawing Views

CommandManager:	Annotations > Geometric Tolerance
Menu:	Insert > Annotations > Geometric Tolerance
Toolbar:	Annotation > Geometric Tolerance

In a shop floor drawing, you need to provide various other parameters, along with the dimensions and dimensional tolerance. These parameters can be geometric condition, surface profile, material condition, and so on. All these parameters are defined using the **Geometric Tolerance** tool. To add the geometric tolerance to the drawing views, choose the **Geometric Tolerance** button from the **Annotations CommandManager**; the **Properties** dialog box will be displayed, as shown in Figure 13-6. Also, a geometric tolerance is attached to the cursor.

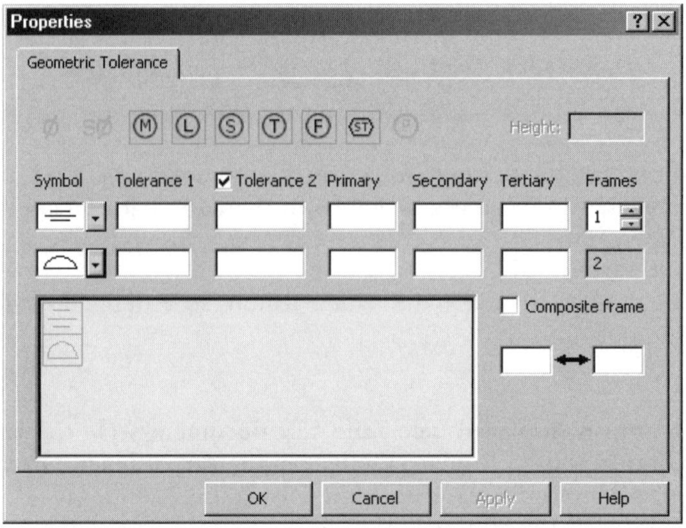

***Figure 13-6** The **Properties** dialog box used to apply geometric tolerance*

Both the rows in this dialog box are separate frames. You can add additional frames using the **Frames** spinner provided on the right of the **Tertiary** edit box. The parameters that can be added to these frames are geometric condition symbols, diameter symbol, value of tolerance, material condition, and datum references. The options available in the **Properties** dialog box, to add the geometric tolerance to the drawing views, are discussed next.

Symbol

The **Symbol** drop-down list is used to define the geometric condition symbol. When you choose the down arrow button, the **Symbols** flyout is displayed, as shown in Figure 13-7. This is used to define the geometric condition symbols in the geometric tolerance. You can select the standard of the geometric condition symbol from this flyout. As soon as you select a symbol, the flyout is closed and the selected symbol is displayed in the **Symbol** edit box. Also, the preview of the geometric tolerance is displayed in the preview area.

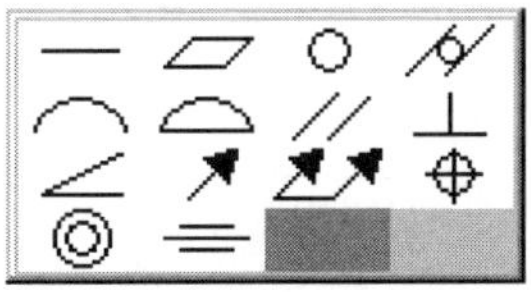

Figure 13-7 The ***Symbols*** *flyout used to define the geometric condition symbols*

Tolerance 1

The **Tolerance 1** edit box is used to specify the tolerance value with respect to the geometric condition defined using the **Symbols** flyout. You can use the buttons available above the rows of the frames, to add symbols such as diameter, spherical diameter, material conditions, and so on.

Tolerance 2

The use of the **Tolerance 2** edit box is the same as that discussed earlier. This edit box is used to define the second geometric tolerance, if required.

Primary

The **Primary** edit box is used to specify the characters to define the datum reference added to the entities in the drawing view using the **Datum Feature Symbol** tool.

Similarly, you can define the **Secondary** datum reference and the **Tertiary** datum reference.

Frames

The **Frames** spinner is used to increase the frames for applying more complex geometric tolerances.

Projected tolerance

The **Projected tolerance** button is the last button available above the frame rows. This button is chosen to define the height of the projected tolerance. When you choose this button, the **Height** edit box will be enabled and you can specify the projected tolerance zone height in this edit box.

Composite frame

The **Composite frame** check box is selected to use a composite frame to add the tolerance.

When you select this check box, the tolerance frame is converted into a composite frame and the preview is modified accordingly.

Between two points

The **Between two points** edit boxes are used to apply geometric tolerance between two points or entities. To do so, specify the tolerance in the edit boxes provided in the **Between two points** area.

Figure 13-8 shows a drawing, after adding annotations to some of the entities in the drawing view.

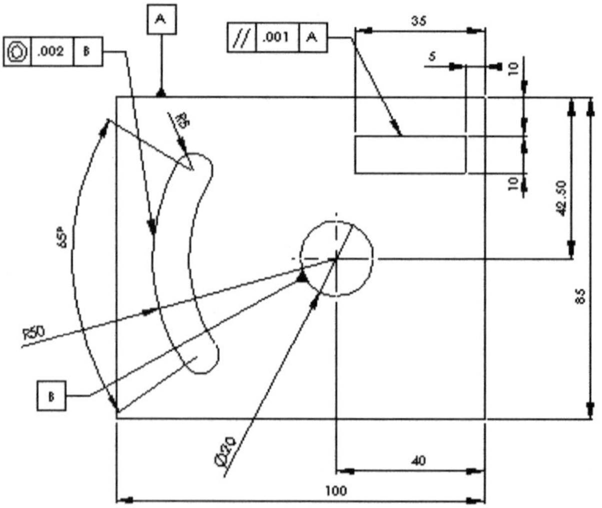

Figure 13-8 A drawing after adding some annotations

Adding Datum Target Symbols to the Drawing Views

CommandManager:	Annotations > Datum Target
Menu:	Insert > Annotations > Datum Target
Toolbar:	Annotation > Datum Target

The **Datum Target** tool is used to add datum targets to the entities in the drawing view. To add the datum targets, choose the **Datum Target** button from the **Annotations CommandManager**. The cursor is replaced by the datum target cursor and the **Datum Target PropertyManager** is displayed, as shown in Figure 13-9. Select a model face, an edge, or a line from the view on which you need to add the datum target and then place the datum target.

This **PropertyManager** is used to define the properties of the datum target symbol. Set the parameters of the datum target symbol in the **Settings** rollout of this **PropertyManager**. You can set the target shape as a point, circle, or rectangle. You can also define the diameter of the target area, if the shapes of the target are selected as a point and circle. If the shape of the target selected is a rectangle, you need to define its width and height. You can specify datum references to the datum target symbol by using the **First reference**, **Second reference**,

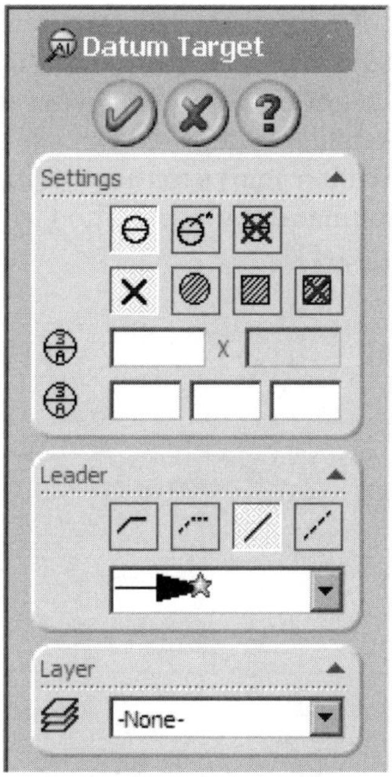

*Figure 13-9 The **Datum Target PropertyManager***

and the **Third reference** edit boxes. Figure 13-10 shows the datum target symbols added to the entities in the drawing view.

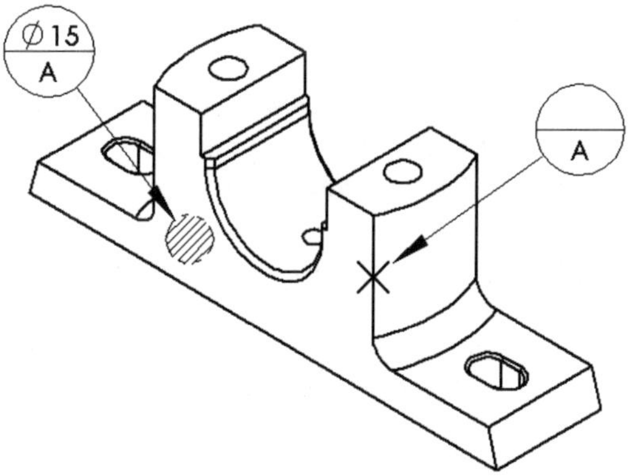

Figure 13-10 Datum target symbols added to the model

Adding Center Marks to the Drawing Views

CommandManager:	Annotations > Center Mark
Menu:	Insert > Annotations > Center Mark
Toolbar:	Annotations > Center Mark

The **Center Mark** tool is used to add center marks to the circular entities. As discussed earlier, center marks are automatically, generated when you generate the model. But if the center marks are not generated while generating the drawing view, or if you have sketched a view, you can use this tool to add the center marks to the drawing views. To add the center marks to the drawing views, choose the **Center Mark** button from the **Annotations CommandManager**. The **Center Mark PropertyManager** will be displayed, as shown in Figure 13-11.

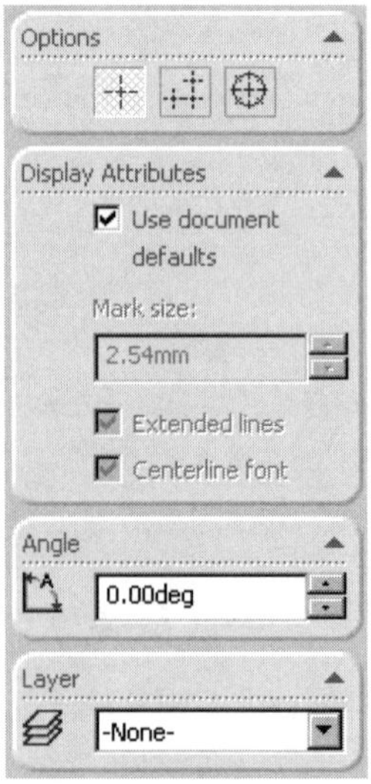

Figure 13-11** Partial view of the **Center Mark PropertyManager

When you invoke the **Center Mark PropertyManager**, the cursor is replaced by the center mark cursor and you are prompted to select a circular edge or an arc for the center mark insertion. By default, the **Single Center Mark** button is chosen in the **Options** rollout of the **Center Mark PropertyManager**. Select the circular edge or arc to add the center mark. Figure 13-12 shows the center marks added using the **Single Center Mark** button.

The **Propagate** button appears on the center marks that are added to holes, which are part of a linear or circular pattern. If you choose this button, the center marks are added to all the remaining instances of the pattern.

The center marks in rectangular and circular patterns can also be applied using the **Linear Center Mark** and **Circular Center Mark** buttons. You can add the center marks in the linear pattern format by using the **Linear Center Mark** button from the **Options** rollout. The center marks will be connected using the centerlines, because the **Connection lines** check box is selected by default, as shown in Figure 13-13.

You can create the center mark in the circular pattern format by using the **Circular Center Mark** button. When you choose this button, the **Circular lines**, **Radial lines**, and **Base center mark** check boxes are displayed. The **Circular lines** check box is used to create a

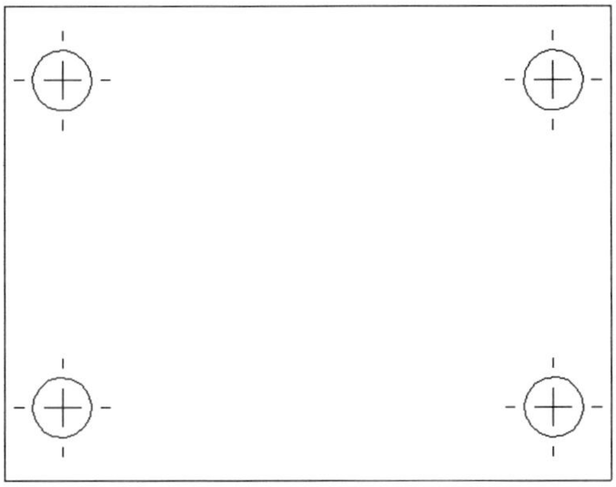

Figure 13-12 *Center marks added using the* ***Single Center Mark*** *option*

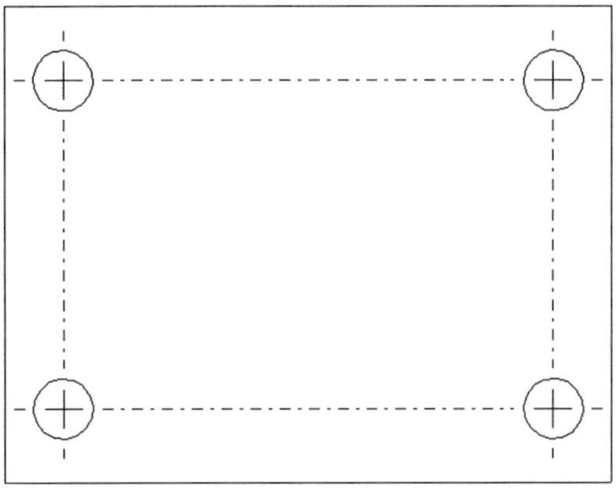

Figure 13-13 *Center marks added using the* ***Linear Center Mark*** *option*

circular line passing through the centers of the circles arranged in the form of a circular pattern. The **Radial lines** check box is used to display the radial lines from the center of the pattern to the center of each instance. The **Base center mark** check box is used to display the center mark at the center of the pattern. Figure 13-14 shows the center mark created with the **Base center mark** check box cleared. Figure 13-15 shows the center mark created with the **Base center mark** check box selected. Figure 13-16 shows the center mark created with the **Radial lines** check box selected. Observe that in all these figures, the **Circular lines** check box is selected.

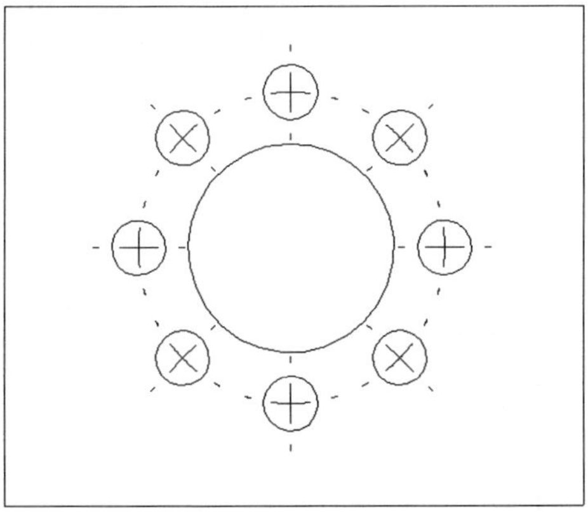

Figure 13-14 *Center marks created with the* ***Base center mark*** *check box cleared*

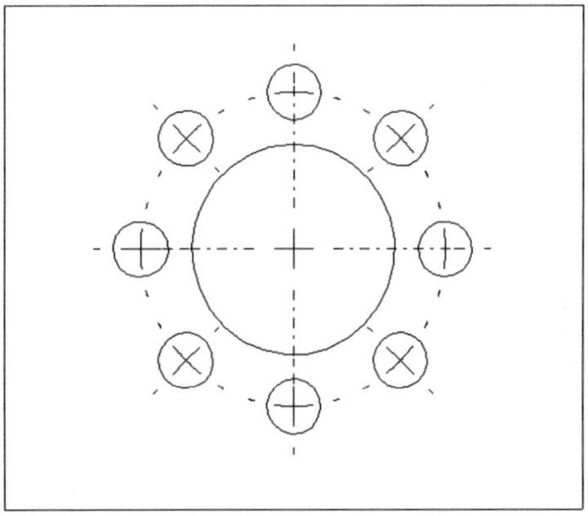

Figure 13-15 *Center marks created with the* ***Base center mark*** *check box selected*

The **Display Attributes** rollout available in the **Center Mark PropertyManager** is used to define the size of the center mark and the extended lines. The **Angle** rollout is used to rotate the center mark at an angle.

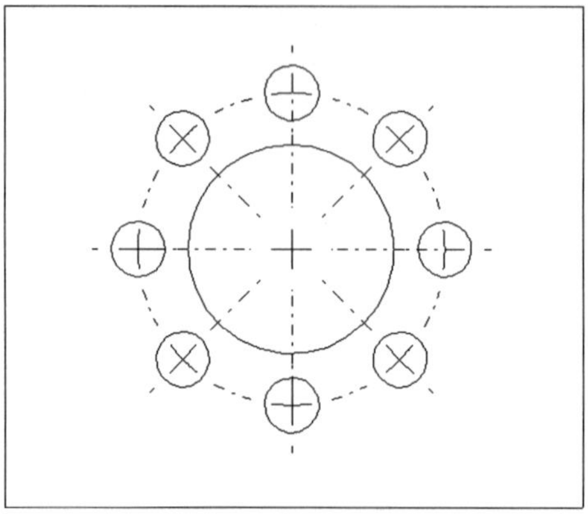

Figure 13-16 *Center marks created with the* ***Radial lines*** *check box selected*

Adding Centerlines to the Drawing Views

CommandManager:	Annotations > Centerline
Menu:	Insert > Annotations > Centerline
Toolbar:	Annotation > Centerline

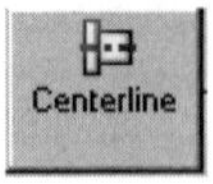

The **Centerline** tool is used to add centerlines to the views by selecting two edges/sketch segments or a single cylindrical/conical face. To add a centerline, choose the **Centerline** button from the **Annotations CommandManager**. The **Centerline PropertyManager** is invoked and it prompts you to select two edges/sketch segments or a single cylindrical/conical face. Select the entity or entities from the view to add the centerline. Figure 13-17 shows the centerlines added to the drawing views by selecting the surface of the cylinder.

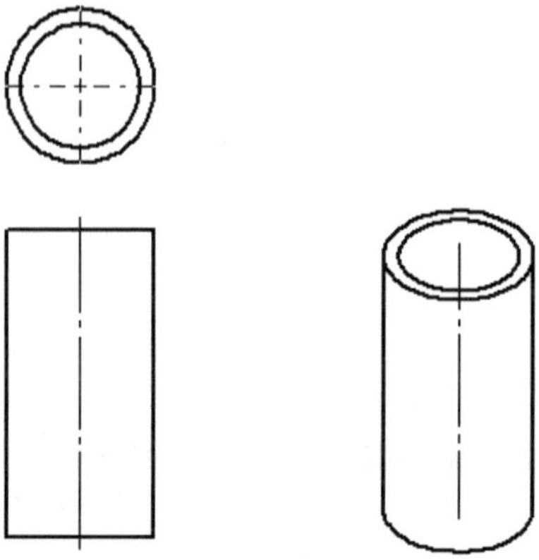

Figure 13-17 *Centerlines added to the front and isometric views using the* ***Centerline*** *tool*

Tip. *You can also set the option for the automatic creation of a centerline, while generating the drawing views. Choose **Tools > Options** from the menu bar to invoke the **System Options- General** dialog box and select the **Document Properties** tab from it. Also, select the **Centerlines** check box from the **Auto insertion on view creation** area and choose **OK** to close the dialog box.*

You can also directly select a view to add the centerlines.

Adding Hole Callout to the Views

CommandManager:	Annotations > Hole Callout
Menu:	Insert > Annotations > Hole Callout
Toolbar:	Annotation > Hole Callout

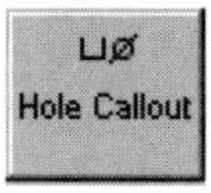

The **Hole Callout** tool is used to generate the hole callouts that are created in the **Part** mode using the **Hole PropertyManager**, the **Hole Wizard** tool, or the cut option. To generate a hole callout, choose the **Hole Callout** button from the **Annotations CommandManager**; the cursor will be replaced by the hole callout cursor. Select the hole from the drawing views; the hole callout will be attached to the cursor. Pick a point on the drawing sheet to place the hole callout. The **Dimension PropertyManager** is displayed, when you place the hole callout. Exit this **PropertyManager**. Figure 13-18 shows a drawing view with hole callouts generated using the **Hole Callout** tool.

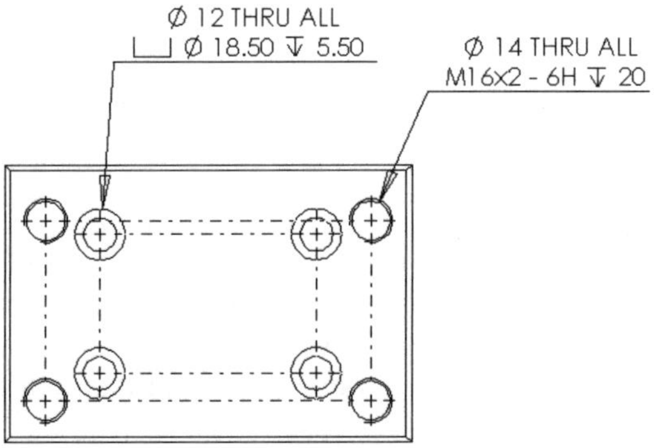

***Figure 13-18** Hole callouts generated using the **Hole Callout** tool*

Adding Cosmetic Threads to the Drawing Views

CommandManager:	Annotations > Cosmetic Thread (*Customize to Add*)
Menu:	Insert > Annotations > Cosmetic Thread
Toolbar:	Annotation > Cosmetic Thread (*Customize to Add*)

The **Cosmetic Thread** tool is used to add the cosmetic threads that will display the thread conventions in the drawing views. To do so, select the circular edge from the drawing view on which you need to apply the cosmetic thread. Now, choose the **Cosmetic Thread** button from the **Annotations CommandManager**;

the **Cosmetic Thread PropertyManager** is displayed, as shown in Figure 13-19.

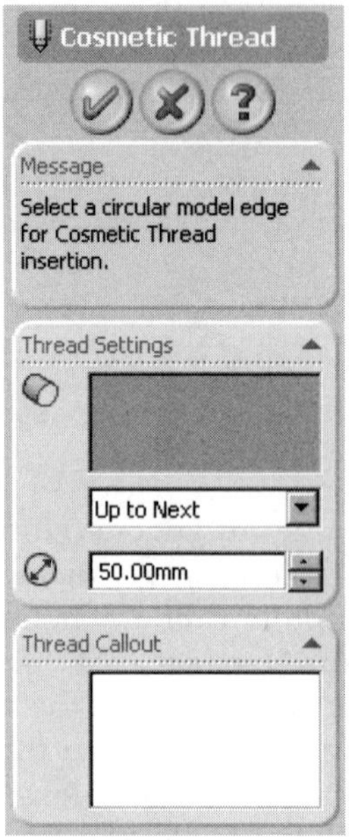

Figure 13-19** The **Cosmetic Thread PropertyManager

The options available in the **Cosmetic Thread PropertyManager** are discussed next.

Thread Settings

The options available in the **Thread Settings** rollout are used to define various parameters of the cosmetic thread. On invoking this tool, you are prompted to select the edges for the threads and set the parameters. Select the required circular edges on which you need to add a cosmetic thread. The name of the selected edge is displayed in the **Circular Edges** selection area. Specify the end condition and set the minor diameter of the thread.

Thread callout

The **Thread Callout** edit box is used to specify the text to be used in the thread callout for the cosmetic thread.

After setting all parameters, choose the **OK** button from the **Cosmetic Thread PropertyManager**. When you add a cosmetic thread to a generated drawing view, the thread convention will be displayed in all the drawing views. Figure 13-20 shows the cosmetic thread added to the drawing views.

Tip. *You can view the cosmetic threads in the part document also, if cosmetic threads are added to the drawing views. You can use any display mode to view the cosmetic threads in the model.*

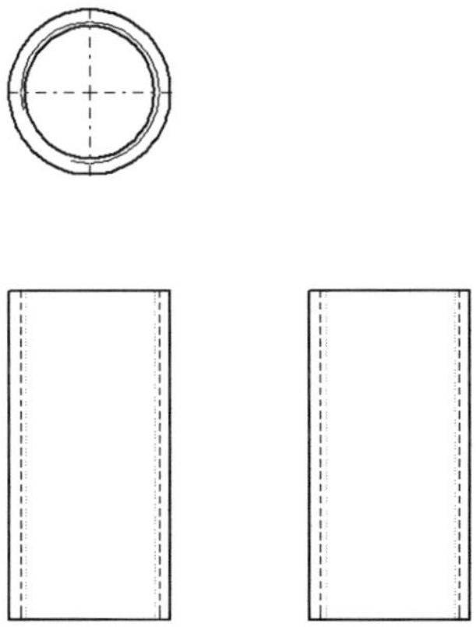

Figure 13-20 *Cosmetic threads added to the drawing views*

Note

*The thread conventions, once added in the drawing view, can be deleted only from the part document. You cannot delete the thread convention from the drawing document. To delete the thread convention in the part document, expand the **Hole** or **Cut** feature in the **FeatureManager Design Tree**. Now, select the thread convention and delete it.*

Adding the Multi-jog Leader to the Drawing Views

CommandManager:	Annotations > Multi-jog Leader (*Customize to Add*)
Menu:	Insert > Annotations > Multi-jog Leader
Toolbar:	Annotation > Multi-jog Leader (*Customize to Add*)

You can add a multi-jog leader line to the drawing views by using the **Multi-jog Leader** tool. A multi-jog leader is basically a leader in which you can add multiple jog lines with arrowheads at both the ends. To add a multi-jog leader, choose the **Multi-jog Leader** button from the **Annotations CommandManager**; the cursor will be replaced by the multi-jog line cursor. Select a point on the sheet, or on an entity from where you need to start the leader. Now, specify the points on the drawing sheet to specify the jogs and then select the second entity, where the end of the leader will be placed. A multi-jog leader will be created. You can also double-click anywhere on the sheet to specify the second end of the multi-jog leader. Figure 13-21 shows a multi-jog leader added to the drawing view. Note that in this figure, the text is written separately.

Adding the Dowel Pin Symbols to the Drawing Views

CommandManager:	Annotations > Dowel Pin Symbol (*Customize to Add*)
Menu:	Insert > Annotations > Dowel Pin Symbol
Toolbar:	Annotation > Dowel Pin Symbol (*Customize to Add*)

The **Dowel Pin Symbol** tool is used to add the dowel pin symbol to the holes in the drawing views. It is also used to confirm the size of the selected hole. To create a dowel pin symbol, select a hole or circular edge from the drawing view and choose the **Dowel Pin Symbol** button from the **Annotations CommandManager**.

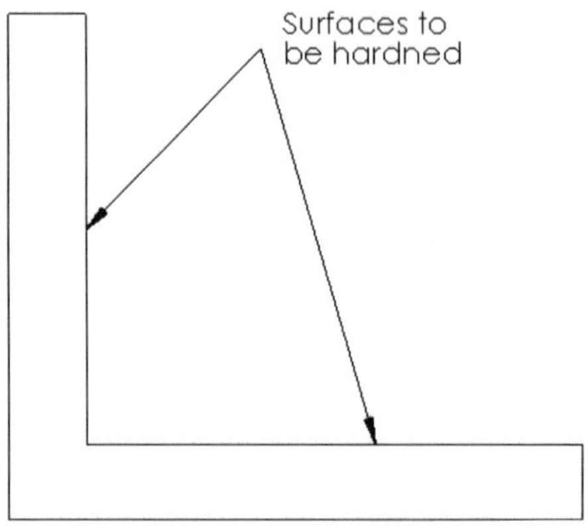

Figure 13-21 *Multi-jog leader added to the drawing view*

The **Dowel Pin Symbol PropertyManager** is displayed. You can flip the direction of the dowel pin symbol by using the **Flip symbol** check box in the **Display Attributes** rollout.

EDITING ANNOTATIONS

You can edit the annotation added to the drawing views by selecting them or by double-clicking them to display their respective **PropertyManager** or dialog boxes in which you can edit the parameters of the selected annotation.

ADDING THE BILL OF MATERIAL (BOM) IN DRAWING

CommandManager:	Annotations > Tables > Bill of Materials
Menu:	Insert > Tables > Bill of Materials
Toolbar:	Annotation > Tables > Bill of Materials

The Bill of Materials (BOM) is a table that displays the list of the components used in the assembly. This table can also be used to provide information related to the number of components in an assembly, their names, quantity, and any other information required to assemble the components. Remember that the sequence of the parts in the BOM depends on the sequence in which they were inserted in the assembly document. The BOM, placed in the drawing document, is parametric in nature. Therefore, if you add or delete a part from the assembly in the assembly document, the change will be reflected in the BOM in the drawing document.

To insert a BOM in the drawing file containing the assembly drawing views, select any one view from the drawing document and choose the **Bill of Materials** button from the **Annotations CommandManager**; the **Bill Of Materials PropertyManager** is invoked, as shown in Figure 13-22. The options available in the **Bill Of Materials PropertyManager** are discussed next.

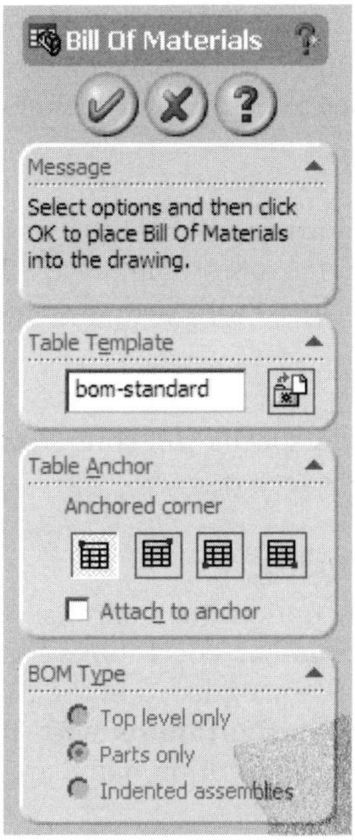

*Figure 13-22 The partial view of the **Bill Of Materials PropertyManager***

Table Template

The **Table Template** rollout is used to define the template to be used in the BOM. By default, the **bom-standard** template is selected. If you choose the **Open table template for Bill of Materials** button, the **Select BOM Template** dialog box is displayed. You can choose any of the default templates available in SolidWorks by using this dialog box.

Table Anchor

The buttons available in the **Table Anchor** rollout are used to select the corner of the BOM, using which, the corner will be attached to the drawing sheet. If the **Attach to anchor** check box is selected, the table will be automatically attached to the anchor point of the drawing sheet.

BOM Type

The **BOM Type** rollout is used to specify the level of the assembly. The options available in this rollout are discussed next.

Top level only

The **Top level only** radio button is used to list only the parts and the subassemblies in BOM. The parts of the subassemblies are not listed in the BOM.

Parts only

If you select the **Parts only** radio button, the subassemblies will not be listed in the BOM. The components of the subassemblies are listed as individual components in the BOM.

Indented assemblies

The **Indented assemblies** radio button is selected to list the components and the subassemblies. The components of the subassemblies are listed as indented, below their respective subassemblies and are not listed with their respective item number.

Configurations

The **Configurations** rollout is used to specify the configuration for creating the BOM. By default, the **Default** configuration is selected.

Part Configuration Grouping

This rollout is used to set the grouping options for a part having more than one configuration. Selecting the **Display as one item number** check boxes ensures that if a component has more than one configurations, all of them are listed in the BOM with the same item number. Selecting the **Display as one item configurations with the same name** ensures that if multiple components with the same configuration name are listed in one row of the BOM.

Keep Missing Items

After creating the BOM, if a component is deleted from the assembly, it will also be deleted from the BOM. To retain the deleted component in the BOM, select the **Keep Missing Items** check box; the **Keep Missing Items** rollout is also invoked. The name of the missing components will be displayed in the BOM with a strikeout by selecting the **Strikeout** check box available in the **Keep Missing Items** rollout.

Zero Quantity Display

The options available in the **Zero Quantity Display** rollout are used to set the option to display the component using -, 0, blank in the BOM if that component is not displayed in a particular configuration.

Item Numbers

You can specify a numeric value from where the sequence of the components starts in a BOM. You can set this value in the **Start at** edit box in this rollout.

After setting all the parameters in the **Bill Of Materials PropertyManager,** choose the **OK** button. A BOM will be attached to the cursor. Specify a point on the drawing sheet to place the BOM. Figure 13-23 shows a BOM added to a drawing sheet.

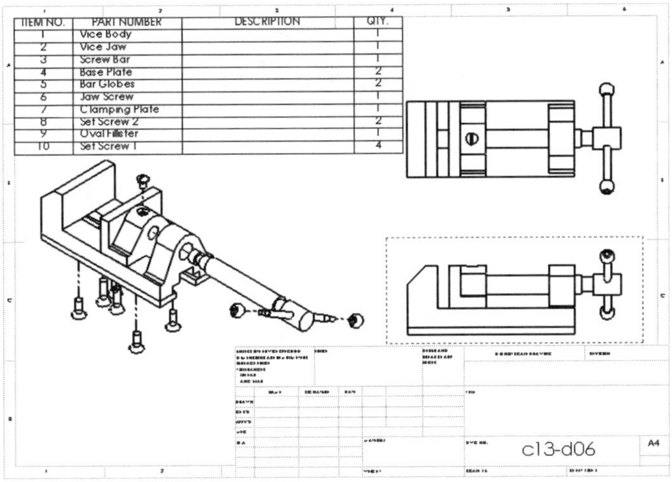

Figure 13-23 BOM added to the drawing sheet

ADDING BALLOONS TO DRAWING VIEWS

CommandManager:	Annotations > Balloon
Menu:	Insert > Annotations > Balloon
Toolbar:	Annotation > Balloon

After adding the BOM, you need to add the balloons to the components in the drawing views. The balloons can be added using the **Balloon** tool or the **AutoBalloon** tool. Both these methods of adding balloons to the components in the drawing views are discussed next.

Adding Balloons Using the Balloon Tool

The **Balloon** tool is used to manually add the balloons to the components of the assembly in the drawing view. The naming of balloons depends on the sequence of the parts in the BOM. To add balloons to the drawing view, choose the **Balloon** button from the **Annotations CommandManager;** the **Balloon PropertyManager** is displayed. The partial view of the **Balloon PropertyManager** is displayed in Figure 13-24.

On invoking the **Balloon PropertyManager**, you are prompted to select one or more locations to place the balloons. Select the components from the assembly drawing view to add the balloons. If you select the face of a component, the balloon will have a filled circle at the attachment point. However, if you select an edge of the component, the balloon will have a closed filled arrow. After adding the balloons to all components, choose the **OK** button from the **Balloon PropertyManager**. Figure 13-25 shows the drawing sheet in which the balloons are added.

*Figure 13-24 Partial view of the **Balloon PropertyManager***

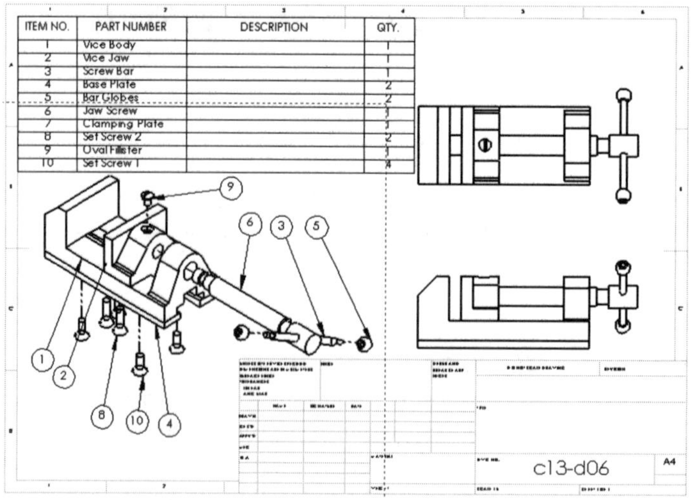

Figure 13-25 Balloons added to the assembly drawing view

Adding Balloons Using the AutoBalloon Tool

CommandManager:	Annotations > AutoBalloon
Menu:	Insert > Annotations > Auto Balloon
Toolbar:	Annotation > AutoBalloon

The **Auto Balloon** tool enables you to add balloons automatically. To do so, select the drawing view in which you need to add the balloons and choose the **AutoBalloon** button from the **Annotations CommandManager**. Balloons are added to the selected drawing view with the default settings and the **Auto Balloon PropertyManager** is displayed, as shown in Figure 13-26. The options available in the **Auto Balloon PropertyManager** are discussed next.

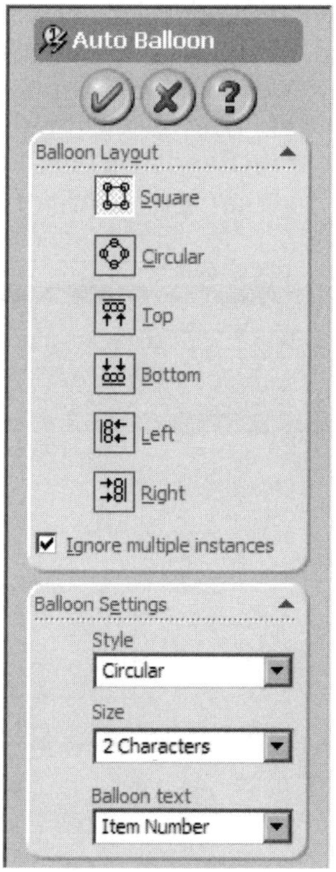

Figure 13-26** Partial view of the **Auto Balloon PropertyManager

Balloon Layout

The options available in the **Balloon Layout** rollout are used to set the layout of the balloons, when they are placed automatically. By default, the balloons are placed in a square form, because the **Square** button is chosen in the **Balloon Layout** rollout. You can also arrange the balloons in circular, aligned to top, aligned to bottom, aligned to left, and aligned to right forms. The **Ignore multiple instances** check box is selected by default to eliminate the creation of multiple instances of a balloon.

Balloon Settings

The **Balloon Settings** rollout is used to set the style, size, and type of the text of balloons.

After setting all parameters, choose the **OK** button from the **AutoBalloon PropertyManager**.

ADDING NEW SHEETS TO DRAWING VIEWS

You can also add new sheets to the drawing document. A multisheet drawing document can be used to generate the drawing views of all the components and the drawing views of the assembly in the same document. You can switch between the sheets easily to refer to the drawings of the different parts of an assembly within the same document. Also, you do not need to open the separate drawing documents. To add a sheet to the drawing document, select **Sheet1** from the **FeatureManager Design Tree** and invoke the shortcut menu. Choose the **Add Sheet** option to add a sheet to the drawing document or else choose **Insert > Sheet** from the menu bar; the **Sheet Properties** dialog box is displayed. You can specify the name, standard sheet format, and scale projection type for the drawing sheet to be added by using this dialog box. Choose the **OK** button from this dialog box; a new sheet will be added in the drawing document. Figure 13-27 shows a drawing document with four drawing sheets added, with the **Sheet 5** as the active sheet.

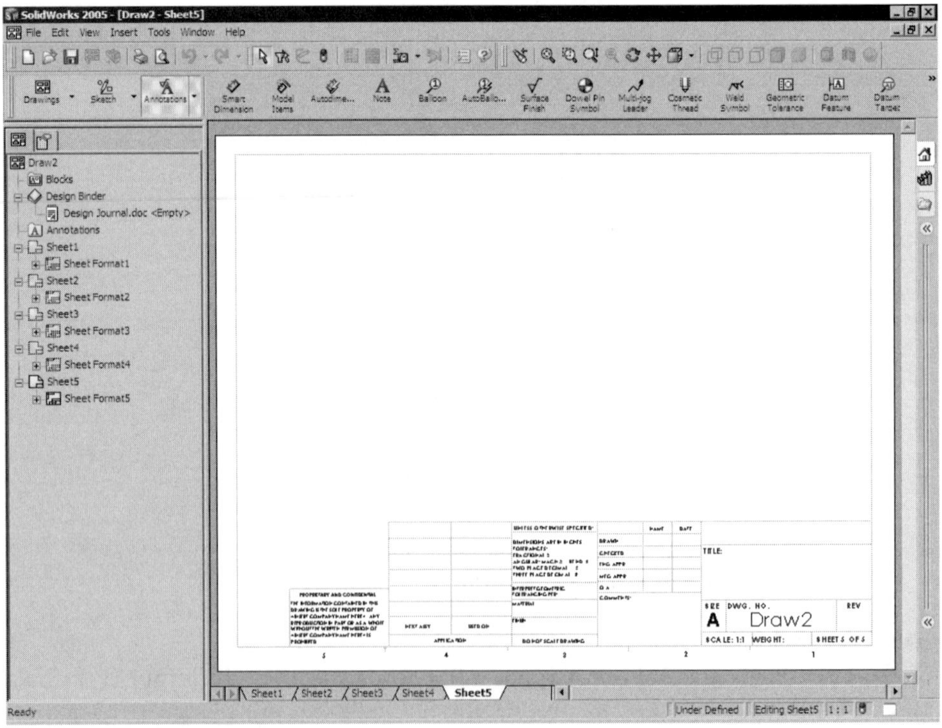

Figure 13-27 Drawing sheets added to the drawing document

To activate a drawing sheet, select the sheet from the **FeatureManager Design Tree,** invoke the shortcut menu and choose the **Activate** option from it. You can also activate the sheet by selecting the sheet tab from the bottom of the drawing document.

EDITING THE SHEET FORMAT

You can edit the standard sheet format according to your requirement. To do so, ensure that the required sheet is active and select it from the **FeatureManager Design Tree**. Right-click on the sheet, invoke the shortcut menu and choose the **Edit Sheet Format** option from it. All the entities, annotations, and views will disappear from the drawing sheet. You can edit the sheet format by using the sketching tools available in the **Sketch CommandManager**. After editing the sheet format, select the active sheet and invoke the shortcut menu. Choose the **Edit Sheet** option from the shortcut menu to switch back to the edit sheet environment.

CREATING A USER-DEFINED SHEET FORMAT

You can also create a user-defined sheet format in SolidWorks. To do so, start a new drawing document with a custom size. On the basis of your design requirements, you can set the size of the sheet in the **Width** and **Height** edit boxes available in the **Sheet Format/Size** dialog box. After defining the sheet, select **Sheet1** from the **FeatureManager Design Tree,** invoke the shortcut menu and choose the **Edit Sheet Format** option from it. The edit sheet format environment is invoked. You can create or modify the sheet format by using the standard sketching tools available in the **Sketch** CommandManager. After creating or modifying the sheet format, switch back to the edit sheet environment. Choose **File > Save Sheet Format** from the menu bar. The **Save Sheet Format** dialog box is displayed. Browse the location where you need to save the sheet format. Specify the name of the sheet format and choose the **Save** button from the **Save Sheet Format** dialog box. Next time when you invoke the **Sheet Format/Size** dialog box, you can retrieve the saved sheet format by selecting the **Standard sheet format** radio button. Using the **Browse** button, browse to the location where you have saved the sheet format. Figure 13-28 shows a user-defined sheet format.

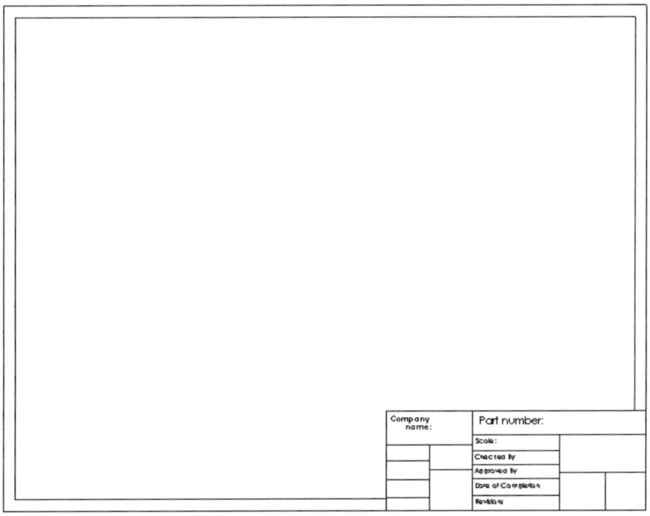

Figure 13-28 *A user-defined sheet format*

TUTORIALS

Tutorial 1

In this tutorial, you will open the drawing created in Tutorial 1 of Chapter 12 and generate dimensions and add annotations to it. Next, you will change the display of the front and the right view to hidden lines visible and the display of the isometric and aligned section views to shaded mode. **(Expected time: 45 min)**

The steps to be followed to complete this tutorial are listed next.

a. Copy the part and the drawing document in the folder of the current chapter.
b. Configure the font settings and generate the dimensions using the **Model Items** tool.
c. Arrange the dimensions and delete the unwanted dimensions, refer to Figures 13-30 and 13-31.
d. Add the datum symbol and geometric tolerance to the drawing views, refer to Figures 13-32 and 13-33.
e. Change the model display state of the drawing views, refer to Figure 13-34.

Copying the Documents in the Folder of the Current Chapter

Before proceeding, you need to copy the model and the drawing document in the folder of the current chapter.

1. Create a folder with the name *c13* in the *SolidWorks* folder and copy *c07-tut02.sldprt* and *c12-tut01.slddrw* from the *\My Document\SolidWorks\c12* folder to this folder.

Opening the Drawing Document

Next, you need to open the drawing document in the SolidWorks window.

1. Invoke the **Open** dialog box and open *c12-tut01.slddrw* document from the folder of the current chapter.

The drawing document, in which you need to add the detailing, is displayed in the drawing area, as shown in Figure 13-29.

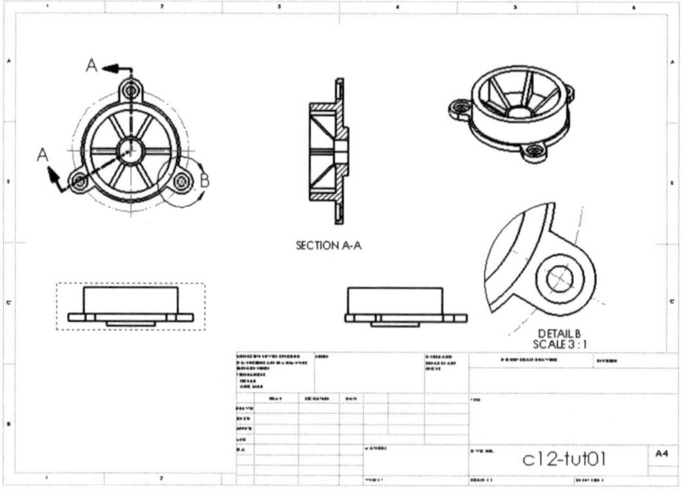

***Figure 13-29** Drawing views generated in Chapter 12*

Applying the Documents Settings

Before generating the model dimensions, you need to configure the document settings. These settings will allow the dimensions and other annotations in the current sheet to be viewed properly.

1. Invoke the **Document Properties - Detailing** dialog box and select the **Annotations Font** option from the area on the left.

2. The **Note** option is selected in the **Annotation type** area. Select this option again. The **Choose Font** dialog box will be invoked.

3. Select the **Points** radio button from the **Height** area and set the value of the font size to **9** from the list box.

4. Choose the **OK** button from the **Choose Font** dialog box. Similarly, set the font size for **Dimension**, **Detail**, and **Section** options to **9**.

5. Now, select the **Arrows** option from the area on the left. The options related to size and shape of the arrows are displayed.

6. Set the value of the **Height** as **1**, **Width** as **3**, and **Length** as **3** in the **Size** area.

7. In the **Section/View size** area, set the value of the **Height** as **2**, **Width** as **4**, and **Length** as **8**.

8. Choose the **OK** button from the **Document Properties - Detailing - Arrows** dialog box.

Generating the Dimensions

Next, you need to generate the dimensions using the **Model Items** tool. As discussed earlier, if you do not select any view, on generating the dimensions using the **Model Items** tool, all the dimensions will be displayed in all the views. Sometimes the dimensions overlap each other. Therefore, you will select the view in which you need to generate the dimension and then you will invoke the **Model Items** tool.

1. Select the top view and choose the **Model Items** button from the **Annotation CommandManager**. The **Insert Model Items** dialog box is displayed. The name of the selected view is displayed in the **Import to views** selection area.

2. Choose the **OK** button from the **Insert Model Items** dialog box.

   The dimensions of the model that can be displayed in the selected view are generated. Any radial dimension attached to the counterbore hole in the top view, needs to be deleted, because you will add a hole callout to this counterbore hole later.

3. Select the radial dimension and press the DELETE key from keyboard to delete the dimension. Similarly, delete the dimension of diameter **20**.

   Note that if SolidWorks 2005 is configured for ANSI system on your machine, the dimension arrowheads are displayed as no filled. You need to change the arrowheads to filled.

4. Drag a window such that all the dimensions are enclosed inside it. Release the left mouse button to select all the generated dimensions. The **Dimension PropertyManager** is displayed.

5. Select the **Filled Arrow** option from the **Style** drop-down list of the **Arrow** rollout. Choose **OK** to close the **Dimension PropertyManager**.

   The arrowheads will be changed to filled arrowheads. Note that the generated dimensions are scattered arbitrarily on the drawing sheet. Therefore, you need to arrange the dimensions by moving them to the required locations.

6. Select the dimension one by one and drag them to the desired location, refer to Figure 13-30. You can reverse the direction of arrowheads by clicking on the control point that is displayed on them.

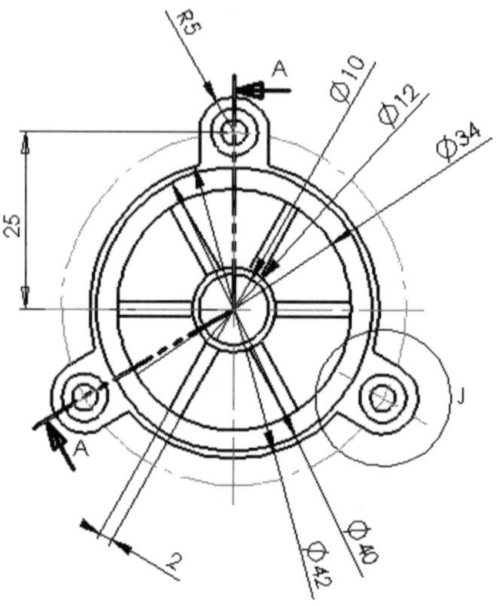

Figure 13-30 *Top view after generating and arranging the dimensions*

7. Select the aligned section view and generate the dimensions using the **Model Items** tool. After placing the dimensions, move them to the appropriate places, refer to Figure 13-31. You may need to change the arrowheads to closed filled, if they are not already filled.

8. Choose the **Hole Callout** button from the **Annotation CommandManager** and select the outer circle of one of the counterbore feature in the front view. The hole callout will be attached to the cursor. Pick a point on the drawing sheet to place the hole callout.

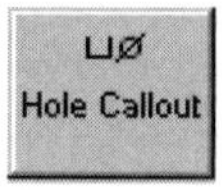

Adding the Datum Feature Symbol to the Drawing View

After generating the dimensions, you need to add the datum feature symbol to the drawing

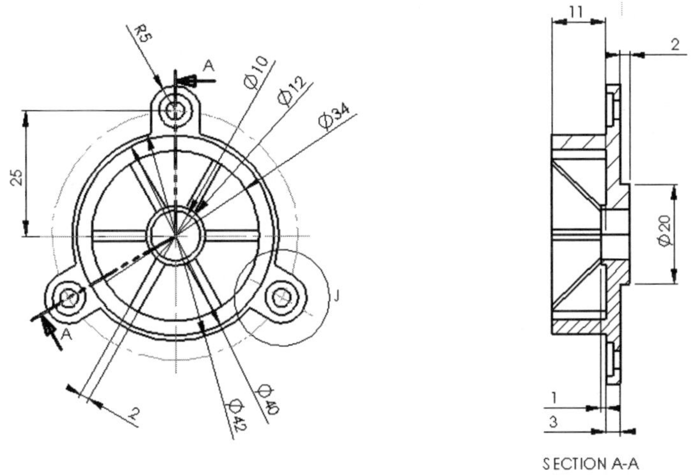

Figure 13-31 *Partial view of the sheet after generating and arranging the dimensions in the aligned section view*

view. Datum feature symbols are used as the datum reference for adding the geometric tolerance to the drawing views.

1. Select the edge of the outer cylindrical feature from the top view and then choose the **Datum Feature** button from the **Annotations CommandManager**. The **Datum Feature PropertyManager** is displayed and a datum callout is attached to the cursor.

2. Place the datum symbol at an appropriate location, refer to Figure 13-32.

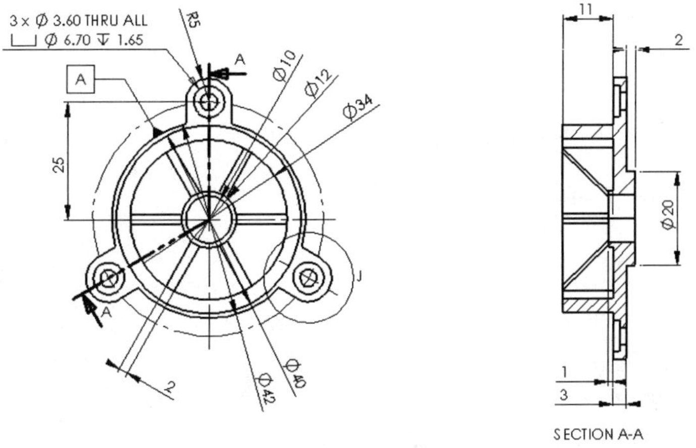

Figure 13-32 *Datum feature symbol hole description added to the top view*

3. Choose the **OK** button from the **Datum Feature PropertyManager**.

Adding the Geometric Tolerance to the Drawing View

After defining the datum feature symbol, you will add the geometric tolerance to the drawing view.

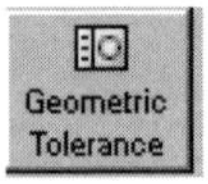

1. Select the circular edge that has a diameter of 12 from the top view and choose the **Geometric Tolerance** button from the **Annotation CommandManager**.

   The **Properties** dialog box is displayed. It is used to specify the parameters of the geometric tolerance.

2. Choose the **GCS** button corresponding to the first row in the **Feature control frames** area. The **Symbols** dialog box is displayed.

3. Select the **Concentricity and coaxiality** option from the list box and choose the **OK** button from the **Symbols** dialog box.

4. Set the value of tolerance in the **Tolerance 1** edit box to **0.002**.

5. Specify **A** in the **Primary** edit box to define the primary datum reference.

6. Choose the **OK** button from the **Properties** dialog box. The geometric tolerance will be placed attached to the selected circular edge. You may need to move the geometric tolerance, if it overlaps the dimensions. Figure 13-33 shows the drawing view, after adding and rearranging the geometric tolerance.

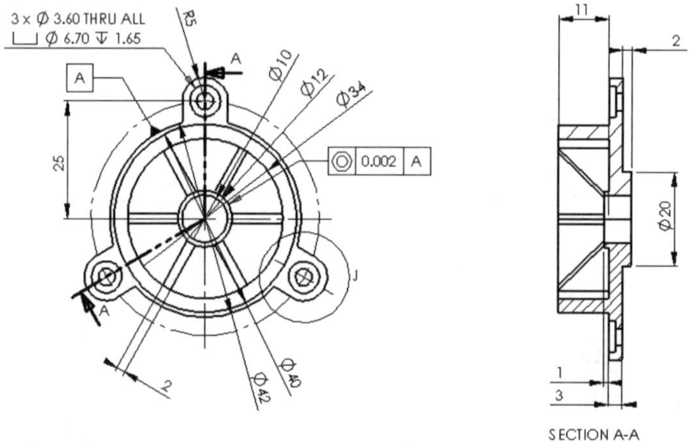

Figure 13-33 Geometric tolerance added to the drawing view

Changing the View Display Options

After adding all the annotations to the drawing views, you need to change the display setting of the drawing views.

1. Press and hold down the CTRL key, and select the front view and the right-side view from the drawing sheet.

2. Choose the **Hidden Lines Visible** button from the **View** toolbar. You can also choose this button from the **Display Style** rollout of the **Multiple Views PropertyManager** that is displayed when you select the two views.

   The hidden lines will be displayed in the selected drawing views.

3. Now, select the isometric view and the aligned section view from the drawing sheet.

4. Choose the **Shaded With Edges** button from the **View** toolbar or the **Multiple Views PropertyManager**. Figure 13-34 shows the final drawing sheet after changing the display view settings.

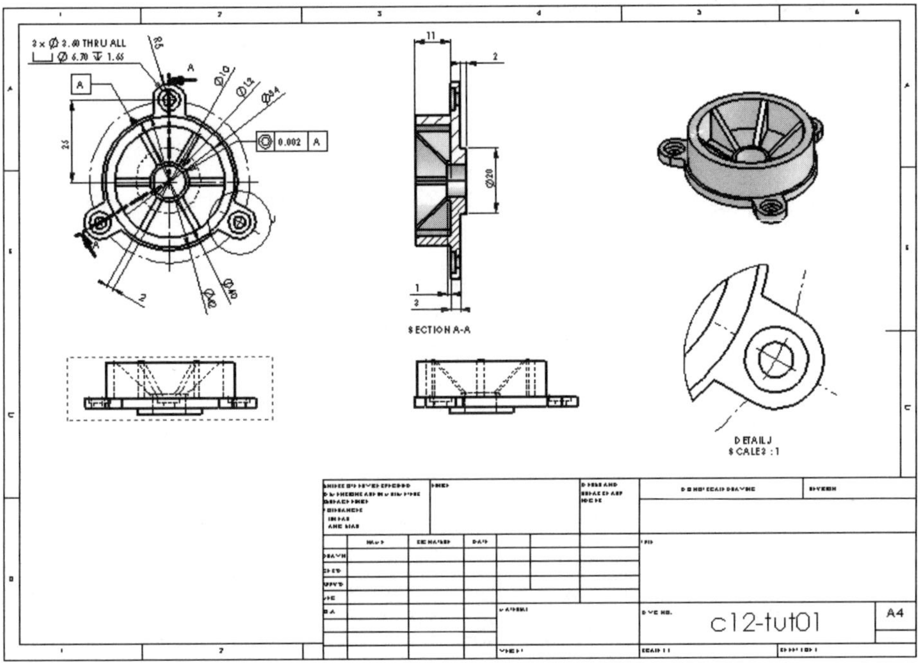

Figure 13-34 *Final drawing sheet*

5. Save and close the drawing document.

Tutorial 2

In this tutorial, you will generate the BOM of the Bench Vice assembly and then add balloons to the isometric view in the exploded state. **(Expected time: 45 min)**

The steps to be followed to complete this tutorial are listed next.

a. Copy the Bench Vice folder that contains parts, assembly, and the drawing document to the folder of the current chapter.
b. Delete the views that are not required in the drawing sheet, refer to Figure 13-36.
c. Move to arrange the views in the drawing sheet, refer to Figure 13-36.
d. Set the anchor on the drawing sheet where the BOM will be attached.
e. Generate the BOM, refer to Figure 13-37.
f. Add balloons to the isometric view, refer to Figure 13-38.

Copying the Bench Vice Assembly Folder to the Current Folder

1. Copy the Bench Vice folder from the *\My Document\SolidWorks\c12* folder to the folder of the current chapter.

Opening the Drawing Document

After copying the folder, you need to open the drawing document in the SolidWorks window.

1. Open *c12-tut02.slddrw* document.

The drawing document, in which you need to generate the BOM and balloons is displayed in Figure 13-35.

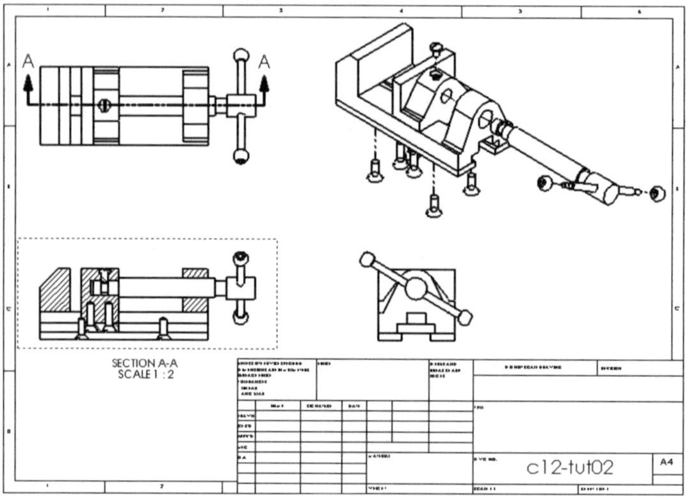

Figure 13-35 *Drawing views generated in Chapter 12*

Deleting the Unwanted View

You need to delete the right view, because it is not required in this tutorial.

1. Select the right-side view and press the DELETE key on the keyboard.

The **Confirm Delete** dialog box will be displayed.

2. Choose the **Yes** button from this dialog box. The view will be deleted from the current drawing sheet.

Moving the Isometric View

You need to move the exploded isometric view, because the BOM will be generated and placed at the top right corner of the drawing sheet.

1. Select the isometric view; the border of the view is displayed in green.
2. Move the cursor to the border; the cursor will be replaced by the move cursor.
3. Drag the cursor to move the drawing view. Place it at the required location, refer to Figure 13-36.

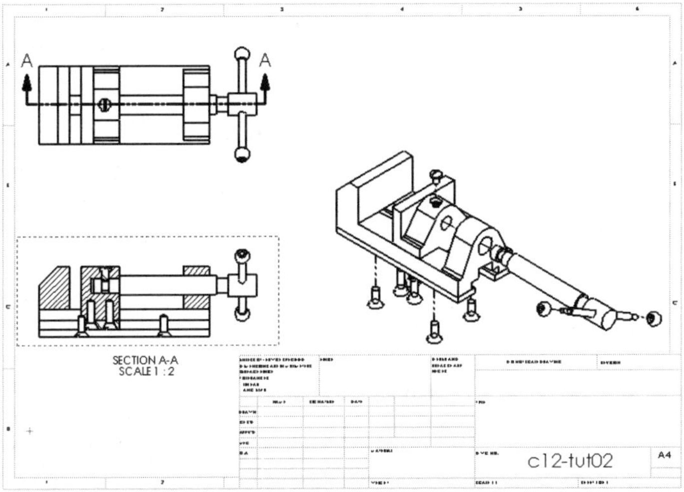

***Figure 13-36** Drawing sheet after moving the drawing view*

Setting the Anchor for the BOM

Before generating the BOM, you need to set its anchor. The anchor is a point on the drawing sheet, on which, one of the corners of the BOM coincides. By default, the anchor is defined at the top left corner of the drawing sheet. But in this tutorial, you need to add the BOM on the top right corner of the drawing sheet. Therefore, you need to set the anchor before generating the BOM.

1. Expand **Sheet1** from the **FeatureManager Design Tree** and then expand **Sheet Format1**.
2. Select the **Bill of Materials Anchor1** option, right-click to invoke the shortcut menu and choose the **Set Anchor** option from it. The drawing views will disappear from the sheet.

3. Specify the anchor point on the inner top right corner of the drawing sheet. A point will be placed at the selected location.

   After setting the anchor, you need to switch back to the edit sheet environment.

4. Select **Sheet1** from the **FeatureManager Design Tree** and right-click to invoke the shortcut menu. Choose the **Edit Sheet** option from the shortcut menu.

Generating the BOM

Next, you need to generate the BOM. As discussed earlier, the BOM generated in SolidWorks is parametric in nature. If a component is deleted or added in the assembly, the change is automatically reflected in the BOM. But before you generate the BOM, you need to set its text parameters.

1. Invoke the **Document Properties - Detailing - Annotation Font** dialog box. Select the **Tables** option from the **Annotation type** area to display the **Choose Font** dialog box.

2. Select the **Points** radio button from the **Height** area and set the value of font size to 9 point. Close both the dialog boxes.

3. Similarly, change the text height of balloons to 12 points.

4. Select the isometric view and choose **Bill of Materials** from the **Annotations CommandManager**. The **Bill Of Materials PropertyManager** is displayed.

5. Select the **Top Right** button from the **Anchored corner** area of the **Table Anchor** rollout. Also, select the **Attach to anchor** check box.

6. Choose the **OK** button from the **Bill Of Materials PropertyManager**.

   The BOM is generated. You will notice that the **Description** column is also displayed in the BOM. But this column is not required and so you need to delete it.

7. Move the cursor over the heading **Description** and right-click to display the shortcut menu. Choose **Delete > Column** from the cascading menu. This column will be deleted. The drawing sheet, after generating the BOM and deleting the **Description** column, is displayed in Figure 13-37.

Note

*If the color of the BOM is gray and is not clearly visible, invoke the **System Options - Colors** dialog box. Select the **Imported Annotations (Driven)** from the **System colors** list box and change its color to black*

Adding the Balloons to the Components

After generating the BOM, you need to add the balloons to the components. Before proceeding further, make sure that you have changed the font height of balloons to 12 point in the last section.

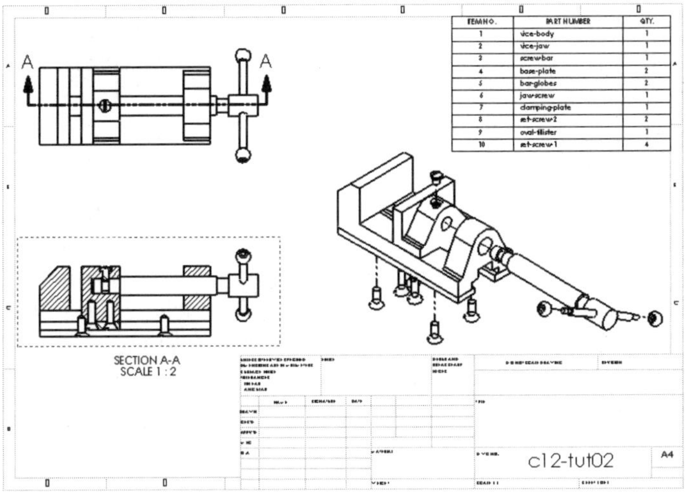

Figure 13-37 Drawing sheet after generating the BOM

1. Select the isometric view and choose the **AutoBalloon** button from the **Annotations CommandManager**. Balloons are automatically added to all the components in the isometric view. The **Auto Balloon PropertyManager** is also displayed.

   The multiple instances of any component are ignored because the **Ignore multiple instances** check box is selected in the **Balloon Layout** rollout.

2. Select **1 Character** from the **Size** drop-down list in the **Balloon Settings** rollout. Choose **OK** to close this **PropertyManager**.

   The balloons are added to all the components, except the Clamping Plate. This is because the clamping plate is not visible in the isometric view. You will notice that the balloons are not properly arranged on the sheet and are placed arbitrarily. Therefore, you need to manually drag each balloon and place it properly.

3. Move the cursor over one of the balloons and when it is highlighted, drag it to place it at a better location, refer to Figure 13-38.

4. Similarly, drag and place the remaining balloons also at the proper locations. The final drawing sheet, after adding and rearranging balloons, is shown in Figure 13-39.

5. Save the drawing and close the document.

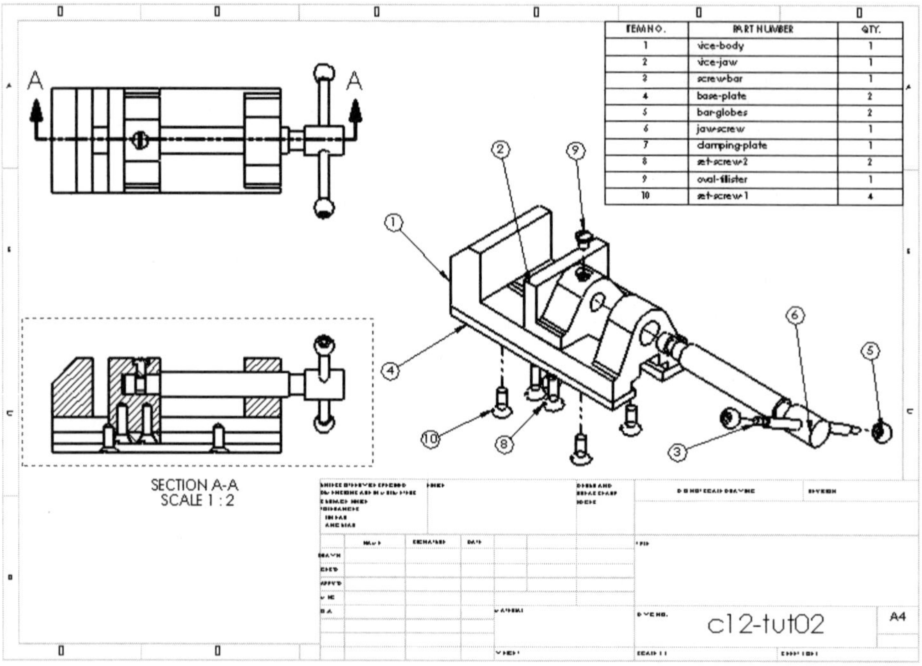

Figure 13-38 Final drawing sheet after adding balloons

SELF-EVALUATION TEST

Answer the following questions and then compare your answers with those given at the end of this chapter.

1. You cannot add annotations while creating the part in SolidWorks. (T/F)

2. You can add the surface finish symbols to the drawing views. (T/F)

3. The **Projected tolerance zone** area is used to define the quality of the projected tolerance. (T/F)

4. You can set the target shape as a point, circle, or rectangle, while adding a datum target. (T/F)

5. You can also set the option for the automatic creation of centerline, while generating the drawing views. (T/F)

6. The __________ spinner is used to define the major diameter of the thread.

7. The __________ tool is used to add the balloons to the components of the assembly in the drawing view.

8. You can also create automatic balloons using the _________ tool.

9. Using the _________ rollout available in the **Bill Of Materials PropertyManager,** you can specify the template needed to create the BOM.

10. The _________ tool is used to create a hole callout.

REVIEW QUESTIONS

Answer the following questions.

1. The _________ tool is used to add the cosmetic threads that are used to display the thread conventions in the drawing views.

2. The _________ tool is used to add reference dimensions to the drawing views.

3. You can change the model display setting from hidden lines removed to hidden lines visible, or wireframe, or shaded using the options available in the _________ toolbar.

4. Select the _________ check box from the **Auto insertion on view creation** area to automatically create centerlines, while generating the view.

5. The _________ tool is used to create the centerlines in the views.

6. Which **PropertyManager** is invoked to add automatic balloons to the selected drawing view?

   (a) **AutoBalloon** (b) **Balloon**
   (c) **Properties** (d) **Center Mark**

7. Which **PropertyManager** is displayed, when you choose the **Cosmetic Thread** button from the **Annotations CommandManager**?

   (a) **Cosmetic Thread Properties** (b) **Cosmetic Thread**
   (c) **Cosmetic Thread Convention** (d) None of these

8. Which **PropertyManager** is used to add center marks to the drawing views?

   (a) **Add Center Mark** (b) **Create Center Mark**
   (c) **Center Mark** (c) **Cosmetic Thread**

9. Which rollout available in the **Cosmetic Thread PropertyManager,** is used to define the depth of the cosmetic thread?

   (a) **Thread Settings** (b) **Thread Depth**
   (c) **Cosmetic Thread** (d) None of these

10. Which **PropertyManager** is used to add balloons in the drawing views?

(a) **Add Balloons** (b) **Balloon Properties**
(c) **Balloons** (d) None of these

EXERCISE

Exercise 1

In this exercise you will generate the isometric view in the exploded view of the assembly created in Tutorial 1 of Chapter 11 on the standard A4 sheet format. The scale of the view will be 1:5. After generating the view, generate the BOM and add balloons to the assembly view shown in Figure 13-39. **(Expected time: 30 min)**

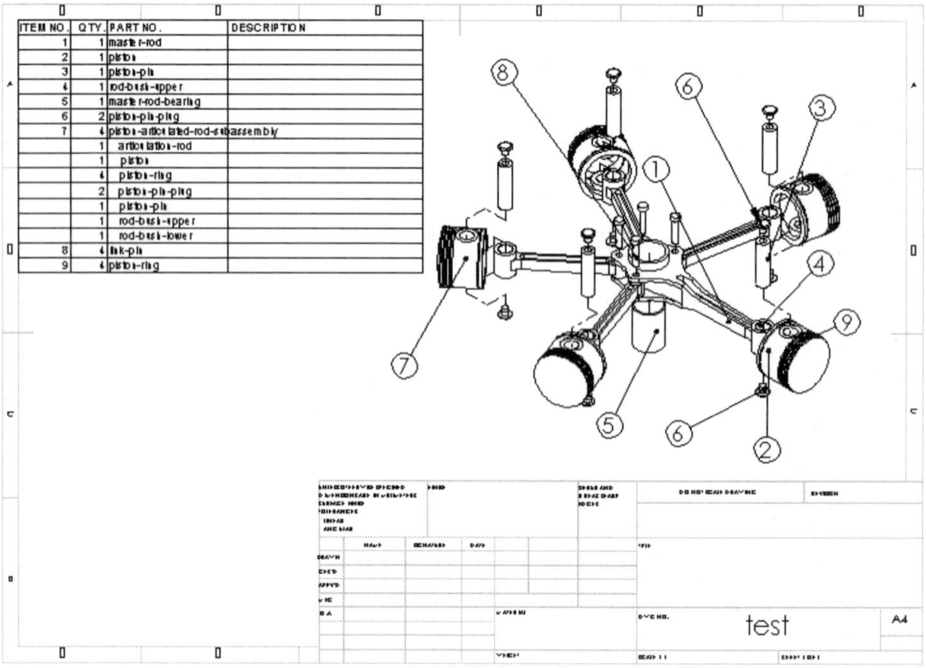

Figure 13-39 Drawing view for Exercise 1

Answers to Self-Evaluation Test

1. T, **2.** T, **3.** F, **4.** T, **5.** T, **6. Minor Diameter**, **7. Balloon**, **8. AutoBalloon**, **9. Table Template**, **10. Hole Callout**

Index

Symbols

A

B

C

D

E

I

L

M

N

O

P

R

S

T

U

V

W

Z

t of other Publications by CADCIM Technologies

he following is the list of some of the publications by Prof. Sham Tickoo and CADCIM Technologies.

CATIA Textbooks

- CATIA for Designers, V5R13
  CADCIM Technologies, USA

Solid Edge Textbooks

- Solid Edge for Designers, Version 15
  CADCIM Technologies, USA

SolidWorks Textbooks

- SolidWorks for Designers, Release 2004
  CADCIM Technologies, USA
- SolidWorks for Designers, Release 2004
  Piter Publishing Press, Russia
- SolidWorks for Designers, Release 2003
  CADCIM Technologies, USA

Autodesk Inventor Textbooks

- Autodesk Inventor for Designers, Release 9
  CADCIM Technologies, USA
- Autodesk Inventor for Designers Release 6 with Release 7 Update Guide
  CADCIM Technologies, USA

Pro/ENGINEER Textbooks

- Pro/ENGINEER Wildfire for Designers, Release 2.0
  CADCIM Technologies, USA
- Pro/ENGINEER Wildfire for Designers
  CADCIM Technologies, USA
- Pro/ENGINEER for Designers, Release 2001
  CADCIM Technologies, USA

Mechanical Desktop Textbook

- Mechanical Desktop Instructor, Release 5
  McGraw Hill Publishing Company, USA

Coming Soon: New Textbooks from CADCIM Technologies

- Solid Edge for Designers Version 16 (November 2004)
- CATIA for Designers, V5R14 (December)
- Autodesk Revit for Architects and Designers (December 2004)